SIGNALS AND COMMUNICATION TECHNOLOGY

For further volumes:
www.springer.com/series/4748

Roberto Verdone • Alberto Zanella
Editors

Pervasive Mobile and Ambient Wireless Communications

COST Action 2100

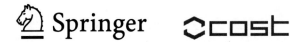

Editors
Roberto Verdone
Dip. di Elettronica, Informatica e Sist.
Università di Bologna
Bologna
Italy

Alberto Zanella
IEIIT
National Research Council of Italy (CNR)
Bologna
Italy

ISSN 1860-4862   Signals and Communication Technology
ISBN 978-1-4471-2314-9   e-ISBN 978-1-4471-2315-6
DOI 10.1007/978-1-4471-2315-6
Springer London Dordrecht Heidelberg New York

British Library Cataloguing in Publication Data
A catalogue record for this book is available from the British Library

Library of Congress Control Number: 2011944421

© Springer-Verlag London Limited 2012
Apart from any fair dealing for the purposes of research or private study, or criticism or review, as permitted under the Copyright, Designs and Patents Act 1988, this publication may only be reproduced, stored or transmitted, in any form or by any means, with the prior permission in writing of the publishers, or in the case of reprographic reproduction in accordance with the terms of licenses issued by the Copyright Licensing Agency. Enquiries concerning reproduction outside those terms should be sent to the publishers.
The use of registered names, trademarks, etc., in this publication does not imply, even in the absence of a specific statement, that such names are exempt from the relevant laws and regulations and therefore free for general use.
The publisher makes no representation, express or implied, with regard to the accuracy of the information contained in this book and cannot accept any legal responsibility or liability for any errors or omissions that may be made.

Printed on acid-free paper

Springer is part of Springer Science+Business Media (www.springer.com)

COST

COST, the acronym for European Cooperation in Science and Technology, is the oldest and widest European intergovernmental network for cooperation in research. Established by the Ministerial Conference in November 1971, COST is presently used by the scientific communities of 35 European countries to cooperate in common research projects supported by national funds.

The funds provided by COST—less than 1% of the total value of the projects—support the COST cooperation networks (COST Actions) through which, with EUR 30 million per year, more than 30,000 European scientists are involved in research having a total value which exceeds EUR 2 billion per year. This is the financial worth of the European added value which COST achieves.

A "bottom up approach" (the initiative of launching a COST Action comes from the European scientists themselves), "à la carte participation" (only countries interested in the Action participate), "equality of access" (participation is open also to the scientific communities of countries not belonging to the European Union) and "flexible structure" (easy implementation and light management of the research initiatives) are the main characteristics of COST.

As precursor of advanced multidisciplinary research COST has a very important role for the realisation of the European Research Area (ERA) anticipating and complementing the activities of the Framework Programmes, constituting a "bridge" towards the scientific communities of emerging countries, increasing the mobility of researchers across Europe and fostering the establishment of "Networks of Excellence" in many key scientific domains such as: Biomedicine and Molecular Biosciences; Food and Agriculture; Forests, their Products and Services; Materials, Physical and Nanosciences; Chemistry and Molecular Sciences and Technologies; Earth System Science and Environmental Management; Information and Communication Technologies; Transport and Urban Development; Individuals, Societies, Cultures and Health. It covers basic and more applied research and also addresses issues of pre-normative nature or of societal importance.

Web: http://www.cost.eu

 ESF provides the COST office through an EC contract

 COST is supported by the EU RTD Framework programme

 The COST 2100 Action was supported by the COST framework from December 2006 to December 2010

Foreword

COST (standing for European COoperation in the field of Scientific and Technical research) Actions in the field of Wireless Communications started from Action 207 Digital Land Mobile Communications, which clarified several open technological issues in Global System for Mobile Communications (GSM)'s specification. COST 207 also paved the way to successful followers, such as COST 231, 259, 273, and, finally, 2100, all characterised by low cost, excellent achievements and openness in disseminating the results.

The scope of COST 2100 has been together broad and ambitious, yet, judging from this book and my lateral experience as Director of NEWCOM++, a network of excellence in wireless communications that cooperated with COST 2100, has been largely achieved.

Following the 1971 statement by Weldon "coding is dead", which proved to be totally wrong in the following 40 years, a recent paper[1] raised the question "Is the PHY layer dead?". The authors of the paper argue that "disconnection between academia and industry has increased the perception of the morbidity of the PHY. Academics largely judge themselves by the theory developed in their papers, ... industry is judged by the stock market on its ability to turn a profit." This book offers the tangible evidence of what can be done to reverse what I also consider a negative trend. The main flavour of this book stems from two main characteristics: a deep, unusual in these days, collaboration between academia and industry and, as a consequence, a constant attention to maintain a difficult equilibrium between rigorous theory and practical constraints.

In its pages, it shows how that convinced cooperation in the area of Wireless Communications Networks (WCN) between major European (and beyond) players in academia and industry can lead to results of interest for both communities.

In today world, a common feature of the main problems the mankind has to face is complexity. Science and technology, in WCN in particular, are no exception to this rule. Overly stressed simplifications, be them in the modelling of the channel or

---

[1] M. Dohler, R.H. Heath, A. Lozano, C.B. Papadias, R.A. Valenzuela. Is the PHY layer dead? *IEEE Comm. Magazine*, vol. 49, n. 4, pp. 159–165, 2011.

transmitter/receiver elements, may lead to elegant results formulation but quite often fail to capture the real essence of the problem and to offer a practical solution. On the other hand, industry should value more the quest of rigour of academia and push the interaction to provide more insight into physical impairments of the channel and the hardware, aiming at a better modelling and more refined theory.

Part I of this book provides an excellent example of how the academia/industry cooperation can advance the knowledge of the physical medium (including antennae) through a joint effort involving measurement, modelling and testing. The width and depth of the results make them a fundamental tool for 4G standardisation bodies.

Almost all the challenges in point-to-point communications have been addressed and solved: the Shannon capacity limits, practical ways to approach them using turbo or low-density parity-check codes, even constructive proofs that they can be attained in some case (polar codes), large increase in spectral efficiency through spatial multiplexing with MIMO.

Moving to a multipoint multiuser paradigm makes the story much different and offers many opportunity of open theoretical and practical problems to researchers and engineers, like cooperation in distributed systems through relaying and network coding, efficient strategies for resource management. These issue are effectively addressed in Parts II and III of the book, again as the result of an evident cooperation between academia and industry.

In the fourth and last part, the book moves one step further in the practical realm by applying the previous results to a few system scenarios, like mobile-to-mobile and body-area communications.

While wishing to all the readers an enjoyable journey through this book, reading the Preface of Roberto Verdone, the Chairperson of COST 2100, I look forward to see the future book that will summarise the results of the next COST action entitled "Cooperative Radio Communications for Green Smart Environments"!

Torino, Italy
Sergio Benedetto
Full Professor, NEWCOM++ Director

Preface

This book contains most of the scientific results achieved within the framework of the European Cooperation in Science and Technology (COST) Action 2100 titled "Pervasive Mobile and Ambient Wireless Communications". The Action (see www.cost2100.org for details) was launched in December 2006 and finished in December 2010. Nearly 600 researchers affiliated to 142 research institutions (Universities, industries and research centres) from 35 countries were active at the end of the project; while most of them were from Europe, companies like Motorola, Azimuth Systems, NTT Docomo and others, were participating from North America and Asia. These researchers presented and discussed 880 scientific articles during the 12 meetings held within four years of activity. This book summarises and presents under a coherent view the contents of a large part of them; those addressing the scientific topics which attracted more interest during the four years brought to larger international visibility the activities of the Action because of the scientific relevance of the researchers involved and the topics.

The COST Action 2100 inherited from the previous Actions on mobile and wireless communications (COST 207, 231, 259, 273) many things: among these, a strong spirit of cooperation, a very large critical mass of researchers and institutions, a deep interaction between academia and industry.

The spirit of cooperation is the engine which allowed the achievement of results that a single research institution cannot reach: the book chapters dedicated to the radio channel present models based on the effort of many researchers, integrating different views and backgrounds; some of these chapters report measurements which were achieved through joint campaigns.

The critical mass around some specific topics, like MIMO or Radio Resource Management of 4G networks, allowed the determination of concerted views and the achievement of results of interest to a wide community of researchers; also, this permitted the creation of working groups addressing some specific topics, like body communications, of innovative nature.

The strong interaction between academia and industry made some of the working groups, like the one dealing with Over-The-Air antenna test methods, a reference point for the scientific world, able to interact and determine the choices made within standardisation bodies like 3GPP.

This book is not only the outcome of four years of activity; it is the intermediate step of a cooperative process started in the 1980s with COST 207, which will progress with the new COST Action born under the auspices of COST 2100: IC1004 "Cooperative Radio Communications for Green Smart Environments". This process contributed in the past to the development of radio systems like GSM and UMTS; with COST 2100, it contributed to the development of 4G networks and their subsystems. Therefore, this book is of interest to researchers, both in academia and industry, and PhD students who are willing to inherit part of the COST 2100 spirit and achievements, and extend them to the study and design of current and future generation mobile and wireless networks.

The book has been contributed by about one hundred researchers, whose names appear in the List of Contributors. I want to thank all of them and all the researchers involved in COST 2100, who made this story a success. The huge effort of summarising and making the contents coherent was in the hands of the Chapter Editors, to whom I am very grateful for their dedication. Their names are reported at beginning of each chapter; in almost all cases they were leading the working groups dedicated to the specific topics in COST 2100. Each chapter of the book has been revised by an external advisor, a scientist not directly involved in COST 2100, who provided comments and suggestions on how to improve the readability of the various chapters; on behalf of the whole group of Chapter Editors, I would like to acknowledge Hanna Bogucka, Carla-Fabiana Chiasserini, Ernst Bonek, Maxime Guillaud, Ove Linnell, Ignacio Llatser, Dirk Manteuffel, Marta Martinez-Vázquez, Andreas Molish, Juan Mosig, Sergio Palazzo, Jordi Perez-Romero, Josep Sole Pereira, Luc Vandendorpe. My warmest thank goes however to the Co-Editor of this Book, Alberto Zanella, who made most of the editorial job. Finally, since the book represents the ultimate outcome of four years of activities, I want to express my eternal gratitude to the people who shared with me all discussions on scientific, administrative, operational aspects of the Action during the four years, meeting regularly the day before each of the twelve COST 2100 events: the Vice-Chairman, Joerg Pamp, the former Chairman, Luis Correia, the three working group leaders, Alister Burr, Narcis Cardona and Alain Sibille, the Chair Assistant, Virginia Corvino, and the COST 2100 secretary, Silvia Zampese. Each of these people contributed somehow to the contents of this book and to a wonderful personal and professional experience. Finally, the Memorandum of Understanding jointly signed with NEWCOM++ (January 2008–April 2011), the FP7 Network of Excellence in Wireless Communications, should be mentioned; under its auspices, COST 2100 and NEWCOM++ shared resources, and jointly organised several successful events (workshops, training schools). NEWCOM++ is publishing for the same publisher, Springer, a Vision Book, which provides a concerted view on the research trends and millennium problems that need to be addressed in the next decade in the area of wireless communications; the NEWCOM++ Vision Book can be considered as a sort of completion of the framework addressed by this COST 2100 book.

Bologna, Italy　　　　　　　　　　　　　　　　　　　　　　　　　Roberto Verdone
　　　　　　　　　　　　　　　　　　　　　　　　　　　　　　COST 2100 Chairperson

Acknowledgements

Authors would like to thank COST office for its valued and continuous support.

Contents

1 Introduction 1
Chapter Editor Roberto Verdone

Part I Radio Channel

2 Channel Measurements 5
Chapter Editor Nicolai Czink, Alexis Paolo Garcia Ariza,
Katsuyuki Haneda, Martin Jacob, Johan Kåredal, Martin Käske,
Jonas Medbo, Juho Poutanen, Jussi Salmi, Gerhard Steinböck, and
Klaus Witrisal

3 Radio Channel Modeling for 4G Networks 67
Chapter Editor Claude Oestges, Nicolai Czink, Philippe De
Doncker, Vittorio Degli-Esposti, Katsuyuki Haneda, Wout Joseph,
Martine Liénard, Lingfeng Liu, José Molina-García-Pardo,
Milan Narandžić, Juho Poutanen, François Quitin, and Emmeric Tanghe

4 Assessment and Modelling of Terminal Antenna Systems 149
Chapter Editor Buon Kiong Lau, Chapter Editor Alain Sibille,
Vanja Plicanic, Ruiyuan Tian, and Tim Brown

5 "OTA" Test Methods for Multiantenna Terminals 197
Chapter Editor Gert F. Pedersen, Chapter Editor Mauro Pelosi,
Jan Welinder, Tommi Jamsa, Atsushi Yamamoto, Miia Nurkkala,
Soon L. Ling, Werner Schroeder, and Tim Brown

6 RF Aspects in Ultra-WideBand (UWB) Technology 249
Chapter Editor Grzegorz Adamiuk, Jens Timmermann,
Christophe Roblin, Wouter Dullaert, Philipp Gentner, Klaus Witrisal,
Thomas Fügen, Ole Hirsch, and Guowei Shen

Part II Transmission Techniques and Signal Processing

7 MIMO and Next Generation Systems 301
Chapter Editor Alister Burr, Ioan Burciu, Pat Chambers,
Tomaz Javornik, Kimmo Kansanen, Joan Olmos, Christian Pietsch,
Jan Sykora, Werner Teich, and Guillaume Villemaud

8 Cooperative and Distributed Systems 341
Chapter Editor Jan Sykora, Vasile Bota, and Tomaz Javornik

9 Advanced Coding, Modulation and Signal Processing 373
Chapter Editor Laurent Clavier, Dejan Vukobratovic, Matthias Wetz,
Werner Teich, Andreas Czylwik, and Kimmo Kansanen

Part III Radio Network Aspects

10 Deployment, Optimisation and Operation of Next Generation Networks 407
Chapter Editor Thomas Kürner, Paolo Grazioso, Andreas Eisenblätter,
Guillaume de la Roche, Fernando Velez, Andreas Hecker, Matias Toril,
Michal Wagrowski, Mario Garcia-Lozano, and Philipp P. Hasselbach

11 Resource Management in 4G Networks 461
Chapter Editor Silvia Ruiz Boqué, Chapter Editor Narcis Cardona,
Andreas Hecker, Mario Garcia-Lozano, and Jose F. Monserrat

12 Advances in Wireless Ad Hoc and Sensor Networks 519
Chapter Editor Paolo Grazioso, Velio Tralli, Pawel Kulakowski,
Andrea Carniani, and Lubomir Dobos

Part IV Broadcasting, Body and Vehicle Communications

13 Hybrid Cellular and Broadcasting Networks 547
Chapter Editor David Gomez-Barquero, Chapter Editor Peter Unger,
Karim Nasr, Jussi Poikonen, and Kristian Nybom

14 Vehicle-to-Vehicle Communications 577
Chapter Editor Christoph Mecklenbräuker, Laura Bernadó,
Oliver Klemp, Andreas Kwoczek, Alexander Paier, Moritz Schack,
Katrin Sjöberg, Erik G. Ström, Fredrik Tufvesson, Elisabeth Uhlemann,
and Thomas Zemen

15 Body Communications 609
Chapter Editor Arie Reichman, Chapter Editor Jun-ichi Takada,
Dragana Bajić, Kamya Y. Yazdandoost, Wout Joseph, Luc Martens,
Christophe Roblin, Raffaele D'Errico, Carla Oliveira, Luis M. Correia,
and Matti Hämäläinen

16 Wrapping Up and Looking at the Future 661
Chapter Editor Luis M. Correia

Index 673

Contributors

Grzegorz Adamiuk Karlsruhe Institute of Technology, Karlsruhe, Germany

Dragana Bajić University of Novi Sad, Novi Sad, Serbia

Laura Bernadó FTW Forschungszentrum Telekommunikation Wien, Vienna, Austria

Vasile Bota Technical University of Cluj-Napoca, Cluj-Napoca, Hungary

Tim Brown University of Surrey, Guildford, UK

Ioan Burciu Laboratory for Analysis and Architecture of Systems (LAAS), Toulouse, France

Alister Burr University of York, York, UK

Narcis Cardona Universidad Politècnica de Valencia, Valencia, Spain

Andrea Carniani University of Bologna, Bologna, Italy

Pat Chambers Université catholique de Louvain, Louvain-la-Neuve, Belgium

Laurent Clavier University of Lille, Lille, France

Luis M. Correia Istituto Superior Técnico (IST), Universidade Técnica de Lisboa, Lisbon, Portugal

Nicolai Czink Forschungszentrum Telekommunikation Wien (FTW), Vienna, Austria

Andreas Czylwik Universität Duisburg-Essen, Duisburg, Germany

Philippe De Doncker Université libre de Bruxelles, Brussels, Belgium

Vittorio Degli-Esposti University of Bologna, Bologna, Italy

Guillaume de la Roche University of Bedfordshire, Bedford, UK

Raffaele D'Errico CEA-LETI, Grenoble, France

Lubomir Dobos Technical University in Kosice, Kosice, Slovakia

Wouter Dullaert Ghent University, Ghent, Belgium

Andreas Eisenblätter atesio, GmbH, Berlin, Germany

Thomas Fügen Karlsruhe Institute of Technology, Karlsruhe, Germany

Mario Garcia-Lozano Universitat Politècnica de Catalunya, Barcelona, Spain

Alexis Paolo Garcia Ariza Ilmenau University of Technology, Ilmenau, Germany

Philipp Gentner Vienna University of Technology, Vienna, Austria

David Gomez-Barquero Universidad Politecnica de Valencia, Valencia, Spain

Paolo Grazioso Fondazione Ugo Bordoni (FUB), Bologna, Italy

Matti Hämäläinen CWC, University of Oulu, Oulu, Finland

Katsuyuki Haneda Aalto University, Espoo, Finland

Philipp P. Hasselbach Institute of Telecommunications, Communications Engineering Lab., Darmstadt University of Technolgy, Darmstadt, Germany

Andreas Hecker Technische Universität Braunschweig, Braunschweig, Germany

Ole Hirsch Ilmenau University of Technology, Ilmenau, Germany

Martin Jacob Technische Universität Braunschweig, Braunschweig, Germany

Tommi Jamsa Elektrobit, Oulu, Finland

Tomaz Javornik Department of Communication Systems, Jozef Stefan Institute, Ljubljana, Slovenia

Wout Joseph Ghent University, Ghent, Belgium

Kimmo Kansanen Norwegian University of Science and Technology, Trondheim, Norway

Johan Kåredal Lund University, Lund, Sweden

Martin Käske Ilmenau University of Technology, Ilmenau, Germany

Oliver Klemp BMW Forschung und Technik, Munich, Germany

Pawel Kulakowski AGH, University of Science and Technology, Krakow, Poland

Thomas Kürner Technische Universität Braunschweig, Braunschweig, Germany

Andreas Kwoczek Volkswagen AG, Wolsburg, Germany

Buon Kiong Lau Dept. of Electrical and Information Technology, LTH, Lund University, Lund, Sweden

Martine Liénard University of Lille, Lille, France

Soon L. Ling Vodafone Group Services Limited, Newbury, UK

Contributors

Lingfeng Liu Université catholique de Louvain, Louvain-la-Neuve, Belgium

Luc Martens Ghent University, Ghent, Belgium

Christoph Mecklenbräuker Vienna University of Technology (VUT), Vienna, Austria

Jonas Medbo Ericsson AB, Stockholm, Sweden

José Molina-García-Pardo Technical University of Cartagena, Cartagena, Spain

Jose F. Monserrat Universidad Politècnica de Valencia, Valencia, Spain

Milan Narandžić Ilmenau University of Technology, Ilmenau, Germany

Karim Nasr National Physical Laboratory, Middlesex, UK

Miia Nurkkala Nokia Corporation, Oulu, Finland

Kristian Nybom Åbo Akademi University, Turku, Finland

Claude Oestges Université catholique de Louvain (UCL), Louvain-la-Neuve, Belgium

Carla Oliveira Istituto Superior Técnico (IST), Universidade Técnica de Lisboa, Lisbon, Portugal

Joan Olmos Universitat Politècnica de Catalunya, Barcelona, Spain

Alexander Paier Vienna University of Technology (VUT), Vienna, Austria

Gert F. Pedersen Aalborg University, Aalborg, Denmark

Mauro Pelosi Aalborg University, Aalborg, Denmark

Christian Pietsch University of Ulm, Ulm, Germany

Vanja Plicanic Sony Ericsson Mobile Communication AB, Lund University, Lund, Sweden

Jussi Poikonen University of Turku, Turku, Finland

Juho Poutanen Aalto University, Espoo, Finland

François Quitin Université libre de Bruxelles, Brussels, Belgium

Arie Reichman Ruppin Academic Center, Kfar Saba, Israel

Christophe Roblin ENSTA Paris-Tech, Paris, France

Silvia Ruiz Boqué Universitat Politècnica de Catalunya, Barcelona, Spain

Jussi Salmi Aalto University, Espoo, Finland

Moritz Schack Technische Universität Braunschweig, Braunschweig, Germany

Werner Schroeder University of Applied Sciences Wiesbaden, Rüsselsheim, Germany

Guowei Shen Ilmenau University of Technology, Ilmenau, Germany

Alain Sibille Département Communications & Électronique, TELECOM ParisTech, Paris, France

Katrin Sjöberg Halmstad University, Halmstad, Sweden

Gerhard Steinböck Aalborg University, Aalborg, Denmark

Erik G. Ström Chalmers University of Technology, Gothenburg, Sweden

Jan Sykora Czech Technical University in Prague, Prague, Czech Republic

Jun-ichi Takada Tokyo Institute of Technology, Tokyo, Japan

Emmeric Tanghe Ghent University, Ghent, Belgium

Werner Teich University of Ulm, Ulm, Germany

Ruiyuan Tian Lund University, Lund, Sweden

Jens Timmermann Karlsruhe Institute of Technology, Karlsruhe, Germany

Matias Toril University of Malaga, Malaga, Spain

Velio Tralli University of Ferrara, Ferrara, Italy

Fredrik Tufvesson Lund University, Lund, Sweden

Elisabeth Uhlemann Halmstad University, Halmstad, Sweden

Peter Unger Technische Universität Braunschweig, Braunschweig, Germany

Fernando Velez University of Beira Interior, Covilhã, Portugal

Roberto Verdone DEIS, University of Bologna, Bologna, Italy

Guillaume Villemaud Institut National des Sciences Appliquées de Lyon, Lyon, France

Dejan Vukobratovic Faculty of Technical Sciences, University of Novi Sad, Novi Sad, Serbia

Michal Wagrowski AGH, University of Science and Technology, Krakow, Poland

Jan Welinder SP Technical Research Institute of Sweden, Borås, Sweden

Matthias Wetz University of Ulm, Ulm, Germany

Klaus Witrisal Graz University of Technology, Graz, Austria

Atsushi Yamamoto Panasonic Corporation, Kadoma, Osaka, Japan

Kamya Y. Yazdandoost National Institute of Information and Communications Technology, Kanagawa, Japan

Thomas Zemen FTW Forschungszentrum Telekommunikation Wien, Vienna, Austria

List of Acronyms

AAS	Advanced Antenna System
ACE	Approximated Cycle Extrinsic Message Degree
ACI	Adjacent Channel Interference
ACK	ACKnowledgment
AcS	Active Set
ADoA	Azimuth Direction of Arrival
ADPP	Azimuth-Delay Power Profile
AF	Amplify and Forward
AGC	Automatic Gain Control
A-GPS	Assisted GPS
AL-FEC	Application Layer FEC
AIR	Antenna Impulse Response
AMC	Adaptive Modulation and Coding
AoA	Azimuth of Arrival
AoD	Azimuth of Departure
AOA	Angle Of Arrival
AOD	Angle Of Departure
AP	Access Point
APP	Application Layer
APPb	A Posteriori Probability
APS	Angular Power Spectrum
ARA	Channel Adaptive Resource Allocation
ARQ	Automatic Repeat reQuest
AS	Angular Spread
ASO	Analogue Switch-Off
ASTM	American Society for Testing and Materials
ATS	Automatic Tuning System
AUT	Antenna Under Test
AWGN	Additive White Gaussian Noise
AZFML	Adaptive Zero Forcing Maximum Likelihood
AZFML-SfISfO-MIMO	Adaptive Zero Forcing Maximum Likelihood Soft Input Soft Output MIMO

BAN	Body Area Network
BANs	Body Area Networks
BAT	Bit Allocation Table
BC	Broadcast Channel
BCH	Bose–Chaudhuri–Hocquenghem
BE	Best Effort
BEC	Binary Erasure Channel
BER	Bit Error Rate
BFC	Block Fading Channel
BICM	Bit Interleaved Coded Modulation
B-IFDMA	Block-Interleaved Frequency Division Multiple Access
BLER	Block Error Rate
BMA	Broadcast/Multiple Access
BP	Broadcast/Point-to-Point
BPr	Belief Propagation
BPSK	Binary Phase Shift Keying
BS	Base Station
BSs	Base Stations
BSC	Binary Symmetric Channel
BSCo	Base Station Controller
BS-MIMO	Beam Steering MIMO
B3G	Beyond 3G
BSPC	BS Power Constraint
CRRM	Common Radio Resource Management
CAC	Call Admission Control
CALM	Communication Architecture for Land Mobiles
CAM	Cooperative Awareness Messages
CAP	Cell Assignment Probability
CAPEX	Capital Expenditures
CART	Classification And Regression Trees
CAS	Cluster Angular Spread
CAST	Compact Antennas Systems for Terminals
CAA	Channel and Application Aware
CBR	Constant Bit Rate
CC	Congestion Control
CCI	Co-Channel Interference
CDD	Cyclic Delay Diversity
CDF	Cumulative Density Function
CDMA	Code Division Multiple Access
CDS	Cluster Delay Spread
CEN	European Committee for Standardization
CF	Cost Function
CFOs	Carrier Frequency Offsets
CINR	Carrier to Interference plus Noise Ratio
CIR	Channel Impulse Response

List of Acronyms

CLD	Cross Layer Design
CMD	Correlation Matrix Distance
CNR	Carrier-to-Noise Ratio
COC	Cell Outage Compensation
COD	Cell Outage Detection
COM	Cell Outage Management
CoMP	Coordinated Multiple Point
COP-STC	Cooperative Overlay Pragmatic Space–Time Code
COST	European Cooperation in Science and Technology
CP	Cyclic Prefix
CPFSK	Continuous Phase Frequency Shift Keying
CPICH	Common Pilot Channel
CPM	Continuous Phase Modulation
CPR	Co-Polar Ratio
CQI	Channel Quality Indicator
CRC	Cyclic Redundancy Check
CS	Carrier Sensing
CSG	Closed Subscriber Group
CSI	Channel State Information
CSMA	Carrier Sense Multiple Access
CSMA/CA	Carrier Sense Multiple Access/Collision Avoidance
CTIA	Cellular Telecommunications & Internet Association
CTIR	Carrier to Interference Ratio
CW	Continuous Wave
C2C CC	Car-to-Car Communication Consortium
DCH	Dedicated Channel
DDF	Dynamic Decode and Forward
DENM	Decentralized Emergency Notification Message
DF	Decode and Forward
DF-DH	DF Dual-Hop
DFMM	Dual-Fed Microstrip Monopole
DFT	Discrete Fourier Transform
DGS	Defected Ground Structure
DL	Down-Link
DT-CDR	Delay-Tolerant Cooperative Diversity Routing
DMC	Dense Multipath Component
DoA	Direction of Arrival
DoD	Direction of Departure
DOSM	Differential Orthogonal Spatial Multiplexing
DP	Dominant Path
DPC	Dirty Paper Coding
DPSK	Differential Phase Shift Keying
DPTCAC	Downlink Power and Throughput based CAC algorithm
DR	Dynamic Range
DRA	Dynamic Resource Allocation

DS	Delay Spread
DSCH	Downlink Shared Channel
DSD	Doppler Spectral Density
DSL	Digital Subscriber Line
DSM	Dynamic Spectrum Management
DSP	Digital Signal Processor
DSRC	Dedicated Short Range Communication
DSSS	Direct Sequence Spread Spectrum
DTT	Digital Terrestrial TV
DTxAA	Double Transmission Antenna Array
DUT	Device Under Test
DVB	Digital Video Broadcasting
DVB-H	DVB-Handheld
DVB-SH	Digital Video Broadcasting-Satellite to Handheld
DVB-S2	Digital Video Broadcasting-Satellite-Second Generation
DVB-T	DVB—Terrestrial
DVB-T2	DVB—Second Generation Terrestrial
EDCA	Enhanced Distributed Channel Access
EDGE	Enhanced Data for Global Evolution
EESM	Exponential Effective Signal-to-Noise Mapping
EFIE	Electric Field Integral Equation
EG	Effective Gain
EGC	Equal Gain Combining
EIRP	Equivalent Isotropic Radiated Power
EM	Electromagnetic
EMA	Exponential Moving Average
eNB	Evolved NodeB
EoA	Elevation of Arrival
EoD	Elevation of Departure
EDoA	Elevation Direction of Arrival
EHF	Extra High Frequency
EKF	Extended Kalman Filter
ER	Effective Roughness
ESNR	Effective Signal-to-Noise Ratio
ESPAR	Electronically Steerable Passive Array Radiator
ESPRIT	Estimation of Signal Parameters via Rotational Invariance Techniques
ETC	Electronic Toll Collection
ETSI	European Telecommunications Standards Institute
EUT	Equipment Under Test
E-UTRA	Evolved UMTS Terrestrial Radio Access
E-UTRAN	Evolved UTRAN
EVM	Error Vector Magnitude
EWF	Expanding Window Fountain
EXIT	Extrinsic Information Transfer

List of Acronyms

EXP	Exponential Scheduling
FACH	Forward Access Channel
FCC	Federal Communications Commission
FCFS	First Come First Served
FDD	Frequency-Division Duplex
FDMA	Frequency Division Multiple Access
FDPF	Finite Difference ParFlow
FDTD	Finite Difference in Time Domain
FEC	Forward Error Correction
FER	Frame Error Rate
FF	Far Field
FFR4	Fractional Frequency Reuse 4
FFT	Fast Fourier Transform
FFR	Fractional Frequency Reuse
FIR	Finite Impulse Response
FMCW	Frequency Modulated Continuous Wave
FOM	Figure Of Merit
FPGA	Field Programmable Gate Array
FRA	Fixed Resource Allocation
FS	Fair Service
FSK	Frequency Shift Keying
FTP	File Transfer Protocol
GBSCM	Geometry Based Stochastic Channel Modeling
GERAN	GSM-EDGE Radio Access Network
GMP	Generalized Matrix-Pencil
GMSK	Gaussian Minimum Shift Keying
GNSS	Global Navigation Satellite System
GO	Geometrical Optics
GOF	Group Of Frequencies
GOP	Group Of Pictures
GoS	Grade of Service
GP	Geometric Programming
GPFS	Group Proportional Fair Scheduling
GPR	Ground Penetrating Radar
GPRS	General Packet Radio Service
GPS	Global Positioning System
GSCM	Geometry-Based Stochastic Channel Model
GSM	Global System for Mobile Communications
GTD	Geometrical Theory of Diffraction
JDRA	Joint Dynamic Resource Allocation
HAC	Hearing Aid Compatibility
H-ARQ	Hybrid ARQ
HCCA	Hybrid Coordination Function Controlled Channel Access
HD	High-Definition

HDTV	High-Definition Television
HNNs	Hopfield Neural Networks
HO	HandOver
HOL	Head-Of-the-Line
HSDPA	High-Speed Downlink Packet Access
HSPA	High-Speed Packet Access
HSPA+	HSPA Evolved
HSUPA	High-Speed Up-link Packet Access
HW	HardWare
MCD	Multipath Component Distance
IARS	Interference-Aware Relay Selection
ICI	Inter-Carrier Interference
ICIC	Inter-Cell Interference Coordination
ICT	Information and Communications Technologies
IDM	Interleave Division Multiplexing
IEEE	Institute of Electrical and Electronics Engineers
IFA	Inverted-F Antenna
ISM	Industrial, Scientific and Medical
IFDMA	Interleaved Frequency Division Multiple Access
IFE	In-Flight-Entertainment
iid	independent identically distributed
IP	Internet Protocol
IPDC	Internet Protocol DataCast
IPv6	Internet Protocol version 6
IQHA	Intelligent Quadrifilar Helix Antenna
IRA	Irregular Repeat Accumulate
IR	Impulse Response
IR-UWB	Impulse Radio Ultra-WideBand
ISI	Inter-Symbol Interference
ISO	International Organization for Standardization
ITRD	International Transport Research Documentation
ITS	Intelligent Transportation Systems
I2I	Indoor-to-Indoor
I2O	Indoor-to-Outdoor
KPI	Key Performance Indicator
LA	Location Area
LAd	Link Adaptation
LB	Load Balancing
LD	Laser Diode
LDGM	Low Density Generator Matrix
LDPC	Low Density Parity Check Code
LLR	Log-Likelihood Ratio
LO	Local Oscillator
LOS	Line-Of-Sight
LMMSE	Linear Minimum Mean Square

List of Acronyms

LMRA	Linear Multiple Regression Analysis
LNA	Low-Noise Amplifier
LSA	Local Service Area
LSF	Local Scattering Function
LTE	3GPP-Long Term Evolution
LTE-Adv	3GPP-LTE Advanced
LU	Lund University
LUD	Location Update
LUC	Land Use Class
LUT	Look-Up-Table
LWDF	Largest Weighted Delay First
L2S	Link-to-System
MAC	Media Access Control
MACh	Multiple Access Channel
MAI	Multi-Access Interference
MANET	Mobile Ad Hoc Network
MAP	Mesh Access Point
MAS	Multiple Antenna System
MaxTP	Maximum Throughput
MBMS	Multimedia Broadcast Multicast Service
MB-OFDM	Multibeam Orthogonal Frequency Division Multiplexing
MC-CAFS	Multicarrier Cyclic Antenna Frequency Spreading
MC-CDMA	MultiCarrier CDMA
MCP	Multi Cell Processing
MCS	Modulation and Coding Scheme
MEG	Mean Effective Gain
MFIE	Magnetic Field Integral Equation
MFN	Multi Frequency Network
MFSK	M-ary Frequency Shift Keying
MGT	Multi-Gigabit Transceiver
MH	Metropolis-Hastings
MI	Mutual Information
MIESM	Mutual Information Effective Signal to Noise Mapping
MIH	Media Independent Handover
MIMO	Multiple-Input Multiple-Output
MISO	Multiple-Input Single-Output
ML	Maximum Likelihood
MLBS	Maximum Length Binary Sequence
M-LWDF	Modified-Largest Weighted Delay First
MMSE	Minimum Mean Squared Error
mm-W	millimeter-waves
m-OCDM	m-Orthogonal Code Division Multiplexing
MODE	Method Of Direction Estimation
MOMENTUM	Models and Simulations for Network Planning and Control of UMTS

MORANS	MObile Radio Access Reference Scenarios
MPC	Multi-Path Component
MPCs	Multi-Path Components
MPE-FEC	Multi-Protocol Encapsulation FEC
MPEG	Moving Picture Experts Group
MPS	Multi Path Simulator
MRC	Maximum Ratio Combining
MS	Mobile Station
MSD	Multiple Symbol Detection
MSE	Mean Square Error
MSK	Minimum Shift Keying
MTC	Mobile Terminated Call
MU	Multiuser
MU-MIMO	MUltiuser Multiple-Input Multiple-Output
MUSIC	MUltiple SIgnal Classification
MWB	Multimedia Web Browsing
NEWCOM++	Network of Excellence in Wireless COMmunications
NC	Network Coding
NLOS	Non-Line-Of-Sight
NLT	Network Lifetime
NMS	Network Management System
NPC	Network Power Constraint
NRW	Normal Random Walk
ODMP	Orthogonal Diversity-Multiplexing Precoding
OFDM	Orthogonal Frequency Division Multiplexing
OFDMA	Orthogonal Frequency-Division Multiple Access
O-LOS	Obstruction-LOS
OLS	Ordinary Least Squares
OPEX	Operational Expenditures
OQAM	Offset Quadrature Amplitude Modulation
ORA	Optimal Relay Assignment
OSTBC	Orthogonal Space–Time Block Code
OTA	Over-The-Air
OTDOA	Observed Time Difference of Arrival
OVSF	Orthogonal Variable Spreading Factor
PARC	Per Antenna Rate Control
PAPR	Peak-to-Average Power Ratio
PAS	Power Angular Spectrum
PC	Personal Computer
PCo	Power Control
PCS	Performance Counter Statistic
PCU	Packet Control Unit
PD	Photo Diode
PDA	Personal Digital Assistant
PDF	Probability Density Function

PDP	Power-Delay Profile
PDPr	Packet Data Protocol
PDU	Protocol Data Unit
PEG	Progressive Edge Growth
PEP	Pairwise Error Probability
PER	Packet Error Rate
PF	Particle Filter
PFS	Proportional Fair Scheduling
PHY	Physical layer
PI	Performance Indicator
PIFA	Planar Inverted F-Antenna
PL	Path Loss
PLP	Physical Layer Pipe
PMI	Precoding Matrix Indicator
PN	Pseudo Noise
PP	Strict Two-Hop
PPM	Pulse Position Modulation
PRB	Physical Resource Block
PSD	Power Spectral Density
PSNR	Peak Signal-to-Noise Ratio
P-STC	Pragmatic Space–Time Code
PTP	Point to Point
P2P	Peer-to-Peer
PTM	Point to Multi-Points
PUSC	Partial Usage of Sub-Channels
QAM	Quadrature Amplitude Modulation
QEF	Quasi Error Free
QHA	Quadrifilar Helix Antenna
QoE	Quality of Experience
QoS	Quality-of-Service
QPSK	Quaternary Phase Shift Keying
RA	Relay Assignment
RAA	Relay-Assignment Algorithm
RAN	Radio Access Network
RARE	RAnk Reduction Estimator
RAT	Radio Access Technology
RBIR	Received Bit Information Rate
RCM	Random Cluster Model
RC-IRA	Rate-Compatible Irregular Repeat-Accumulate codes
RCS	Radar Cross Section
RF	Radio Frequency
RFID	Radio Frequency Identification
RI	Rank Indicator
RIMAX	Iterative Gradient-Based ML Parameter Estimation Algorithm

RLC	Radio Link Control
RMS	Root-Mean-Square
RMSE	Root Mean Square Error
RN	Relay Node
RNC	Radio Network Controller
RP-ACF	Received Pulse AutoCorrelation Function
RP-CCF	Received Pulse CrossCorrelation Function
RR	Round Robin
RRA	Random Resource Allocation
RRM	Radio Resource Management
RS	Reed–Solomon
RSCP	Received Signal Code Power
RSRP	Receive Signal Reference Power
RSS	Received Signal Strength
RSSI	Received-Signal Strength Indicator
RSU	Road Side Unit
RT	Ray Tracing
RTT	Round Trip Time
RTTT	Road Transport and Traffic Telematics
RU	Resource Unit
RV	Random Variable
Rx (or RX)	Receiver
SAGE	Space-Alternating Generalized Expectation-maximization
SAM	Specific Anthropomorphic Mannequin
SAR	Specific Absorption Rate
SC	Selection Combining
SCM	Spatial Channel Model
SCME	3GPP Spatial Channel Model Extended
S-CDR	Synchronous Cooperative Diversity Routing
SEM	Singularity Expansion Method
SER	Symbol Error Rate
StCM	Stochastic Channel Modeling
SC-FDMA	Single Carrier Frequency Division Multiple Access
SDCCH	Stand-alone Dedicated Control CHannel
SDMA	Space Division Multiple Access
SDRA	Semi-Distributed Relay Assignment
SEACORN	Simulation of Enhanced UMTS Access and Core Networks
SF	Shadow Fading
SFBC	Space-Frequency Block-Code
SFC	Space-Frequency Coding
SFN	Single-Frequency Network
SFE	Spatial Fading Emulator
SHB	Shadowing by Human Bodies

List of Acronyms

SHO	Soft HandOver
SIC	Successive Interference Cancellation
SIMO	Single-Input Multiple-Output
SINR	Signal-to-Interference-plus-Noise Ratio
SISO	Single-Input Single-Output
SIR	Signal-to-Interference Ratio
SLOS	Semi-Line-Of-Sight
SM	Spatial Multiplexing
SMEM	Spherical Mode Expansion Method
SNR	Signal-to-Noise Ratio
SoA	State of the Art
SON	Self Organization
SOTIS	Self-Organizing Traffic Information System
SPA	Self-Positioning Algorithm
SPrA	Sum-Product Algorithm
SRA	Sequential Relay Assignment
SiRA	Simple Resource Allocation
SSV	Security Service Vector
STC	Space–Time Coded
STBC	Space–Time Block-Code
STD	Short-Term Dynamic
STDMA	Self Organising Time Division Multiple Access
SU	Single User
SUMO	Simulation of Urban Mobility
SURA	Sequential Unique Relay Assignment
sub-mm-W	sub-millimeter-waves
SVC	Scalable Video Coding
SW	SoftWare
TB	Transport Block
TC	Technical Committee
TCAC	Throughput-Based CAC algorithm
TD	Transmit Diversity
TDD	Time Division Duplex
TF	Transfer Function
TDMA	Time-Division Multiple Access
TDoA	Time Delay of Arrival
TDR	Transmission Data Rate
TGad	IEEE P802.11 Task Group AD in Very High Throughput in 60 GHz
TIS	Total Isotropic Sensitivity
TH	Time Hopping
THP	Tomlinson–Harashima Precoding
TLM	Transmission Line Method
TLP	Traffic Loading Platform
ToA	Time of Arrival

TR	Transition Region
TRP	Total Radiated Power
TRS	Total Radiated Sensitivity
TS	Transport Stream
TTI	Transmission Time Interval
TTL	Time-To-Live
TTT	Time-To-Trigger
TV	Television
TVSHB	Time-Varying Shadowing by Human Bodies
Tx (or TX)	Transmitter
TxAA	Single-stream Transmit Antenna Array
UCA	Uniform Circular Array
UE	User Equipment
UEP	Unequal Error Protection
UF	User Fairness
UK	United Kingdom
UL	Up-Link
ULA	Uniform Linear Arrays
UMTS	Universal Mobile Telecommunications System
US	Uncorrelated Scattering
USB	Universal Serial Bus
USF	User's Simultaneous Factor
UT	User Terminal
UTA	Uniform Triangular Array
UTD	Uniform Theory of Diffraction
UTRAN	UMTS Terrestrial Radio Access Network
UWB	Ultra-WideBand
VANET	Vehicular Ad Hoc Network
VBR	Variable Bit Rate
VHO	Vertical HandOver
VHP	Visual Human Project
VNA	Vector Network Analyzer
VoIP	Voice over Internet Protocol
VR	Visibility Region
VSC	Vehicular Security Communication
VSG	Vector Signal Generator
VTE	Video Telephony
V2I	Vehicle-to-Infrastructure
V2V	Vehicle-to-Vehicle
V2X	Vehicle-to-X
WAVE	Wireless Access in Vehicular Environments
WCN	Wireless Communications Networks
WCDMA	Wideband Code Division Multiple Access
WFS	Wireless Fair Service
WiMAX	Worldwide Interoperability for Microwave Access

WINNER	Wireless World Initiative New Radio
WLAN	Wireless Local Area Network
WLF-CG	Worst-Link-First Coding-Gain
WLF-PL	Worst-Link-First Path Loss
WG	Working Groups
WiFi	Wireless Fidelity
WMN	Wireless Mesh Network
WMNs	Wireless Mesh Networks
WNC	Wireless Network Coding
WPAN	Wireless Personal Area Network
XPD	Cross-Polar Discrimination
WPT	Wireless Planning Tool
WSAN	Wireless Sensor and Actuator Network
WSNs	Wireless Sensor Networks
WSS	Wide-Sense Stationary
WSSUS	Wide-Sense Stationary Uncorrelated Scattering
WWAN	Wireless Wide Area Network
XPR	Cross-Polarization Ratio
ZF	Zero Forcing
2D	Two-Dimensional
2G	Second Generation
2WRC	2-Way Relay Channel
3D	Three-Dimensional
3G	Third Generation
3GPP	3rd Generation Partnership Project
3GPP-SCM	3GPP Spatial Channel Model
3GPP2	3rd Generation Partnership Project 2
4G	Fourth Generation

Chapter 1
Introduction

Chapter Editor Roberto Verdone

Mobile radio communications, since the first deployments over wide areas during the 1980s, have helped society in increasing self-awareness, social and individual security. After the provision of television broadcasting, and before the success of the World Wide Web, cellular networks have represented a step forward in the progress of humanity, when GSM (the Global System for Mobile Communications) was launched during the 1990s. More recently, cellular networks have started integration with TV broadcasting systems and the Web.

Mobile radio communications faced an important evolution during the years of COST2100 (December 2006–December 2010), with the standardization of 3GPP-Long Term Evolution (LTE): new air interfaces based on multi-carrier techniques at the physical layer were defined, with strong implications on the algorithms for radio resource management. Moreover, Multiple-Input Multiple-Output (MIMO) systems, formerly relegated to laboratory level, have became a reality and among the key technology enablers for broadband radio access in many standardized systems. The radio access has therefore been significantly improved in terms of speed, allowing the provision of new services to mobile users.

In the meanwhile, the pervasiveness of radio communications, determined by the availability of small, low-cost and energy-efficient radio devices, has raised the interest towards new applications of wireless communications, ranging from sensor networks for environmental monitoring to body communication systems. After Bluetooth was launched several years ago, a new generation of radio standards, like IEEE802.15.4, 4a and 16, has made the field of short-range communication systems and Personal Area Networks very attractive for industries.

COST2100 addressed these technology trends, both at the large scale of mobile radio communications and at the small scale of Personal Area Networks, providing support to the evolution of mobile and wireless communications.

In this field, proper design of transmission technologies and network algorithms require accurate knowledge of the radio channel characteristics, as it is the core of

R. Verdone (✉)
DEIS, University of Bologna, Bologna, Italy

all wireless communication systems. On the other hand, many among the worldwide recognized experts of radio channel modeling and characterization, were involved in COST2100. So, the Action could play a key role in the definition of the techniques implemented in recent standards.

The book, which addresses all types of radio communication systems, is basically organized in four parts.

The first one deals with the radio channel and comprises Chap. 2, dedicated to channel measurements, Chap. 3, which provides modeling of the radio channel for 4G networks, Chap. 4, addressing terminal antenna systems, Chap. 5, reporting on the activities related to "Over-The-Air" test methods for multiantenna terminals, and Chap. 6, addressing Radio Frequency (RF) aspects of Ultra-WideBand (UWB) technologies.

The second part is related to signal processing and transmission techniques: it includes Chap. 7, dedicated to MIMO technologies, Chap. 8, which addresses cooperative technologies, and Chap. 9, dealing with advanced modulation and coding techniques.

The third part is more oriented to network aspects: Chap. 10 describes the potentials of next generation cellular networks, Chap. 11 addresses measurement-based optimization of wireless networks, Chap. 12 is dedicated to radio resource management issues for 4G networks, and Chap. 13 to ad hoc and sensor networks.

Finally, Chaps. 14, 15 and 16 address the application of the models and the techniques discussed in the other parts of the book to specific environments/application contexts: hybrid cellular and broadcasting networks, vehicle-to-vehicle and body communications.

The very final chapter of the book shows a vision on the future of research in mobile and wireless communications.

The book is the outcome of the research effort provided by hundreds of institutions and researchers during four years. It is in the belief of all those who participated to COST2100 that the book will represent an other humble contribution to the progress of humanity.

Part I
Radio Channel

Chapter 2
Channel Measurements

Chapter Editor Nicolai Czink, Alexis Paolo Garcia Ariza, Katsuyuki Haneda, Martin Jacob, Johan Kåredal, Martin Käske, Jonas Medbo, Juho Poutanen, Jussi Salmi, Gerhard Steinböck, and Klaus Witrisal

Radio channel measurements and channel modeling have always been a driving force of the wireless COST actions. These high spirits have been kept up also in COST 2100, where significant advances, particularly in channel measurements, have been made.

Channel measurements are indispensable for wireless system design. It is the wireless channel that determines the ultimate performance limits of any communication system. In the beginnings of cellular communications, fading and path loss of the narrowband channel were the key figures of merit. This has changed with wideband multiantenna, multiuser systems. New important features of the radio channel became obvious: the channels' frequency selectivity, directivity, polarimetric properties, and their relation to channels of the other users.

Also for emerging systems, the channel properties need to be identified and modeled: Peer-to-peer systems, which employ distributed nodes indoors or outdoors, show surprising fading characteristics. The propagation channels between vehicles suffer from severe fading under high Doppler shifts. Ultrawideband systems provide an extremely high delay resolution, which needs to be captured. Finally, radio systems using millimeter and submillimeter waves experience strong shadowing by human bodies. All these effects need to be measured and modeled accordingly.

Once measurements have been done, and researchers obtained an abundance of data, these measurements need to be analyzed in a sound way. For modeling purposes it is often advantageous to separate the effects of the antenna from the radio channel. This can be achieved by high-resolution parameter estimation, where the channel is characterized as a superposition of multiple wave fronts. Using this data along with maps or photographs of the environment, scattering objects can be automatically located, which provides a deep insight into the processes going on in the radio channel. However, high-resolution parameter estimation methods have several pitfalls, depending on the calibration of the measurement equipment. Only with the knowledge of possible pitfalls, these can be avoided.

N. Czink (✉)
Forschungszentrum Telekommunikation Wien (FTW), Vienna, Austria

Due to the large number of successful measurement campaigns in all these areas, and the consequential results, it is impossible to describe all the findings in depth. This chapter rather presents a summary of the measurements methods including respective examples and provides references for further reading.

The first part of this chapter gives an overview of recent advances in channel measurements in Sect. 2.1. Starting with measuring distributed, cooperative communication systems in Sect. 2.1.1, the discussion continues with the specificities of polarimetric channel measurements in Sect. 2.1.2. Next, the challenges of measuring channels with high mobility are described in Sect. 2.1.3. Subsequently, the methods of ultrawideband channel sounding are shortly outlined in Sect. 2.1.4. Finally, the measurements and channel characteristics of millimeter and submillimeter channels are discussed in depth in Sect. 2.1.5.

The second part of this chapter in Sect. 2.2 is devoted to recent methods of measurement analysis. First, different approaches for high-resolution parameter estimation are discussed in Sect. 2.2.1. Subsequently, Sect. 2.2.2 describes how to localize and visualize scattering in the environment. Finally, the pitfalls of high-resolution parameter estimation are pointed out in Sect. 2.2.3. Finally, Sect. 2.3 draws the conclusions from this chapter by walking through the lessons learned.

The evaluation of the measurement data and consecutive radio channel modeling is not part of this chapter. These aspects are addressed in the following chapters, specific to the considered systems.

2.1 Advances in Channel Measurements

This section discusses the latest methods to measure the radio channels of different kinds of systems. Due to the broad nature of the research in recent years, a number of different topics is addressed: Distributed cooperative systems, the polarimetric channel, vehicular channels under high mobilities, ultrawideband channel sounding, and millimeter and submillimeter wave systems.

2.1.1 Measuring Distributed Systems

We use the term "distributed systems" for the whole range from distributed ad hoc networks to cooperating base stations in a cellular system. All these systems have in common that they are typically wideband, make use of multiple antennas at each node and suffer from interference. These properties need to be measured and modeled accordingly. In this section, we focus on the different ways to measure distributed systems.

Generally, the techniques for measuring distributed systems can broadly be split into three methods:

Single-sounder sequential measurements In these measurements, a single-channel sounder equipment is used sequentially along multiple routes to mimic multiple

2 Channel Measurements

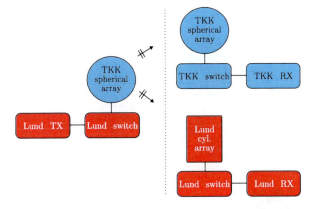

Fig. 2.1 Setup of the WILATI indoor multiuser MIMO measurements [KAS+] (©2010 IEEE, reproduced with permission)

users in a sufficiently static scenario. Another possibility is to measure the same route sequentially with different base station positions. For this reason, this kind of measurement is sometimes also called "multipass measurements."

Single-sounder multinode measurements Again, only a single-channel sounder is used, however multiple nodes are connected to the sounder by long RF cables. In this way, the nodes can be distributed across multiple rooms in a building (or even farther when using optical RF cables).

Multi-sounder measurements This type of measurements employs multiple channel sounders, where typically a single transmitter and multiple receivers are used.

Since there are multiple ways to implement these three methods, this section will provide an overview of recent measurement activities in *different scenarios* explaining the key ideas, the measurement method, and an exemplary result. The following paragraphs are organizing the scenarios into *multinode* (multiuser) measurements and *multibase-station* measurements.

2.1.1.1 Multinode Measurements

Multinode measurements comprise both measurements with a single base station and multiple users, and also peer-to-peer measurements.

Indoor Multiuser MIMO With huge technological effort, Aalto University and Lund University (LU) married their fundamentally different channel sounders to do multiuser Multiple-Input Multiple-Output (MIMO) measurements [KAS+, RTR+08] in the WILATI project. The general system setup is shown in Fig. 2.1.

The LU sounder is a commercial RUSK channel sounder produced by MEDAV GmbH [cha10]. It uses fast Radio Frequency (RF) switching and periodic multifrequency signals. The transmitter of the LU sounder has an arbitrary waveform generator unit. The receiver equipment is down converting, sampling, and processing the signal.

Table 2.1 Lund University and Aalto University sounder specifications

Parameter	LU sounder	Aalto sounder
Center frequency	5150–5750 MHz	5300 MHz
Bandwidth	10–240 MHz	120 MHz
Tx-code length	1.6 µs	N/A
Sampling rate at Tx	320 MHz	N/A
Rx-element switching interval	3.2 µs	3.2 µs
Time between MIMO snapshots	39.3216 ms	39.3216 ms

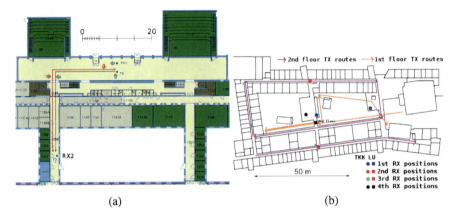

Fig. 2.2 Scenarios of the WILATI measurements at (**a**) Lund University [KAK+07] (©2007 EurAAP, reproduced with permission). (**b**) Aalto University [RTR+08]

The receiver of the Aalto sounder consists just of a down converter and sampling unit. All the signal processing is done in a postprocessing step and was adjusted to cope with the LU transmitter's signals.

A special procedure for achieving synchronization between the different sounder architectures was necessary for this specific setup. A comparison of the sounder specifications is provided in Table 2.1. It is noteworthy that both sounders used calibrated antenna arrays, which enabled high-resolution estimation of the propagation paths in the channel, as discussed in Sect. 2.2.1. The indoor scenarios included dynamic measurements along routes on corridors (see Fig. 2.2a), and in big multistorey halls (see Fig. 2.2b).

A number of interesting findings resulted from these measurements. Most prominently, the analysis of the location of scatterers (see Fig. 2.30) lead to the investigation of common clusters (see Sect. 3.5.3.5). Additionally, interlink correlations were analyzed from the measurements in [KHH+10].

Outdoor Multiuser Measurements Eurecom employed their own equipment, the Eurecom MIMO Open-Air Sounder (EMOS), for outdoor measurements of multiuser MIMO channels [KKC+08]. The platform consists of a BS that continuously

Table 2.2 Eurecom EMOS sounder specification

Parameter	Eurecom EMOS
Center frequency	1917.6 MHz
Bandwidth	4.8 MHz
BS transmit power	30 dBm
Number of antennas (BS)	4 (2 cross-polarized)
Number of antennas (UE)	2
Number of OFDM subcarriers	160

sends a signaling frame, and one or more User Equipments (UEs) receive these frames to estimate the channel. For the BS, a PC with four PLATON data acquisition cards is used along with a Powerwave 3G broadband antenna. The UEs are ordinary laptop computers with Eurecom's dual-RF CardBus/PCMCIA data acquisition cards fed by two clip-on 3G Panorama Antennas (all antennas uncalibrated). While the platform is designed for a full software-radio implementation, it is also well suited for channel sounding. The parameters of this system are provided in Table 2.2.

The sounder is based on over-the-air synchronization and channel estimation by an Orthogonal Frequency Division Multiplexing (OFDM) sounding sequence. As shown in Fig. 2.3, one transmit frame is 2.667-ms long and consists of a synchronization symbol (SCH), a broadcast data channel (BCH) comprising 7 OFDM symbols, a guard interval, and 48 pilot symbols from a random QPSK sequence used for channel estimation. The subcarriers of the pilot symbols are multiplexed over the four transmit antennas to ensure orthogonality in the spatial domain.

A sample scenario investigated in these measurements was outdoor in a semiurban hilly terrain in Sophia-Antipolis, France. The base station was mounted on the rooftop of a building and four mobile stations were placed in passenger cars (see Fig. 2.4a). As an exemplary result, a comparison of the sum rate capacity of Single User (SU)-MIMO and Multiuser ((MU))-MIMO is given in Fig. 2.4(b). The results show that, both for measured and simulated channels, MU-MIMO provides a higher sum rate capacity than SU-MIMO Time-Division Multiple Access (TDMA).

These measurements also provide the basis for the investigations presented in Sect. 3.5.2.4. The EURECOM sounder was also extended to the 800 MHz band,

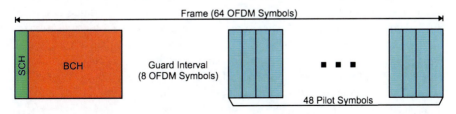

Fig. 2.3 Frame structure of the Eurecom MIMO OpenAir Sounder (EMOS) [KKC+08] (©2008 IEEE, reproduced with permission)

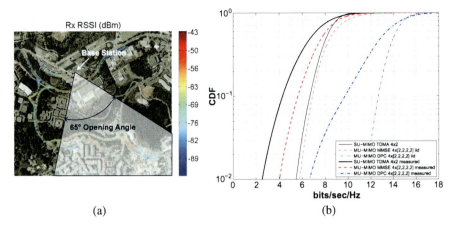

Fig. 2.4 Outdoor multiuser MIMO measurements using the Eurecom sounder; (**a**) received signal strength in the scenario (©2008 IEEE, reproduced with permission). (**b**) Capacity of different multiuser MIMO communication schemes in this scenario [KKC+08]

in which it was used for the comparison of LTE transmission modes in rural areas [KGL+10].

Outdoor-to-Indoor Multiuser MIMO In this multiuser scenario, FTW and Stanford University used sequential measurements using a single-channel sounder [CBVV+08b]: They demonstrated that, given the *scenario* is static, multiuser measurements can be performed sequentially, even if the users themselves are moving [CBVV+08a].

As equipment, the RUSK Stanford Channel sounder produced by MEDAV GmbH [cha10] was used. The sounder architecture is similar to the Lund University equipment presented before. Additionally, a distance wheel was used to trigger the acquisition of MIMO snapshots every 1.6 cm of receiver movement. At the base station, two dual-polarized WiMAX base-station antennas were used as transmitters. At the receiver side, three types of commercial WiMAX dual-antenna MIMO arrays were used: a customer-premises equipment (CPE) array, a PC-card array, and a USB-stick array. Additionally, two discone antennas were used as reference array. The parameter settings of the measurements are provided in Table 2.3.

Figure 2.5(a) shows the map of the outdoor scenario with two Tx positions and the corresponding Tx locations. At every position, the BS was rotated into three different directions. Indoors, the measurements were done along five routes for each transmitter location and orientation as indicated in Fig. 2.5(b). These routes were maintained with meticulous precision (maximum deviation of 2 cm) to ensure the repeatability of the measurements. The indoor environment is a cubicle-style office, where the cubicles consisted of cardboard walls with metal frames. A glass wall was along Route 3, supported by a metallic structure.

It should be noted that the measurement data of this campaign is publicly available upon request [CBVV+08b].

2 Channel Measurements

Table 2.3 RUSK Stanford sounder specification for multiuser MIMO measurements

Stanford sounder parameters	MU-MIMO	O2I distr.	I2I static	I2I mobile
Center frequency	2.45 GHz			
Bandwidth	240 MHz (70 MHz used)			
Tx code length	3.2 µs			
MIMO snapshots spacing	1.6 cm = 0.13λ	250 ms	250 ms	9.8 ms
Number of antennas (BS)	4 (2 x-pol)	4 (2 x-pol)	8	8
Number of antennas (MS)	4 arrays × 2 ant.	8	8	8

Fig. 2.5 Outdoor-to-indoor multiuser MIMO measurements [CBVV+08b]; (**a**) outdoor scenario (©2011 Google-Imagery ©DigitalGlobe, USDA Farm Service Agency, GeoEye, U.S. Geological Survey, MapData ©2011 Google), (**b**) indoor routes

To demonstrate the feasibility of the single-sounder approach, the channel matrices from two consecutive measurement runs along the same route were compared using the matrix collinearity measure (3.22).

Figure 2.6 demonstrates that the measured channels showed a strong collinearity in the distance range between 5 and 10 meters when the measurements of the two runs were properly aligned with each other. When using the collinearity measure on the channel *correlation matrices* (instead of the channel matrices), a high collinearity (larger than 0.9) is achieved throughout the whole measurement route. This demonstrated that, by appropriate measures, the environment was sufficiently static to perform multiuser channel measurements with just a single-channel sounder.

These measurements were used to derive and analyze a subspace-based model of the multiuser MIMO channel (see Sect. 3.5.2.1).

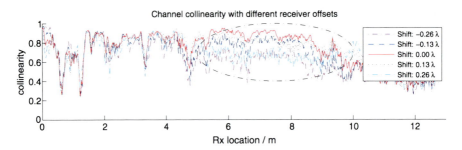

Fig. 2.6 Validation of doing multipass measurements: The channels matrices are collinear over a longer range when the environment does not change between runs [CBVV+08a]

Outdoor-to-Indoor and Indoor-to-Indoor Distributed Nodes Two similar measurement campaigns of the radio channel between indoor nodes distributed throughout a building and between the nodes and an outdoor base station were first presented in [CBVV+08b] (US-style office building with cubicles) and were succeeded by a follow-up campaign [CCL+10] (European-style office building with brick-walls).

The basic idea in both campaigns was to use a single switched-array channel sounder but connect long low-loss RF cables to the RF switches. In this way, the measured antenna "array" consisted of the antennas at the distributed nodes. Naturally, this approach results in a lower measurement SNR due to the cable losses, which needs to be compensated for in a postprocessing step.

The measurements in [CBVV+08b] were performed with the RUSK Stanford channel sounder produced by MEDAV GmbH [cha10]. Three different scenarios were measured: (i) outdoor-to-indoor (O2I) static, (ii) indoor-to-indoor (I2I) static, and (iii) I2I mobile. The sounder parameters for each of the measurements are provided in Table 2.3.

The scenario is shown in Fig. 2.7(a). Both static and mobile measurements were performed. In the static measurements, the antennas were at fixed positions while fading was generated by people that were moving throughout the room. For the mobile measurements, specific antennas were moved locally. From these measurements a comprehensive O2I and I2I model for distributed channels was designed [OCB+10].

The follow-up campaign presented in [CCL+10] was using an Elektrobit Propsound CS channel sounder [Ele10, HKY+05]. This sounder is also a switched-array MIMO channel sounder but using pseudorandom sequences for probing the channel. At the receiver, a matched-filtered impulse response can be obtained by postprocessing.

In this campaign a larger number of static and mobile scenarios were measured, both outdoor-to-indoor and indoor-to-indoor. Additionally, indoor nodes were occasionally equipped with multiple antennas. An exemplary map of indoor measurements is depicted in Fig. 2.7(b).

From these measurements a statistical model for outdoor-to-indoor and indoor-to-indoor distributed links was developed [OCB+10] (cf. Sect. 3.5.4.2).

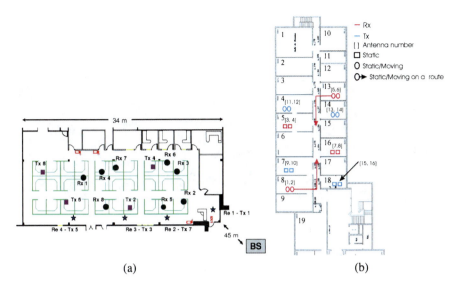

Fig. 2.7 Distributed node measurement scenarios: (**a**) Stanford campaign—*black dots* denote receivers (O2I & I2I), *stars* denote "relays" (Rx for O2I, Tx for I2I), and *squares* denote transmitters (I2I) [OCB+10] (©2010 IEEE, reproduced with permission). (**b**) UCL campaign [CCL+10]—exemplary measurement

Outdoor Relay Channel An intermediate step toward the measurements of multiple base stations was performed by Bristol University [WWW+07]. In their study, they considered a realistic environment of outdoor BS-to-relay and relay-to-UE channels, as addressed in the IEEE 802.16j standard for WiMAX relaying. Both at the BS and at the Relay station, two signal generators were used, transmitting a CW at 3.59 GHz and 3.47 GHz, respectively. As UE, a spectrum analyzer was used to quantify the propagation loss for both frequencies. The UE was mounted on a trolley and pushed along several routes to collect samples for estimating the path loss. For the path loss on the BS-UE link, the COST 231 Walfish–Ikegami model was found to provide the closest match to the measured data. For the relay-UE link, the conventional models are farther off, which implies that the development of models with more localized parameters is called for. Finally, the BS-relay link showed a tight fit to the IEEE 802.16d [IEE04] model, with deviations due to the antenna heights.

Orange Labs [CCC10a, CCC10b] performed two extensive studies where they included the outdoor-to-indoor link. In their first study [CCC10a], a narrowband signal was used as well, but at three carrier frequencies (900 MHz, 2.1 GHz, and 3.5 GHz). The signal was transmitted from a roof of a 21-m high building in an urban environment. The receiver was placed at different floors (outdoors and indoors) of a close-by 5-floor building. The distance between Tx and Rx was about 150 meters.

Their second study [CCC10b] focused on the path loss between base station and relay station. The base station was mounted on the roof of a 20-m high building. An extensible antenna was mounted on a van to simulate relay stations with different

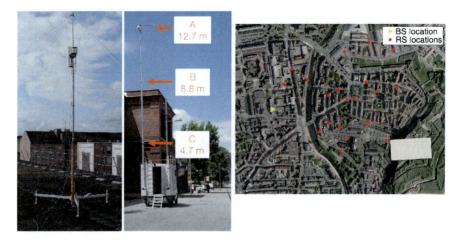

Fig. 2.8 Relay measurements in Belfort, France [CCC10b]; *Left*: Photo of base station and relay station with 3 antenna heights; *Right*: Map of the environment

height. The measured points for the relays were distributed throughout the city center of Belfort, France. The antennas and measurement map are shown in Fig. 2.8. In both studies, narrowband transmitters and receivers were used.

The measurements demonstrated that in general, a higher floor provided more power for all considered frequencies. It was found that the corresponding WINNER feeder-link models for LOS (B5a) was consistent with the measurements. For the NLOS model (B5f), higher deviations are observed for the receiver at lower floors. However, the model was based on the assumption that NLOS receivers are on high floors, which explains the deviations.

More details on the path loss results and fitting path loss models are provided in Sect. 3.1.

Outdoor Peer-to-Peer Channel In [ETM07], the authors measured an outdoor peer-to-peer channel at 300 MHz. MIMO at such a low carrier frequency is attractive for peer-to-peer communications because it combines the good coverage of a low-carrier-frequency system with the high data rates enabled by MIMO.

The measurements were performed using the RUSK LUND channel sounder using uniform circular dipole arrays mounted to the roof top of two cars. The measurements took place in a rural/semi-suburban area near Linkoping, Sweden, showing groups of trees and buildings in the environment.

The evaluation of these measurements are presented in Sect. 3.5.4.1.

2.1.1.2 Multibase-Station Measurements

Although single-base-station channel measurements are commonplace today, they involve an appreciable effort to accomplish. The increased complexity of perform-

Table 2.4 TU-Ilmenau sounder specifications

Parameter	TU-Ilmenau RUSK sounder
Center Frequency	2.53 GHz
Bandwidth	2 × 40 MHz
CIR length	6.4 μs
Transmit power	46 dBm
MIMO Snapshot rate	75 Hz
Number of BS	3
Inter Site Distance	580–680 m
BS antenna	16 ULA (8 cross-polarized)
UE antenna	48 cylinder (24 cross-polarized) + 10 (MIMO-Cube)

ing multibase-station channel measurements implies an extraordinary challenge explaining the fact that those campaigns are relatively rare.

One straightforward way to accomplish multibase-station measurements is to repeat a specific UE measurement route using different single sites in sequence. The drawback is that the work load increases proportionally to the number of base station sites. Nevertheless, this method has been employed in MIMO multilink campaigns in three cities in Germany using Medav RUSK channel sounders. GPS in combination with odometers are used to make sure that sufficient UE position accuracy is kept when the same route is repeated for the different sites. In a campaign performed by the Technical University of Ilmenau [NKS[+]09] within the EASY-C project, three base station sites (see Table 2.4) were placed in the center of Ilmenau at different heights using a sky lift. First investigations were done on the large-scale parameters with the results that multipass measurements are possible even in suburban environments under certain precautions [NKJ[+]10].

The campaigns in Berlin and Dresden were also performed within the EASY-C EU project. Different scenarios (see Table 2.5) with intersite distances between 100 m and 700 m were employed. The gain of cooperation between the sites has been evaluated in a scenario with five mobile users and four base station cites at the Technical University of Berlin (TUB) campus [JJT[+]09]. A five-fold gain in cell throughput is obtained as shown in Fig. 2.9.

In a measurement campaign performed by Ericsson Research [LMF10] simultaneous coherent transmission from three different base station sites was accomplished by means of fiber optical distribution on the RF signal to each antenna at each site. A map of the measurements is shown in Fig. 2.10. The Ericsson channel sounder is based on their own developed MIMO testbed for LTE. In this campaign a single Kathrein base station antenna was used at each site. Details of the setup is provided in Table 2.6. The measured channel data were used to investigate the gain of multibase-station coherent cooperation in a single-user scenario. A convenient measure of the MIMO performance of the channel is the eigenvalue dispersion σ_λ according to

Table 2.5 EASY-C project sounder specifications

Parameter	Berlin	Dresden	TUB (Berlin)
Center Frequency	2.53 GHz	2.53 GHz	5.2 GHz
Bandwidth	20 MHz	21.25 MHz	120 MHz
BS/Sectors	3/3	3/3	4/1
Transmit power	41 dBm	41 dBm	40.3 dBm
MIMO Snapshot rate	500 Hz	1150 Hz	500 Hz
BS antennas	ULA 16	Kathrein 2	Star 16
UE antennas	Cube 10	CUBA 16	Cube 10
Inter Site Distance	500 m	750 m	100–200 m

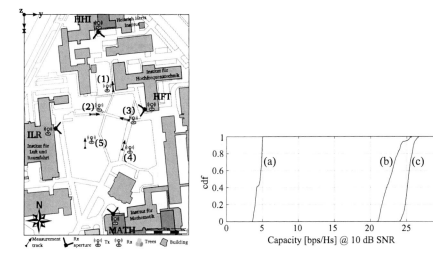

Fig. 2.9 *Left*: Campus map of the TUB. BS locations are indicated by the short names of the institute buildings. The five short measurement tracks are indicated by *arrows* [JJT+09] (©IEEE 2009, reproduced with permission). *Right*: Distributions of capacity for the traditional deployment with different resources in each cell (*a*), mean capacity of isolated cells (*b*) and capacity with base station cooperation using a common resource (*c*), all for SNR = 10 dB [JJT+09] (©IEEE 2009, reproduced with permission)

$$\sigma_\lambda = \left(\prod_{i=1}^{N} \lambda_i\right)^{1/N} \bigg/ \left(\frac{1}{N}\sum_{i=1}^{N} \lambda_i\right). \tag{2.1}$$

σ_λ may take values between 0 and 1 where 0 corresponds to fully correlated antennas and 1 to the best possible MIMO channel in the sense that all eigenvalues are equal. In the case of a 2×2 system, $\sigma_\lambda \approx 0.6$ for an i.i.d. Rayleigh fading channel. Figure 2.10 shows the distributions of σ_λ together with the average channel gain for the three base station sites. It is clear that the best MIMO performance is obtained in areas between the three sites where the signal strengths are similar.

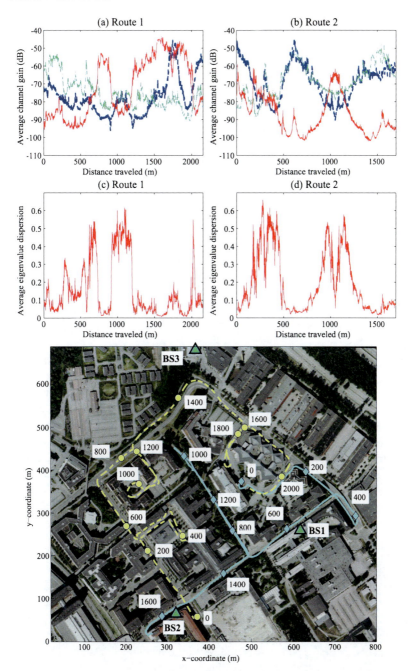

Fig. 2.10 Simultaneous channel measurements with multiple base stations. *Top*: Average channel gain per receive antenna over 20 MHz for BS 1 (*blue*), BS 2 (*green*), BS 3 (*red*) for (**a**) route 1 and (**b**) route 2. Average eigenvalue dispersion for (**c**) route 1 and (**d**) route 2 [LMF10] (©IEEE 2010, reproduced with permission). *Bottom*: Location of BSs and route 1 (*yellow*) and route 2 (*cyan*) of MS. Distances from starting points are indicated in meters

Table 2.6 Ericsson sounder specifications

Parameter	Ericsson sounder
Center Frequency	2.66 GHz
Bandwidth	19.4 MHz
CIR length	22.2 µs
Transmit power	36 dBm
MIMO Snapshot rate	190 Hz
Number of BS	3
Inter Site Distance	580–680 m
BS antenna	1 Kathrein (18 dBi 45 deg polarized)
UE antenna	2 dipoles + 2 magnetic loops

Additionally, it was shown that—using the coherent measurements—the mobile nodes' location could be well estimated [MSKF10].

2.1.2 Polarimetric Channel Measurements

With increasing interest in multiantenna communications over the past decade, the polarimetric behavior of the radio channel is investigated intensively. A lot of experiments have been conducted in order to improve the understanding of the polarized radio channel. In order to determine the polarized electromagnetic field fully, its vector components have to be measured. This may be achieved by means of a tripole antenna which measures all three field components at one point in space. For radio channel modeling, it is, however, convenient to decompose the electromagnetic field into a set of plane waves. A plane electromagnetic wave is fully characterized by the two field components which are perpendicular to its direction of propagation (the corresponding longitudinal component vanishes). In channel measurements each plane wave is typically characterized with the horizontal and polar (vertical in the case of propagation in the horizontal plane) field components.

There are two basic metrics for statistical characterization of the polarization properties of measured channels. The first is the cross polar discrimination (XPD), i.e., the power ratio between co-polar and cross-polar transmission and reception (P_{VV}/P_{HV}, P_{HH}/P_{VH}). The second is the copolar power ratio (CPR), i.e., the power ratio between the two copolar transmission and reception cases (P_{VV}/P_{HH}).

In [QCOD07, QOHD08b] the polarization characteristics in an indoor corridor-to-room scenario have been measured. At the transmitter location in the corridor two log periodic vertically and horizontally polarized antennas were used. At the receiver locations in the room a tripole antenna was used. To determine the direction of each incoming wave at each receiver location a Bartlett beamformer, based on a two-layer horizontal planar virtual array each of 10 × 10 elements, was used. Here the observed XPD was around 10 dB. Interestingly, the authors derived in

Table 2.7 Sounder specifications and channel polarization XPD and CPR for the different polarization campaigns

Parameter	[DC08]	[QOHD08b]	[KMGT09]	[PQD+09]	[MGPRJL08]
Center Freq.	2.2 GHz	3.6 GHz	4.5 GHz	3.5 GHz	2.45 GHz
Bandwidth	62.5 MHz	200 MHz	120 MHz	CW	200 MHz
BS antenna	VH	VH Log-periodic	VH planar $2 \times 4 \times 2$	VH	4 monopoles
UE antenna	VH Omni	Tripole	VH circular $(2 \times 24 \times 2)$	Tripole	4 monopoles
Scenario	Urban Macro	Corridor-to-room	Urban Macro	Indoor + Outdoor-to-indoor	indoor
XPD	5 dB	10 dB	9 dB	5 dB	8–16 dB
CPR	3 dB	NA	−1 dB	−2 dB	NA

[QOHD08a] analytically that the XPD should fit an F-distribution for Rayleigh and Ricean fading, which is also supported by measurements.

A somewhat lower observed channel XPD around 5 dB was observed in [PQD+09] in an indoor scenario and outdoor-to-indoor scenario. The CPR was around −2 dB.

In an urban macro-cellular campaign in Mulhouse [DC08] the polarized radio channel was measured at 47 UE locations. The average XPD of the measured channel is around 5 dB. Comparing the gains between the copolar transmission modes, a 3 dB effect of lower loss for vertical polarization is observed (CPR = 3 dB). These findings are also supported by the Ilmenau multibase campaign [SSN+09] where the XPD and CPR were measured (in the same type of scenario) to about 7 dB and 3.5 dB, respectively.

Another urban macrocellular campaign was performed by Tokyo Institute of Technology [KMGT09]. Here a circular stacked double polarized array was used at the UE. For the polarimetric analysis, the RIMAX superresolution method was used.

The vertical to horizontal cross-polar discrimination (XPD) of the radio channel ranged from about 5 dB to about 10 dB. The corresponding CPR was −1 dB, which is in contrast to the other two macrocellular campaigns [DC08, SSN+09].

A polarized MIMO indoor campaign was carried out by [MGPRJL08] using a 4×4 system of monopole antennas. To capture different polarizations, the antennas were rotated between measurement runs. Depending on the environment, the authors observed XPDs between 8 and 16 dB.

Investigations of polarized MIMO in a tunnel [LSD+09] showed that the XPD increases, as soon as the distance between Tx and Rx becomes larger than few times the tunnel width (or height). An overview of the sounder specifications for all the measurements presented are provided in Table 2.7.

2.1.3 Vehicular Channel Measurements

2.1.3.1 Measurement Practice

Vehicular propagation channels can be separated into Vehicle-to-Vehicle (V2V) and Vehicle-to-Infrastructure (V2I) channels, and are typically highly dynamic since one or both link ends can move at very high speeds. A lot of the research that has been conducted in this area was sparked by international standardization work envisioning intelligent transport systems in the 5.9 GHz band (e.g., IEEE 802.11p [WAV]).

Measurements of propagation channels for vehicular systems put high requirements on the measurement equipment. The mobility of the link end(s) in such systems implies high Doppler shifts which the channel sounder needs to cope with by means of very fast sampling. V2V channels are particularly challenging, since, in contrast to cellular systems, Tx and Rx are at the same height, which results in multiple scattering around both link ends. Furthermore, there are many important scatterers moving at high speeds, which combined with the mobility of Tx and Rx even can result in propagation channels that are nonstationary [PZB$^+$08]. The high mobility and challenges of measuring V2V propagation channels can be demonstrated by considering a Tx and Rx that are driving in convoy at a speed v. A multipath component that is generated by a single bounce off another vehicle driving at the same speed toward the Tx/Rx convoy will then have a Doppler shift of $v = 4 f_c v/c$ (where c is the speed of light). For a center frequency $f_c = 5.9$ GHz and a speed $v = 110$ km/h, this results in a Doppler shift of 2.4 kHz.

The vehicle-to-vehicle measurement campaigns within COST 2100 have used different parameter settings in their measurement setups, particularly concerning the size of the MIMO system and the vehicle speed. In [RKVO08] and [RKVO10], large MIMO systems were measured (30×30 and 30×4, respectively) while using a low driving speed of the Tx and Rx cars (5–20 km/h). The measurements in [PKC$^+$07] and [PKZ$^+$08], on the other hand, were conducted at speeds up to 110 km/h but involved a much smaller MIMO system (4×4).

The V2I measurements in [PKH$^+$08], which characterized the propagation channel between a base station and a high-speed train (260–290 km/h), deployed a 16-element array at one link end and a single antenna element at the other. These measurements were performed in two steps: First the antenna array acted as base station, and the single antenna was used as mobile unit. Then, the measurement was repeated with the reversed situation. In this way, directional information could be extracted at both sides (though, strictly speaking, double-directional data is neglected).

The parameter settings for the V2V measurement activities within COST 2100 are summarized in Table 2.8. It is noteworthy that no campaign has been conducted at 5.9 GHz as suggested by standards [WAV], usually for regulatory reasons or due to limitations in the measurement equipment. The difference between channel characteristics at the measured frequency bands and 5.9 GHz are commonly expected to be negligible [PKC$^+$07, RKVO08, PBK$^+$10, RKVO10].

One aspect that requires special attention in V2V channel measurement campaigns is their documentation, most importantly of the Tx and Rx (spatial) coordinates. Whereas it is simple to obtain these coordinates for measurements where

Table 2.8 Measurement parameters in vehicle-to-vehicle measurement campaigns within COST 2100

	[PKC+07]	[MFP+08]	[RKVO08]	[PBK+10]	[RKVO10]
Center freq. [GHz]	5.2	5.2	5.3	5.6	5.3
Bandwidth [MHz]	240	120	120	240	60
Snap. rep. [ms]	0.3072	1/10	8.4	N/A	15
Tx power [dBm]	27	33	36	27	36
Tx antennas	4	1	30	1/4	4
Rx antennas	4	1	30	1/4	30
Vehicle speed [km/h]	30–110	10/90	5–15	N/A	5–20
Snapshots	32500	N/A	967–2006	N/A	N/A
Recording time [s]	10	10	67–140	N/A	N/A
Tx ant. height [m]	2.4	roof + 0.2	2.12	N/A	N/A
Rx ant. height [m]	2.4	roof + 0.2	2.32	N/A	N/A

both link ends are static, or one is moving at a moderate speed, the fast sampling and high speeds make the acquisition less feasible in V2V measurements. For directional estimation analysis of measurement data, this is especially important. Even though exact knowledge of the Tx and Rx coordinates is not necessary for extracting directional estimates, it is still essential for their interpretation; without such knowledge the mapping of the extracted results to physical scattering objects is impossible. The usual way of conduct is to use GPS receivers at Tx and Rx in order to log their time-varying coordinates [PKC+07, PBK+10, RKVO10]. GPS data, however, can suffer from inaccuracies due to multipath propagation, especially in urban environments, and may for this reason be supplemented by video documentation [PKC+07]. An alternative approach was used in [MFP+08], where the Tx car was kept static while the Rx car was moved at a low speed (10 km/h). The aim here was to verify a raytracing model and simultaneous track-keeping of two vehicles was deemed to inaccurate for such purposes.

The characteristics of propagation channels depend on the particular propagation environment and measurement data is therefore usually classified into several well-known environments. For V2V channels, the most common are highway, urban, rural, campus, and suburban [PKC+07, MFP+08, RKVO08], which differ in, inter alia, scatterer densities, (road) geometry (lane width, number of lanes etc.), and roadside objects (houses, sound abatement walls, metal fences, etc.). However, there are also some additional classification concerns that are of importance for V2V propagation channels. First, the distinction between LOS and NLOS is not straightforward since shifts between the two situations may occur rapidly due to the sudden obstruction by, e.g., a truck. One measurement method is to let the Tx and Rx drivers apply a "realistic" driving behavior [RKVO08] and let the channel conditions shift naturally between LOS and NLOS, but dedicated measurements where the LOS is deliberately obstructed by letting a truck drive between Tx and Rx have also been conducted in order to isolate the effect of LOS obstruction [PBK+10]. Second, it is common to distinguish between measurements where the Tx and Rx cars are driving in the same direction [PKC+07, RKVO08, MFP+08] and measurements where the direction of travel is the same [PKC+07, RKVO10] since the channel properties can be significantly different in those situations (e.g., in terms of Doppler shifts). Third, in the context of traffic safety applications, it is of interest to characterize propagation channels for the particular traffic situations that are important for such applications, e.g., precrash applications such as intersection collision avoidance. For such purposes, the regular environment classification (highway, urban, etc.) is not sufficient, and therefore measurements have also been conducted in intersections [PBK+10], traffic congestion [PBK+10], and overtaking situations [RKVO10].

2.1.3.2 Choice of Vehicles and Antennas

In any measurement campaign, the choice of antennas (or antenna arrays) is important and will impact the end result. For vehicular channel characterization, the type of vehicle that is used and the location of the antenna on the vehicle will also largely impact the result. The antenna location on a given vehicle not only affects the influence from the subject vehicle, but also how much influence other vehicles and scatterers will have on the received signal. The impact of antenna location on system performance was investigated through simulations in [RPZ09] but has not yet been analyzed by measurements.

In vehicular measurements, the most common approach is to use a "regular" antenna array (i.e., an antenna array specially designed for channel characterization) mounted at an elevated position on the vehicle (usually on/above the vehicle roof) [PKC+07, MFP+08, RKVO08, RKVO10] or inside the vehicle [PKH+08]. Examples of regular antennas include a monopole [MFP+08], a uniform circular array [PKC+07, PKH+08], or a semi-spherical array [RKVO08, RKVO10]. Such an approach leads to a straightforward characterization of the propagation channel, but the elevated position reduces the possibility of obstruction of the LOS and other important propagation paths, and may thus lead to an overly beneficial channel. An alternative is to use an antenna array specially designed for vehicular applications, thus constituting a realistic example design that could be used by a commercial V2V communications system [PBK+10]. This leads to a channel that is realistic from an application point-of-view, though the generality of the results can be limited by the particular antenna arrangement that is used.

2.1.3.3 Performance Analysis of 802.11p Vehicle-to-Infrastructure Links

The usual goal of channel measurements is to obtain a description of the channel transfer function, or conversely, the channel impulse response. However, channel measurements can also be conducted in order to assess the performance of a particular wireless system. In [PTA+10], the performance of the physical layer in the IEEE 802.11p standard was investigated by means of V2I measurements. The goal was to characterize the average downstream packet broadside performance for a vehicle passing two road side units. The road side units were mounted on top of a highway gantry and the van had an onboard unit mounted on top of it. The measurements were conducted by disabling the retransmission function of the MAC layer, filling the MAC service data unit of the transmitting road side units with random data (of a specific length) and then continuously transmitting these data while the van was driving along the highway at 80 km/h or 120 km/h. After reception, it was determined whether the frames where correctly decoded by the receiving onboard unit.

2.1.4 UWB Channel Sounding

Ultra-WideBand (UWB) channel sounding requires equipment that is capable of handling the ultrawide frequency bands under consideration. This section describes measurement efforts in the 3.1–10.6 GHz band, the frequency range allocated in the US by the FCC for unlicensed use of UWB systems [FCC02]. Only subsets of this band may be available in other parts of the world, or of interest for certain applications, therefore some measurement campaigns address only parts of these frequencies.

Vector Network Analyzer (VNAs) that cover this frequency band are available in many labs and have therefore been used in many UWB measurement campaigns [Mol09, ASQ09]. VNA-based measurements have also been presented to this COST action by various groups (see Sect. 6.3.1) and already to the predecessor action COST 273 (see [Cor06]). The VNA's key advantages are a large dynamic range and a (rather) free and flexible choice of the measurement band. Its key disadvantages are the required RF cables to connect the VNA to the TX antenna *and* the RX antenna, and the relatively slow measurement speed due to the sequential scanning of the frequency points. The latter is fortunately less critical in many cases, since UWB communications is mostly limited to indoor applications by regulation, where one can often make sure that there are no moving objects in the environment. But it prevents—on the other hand—characterization of time variations of the channel.

This section describes alternative measurement equipment that has been employed in UWB channel measurements presented within COST 2100. We first address a Pseudo Noise (PN) Direct Sequence Spread Spectrum (DSSS) channel sounder that has been used by the authors of [CPB07, CTB08]. The channel sounder better supports dynamic channels, as it simultaneously measures the complete frequency band. The second topic of this section is a demonstrator system for the standardized IEEE802.15.4a UWB signaling scheme that has been discussed in [GBG$^+$09, GBA$^+$09, AAM09]. The IEEE802.15.4a standard defines preamble sequences that are designed for estimating a channel impulse response at fine time resolution in order to support robust ranging and positioning in indoor applications [IEE07]. Its signal scheme is thus well suited for UWB channel sounding.

Note that the results of the measurement campaigns are described in Sect. 6.3.1; this section only details the measurement technology. For an overview of the measurement campaigns, please refer to Table 6.6.

2.1.4.1 Time-Domain Channel Sounding

Two methods are commonly used for time-domain channel sounding, which differ by the type of RF signal used. One method is based on transmission of short "isolated" pulses that yield at the receiver directly the channel's impulse response (convolved with the pulse shape of the probing signal). The time resolution is determined by the bandwidth of the transmitted pulse, i.e., the pulse duration. Unfortunately, the pulse energy decreases with increased bandwidth such that the resulting sounding signal—a low-duty-cycle pulse stream—has very low power. As a consequence, high pulse voltages are needed at the transmitter, and demanding specifications result for the receiver to offer sufficient dynamic range and sensitivity.

The second method is based on PN sequences as sounding signals, and it bears the advantages of DSSS systems. In stead of isolated pulses, PN sequences of digital symbols are transmitted. The continuous transmission increases the average power of the transmitted signal, while the bandwidth is still determined by the bandwidth (duration) of the (baseband) pulse used for modulating the PN sequence. The receiver performs correlation processing to determine the channel impulse response in time domain. It thereby exploits processing gain proportional to the time-bandwidth product of the PN sequence [BLM04, Pro95], which quantifies the energy gain with respect to the method using isolated pulses. Correlation processing also suppresses (narrowband) interference signals, another important advantage of this technique.

2.1.4.2 M-Sequence UWB Channel Sounding

A channel sounder based on an Maximum Length Binary Sequence (MLBS) has been used in the measurements presented in [CPB07]. The channel sounder [KSP+05, SHK+07] generates a length-4095 MLBS by using a shift register clocked at 7 GHz. Up-converting the obtained baseband signal to a 7 GHz (synchronous) carrier provides a sounding signal covering the 3.5–10.5 GHz band, which agrees reasonably well with the frequencies allocated by the FCC. The sequence repeats after roughly 600 ns, which determines the maximum channel excess delay that can be measured unambiguously. This is sufficient for common indoor UWB channels. On the other hand, the short sequence length allows for subsampling to simplify the receiver hardware. A subsampling by a factor of 512 still allows for the acquisition of more than 3000 CIRs per second, while the sampling rate is reduced to a moderate speed of 13.7 MHz. Of course, subsampling means that a large amount of signal power is lost. Further details on the sounding architecture can be found in [KSP+05].

Calibration is a key step for ensuring accurate channel measurements. In case of the MLBS channel sounder, the data post processing includes suppression of

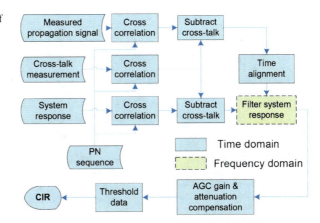

Fig. 2.11 Postprocessing of measurement data [CPB07] (©2007 IEEE, reproduced with permission)

crosstalk between the transmitter and the receiver, and calibration for the system response of the measurement equipment. Figure 2.11 illustrates the required signal processing steps [CPB07]. Cross-correlation with the MLBS is the first operation performed upon the acquired raw data. The crosstalk is measured with the receive port disconnected. It is also cross-correlated with the MLBS and then subtracted from the input data and the system response. The system response is measured with the transmit and receive ports connected directly. Finally, the system response is removed from the channel data by division in the frequency domain, where one has to take care to avoid noise amplification [CPB07].

The accuracy of the captured channel data was verified by comparison to measurements performed with a VNA. For that purpose a completely static, artificial multipath environment has been created inside an anechoic chamber [CPB07]. A close match has been observed within the frequency band of interest. In summary, the paper [CPB07] concludes, that "the time-domain sounder offers rapid data acquisition of real-time channel data at the expense of increased processing complexity and reduced dynamic range."

2.1.4.3 Channel Sounding Using IEEE 802.15.4a Preamble Signals

The preamble of the standardized IEEE 802.15.4a UWB signaling scheme [IEE07] starts with a code sequence that has ideal properties for channel sounding. Its autocorrelation shows a single peak at zero-shift and zeros otherwise, i.e., it is perfectly flat in the frequency domain, which means that it can probe any frequency (occupied by the signal) equally well.

In the IEEE 802.15.4a preamble ternary codesequences of length 31 or 127 are used to achieve such properties. That is, the chips of the preamble are taken from the set $\{-1, 0, +1\}$. Each chip is modulated by a baseband pulse with a bandwidth between 500 MHz and 1.5 GHz and then modulated on an RF carrier that sits at predefined channels within the UWB band. The spacing between consecutive preamble pulses is extended by a factor of 4, 16, or 64 with respect to the nominal pulse repetition frequency of 499.2 MHz, which stretches the period of the ternary code to

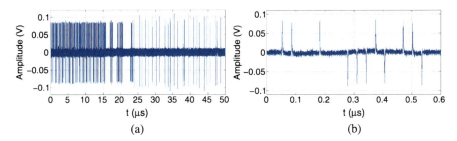

Fig. 2.12 (**a**) IEEE 802.15.4a signal in baseband. The cyclically repeated preamble is terminated by a start-of-frame delimiter, which is followed by pulse-position-modulated data symbols [GBA+09] (©IEEE 2009, reproduced with permission). (**b**) Close-up on the ternary preamble sequence [GBA+09] (©IEEE 2009, reproduced with permission)

about 1 or 4 µs, depending on the parameter settings. Finally, the ternary sequences are periodically repeated up to 4096 times, to increase their energy and to simplify code acquisition. The allowed parameter combinations can be found in the standard [IEE07]. Figure 2.12 shows an IEEE 802.15.4a signal with a close-up on the preamble sequence. Sixteen cyclic repetitions of a length-31 sequence can be seen with a pulse spacing extended by a factor of 16 to 32 ns.

The ternary code has the special property that its absolute value shows perfect cross-correlation properties with the original code sequence. Therefore simple energy detection receivers can also perform channel estimation [LCK06] and thereby exploit processing gain, albeit suffering from so-called noncoherent combining loss [GTPPW10].

Due to these code properties, IEEE 802.15.4a nodes are suitable for channel sounding within the frequency band they are operated at. This has been presented in [GBG+09, GBA+09, AAM09], using a standard-compliant demonstrator transmitter. The demonstrator, see Fig. 2.13(a), uses fast serial IO-ports of a Xilinx Field Programmable Gate Array (FPGA), so-called Multi-Gigabit Transceivers (MGTs), to generate the baseband sequence. Two MGT outputs are combined to obtain the ternary code symbols. The baseband signal is lowpass filtered for pulse shaping [AFG+09] and then up-converted to the carrier frequency, using either a Vector Signal Generator (VSG) as shown in the figure or a custom-designed up-converter [AGFW09, AAM09]. At the receiver side, the demonstrator acquires the received UWB signal directly with a real-time scope, a suitable bandpass filter, and an Low-Noise Amplifier (LNA). Any post processing is done offline. Clock synchronization of the FPGA, the VSG, and the scope simplifies this postprocessing, as the chip and sampling clocks then have a fixed phase relation, avoiding clock offset estimation and correction. Coherent processing of the UWB data yields channel impulse responses at high dynamic range as illustrated in Fig. 2.13(b). The real-time sampling supports measurements in time-variant environments. However, transferring the data to the PC is rather time consuming, therefore a real-time observation of channel variations is hard to achieve.

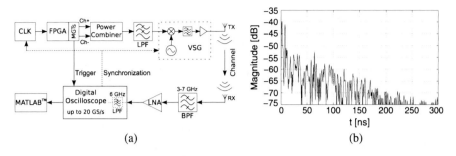

Fig. 2.13 (**a**) Demonstrator transmitter for the IEEE802.15.4a standard [GBA+09] (©IEEE 2009, reproduced with permission). (**b**) Channel impulse response captured in an industrial environment [GBA+09] (©IEEE 2009, reproduced with permission)

2.1.5 Millimeter and Submillimeter Waves

Wireless systems at millimeter-waves (mm-W) and sub-millimeter-waves (sub-mm-W) are promising solutions to reach broadband data rates for short-range communications [Smu02]. However, the realization of multi-Giga-bit/s data transmission at such frequencies becomes a challenge: about 3 Gb/s will be required for High-Definition (HD) uncompressed video [SOK+08], and similar rates are foreseen for PC connectivity and peer-to-peer communications. If this goal is achieved, the mm-W/sub-mm-W solutions will reduce the number of wires over our desks, TV rooms, and also within cars and aircraft cabins.

These new technologies are being developed in parallel with IEEE standards, i.e., 802.15.3c [IEE09] and 802.15.11ad [IEE10]. In Europe, the standard ECMA-387 has been defined for mm-W [ECM08]. Globally, about 7 GHz of unlicensed bandwidth (in some cases 9 GHz from 57 to 66 GHz) have been allocated, e.g., [FCC01] and [ECC03], which has increased the industrial interest. This is the highest amount of free spectrum assigned so far for short-range applications. Nowadays wireless multi-Giga-bit/s systems are being developed in some industrial projects in Europe, e.g., the *EASY-A* project[1] (Enablers for Ambient Services and Systems, Part A—60 GHz Broadband Links) or the MEDEA+ project *Qstream* (Ultrahigh-data-rate wireless communication). Massive markets include HD-TV home-cinema [SOK+08], office/meeting room point-to-point high-data-rate communications [JK09, JMK10], in-car/cabin point-to-multipoint video streaming [GKZ+10], real-time video streaming in medical applications [KSH+10, KHS+10], and fixed point-to-point radio links [GKKV10] (see some scenario examples in Fig. 2.14). Recently, public transportation scenarios have also been proposed for IEEE standardization at 60 GHz [Gar10].

In this context, this section addresses measurements and characterization of mm-W and sub-mm-W channels. There are still many open questions about the

---

[1] URL: http://www.easy-a.de/.

Fig. 2.14 Scenarios envisioned for mm-W/sub-mm-W: (*top-left*) kiosk, living room, office cubicle, and meeting room; (*top-right*) public transportation, and (*bottom*) medical and fix point-to-point links [GKZ+10] (©2010 IEEE, reproduced with permission)

channel at such Extra High Frequency (EHF) bands (30–300 GHz), mainly related with the confined propagation scenarios envisioned for Wireless Local Area Network (WLAN) and Wireless Personal Area Network (WPAN) [CTT+09]. Besides, some applications are included within the public transportation scenarios for multiple access, e.g., for In-Flight-Entertainment (IFE), see Fig. 2.14 (top-right), [Gar10, GKZ+10]. These environments differ from others due to the higher density of users, human behavior, and usual metal cabins. Indeed, mm-W/sub-mm-W radio channels (with wavelengths between 1 and 5 mm) do not follow the usual 2–5 GHz channel characteristics (e.g., see [YSH05] for 60 GHz channels), showing smaller delay spread [GKT+09], remarkable impact of shadowing due to the human activity [GKT+10a], important diffraction effects due to human bodies [JMK10, GKT+10a], high water vapor absorption, and different directional characteristics due to the high path-loss.

Thus, investigations about channel characteristics for different scenarios and antenna patterns, including directional and polarimetric information, are important issues for designing multi-Giga-bit/s access. Besides, new channel sounders, extensive measurement campaigns, analysis, and modeling will be required for proposing possible enhancement techniques, e.g., beamforming, macro-diversity, and seamless handover, among others.

The section is structured as follows. At first, the propagation phenomena are described to highlight the peculiarity of mm-W and sub-mm-W. Then, different channel sounder architectures and their characteristics are presented. At the end of the section, radio propagation channel analyses are summarized for different scenarios, namely: indoor, in-cabin, in-car, hospital, and outdoor.

2.1.5.1 Propagation Phenomena

To derive accurate deterministic mm-W and sub-mm-W channel models, the knowledge of the significant propagation mechanisms is very important. Propagation phe-

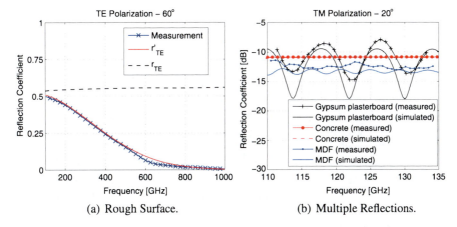

Fig. 2.15 Calculated and measured reflection coefficient for a rough surface [PSKK07] and different optically thin materials [PSKK07]

nomena are significantly different than at lower frequencies due to the small wavelengths at 60 GHz and beyond 100 GHz. In the following, these phenomena and the possibilities to model them are presented. At first, scattering and reflection measurements are introduced, then the path loss at mm-W and sub-mm-W is analyzed, and at the end the effects of shadowing due to human activity are summarized.

The scattering from rough surfaces and the effect of multiple reflections have been investigated during the last years up to THz frequencies [PSKK07, PJKK08, JPJ+09]. In [PSKK07] the specular reflection from rough concrete plaster has been measured with terahertz time-domain spectroscopy. In this work the reflection coefficient r has been calculated from measurements of absorption coefficient and refractive index using the conventional Fresnel equations. Following Kirchhoff's Scattering theory [BS87], these coefficients need to be multiplied by the so-called Rayleigh factor, taking into account the surface roughness in order to obtain a modified reflection coefficient r'. A comparison between measurement and simulation results using this procedure is shown in Fig. 2.15(a) for frequencies between 100 GHz and 1 THz. The simulated reflection coefficient r fits with the measured data, but only for longer wavelengths. Except of these limited cases, the simulated conventional reflection coefficient deviates significantly from the measured one. The difference becomes more and more significant as the frequency increases. This is a consequence of the effective roughness of the material, which grows along with the scattering losses in the specular direction if the frequency increases. However, the modified reflection coefficient r' fits well with the measurement data in all the spectrum.

On the other hand, multiple reflections within materials significantly influence the radio wave propagation, especially in case of optically thin or layered materials. This propagation mechanism can be modeled by the so-called transfer matrix method [PJKK08]. Figure 2.15(b) shows measurement and simulation results for concrete, medium density fiberboard (MDF) and gypsum plasterboard samples. In case of the concrete slab, a flat characteristic is observed. This behavior is due to the large thickness of the concrete slab compared to the wavelength of the propagated waves, and also due to the relatively high absorption losses of concrete. The reflection pattern of the MDF sample is similar to that of the concrete slab. However, it does differ in some details. The most significant difference is the influence of multiple reflections within the material, which can be clearly recognized as an oscillation in the frequency pattern. The multiple reflection effect is even more pronounced for the gypsum plasterboard sample, caused by the small absorption losses and the small sample thickness.

Other relevant propagation mechanism at 60 GHz and THz is the path-loss, especially if we consider that a likely application is the wireless interconnection of different electronic devices for ultrafast file transfer on a desktop or within indoor environments with TX-RX ranging from few centimeters to 10 m. In [JPJ+09] the path-loss for the desktop case has been characterized experimentally and compared to the two-ray model [HBH03] for frequencies between 290 and 310 GHz. Figure 2.16(a) illustrates the simulated distance dependent path-loss for 304.27 GHz. Besides, the two-ray model results and the free space loss are shown as reference. In addition, the same data are depicted for a frequency of 2.4 GHz, where current systems like Wi-Fi or Bluetooth are deployed. For 304.27 GHz, the combined direct and reflected rays cause alternating reinforcement and reduction of the signal power, with peak levels reaching up to 6 dB below free space loss. Regarding 2.4 GHz, the combined path-loss is always higher than free space loss (2–26 dB), whereas no fading dips can be observed. Figure 2.16(b) depicts the comparison between the two-ray model and measurements around the fading dip at 67.5 cm. The fitting between simulation and measurement shows that the two-path theory is well suited to describe the analyzed scenario. Moreover, Fig. 2.16(c) compares the empirical path-loss fitting obtained for different 60 GHz indoor environments, i.e., residential and hospital environments [KHS+10]. It was found that the empirical path-loss model for hospital environments fell between small rooms and residential Line-Of-Sight (LOS) environments, with path-loss exponents between 1.34 and 2.23 (for small rooms and residential between 1.53 and 2.44; shielded room 4.18). This is reasonable since the hospital environments have more metallic scattering objects than residential environments, but these environments are more spacious than a small room where wave scattering could be very rich.

Other relevant propagation phenomena at mm-W and beyond are the obstruction losses and the dynamics from LOS to Obstruction-LOS (O-LOS) due to the human activity, which is denoted by Garcia et al. as Shadowing By Human Bodies (SHB) [GKT+10a]. This is specially relevant for public transportation scenarios (see Fig. 2.14, top-right). More precisely, the Time-Varying Shadowing by Human Bodies (TVSHB) [GKT+10a] shows peculiar characteristics at mm-W compared to other frequencies (see Fig. 2.17). This is a consequence of the small Fresnel zones and, due to the body skin, reflects partially the waves, and the through body propagated waves are strongly attenuated. Therefore, short-term oscillations due to diffracting waves appear during the partial obstruction of the Fresnel zones (see Fig. 2.17, right and top-left). Besides, strong and long shadow fading events appear

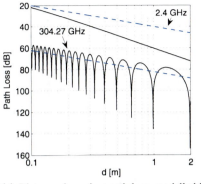

(a) Distance dependent path-loss, modelled by two-ray theory (solid line) and Friis' formula (dashed line).

(b) Distance dependent path-loss, measured and simulated by two-ray theory (f=304.27 GHz).

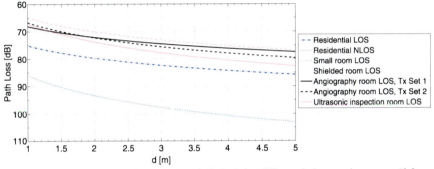

(c) Distance dependent path-loss, measured (fitting) for different indoor environments (f from 61 to 65 GHz).

Fig. 2.16 Measurement and simulation results of the path-loss for 60 GHz and THz applications [JPJ+09, KHS+10] (©2009 IEEE, reproduced with permission)

due to the total obstruction of the first Fresnel zone, including some contributions from reflected, diffracted, and wake through body propagated waves (more details can be found in [GKT+10a]).

Therefore, the main challenge for the statistical characterization of TVSHB is the real-time measurement (wideband) of the duration/depth of each obstruction event. For instances, for the isolation of the events in Fig. 2.17, right, the relative receiver power and its gradient were classified using thresholds of -2 dB and ± 0.3 dB/100 ms, respectively (see also Fig. 2.17, bottom left). From preliminary results within an Airbus-340 cabin [GKT+10a], the analyzed TVSHB events indicated fading dynamics up to about ± 10 dB/100 ms, with maximum depth durations of 8 s. These are relevant channel characteristics for a proper link budget and MAC simulations in 60 GHz WLAN/WPAN.

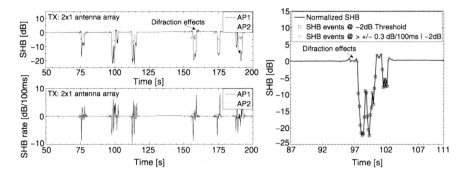

Fig. 2.17 TVSHB characteristics during obstruction LOS using 2 × 1 antenna arrays for two distributed access points (AP): (*top-left*) SHB, (*bottom-left*) SHB rate, (*right*) SHB around isolated human events [GKT+10a] (©2010 EurAAP, reproduced with permission)

2.1.5.2 Channel Sounding

Channel sounding at mm-W/sub-mm-W is a challenging task, and nowadays real-time implementations have been realized at mm-W. Some sounders based on VNA and real-time hardware have been introduced in different publications, e.g., see [GKT+09, JK09, FS10, GWF+10, KSH+10, Kiv07, ZBN05]. One of these solutions, developed by MEDAV GmbH and TU-Ilmenau, fulfill both MIMO and UWB characteristics in real-time [GKT+09]. This sounder is based on a commercial UWB channel sounder and 60 GHz up/down-converters as described in Fig. 2.18 for a Single-Input Single-Output (SISO) case. This solution is based on M-Sequence UWB radar chip-sets for baseband (0 to 3.5 GHz), and two stages of frequency conversion from/to the UWB band (3.5 to 10.5 GHz) and to/from the 60 GHz band (59.5 to 66.5 GHz). The reference LO is a 7-GHz signal which is distributed within the system including different multiplication stages. Distributed MIMO capability is supported at the UWB stage with multiple modules. This modular concept allows one to replace the 60-GHz stage for different single and dual-polarized frontends. Note that the system has a common chain for obtaining I and Q signals at baseband. Thus, for calibration purposes, a serial I/Q concept was implemented based on a 90° LO phase sifting and a switch between 0° and 90° LO at the receiver chain. During I/Q and frequency response calibrations (using a high-precision step attenuator and a delay line at the UWB stage), the PC-based system saves the calibration vectors, which can be used in postprocessing for data analysis. Back-to-back calibration at 60 GHz (including a high-precision attenuator) and cross-talk calibration (at UWB with 50 Ohm terminations) are performed for this channel sounder architecture.

The performance of this sounder is presented in Fig. 2.19, based on a 3-GHz bandwidth implementation at 60 GHz, with real-time Single-Input Multiple-Output (SIMO) capability (1 TX and 2 RX), and linear polarization. The frequency/phase responses and impulse response from back-to-back measurements using a high-precision coaxial attenuator (59 dB) after calibrations for both TX-RX pairs (CH1 and CH2) are presented. Once the calibration was performed and the frequency

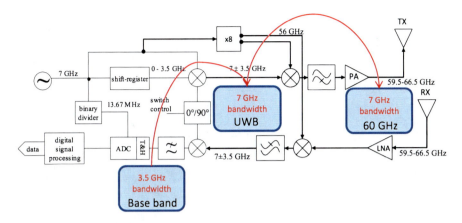

Fig. 2.18 60-GHz-UWB real-time channel sounder architecture [GKZ+10] (©2010 IEEE, reproduced with permission)

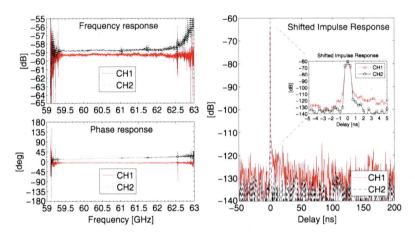

Fig. 2.19 60-GHz real-time channel sounder performance [GKZ+10] (©2010 IEEE, reproduced with permission)

span set (in postprocessing from 59.5 to 62.5 GHz), the dynamic range of the system reaches up to approximately 50 dB, with maximum detectable paths at ∼ 585 ns and a resolution better than 1 ns. These characteristics guarantee sufficient dynamic range for TVSHB analysis and present the highest sounding resolution reported in literature based on a real-time SIMO configuration. This configuration is also useful for macro-diversity analysis at mm-W. Measurements campaigns based on this channel sounder have been reported in [Gar10, GKT+10a, GKZ+10, GKBT09]. Examples of measured Power Delay Profile (PDP) (TX-RX in alignment), and both RMS delay spread and coherence bandwidth (at 0.9 correlation) under TVSHB (dynamics) are presented in Fig. 2.20.

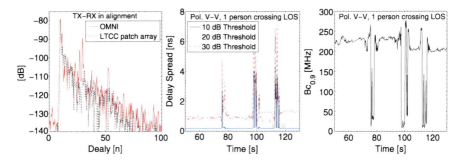

Fig. 2.20 60-GHz real-time channel results: (*left*) PDPs with two different antennas within the Airbus-340, (*center*) RMS delay spread under TVSHB, and (*right*) coherence bandwidth, Bc, at 0.9 correlation under TVSHB [GKBT09]

Another real-time sounder implementation, based on Frequency Modulated Continuous Wave (FMCW) techniques, has been presented in [FS10], and some channel results in [GWF+10]. In this case, the channel is repeatedly excited with a Continuous Wave (CW) signal (chirp) that is swept across the channel in the frequency range from 59 GHz to 67.5 GHz. Besides, this sounder allows frequency scaling for the final bandwidth at 60 GHz up to 300 MHz × 4 for the results presented in [GWF+10]. The phase noise for this sounder was predicted to be about −45 dBc, the dynamic range about 40 dB (at 260 MHz sweep), and the Doppler/delay profile demonstrated a small degradation [FS10].

On the other hand, channel sounders using VNAs and up/down-converters, e.g., [JK09] and [KSH+10], have been implemented with the advantage of high dynamic range and huge bandwidth. The drawbacks of these solutions are the time consumption for performing measurements and the feasibility of real-time measurements only when using single tones. These channel sounders are useful for office and home environments as described in Fig. 2.14 (top-left), where channels are mainly LOS with very long coherence.

Finally, it is worth mentioning that the performance analysis of new enhancement techniques for 60 GHz WLAN/WPAN, for instance, pixel-partitioning and multiantenna-beam methods [SOK+08, PGR09], require real-time sounding at 7 to 9 GHz bandwidth. Besides, multiantenna and dual-polarization capabilities (the first 60 GHz dual-polarized architecture with phase adjustment was presented in [GM+10]) are also required to investigate macro-diversity, seamless handover, beamforming, and polarization mismatch. Moreover, PHY channelization is important for channel sounding at mm-W to cover most of the 60 GHz spectrum for analyzing Signal-to-Noise Ratio (SNR) and frequency reuse of different subbands [GKZ+10, GM+10].

2.1.5.3 Channel Analysis for Different Scenarios

Besides the pure characterization of the radio propagation channel, channel measurements are used to build statistical channel models. In [CAS08] a new approach

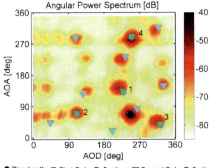
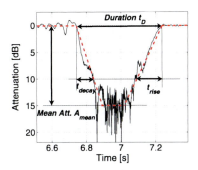

(a) Measured angular power spectrum and AoA/AoD predicted by ray-tracing [KJ09] (©2009 IEEE, reproduced with permission).

(b) Example for temporal characteristics of human induced shadowing events with parameter definitions [JMK10] (©2010 IEEE, reproduced with permission).

Fig. 2.21 Double directional channel measurement results and example for temporal characteristics of human induced shadowing events

based on α-stable distributions is proposed as an alternative to the widely used Saleh–Valenzuela model [SV87]. The model is based on a thorough study of the second-order statistical properties of the indoor channel in the 60-GHz band based on 2-GHz bandwidth measurements. In order to show that the model is appropriate, observed impulse responses have been tested against model results. Besides, the visible closeness between simulated and real data has indicated that the model is able to detect the time of arrival of multipath components with a great precision. In addition to the comparison between single impulse responses, the cumulative distribution functions of RMS delay spread from observed impulse responses and those generated by the model have shown a good fit.

On the other hand, the knowledge of angular dispersive channel characteristics is important for modeling, when beam steering [IEE09, IEE10] should be applied. In [KJ09, JK09], double directional mm-W channel measurements and channel characteristics derived by ray tracing in a fully furnished conference room have been compared (see Fig. 2.21a). In this figure the measured angular power spectrum depending on AoA and AoD is shown and compared to ray tracing results. For first- and second-order reflections, the corresponding paths, derived by ray tracing, have been identified and have shown a good agreement with the measurement data. Although material parameters have been used from literature instead of performing a site-specific calibration, the path-loss predicted for the different rays is in reasonable agreement with the measurements (mean error of 0 dB and a standard deviation of 4 dB). Another directional channel analysis has been performed in a conference room [GTWH10], where the Space-Alternating Generalized Expectation-maximization (SAGE) algorithm has been used. This work concluded that in LOS scenarios the strongest component contributes to 95% of the available power, and the strongest component in Non-Line-Of-Sight (NLOS) scenarios is on the order of 20 dB lower than the LOS component. Besides, the authors concluded that none of the channels can be regarded as having rich scattering.

Another important issue is the time variability of the channel in environments with stationary devices. The nonstationarity of the 60-GHz channel mainly appears from moving people that may attenuate the communication link by 20 dB and more [CZZ04, GKT+10a, JMK10]. Figure 2.21(b) shows an exemplary human-induced shadowing event obtained from experimental measurements (solid line) as well as the shape of a model (dashed line) proposed by Jacob et al. to describe this event [JMK10]. Four parameters have been chosen to describe a shadow fading event: the duration t_D, the decay time t_{decay}, the rising time t_{rise}, and the mean attenuation A_{mean} calculated in the interval $\frac{1}{3}t_D < t < \frac{2}{3}t_D$. The above-mentioned parameters have been statistically analyzed based on several hundreds of measurements. During the measurements, the received power of the LOS path was observed while a person was moving around within a living room scenario. The investigations have shown that the drop of signal level happens in the order of tens of milliseconds. On average the signal decreases by 20 dB in 230 ms, whereas it takes 61 ms for a reduction of 5 dB. In 90% of the cases, the signal decrease took at least 4 ms for a 1 dB, 27 ms for a 5 dB, and 101 ms for a 20 dB threshold. The duration of a single fading event reach up to 550 ms on average, and the mean attenuation A_{mean} ranges between 6 and 18 dB, whereas the maximum attenuation can reach up to 36 dB. These parameters can be statistically modeled by well-known distributions and serve as guidelines for 60 GHz WLAN MAC layer development.

For aircraft cabin applications [Gar10, GKT+09, GKBT09, GKT+10a, GKZ+10], redundancy (macro-diversity or handover) must be addressed as a possible system enhancement against the channel dynamics, mainly due to the high SHB[2] margins for network planning. In [GKT+09, GKT+10a, GKZ+10], two measurement campaigns were performed within an Airbus-340 cabin, taking into account different redundant cell configurations (distributed Access Point (AP); AP1 and AP2), human activities, and TX antennas (horns, open-waveguides, and cavity/patch arrays). The TVSHB when passengers were walking, blocking, and standing up has been considered for the analysis of shadow fading in dB (denoted by SHB in Fig. 2.22, left), fading-rate in dB/ms (denoted by SHB rate in Fig. 2.22, center), depth-duration, SHB auto-correlation, and cross-correlation between APs. The TVSHB events showed up to about ±10 dB/100 ms of fading-rate, with maximum depth durations of ∼8 s, median depth durations of ∼0.5 s, shadow fading up to ∼−20 dB at 10% outage, and stationary intervals between 112 ms and 1 s (see also Fig. 2.23).

These SHB results need to be considered for in-cabin coverage analysis. For instance, Fig. 2.22 (right) shows the SNR outage results plotted for different antennas (horn and open waveguide), taking into account an Airbus-340 cabin in static conditions [GKZ+10]. Here, 50% of the in-cabin coverage area fulfill the SNR requirements for compressed video, but if SHB is considered, redundancy must be addressed.

---

[2]SHB refers to the normalized receiver power with respect to LOS conditions during each human event.

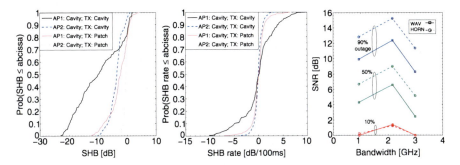

Fig. 2.22 Examples of the CDFs of SHB [GKT+10a] (©2010 EurAAP, reproduced with permission) (*left*), CDFs of SHB rate (*center*), and SNR coverage (*right*) for the Airbus-340 in-cabin channel at mm-W [GKZ+10] (©2010 IEEE, reproduced with permission)

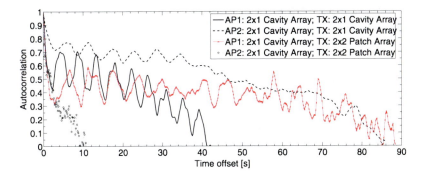

Fig. 2.23 Real-time autocorrelation of the SHB for the in-cabin channel at mm-W for a 1×2 AP configuration and different antennas [GKT+10a] (©2010 EurAAP, reproduced with permission)

Other analyses claim that the best configuration against TVSHB is obtained combining wide and narrow beam patterns at TX and RX, respectively [GKT+10a]. This can reduce both the shadowing margins at 10% outage and the fading depths during each event. Moreover, based on another TVSHB characterization described in [GKT+09], SHB margins can be reduced by 5 dB (mean) using wider beam patterns, obtaining more than 50% relative power increments for link budgets. For the same configurations, the SHB rate can be improved up to ± 3 dB/100 ms at 99% reliability.

Results from dynamic mm-W channel measurements and modeling in a shielded room and hospital environments were presented in [KSH+10] and [KHS+10]. The goals of these works were to investigate potential and feasibility of 60-GHz radios for multi-Giga-bit/s short-range communications in indoor medical applications. The results revealed that shadowing by human activity can reach up to 10 dB (for fixed TX/RX positions). With a small movement of 20 cm, the normalized received power has changed by as much as 6 dB. Besides, for indoor propagation modeling purposes, the empirical path-loss exponent for a shielded room was estimated close to 4, and for other environments, between 1.54 and 2.23.

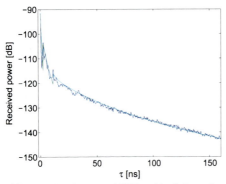 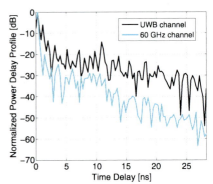

(a) Example of measured PDP and its fitting using the power law model for the shielded room environment [KSH+10] (©2010 IEEE, reproduced with permission).

(b) Normalized PDPs for a 60 GHz and a UWB in-car-channel [SJK10](©2010 EurAAP, reproduced with permission).

Fig. 2.24 Power delay profile for a shielded room environment [KSH+10] and an in-car environment [SJK10]

The main difference between propagation in a shielded room and other indoor environments considered in this section is the occurrence of clustering in the PDPs (see, e.g., Fig. 2.24a and Fig. 2.20, left). In the investigated shielded room, the PDPs consisted of 3 clusters in many cases. It turned out that the PDPs were better modeled by the power law of the paths, rather than conventional exponential decay. Additionally, a strong correlation between model parameters, e.g., between peak power and power decay [KSH+10], was observed.

The RMS delay spread was estimated for all indoor environments to be lower than 15 ns for thresholds of 10, 20, and 30 dB (see, e.g., Fig. 2.20, center).

The 60-GHz band is also of interest for applications where distances of less than 1 or 2 meters have to be linked, like the in-car scenario. In order to save installation costs, particularly in luxury class cars which are equipped with infotainment and entertainment devices, replacing wired by wireless communication is of great interest. Figure 2.24(b) shows the normalized PDPs for a 60-GHz and a UWB in-car channel [SJK10]. The transmitting antenna has been installed below the interior rear view mirror. The receiver was placed on the height of the rear seat armrests, assuming passengers accessing the car network with their own handheld devices. Both PDPs exhibit a very strong direct path, which reflects the LOS situation of this Rx position. It appears that the power decay time constant of the multipath components for the 60-GHz channel is in general much higher than for the UWB channel. The possible explanation is that the multipath components of the 60-GHz frequency band are strongly attenuated by the leather-cladding of the doors and seats in contrast to the UWB frequency band where a reflection at the metallic structure is possible. Other investigations lead to the conclusion that the antenna alignment, particularly when the main lobe of the antenna pattern is wide enough, does not play a significant role for UWB and 60-GHz system performance. Furthermore, a feasibility study which was based on the conducted measurements has shown that both systems can be realized even in such a multipath-rich environment. In these scenarios, inter-symbol interference becomes a limiting factor at increased rates.

Finally, the mm-W band is also of interest for long range outdoor point-to-point links. An example about channel investigations for this kind of links is addressed in [GKKV10]. The authors have found that the LOS path is up to about 30 dB stronger than other MPCs. This is caused by the high directivity of the parabolic antennas (45 dBi gain) used during the measurement campaigns. Therefore, in E-band point-to-point channels, the plane waves propagate like in free space. For link budgets including large-scale channel fading, the authors have proposed to introduce a fading margin of 2 dB considering both the high channel bandwidth and the focus ability of high-gain antennas. Besides, a model of rain attenuation for E-band at 80 GHz is proposed in this work based on rain attenuation models from ITU-R [IR03]. Based on the proposed model, rain attenuations were estimated to be about 8.7 dB/km and 19 dB/km, for medium and downpour rainfalls, respectively.

2.2 Channel Parameter Estimation

2.2.1 High-Resolution Parameter Estimation

High-resolution parameter estimation is an important processing step for many propagation (channel sounding) measurements. The main objective of the procedure is to provide a description of the radio propagation channel that is—with certain limitations—independent of the employed measurement equipment. The obtained model parameters can be later used, for example, to (i) simulate a MIMO radio link with an arbitrary array configuration and other system parameters or (ii) derive parameters for stochastic channel models, such as cluster-based models.

This section describes the current state-of-the-art in high-resolution parameter estimation, reflecting the latest developments in the COST 2100 framework. First, the underlying modeling assumptions and their alternatives are summarized. Then, brief descriptions of available algorithms are given. Finally, the clustering of the resulting propagation path parameters is discussed.

2.2.1.1 On the Model Assumptions and Initialization

Let us have a sampled realization of the multidimensional radio channel $\mathcal{H} \in \mathbb{C}^{M_f \times M_t \times M_r}$ with M_f frequency points, M_t transmit antennas, and M_r receive antennas. In the following, a vector form $\mathbf{h} = \text{vec}(\mathcal{H}) \in \mathbb{C}^{M \times 1}$ with $M = M_f M_t M_r$ is used for clarity. High-resolution parameter estimation methods are based on modeling the channel as a superposition of P Multi-Path Components (MPCs) as

$$\mathbf{h}_S = \sum_{p=1}^{P} \mathbf{h}_s(\Theta_p), \qquad (2.2)$$

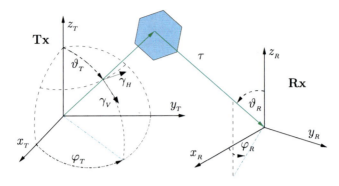

Fig. 2.25 Illustration of the double-directional propagation path parameters [Sal09]

where $\mathbf{h}_s(\Theta_p)$ denotes the contribution of the pth specular-like (hence subscript s) propagation path. Typically, the MPCs are modeled as planar wavefronts at both link ends. Also, the narrowband assumption of the propagation channel is commonly made, i.e., the complex response of the antennas, along with the interactions that the radio waves undergo, are assumed to be constant over the frequency band occupied by the employed RF signal. Hence, a single path is parameterized by

$$\Theta_p = \left[\tau^{(p)} \phi_t^{(p)} \theta_t^{(p)} \phi_r^{(p)} \theta_r^{(p)} \gamma_{HH}^{(p)} \gamma_{HV}^{(p)} \gamma_{VH}^{(p)} \gamma_{VV}^{(p)} \right]^T, \qquad (2.3)$$

where τ denotes the Time Delay of Arrival (TDoA), $\phi_{r/t}$ and $\theta_{r/t}$ denote the azimuth and elevation angles at Rx/Tx, respectively, and $\gamma_{ij}^{(p)}$ denote the complex path weights of each $ij \in \{HH, HV, VH, VV\}$ polarization component. The propagation path parameters are illustrated in Fig. 2.25.

The contribution of a single Multi-Path Component (MPC) in the observed radio channel, including the antennas, is given by

$$\mathbf{h}_s(\Theta_p) = \sum_i \sum_j \mathbf{b}_{r,j}\left(\phi_r^{(p)}, \theta_r^{(p)}\right) \otimes \mathbf{b}_{t,i}\left(\phi_t^{(p)}, \theta_t^{(p)}\right) \otimes \mathbf{b}_f\left(\tau^{(p)}\right) \cdot \gamma_{ij}^{(p)}. \quad (2.4)$$

The vectors $\mathbf{b}_{r,j/t,i}(\phi_{r/t}^{(p)}, \theta_{r/t}^{(p)}) \in \mathbb{C}^{M_r/M_t \times 1}$ denote the angle-dependent responses (narrowband steering vectors) of the receive or transmit (r/t) antenna arrays for the horizontal ($i/j = H$) or vertical ($i/j = V$) polarization, and $\mathbf{b}_f(\tau^{(p)}) \in \mathbb{C}^{M_f \times 1}$ is the frequency response resulting from the delay of arrival $\tau^{(p)}$.

The propagation path model (2.2) has the ability to separate the influence of the antenna array response, as the parameters in (2.3) depend only on the propagation channel itself. Hence, it is a very intriguing approach to try to model the complete measured propagation channel as a superposition of propagation paths along with the assumption of i.i.d. complex white normal distributed measurement noise $\mathbf{n}_w \sim \mathcal{CN}(\mathbf{0}, \sigma_n^2 \mathbf{I})$ as

$$\tilde{\mathbf{h}} = \mathbf{h}_S + \mathbf{n}_w. \qquad (2.5)$$

This model can be reasonably accurate especially in outdoor scenarios. This was indicated, e.g., in [EWLT10], where the goodness of fit was compared in terms of both

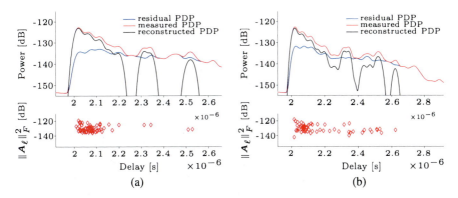

Fig. 2.26 Measured, reconstructed, and residual PDP using (**a**) SIC-based, and (**b**) ADPP-based initialization [SCPF09]

power and capacity from measurements at 285 MHz. However, it has been shown that this approach comes with several deficiencies including (i) a huge number of MPCs is required to obtain a decent fit to a given channel realization [SCPF09, MRAB05], (ii) depending on the initialization technique, the path estimates may be concentrated only on the early, high-power part of the delay profile [SCPF09], and (iii) inference based on describing the channel with propagation paths only (2.5) may result in false conclusions on the performance measures such as channel capacity [RSK06].

In [SCPF09], the authors propose an alternative approach for initializing high-resolution propagation path estimation algorithms with the aim of extracting channel features over the whole delay range of the channel response. The late arriving parts have low power but still contain information which may be useful for, e.g., localization purposes. The contribution [SCPF09] considers the extraction of such features by designing initialization steps which do not only consider the minimization of the residual power. Instead of using conventional Successive Interference Cancellation (SIC) methods for detecting paths from the residual signal based on the l_2-norm, the algorithms are initialized by either (i) dividing the effective delay range into a number of equidistant bins and using a fixed number of starting values for path estimates within each bin or (ii) detecting the maxima in the PDP or Azimuth-Delay Power Profile (ADPP) and assigning a number of initial path estimates for each of such maxima. A simple measure such as the widely used ratio of the residual power versus the measured power is inadequate to evaluate the channel estimates. As can be seen in Fig. 2.26, the lower residual power in the early part of the channel response would lead with this measure to the conclusion that the SIC-based method provides better channel parameter estimates. However, it neglects the MPCs in the later part of the channel response almost completely. The proposed approaches of [SCPF09] provide path estimates for a wider delay range, yielding a more realistic estimate of the overall radio channel. Figure 2.26 shows an estimation example comparing SIC and ADPP-based initialization.

While leading to a better extraction of path parameters in the late part of the response, the estimation method suggested in [SCPF09] is not concerned with catching *all* the power in the full response. Thus a significant portion of the power in the channel remains unaccounted for, see the residual PDPs in Fig. 2.26. Another approach introduces a so-called Dense Multipath Component (DMC) as a stochastic part $\mathbf{h}_D \sim \mathcal{CN}(\mathbf{0}, \mathbf{R}_D)$ in the channel model [Ric05, Sal09], yielding

$$\mathbf{h} = \mathbf{h}_S + \mathbf{h}_D + \mathbf{n}_w. \tag{2.6}$$

The DMC describes the contribution in the radio channel from diffuse scattering, which is difficult and computationally exhausting to capture using a superposition of deterministic propagation paths (2.2) only. From the point of view of high-resolution propagation path parameter estimation, the model (2.6) can be simplified as

$$\mathbf{h} = \mathbf{h}_S + \mathbf{n}_D, \tag{2.7}$$

where $\mathbf{n}_D = \mathbf{h}_D + \mathbf{n}_w \sim \mathcal{CN}(\mathbf{0}, \mathbf{R}_D + \sigma_n^2 \mathbf{I})$. Furthermore, [Ric05, Sal09] suggest the use of a Kronecker model to describe the DMC covariance matrix as $\mathbf{R}_D = \mathbf{R}_R \otimes \mathbf{R}_T \otimes \mathbf{R}_f$. The Kronecker model provides computational benefits and limits the number of free parameters, which results in good identifiability of the model. However, due to the presence of paths $\mathbf{h}_S$ (2.2), the overall channel (2.7) is not a Kronecker model.

To conclude, the benefits of having the DMC in the channel model (2.7) include: (i) the detection of significant propagation paths over the whole parameter range of interest is improved, (ii) a better fit of the overall channel is obtained, and (iii) computational complexity of the resulting parameter estimator is reasonable. A downside of the currently employed DMC models is that they do not directly support the separation (deembedding) of the radio propagation channel from the antenna arrays. However, active research to overcome such limitations is currently taking place, see [KLT09, SPH+10, QOHDD10]. For more discussion on DMC, see Sect. 3.3.

In [SSM10], the MPC model (2.4) was extended to a frequency-dependent, wideband antenna array model. A wideband antenna array model is necessary if the frequency response of the antennas varies over the signal band, or if the array size becomes significant.[3] The frequency dependence of the wideband antenna array model invalidates the Kronecker separability of (2.4). However, one of the benefits of the wideband modeling is that the commonly assumed $\lambda/2$ restriction on the antenna spacing may be relaxed. This is illustrated in Fig. 2.27 for a 10-element Uniform Linear Arrays (ULA) with 5-cm antenna spacing. The ambiguity function (correlation of the antenna array response at different angles) is defined as

$$C(\phi_i, \phi_j) = \frac{\text{vec}(\mathbf{B}_r(\phi_i))^H \text{vec}(\mathbf{B}_r(\phi_j))}{\|\text{vec}(\mathbf{B}_r(\phi_i))\|_F \|\text{vec}(\mathbf{B}_r(\phi_j))\|_F}, \tag{2.8}$$

where $\mathbf{B}_r(\phi_i) \in \mathbb{C}^{M_r \times M_f}$ denotes the wideband "steering matrix." The wideband model does not suffer from the angular ambiguity present at the highest frequency

---

[3]The antenna array size can be considered insignificant while $D \ll c/B$, where D is the diameter of the smallest sphere enclosing the array, c is the speed of light and B is the signal bandwidth.

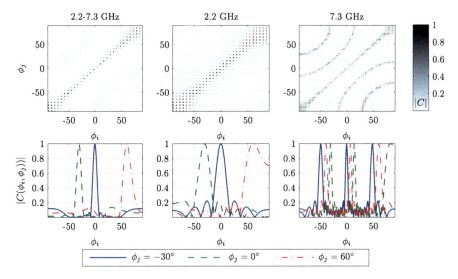

Fig. 2.27 Ambiguity function of a (ULA-10, 5 cm antenna separation) for UWB (2.2–7.3 GHz) antenna array model compared to narrowband model at 2.2 GHZ ($\lambda_l \approx 13.6$ cm) and 7.3 GHz ($\lambda_h \approx 4.1$ cm) [SSM10] (©2010 IEEE, reproduced with permission)

and has also improved resolution (narrower mainlobe) than at the lowest frequency [SSM10].

2.2.1.2 Algorithms

The notion of high-resolution parameter estimation implies the ability of an algorithm to distinguish between multiple closely separated and/or correlated propagation paths. Figure 2.28 provides a classification of available parameter estimation techniques. The methods that are currently widely used and can be classified as high-resolution techniques include subspace-based techniques such as Estimation Of Signal Parameters via Rotational Invariance Techniques (ESPRIT) [PRK85] and its variations [HZMN95], RAnk Reduction Estimator (RARE) [PMB04], Maximum Likelihood (ML)-based methods including SAGE [FTH+99] and Iterative Gradient-Based ML Parameter Estimation Algorithm (RIMAX) [Ric05], as well as sequential estimation (Bayesian filtering) based Extended Kalman Filter (EKF) [SRK09] and particle filter [YSK+08]. Recent summaries of the available high-resolution parameter estimation techniques can be found in [Yin06, Sal09].

The subspace-based methods rely on the model (2.5), with the additional assumption of $\mathbf{h}_s(\Theta_p) \sim \mathcal{CN}(\mathbf{0}, \mathbf{R}_S)$. The eigenvalue decomposition of the data covariance matrix is then given by

$$\tilde{\mathbf{R}} = \mathcal{E}\{\tilde{\mathbf{h}}\tilde{\mathbf{h}}^H\} = \mathbf{U}\boldsymbol{\Lambda}\mathbf{U}^H = \mathbf{U}_s\boldsymbol{\Lambda}_s\mathbf{U}_s^H + \mathbf{U}_w\boldsymbol{\Lambda}_w\mathbf{U}_w^H, \qquad (2.9)$$

where $\mathbf{U}_s \in \mathbb{C}^{\tilde{M} \times P}$ and $\boldsymbol{\Lambda}_s \in \mathbb{R}^{P \times P}$ denote the eigenvectors and eigenvalues associated with the signal subspace with P sources, whereas $\mathbf{U}_w \in \mathbb{C}^{\tilde{M} \times (\tilde{M} - P)}$ and

Static Model		Dynamic Model
Spectral-based	Maximum likelihood	Kalman filters
- Beamforming (Bartlett, MVDR)	Stochastic ML	- Extended Kalman filter (EKF)
Subspace-based	Deterministic ML	- Unscented Kalman filter (UKF)
- MUSIC (spectral)	- SAGE	Sequential Monte Carlo
- ESPRIT, RARE (parametric)	- RIMAX	- Particle filter

Fig. 2.28 Classification of parameter estimation techniques [Sal09]

$\Lambda_w \in \mathbb{R}^{(\tilde{M}-P) \times (\tilde{M}-P)}$ denote those of the noise subspace, which is orthogonal to the signal subspace. The symbol $\tilde{M} \leq M$ denotes the length of vector $\mathbf{h}$, which may be only a subset of the complete data model. A limitation of the subspace-based techniques is the fact that they require a sufficient number of realizations to obtain reliable estimates of the subspaces. ESPRIT is further applicable only to array geometries, where the array may be divided into a number of equidistantly spaced and identical subarrays.

The multidimensional Unitary ESPRIT algorithm was recently applied in [KH08, KdJBH08] for nonstationary SIMO measurements. The measurement setup employed a 31-element tilted cross array with three 11-element subarrays. An estimate $\hat{\mathbf{R}}$ of (2.9) was obtained from 10 consecutive snapshots. This allowed one to identify a maximum of 10 Direction of Arrival (DoA) estimates for each delay bin. In other words, the delay estimates $\hat{\tau}$ were in fact not high-resolution, as they were based on the Channel Impulse Response (CIR) delay bins.

Other contributions in subspace-based high-resolution parameter estimation methods include [GR07b], where a Root-MUSIC-based 2-D estimation algorithm was proposed for cylindrical antenna array configuration, as well as [GR07a], where RARE [PMB04] was employed for DoA estimation.

The ML-based methods, i.e., SAGE and RIMAX are perhaps currently the most commonly applied methods for high-resolution propagation parameter estimation. They are based on optimizing the initial parameter estimates, typically obtained by a SIC-based global grid search procedure, by assuming either the data model (2.7) or (2.5), and the probability density $\mathbf{h} \sim \mathcal{CN}(\mathbf{h}_S(\boldsymbol{\Theta}), \mathbf{R})$. The parameter estimates are found by maximizing the corresponding likelihood function

$$p(\mathbf{h}|\boldsymbol{\Theta}, \mathbf{R}) = \frac{1}{\pi^M \det(\mathbf{R})} e^{-(\mathbf{h}-\mathbf{h}_S(\boldsymbol{\Theta}))^H \mathbf{R}^{-1}(\mathbf{h}-\mathbf{h}_S(\boldsymbol{\Theta}))}. \qquad (2.10)$$

Both SAGE and RIMAX are iterative procedures for finding the optimal parameters. In SAGE, the likelihood function (2.10) is maximized w.r.t. a subset of (or even a single) parameters at a time, keeping the rest of the parameters fixed. Each complementary subset is optimized in alternating fashion until convergence is reached. The beauty of SAGE is that the computational complexity of a single update can be controlled by the choice of the size of the parameter subset. On the other hand, a poor choice of subsets may reduce the convergence speed. RIMAX relies on a gradient-based optimization algorithm, which typically leads to faster convergence.

The parameters may be updated jointly or in subsets to reduce computational complexity. However, the parameter subsets, e.g., groups of paths, should be arranged so that correlated MPCs are optimized jointly. Also the introduction of the DMC (2.7) to support the high-resolution estimation of propagation path parameters was originally proposed for RIMAX [Ric05]. Both SAGE and RIMAX have been widely used within the COST 2100 framework, see, e.g., [CTW+07, SCPF09, KST08, ZTE+10, NKJ+10, BDHZ10, EWLT10].

A parametric study comparing the DoA estimation performance of beamforming, MUltiple SIgnal Classification (MUSIC), ESPRIT, and SAGE was conducted in [SGL07]. Simulations using multiple paths with different power levels resulted in the conclusion that SAGE outperforms the other compared methods, especially in terms of estimation accuracy for the paths other than the strongest one. In [GTS+10], an attempt was made to compare the performance of ESPRIT and RIMAX in estimating the MPC parameters in the presence of DMC. Although RIMAX clearly outperforms ESPRIT as expected, it is also evident that there would be much room for improvement in terms of correct pairing of the simulated parameters with the estimated ones.

The contribution [SKI+10a] proposes an approach, where initial MPC estimates are found using SAGE, after which they are clustered. The MPCs of each cluster are then refined using an extension of the subspace-based Method Of Direction Estimation (MODE) [SS90] algorithm. The approach in [SKI+10a] is motivated by the degraded performance of SAGE while estimating coherent signals.

The algorithms presented so far assume that the propagation path parameters remain constant during the observation period and are more or less independent from an observation to another. However, as channel sounding measurements are often performed in a dynamic, mobile environment, a state-space model along with a sequential estimation technique may be employed. Two such Bayesian filtering approaches have been proposed recently, one relying on the EKF [SRK09] and another utilizing the particle filter [YSK+08]. In both methods, the propagation path parameters $\boldsymbol{\Theta}$ (2.3), as well as their rate of change $\Delta\boldsymbol{\Theta}$, are modeled as the state $\boldsymbol{\theta} = [\boldsymbol{\Theta} \ \Delta\boldsymbol{\Theta}]^T$ of the system. The time evolution is then governed by a linear state transition equation

$$\boldsymbol{\theta}_k = \mathbf{F}_k \boldsymbol{\theta}_{k-1} + \mathbf{v}_k, \qquad (2.11)$$

where $\mathbf{v}_k \sim \mathcal{N}(\mathbf{0}, \mathbf{Q}_k)$ is the state noise vector at time k. The measurement equation is defined as the nonlinear mapping of the state variables to the observation, i.e., the measured channel as

$$\mathbf{y}_k = \mathbf{h}(\boldsymbol{\theta}_k) + \mathbf{n}_k, \qquad (2.12)$$

where $\mathbf{n}_k$ is the measurement noise vector. Both [SRK09] and [YSK+08] assume complex Gaussian entries for $\mathbf{n}_k$. However, in [SRK09], the measurement noise was colored due to the inclusion of the DMC (2.7), whereas [YSK+08] assumed white complex Gaussian noise (2.5).

The EKF relies on a Taylor series approximation for linearizing the nonlinear measurement equation (2.12) about the current parameter estimates. The computational complexity of the EKF solution is comparable to that of a single iteration of the RIMAX algorithm, typically leading to about an order of magnitude improvement in terms of the overall computation time [SRE+06]. The EKF method has been applied for processing vast channel sounding data sets, and the results have been used for example in analyzing indoor scattering mechanisms in [PHS+09b].

The particle filter works by using a set of particles, drawn from the assumed probability densities of the parameter space, to approximate the joint distribution that best fits the observed measurements. The advantage of the method is that it is not restricted to Gaussian probability densities, and it allows the propagation paths to have some spread in the respective parameter domains. A method was presented in [YSK+08] to avoid a large number of closely separated particles due to typically highly concentrated paths. This method achieves a computational complexity lower than the SAGE algorithm.

Both EKF and particle filter-based sequential estimation approaches have been reported to outperform single-snapshot ML-based algorithms in terms of parameter estimation accuracy in simulations [YSK+08, SRE+06].

2.2.1.3 Clustering of Propagation Path Parameters

It is a widely accepted assumption that the propagation paths typically appear in so-called clusters. Clustering is also a convenient approach from the channel simulation point of view, as it provides a well-defined framework for drawing time-evolving channel realizations. Clustering has formed a basis for several standardized models, including COST 273 [Cor06].

In order to parameterize cluster-based channel models, it is necessary to perform clustering of the high-resolution propagation path parameter estimates and track them over time. Such an approach was proposed in [CTW+07], where a KPower-Means clustering algorithm using the Multipath Component Distance (MCD) distance metric [CCS+06] was suggested for clustering windows of data, and a Kalman filter was applied to predict and track the cluster positions over time. The authors in [CTW+07] also propose to use a so-called *closeness function* as a metric while associating the clusters over time. A 3D-illustration of the clustering results for a single time window is shown in Fig. 2.29. This clustering method was also employed in [ZTE+10] in order to parameterize a cluster-based channel model. A similar approach, but using particle filters for cluster tracking, was described in [SKI+10b].

An alternative clustering approach using the so-called *hierarchical whole clustering method* [JD88] was proposed in [KH08]. Comparing the results based on the two clustering approaches [CTW+07] and [KH08], it is evident that the definition of path clustering is somewhat ambiguous. Namely, the results in [KH08] indicate that the clusters have Root-Mean-Square (RMS) angular spreads in the order of one degree or even less. One can then argue whether such results are in fact uncertainty resulting from estimating and tracking a single propagation path instead of a cluster of paths, as the approach yields similar results as the sequential estimation algorithms (single-path tracking) [SRK09, YSK+08], where the association of the paths over time is implicitly obtained.

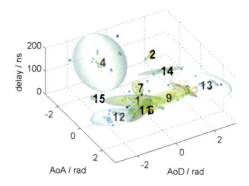

Fig. 2.29 An example of propagation path clustering in an indoor scenario [CTW+07] (©2007 IEEE, reproduced with permission)

As clustering is based on the estimated propagation paths which typically represent only a portion of the propagation channel, see, e.g., [RSK06, SCPF09], it can be hard to quantify in retrospect how much of the overall channel remains unaccounted for. Recently, a cluster-based model for the DMC was proposed in [SPH+10] to overcome this limitation. However, no method for automatic extraction of the parameters of such a model has been reported so far.

2.2.2 Scatterer Localization and Visualization

The development of radio channel models should always be based on thorough understanding of the physical propagation phenomena. To obtain this understanding, it is essential to identify underlying propagation mechanisms, such as reflections, diffraction, or scattering from physical objects of propagation environments. However, understanding all the relevant propagation mechanisms by manual comparison of the measurement results with the maps or photos of the environment is a cumbersome task. Therefore it is helpful to develop simple and efficient scatterer localization and visualization tools and methods. In this subsection four examples of such tools and methods are introduced [KH07, PHS+09a, CPD10, SKA+08, MB10]. Fundamental ideas of those tools and methods are to relate the data obtained from channel measurements to the physical scattering objects of the surrounding environment.

Poutanen et al. [PHS+09a] developed a so-called measurement-based ray tracer. The ray tracer combines a digital map of the environment with high-resolution directional parameter estimates obtained from the measurement data. The environment is described by representing surfaces of the objects by straight lines, and rays are plotted on top of the map from the locations of transmit and receive terminals. The directions of the rays are based on the measured Direction of Departure (DoD) and DoA, and the rays hit a surface of a certain object. In this way the scattering points and propagation paths of each multipath component can be identified. Figure 2.30 shows an example of identified scattering points from a double-directional propagation measurement in an indoor hall environment. The Transmitter (Tx) was

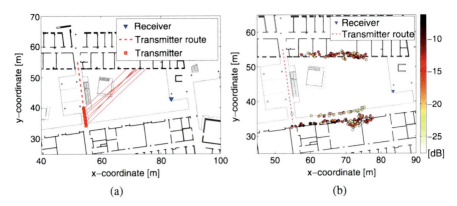

Fig. 2.30 Example of identifying propagation paths of rays in an indoor hall scenario using the measurement-based ray tracer: (**a**) Rays are launched from Tx to the direction of DoD, and a specular reflection is assumed to update the signal direction after the ray hits the wall. (**b**) Scattering points of single bounce paths on two walls. They are calculated by intersection of rays launched from Tx and Rx

moved along a route shown in a red dashed line, while the Receiver (Rx) was fixed. Figure 2.30(a) shows that rays are launched from consecutive Tx locations along a measurement route to the direction of the DoD, and hit a surface of a wall defined in an AutoCAD drawing of the environment. By calculating a specular reflection on the wall surface and by using the delay information of each multipath component, the rays arrive approximately at the Rx. Figure 2.30(b) shows scattering points calculated by the intersection point of rays launched from Tx and Rx for single-bounce paths from the two walls. The color of scattering points represents the relative power level of the rays. It is clearly seen on the map that the rays from scattering points between $y = 30$ and 40 m stem from both wall reflections and reflections from a fence along the corridor. In this way the measurement-based ray tracer enables us to analyze contributions of different propagation mechanisms and also to group multipaths originating from same scattering processes to form physically motivated clusters.

Kwakkernaat and Herben [KH07], Cornat et al. [CPD10], and Medbo [MB10] developed tools to plot measurement results on the photograph of the environment. The tool is a convenient way to physically analyze the channel behavior. In addition, since complex channel measurements are prone to errors, the tool can be used to verify the soundness of the measured data. The basic idea of the proposed approach is to generate a 3-D panoramic photograph by taking several photos that slightly overlap and stitch them together. A 3-D panoramic photograph covers 180 degrees in elevation and 360 degrees in azimuth, so it can be combined with the measured power angular spectrum and angles of propagation path parameter estimates that cover the same range. The combination of the 3-D panoramic photograph and the measured power angular spectrum is called a radiophoto in [CPD10]. Such a photograph was first reported in [KH07] for their macrocellular outdoor measurements in Eindhoven, the Netherlands, to plot parameter estimates from the 3-D Unitary ESPRIT as shown in Fig. 2.31(a). The photo was taken by a camera located directly

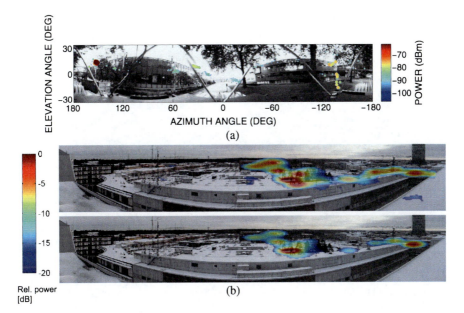

Fig. 2.31 (a) A photograph of a macrocellular environment seen from the MS, Eindhoven, the Netherlands. Propagation path estimates from the 3-D Unitary ESPRIT are overlaid on the photo. (b) Aggregate angular spectrum overlaid on a panoramic photo taken from the BS. The measurement was in a macrocellular scenario, Kista, Sweden

underneath a mobile antenna array that was mounted on a rooftop of a measurement vehicle. Similarly, a radiophoto obtained in an outdoor macrocellular scenario in the city center of Mulhous, France, is shown in [CPD10]. The photos clearly highlight the main propagation mechanisms, such as diffraction over the rooftops, reflections from surrounding buildings, and street canyons toward the direction of base station. The photo on the Base Station (BS) side was introduced in [MB10] for outdoor macrocellular scenario in Kista, Sweden, at 5.25 GHz as shown in Fig. 2.31(b). The angular power spectrum overlaid on the photo was derived by aggregating all the propagation paths estimated from 10 different Mobile Station (MS) locations. The power spectrum in the upper figure was obtained after removing the pathloss difference between BS-MS links, while that in the lower figure was derived with the link pathloss difference preserved. The photo revealed that propagation paths between the BS and the MSs were clustered in directions where building walls or roof edges have LOS conditions to both the BS and the MSs.

Santos et al. [SKA+08] established a method which identifies scatterers using successive interference cancellation (a detect-and-subtract method) for ultrawideband channel measurements. The novelty of the work is to map peaks of channel impulse responses detected in the delay domain to scatterers on the two-dimensional spatial domain. To this end, the channel measurement was carefully designed; a single Tx antenna was put on a fixed position, while a quite large array which stretched beyond the channel stationarity region was formed by moving a single antenna on

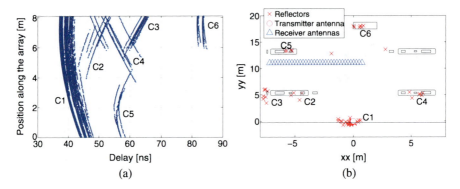

Fig. 2.32 Propagation path detection and scatterer identification in outdoor petrol station by ultrawideband channel sounding: (**a**) Detected scatterer peaks in the delay domain. (**b**) Map of the measurement scenario and the calculated spatial positions of scatterers. Petrol pumps and shop building are also shown [SKA[+]08] (©2008 IEEE, reproduced with permission)

the Rx side. The scatterer identification was made possible by assuming a single bounce of each scattered wave. An example of such measurements and scatterer identification is reported for outdoor measurements in a petrol station in Lund, Sweden. The scatterers were detected in the delay domain for successive Rx positions as shown in Fig. 2.32(a) and were related to the physical objects in the spatial domain as shown in Fig. 2.32(b). The position of Tx and Rx antennas as well as the most significant objects in the channel, such as the petrol pumps and the shop wall, are also shown. Almost all the scatterers identified in the measurements corresponded to physical objects, which was a strong indication that single-bounce reflections were dominating in this scenario. The method is, therefore, of great value for ultrawideband channel modeling and for automated generation of environment maps.

2.2.3 Impact of System Calibration

When conducting measurements of the mobile radio channel, it is important to keep in mind that the measurement data is always influenced by the measurement system. This means the data always contains both the actual channel transfer function $H(f,t)$ and the characteristics of the measurement system. Another very important aspect is the impact of the antenna arrays used during measurements. In general, the beampatterns of the arrays lead to a spatial filtering of the mobile radio channel. Thus, the measured channel transfer function can differ depending on the characteristics of the antennas. This influence can only be removed by high-resolution parameter estimation or to some extent by beamforming [WHR[+]10]. In order to perform these procedures, a precise knowledge of the antenna beampatterns is necessary.

2.2.3.1 Calibration of Receiver/Transmitter

In general, both the receiver and transmitter possess a frequency response that is not flat. This can be caused by, e.g., power amplifiers at the transmitter or low-noise amplifiers at the receiver. The impact on the measured transfer function of the radio channel is that it is multiplied with the transfer function of the receiver and transmitter, respectively:

$$\widehat{H}(f,t) = G_{sys,Rx}(f) \cdot H(f,t) \cdot G_{sys,Tx}(f), \tag{2.13}$$

where $G_{sys,Rx/Tx}$ are the time-independent transfer functions of the receiver and transmitter, respectively. Since the transfer functions of the measured device are assumed to be constant over time, it is possible to remove their impact by a single Back-2-Back system calibration. This procedure is performed by directly connecting the receiver and transmitter and conducting a measurement. The resulting transfer function $C(f)$ is then stored and later on used to get the estimate of the actual channel transfer function:

$$C(f) = G_{sys,Rx}(f) \cdot G_{sys,Tx}(f), \tag{2.14}$$

$$\widetilde{H}(f,t) = \frac{\widehat{H}(f,t)}{C(f)}. \tag{2.15}$$

Some measurement devices are equipped with an Automatic Gain Control (AGC) which controls the input power of the internal components of the system. This is useful in order to avoid any nonlinear effects of, e.g., amplifiers or analog-digital-converters. It is important to note that in this case it is necessary to perform the Back-2-Back calibration for each of the different AGC settings. The necessity to perform a Back-2-Back system calibration stems from the purpose of a channel sounder to measure the transfer function of the radio channel without the influence of the measurement system as accurately as possible. In an actual communication system, on the other hand, there is no need to separate the radio channel from the receiver or transmitter system; instead they can be treated jointly. For the communications systems, it is only important that there is a frequency-selective transfer function, but it is not important whether the frequency selectivity arises from the multipath propagation in the environment or is induced by the RF-components of the receiver or transmitter.

Nevertheless, it is still necessary to measure the (joint) channel transfer function. This information (channel state information) is later used to equalize the influence of the total channel (e.g., in an OFDM-System) or is used to exploit certain properties of the channel (like water pouring in a MIMO system). Since the RF-components of the transmitter and receiver are in general not identical, it is necessary to measure both the downlink (transmitter to receiver) and the uplink (receiver to transmitter). The information of, e.g., the uplink is then transferred back to the transmitter to get the full CSI. If the uplink and downlink are reciprocal, the knowledge about either of them can be obtained by a procedure called relative calibration [KJGK10].

As stated above, the transfer function of the measurement device is presumed to be time invariant. While this is true for the magnitude, it might not be correct for the phase. Each system is affected by phase noise and phase drift of the receiver and transmitter, respectively. If this is not accounted for by different means, for example, by using a reference channel or highly stable reference signals, it might *pretend* a variation of the radio channel, thus leading to overestimation of metrics like channel capacity [CO09]. If no reference signal can be used, the impact on channel analysis depends on the magnitude of the phase noise. Since channel sounding measurements are conducted with highly specialized equipment, it can be assumed that the strength of the phase noise is low. In [GKT10b] Ghoraishi et al. used an OFDM-based channel sounder with the phase noise magnitude lower than −60 dBc/Hz to investigate the influence of phase variations. They used simulations and measurements to show that the impact of the phase noise on the channel impulse response can be neglected when the magnitude is below −50 dBc/Hz.

2.2.3.2 Calibration of Antenna Arrays

In the previous section the influence of the RF-components (of both the receiver and transmitter) on the frequency response of the system was discussed. Its impact can easily be equalized by applying some sort of system calibration. Another important aspect of channel measurements is the influence of the antenna arrays, particularly the influence of the spatial directivity of the individual array elements. In order to compensate for the effects of the antennas, it is common to perform high-resolution parameter estimation (see Sect. 2.2.1). This procedure is able to "deembed" the antennas from the channel measurements. One prerequisite of any estimator is the availability of the beam patterns of the antenna arrays. It is apparent that the quality of any subsequent parameter estimate is limited by the accuracy of the beam pattern. The beam pattern is usually obtained by measuring the response of an array inside an anechoic chamber. The accuracy of the beam patterns is, however, limited by the actual calibration procedure. Although the measurement is performed in an anechoic chamber, multipath propagation can still occur, although it is only attenuated to a certain level. Depending on the nature of these multipath (e.g., reflections close to the antenna under test), the measurements can be irreversibly distorted. In order to quantify the impact of the distortion, a measure about the quality of the data is required. In [KT10] it is suggested to use the measured beam patterns during a parameter estimation of a reference measurement—such as the calibration measurement itself—and to evaluate the accuracy of the estimation.

Besides these inherent limitations, the beam patterns are, however, often deliberately restricted in order to reduce the computational complexity of the estimation process or because of assumptions that are made about the radio channel or about the properties of the antenna arrays. Some of these assumptions are:

1. nominal single-polarized antennas with high Cross-Polar Discrimination (XPD) in the main-beam direction will keep the same XPD for different angle of arrival

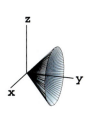
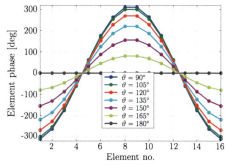

(a) DoAs that result in the same phase constellation at an ideal ULA (elements placed along y-axis, $\Delta\alpha = 45°$)

(b) Phase over 16 element UCA, ⌀ 10.5 cm at 5.2 GHz for incidence at different co-elevations ϑ.

Fig. 2.33 Phase over 16-element UCA (©2007 EurAAP, reproduced with permission)

2. nominal uniform arrays exhibit equal beam patterns for all elements, thus ignoring mutual coupling
3. if only azimuthal properties are to be examined, it is sufficient to use only azimuthal "cuts" of the full-3D beam pattern
4. if an array is used that cannot resolve the Elevation of Arrival (EoA), the estimates of the AoA are still correct

It must be said that the last assumption is different from the others in the sense that it cannot be resolved by using the full polarimetric 3D beam pattern; in fact, it is a consequence of the choice of the array used. If these assumptions are violated—which is nearly always the case with practical arrays—and still the restricted beam patterns are used, it can lead to biased estimates or even cause artifacts that do not reflect the radio channel [LKT07]. Especially the emergence of artifacts is harmful since it may pretend clusters in the propagation environment, which are not physically correct.

The ULA is a type of antenna array that reveals the consequences of the fourth assumption. Since it is ambiguous with respect to the elevation angle, it is often only used to examine the AoA, thus, using only an azimuthal cut of the beam pattern. The ambiguity of the ULA lies in its inability to resolve any paths in the elevation domain. Nevertheless, that does not mean that the beam pattern of the ULA will not change with elevation. This is particularly true for the phase variation over the array elements. This phase variation is the most important information for an estimator to determine the direction of the impinging wave. If the phase variation over the elements is unique for all pairs of angles in azimuth and elevation, the array is unambiguous. If there are multiple pairs of angles that lead to the same phase pattern, it is impossible to reliably decide which angle is the correct one. In Fig. 2.33 one can see the "ambiguity-cone" of the ULA. It shows that any impinging wave whose direction of arrival is on this cone will lead to the same phase pattern on the array elements. If an azimuthal cut of the beam pattern is used and the cut is not performed for the correct elevation angle, this will lead to biased estimates for the azimuth angle. Note that this problem can only be solved for the ULA by suppressing waves

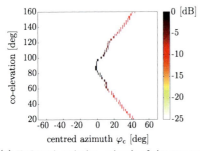 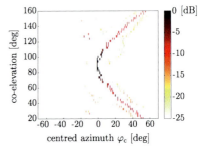

(a) Horizontal excitation, azimuth of the true path $\varphi = -70°$.

(b) Vertical excitation, azimuth of the true path $\varphi = -70°$.

Fig. 2.34 Power distributions of the estimated paths as function of the centered azimuth (estimation bias) around the true path and co-elevation using the full polarimetric 1D data model of a 8-element ULA for $\varphi = -70°$ of the true path from top to bottom and for horizontal (*left*) and vertical excitation (*right*) [LKT07] (©2007 EurAAP, reproduced with permission)

Fig. 2.35 Estimation artifacts (bias) around the true path using the azimuth-only data model of a UCA [LKT07] (©2007 EurAAP, reproduced with permission)

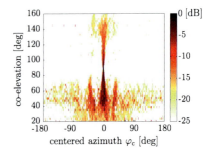

with other elevation angles already during the measurements. A better approach is to use a Uniform Circular Array (UCA) where the phase variation is unique, at least for one hemisphere of the spherical coordinate system, or to use a stacked UCA or other three-dimensional array configurations. Figure 2.34 shows some examples of estimation results using a ULA where estimations where performed with azimuth-only beam patterns that where derived for an elevation angle (90° coelevation in this case) that do not match the elevation angles of the actual propagation paths. It can be seen that the AoA is only estimated correctly for a coelevation of 90°, which is the angle where the EoA of the impinging wave matches the EoA of the azimuth-only beam pattern. It can be seen that although the bias of the estimates is sometimes high, the spread, i.e., the estimation artifacts around the true path, is relatively small. This is caused by the ambiguity of the ULA. In other words, within the 1D-beam pattern, there exists a phase pattern that, although belonging to a different combination of AoA and EoA (compared to the true parameters) allows the estimator to find a proper match with the received signal. If an array type is used that does not have this property and still a 1D-beam pattern is used, the situation is completely different. Figure 2.35 shows the estimation artifacts that arise when a UCA is used with only a 1D-beam pattern. In this case it is not possible for the estimator to find a suitable phase pattern within the 1D-beam pattern if the EoA of the impinging wave is different from the EoA the 1D-beam pattern was created for. This causes the estimator to introduce additional artificial paths to match the received signal.

Concluding, the impact of both the array calibration accuracy and the simplification of the data model depend heavily on the types of analyses or simulations that are to be performed using the parameter estimates. In [EWLT10] the authors investigated the reliability of estimation results by synthesizing channels and comparing them to actual measurement data. It was found that a good agreement w.r.t. to power and MIMO channel capacity can be achieved. However, using this approach, it is not always possible to determine the physical relevance of the estimated propagation parameters, since the authors used the same antenna array (array calibration data) for the synthesis that was used for the estimation. The estimation procedure aims to describe the measured data as accurately as possible, and if the accuracy of the array calibration data is limited, the estimator will compensate for this by introducing artifacts. Thus, a synthesis using the same (limited) calibration data will again lead to an accurate-as-possible description of the measurement data. Therefore, the estimation results in [EWLT10] are also validated using visual inspection showing some possibilities of estimation artifacts. In [KSKT10] measurement data is used to derive an antenna-independent description of the radio channel and to synthesize data for a different antenna. The accuracy of the synthesis is again verified using the MIMO channel capacity. Since a different array is used for the synthesis, the impact of estimation artifacts can be observed in the results. It is furthermore shown that the impact of the estimation errors increases with rising SNR of the simulations. Therefore, it can be concluded that the calibration of the measurement devices and the validity of the data model ultimately limit the usefulness of any subsequent processing of the data.

2.3 Lessons Learned

The previous paragraphs presented a technical overview of the wide range of activities done in the field of radio channel measurements. This last part shall give a short insight into the practical problems when planning new field measurements. No matter for which kind of technology the radio measurements shall be done, there are a number of basic points to be always considered before starting out.

- *Focus of the measurements*—The most important thing that one should ask oneself before a measurement campaign is: What do I expect to learn from the measurements? Only after having a clear and concise answer to this question, meaningful measurements can be designed. This may sound straightforward, however, it happened in the past that measurements have been performed without a defined focus, which obviously lead to quite limited results.
- *Equipment*—Given the focus, one needs to decide which equipment can fulfill the needs. For already existing specialized measurement equipment, it might turn out that it does not fit the purpose. In this case, either the focus or the equipment need to be updated.

Furthermore, recent experience showed that combining specialized channel sounding equipment of different makes is extremely challenging and time consuming. It is of great importance to test the equipment thoroughly before going to the field trials. Every glitch that is not captured before the measurements multiplies time and cost of postprocessing.

It is also necessary to plan for additional tools and items, like connectors, converters, extension cords, but also for cars, gas, power generation, or simply for having a great supply of duct tape.

- *Schedule*—Allow and plan for enough time. In the preparations, the exact measurement locations, equipment settings, and measurement parameters should be defined. This may also involve doing some sample measurements before the campaign starts to adjust the parameters. No doubt, the planning of a campaign takes much longer than the measurement campaign itself.

  For the period of the campaign also include some extra time in your planning for solving unforeseen problems. Measurement equipment is generally error-prone since it is usually experimental equipment. Having a specialized channel sounder at hand usually helps a lot but does not completely prevent faults from happening.

  Finally, the large amounts of data collected need to be evaluated. One should already have a time schedule for the evaluations agreed as well.

- *Documentation*—Meaningful documentation is the most essential part of every measurement campaign. What does an abundance of data help if one does not know where (and with which parameter settings) it was collected? Using so-called "field note sheets" is encouraged, where the measurement location, settings, possible photographs, and other important information can be written during the measurement.

- *People*—Having enough helpers makes life much easier. Distributing tasks (coordination, documentation, measuring, assisting the measurement team) allows everybody to focus on their part and leaves the coordinator to oversee and guide the action.

As a final comment, radio channel measurements are always a challenge, but when keeping to a well-prepared plan, the results merit the efforts.

References

[AAM09] A. Adalan, H. Arthaber, and C. Mecklenbraeuker. On the potential of IEEE 802.15.4a for use in car safety and healthcare applications. Technical Report TD-09-865, Valencia, Spain, May 2009.

[AFG+09] A. Adalan, M. Fischer, T. Gigl, K. Witrisal, A. L. Scholtz, and C. F. Mecklenbraeuker. Ultra-wideband radio pulse shaping filter design for IEEE 802.15.4a transmitter. In *Proc. WCNC 2009—IEEE Wireless Commun. and Networking Conf.*, Budapest, Hungary, April 2009.

[AGFW09] A. Adalan, T. Gigl, M. Fischer, and K. Witrisal. A modular impulse radio ultra-wideband research & development platform for IEEE 802.15.4a. In *European Conference on Wireless Technologies, ECWT*, pages 116–119, Rome, Italy, September 2009.

[ASQ09] J. Ahmadi-Shokouh and R. C. Qiu. Ultra-wideband (UWB) communications channel measurements—a tutorial review. *Int. J. Ultra Wideband Communications and Systems*, 1(1):11–31, 2009.

[BDHZ10] R. J. C. Bultitude, G. S. Dahman, R. H. M. Hafez, and H. Zhu. Double directional radio propagation measurements and radio channel modelling pertinent to mobile MIMO communications in microcells. Technical Report TD-10-11019, Aalborg, Denmark, June 2010.

[BLM04] J. R. Barry, E. A. Lee, and D. G. Messerschmitt. *Digital Communication*, 3rd edition. Kluwer, Boston, USA, 2004.

[BS87] P. Beckmann and A. Spizzichino. *The Scattering of Electromagnetic Waves from Rough Surfaces*. Artech House, Norwood, 1987.

[CAS08] L. Clavier, N. Azzaoui, and W. Sawaya. UWB and 60 GHz channel model as an α-stable random process. Technical Report TD(08)634, Lille, France, October 2008.

[CBVV+08a] N. Czink, B. Bandemer, G. Vazquez-Vilar, L. Jalloul, and A. Paulraj. Can multi-user MIMO measurements be done using a single channel sounder? Technical Report TD-08-621, Lille, France, October 2008.

[CBVV+08b] N. Czink, B. Bandemer, G. Vazquez-Vilar, L. Jalloul, and A. Paulraj. Stanford July 2008 radio channel measurement campaign. Technical Report TD-08-620, Lille, France, October 2008.

[CCC10a] Q. Chu, J.-M. Conrat, and J.-C. Cousin. On the impact of receive antenna height in a LTE-advanced relaying scenario. Technical Report TD-10-11005, Aalborg, Denmark, June 2010.

[CCC10b] Q. Chu, J.-M. Conrat, and J.-C. Cousin. Path loss characterization for LTE-advanced relaying propagation channel. Technical Report TD-10-12019, Bologna, Italy, November 2010.

[CCL+10] P. Chambers, P. Castiglione, L. Liu, F. Mani, F. Quitin, O. Renaudin, F. Sanchez-Gonzales, N. Czink, and C. Oestges. PUCCO radio measurement campaign. Technical Report TD-10-11015, Aalborg, Denmark, June 2010.

[CCS+06] N. Czink, P. Cera, J. Salo, E. Bonek, J.-P. Nuutinen, and J. Ylitalo. A framework for automatic clustering of parametric MIMO channel data including path powers. In *Proc. VTC 2006 Fall—IEEE 64th Vehicular Technology Conf.*, Montreal, Canada, September 2006.

[cha10] MEDAV RUSK channel sounder, 2010. http://www.channelsounder.de.

[CO09] N. Czink and C. Oestges. Impacts of channel sounder phase noise (forum discussion). Technical Report TD-09-738, Braunschweig, Germany, February 2009.

[Cor06] L. M. Correia, editor. *Mobile Broadband Multimedia Networks*. Academic Press, San Diego, 2006.

[CPB07] R. Cepeda, S. C. J. Parker, and M. Beach. The measurement of frequency dependent path loss in residential LOS environments using time domain UWB channel sounding. In *IEEE Intern. Conf. on Ultra-Wideband, ICUWB*, Singapore, September 2007. [Also available as TD(07)306].

[CPD10] J.-M. Conrat, P. Pajusco, and A. Dunand. On the use of panoramic photography for understanding propagation channel physical phenomena. Technical Report TD(10)12024, Bologna, Italy, November 2010.

[CTB08] R. Cepeda, W. Thompson, and M. Beach. On the mathematical modelling and spatial distribution of UWB frequency dependency. In *2008 IET Seminar on Wideband and Ultrawideband Systems and Technologies: Evaluating Current Research and Development*, pages 1–5, November 2008. [Also available as TD(08)456].

[CTT+09] C. Cordeiro, S. Trainin, J. Trachewsky, S. Shankar, Y. Liu, G. Basson, and J. Yee. Implications of usage models on TGad network architecture. In *IEEE Doc. 802.11-09/0391r0*, Vancouver, BC, Canada, March 2009.

[CTW+07] N. Czink, R. Tian, S. Wyne, F. Tufvesson, J.-P. Nuutinen, J. Ylitalo, E. Bonek, and A. F. Molisch. Tracking time-variant cluster parameters in MIMO channel measurements. In *The 2nd International Conference on Communications and Networking*

	in China (CHINACOM 2007), pages 1147–1151, August 2007. [Also available as TD(07)336].
[CZZ04]	S. Collonge, G. Zaharia, and G. E. Zein. Influence of the human activity on wideband characteristics of the 60 GHz indoor radio channel. *IEEE Transactions on Wireless Communications*, 3(6):2396–2406, 2004.
[DC08]	A. Dunand and J.-M. Conrat. Polarization behaviour in urban macrocell environments at 2.2 GHz. Technical Report TD-08-406, Wroclaw, Poland, February 2008.
[ECC03]	ECC. The european table of frequency allocations and utilizations covering the frequency range 9 kHz to 275 GHz. In *ERC Report 25*, pages 1–268, Dublin, Ireland, January 2003.
[ECM08]	ECMA. High rate 60 GHz PHY, MAC and HDMI PAL. In *ECMA-387*, 1st edition, pages 1–344, Geneva, Switzerland, December 2008.
[Ele10]	Elektrobit EB Propsim Homepage. 2010. http://www.propsim.com.
[ETM07]	G. Eriksson, F. Tufvesson, and A. F. Molisch. Characteristics of MIMO peer-to-peer propagation channels at 300 MHz. Technical Report TD-07-376, Duisburg, Germany, September 2007.
[EWLT10]	G. Eriksson, K. Wiklundh, S. Linder, and F. Tufvesson. Directional channel estimates from urban peer-to-peer MIMO measurements at 285 MHz. Technical Report TD-10-12099, Bologna, Italy, November 2010.
[FCC01]	FCC. Part 15-radio frequency devices section 15.255: operation within the band 57.0–64.0 GHz. In *Code of Federal Regulations*, pages 1–762, USA, January 2001.
[FCC02]	FCC. Revision of part 15 of the commission's rules regarding ultra-wideband transmission systems. First Report and Order, ET Doc. 98-153, FCC 02-48, Adopted: February 14, 2002, Released: April 22, 2002.
[FS10]	S. Feeney and S. Salous. Implementation of a channel sounder for the 60 GHz band. Technical Report TD-10-10043, Athens, Greece, February 2010.
[FTH$^+$99]	B. H. Fleury, M. Tschudin, R. Heddergott, D. Dahlhaus, and K. I. Pedersen. Channel parameter estimation in mobile radio environments using the SAGE algorithm. *IEEE J. Select. Areas Commun.*, 17(3):434–450, 1999.
[Gar10]	A. Garcia. The 60 GHz in-cabin channel. In *IEEE Doc. 802.11-10/0027r0*, pages 1–32, Los Angeles, CA, USA, January 2010.
[GBA$^+$09]	T. Gigl, T. Buchgraber, A. Adalan, J. Preishuber-Pfluegl, M. Fischer, and K. Witrisal. UWB channel characterization using IEEE 802.15.4a demonstrator system. In *IEEE Intern. Conf. on Ultra-Wideband, ICUWB*, pages 230–234, Vancouver, Canada, September 2009.
[GBG$^+$09]	T. Gigl, T. Buchgraber, B. Geiger, A. Adalan, J. Preishuber-Pfluegl, and K. Witrisal. Pathloss and delay spread analysis of multipath intensive environments using IEEE 802.15.4a UWB signals. Technical Report, Vienna, Austria, September 2009. [TD(09)965].
[GKBT09]	A. P. Garcia, W. Kotterman, D. Brückner, and R. S. Thomä. 60 GHz in-cabin channel characterisation and human body effects. Technical Report TD-09-756, Braunschweig, Germany, February 2009.
[GKKV10]	S. Geng, M. Kyrö, V.-M. Kolmonen, and P. Vainikainen. Feasibility Study of E-band Radio for Gigabit Point-to-Point Wireless Communications. Technical Report TD(10)10076, Athens, Greece, February 2010.
[GKT$^+$09]	A. P. Garcia, W. Kotterman, R. S. Thomä, U. Trautwein, D. Brückner, W. Wirnitzer, and J. Kunisch. 60 GHz in-cabin real-time channel sounding. In *Proc. Fourth Int. Conf. on Commun. and Networking in China (ChinaCOM2009)*, pages 1–5, Xi'an, China, 2009. [Also available as TD(09)877].
[GKT$^+$10a]	A. P. Garcia, W. Kotterman, U. Trautwein, D. Brückner, J. Kunisch, and R. S. Thomä. 60 GHz time-variant shadowing characterization within an Airbus 340. In *Prod. 4th European Conf. on Antennas and Propagation (EuCAP 2010)*, pages 1–5, Barcelona, Spain, 2010. [Also available as TD(09)970].

[GKT10b]   M. Goraishi, M. Kim, and J. Takada. Influence of phase noise on the frequency division multiplexing channel sounding. Technical Report TD-10-12044, Bologna, Italy, November 2010.

[GKZ+10]   A. P. Garcia, W. Kotterman, R. Zetik, M. Kmec, R. Müller, F. Wollenschläger, U. Trautwein, and R. S. Thomä. 60 GHz-ultrawideband real-time multi-antenna channel sounding for multi giga-bit/s access. In *IEEE 72nd Vehicular Technology Conference (VTC-Fall 2010)*, pages 1–6, Ottawa, Canada, September 2010. [Also available as TD(10)11090].

[GM+10]   A. P. Garcia, R. Müller, F. Wollenschläger, L. Xia, A. Schulz, M. Elkhouly, Y. Sun, U. Trautwein, and R. S. Thomä. Dual-polarized architecture for ultrawideband channel sounding at 60 GHz with digital/analog phase control based on 0.25 mm SiGe BiCMOS and LTCC technology. Technical Report TD-10-12015, Bologna, Italy, November 2010.

[GR07a]   R. Goossens and H. Rogier. 2-D direction-of-arrival estimation in the presence of mutual coupling by exploiting the symmetry in uniform circular array. Technical Report TD-07-214, Lisbon, Portugal, February 2007.

[GR07b]   R. Goossens and H. Rogier. Improved root-music based 2-d DOA estimation algorithm by considering a cylindrical antenna array configuration. Technical Report TD-07-370, Duisburg, Germany, June 2007.

[GTPPW10]   T. Gigl, F. Troesch, J. Preishuber-Pfluegl, and K. Witrisal. Maximal operating distance estimation using IEEE 802.15.4a ultra wideband. In *Workshop on Positioning, Navigation, and Communication, WPNC*, Dresden, Germany, March 2010.

[GTS+10]   D. P. Gaillot, E. Tanghe, P. Stefanut, W. Joseph, M. Lienard, P. Degaque, and L. Martens. Accuracy of specular path estimates with ESPRIT and RiMAX in the presence of diffuse multipath. Technical Report TD-10-12022, Bologna, Italy, November 2010.

[GTWH10]   C. Gustafson, F. Tufvesson, S. Wyne, and K. Haneda. Directional analysis of measured 60 GHz indoor radio channels in a conference room. Technical Report TD-10-12077, Bologna, Italy, November 2010.

[GWF+10]   A. P. Garcia, F. Wollenschläger, S. M. Feeney, S. Salous, and R. S. Thomä. 60 GHz channel sounding based on frequency-modulated-continuous-wave techniques. Technical Report TD-10-10009, Athens, Greece, February 2010.

[HBH03]   M. P. M. Hall, L. W. Barclay, and M. T. Hewitt. *Propagation of Radiowaves*, 2nd edition. Institution of Electrical Engineers, Inspec/IEE, 2003.

[HKY+05]   L. Hentilä, P. Kyösti, J. Ylitalo, X. Zhao, J. Meinilä, and J.-P. Nuutinen. Experimental characterization of multi-dimensional parameters at 2.45 and 5.25 GHz indoor channels. In *WPMC2005*, Aalborg, Denmark, 2005. See also http://www.propsim.com/.

[HZMN95]   M. Haardt, M. D. Zoltowski, C. P. Mathews, and J. Nossek. 2D unitary ESPRIT for efficient 2D parameter estimation. In *Proc. ICASSP 1995—IEEE Int. Conf. Acoust. Speech and Signal Processing*, vol. 3, pages 2096–2099. Detroit, MI, May 1995.

[IEE04]   IEEE standard for local and metropolitan area networks part 16: Air interface for fixed broadband wireless access systems, 2004.

[IEE07]   IEEE P802.15.4a-2007 (Amendment 1). 802.15.4: Wireless medium access control (MAC) and physical layer (PHY) specifications for low-rate wireless PANs, 2007.

[IEE09]   IEEE. Part 15.3: wireless medium access control MAC and physical layer PHY specifications for high rate wireless personal area networks WPANs: Amendment 2: millimeter-wave based alternative physical layer extension. In *IEEE P802.15.3c/D02*, pages 1–194, New York, USA, July 2009.

[IEE10]   IEEE. Part 11: wireless LAN medium access control (MAC) and physical layer (PHY) specifications, amendment 6: Enhancements for very high throughput in the 60 GHz band. In *IEEE P802.11ad/D0.1, Unapproved Draft*, pages 1–357, New York, USA, June 2010.

[IR03]   ITU-R. Specification attenuation model for rain for use in prediction methods. In *Recommendation ITU-R P.838-2*, 2003.

[JD88] A. K. Jain and R. C. Dubes, editors. *Algorithms for Clustering Data*. Prentice Hall, New York, 1988.

[JJT+09] V. Jungnickel, S. Jaeckel, L. Thiele, L. Jiang, U. Kruger, A. Brylka, and C. von Helmolt. Capacity measurements in a cooperative MIMO network. *IEEE Trans. Veh. Technol.*, 58(5):2392–2405, 2009. [Also available as TD(09)730].

[JK09] M. Jacob and T. Kürner. Radio channel characteristics for broadband WLAN applications between 67 and 110 GHz. In *Proc. The Third European Conference on Ant. and Prop. (EuCAP)*, pages 1–5, Berlin, Germany, March 2009. [Also available as TD(09)745].

[JMK10] M. Jacob, C. Mbianke, and T. Kürner. A dynamic 60 GHz radio channel model for system level simulations with MAC protocols for IEEE 802.11ad. In *Proc. IEEE International Symposium on Consumer Electronics ISCE*, Berlin, Germany, June 2010. [Also available as TD(10)1109].

[JPJ+09] M. Jacob, S. Priebe, C. Jastrow, T. Kleine-Ostmann, T. Schrader, and T. Kürner. An overview of ongoing activities in the field of channel modeling, spectrum allocation and standardization for mm-Wave and THz indoor communications. In *Proc. Globecom 2009—IEEE Global Telecommunications Conf.*, pages 1–5, Honolulu, HI, USA, December 2009. [Also available as TD(10)10055].

[KAK+07] J. Koivunen, P. Almers, V.-M. Kolmonen, J. Salmi, A. Richter, F. Tufvesson, P. Suvikunnas, A. Molisch, and P. Vainikainen. Dynamic multi-link indoor MIMO measurements at 5.3 GHz. In *Proc. of EuCAP 2007*, 2007.

[KAS+] V.-M. Kolmonen, P. Almers, J. Salmi, J. Koivunen, K. Haneda, A. Richter, F. Tufvesson, A. F. Molisch, and P. Vainikainen. A dynamic dual-link wideband MIMO channel sounder for 5.3-GHz. *IEEE Trans. Instrum. Meas.*, 59(4):873–883.

[KdJBH08] M. Kwakkernaat, Y. de Jong, R. Bultitude, and M. Herben. High-resolution angle-of-arrival measurements on physically-nonstationary mobile radio channels. *IEEE Trans. Antennas Propagat.*, 56(8):2720–2729, 2008.

[KGL+10] F. Kaltenberger, R. Ghaffar, I. Latif, R. Knopp, D. Nusbaum, and H. Callewaert. Comparison of LTE transmission modes in rural areas at 800 MHz. Technical Report TD-10-12080, Bologna, Italy, November 2010.

[KH07] M. Kwakkernaat and M. Herben. Analysis of clustered multipath estimates in physically nonstationary radio channels. In *Proc. PIMRC 2007—IEEE 18th Int. Symp. on Pers., Indoor and Mobile Radio Commun.*, pages 1–5, Athens, Greece, September 2007. [Also available as TD(07)324].

[KH08] M. R. J. A. E. Kwakkernaat and M. H. A. J. Herben. Analysis of scattering in mobile radio channels based on clustered multipath estimates. *Int. J. Wireless Inf. Networks*, 15(3–4):107–116, 2008. [Also available as TD(07)324].

[KHH+10] V.-M. Kolmonen, K. Haneda, T. Hult, J. Poutanen, F. Tufvesson, and P. Vainikainen. Measurement-based evaluation of interlink correlation for indoor multi-user MIMO channels. Technical Report TD-10-10070, Athens, Greece, February 2010.

[KHS+10] M. Kyrö, K. Haneda, J. Simola, P. Vainikainen, K. Takizawa, and H. Hagiwara. 60 GHz radio channel measurements and modelling in hospital environments. Technical Report TD-10-11049, Aalborg, Denmark, June 2010.

[Kiv07] J. Kivinen. 60 GHz wideband radio channel sounder. *IEEE Trans. Instrum. Meas.*, 56(5):1266–1277, 2007.

[KJ09] T. Kürner and M. Jacob. Application of ray-tracing to derive channel models for future multi-gigabit-systems. In *Proc. International Conference on Electromagnetics in Advanced Applications (ICEAA)*, pages 1–4, Torino, Italy, September 2009.

[KJGK10] F. Kaltenberger, H. Jiang, M. Guillaud, and R. Knopp. Relative channel reciprocity calibration in MIMO/TDD systems. In *Proc. ICT Future Network and Mobile Summit*, Florence, Italy, June 2010. [Also available as TD(09)950].

[KKC+08] F. Kaltenberger, M. Kountouris, L. S. Cardoso, R. Knopp, and D. Gesbert. Capacity of linear multi-user MIMO precoding schemes with measured channel data. In *Proc. IEEE Intl. Workshop on Signal Processing Advances in Wireless Communications (SPAWC)*, Recife, Brazil, July 2008. [Also available as TD(08)407].

2 Channel Measurements

[KLT09] M. Käske, M. Landmann, and R. Thomä. Modelling and synthesis of dense multipath propagation components in the angular domain. In *3rd European Conference on Antennas and Propagation (EuCAP 2009)*, pages 2641–2645, Berlin, Germany, March 2009. [Also available as TD(09)762].

[KMGT09] Y. Konishi, L. Materum, M. Ghoraishi, and J.-i. Takada. Multipath cluster polarization characteristics of a small urban MIMO macrocell. Technical Report TD-09-737, Braunschweig, Germany, February 2009.

[KSH+10] M. Kyrö, J. Simola, K. Haneda, S. Ranvier, P. Vainikainen, and K. Takizawa. 60 GHz radio channel measurements and modeling in a shielded room. In *IEEE 71st Vehicular Technology Conference (VTC-Spring 2010)*, pages 1–5, Taipei, Taiwan, May 2010. [Also available as TD(09)980].

[KSKT10] M. Käske, C. Schneider, W. Kotterman, and R. Thomä. Solving the problem of choosing the right MIMO sounding antenna: embedding/de-embedding. Technical Report TD-10-12081, Bologna, Italy, November 2010.

[KSP+05] M. Kmec, J. Sachs, P. Peyerl, P. Rauschenbach, R. Thomae, and R. Zetik. A novel ultra-wideband real-time MIMO channel sounder architecture. In *28th General Assembly of the International Union of Radio Sciences (URSI)*, New Delhi, India, October 2005.

[KST08] W. Kotterman, G. Sommerkorn, and R. Thomä. Ilmenau measurement data for SIG A. Technical Report TD-08-446, Wroclaw, Poland, February 2008.

[KT10] M. Käske and R. Thomä. Validation of estimated dense multipath components with respect to antenna array calibration accuracy. In *Proc. 4th European Conference on Antennas and Propagation (EuCAP 2010)*, Barcelona, Spain, April 2010. [Also available as TD(10)10075].

[LCK06] Z. Lei, F. Chin, and Y.-S. Kwok. UWB ranging with energy detectors using ternary preamble sequences. In *Proc. WCNC 2006—IEEE Wireless Commun. and Networking Conf.*, pages 872–877, 2006. doi:10.1109/WCNC.2006.1683585.

[LKT07] M. Landmann, W. Kottermann, and R. Thomä. On the influence of incomplete data models on estimated angular distributions in channel characterisation. In *Proc. 2nd European Conference on Antennas and Propagation (EuCAP 2007)*, Edinburgh, UK, November 2007. [Also available as TD(07)321].

[LMF10] B. K. Lau, J. Medbo, and J. Furuskog. Downlink cooperative MIMO in urban macrocell environments. In *Proc. IEEE Int. Symp. Antennas Propagat. (APS'2010)*, 2010.

[LSD+09] M. Lienard, E. Simon, P. Degauque, J.-M. Molina-Garcia-Pardo, and L. Juan-Llacer. Polarization diversity and MIMO capacity in tunnels. Technical Report TD-09-802, Valencia, Spain, May 2009.

[MB10] J. Medbo and J.-E. Berg. Directional propagation characteristics at the base station. Techical Report TD(10)12096, Bologna, Italy, November 2010.

[MFP+08] J. Maurer, T. Fugen, M. Porebska, T. Zwick, and W. Wiesbeck. A ray-optical channel model for mobile to mobile communications. Technical Report TD(08)430, Wroclaw, Poland, February 2008.

[MGPRJL08] J.-M. Molina-Garcia-Pardo, J.-V. Rodrıguez, and L. Juan-Llacer. Polarized indoor MIMO channel measurements at 2.45 GHz. Technical Report TD-08-605, Lille, France, October 2008.

[Mol09] A. F. Molisch. Ultra-wide-band propagation channels. *Proc. IEEE*, 97(2):353–371, 2009. doi:10.1109/JPROC.2008.2008836.

[MRAB05] J. Medbo, M. Riback, H. Asplund, and J. Berg. MIMO channel characteristics in a small macrocell measured at 5.25 GHz and 200 MHz bandwidth. In *Proc. VTC 2005 Fall—IEEE 62nd Vehicular Technology Conf.*, pages 372–376, Dallas, TX, September 2005.

[MSKF10] J. Medbo, I. Siomina, A. Kangas, and J. Furuskog. Propagation channel impact on LTE positioning accuracy—a study based on real measurements of observed time difference of arrival. Technical Report TD-10-11079, Aalborg, Denmark, June 2010.

[NKJ+10]   M. Narandžić, M. Käske, S. Jäckel, G. Sommerkorn, C. Schneider, and R. S. Thomä. Variation of estimated large-scale MIMO channel properties between repeated measurements. Technical Report TD-10-11088, Aalborg, Denmark, June 2010.

[NKS+09]   M. Narandzic, M. Käske, C. Schneider, G. Sommerkorn, A. Hong, W. A. Th. Kotterman, and R. S. Thomä. On a characterisation of large-scale channel parameters for distributed (multi-link) MIMO—the impact of power level differences. Technical Report TD-09-981, Braunschweig, Germany, February 2009.

[OCB+10]   C. Oestges, N. Czink, B. Bandemer, P. Castiglione, F. Kaltenberger, and A. J. Paulraj. Experimental characterization and modelling of outdoor-to-indoor and indoor-to-indoor distributed channels. *IEEE Trans. Veh. Technol.*, 59(5):2253–2265, 2010. doi:10.1109/TVT.2010.2042475.

[PBK+10]   A. Paier, L. Bernadó, J. Karedal, O. Klemp, and A. Kwoczek. Overview of vehicle-to-vehicle radio channel measurements for collision avoidance applications. In *Proc. VTC 2010 Spring—IEEE 71st Vehicular Technology Conf.*, Taipei, Taiwan, May 2010. [Also available as TD(09)928].

[PGR09]   M. Park, P. Gopalakrishnan, and R. Roberts. Interference mitigation techniques in 60 GHz wireless networks. *IEEE Commun. Mag.*, 47(12):34–40, 2009.

[PHS+09a] J. Poutanen, K. Haneda, J. Salmi, V.-M. Kolmonen, F. Tufvesson, T. Hult, and P. Vainikainen. Development of measurement-based ray tracer for multi-link double directional propagation parameters. In *Proc. 3rd European Conf. Antennas and Propagation (EuCAP 2009)*, pages 2622–2626, Berlin, Germany, March 2009. [Also available as TD(09)771].

[PHS+09b] J. Poutanen, K. Haneda, J. Salmi, V.-M. Kolmonen, F. Tufvesson, and P. Vainikainen. Analysis of radio wave scattering processes for indoor MIMO channel models. In *Proc. PIMRC 2009—IEEE 20th Int. Symp. on Pers., Indoor and Mobile Radio Commun.*, Tokyo, Japan, September 2009. [Also available as TD(09)839].

[PJKK08]   R. Piesiewicz, C. Jansen, M. Koch, and T. Kürner. Measurements and modeling of multiple reflections effect in building materials for indoor communications at THz frequencies. In *Proc. German Microwave Conference, GEMIC 2008)*, pages 3089–3092, Hamburg, Germany, March 2008. [Also available as TD(07)427].

[PKC+07]   A. Paier, J. Karedal, N. Czink, H. Hofstetter, C. Dumard, T. Zemen, F. Tufvesson, C. F. Mecklenbräuker, and A. F. Molisch. First results from car-to-car and car-to-infrastructure radio channel measurements at 5.2 GHz. In *Proc. PIMRC 2007—IEEE 18th Int. Symp. on Pers., Indoor and Mobile Radio Commun.*, Athens, Greece, September 2007. [Also available as TD(07)303].

[PKH+08]   R. Parviainen, P. Kyösti, Y.-T. Hsieh, P.-A. Ting, J.-S. Chiou, and M. Yang. Results of high speed train channel measurements. Technical Report TD(08)646, Lille, France, October 2008.

[PKZ+08]   A. Paier, J. Karedal, T. Zemen, N. Czink, C. Dumard, F. Tufvesson, C. F. Mecklenbräuker, and A. F. Molisch. Description of vehicle-to-vehicle and vehicle-to-infrastructure radio channel measurements at 5.2 GHz. Technical Report TD(08)636, Lille, France, October 2008.

[PMB04]   M. Pesavento, C. F. Mecklenbräuker, and J. F. Böhme. Multidimensional rank reduction estimator for parametric MIMO channel models. *EURASIP J. Appl. Signal Process.*, 2004(1):1354–1363, 2004.

[PQD+09]   A. Panahandeh, F. Quitin, J. M. Dricot, F. Horlin, C. Oestges, and P. De Doncker. Cross-polar discrimination statistics for outdoor-to-indoor and indoor-to-indoor channels. Technical Report TD-09-815, Valencia, Spain, May 2009.

[PRK85]   A. Paulraj, R. Roy, and T. Kailath. Estimation of signal parameters via rotational invariance techniques—ESPRIT. In *Proc. 19th Asilomar Conference on Circuits, Systems and Computers*, pages 83–89, Pacific Grove, CA, November 1985.

[Pro95]   J. G. Proakis. *Digital Communications*, 2nd edition. McGraw-Hill, New York, USA, 1995.

[PSKK07]	R. Piesiewicz, J. Schoebel, M. Koch, and T. Kürner. Propagation measurements and modeling for future indoor communication systems at THz frequencies. In *Proc. Wave Propagation in Communication, Microwave Systems and Navigation (WFMN 2007)*, Chemnitz, Germany, July 2007. [Also available as TD(07)367].
[PTA+10]	A. Paier, R. Tresch, A. Alonso, D. Smely, P. Meckel, Y. Zhou, and N. Czink. Average downstream performance of measured IEEE 802.11p infrastructure-to-vehicle links. In *Proc. ICC 2010—IEEE Int. Conf. Commun.*, Cape Town, South Africa, May 2010. [Also available as TD(10)014].
[PZB+08]	A. Paier, T. Zemen, L. Bernadó, G. Matz, J. Karedal, N. Czink, C. Dumard, F. Tufvesson, A. F. Molisch, and C. F. Mecklenbräuker. Non-WSSUS vehicular channel characterization in highway and urban scenarios at 5.2 GHz using the local scattering function. In *Proc. Int. Workshop Smart Antennas (WSA)*, pages 9–15, Darmstadt, Germany, 2008.
[QCOD07]	F. Quitin, F. H. C. Oestges, and P. De Doncker. Cross-polarized MIMO channel measurements for indoor environments. Technical Report TD-07-388, Duisburg, Germany, September 2007.
[QOHD08a]	F. Quitin, C. Oestges, F. Horlin, and P. De Doncker. Small-scale variations of cross-polar discrimination in polarized MIMO systems. Technical Report TD-08-603, Lille, France, October 2008.
[QOHD08b]	F. Quitin, C. Oestges, F. Horlin, and P. De Doncker. Spatio-temporal characterization of polarized MIMO channels. Technical Report TD-08-602, Lille, France, October 2008.
[QOHDD10]	F. Quitin, C. Oestges, F. Horlin, and P. De Doncker. A spatio-temporal channel model for modeling the diffuse multipath component in indoor environments. In *Proc. 4th European Conference on Antennas and Propagation (EuCAP 2010)*, Barcelona, Spain, April 2010. [Also available as TD(10)003].
[Ric05]	A. Richter. *Estimation of radio channel parameters: models and algorithms*. PhD dissertation, Technischen Universität Ilmenau, Ilmenau, Germany, May 2005.
[RKVO08]	O. Renaudin, V.-M. Kolmonen, P. Vainikainen, and C. Oestges. Wideband MIMO car-to-car radio channel measurements at 5.3 GHz. In *Proc. VTC 2008 Fall—IEEE 68th Vehicular Technology Conf.*, Calgary, Canada, September 2008. [Also available as TD(08)510].
[RKVO10]	O. Renaudin, V.-M. Kolmonen, P. Vainikainen, and C. Oestges. Description of the August 2009 car-to-car radio channel measurement campaign. Technical Report TD(10)013, Athens, Greece, February 2010.
[RPZ09]	L. Reichardt, J. Pontes, and T. Zwick. Performance improvement using multiple antenna systems for car-to-car communications in urban environment. Technical Report TD(09)966, Vienna, Austria, September 2009.
[RSK06]	A. Richter, J. Salmi, and V. Koivunen. On distributed scattering in radio channels and its contribution to MIMO channel capacity. In *Proc. 1st European Conference on Antennas and Propagation (EuCAP2006)*, Nice, France, November 2006.
[RTR+08]	A. Richter, F. Tufvesson, P. S. Rossi, K. Haneda, J. Koivunen, V.-M. Kolmonen, J. Salmi, P. Almers, P. Hammarberg, K. Pölönen, P. Suvikunnas, A. F. Molisch, O. Edfors, V. Koivunen, P. Vainikainen, and R. R. Müller. Wireless LANs with high throughput in interference-limited environments—project summary and outcomes. Technical Report TD-08-432, Wroclaw, Poland, February 2008.
[Sal09]	J. Salmi. *Contributions to measurement-based dynamic MIMO channel modeling and propagation parameter estimation*. PhD dissertation, Helsinki University of Technology, Dept. of Signal Processing and Acoustics, Espoo, Finland, August 2009. [Also available as TD(08)471].
[SCPF09]	G. Steinböck, J.-M. Conrat, T. Pedersen, and B. H. Fleury. On initialization and search procedures for iterative high-resolution channel parameter estimators. Technical Report TD-09-956, Vienna, Austria, September 2009.

[SGL07]   P. Stefanut, D. P. Gaillot, and M. Liénard. Parametric study of the performance of high-resolution estimation algorithms. Technical Report TD-09-763, Braunschweig, Germany, February 2007.

[SHK+07]   J. Sachs, R. Herrmann, M. Kmec, M. Helbig, and K. Schilling. Recent advances and applications of m-sequence based ultra-wideband sensors. In *IEEE Intern. Conf. on Ultra-Wideband, ICUWB*, pages 50–55, September 2007. doi:10.1109/ICUWB.2007.4380914.

[SJK10]   M. Schack, M. Jacob, and T. Kürner. Comparison of in-car UWB and 60 GHz channel measurements. In *Proc. The Fourth European Conference on Ant. and Prop. (EuCAP)*, pages 1–5, Barcelona, Spain, March 2010. [Also available as TD(10)11037].

[SKA+08]   T. Santos, J. Karedal, P. Almers, F. Tufvesson, and A. Molisch. Scatterer detection by successive cancellation for UWB method and experimental verification. In *Proc. VTC 2008 Spring—IEEE 67th Vehicular Technology Conf.*, pages 445–449, Marina Bay, Singapore, May 2008. [Also available as TD(08)411].

[SKI+10a] K. Saito, K. Kitao, T. Imai, Y. Okano, and S. Miura. The DoA estimation method using EM/SAGE algorithm and extended MODE algorithm with array interpolation. Technical Report TD-10-12040, Bologna, Italy, November 2010.

[SKI+10b] K. Saito, K. Kitao, T. Imai, Y. Okano, and S. Miura. The modeling methods of time-correlated MIMO channels using the particle filter. Technical Report TD-10-11086, Aalborg, Denmark, June 2010.

[Smu02]   P. F. M. Smulders. Exploiting the 60 GHz band for local wireless multimedia access: prospects and future directions. *IEEE Commun. Mag.*, 40(1):140–147, 2002.

[SOK+08]   H. Singh, J. Oh, C. Kweon, X. Qin, H.-R. Shao, and C. Ngo. A 60 GHz wireless network for enabling uncompressed video communication. *IEEE Commun. Mag.*, 46(12):71–78, 2008.

[SPH+10]   J. Salmi, J. Poutanen, K. Haneda, A. Richter, V.-M. Kolmonen, P. Vainikainen, and A. F. Molisch. Incorporating diffuse scattering in geometry-based stochastic channel models. In *Proc. 4th European Conference on Antennas and Propagation (EuCAP 2010)*, Barcelona, Spain, April 2010. [Also available as TD(10)047].

[SRE+06]   J. Salmi, A. Richter, M. Enescu, P. Vainikainen, and V. Koivunen. Propagation parameter tracking using variable state dimension Kalman filter. In *Proc. VTC 2006 Spring—IEEE 63rd Vehicular Technology Conf.*, pages 2757–2761, Melbourne, Australia, May 2006.

[SRK09]   J. Salmi, A. Richter, and V. Koivunen. Detection and tracking of MIMO propagation path parameters using state-space approach. *IEEE Trans. Signal Processing*, 57(4):1538–1550, 2009.

[SS90]   P. Stoica and K. C. Sharman. Novel eigenanalysis method for direction estimation. *IEE Proc. Radar and Signal Processing*, 137(1):19–26, 1990.

[SSM10]   J. Salmi, S. Sangodoyin, and A. F. Molisch. High resolution parameter estimation for ultra-wideband MIMO radar. In *The 44th Asilomar Conference on Signals, Systems, and Computers*, Pacific Grove, CA, November 2010. [Also available as TD(10)2042].

[SSN+09]   C. Schneider, G. Sommerkorn, M. Narandzic, M. Käske, A. Hong, V. Algeier, W. A. T. Kotterman, and R. S. Thomä. Part I: Reference campaign—description and application. Technical Report TD-09-776, Braunschweig, Germany, February 2009.

[SV87]   A. Saleh and R. Valenzuela. A statistical model for indoor multipath propagation. *IEEE J. Select. Areas Commun.*, 5(2):128–137, 1987.

[WAV]   IEEE P802.11p/D4.0: Part 11: Wireless LAN Medium Access Control (MAC) and Physical Layer (PHY) Specifications: Amendment: Wireless Access in Vehicular Environments (WAVE), Draft 4.0, March 2008.

[WHR+10]   S. Wyne, K. Haneda, S. Ranvier, F. Tufvesson, and A. Molisch. Beamforming effects on measured mm-wave channel characteristics. Technical Report TD-10-10023, Athens, Greece, February 2010.

[WWW+07] M. Webb, G. Watkins, C. Williams, T. Harrold, R. Feng, and M. Beach. Mobile multihop: Measurements vs. models. Technical Report TD-07-322, Duisburg, Germany, September 2007.

[Yin06] X. Yin. *High-resolution parameter estimation for MIMO channel sounding*. PhD dissertation, Department of Electronic Systems, Aalborg University, Aalborg, Denmark, 2006.

[YSH05] H. Yang, P. F. M. Smulders, and M. H. A. I. Herben. Indoor channel measurements and analysis in the frequency bands 2 GHz and 60 GHz. In *Proc. PIMRC 2005— IEEE 16th Int. Symp. on Pers., Indoor and Mobile Radio Commun.*, Dublin, Ireland, January 2005.

[YSK+08] X. Yin, G. Steinböck, G. E. Kirkelund, T. Pedersen, P. Blattnig, A. Jaquier, and B. H. Fleury. Tracking of time-variant radio propagation paths using particle filtering. In *IEEE International Conference on Communications (ICC 2008)*, pages 920–924, Beijing, China, May 2008. [Also available as TD(07)380].

[ZBN05] T. Zwick, T. J. Beukema, and H. Nam. Wideband channel sounder with measurements and model for the 60 GHz indoor radio channel. *IEEE Trans. Veh. Technol.*, 54(4):1266–1277, 2005.

[ZTE+10] M. Zhu, F. Tufvesson, G. Eriksson, S. Wyne, and A. F. Molisch. Parameterization of 300 MHz MIMO measurements in suburban environments for the COST 2100 MIMO channel model. Technical Report TD-10-11071, Aalborg, Denmark, June 2010.

Chapter 3
Radio Channel Modeling for 4G Networks

Chapter Editor Claude Oestges, Nicolai Czink, Philippe De Doncker, Vittorio Degli-Esposti, Katsuyuki Haneda, Wout Joseph, Martine Liénard, Lingfeng Liu, José Molina-García-Pardo, Milan Narandžić, Juho Poutanen, François Quitin, and Emmeric Tanghe

This chapter is dedicated to radio channel modeling for 4G networks. In addition to recent results in the area of 4G channel modeling at large, including complex environments such as aircrafts, COST 2100 has dealt with a number of specific topics which are presented in this chapter.

- Improved deterministic methods, including models of diffuse components, are detailed in Sect. 3.2. The developed models namely propose new methods to account for macroscopic diffuse scattering in ray-tracing tools.
- The measurement-based modeling of the same diffuse or dense multipath components has represented one major research topic in COST 2100, especially regarding the extraction of such components from experimental data (Sect. 3.3).
- Another important topic of research has concerned the modeling of the polarization behavior of wireless channels (see Sect. 3.4), as multipolarized antenna arrays appear more and more as a realistic implementation of MIMO systems.
- The specific representation and modeling of multilink scenarios is dealt with in Sect. 3.5. Considering the correlations between multiple links is a significant requirement to design robust schemes in cooperative, relay or multihop networks.
- Finally, the COST 2100 model is presented in Sect. 3.6. Starting from a single-link implementation based on the COST 273 model (available online), it builds upon the work described in this chapter to propose an updated version of the COST 2100 model including enhancements such as diffuse and cross-polar components, as well as multilink aspects.

3.1 General Aspects of Channel Characterization and Modeling

Before detailing COST 2100 achievements in specific aspects of 4G channel characterization and modeling, this first section reviews general results not covered

---

C. Oestges (✉)
Université catholique de Louvain (UCL), Louvain-la-Neuve, Belgium
e-mail: claude.oestges@uclouvain.be

Fig. 3.1 Excess loss for a vertical polarization wave at the top of the houses/trees for different tree widths [VQAR07]

elsewhere in this chapter. In particular, updated path loss models are provided for a number of classical environments (office indoor, outdoor, and outdoor-to-indoor). A large part is also dedicated to stochastic approaches. In this context, one of the main objectives of the efforts was to increase the ability of stochastic models to describing complex propagation phenomena in challenging environments (such as aircrafts).

3.1.1 Updated Path Loss Models in Challenging Environments

3.1.1.1 Hilly Terrains and Roadside Tree Residential Environments

An improved theoretical model is proposed in [TL08] for wooded residential environments to include incoherent fields and quasi-coherent fields when computing the excess path loss. Analogous to classical path loss models, the row of houses/buildings is viewed as diffracting cylinders lying on the earth, and the trees are located adjacent to the houses/buildings. Because of the small grazing angle between the incident field and the rooftops of the buildings, the buildings are modeled as perfect absorbing screens, whereas the tree canopies are modeled as elliptical cylinders. The fields at the aperture of each half-screen depend on the mean field passing through a tree and on the incoherent fields produced by the tree canopy. The mean field in the tree canopy is evaluated by Foldy–Lax multiple scattering theory. Figure 3.1 shows the propagation loss relative to free-space loss at the top of successive houses, with and without trees. Figure 3.1 shows that, were the trees not considered, the propagation loss would be underestimated, especially for the first few houses/trees.

3 Radio Channel Modeling for 4G Networks

Table 3.1 Parameters for the path loss model in [LDMGP08]

Parameters	L_0 [dB]	n_{L_0}	n	σ [dB]
	68	0.82	0.57	2.7

3.1.1.2 Confined Areas

Wireless communications in confined environments, such as tunnels, have been widely studied for a number of years, and a number of experimental results have been presented in the literature. However, for the time being, there is a lack of experimental data dealing with (ultra)wideband propagation characteristics in environments such as train tunnels. In [LDMGP08], the authors investigate the statistics of the electric field in a frequency range extending from 2.8 GHz to 5 GHz in a tunnel environment. A simple empirical expression is given to predict the mean path loss L as a function of frequency f and distance d:

$$L = \left(L_0 + 10 n_{L_0} \log_{10}(f(\text{GHz}))\right) + 10 n \log_{10}(d) + X_\sigma. \tag{3.1}$$

Using an iterative procedure, the parameters minimizing the mean square error are given in Table 3.1.

Regarding the small-scale fading, a Rice distribution fits quite well the experimental E-field amplitude. The phase of the E-field in the transverse plane is found to be uniformly distributed, with a range of values that decreases with longitudinal distance and increases with frequency.

A similar study was performed by [LP09] considering in-room communications. The spectral statistics are found to match Rayleigh and Rice distributions in the 2–18 GHz band, although alternative distributions (Nakagami and Log-Normal) are sometimes required in the low and high ends of the band. It is also shown that the K-factor depends on the distance d between transmitter and receiver and is modeled in the 3–12 GHz range by

$$K(f_c, d) = \left(\frac{\Theta f_c + \Omega}{d}\right)^n, \tag{3.2}$$

where Θ, Ω and n are parameters set to fit the results (see Table 3.2).

In addition to this study, a model for the power delay spectrum of an in-room reverberant channel is proposed in [SPF10]. It includes both a dominant component following a standard inverse distance power law and a reverberant component exponentially decaying with distance. This model allows for the prediction of path gain, mean delay, and RMS delay spread.

Table 3.2 Parameters extracted from K-factor results. The last column (denoted as RMS) is the root mean square error between the model and the measured results

	Θ [$\frac{m}{GHz}$]	Ω [m]	n	RMS [ns]
big room	0.70	1.03	1.50	1.70
small room	0.25	1.19	1.92	1.71

3.1.2 Channel Characterization in Various Outdoor-to-X Scenarios

3.1.2.1 Outdoor-to-Outdoor

For wide outdoor area scenarios, 3GPP has defined new spatial channel models (SCM) based on tapped delay-lines with fixed values for angular parameters or correlation matrices. On the one hand, these models simplify link-level simulations and reduce the amount of simulation time, but on the other hand, the great variability of MIMO propagation channels is not taken into account.

In [CP05], a directional wideband measurement campaign was performed in urban macrocells at 2 GHz using a channel sounder and an 8-sensor linear antenna array at the base station. Directions of arrival at the Base Station (BS) and the Mobile Station (MS) were estimated by a beamforming technique. Global parameters such as delay spread and azimuth spread at BS were processed from the Azimuth-Delay Power Profiles (ADPP) at BS and MS. Furthermore, a maximum factor (max_R) defined by (3.1) was introduced to quantify the spatial diversity at the MS. A maximum factor close to 0 tends to indicate a uniform distribution of the power around the mobile and thus a high potential spatial selectivity at MS,

$$max_R = \frac{\max(P_{MS\text{-}cluster}(k))}{\sum_k (P_{MS\text{-}cluster}(k))}, \qquad (3.3)$$

where $P_{MS\text{-}cluster}(k)$ is the power of the kth mobile cluster (MS-Cluster). To identify different groups of environments, a hybrid method combining hand-made filtering and K-means algorithm was used. The K-means method partitions the MIMO points into K mutually exclusive groups such that MIMO points within each group are as close to each other as possible and as far from MIMO points in other groups as possible. Three typical and three atypical measurement files have been identified with the following procedure. The hand-made filtering was applied to extract atypical groups, and the K-means algorithm was applied to identify typical groups. An example is shown in Fig. 3.2, where the BS Angular Spread (AS) is plotted versus the Delay Spread (DS). Six groups of files are identified on the graph. The main characteristics of typical groups are the following.

- Typical Group 1: the vast majority of channels of this group have characteristics similar to those extracted from LOS measurements (low spatial diversity, low frequency diversity) even if there is no BS-MS visibility.
- Typical Group 2: this group differs from group 1 because the median value of max_R is 0.3 instead of 0.7, which indicates a relatively higher diversity at MS.
- Typical Group 3: the features of this group indicate a relatively high frequency diversity and a relatively high spatial diversity at BS and MS.

Atypical groups correspond either to High DS, or High BS-AS, or Low BS-AS.

Table 3.3 summarizes the statistical channel parameters. The mean values of the azimuth spread and delay spread at BS are equal to 9.5 degrees and 0.25 µs, respectively. Others directional wideband results are also described in [KH07]. In addition

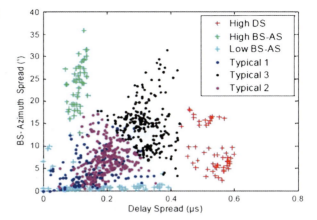

Fig. 3.2 Selection of typical and atypical files

to those results, authors in [RO07] investigate wideband propagation characteristics of a 1.87 GHz pedestrian radio channel. Table 3.4 summarizes the channel parameters for different Rx locations. Delay spreads are in accordance with results presented in Table 3.3. It should be noticed that there is no clear correlation between the RMS delay spread and the coherence bandwidth. The influence of the filtering bandwidth on the channel parameters is also emphasized, in particular on the apparent RMS delay spread.

In [ALO07], SISO urban radio channels are revisited and delay-spreads derived from well-known models are compared with those obtained from new measurement data. As illustrated in Fig. 3.3, the measured median RMS delay spreads are around 100 ns, i.e., in the same range as those of [CP05] and the 3GPP Rural Area (RA) channel model. They are however somewhat higher than the ITU Pedestrian A (PedA) model and well below the time dispersion modeled in the 3GPP Typical Urban (TU) model.

Finally, [BDHZ10] reports also some results of double directional sounding measurements made in downtown Ottawa at 2.25 GHz. Based on the fact that multiple runs over the same measurement routes resulted in very similar delay and angular time series, the possibility that features of the built-environment dominate and that variations due to different traffic situations are less significant than might be expected is considered worthy of further investigation.

Table 3.3 Statistics of outdoor channel parameters [CP05]

	Atypical			Typical		
	High DS	High AS	Low BS-AS	Type 1	Type 2	Type 3
Occurrence %	5%	5%	10%	30%	30%	20%
Distance (m)	327	33	530	382	411	316
MS-Angle (deg.)	30	47	5	32	45	54
DS (ns)	540	110	123	159	195	319
BS-AS (deg.)	7.7	24	0.7	3.62	6.6	15

Table 3.4 RMS delay/Doppler spread and coherence bandwidth/time values for different measurement location [RO07]

Measure. location	rms delay spread [ns]	rms Doppler spread [MHz]	coherence bandwidth [MHz]	coherence time [s]
1	150	94.1	10.1	0.079
2	162	94.1	2.2	0.070
3	107	114.1	24.4	0.047
4	245	77.4	10.4	0.023
5	89	82.2	29.7	0.036
6	180	69.7	6.2	0.026
7	207	59.0	15.5	0.036
8	174	85.1	18.5	0.058
9	200	65.0	2.9	0.060

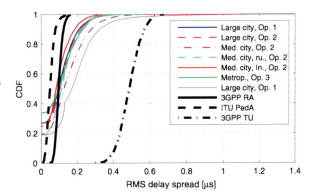

Fig. 3.3 Cumulative distribution functions for the instantaneous RMS delay spread measured in 3G networks in three different cities. Simulated distributions for three common channel models are plotted for comparison [ALO07]

3.1.2.2 Outdoor-to-Relay

Relaying technology is a key technical enhancement in 3GPP LTE-Advanced. In [CCC10], a measurement-based study analyzes the impact of relay antenna height on path loss channel models. Path loss prediction of 3GPP Urban Macro and COST-231 WI model is given with street width $w = 10$ m, average building height $h_{Roof} = 15$ m, and average building separation $b = 20$ m. A comparison between the prediction of these models and measurement is presented in Fig. 3.4. The 3GPP Urban Macro model slightly underestimates the measurements. The path loss reduction when RS antenna height is raised from 4.8 m to 8.8 m is smaller than that obtained when the antenna height increase from 8.8 m to 12.7 m. However, 3GPP Urban Macro model uses the logarithm fit under the form $M \log_{10}(h_{RS})$ to describe antenna height correction. As a consequence, the prediction of this model is not coherent with the measurements. The predicted path loss reduction from 4.8 m to 8.8 m is greater than that from 8.8 m to 12.7 m. COST-231 WI prediction results presented in Fig. 3.4 obviously provide the closest agreement with the measurement

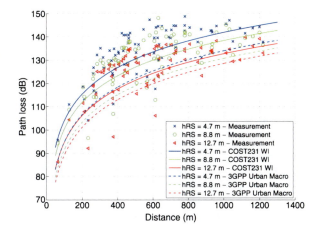

Fig. 3.4 Comparison between COST-231 WI and 3GPP Urban Macro models and measurements [CCC10]

even though this model is neither designed for relaying systems nor formally specified to use at frequencies beyond 2 GHz. The disadvantage of COST-231 WI model is that it requires a detailed knowledge of the propagation environment. A path loss model for BS to Relay Station (RS) is deduced from measurements and can be expressed as follows:

$$PL(\text{dB}) = 34\log_{10}(d) + 11.7 + 25.5\log_{10}(20 - h_{\text{RS}}), \qquad (3.4)$$

where d is range in meters, and h_{RS} is the relay antenna height in meters. The model is valid only for the NLOS BS-RS condition and with d in the range from 20 m to 1000 m. The RS antenna height is limited to 15 m, which is suitable for typical relaying deployments.

3.1.2.3 Outdoor-to-Indoor

Outdoor-to-indoor communications have been widely studied, but most measurements were performed in a specific configuration. Typically, the transmitter faced the building where the receiver was located. The characterization therefore concerns primarily penetration inside buildings, rather than typical outdoor-to-indoor propagation in an urban environment. In [CDNL09], measurements were performed to characterize such a channel where the line BS-building is obstructed by other buildings. The carrier frequency was 3.7 GHz. The study is based on a SIMO approach, performed on the campus of the University of Lille, France. The measurement equipment included a channel sounder and a 12×12 virtual antenna array at the MS. Directional impulse responses at MS were estimated by beamforming. The mean value of AS is 66 degrees, and the mean value of DS is between 75 ns and 160 ns, depending on the visibility conditions between the building and the base station. The results presented in Table 3.5 can be compared to values obtained by other authors and for different environments. Regarding the AS statistics, an acceptable

Table 3.5 Summary of measurement results for outdoor-to-indoor scenario [CDNL09]

	Place	R&D Institution	Bandwidth (MHz)	Frequency (GHz)	MS-AS	DS (ns)
Outdoor to indoor measurement campaign	Lille Building P3	Lille Univ.& FTR&D	250	3.6	66	160
	Lille Building P4	Lille Univ.&FTR&D	250	3.6		75
	Lille Outdoor	Lille Univ.&FTR&D	250	3.6	50	156
	Lund	Lund Univ.	120	5.2		10
	Helsinki	Helsinki Univ.	60	5.3	40	
	Oxford	Multiple	80	2.5/5.7		37
	Stockholm	KTH	Narrowband	1.8	70	
	Stockholm	Ericsson	200	5.2	70	250
	Oulu	Oulu Univ.	100	5.2	61	40
	Ottawa	Oulu Univ.	100	4.9		16
	Cantabria	Cantabria Univ.	250	3.5		25
	Beijing	Beijing Univ.	100	5.2		36
Winner Phase II O2I Channel models	Model Winner II microcell O2I B4		100	from 2 up 6	58	49
	Model Winner II microcell O2I C4					240
Outdoor environment	Mulhouse	FTR&D	62.5	2.2	62	167
	Model Winner II microcell C2		100	from 2 up 6	53	234

agreement between results described in [CDNL09] and values defined in the WINNER phase II channel models was found. Furthermore, one can note in this table a large dispersion of DS values ranging from 10 to 250 ns.

Finally, let us mention that further indoor penetration losses at 3.5 GHz WIMAX frequency are provided in [BCG+09].

3.1.2.4 Outdoor-to-Aircraft

Outdoor to indoor scenario also includes the penetration of signal inside an aircraft. In [MPC10], an outdoor-to-aircraft measurement campaign was conducted in Airbus A340-300 aircraft to evaluate the attenuation of the hull. This study is useful to perform an interference study inside the aircraft and to assess whether the aircraft mobile terminal interferes to the ground station. The minimum aircraft attenuation was 15.5 dB. This value increases by 5.5 dB when the aircraft flies above the base station. Taking into account the required margins to prevent interference, it is found that from a ground base station to an on-board mobile terminal, an additional attenuation of 12.3 dB is required.

Table 3.6 Summary of correlation coefficient for distance between elements array of λ in indoor environment

f (GHz)	2	2.4	6	12
LOS	0.326	0.452	0.785	0.979
NLOS	0.283	0.425	0.780	0.870

3.1.3 Solving New Issues in Stochastic Channel Models

3.1.3.1 Frequency Dependence of Channel Parameters in Indoor Environments

Since the frequency dependence of channel parameters is relevant for future radio systems, such as cognitive networks, it has attracted recent attention in the propagation community. In [GRC07], channel characteristics between 2 and 12 GHz were extracted from MIMO wideband measurements in an office environment (in LOS and NLOS scenarios, for ranges below 12 m). The spatial correlation was computed for four carrier frequency bands, namely around 2, 2.4, 6, and 12 GHz. Table 3.6 gives the Rx correlation coefficient averaged over four and five Tx locations. It clearly appears that, for a distance between array elements equal to one wavelength, the correlation coefficient increases with frequency. A possible explanation is related to changes in both the scatterers and the angles of departure and arrival at both ends of the link and to possible reductions in the antenna vertical beamwidth.

Measurements at three frequencies (2.5, 3.5, and 5.7 GHz) [Sal07] were also performed in indoor and indoor-to-outdoor environments. The RMS delay spread was subsequently computed, and differences between the three frequencies were within 10 ns.

3.1.3.2 Stationarity Distance

For system level simulations, mobile radio channels are often assumed to be wide-sense stationary, i.e., with unchanging stochastic parameters. Measurements, however, show that the mobile channel characteristics may change rapidly. In [Bul08, ST10], different approaches are suggested to estimate the rate of change of mobile radio channel statistics. In [Bul08], the detection of changes in measured outdoor radio propagation is investigated by comparing cumulative distributions of angle of arrival and excess delay in consecutive time intervals.

When studying the temporal stationarity of the time-frequency properties only, [ST10] uses the collinearity of the Generalized Local Scatter Function (GLSF) between different time instances for all MIMO sublinks. The goal was to analyze the dependence of the quasi-stationarity regions upon the orientation of the mobile terminal (MT) and to compare the quasi-stationarity of the time-frequency with the spatial properties in a specific urban macro cell scenario in the 3GPP LTE band (2.53 GHz). An overview of the Tx location and tracks is shown in Fig. 3.5. Four orientations of the three neighboring elements of the 2-rings array are considered,

Fig. 3.5 Overview of the MT reference tracks

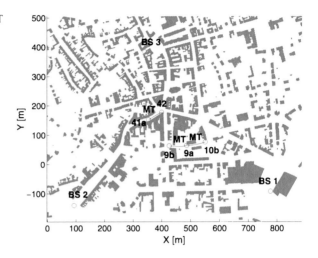

Table 3.7 Average stationarity distances using collinearity of the GLSF [ST10]

Track	$d_{\text{stat,avg}}$ per orientation [m]			
	front	right	back	left
10b-9a	1.80	1.34	2.20	2.75
9a-9b	0.96	0.64	0.93	1.21
41a-42	0.80	1.41	0.23	0.42

and Table 3.7 shows the average stationarity distances $d_{\text{stat,avg}}$ per orientation for some of the tracks. They are rather short, ranging between 0.23 m and 2.75 m. Longer average stationarity distances are observed when main reflections are identified: for track 10b-9a, a component is expected to occur partly for the back and left orientation. For track 41a-42, a strong constant reflection is expected for the front and right orientation due to a high building in the direct vicinity.

The above results have considered stationarity in time and frequency. A detailed analysis of stationarity in the space–time dimension is discussed in Sect. 3.5.

3.1.3.3 MIMO Correlation in Static Scenarios

Correlation matrices are popular when modeling MIMO channels. The essence of such models is that received signals are inherently stochastic, with correlation of any degree between transmitters, between receivers, and between transmit and receive sides being expressed by the channel correlation matrix. Probably, the best-known model in this category is the Kronecker model, stating that the transmit- and receive-sided correlation matrices ($\mathbf{R}_t$ and $\mathbf{R}_r$, respectively) are fully independent and that the full-correlation matrix can be expressed as the Kronecker product of lower size matrices. This means that, writing the full correlation matrix $\mathbf{R}_\mathbf{H}$ in the usual way,

$$\mathbf{R}_\mathbf{H} = E\{\text{vec}(\mathbf{H})\,\text{vec}(\mathbf{H})^H\}, \tag{3.5}$$

we can write the full correlation matrix of a Kronecker-separable channel as

$$\mathbf{R} = \mathbf{R}_t^T \otimes \mathbf{R}_r. \tag{3.6}$$

Such a separation is attractive as the full correlation matrix grows rapidly with the size of the antennas array. Among propagation experts, some form of consensus has been reached that for small MIMO systems, e.g., 2×2, the Kronecker separability holds reasonably well, due to the limited resolution of small antenna arrays. However, for larger systems, discrepancies might become significantly large. Authors in [KT07] illustrate this point by considering a MIMO system of size 8×16. In this case, 1000 independent realizations of $\mathbf{H}$ are needed to correctly estimate $\mathbf{R_H}$ and to validate the full correlation model. Therefore, for large MIMO configurations, the amount of independent data needed for such estimations could exceed what is practically available from measurements.

For multiband and multiuser applications, [Sib01] suggests a "semi-Kronecker" approximation to model the channel matrix as a frequency-dependent correlation matrix on the one hand and a user-dependent frequency correlation matrix on the other hand. The channel matrix $\mathbf{H}$ can be written as

$$\mathbf{H} = \mathbf{R}_f^{\frac{1}{2}} \mathbf{X} \mathbf{R}_u^{\frac{T}{2}}, \tag{3.7}$$

where $\mathbf{R}_f$ and $\mathbf{R}_u$ are the frequency and user correlation matrices of dimensions $n_f \times n_f$ and $n_u \times n_u$, respectively, and $\mathbf{X}$ is a matrix of zero mean and unit variance, n_u and n_f being the numbers of users and radio link frequencies, respectively. It can be shown that the semi-Kronecker approximation improves the model performance for channels exhibiting both frequency-dependent spatial correlations and position-dependent frequency correlations, at a low computational cost.

3.1.3.4 Channel Dynamics in Geometry-Based Channel Models

Geometry-based channel models parameterized from measured data, such as the 3GPP SCM, the WINNER Channel Model have been widely used for evaluating the performance of MIMO transmission technologies in the next generation cellular systems such as LTE and IMT-Advanced. The inability of these models to evaluate the temporal evolution of MIMO transmission throughput has led to demands for the enhanced channel models that cover not only the static and time-averaged channel behavior but also the dynamic and time-variant channel behavior as already discussed in [Bul08] and [ST10]. While this issue is also solved with the COST 2100 model (see Sect. 3.6), alternative works have been conducted within COST 2100.

In particular, results in [Sai10] (for outdoor scenarios) and [CBYZ08, Czi07] (for indoor scenarios) suggest new methods to model the time-variant channel. The dynamic channel model is created by generating clusters according to the distributions of cluster parameters. The method proposed by [SKI+11] uses particle filter to parameterize the dynamic properties of rays in urban environments.

The Random Cluster Model (RCM) proposed by [CBYZ08] uses also multipath clusters to represent the channel. For the description of the environment, [Czi07] considers the multivariate distribution of the cluster parameters. This approach is advantageous over methods using only marginal distributions, which cannot model correlations between the cluster parameters. Small-scale time variations are implemented by linearly changing the parameters of the propagation paths. In every sampling time instance, the increments determined in the cluster parameters are added to the respective path parameters. In this way, clusters are moving in delay (causing Doppler shifts), as well as in angles, and smoothly change their power. To introduce large-scale changes in the propagation conditions, a cluster birth/death process is also included. In every cluster lifetime interval, the lifetime of each cluster is decreased. Dying clusters are fading out during the next cluster lifetime interval. New clusters are then drawn.

3.1.3.5 Multiple Bounces in Street Microcells

For street microcells, geometrical channel models have the main drawback that multiple bounces are generally not considered. The single-bounce assumption is rather restrictive, since the street width is not sufficient to match the ellipse of the maximum delay. In fact, one of the most critical point is the scattering power distribution or path gain.

To overcome this issue, the concept of effective street width can be introduced. In [GT07b] and [GT07a], a Geometrical Stochastic Channel Model (GSCM) is proposed for the dense urban areas, according to the physical propagation mechanisms in the street microcell. This model is site-specific and deterministically outlines specific zones. Multiple scattering is taken into account, yet such process still appears as single-bounce scattering when computing the field.

3.2 Improved Deterministic Channel Models

Deterministic channel models aim at simulating the physical radio-propagation process that takes place in a given environment with a given system setup and therefore at reproducing the actual radio coverage and channel transfer-function in specific cases. Not only are deterministic models useful to deploy or optimize a radio system in specific environments, but they can also be adopted in the system-design stage to assess channel or system behavior in important, reference cases. Since deterministic models in general and ray models in particular also correspond to some extent to mental visualizations of the actual propagation mechanisms, they also represent very valuable study tools.

Deterministic models require as input a detailed geometric and electromagnetic description of the environment. The geometric description is usually provided with a digital file called *environment database*. Urban environments are usually stored in 3D vectorial form where building walls and edges are represented as sets of points or vectors in a 3D Cartesian reference systems. Once expensive and hardly available, urban and building databases are now relatively cheap and of widespread use. Terrain profile in both urban and rural environment is generally discretized and stored in *raster data* form. The electromagnetic description requires that every object in the environment (building wall, terrain, etc.) be associated with its electromagnetic properties (e.g., electrical permittivity and conductivity) so that its electromagnetic behavior (reflection coefficient, penetration loss, etc.) be defined.

The output of deterministic models can be anything between mere field-strength and the complete description of the multidimensional channel transfer function over a spot, a route, a cell, or the entire service-area of the system.

Although physical and deterministic methods are traditionally regarded as opposed to empirical or statistical models, which provide a simplified, but more general, stochastic description of the radio channel, there is indeed a continuum of hybrid models in between the two extremes. Since environment description can never be very detailed, statistical elements must be introduced in deterministic models to account for the effect of the missed details on the propagation channel. If, for example, surface roughness or building details, such as decorations, widows, indentations, etc., are missing in the database, then a statistical modeling of scattering generated by those details should be introduced in the deterministic model. Similarly, if the geometric database definition is poor compared to the wavelength, then the deterministic model cannot accurately predict fast-fading, and a statistical description of it should (and could) be provided.

Modern deterministic channel models are essentially based on two different approaches, the electromagnetic approach and the ray approach. In the former case, Maxwell's equations or other analytical formulations derived from them are discretized and solved through numerical methods. In the latter case, the propagating field is computed as a set or rays whose interactions with the objects in the environment follow Geometrical Optics (GO) theory and its extensions to treat edge-diffraction and diffuse scattering.

Due to their adherence to the physical, multipath propagation phenomenon, deterministic models potentially allow for a multidimensional characterization of propagation, i.e., they could give information on the distribution in time and space (propagation delay, ray paths, angles of departure/arrival $(\theta_t, \phi_t)/(\theta_r, \phi_r)$) of the multipath field. This would be a valuable quality for the design and planning of Fourth Generation (4G) (and beyond) systems adopting Multiple-Input Multiple-Output (MIMO) schemes with array antennas and/or space–time coding techniques. There are however still shortcomings due to lack of database accuracy, deficiencies or problems in diffuse scattering modeling, and excessive computation time that still make deterministic models fall short of widespread use despite their potential. This is were improvements are still needed and were the COST 2100 researchers working on deterministic channel models put much of their effort.

Similarly to what happened in the last decade in the GPS navigation-system field, where the increasing availability and accuracy of digital maps and technology advances spurred the widespread diffusion of GPS navigators in vehicles and portable devices, deterministic radio propagation and channel models will most likely become of common use in the future for both research and commercial purposes.

3.2.1 Electromagnetic Models

Electromagnetic methods are aimed at solving Maxwell's equations, or other sets of equations derived from them, on a computer using some sort of domain discretization. In particular, the space domain must be discretized with a sufficiently small step, usually a fraction of the wavelength. Therefore, electromagnetic methods are not commonly used for propagation modeling at radio frequencies, because a huge space grid would be necessary to discretize an entire radio propagation environment, which is usually orders of magnitude larger than the wavelength. However, with the dramatic growth rate in memory capacity and CPU power of modern computers, the use of electromagnetic models for radio propagation prediction is becoming every year more and more feasible.

In [dlRJRG07] and [VRZ08], efficient finite-difference methods for radio propagation simulation are presented. The classic finite-difference electromagnetic method is Finite Difference in Time Domain (FDTD). In its basic form, the FDTD method consists of a discretization of time-domain Maxwell's equations in both time and space. The electromagnetic field is then propagated from the source over the discrete space grid in subsequent time steps. Proper absorbing boundary conditions at the border of the considered domain must be enforced to simulate free-space behavior around it. A different version of the FDTD method is the Transmission Line Method (TLM) method, where the domain is discretized into strips, and the field is propagated along them with proper matching conditions on the borders. Among other finite-difference methods, there are also frequency-domain versions of the TLM method.

In [VRZ08], the standard 2D FDTD algorithm is applied to an Non-Line-Of-Sight (NLOS) domain in which diffraction on corners is one of the main propagation phenomena. In order to reduce the memory requirements and the computation time, the FDTD method is run here at a lower simulation frequency. This way, the discretization steps grow, thus also reducing the computational load. To cope with the behavior of the materials at different frequencies, lower-frequency FDTD simulations require a calibration of the material parameters. However, it is shown that the results of this frequency reduction method are also subject to an error which is not related to the material properties, but to the dependence of wedge diffraction on the geometry of the problem and on the wavelength. Then, by comparing lower-frequency FDTD predictions with a single knife-edge diffraction model the author is able to evaluate the diffraction error, which is shown to depend on the following ratio:

$$k = \frac{h}{R_1}, \qquad (3.8)$$

where h is the intercept of the obstacle with the Line-Of-Sight (LOS) line, and R_1 is the radius of the first Fresnel's ellipsoid at the obstacle's location. The diffraction error vs. the frequency reduction factor for different values of k is shown in Fig. 3.6.

According to the author, this diffraction error bounds the achievable accuracy of lower-frequency simulations of finite-difference methods.

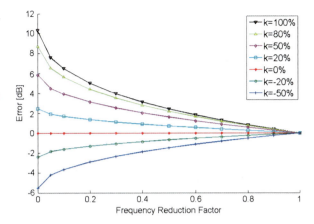

Fig. 3.6 Diffraction error vs. frequency reduction factor [VRZ08] (©2008 IEEE, reproduced with permission)

In [dlRJRG07], a method similar to TLM and called Finite Difference ParFlow (FDPF) is adopted. The method is able to simulate steady-state propagation of a narrowband field resorting to a linear inverse problem in the frequency domain [GJRR07]. The basic FDPF method is extended into a Multi-Resolution version, which is very efficient to compute multiple-source propagation, and tested against measurements in both indoor and outdoor environments. By using a very fine spatial step over limited areas it is possible to predict the statistics of fading. For use in outdoor environment, due to the large domain size, a reduced-frequency technique is adopted similarly to what done in [VRZ08]. The method is shown to predict with good accuracy the Rice factor of the fading, and the computation time is of only a few seconds, once a proper preprocessing of the environment is performed. The preprocessing, which is the most time consuming task, can be done only once for a given environment. Hence, this model is very efficient for multiple simulations in the same environment and is very fast compared to other finite difference models in the time domain. However, a disadvantage of this approach is that, unlike FDTD, it is not possible to simulate polarization, since the field in TLM is represented by scalar values. 3D and 2.5D extensions of this model have also been implemented. Moreover, in scenarios like outdoor-to-indoor where the size of the environment is too large, this model has been combined with a Ray Launching model [dlFL+10], providing a good accuracy with respect to measurements in terms of field amplitude and low computing time on a 2008-average performance personal computer, as summarized in Table 3.8.

Other kinds of electromagnetic methods based on boundary integral equations are aimed at solving radiation problems by directly calculating a solution to Maxwell's equations. The Electric Field Integral Equation (EFIE) method, frequently used for calculating radiation and scattering from perfect conductors is considered in [BLdC09]. This method, a frequency-domain approach as opposed to the time-domain finite-difference methods considered above, suffers from a lower bound in frequency, i.e., fails to provide accurate results for fields radiated by scatterers small with respect to the wavelength. In the paper, the authors examine the cause of this limitation by analyzing the asymptotic behavior of the method for

Table 3.8 Model computation time and RMS Error between measured and simulated field amplitudes

	Outdoor points	Indoor points	All the points
Number of points	72	32	104
Preprocessing time	0 s	41 s	41 s
Simulation time	58 s	57 s	115 s
RMSE	7.9 dB	2.4 dB	6.2 dB

decreasing frequency. This low frequency limit is compared to the equation of electrostatics and to that of the Magnetic Field Integral Equation (MFIE) in order to determine precisely what aspect of the equation fails. The goal of the study is to derive a better-conditioned formulation for use in Ground Penetrating Radar (GPR) simulations, where objects are of nonstandard shape and can be both larger and smaller than a wavelength.

3.2.2 Ray-Tracing Models

Ray Tracing (RT) models are based on GO and its extension to treat diffraction, namely the Geometrical Theory of Diffraction (GTD) or the Uniform Theory of Diffraction (UTD) and are usually adopted for field prediction in man-made environments such as the indoor and the urban environments. Recently, extensions to include diffuse scattering have also been proposed [DEFVF07, KH08, DEFVF08, MO10, MQO10]. RT models are based on the representation of the propagating field through a set of optical rays experiencing *interactions* along their paths such as reflection on plane surfaces, diffraction on straight edges, etc. Curved surfaces and edges are usually not present in environment databases. All rays experiencing up to a maximum number of interactions are computed, where such a maximum number can be different for different interactions. For example, if an RT simulation is performed with a maximum of four reflections and two diffractions, it usually means that all rays experiencing from zero (direct path) to six interactions are considered and among them, only those with a maximum of four reflections and two diffractions. Rays have a zero transverse dimension and therefore can describe the field with infinite resolution. Beams (or *tubes of flux*) have a finite transverse dimension because an angle-discretization is adopted. In models adopting beams, a limit to space resolution is set, and the RT model is more specifically called Beam-Launching or Ray-Launching model. RT models *naturally* simulate multipath propagation and therefore can potentially reproduce the distribution of the radio signal in time, space, and polarization, i.e., can perform a *multidimensional* radio channel simulation. Multidimensional channel modeling is now very important as performance of upcoming technologies such as 3GPP-Long Term Evolution (LTE) and Worldwide Interoperability for Microwave Access (WiMAX) will no longer depend only on Signal-to-Noise Ratio (SNR) but also on the time- and angle-dispersion of the radio channel. However, due to limits in the accuracy of the input database and in the capabilities of the today's available models, multidimensional simulation performance of RT is still object of debate [MBvDK09, MCBH10], and there is still a wide margin for improvements and extensions [Man07, KH08, BLdC08].

3.2.2.1 Ray-Tracing Performance Evaluation

In [MBvDK09] and [MCBH10], the multidimensional prediction performance of the same commercially available RT model has been evaluated through comparison with measurements. The adopted 3D prediction tool is based on a Ray-Launching method described in [HWW03].

In [MBvDK09], performances in terms of path loss L, DS σ_τ, and AS ϕ_τ have been considered, where the definition of such parameters are compliant with the 3rd Generation Partnership Project (3GPP) definition. Measurements were carried out at 2.2 GHz over six routes in Rotterdam, The Netherlands, with three different directional base stations at a height of about 27 m. Multidimensional measurements were done with the 3D high-resolution channel sounder available at the Eindhoven University of Technology. The multidimensional channel parameters were extracted from the measurements using the 3D Unitary ESPRIT algorithm. Then the prediction algorithm has been run with a maximum number of reflections of 2 and 1 diffraction in two different cases: with physical reflection and diffraction coefficients (derived from Fresnel's coefficients and UTD) and with empirical, effective reflection and diffraction coefficients. No tuning over such coefficients was performed. By comparing measurement and simulation the mean error and the standard deviation of the error are computed for L, σ_τ, and ϕ_τ over the six routes. Results show that while L is predicted with a very good accuracy, σ_τ is generally overestimated by the model, while the mean ϕ_τ is fairly well predicted, but its variations along the routes are not, as the mean error is of only 2 degrees, while the standard deviation of the error is of 16 degrees. Generally the model performs better with empirical coefficients than with physical coefficients. The limited accuracy of the model can be explained by the fact that conventional ray-tracing models provide only a limited modeling of the propagation environment, and therefore for example scattering from rough surfaces and trees is not included.

A similar comparison between the same Ray-Launching program and measurements is reported in [MCBH10]. Performance metrics are again the mean error and the standard deviation of the error for L, σ_τ, and ϕ_τ over one route in macrocellular environment. Here however the angular spread is divided into azimuth spread and elevation spread. The measurement campaign was carried out in Mulhouse, France, at a frequency of 2.2 GHz along a route with the Rx antenna on a car at 2.5 m of height and the Tx antenna on the roof of a building at a height of 25 m. The Rx antenna is a planar virtual array composed of 21×21 omnidirectional antennas. Measured impulse responses from all the Rx antennas were postprocessed using a beamforming algorithm and a multidimensional maximum detection algorithm in order to derive propagation parameters. The prediction tool was run with both physical and empirical coefficients and with a maximum of six reflections and two diffractions.

Results show that mean errors are small for all parameters, including σ_τ and ϕ_τ, but the standard deviations of the error are of the same order of magnitude as measured standard deviations of the same parameter over the route, meaning that the model is not accurate at describing the local value of propagation parameters. Also, empirical reflection and diffraction coefficients yield better results than physical ones, probably because such parameters are more appropriate at describing the effective behavior of building walls and edges in the simplified scenario provided by the urban database. The small large-scale dynamic range over the chosen route makes the measured and simulated azimuth spread curves appear as random. If a route with great variability of propagation conditions were chosen, such as a transition from LOS to NLOS, probably the trend would appear more clear, and the model more adherent to the measurements.

Similarly to the two previous works, an RT model is checked against multidimensional measurements along four routes in Tokyo, Japan, in [KIS+08]. Carrier frequency is 2.2 GHz, base station height is 38 m, and a sophisticated channel sounder is adopted with an 8-element ULA antenna at the base station (Tx) and a 96-element cylindrical antenna at the mobile station. Here the goal is to check RT reliability for prediction of dynamic channel properties, which are important to evaluate dynamic control techniques such as rank adaptation for LTE systems. Also in this case diffuse scattering is not included in the model. Results show that the mean delay and the mean angle of departure at the base station are well predicted by the model, while delay spread, angle of arrival, and azimuth spread at the mobile station are not very well reproduced, as stated also by the low correlation coefficients between simulated and measured parameters in the latter cases.

In [BCE+10], the path-loss prediction capabilities of an advanced RT model including Effective Roughness (ER)-scattering are checked against both measurements and other empirical-statistical models in microcellular, WiMAX environment. Measurement and simulation refer to a dense urban environment and an indoor environment in the city of Bologna, Italy. The RT model performs far better than the COST 231 and the Stanford University Interim (SUI) models over all the considered routes, except those were the environment description is failing. Typical performance metrics include a mean error close to zero in all cases and a standard deviation of the error of about 10 dBs in outdoors and 8 dBs in indoor cases. Such performances are achieved with few interaction events (two or three, including one diffraction) and computation time of the order of a few minutes.

3.2.2.2 Ray-Tracing Enhancement

A great deal of work has been addressed at enhancing RT performance, especially with the aim of improving the prediction of multidimensional propagation parameters and of reducing computation time. One possible way of attaining both goals is represented by the introduction of diffuse scattering into the model, but this topic is covered in the next section.

Fig. 3.7 Simplification of a piecewise line

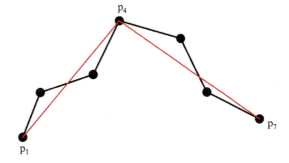

A method for computation time reduction through the simplification of the environment database is presented in [Man07]. The method, based on the recursive Douglas–Peuker algorithm, operates on the polygon representing the footprint of a building by reducing the number of its facets with a given *tolerance-distance* while preserving as much as possible its shape. The simplification of a line is illustrated in Fig. 3.7. The method is shown to be capable of reducing computation time to about one-third with a negligible prediction accuracy degradation.

Another important limitation of RT simulators is the limited number of reflections they can handle with a reasonable computation time. Unfortunately, when computing reflection from multiple-layer compound materials an infinite number of reflections would be needed due to the resonant nature of the field within each layer. This problem is often encountered when applying RT to GPR problems. A solution to this problem is presented in [BLdC08], where a compact, recursive formulation for reflection for multilayer material is derived. The formulation, not reported here for brevity, can be embedded in RT simulators. Since multiple reflection within a multilayer structure also cause a time spreading of the field, the proposed formulation is also useful to enhance the time-domain performance of an RT simulator.

3.2.3 Diffuse Scattering Modeling

Conventional ray models only account for rays that undergo specular reflections or diffractions, but neglect diffuse scattering phenomena which, according to recent studies, can have a significant impact on propagation and especially on multidimensional radio channel characteristics. The modeling of diffuse scattering, or of the Dense Multipath Component (DMC) component, is also treated in more general terms in Sect. 3.3. Diffuse scattering is intended here as the signal scattered in other than the specular direction as a result of deviations (surface or volume irregularities) in a building wall from a uniform flat layer.

In [DE01] and [DEFVF07], the so-called ER approach for the integration of diffuse scattering contribution into RT models is proposed. According to the ER approach, diffuse scattering, although due to a number of surface and volume details (windows, decorations, internal partitions, pipes, cables etc.) is assumed originating

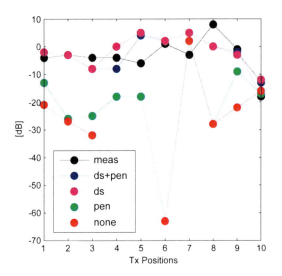

Fig. 3.8 Comparison with Measurements (meas) of Cross-polarization Ratio for Different Implementations: Diffuse Scattering and Penetration (ds + pen), Diffuse Scattering Only (ds), Penetration Only (pen), and Neither Diffuse Scattering nor Penetration (none) (warehouse scenario)

from the surface. An *effective roughness* is associated to each wall, the wall is discretized into surface elements and a given scattering pattern (Lambertian, i.e., with a cosine-shaped lobe directed toward the surface's normal, or with a directive lobe steered toward the specular reflection direction) is ascribed to each element, then the scattered rays are computed whose intensity must satisfy a power-conservation balance at the surface.

This model is adopted in [MO10] were it is embedded in an RT tool. A warehouse scenario is simulated were also measurements were carried out and measurement and simulation are compared in terms of σ_τ and Cross-Polarization Ratio (XPR), with the latter defined as

$$\text{XPR} = 10\log_{10}\left[\frac{P_{xp}}{P_{co}}\right], \tag{3.9}$$

where P_{xp} and P_{co} are the powers received in cross-polarization and copolarization, respectively. Measurements were performed with a channel sounder working at 1.9 GHz and with a bandwidth of 80 MHz using omnidirectional antennas. Results show that the implementation of diffuse scattering is very important to obtain a good accuracy of the XPR prediction in the considered scenario (see Fig. 3.8). Although originally the ER model does not include any depolarization model, the depolarization of the received field is given by the fact that the presence of diffuse scattering tends to move the propagation away from the quasi-horizontal plane containing the antennas, where most of the reflected and diffracted rays lay, therefore producing depolarization due to the inclination of the scattered rays. As regards delay spread σ_τ, diffuse scattering improves the prediction accuracy mainly in NLOS regions, but σ_τ is generally underestimated by RT simulations.

Similar studies are carried out in [MQO10] and [MO11]. In [MQO10], an office scenario is considered. Measurements are here at 3.6 GHz and were carried out with a virtual array of $10 \times 10 \times 2$ antennas. Results are similar to the warehouse case,

3   Radio Channel Modeling for 4G Networks

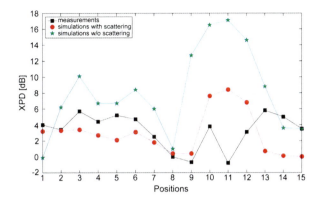

Fig. 3.9 Comparison with measurements of the XPR simulated with and without the diffuse scattering component (office scenario)

except that the directive diffuse scattering pattern performs better here than the Lambertian one. In [MO11], a microcellular outdoor scenario in a university campus is considered, where measurements have been carried out with an Elektrobit MIMO Channel Sounder with linear arrays of four dual-polarized patch antennas. Comparison between RT simulation and measurement yields interesting results. First of all, a calibration of the ER model is performed, and parameter values surprisingly similar to what reported in previous work [DEFVF07] are found. Then, again the inclusion of the diffuse component is shown to be very important to improve both Path Loss (PL) and XPR results, as shown in Fig. 3.9.

Therefore, diffuse scattering is definitely important for the correct RT multidimensional simulation of both the indoor and outdoor radio channels.

In [KH08] and [DEFVF08], similar advanced diffuse scattering models for the use in RT simulators are derived. The method described in [DEFVF08] is derived from the ER approach, while the one described in [KH08] is derived from an analytic approach where the stochastic properties of the surface of the wall convert into a random direction of the scattered power at the receiver [DDGW03]. The proposed methods can be used to model the dispersive effects of rough surface scattering in a manner similar to using the reflection reduction factor for Gaussian surfaces, except that the reduced power in the specular direction is distributed in the angular domain. The novelty and the value of the two new methods is that the angle distribution of the power scattered from a wall at a receiving point is computed analytically through relatively simple formulas.

The method presented in [KH08] has been embedded into a ray tracing tool, calibrated and validated against measurements in [KH09]. For a more complete coverage of this work, refer to [Kwa08]. Results show that the method can reproduce the angular distribution of the signal at the Rx with good accuracy, as shown in Fig. 3.10.

At sub-mm wave frequencies diffuse scattering is also very important. Due to the small wavelength with respect to the size of objects in both indoor and outdoor scenarios, almost every interaction, even with the smallest objects and details, could be treated as reflection or diffraction. However, the effect of rough surfaces, which must be considered as random, can only be modeled as diffuse scattering adopting

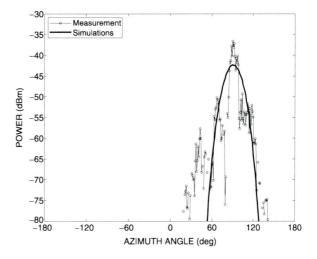

Fig. 3.10 Measured and estimated (from calibrated simulation) power-angular profile in a reference microcellular scenario [KH09] (©2010 EurAAP, reproduced with permission)

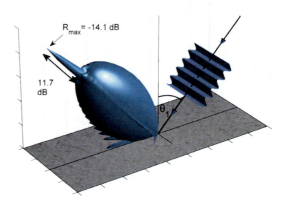

Fig. 3.11 Angular dependent scattering behavior of a rough surface element at THz frequencies, where surface roughness standard deviation is much greater than the wavelength

appropriate approaches. Rough surface scattering for THz propagation modeling is studied in [PJK] and [PJJK]. In [PJJK], nonspecular rough surface scattering is investigated at 300 GHz using the Kirchhoff scattering theory. Figure 3.11 gives an example of the angular-dependent scattering behavior of a rough surface element.

The practical implementation of the Kirchhoff model into a self-developed ray tracing algorithm is discussed. Up to three reflections and second-order scattering processes are included in the full-3D, image method-based algorithm. Ray tracing simulation results are compared with measurements in a simple indoor environment for the purpose of validation. Good agreements between simulations and measurements are observed. However, it has shown that including depolarization may help to further improve the accuracy of the results, which has been done in a next step. Geometrical depolarization and incoherent scattering according to a perturbation approach is implemented additionally in [PJK], whereas the Kirchhoff theory is still used for the coherent contribution. Ray tracing simulations are performed in small indoor scenario in order to study the impact of scattering in ultrabroadband THz propagation channels. These investigations show that a nonnegligible cross-polarization coupling is expected in THz propagation channels due to scattered path components. The most relevant scattering influence is found especially if the Rx is close to a rough surface and far from the Tx. Regarding the broadband channel transfer functions, an increased frequency selectivity may occur due to scattering. From the results it becomes clear that scattering needs to be considered for propagation simulations at THz frequencies in indoor environments as long as rough materials are present.

Also diffuse scattering due to vegetation is of importance in both rural and urban radio propagation. The main effects of scattering in random sparse media, such as the vegetation canopy, are the diffusion of power in all directions (incoherent component) and the attenuation of the main forward, coherent component. Attenuation of the forward component in vegetated areas is studied in detail in [TL] and [CTK10]. In [TL], a theoretical model based on the *Radiative Transport Theory* is developed to compute forward-component attenuation at millimeter-wave frequencies in a vegetated area, which is modeled as a layer of thick dielectric discs and cylinders. The radiative transport equation is simplified and then evaluated by using the forward scattering approximation. Results show that considering the coherent intensity only, the excess loss is overestimated. As a matter of fact, at a low penetration distance into the vegetation, with only few trees on path, the attenuation increase with distance is greater than at larger vegetation depths where a transition occurs from the strongly attenuated direct-path mode to a multiple scatter mode, which has a lower attenuation rate. In [CTK10], a method previously developed by the authors of [TL], based on the Foldy–Lax multiple scattering theory, is applied to the computation of vegetation attenuation for broadband access systems at 3.5 GHz. Three measurement campaigns were performed in a rural area using a mobile WiMAX system (IEEE 802.16e) during winter, spring, and midsummer. Foliage attenuation for different degrees of foliation in different seasons is studied. The derived foliage loss is then verified and compared with an empirical exponential decay mode. This excess foliage attenuation due to foliation, which peaks in summer, is found to be about 0.7 dB/m under the assumption that the primary trunks of trees along the propagation paths remain in place throughout various defoliation processes.

3.2.4 Other Issues Related to Deterministic Channel Models

Other studies have been carried out within the COST 2100 action that are related to deterministic channel modeling. One very attractive application of RT models is the possibility of replacing measurements to parameterize statistical or geometric-stochastic propagation models. In [ZK10] the RT model developed at the University of Bologna [DEG+04] is used to analyze the dynamics of change of major multipath components along a route in urban environment in order to determine their

Fig. 3.12 Rays diffracted on vertical edge belong to the Keller's cone whose vertex is the VTX. The VTX is almost stationary in the vicinity of *point C* (route tangent to the cone), while in other points, where route cuts the cone (e.g., *points A, B, D*), the VTX slides over the edge as the Rx moves along the route [ZK10] (©2010 EuMA, reproduced with permission)

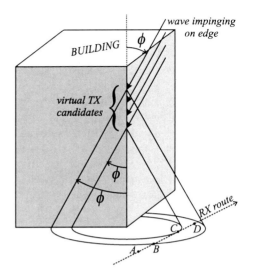

visibility region-size. The "*visibility region*" of a multipath component or, better, of a *multipath cluster*, is a key concept of geometric-stochastic propagation models and represents the space- or time-span over which the multipath component is present at the generic radio terminal antenna. As stated by Geometrical Optics, it is well known that a ray undergoing a number or reflections appears as generated from a Virtual Tx (VTX), whose fixed position corresponds to the multiple-reflected image of the actual Tx. Using the RT program, in [ZK10], the authors identify a ray (or better a *ray entity*) with a fixed VTX as the mobile terminal moves along the route and assume the ray entity "dead" when no ray comes from that virtual Tx anymore. With this procedure, the visibility region, i.e., the space duration of the ray entity, is determined. If however the VTX is not fixed, as for diffracted rays, where the VTX can slide up and down over the corresponding vertical building edge (see Fig. 3.12), then the ray life appears as very short.

Therefore, in diffraction-dominated environments, such as the one considered in [ZK10], the estimated visibility region can be as low as 1 m and therefore much shorter than what commonly expected, i.e., 10 to 100 m. It must be pointed out however that the visibility region is usually intended for clusters of rays rather than single rays, which on the whole should have a greater lifespan than the single ray components.

In [MK07], a study on the influence of geographical input database resolution on prediction accuracy is presented. The considered kind of prediction model is a semi-deterministic one for field-strength prediction in macrocells [KFW96], and the input database includes terrain-height data and land-usage with three different pixel dimensions (and therefore resolutions): 50 m, 100 m, and 200 m. Results refer to simulation and measurement in urban, rural-open, and rural-forested environments at 900 MHz and 1800 MHz. Results in terms of mean error and standard deviation of the error are reported in Table 3.9 for different resolution levels and land usage.

Table 3.9 Performance of field prediction vs. resolution level for different environments

Environment	μ (50 m)	σ	μ (100 m)	σ	μ (200 m)	σ
urban	−0.5	7.7	−0.2	8.0	0.4	8.3
open	0.6	8.3	0.4	8.5	0.7	8.8
forested	−0.9	9.7	−0.6	9.6	0.5	9.7

It is evident that the higher the resolution, the lower the error, especially the standard deviation of the error, but of course computation time is much higher with higher resolutions. Also, the accuracy of the prediction is shown to be lower at the higher frequency of 1800 MHz.

In [CL08], a new method for the extrapolation of the 3D antenna pattern from 2D diagrams is presented. It is well known that antenna manufacturers do not usually provide complete 3D antenna patterns in data sheets but only two 2D antenna diagrams in two perpendicular planes (e.g., the vertical and the horizontal plane). In the last two decades however, with the diffusion of mobile radio systems the knowledge of the complete radiation properties of antennas has become necessary even in direction far apart from the main radiation axes, especially for system simulation and interference evaluation purposes. Traditional extrapolation methods do not give good results outside of the main radiation lobe. Therefore a new method is proposed which is based on the following expression. Given the two partial radiation diagrams $G_v(\theta)$ and $G_h(\phi)$ expressing antenna gain in dB, the 3D antenna pattern is

$$G(\theta, \phi) = G_v(\theta) + \lambda G_h(\phi), \tag{3.10}$$

where

$$\lambda = \frac{G_v(\theta') - G_v(\theta)}{G_v(\pi) - G_v(0)}, \tag{3.11}$$

and $\theta' = \pi - \theta$ if $\theta < \pi/2$ or $\theta' = 3\pi - \theta$ if $\theta \geq \pi/2$. The method is shown to give more realistic results with respect to traditional extrapolation methods for directions far apart from the main radiation lobe.

In [LwDBTdC09] an approximate method for computing the RF field in presence of moving objects, such as blades of wind turbines, is presented. By using simple formulas for the reflected fields of objects that are large enough with respect to the wavelength, it is possible to gain insight in the scattering mechanisms, in particular from moving objects, e.g., wind turbines. To obtain a reasonably accurate representation of the near-field cross-section of the object, it has to be limited to its coherent part. For a flat plate, this can be deduced from the far-field distance phase requirement formula. For a curved object, the radii of curvature have to be taken into account. It was shown that for a cylinder, the formula was very close to the ones that can be obtained from the ideal flat wedge diffraction formulas. They also predict a $1/d$ behavior for the very close but practically unrealistic near-field ($d \ll \lambda$), over a $1/d^{1.5}$ for the near-field and then the $1/d^2$ behavior in the far-field.

Table 3.10 Summary of studies about DMC modeling

Study	Measurement campaign	Portion of channel power in DMC	Correlation with specular
[KLT09]	urban Tokyo, 4.5 GHz	same order as specular	–
[PSH$^+$11]	modern office TKK 5.3 GHz	specular 5 to 15 dB below DMC	PDP: 0.6–0.8
[QOHD10b, QBR$^+$10]	indoor office UCL 3.6 GHz	mean: 30%	0.99 (azimuth) 0.98 (elevation) corr. delay
[SPH$^+$10]	modern office TKK 5.3 GHz	indoor: DMC 50–90% outdoor: DMC 20–80%	corr. angular/delay
[CKZ$^+$10]	vehicle-to-vehicle 5.9 GHz	–	–
[PTH$^+$10b]	based on [PSH$^+$11, QOHD10b]	–	–

3.3 Modeling Dense Multipaths

3.3.1 Introduction

As already introduced in Sect. 3.2.3, the radio channel cannot be characterized by specular components alone, but one has to take into account the distributed random scattering, i.e., the DMC (also known as dense or diffuse multipaths). When extracting specular propagation paths using high-resolution algorithms, part of the power is not captured. This residual portion of received power is defined as DMC and can be remarkably substantial. One possibility is to model the DMC within ray-tracing tools, as outlined in Sect. 3.2.3. This section investigates alternative solutions in the framework of Geometry-Based Stochastic Channel Modeling (GBSCM).

Only a few studies are available concerning the characterization and modeling of the directional properties of the DMC. A first model of DMC in the time delay domain is included in [Ric05]. The DMC was modeled with an exponential power delay profile. First attempts to model DMC ignored the spatial distribution of the DMC and modeled it as an angular-white contribution. In [SRK09] and [KLT09, PSH$^+$11, QOHD10b, CKZ$^+$10, SPH$^+$10, QBR$^+$10], it is shown that the angular-white assumption cannot be used for modeling the angular spectrum of the DMC. A number of contributions [KLT09, PSH$^+$11, QOHD10b, CKZ$^+$10, SPH$^+$10, PTH$^+$10b, QBR$^+$10] deal with modeling of DMC. A cluster-based Kronecker modeling approach is proposed in [KLT09, PSH$^+$11, SPH$^+$10], a spatio-temporal model using von Mises distributions for the angular power spectrum is discussed in [QOHD10b, KLT09, QBR$^+$10], and a low-complexity geometry-based modeling approach is presented in [CKZ$^+$10]. Finally, [PTH$^+$10b] proposes a method to include DMC in the COST 2100 channel model (see Sect. 3.6). Table 3.10 summarizes the different contributions about DMC modeling.

3.3.2 Modeling the DMC

As already mentioned in Sect. 3.1.3, MIMO radio propagation channels are classically modeled by means of correlation matrices. When including DMC, the channel matrix consists of the superposition of two components, the specular component and the dense multipath component:

$$\mathbf{H} = \mathbf{H}_{SC} + \mathbf{H}_{DMC}, \quad (3.12)$$

where tensors $\mathbf{H}$, $\mathbf{H}_{SC}$, and $\mathbf{H}_{DMC} \in \mathbb{C}^{n_r \times n_t \times n_f}$ represent the "total" impulse response, the impulse response resulting from specular interactions, such as reflections, and the impulse response from the diffuse scattering, respectively, and n_r, n_t, and n_f represent the numbers of receiver antennas, transmitter antennas, and frequency (delay) samples, respectively. The stochastic model of the specular component is well known and relies, e.g., on the Kronecker or Weichselberger models [Cor06].

To model the DMC in a stochastic fashion, the first assumption is that dense multipaths can be modeled by a complex-valued, zero-mean, normally distributed, circularly symmetric, multivariate random variable $\text{vec}(\mathbf{H}_{DMC}) \sim \mathcal{CN}(0, \mathbf{R}_{DMC})$ [KLT09, SPH+10]. The second assumption is that the covariance matrix $\mathbf{R}_{DMC}$ of the DMC can be written as follows:

$$\mathbf{R}_{DMC} = \mathbf{R}_r \otimes \mathbf{R}_t \otimes \mathbf{R}_f, \quad (3.13)$$

where $\mathbf{R}_r$ and $\mathbf{R}_t$ denote the covariance matrix at the receiver and transmitter, respectively, and $\mathbf{R}_f$ denotes the covariance matrix in the frequency domain. Thus in (3.13) a Kronecker structure is assumed, implying for the DMC, uncorrelated scattering between receive angles (azimuth and elevation), transmit angles (azimuth and elevation), and delay [Ric05, SPH+10].

As a model for $\mathbf{R}_f$, the following is proposed. In the delay domain, one can observe an exponentially decaying function for the DMC power delay profile (PDP) (see also Fig. 3.13). The exponential power profile $\psi_d(\tau)$ can thus be described by the parameters peak power α_d, base delay τ_d, and slope B_d in (3.14) [QOHD10b]:

$$\psi_d(\tau) = \begin{cases} 0, & \tau < \tau_d, \\ \alpha_d/2, & \tau = \tau_d, \\ \alpha_d e^{-B_d(\tau - \tau_d)}, & \tau > \tau_d. \end{cases} \quad (3.14)$$

The power spectrum density is then calculated as the Fourier transform of the DMC PDP. The covariance matrix $\mathbf{R}_f$ in the frequency domain can be computed using (5) of [KLT09]. $\mathbf{R}_f$ is a Toeplitz matrix based on the sampled version of the power spectrum density [KLT09, SPH+10].

For the DMC angular distribution, a double von Mises distribution is proposed, with individual mean and concentration parameter for both azimuth and elevation (thus four parameters). The spatial domain covariance matrices $\mathbf{R}_{Rx}$ and $\mathbf{R}_{Tx}$ at the receiver and transmitter can be obtained from the DMC angular distributions and the steering matrices (using (7) in [KLT09] and (6)–(8) in [SPH+10]).

The above modeling can be extended to account for DMC clusters, where the DMC parameters are determined for each cluster separately [PSH+11, QOHD10b, SPH+10, QBR+10]. Geometry-based stochastic channel models (GSCMs) can incorporate these models to take the influence of the diffuse scattering in the MIMO radio propagation channel into account.

The approach to include DMC in the COST 2100 MIMO channel model is the following [PTH+10b]. In angular domain, the DMC is included by distributing the multipath components (MPCs) for the DMC around the cluster centroids within an area larger than that of used for the specular paths. In the delay domain, each MPC of the DMC is assigned an extra delay time in addition to the delay determined by the geometry in order to achieve an exponentially decaying PDP for the DMC. Hence, the delay time of the MPCs of the DMC can be expressed by

$$\tau_{DMC} = \tau_{DMC,geometrical} + \tau_{DMC,additional}, \qquad (3.15)$$

where $\tau_{DMC,geometrical}$ is the delay coming from the geometry, and $\tau_{DMC,additional}$ is a random additional delay that is added to each MPC of the DMC.

3.3.3 DMC Extraction from Measured Data

In [PTH+10b], three possible procedures for the extraction of the parameters for the DMC from measurement data are proposed. In the angular spectrum-based approach, the DMC part of the clusters can be identified from the measurement data based on the power-angular-profile (PAP) calculated through beamforming. In the physical scattering point-based approach, scattering points of physical objects can be identified for the MPCs of the specular component by combining the measured signal directions with a map of the environment. Finally, in the trial-and-error-based approach, parameter values are extracted by obtaining a good match between a measurement and a simulation based on validation metrics such as delay spread, angular spread, etc. In the delay domain three parameters are needed in order to include the DMC to the COST 2100 channel model, namely, the base delay, the peak power of the DMC, and the DMC cluster factor [PTH+10b].

As an illustration of the angular spectrum-based approach, the DMC in (3.12) is obtained by subtracting the impulse response formed by the specular components as estimated by a high-resolution algorithm (e.g., SAGE [QOHD10b, QBR+10], RiMAX [KLT09], or EKF [PTH+10b]) from the original total impulse response. Figure 3.13 illustrates this principle with an example of the measured power delay profile of [QOHD10b]. The spatio-temporal spectrum for DMC at the receive side or at the transmit side can then be obtained by applying a beamforming technique ((10) in [KLT09], (2) in [QOHD10b]).

Käske et al. [KLT09] provide a first step to discard the assumption of an angular white spectrum for DMC. The nonwhite angular spectrum of DMC is later also shown in [PSH+11, QOHD10b, SPH+10]. The extension of the DMC to the angular domain yields an improved description of a MIMO-channel both in terms

Fig. 3.13 Measured PDP, SAGE-reconstructed PDP, and resulting DMC [QOHD]

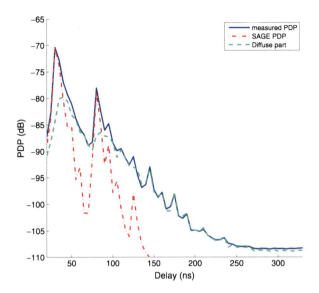

of channel capacity and angular distribution of power [KLT09]. The model in the angular domain is comprised of a von Mises distribution and an additional uniform distribution.

An important contribution of [SPH+10] is the proposition of a DMC model using a superposition of so-called DMC clusters, which consider both delay and angular characteristics, and polarization (Table 3.10). The DMC is modeled as a multivariate zero-mean normally distributed random variable with specific correlation properties, hence yielding Rayleigh-like fading, whereas the specular contribution is a superposition of deterministic propagation paths. The angular and delay parameters of the DMC should have correlation with those of the specular parameters. Each DMC cluster is modeled by an exponentially decaying power delay profile. For both azimuth and elevation for each DMC cluster, a double von Mises angular distribution with individual mean and concentration parameter is proposed. The model also enables time-evolution modeling.

In [PSH+11], the angular characteristics of the DMC have been studied by analyzing data from a dynamic dual-link indoor channel measurement at 5.3 GHz (Table 3.10). The DMC is assumed to obey a Kronecker model with no specific structural model for the individual covariance matrices. The advantage of the estimation approach is that the power-angular-delay profiles (PADPs) of both the specular and the dense multipath components can be analyzed. The specular power is 5–15 dB below the power of the DMC for the most of the time in the considered environment. The correlation between the power-angular profiles of specular and dense MPCs was generally in the order of 0.6–0.8.

In [QOHD10b], the spatio-temporal characteristics of the DMC are extracted from an extensive measurement campaign (Table 3.10). It is observed that the angular power spectrum of the DMC is not white and that the angular properties of the DMC are significantly correlated with the angular properties of the specular part of the channel. The diffuse part of the channel is modeled as the sum of several diffuse clusters. Each diffuse cluster has a von Mises distribution in the azimuth and in the co-elevation domain and an exponentially decaying power delay profile. In [QBR+10], the methodology for clustering specular-diffuse channels is based on joint clustering of the specular MPC and the spatio-temporal diffuse bins. The specular and diffuse cluster spatio-temporal location show to be highly correlated to the global cluster location. Spreads are slightly higher for the diffuse component. The specular angular spread is always lower than the global cluster angular spread, while the diffuse angular spread is always higher.

3.3.4 Low-Complexity Geometry-Based Modeling Approach

In Sect. 3.3.2, DMC is modeled as a continuous power distribution in the delay and angular domains. Alternatively, DMC can also be modeled as the summation of a large number of discrete paths, placed in the propagation environment by geometrical considerations (much like specular paths in GSCMs). This last approach is usually computationally very complex and time consuming, as it requires to add a large number of complex exponentials. To overcome this complexity constraint, the simulation method from [KZU07] is proposed [CKZ+10]. The modeling of the diffuse components is demonstrated by the example of a vehicle-to-vehicle channel, where the proposed low-complexity approach reduced the simulation time by a factor of 30.

3.4 Multipolarized Channel Modeling

MIMO systems show tremendous performance improvement in terms of outage capacity or system diversity. However, in some circumstances, it is impossible to design a MIMO system where the antennas are separated by half a wavelength or more. Polarized MIMO systems have been proposed as a solution to resolve this problem. By using perpendicularly polarized, colocated antennas, one can benefit from the polarization diversity to obtain low inter-antenna correlation while still maintaining a compact equipment size.

3.4.1 Polarized MIMO Systems

When considering polarized MIMO systems, the polarization of the channels is not perfectly maintained. Some power leaks from one polarization to another polarization (e.g., from the vertical transmit antenna to the horizontal receive antenna), mainly due to three mechanisms. The first one is the antenna polarization patterns; although an antenna is usually designed to capture a given polarization, it will always capture some of the power of the perpendicular polarization. This can easily

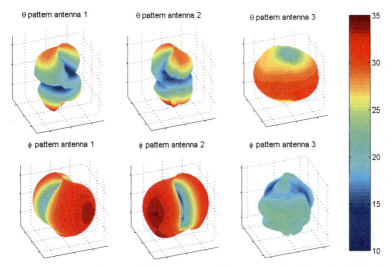

Fig. 3.14 θ- and ϕ-polarization radiation patterns of three perpendicularly polarized co-located antennas [QOHD09b] (©2009 IEEE, reproduced with permission)

be accounted for by considering the θ- and the ϕ-polarization patterns of the antennas. Figure 3.14 shows an example of the radiation patterns of three perpendicularly polarized colocated antennas. Antennas 1 and 2 are designed to capture two perpendicular horizontal polarizations, and antenna 3 is designed to capture only vertical polarization. However, antennas 1 and 2 will also capture some power of the θ-polarization in certain directions, causing leakage from a vertical antenna. Antenna 3 will mainly capture the power of the θ-polarization, but the imperfect cross-polar isolation of the antenna causes it to capture some power (although very low) in the ϕ-polarization, leading to interference from a horizontal transmitter.
The second mechanism is the tilting of the antenna array. The mismatch between the transmit antenna array tilting and the receive antenna array tilting will result in polarization crosstalk. This is especially critical when considering mobile handsets that can hardly be kept in a fixed orientation. The third mechanism is due to the channel itself. The interaction of the electromagnetic wave with its surroundings will cause the wave to be depolarized, causing polarization leakage between the antennas of the transmitter and the antennas of the receiver.

3.4.2 Polarization in Stochastic Channel Models

Polarization has two effects on the MIMO channel matrix. The first one is to reduce interantenna correlation by exploiting the polarization diversity. In [MGPLD+09], measurements are performed in a tunnel environment, and cross-polarized and copolarized systems are compared in terms of capacity and correlation. The results show that the interantenna correlation of a cross-polarized system is significantly lower than the interantenna correlation of a copolar system (for an identical total equipment size), thereby validating the idea of using cross-polarized systems to reduce correlation.

The second effect of polarized MIMO systems is to cause a power imbalance between the different branches of the MIMO channel matrix; for example, most of the power transmitted by a vertically polarized antenna will remain in the vertical polarization, but some of the power will leak into the horizontal polarization because of the three mechanisms described previously. In statistical channel models, the power imbalance between polarizations is usually represented by the Cross-Polar Discrimination (XPD), which is defined as the power ratio between the different branches of the MIMO channel matrix. The XPD can be defined in several ways. In the following, dual-polarized arrays with vertically and horizontally polarized antennas will be considered. The formalism that is developed can be extended to other types of polarized arrays (e.g., $+45°/-45°$ polarized arrays). At the receiver side, the XPD is defined as

$$\text{XPD}_V^{Rx} = \frac{\text{E}\{|h_{VV}|^2\}}{\text{E}\{|h_{HV}|^2\}}, \qquad \text{XPD}_H^{Rx} = \frac{\text{E}\{|h_{HH}|^2\}}{\text{E}\{|h_{VH}|^2\}}, \qquad (3.16)$$

where $\text{E}\{\ldots\}$ represents the expectation operator, h_{XY} is the channel between antenna with receive polarization X and the antenna with transmit polarization Y, V represents the vertical polarization, and H represents the horizontal polarization. At the transmitter side, the XPD is defined as

$$\text{XPD}_V^{Tx} = \frac{\text{E}\{|h_{VV}|^2\}}{\text{E}\{|h_{VH}|^2\}}, \qquad \text{XPD}_H^{Tx} = \frac{\text{E}\{|h_{HH}|^2\}}{\text{E}\{|h_{HV}|^2\}}. \qquad (3.17)$$

Finally, the Co-Polar Ratio (CPR) between the VV-link and the HH-link is defined as

$$\text{CPR} = \frac{\text{E}\{|h_{VV}|^2\}}{\text{E}\{|h_{HH}|^2\}}. \qquad (3.18)$$

The behavior of XPD and CPR show no significant dependence on the distance between the transmitter and the receiver. In [MGPRJL08], an extensive indoor measurement campaign at 2.45 GHz shows that the path loss coefficients of the different polarizations combinations show no significant differences and that wall attenuation is identical for all combinations of polarizations (see Fig. 3.15). The XPD values range between 10 and 20 dB, and a CPR around 0 dB was measured. Similar conclusions have been drawn in [PQD+10] for outdoor-to-indoor and indoor-to-indoor measurements at 3.5 GHz. The evolution of XPD and CPR with distance are rather low and show no significant trends. Mean XPD values are around 5 dB for the outdoor-to-indoor case and between 5 and 10 dB for the indoor-to-indoor case. The measured CPR shows slightly negative values (around -2 dB).

The large-scale variations of the XPD and CPR around their mean can be modeled with a log-normal distribution, with typical spreads between 2 and 4 dB.

Finally, if one is interested in the small-scale variations of the XPD, that is, if one considers the XPD of a MIMO channel without applying the expectation operator

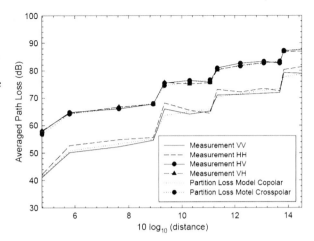

Fig. 3.15 Path-loss of different polarization combinations. It can be observed that the power ratio between the different polarizations does not change significantly with distance, nor with wall attenuation [MGPRJL08] (©2008 IEEE, reproduced with permission)

in (3.16)–(3.18), it has been shown in [QOHD09a] that the small-scale variations of the XPD can be modeled with a doubly noncentral F-distribution:

$$\alpha^2 \chi_V \sim F''(2, 2, 2K_{VV}, 2K_{HV}), \quad (3.19)$$

where $\chi_V = |h_{VV}|^2/|h_{HV}|^2$ is the instantaneous XPD, $\alpha = 1/\sqrt{\text{XPD}_V}$ is the inverse of the square root of the XPD as defined in (3.16), and K_{VV} and K_{HV} are the Ricean K-factors of the VV-link and the HV-link, respectively. Equation (3.19) describes the behavior of the instantaneous XPD around its mean value. Measurement were performed to confirm this theoretical result, and excellent agreement was obtained between the measurements and the model.

The effects of polarization on capacity is twofold. On one hand, the interantenna correlation is strongly reduced, increasing the diversity and thereby increasing the MIMO channel capacity. On the other hand, when considering polarized MIMO systems, the channel coefficients between crossed polarizations have lower power than the channel coefficients between identical polarizations. When considering a fixed transmit power, this will reduce the received SNR, which will finally result in reducing the MIMO channel capacity. The comparison of a cross-polar and a copolar system in a tunnel environment in [MGPLD+09] determines that the capacity improvement due to lower correlation is more or less balanced out by the capacity penalty due to lower SNR. Figure 3.16 show the measured capacities, for a fixed transmit power, between different cross-polar and copolar configurations. It can be deduced that most systems perform equally well, except for the VH configuration (all transmit antennas vertical and all receive antennas horizontal).

3.4.3 Polarization in Geometrical Channel Models

Another way to consider polarization in channel models is to include a per-wave polarization in geometrical channel models. In that case, the interaction between

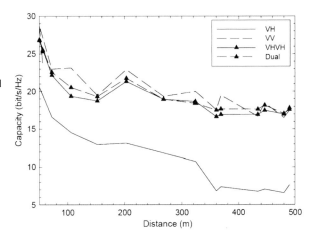

Fig. 3.16 Capacity of different cross-polarized or copolarized MIMO configurations. VH is all Tx antennas vertical and all Rx antennas horizontal, VV is all Tx antennas vertical and all Rx antennas vertical, VHVH is half of the Tx and Rx antennas vertical and half of the Tx and Rx antennas horizontal, and Dual is dual-polarized Tx and Rx antennas

polarized wave and radiation pattern is straightforward. The antenna effects, array tilting effects, and channel effects are perfectly decoupled. The contribution of a single wave k, with angles of departure $(\theta_{t_k}, \phi_{t_k})$ and angles of arrival $(\theta_{r_k}, \phi_{r_k})$, to the MIMO channel matrix can be described as

$$\mathbf{H}(t, \tau) = \underbrace{\exp(j 2\pi \nu_k t)}_{\text{Doppler shift}} \underbrace{\mathbf{C_r}(\theta_{r_k}, \phi_{r_k})}_{\text{project. on Rx array}} \begin{bmatrix} \gamma_{\theta\theta,k} & \gamma_{\phi\theta,k} \\ \gamma_{\theta\phi,k} & \gamma_{\phi\phi,k} \end{bmatrix} \underbrace{\mathbf{C_t}(\theta_{t_k}, \phi_{t_k})^T}_{\text{project. on Tx array}} \delta(\tau - \tau_k), \tag{3.20}$$

where ν_k is the Doppler shift of wave k, and τ_k is the delay of wave k. The matrix $\boldsymbol{\gamma}_k$ is the polarization matrix of wave k. Note that in literature, the elements of γ are often referred to as γ_{VV}, γ_{HV}, γ_{VH}, and γ_{HH}. These elements represents the θ- and ϕ-components of the wave k, not to be confused with the polarization V and H of the antenna systems! If we consider N receive antennas, the matrix $\mathbf{C_r}$ is an $N \times 2$ matrix with the first column representing the array response for the θ-polarization in the direction $(\theta_{r_k}, \phi_{r_k})$, and the second column representing the array response for the ϕ-polarization in the direction $(\theta_{r_k}, \phi_{r_k})$. Similarly, for M transmit antennas, the matrix $\mathbf{C_t}$ is an $M \times 2$ matrix with the first column containing the array response for the θ-polarization in direction $(\theta_{t_k}, \phi_{t_k})$, and the second column contains the array response for the ϕ-polarization in direction $(\theta_{t_k}, \phi_{t_k})$. The array response accounts for the phase shift between the different antenna elements and for the polarized antenna pattern of each antenna element. It is clear from Eq. (3.20) that including polarization in geometrical channel models is only a matter of properly parameterizing matrix $\boldsymbol{\gamma}_k$.

The model of (3.20) implicitly includes two mechanisms for transferring the power from one polarized antenna to another polarized antenna. The first mechanism is the depolarization of the wave itself, causing the antidiagonal elements of $\boldsymbol{\gamma}_k$ to be nonnegligible. The second mechanism is due to the propagation mechanisms: rays that have strong diffraction or diffuse scattering can be incident on the receiver with skew angles that will cause the θ-polarization to be projected mostly on a horizontal antenna, causing global channel depolarization.

3 Radio Channel Modeling for 4G Networks

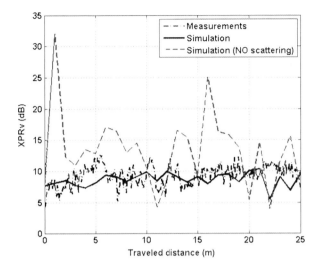

Fig. 3.17 Comparison of measured XPD, simulated XPD with diffuse scattering, and simulated XPD without diffuse scattering

One of the goals of ray-tracing models has been to correctly model polarization behavior. In [SA08, SA09], ray-tracing results were presented for canonical scenarios using only reflection and diffraction mechanisms. In this case, the channel depolarization is due to wave depolarization, which is rather low when considering Fresnel reflection coefficients, and to oblique incidence of the waves upon the receiver, which is only possible when considering three-dimensional propagation and a large number of reflections (such as wall-to-ground, ground-to-wall, etc.). Capacity values are presented to show the potential improvement of polarized MIMO systems. In [VKDEV08, VKDEV09], diffuse scattering has been added to the ray-tracing model. This will cause the two depolarization mechanisms to contribute to the channel depolarization: some diffuse rays will contribute to an antenna they are not supposed to contribute to due to their skew angle of arrival. Additionally, the diffuse waves are introduced with a cross-polarization parameter K_{XPOL} that indicates the amount of power that is scattered in each polarization (antidiagonal elements of γ_k). The parameter K_{XPOL} serves as a tuning parameter that can be adjusted by comparison with experimental results. The optimum value of K_{XPOL} is found to be 1%, indicating that wave depolarization is rather minor and that skew angle of arrival is the main cause of channel depolarization. Comparison with experimental result is shown in Fig. 3.17. It is obvious that diffuse scattering is fundamental for modeling polarization behavior properly in ray-tracing models. Measurements were used in combination with ray-tracing tools in [VDEMO10] to determine the polarimetric properties of diffuse scattering in the case of a simple reflection on a building. Results suggested that the diffuse scattering contains around 30% of the power and that the cross-polar ratio of the diffuse component was low (around 5 dB), indicating that the cross-polarization coupling due to the diffuse scattering is high.

Extensive measurements have been conducted to experimentally identify the waves cross-polarization parameters. In [DC07a, DC07b] and [QOHD09c, QOHD10a], polarized Single-Input Multiple-Output (SIMO) measurements were performed in an outdoor environment and an indoor environment, respectively. The measurements were performed with virtual arrays, and propagation paths were detected by applying Bartlett beamforming and finding the local maxima of $h(\tau, \theta, \phi)$. The ray XPD is defined as follows:

$$\text{XPD}_V^{ray} = \frac{P_{VV}^{ray}}{P_{HV}^{ray}}, \qquad \text{XPD}_H^{ray} = \frac{P_{HH}^{ray}}{P_{VH}^{ray}}, \qquad \text{CPR}^{ray} = \frac{P_{VV}^{ray}}{P_{HH}^{ray}}, \qquad (3.21)$$

where P_{XY}^{ray} is the path power of polarization XY. The ray XPD are found to be independent of delay or azimuth of arrival, but to slightly increase with coelevation. However, this latter is probably due to imperfect antenna deembedding. The ray XPD shows a log-normal distribution with means between 5 and 6 dB, and spreads between 5 and 6 dB. The ray CPR also has a log-normal distribution, with means around 2 dB and a spread around 6 dB. Finally, in [QOHD09c], LOS rays show significantly higher XPD (ray XPD between 15 and 20 dB).

Finally, in [KMT+09] and [QOHD09b], polarization of the directional channel was not expressed in terms of ray polarization, but in terms of cluster polarization, for outdoor and indoor SIMO channels, respectively. The cluster XPD is defined as in (3.21), except that one has to consider the power of all the rays of the cluster for one polarization. In the outdoor case, it is investigated if cluster polarization can be expressed as a function of the physical propagation parameters (angles of arrival, delay, etc.). No significant trends are observed, confirming the fact that polarization is independent of the physical propagation parameters. However, in both environments, it is observed that the cluster XPD is log-normally distributed, with mean values around 10 dB for the outdoor channel and around 5 dB for the indoor channel. The cluster XPD spread is around 5–6 dB. In both environments, the cluster CPR is shown to have slightly negative values (between -1 and -2 dB) and a spread around 5 dB.

3.5 Multilink Channel Characterization

The joint characterization of multiple radio links has become important for deployment scenarios involving common terminals, such as relay networks, distributed cooperative communications, virtual MIMO schemes, etc. The design and performance evaluation of such techniques require reliable models of such radio channels. Additionally, multilink scenarios range from cellular multi-BS cooperation to peer-to-peer links (where terminals on both link-ends can be mobile) and relay feeder links (with stationary terminals on both ends). These scenarios significantly differ from the classical cellular concept: changes in terminal deployment (height) and mobility pattern result in considerably different propagation conditions.

3 Radio Channel Modeling for 4G Networks

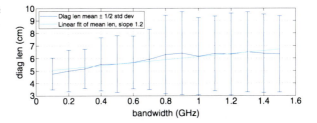

Fig. 3.18 Dependence of the average path visibility region and its standard deviation on bandwidth [PHKU09]

Multilink channel models can be distinguished by their consistency in representing the proper space–time evolution of multiple radio links. The most consistent environmental description is provided by detailed databases, such as those employed in ray-tracing (see Sect. 3.2). Geometry-based approaches (such as Geometric Stochastic Channel Model (GSCM)) aim at maintaining a high consistency of the environment by introducing the concept of visibility regions. Representatives of this approach are the COST 273 [Cor06] and the COST 2100 model (see Sect. 3.6). When a consistent description of the environment is not available, the knowledge of joint link properties is required to emulate *realistic* space–time relations. This is the case with models that place multipaths directly in the parameter space (angle/delay/Doppler). Examples for such models are the 3GPP SCM [3GP09], the WINNER channel model [KMH+07], and the ITU model [ITU08].

3.5.1 Space–Time Stationarity

An important concept for multilink channel modeling is the (non)stationarity of the channel in space. One way to approach the space–time (ST) variability is to divide the analyzed space–time domain into small regions with high similarity between channel realizations, the local stationarity regions [Ste01]. This similarity can be quantified by the stationarity distance, which has been discussed in Sect. 3.1.3. This concept can be applied for channel characterization and also for channel simulation with low complexity. Local stationarity regions are implemented in many current GSCM, but often under different names ("drops", "channel segments", etc.).

3.5.1.1 Similarity Regions

The spatial stability of wideband and ultawideband indoor channel responses was analyzed at a frequency range of 2–17 GHz [PHKU09]. The average number of simultaneously present MPCs was 41 at a delay resolution of 28 ps. An MPC visibility region corresponded to consecutive Rx positions having a power above a predefined threshold. It was found that the average extent (per measurement) of a visibility region ranges from 7 to 48 cm, which agrees with other reported results [PCRH02]. The average path visibility over the entire set of measurements is 14.4 cm with a standard deviation of 8.6 cm. The spatial stability of measured responses, i.e., the size of the typical area of visibility of each multipath component, increases with the signal bandwidth, as demonstrated in Fig. 3.18. An investigation of the path stabil-

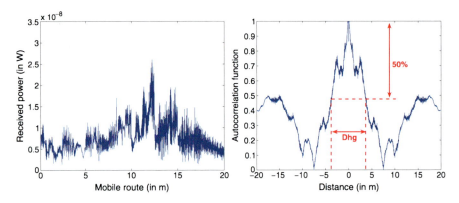

Fig. 3.19 Received power: (**a**) along trajectory (**b**) associated autocorrelation function [DPP+07]

ity w.r.t. its path delay has shown that early paths are more stable, i.e., they have a larger ST visibility regions than late paths.

The similarity of the received power levels for ad hoc (mobile) nodes in outdoor environment is investigated in [DPP+07]. Ray-tracing simulations were performed within circular zones with a radius of 20 m along radial tracks with a resolution of 0.1 m (comparable to the wavelength). The extent of the homogeneous zone is defined by the radius for which the autocorrelation function of the measured realizations drops below 50% (see Fig. 3.19). The measurement analysis showed that the radius of the homogeneous zone equals 13.2 m in a suburban scenario (Poitiers) and 17.7 m for an urban scenario (Munich).

The authors of [KHH+10] point out that the correlation between eigenvectors of different MIMO links can be high under similar *dominant propagation mechanisms*[1] (DPM). In this concept, the relevance of existing propagation mechanisms (and related MPCs/clusters) is evaluated according to the level of received power. The identification of the DPM similarity regions leads to a segmentation of the ST domain and by that reveals the large-scale evolution of the radio channel.

3.5.1.2 Drop-Based Simulations

The computational complexity of channel simulation grows with increasing bandwidth, number of antennas, and number of users, as well as with advanced network topologies. The knowledge about the space–time channel variability can be exploited to reduce the simulation complexity as will be discussed in this subsection.

The authors of [KJBN08] investigated the complexity of *drop-based* MIMO simulations, where a number of user terminals are dropped into a network topology, and short segments of the radio channel are simulated. Two major aspects of GSCM are

---

[1]The concept of dominant propagation mechanisms is often applied in channel characterization, see, e.g., [KHH+10, PHS+09, ETM07].

3 Radio Channel Modeling for 4G Networks

Table 3.11 Computational complexity of synthesizing channels from the WINNER GSCM (×1000 real operations) for 4×4 MIMO, 20 clusters, 20 path per cluster, and different number of considered polarizations P [KJBN08]

Operation	Processing stage	$P = 1$	$P = 2$
Generation of MPC param.	Drop preprocessing	46.3	46.3
Generation of chan. coeff.		467.9	1021.1
	Processing per time sample	140.8	N/A

analyzed: generation of the MPC parameters and generation of channel coefficients, as displayed in Table 3.11. The computational complexity is quantified by the number of *real operations* [KJ07], where each of the following operations is counted individually: real multiplication, division, addition, and table lookup.

The idea behind drop-based simulations is to keep the multipath structure constant in the local stationarity region ("drop"). This has the advantage that all time-independent operations can be performed only once per drop in a preprocessing phase. In this way, the recomputation of all MPC parameters and the complete generation of channel coefficients are avoided in consecutive simulation time instants. For GSCM, using only a single polarization ($P = 1$), this saves 514 200 real operations and reduces complexity by 78.5% per time sample. According to Table 3.11, the generation of four or more time samples within a drop during a simulation requires more computation than the complete drop-related preprocessing.

The reduction of ray-tracing complexity by a drop-like concept is also analyzed in the context of 802.11g-based ad hoc networks [DPP+07]. For the studied suburban and urban (outdoor) environments, the received power level was kept constant within a circular area (*homogeneous zone*) with radii of 13 m and 18 m, respectively. According to the time that the mobile spends within the same homogeneous zone, and due to a dropout of unfeasible connections, a reduction of the computation time of $\approx 50\%$ is observed.

3.5.2 Channel Similarity Measures

One major question in multilink propagation is how similar channels of different links are to each other. Depending on the channel similarity, algorithms show different performance.

In the majority of the presented cases, a correlation-like measure is introduced to characterize the similarity: interlink eigenvalue correlations, and shadowing correlation are two examples. These correlations are equally applicable to link-dependencies with and without embedded space–time evolution. Alternatively, stochastic characterization of the ratios/differences between corresponding link properties can be applied, like in [NKK+].

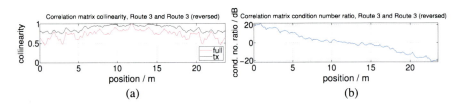

Fig. 3.20 (a) Collinearity and (b) condition number ratio of the spatial correlation matrices for selected route and its inverse run [CBVV+]

3.5.2.1 Matrix Collinearity

The distance between two matrices of same dimensions can be quantified by their collinearity [GvL96]:

$$c(\mathbf{A}, \mathbf{B}) = \frac{|\operatorname{tr}(\mathbf{A}\mathbf{B}^H)|}{\|\mathbf{A}\|_F \|\mathbf{B}\|_F}, \quad (3.22)$$

where $\mathbf{A}$ and $\mathbf{B}$ are (complex-valued) matrices, $\|\cdot\|_F$ denotes the Frobenius norm, and $(\cdot)^H$ is the matrix conjugate transpose operation. In general, the collinearity describes how similar the subspaces of the compared matrices are. This measure ranges from zero (no collinearity, i.e., matrices are orthogonal to each other) to one (full collinearity). For real-valued matrices matrix collinearity has the geometric interpretation as being equivalent to the cosine between vectors obtained by matrix unfolding: $c(\mathbf{A}, \mathbf{B}) = |\cos \angle (\operatorname{vec}(\mathbf{A}), \operatorname{vec}(\mathbf{B}))|$.

The suitability of this measure for quantification of the separability of multiuser MIMO channels is demonstrated in [CBVV+], by calculating the matrix collinearity along single route and its time (i.e., position) reversal, Fig. 3.20(a). The resulting curves are, of course, symmetric and show that the similarity increases, the closer the MSs get (central position in Fig. 3.20).

Note that the *Correlation Matrix Distance* [HCOB05] can be considered as a special case of this measure, being applicable only to Hermitian matrices (e.g., correlation matrices):

$$d(\mathbf{R}_1, \mathbf{R}_2) = 1 - c(\mathbf{R}_1, \mathbf{R}_2). \quad (3.23)$$

3.5.2.2 Ratio of Condition Numbers

This measure is given by

$$\chi(\mathbf{A}, \mathbf{B}) = 10 \log_{10} \left(\frac{\lambda_{\max}(\mathbf{A})}{\lambda_{\min}(\mathbf{A})} \Big/ \frac{\lambda_{\max}(\mathbf{B})}{\lambda_{\min}(\mathbf{B})} \right), \quad (3.24)$$

where $\lambda_{\max}(\mathbf{A})$ denotes the largest singular value of the matrix $\mathbf{A}$. In this measure, similarity between the condition numbers is indicated by values close to 0 dB.

Figure 3.20(b) shows an example, where similar condition numbers (0 dB) are associated with high collinearity, although [CBVV+] gives cases where this does not apply. Therefore, the condition number ratio and matrix collinearity provide a different notion of the (dis)similarity of the spatial structure, while the condition number ratio tells whether some channels are more directive. The collinearity measure is sensitive to the alignment or nonalignment of the preferred directions.

3.5.2.3 MIMO Capacity Under Interference

Another measure that aims directly at the capacity of multiuser communications is the MIMO capacity under interference [Blu03]. This measure is defined as

$$I = \log_2 h \det\left(\mathbf{I} + \mathbf{H}_0 \mathbf{H}_0^H \left(\sigma^2 \mathbf{I} + \sum_{k=1}^{K} \mathbf{H}_k \mathbf{H}_k^H\right)^{-1}\right), \quad (3.25)$$

where $\mathbf{H}_0$ is the MIMO channel matrix for the considered link, and $\mathbf{H}_k$ are the channel matrices of the interfering links.

A convenient property of this measure is that its value depends both on the eigenvalues and on the eigenstructure of the channels. For this reason, the authors of [CBOP10] applied this measure to statistically model interlink correlations (see Sect. 3.5.3.6).

3.5.2.4 Spectral Divergence

The *Spectral Divergence* (SD) [Geo06] measures the distance between strictly positive, nonnormalized spectral densities, such as the power delay profile. The SD between two different links l_1 and l_2 is given by

$$\gamma_{l_1,l_2}(t) = \log \frac{1}{T^2} \sum_{\tau} \frac{P_{l_1}(t,\tau)}{P_{l_2}(t,\tau)} \sum_{\tau} \frac{P_{l_2}(t,\tau)}{P_{l_1}(t,\tau)}, \quad (3.26)$$

where $P(t,\tau)$ denotes the (time-variant) power delay profile at time t and propagation delay τ, and T is the number of samples in the delay domain. This measure is applied in [KBZ09] to characterize the (dis)similarity of different links in wideband multiuser multiple-input multiple-output (MU MIMO) channels.

Figure 3.21 shows that the spectral divergence is not strongly related to distance, since small spatial divergence (high similarity) may appear for large user separation.

3.5.3 Modeling Interlink Dependencies

The dependency of different links on each other is a significant aspect in multiuser channel modeling, as already indicated by the previous metrics. For proper channel simulation, such dependencies need to be modeled accordingly. The following paragraphs will discuss these dependencies starting from large-scale interactions, followed by the mechanism of a shared multipath structure, and finally elaborate on correlation-based models.

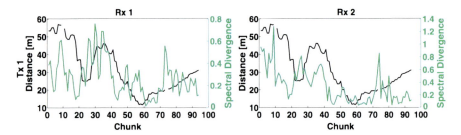

Fig. 3.21 Spectral divergence between MIMO sublinks for different users and distance between them (*black line*) [KBZ09]

3.5.3.1 Shadowing Correlation due to Nearby Obstacles

Considering an exemplary indoor communication system with two access points and one mobile station, the large-scale fading behavior of the two links can be strongly correlated, even when the separation between the access points is large.

The cause of such correlation in an indoor large-hall environment is analyzed in [PHK$^+$09] by looking into the physical propagation mechanisms. The classification of the specular components (representing 5–20% of the total power) have revealed different dominant propagation mechanisms for each of the access points. However, it is also found that, due to shadowing effect of pillars along the track of the mobile, relevant MPCs for both links have similar propagation directions at the mobile side. Thus, a significant correlation of the received power was observed.

3.5.3.2 Generation of Correlated Shadowing in Macrocell Scenarios

Under the assumption that the dominant propagation mechanism for NLOS conditions is over-obstacle propagation (i.e., over roof-top diffraction), a *virtual environment* may be used to generate correlated shadowing between different links. The model presented in [FMC08] offers the adaptability to different environments, ranging from rural environments to small urban macrocells, and to different antenna heights, what is required for multilink/multihop representation.

This shadowing model assumes that increasing power loss under NLOS is well approximated by the sum of diffraction losses on multiple obstacles. The model associates the diffraction loss variability (i.e., shadowing) with height variations of obstacles in a virtual scenario:

$$L_{\text{diff}} \approx \sum_{i=1}^{m} L_i|_{h_i=0} + \sum_{i=1}^{m} h_i \left.\frac{\delta L_i}{\delta h_i}\right|_{h_i=0}, \quad (3.27)$$

where h_i is the height of the ith obstruction, and L_i is corresponding knife-edge diffraction loss determined according to Deygout's model [Par01]. The main adjustable parameters of the described model are the standard deviation of obstacle heights and the sampling distance (grid) of the virtual environment.

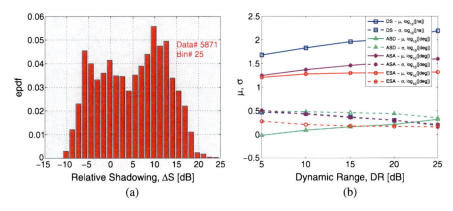

Fig. 3.22 (a) Empirical PDF of relative (pairwise) shadowing, (b) dependence of log-normal model parameters of power spreads from effective dynamic range [NKK$^+$] (©2010 EurAAP, reproduced with permission)

It turns out that decorrelation distance is significantly affected by the choice of the grid and that standard deviation of shadowing is proportional to the obstacle heights. Therefore, for the targeted statistic of shadowing, the model parameters can be adjusted in the following order: first select the grid in order to obtain the desired decorrelation distance and, afterwards, choose the value of obstacle heights that provides the desired standard deviation of shadowing. Both adjustments are expected to be fairly straightforward due to the linear nature of dependencies.

3.5.3.3 Dependence of Large-Scale Parameters of Cooperative Links on Absolute Power Levels

Many current models use the statistics of the measured power spread to describe large-scale channel evolution [3GP09, KMH$^+$07, ITU08]. They, however, do not consider downlinks in a cooperative mode, i.e., a common receiver and several spatially distributed transmitters. Due to the reception of signals with different power levels and the limited dynamic range of the receiver, a perception of spreads on any particular link will be changed by other existing links.

The joint characterization/modeling of cooperative links can be based on appropriate relationships between *relative peak-power differences* and *perceived power spreading* [NKK$^+$]. The differences between link power levels can be represented by the PDF of peak power-level differences $\Delta P = |P_1 - P_2|$. Alternatively, by accounting for the peak level dependence over distance, $\hat{P}(d) = A \log_{10}(d) + B$ (which is similar to path loss modeling), the statistics of *relative shadowing*, $\Delta S = \Delta P - |\hat{P}(d_1) - \hat{P}(d_2)|$, can be used (see Fig. 3.22a). Additionally, the Dynamic Range (DR) of the receiver is introduced as model parameter. For a given realization of the peak power differences (random process) and a dynamic range of the receiver, the effective dynamic ranges for each of cooperative links can be determined. As a consequence of the mutual interaction between links, the effective DRs will impact the distributions of perceived power spreads. Assuming a log-normal model for the measured delay and angular spreads, a dependence of their distribution parameters (μ, σ) from effective DRs is observed as shown in Fig. 3.22(b). In this manner, a representation of cooperative links is adjusted according to the receiver perception.

3.5.3.4 Intercell Correlation Properties of Large-Scale Parameters

The work in [ZTM10] provided the first available analysis of intercell large-scale parameter correlations. The work is based on simultaneous multibase-station measurements conducted by Ericsson (see Sect. 2.1.1, with the measurement map in Fig. 2.10).

The work investigated the parameter correlations between all combinations of shadow fading and delay spread of two links, in both NLOS and LOS situations.

The (conventional) intrasite parameter correlations and their autocorrelation values were found to be similar to the established parameters in COST 2100 (see Sect. 3.6.2) and in the WINNER II C2 model.

Intersite large-scale parameter correlations are also an important property because their existence affects system performance [Jal10]. It is demonstrated in [ZTM10] that careful selection of spatial regions where correlations are calculated easily boosts intracorrelation levels above $1/e$. Generally, when the MS is located between two BSs, and also the main lobes of the respective BSs point toward the MS's route, a high intersite correlation exists—even if the BSs are far away from each other. If the MS is moving toward or away from both BSs, positive intersite correlations are obtained; otherwise negative correlations are obtained.

3.5.3.5 Interlink Correlation Modeling by Common Scatterers

For the development of multilink channel models, it is essential to have understanding about the physical phenomena that increase correlation between different links, such as scatterers that are common for two or more links, i.e., common scatterers (or clusters of scatterers). Common scatterers are especially harmful for systems that depend on the spatial characteristics of the channel, since underestimating the significance of common scatterers in simulations would result in overestimating the system performance. For instance, mitigation of the interference from other users by beamforming might become difficult, or the desired performance improvement of a relay scheme deteriorated, if different links are correlated due to similar scatterers. In order to quantify the amount of energy that propagates via the same scatterers in different links, a measure called the significance of common scatterers was introduced in [PHS$^+$10] as follows. In a dual-link case, the significance of the nth common scatterer is denoted as a function of a measurement time instant k by

$$S_{\text{common}}^n(k) = \sqrt{s_{\text{common}}^{(1),n}(k) \cdot s_{\text{common}}^{(2),n}(k)}, \qquad (3.28)$$

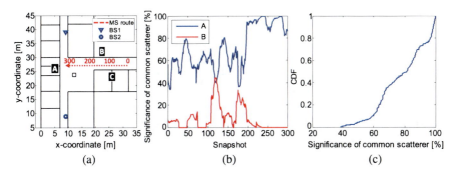

Fig. 3.23 (**a**) The floor plan of the environment where the significance of common scatterers was investigated. (**b**) Significance of common scatterers A and B. (**c**) CDF of the total significance of common scatterers [PHS+10] (©2009 EurAAP, reproduced with permission)

where $s_{common}^{(i),n}(k)$ is the proportion of the nth common scatterer in total power in the ith link. If the number of scatterers that are common for the different links is denoted by $N(k)$, the total significance of the common scatterers can be expressed by the sum of the significances of the individual common scatterers $S_{common}^n(k)$ by

$$S_{common,tot}(k) = \sum_{n=1}^{N(k)} S_{common}^n(k). \tag{3.29}$$

The total significance of the common scatterers $S_{common,tot}(k)$ gets values between 0 and 1, where 0 means that the different links have no common scatterers, and 1 indicates that all scatterers are common for the two links. This concept is embedded into the COST 2100 model presented in Sect. 3.6.

The significance of common scatterers was studied in the corridor of the Department of Radio Science and Engineering, in Aalto University School of Science and Technology [PHS+10]. The floor plan of the environment is shown in Fig. 3.23(a). Figure 3.23(b) shows the significance of the common scatterers A and B. In Fig. 3.23(b), "A" corresponds to the propagation along the corridor, and "B" to the interaction points calculated on wall B. Figure 3.23(c) shows the CDF of the total significance of common scatterers. The propagation along corridor (i.e., "A") is a dominating propagation mechanism in both links and thus forms a significant common scatterer. The total significance of common scatterers is typically between 60% and 90% in this scenario, as can be seen in Fig. 3.23(c).

The impact of common clusters on interlink correlations is evaluated in [PTH+] by the collinearity of generated correlation matrices (see Sect. 3.5.2.1). The simulated scenario considers links from two BSs to a single (fixed) mobile. These links share a single common cluster and have up to five additional clusters per link, all of which are randomly positioned. The resulting CDFs of the collinearity are shown in Fig. 3.24 for the significance levels $S_{CC} \in \{0, 1, 10, 50, 90, 99, 100\}$. In both cases in Fig. 3.24, higher common cluster significance leads to increased collinearity (CDF curves are shifted to the right). Note that increasing the number of uncommon clusters also leads to the effect of higher collinearity.

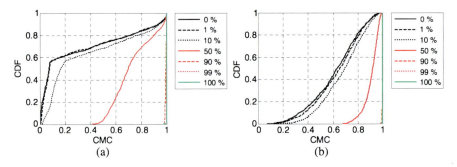

Fig. 3.24 Effect of common clusters on correlation matrix collinearity (CMC) in the simulated random scenarios with one common cluster: (**a**) 1 uncommon cluster, (**b**) 5 uncommon clusters [PTH[+]] (©2011 IEEE, reproduced with permission)

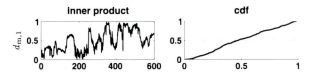

Fig. 3.25 Interlink correlation of eigenvectors that correspond to dominant eigenvalues [KHH[+]10] (©2010 IEEE, reproduced with permission)

3.5.3.6 Analytical Dual-Link MIMO Channel Models

Analytical models represent MIMO channels by the correlation matrix between transmit and receive antennas. The multilink extension of an analytical channel model introduces additional analytical relations between correlation matrices describing individual links. Therefore, the relations between different links from multiple users need to be analyzed and modeled accordingly.

Reference [KHH[+]10] compares the eigenvectors of correlation matrices that describe different links. The normalized correlation matrices $\mathbf{R}^{(i)} \in \mathbb{C}^{4\times 4}$, $i = 1, 2$, on the mobile side are determined for each position of the mobile toward two fixed BS positions. The eigenvectors corresponding to these matrices are determined and ordered. The inner products between corresponding ordered eigenvectors are used to express the link similarity. The empirical CDFs of inner products (given separately for every pair of matched eigenvectors) provide a statistical description of the measurement, as demonstrated in Fig. 3.25.

In [HTK[+]], the authors both analyzed and modeled multiuser channels. First, the Correlation Matrix Distance (CMD) metric (Eq. (3.23)) is used to quantify the dependence between multiple analytically modeled links. The correlation matrices and CMD, $d_{\mathbf{R}_1 \mathbf{R}_2}(s)$, change with propagation conditions, which are indicated by a dependence upon a parameter s. By analyzing correlations on the Tx side, it was noticed that the corresponding elements of the correlation matrices $\mathbf{R}_1(s)$ and $\mathbf{R}_2(s)$ toward two different receivers could experience high correlation under similar propagation condition.

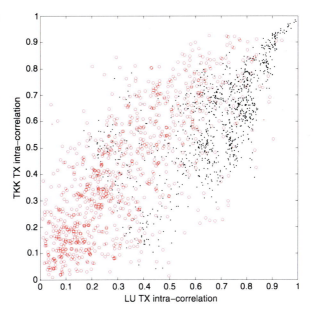

Fig. 3.26 Scatter plots of the intralink Tx-correlation matrix elements of the two links: measured—*black dots*, simulated—*red circles* [HTK$^+$] (©2010 EurAAP, reproduced with permission)

From these findings the following model was derived: Starting from a reference link, a correlation matrix with a specific CMD with respect to the first channel can be generated by

$$\mathbf{R}_2(s) = \left(1 - \sqrt{d_{R_1 R_2}(s)}\right)\mathbf{R}_1(s) + \sqrt{d_{R_1 R_2}(s)}\mathbf{R}_1^\perp(s), \tag{3.30}$$

where $\mathbf{R}_1^\perp(s) = \frac{\mathbf{R}_1^{-1}(s)}{\|\mathbf{R}_1^{-1}(s)\|_F}$. Note that this equation holds exactly only for singular $\mathbf{R}_1$, while it provides an approximation for nonsingular correlation matrices. A demonstration of the model is shown in Fig. 3.26. The scatter plot of the simulated correlation matrix elements (indicated by red circles) presents a similar trend as the selected measured data. This model was subsequently extended in [HTK$^+$10] to fit the CMD metric (3.23) by a deterministic rotation of an eigenmatrix.

[CBOP10] presented a more complete model that enables the use of nonsingular correlation matrices. The authors use the concept of the MIMO capacity under interference [Blu03] (see Sect. 3.5.2.3) to show the impact of (multiple) interfering links on a desired link. The MIMO capacity under interference can be approximated by the Rx correlation matrices of the considered link $\mathbf{R}_0 = \mathcal{E}\{\mathbf{H}_0 \mathbf{H}_0^H\} = \mathbf{U}\boldsymbol{\Lambda}\mathbf{U}^H$ and of the interfering link(s) $\mathbf{R}_I = \mathcal{E}\{\sum_{i=1}^{N_I} \mathbf{H}_i \mathbf{H}_i^H\} = \mathbf{V}\boldsymbol{\Gamma}\mathbf{V}^H$.

For given $\boldsymbol{\Gamma}$ and $\mathbf{R}_0$, this metric only depends on the eigenvalues of the interference $\mathbf{V}$, having a maximum for $\mathbf{V} = \mathbf{U}$ and minimum for $\overleftarrow{\mathbf{U}}$ (having reverse order of column vectors in $\mathbf{U}$).

[CBOP10] propose to analytically model multiuser MIMO channels by controlling the severity of interference by smoothly rotating the eigenspace of the interference correlation matrix. In this model, all possible values between the two extremes are achieved by proper rotation of the unitary matrix. The unknown unitary matrix

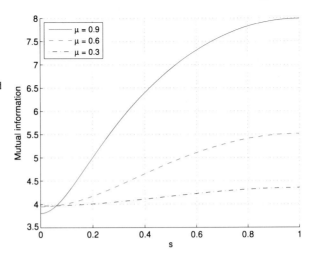

Fig. 3.27 MIMO capacity under interference metric for different alignments of signal and interference subspace; μ denotes the antenna correlation of both the desired and the interference channels [CBOP10]

$\mathbf{V}(s)$ can be parameterized by a single scalar parameter s by letting

$$\mathbf{V}(s) = \mathbf{U}\mathbf{W}e^{js\boldsymbol{\Phi}}\mathbf{W}^H, \qquad (3.31)$$

where $\mathbf{W}$ and $\boldsymbol{\Phi}$ are defined from the eigenvalue decomposition $\mathbf{U}^{-1}\overleftarrow{\mathbf{U}} = \mathbf{W}e^{j\boldsymbol{\Phi}}\mathbf{W}^H$. This transformation describes the shortest path of rotation from $\mathbf{U}$ to $\overleftarrow{\mathbf{U}}$. An extension of the model to describe any possible path between the two extreme values is also indicated in [CBOP10].

In the special case where $\mathbf{R}_0$ is 4×4 Toeplitz matrix with elements $\{\mu^0, \mu^1, \mu^2, \mu^3\}$ and $\boldsymbol{\Gamma} = \boldsymbol{\Lambda}$, the impact of the parameter s on the MIMO capacity under interference metric is shown in Fig. 3.27. The presented results demonstrate that for given singular values, correlation-based metrics solely depend on the alignment of the correlation matrix eigenspaces. Therefore, the alignment of the eigenmodes of the receive correlation matrices significantly influences the capacity (up to 100%) of the 4×4 link under interference.

3.5.4 Peer-to-Peer Links

The material presented in this section studies important characteristics of links in a (cooperative) multihop network.

3.5.4.1 Peer-to-Peer MIMO Channel Characteristics at 300 MHz

Using outdoor peer-to-peer measurements in the 300-MHz band introduced in Sect. 2.1.1.1, the authors of [ETM07] evaluated a large number of model parameters: the propagation loss, dominant propagation mechanisms, fading characteristics, delay properties, directional properties, channel capacity, and antenna element correlation.

3 Radio Channel Modeling for 4G Networks

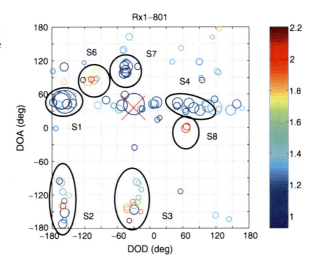

Fig. 3.28 Identified propagation paths (clustered) from the measurements in the direction of arrival/departure domain, fitting scatterers in the environment. The size of the circles indicates the power; normalized to the strongest MPC. The delay is color-coded and given in μs

As an example, Fig. 3.28 shows the dominant propagation mechanisms evaluated by high-resolution parameter estimation. Groups of scatterers could be attributed to specific propagation mechanisms, like LOS, single-bounce scattering, and double-bounce scattering.

Further results show that vegetation (groups of trees) as well as buildings can act as efficient reflectors (scatterers). A capacity analysis confirmed that the presence of a LOS component, leads to reduced diversity, but simultaneously to smaller fading depth (higher Ricean K-factor). Another important insight is that the channel characteristics at Tx and Rx depend on each other, which leads to three interconnected effects: (i) the joint DOA/DOD spectrum cannot be factorized into the product of the marginal DOA and the marginal DOD spectra; (ii) the correlation matrix at the Rx is different for different Tx antenna elements, and (iii) a Kronecker-type channel model underestimates the true (measured) channel capacity.

3.5.4.2 Stochastic Channel Models for Distributed Networks

Modeling distributed networks may seem straight forward, because it usually means to model the radio channel between nodes that are rather far from each other. However, the fading statistics, both large-scale and small-scale, have astonishing features when considering distributed channels, i.e., channels of distributed transceiver nodes.

Generally, peer-to-peer channels can be classified by their mobility: *Nomadic* channels are characterized by fixed nodes, where only the environment changes, e.g., moving people or cars driving by. In *single-mobile* channels, one of the peers is moving, while the other is static. *Double-mobile* channels occur between two moving peers.

When modeling peer-to-peer channels, the kind of mobility needs to be taken into account. A theory put forward by [OCB+10] models the channel as superposition of the following propagation effects (expressed in logarithmic scale):

$$\text{channel} = \text{path loss} + \text{static shadowing} + \text{dynamic shadowing} + \text{fading}. \quad (3.32)$$

The path loss is defined as the propagation loss depending on the distance. The investigation in [OCC10] showed that for indoor-to-indoor links, the path loss also depends on the number of walls in between transmitter and receiver. One can even distinguish between thick (brick) walls and thin (plasterboard) walls. Dynamic shadowing reflects the well-known large-scale variations around the path loss, which are typically log-normal distributed. Fading accounts for small-scale variations of the signal power. *Static shadowing* is due to two effects: (i) time-invariant obstructions of the paths due to large shadowing objects, and (ii) constructive and destructive interference of coherent multipath. The latter effect is *specific to nomadic links*, where temporal and spatial (or frequency) fading are unrelated.

Another feature of peer-to-peer channels is that dynamic shadowing may be correlated between the links. The authors of [FWE10] suggest to use the model of Gudmundson [Gud91] to model correlated shadowing. However, this approach is able to model strictly positive shadowing correlations, only. On the other hand, [OCB+10] showed from measurements (see the measurement map in Fig. 2.5) that shadow fading correlations can also have negative values. Hence, they suggested a statistical model of the correlation coefficients that depends on the node mobility. It turned out that shadow fading correlations are stronger in directional environments (like a large office room having cubicles) [OCB+10] than in European-style offices with small, separated rooms [OCC10].

Small-scale fading is heavily influenced by the different mobility cases. Nomadic channels are rather static over time and can thus be modeled by a Ricean distribution. Single-mobile and double-mobile channels can show any distribution between Ricean fading, Rayleigh fading, and Double-Rayleigh fading. The higher the mobility, the more severe the fading becomes. Reference [OCB+] introduced the *second-order scattering function* (SOSF), where a combination of Ricean, Rayleigh, and double-Rayleigh fading can be easily modeled and parameterized. An example of estimated SOSF parameters from single-mobile measurements are shown in Fig. 3.29. Additionally, it turned out that the statistics of small-scale fading are time variant. Reference [GCC+10] devised a model to capture the time-variance of the parameters of the SOSF. Fading is categorized into three sorts: Ricean fading, Rayleigh/double-Rayleigh fading, and a mixture between them. Using a hidden Markov model, the model is switching between these three states. The parameters of the distributions are in turn modeled using the Beta-distribution. In that way, the overall statistics of the small-scale fading are fitted well.

Another approach that enables to model frequency-selective channels is presented in [FWE10], where the authors suggests to model the channel by a combination of a coherent LOS path and diffuse multipath with an exponentially decaying power delay profile.

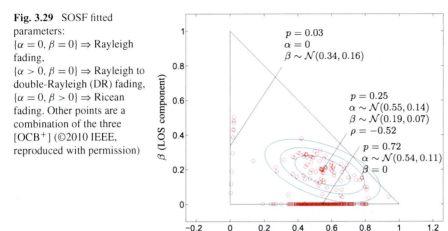

Fig. 3.29 SOSF fitted parameters: $\{\alpha = 0, \beta = 0\} \Rightarrow$ Rayleigh fading, $\{\alpha > 0, \beta = 0\} \Rightarrow$ Rayleigh to double-Rayleigh (DR) fading, $\{\alpha = 0, \beta > 0\} \Rightarrow$ Ricean fading. Other points are a combination of the three [OCB+] (©2010 IEEE, reproduced with permission)

Finally, [FWE10] discusses simplifications of such peer-to-peer models to make them suitable for packet-level simulations. Should only the statistics of the channel *power gain* be necessary, this can be simply modeled by a χ^2 distribution of independently fading (frequency-flat) subchannels.

3.6 COST 2100 Multilink MIMO Channel Model

A significant achievement of COST 2100 builds on the previous COST 273 MIMO channel model [Cor06, LCO09, CO08a, LOC08], to consider recent developments highlighted in the previous sections, namely: polarization (see Sect. 3.4), dense multipaths (see Sect. 3.3) and multilink aspects (see Sect. 3.5). Its goal is to describe radio propagation in various cellular scenarios, including macro/micro/picocells, using a generic and flexible structure. As a GBSCM, the COST 2100 channel model relies on the assumption that the wideband propagation channel can be described in the delay and the direction domains at both Transmitter (Tx) and Receiver (Rx) sides via physical clusters, i.e., groups of Multi-Path Component (MPC)s. As detailed in [MAH+06, OC07], the double-directional (angular) domains, when combined with the Tx/Rx array steering vectors, can be easily transformed into spatial dimensions for MIMO channel simulations. By contrast to WINNER II model, which describes the channel based on the system-level parameters [NST08, JKM08], the COST 2100 channel model characterizes the channel at the level of individual clusters. Such cluster-oriented structure is inherently more flexible when characterizing channels in multilink scenarios.

The COST 2100 channel model adopts a generic structure of the clusters for both local scattering and remote (far) scattering with single bounce and multiple bounces, which is described in Sect. 3.6.1. Recent works have improved this structure by including enhanced polarization and DMC modeling [QOHD09a, PTH+10a], as

Fig. 3.30 General description of the COST 2100 channel model in a single-link scenario

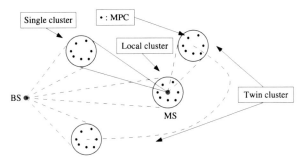

detailed in Sects. 3.4 and 3.3. Additionally, [VQAR07] has extended the cluster generation model by introducing terrain and clutter information.

Another important feature of the updated COST 2100 channel lies in its generic multilink extension for modeling transmissions between multiple Base Station (BS)s and multiple Mobile Station (MS)s. While keeping a consistent channel modeling for each individual link, the COST 2100 approach maintains a controlled complexity to describe the essential properties of multilink channels, such as interlink correlation.

3.6.1 COST 2100 Single-Link MIMO Channel Model

The general philosophy of the single-link COST 2100 MIMO channel model is described in Fig. 3.30. The signal waves propagate through the physical environment from a static BS to an MS via different MPCs resulting from the interaction with scatterers. An MPC is characterized in delay and angular domains by its delay (τ), angle of departure (Azimuth of Departure (AoD), Elevation of Departure (EoD)), and angle of arrival (Azimuth of Arrival (AoA), Elevation of Arrival (EoA)). The MPCs with similar delay and angles are grouped into clusters. Such clustering enables a description of the channel with a reduced number of parameters.

In the COST 2100 modeling approach, there are two kinds of clusters. *Local clusters* are located around MS or BS, and *far clusters* (remote clusters) are located away from both the BS and MS sides. Far clusters include clusters with single-bounce scatterers and clusters with multiple-bounce scatterers. Single-bounce clusters can be explicitly described at a certain position by matching their delay and angles through a geometric approach. On the contrary, multiple-bounce clusters have to be described by two representations viewed from the BS and MS sides, respectively, to reflect their geometric mismatch. Multiple-bounce clusters are also called *twin clusters* as in Fig. 3.30.

The visibility of a far cluster depends is determined by its Visibility Region (VR). A connection between the BS and the MS through a far cluster is established only when the MS is inside the corresponding VR. This determines the number of active far clusters. By definition the local clusters are always visible.

The COST 2100 model calculates the time-varying Channel Impulse Response (CIR) in delay and direction domain as

$$h(t, \tau, \mathbf{\Omega}^{BS}, \mathbf{\Omega}^{MS})$$
$$= \sum_{n \in \mathscr{C}} \sum_{p} \alpha_{n,p} \delta(\tau - \tau_{n,p}) \delta(\mathbf{\Omega}^{BS} - \mathbf{\Omega}^{BS}_{n,p}) \delta(\mathbf{\Omega}^{MS} - \mathbf{\Omega}^{MS}_{n,p}), \quad (3.33)$$

where $\mathscr{C}$ is the set of visible cluster indexes, $\alpha_{n,p}$ is the complex amplitude of the pth MPC in the nth cluster, $\mathbf{\Omega}^{BS}_{n,p}$ is the direction of departure (AoD, EoD), and $\mathbf{\Omega}^{MS}_{n,p}$ is the direction of arrival (AoA, EoA) of the MPC.

For a MIMO system using V and U multiple antennas arrays at BS and MS, respectively, we may express the $U \times V$ MIMO channel matrix $\mathbf{H}(t, \tau)$, under the plane wave and balanced narrowband array assumptions, as

$$\mathbf{H}(t, \tau) = \iint h(t, \tau, \mathbf{\Omega}^{BS}, \mathbf{\Omega}^{MS}) s_{MS}(\mathbf{\Omega}^{MS}) s_{BS}^T(\mathbf{\Omega}^{BS}) d\mathbf{\Omega}^{MS} d\mathbf{\Omega}^{BS}$$
$$= \sum_{n \in \mathscr{C}} \sum_{p} \alpha_{n,p} s_{MS}(\mathbf{\Omega}^{MS}) s_{BS}^T(\mathbf{\Omega}^{BS}). \quad (3.34)$$

3.6.1.1 Visibility Region

Visibility regions are identically sized circular regions on the azimuth plane. Each VR is connected to only one cluster, so that the visibility level of the cluster is controlled by the visibility gain, A_{VR}, which smoothly increases as the MS is approaching the VR center,

$$A_{VR}(r_{MS}) = \frac{1}{2} - \frac{1}{\pi} \arctan\left(\frac{2\sqrt{2}(L_C + d_{MS,VR} - R_C)}{\sqrt{\lambda L_C}}\right), \quad (3.35)$$

where $d_{MS,VR}$ is the distance from the MS to the center of VR, R_C is the VR radius, and L_C is the transit region Transition Region (TR) width. This concept is also depicted in Fig. 3.31(a).

The VRs are uniformly distributed in the cell, as shown in Fig. 3.31; the density corresponding to the average number of visible far clusters (N_C) is given by

$$\rho_C = \frac{N_C}{\pi (R_C - L_C)^2} [m^{-2}]. \quad (3.36)$$

The quantity N_C is modeled as a Poisson-distributed random variable. Different VRs may overlap, which allows multiple visible clusters to be simultaneously active when the MS lies in the overlapping region.

3.6.1.2 Clusters

A cluster is depicted as an ellipsoid in space, its size being determined by the axes along different directions. These correspond to the maximum Cluster Delay Spread

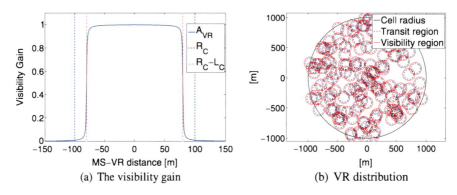

Fig. 3.31 Examples of visibility gain and VR distribution in a macrocell

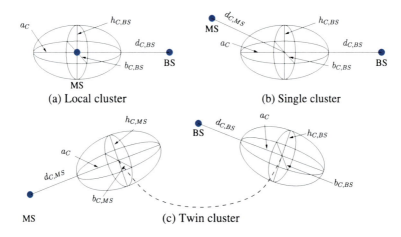

Fig. 3.32 Three kinds of clusters

(CDS), denoted as δ_C, and the Cluster Angular Spread (CAS), denoted as $\phi_{C,BS}$ and $\phi_{C,MS}$ in the azimuth plane, and as $\theta_{C,BS}$ and $\theta_{C,MS}$ in the elevation plane. For twin clusters, the position and spread of the cluster are specified at both the BS and MS sides, whereas local and single-bounce clusters only need one representation in space. Before detailing the characteristics of clusters, we introduce the following notations related to the quantities used in Fig. 3.32:

- Δ_C, $\phi_{C,BS}$, $\phi_{C,MS}$, $\theta_{C,BS}$, and $\theta_{C,MS}$ are the delay, AoD, AoA, EoD, and EoA spread of the cluster,
- a_C, $b_{C,BS}$, $b_{C,MS}$, $h_{C,BS}$, and $h_{C,BS}$ are the corresponding extent of the cluster in space that leads to the delay, AoD, AoA, EoD, and EoA spread of the cluster.
- $d_{C,BS}$ and $d_{C,MS}$ are the cluster-to-BS distance and cluster-to-MS distance.

1. **Local cluster:** the local cluster contains single-bounce MPCs uniformly distributed in azimuth. The spatial spread of the local cluster is only determined

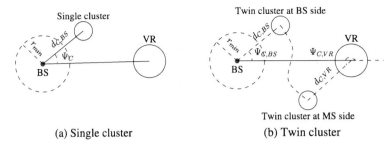

Fig. 3.33 Distributions for single and twin clusters

by its delay and elevation spreads, thanks to a simple geometry relationship [Cor06]:

$$a_{C,BS} = \frac{1}{2}\Delta_C c_0, \quad b_{C,BS} = a_{C,BS}, \quad h_{C,BS} = d_{C,BS}\tan\theta_C, \quad (3.37)$$

where c_0 is the free-space wave speed.

2. **Single-bounce cluster:** a single-bounce cluster has independent delay and azimuth spreads and is depicted as a rotated ellipsoid with respect to BS in order to match the spatial spread with the delay and angular spreads viewed from the BS.

    The single-bounce cluster is always distributed relative to the BS and the VR, as shown in Fig. 3.33(a): its position is determined by a random vector originating from the BS and rotated with a Gaussian distributed angle Φ_C relative to the imaginary line between the BS and the VR, $\Psi_C \sim \mathcal{N}(\mu_{\Psi_C}, \sigma_{\Psi_C})$. The distance to the BS follows a given nonnegative distribution (e.g., an exponential distribution in macrocells) bounded by a minimum distance $r_{\min}$.

3. **Twin cluster:** a twin cluster consists of two representations of an identical ellipsoid, corresponding to the BS and MS sides. The ratio of twin clusters to the total number of clusters is determined by the selection factor K_{sel} [Cor06, CO08b]. To determine the twin cluster distribution, the method applied for single-bounce clusters is performed twice: first, from the BS side, then from the VR, as in Fig. 3.33(b). The distance between the cluster at MS side and the VR is computed by

$$d_{C,BS}\tan\phi_{C,BS} = d_{C,VR}\tan\phi_{C,MS}, \quad (3.38)$$

in order to maintain the angular spreading consistency as viewed from BS and MS sides.

An additional cluster-link distance d_C and delay $\tau_{C,\text{link}}$ are introduced to compensate for the delay mismatch in the twin cluster; d_C is a geometrical distance between twin cluster centers, and $\tau_{C,\text{link}}$ is a nonnegative random variable with a minimum delay when single bounce between the twin cluster centers occurs. Note that local and single-bounce clusters can actually be regarded as special twin clusters with $d_C = 0$ and $\tau_{C,\text{link}} = 0$. Hence, the cluster is eventually expressed for all clusters as

$$\tau_C = (d_{C,BS} + d_{C,MS} + d_C)/c_0 + \tau_{C,\text{link}}. \quad (3.39)$$

4. **Cluster parameterization:** cluster parameters Δ_C, $\phi_{C,BS}$, $\theta_{C,BS}$, $\phi_{C,MS}$, $\theta_{C,MS}$ and the shadow fading factor σ_S are considered as correlated nonnegative random variables. The actual correlation coefficients depend on the scenarios, and some are tabulated in [Cor06]. The cluster power attenuation A_C is assumed to exponentially decay when the difference between the cluster delay τ_C and the Line-Of-Sight (LOS) delay τ_0 increases:

$$A_C = \max\bigl(\exp\bigl[-k_\tau(\tau_C - \tau_0)\bigr], \exp\bigl[-k_\tau(\tau_B - \tau_0)\bigr]\bigr), \quad (3.40)$$

where k_τ is the decaying parameter, and τ_B is the cut-off delay, introduced to avoid an extremely low cluster power.

3.6.1.3 Multipath and LOS Components

Within a cluster, MPCs are spatially Gaussian distributed. The distribution of MPCs for a twin clusters is identical at BS and MS sides to guarantee a consistent delay and angular spreads. Rayleigh fading is assumed for the MPC fading in each cluster.

The complex amplitude of the nth MPC in the mth cluster is eventually obtained as

$$\alpha_{m,n} = \sqrt{P}\, A_{VR,m}\, A_{MPC,m,n} \sqrt{\sigma_{S,m} A_{C,m}}\, e^{-2j\pi \tau_{m,n} f_c}, \quad (3.41)$$

where P is the pathloss of the channel, $\tau_{m,n}$ is the delay of the MPC, which is determined by

$$\tau_{m,n} = (d_{MPC_{m,n},BS} + d_{MPC_{m,n},MS})/c_0 + \tau_{m,C,link}. \quad (3.42)$$

The LOS is a special cluster containing a single MPC in the COST 2100 channel model. Its VR is determined by a cut-off distance from the BS. The power of the LOS is determined by a log-normal distributed power factor over the sum of the power from the other visible clusters [Cor06].

3.6.1.4 Temporal Evolution

The COST 2100 channel model assumes that the environment remains static so that the MPCs maintain a constant fading and deterministic positions throughout one simulation instance. The time variation of the channel is therefore only caused by the movement of the MS. Such motion will change the visibility of the clusters as the MS enters and leaves different VRs. It will also cause the motion of the local cluster around the MS, resulting in necessary updates of the local MPCs.

3.6.1.5 Polarization

The COST 2100 channel model structure enables to include polarization characteristics. Instead of generating a single channel for each path which will be projected on the transmit and receive antenna coordinate system, four channels are created for each path, as shown in (3.20), and are then projected on the transmit and the receive antenna system with the according polarization characteristics.

The four-channel polarization matrix of path n belonging to cluster m is defined as follows:

$$\gamma_{m,n} = \begin{bmatrix} e^{j\varphi_{11,m,n}} & \frac{1}{\sqrt{CPR_{m,n} \cdot XPD_{H,m,n}}} e^{j\varphi_{12,m,n}} \\ \frac{1}{\sqrt{XPD_{V,m,n}}} e^{j\varphi_{21,m,n}} & \frac{1}{\sqrt{CPR_{m,n}}} e^{j\varphi_{22,m,n}} \end{bmatrix}, \quad (3.43)$$

where XPD_V, XPD_H, and CPR are defined as the power ratio of the VV to the HV channel, the HH to the VH channel, and VV channel to the HH channel, respectively. These power ratios are defined for each propagation path m, n. The phase of the four channels of each path is defined as an i.i.d. uniform random variable between 0 and 2π. The four-channel path matrix is then normalized and multiplied by the complex path amplitude coefficient $\alpha_{m,n}$ in (3.41) containing path loss, shadowing, cluster attenuation, etc.:

$$\gamma_{m,n} = \alpha_{m,n} \gamma_{m,n} / \|\gamma_{m,n}\|_F. \quad (3.44)$$

The three cross-polar ratios are defined per path and per cluster. The cross-polar ratios of the different paths of one cluster are log-normally distributed, with mean μ_{XPD_m} and standard deviation σ_{XPD_m} defined per cluster. The mean and standard deviation of the cluster cross-polar ratios are also log-normally distributed, with parameters $(m_{\mu XPD}, S_{\mu XPD})$ and $(m_{\sigma XPD}, S_{\sigma XPD})$ for the mean and the standard deviation, respectively. Their values can be extracted from measurements and are tabulated in Table 3.17.

3.6.1.6 Dense and Diffuse Multipaths

There are two approaches of including the dense or diffuse multipaths into the COST 2100 channel model, according to the DMC modeling results detailed in Sect. 3.3.1: (1) in a parameter space and (2) in a geometrical simulation space. Both approaches generate the diffuse multipaths around the specular components, and their differences are complexity in implementation, applicability to dynamic channel simulation, and scalability; the parameter space approach gives lower model complexity while scalability is not considered. On the other hand, the geometry space approach provides flexible and scalable simulation environment at the expense of increased simulation complexity. The following summarizes the two approaches.

1. **Parameter space approach:** in [QOHD09a], a parameter space approach is investigated, where the DMC is modeled by constructing the desired spatio-temporal power spectrum. The spatio-temporal spectrum $\psi(t, \tau, \Omega^{BS}, \Omega^{MS})$ is constructed in the angular and temporal domain, with a delay resolution at least as high as the system's bandwidth, and an angular resolution as least as high as the MIMO system's angular resolution.

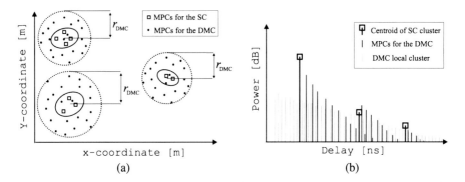

Fig. 3.34 Inclusion of the DMC on the geometry of the simulation. (**a**) In spatial domain, the MPCs of the DMC are placed around the SC clusters with a distribution wider than that of the SC. The resulting angular distribution of the DMC is wider than that of the SC. (**b**) Each MPC of the DMC is associated with an additional delay in order to achieve exponential decaying for the DMC part of each cluster (taken from [PTH+10a])

The diffuse clusters are generated in the angular domains to match the generated cluster's diffuse angular spreads. The distribution used for the cluster's shape in the angular spectrum can be either a multivariate normal distribution, or a distribution more accurate for modeling spherical data (e.g., a Fisher–Bingham (FB5) distribution). In the delay domain, the clusters are modeled with an exponential decaying power profile.

The spatio-temporal impulse responses are obtained then obtained as follows:

$$\mathbf{h}(t, \tau, \boldsymbol{\Omega}^{BS}, \boldsymbol{\Omega}^{MS}) = \left[\psi(t, \tau, \boldsymbol{\Omega}^{BS}, \boldsymbol{\Omega}^{MS})\right]^{1/2} \odot \mathbf{h}_\omega^{iid}, \quad (3.45)$$

where $\mathbf{h}_\omega^{iid}$ is a matrix of complex-normal i.i.d. elements with the same dimensions as $\psi(t, \tau, \boldsymbol{\Omega}^{BS}, \boldsymbol{\Omega}^{MS})$. For each channel realization, a new realization of h_ω^{iid} is generated. The wideband MIMO channel matrix $\mathbf{H}(t, \tau)$ is then obtained by integrating the spatio-temporal impulse response over all solid angles as in (3.34).

2. **Geometry-based approach:** another method to consider the DMC is developed in [PTH+10a]; MPCs representing the DMC are dropped around clusters on a geometrical map within a larger area than the MPCs representing the Selection Combining (SC), as shown in Fig. 3.34(a). The resulting response in the angular domain is spread over a larger range than the SC. The parameter that needs to be adjusted based on measurements is the radius of the area within which the DMC are distributed, i.e., r_{DMC}. In order to achieve the desired behavior of the DMC in the delay domain, each MPC of the DMC is associated with an additional delay on top of the delay determined by the geometry, i.e., the cluster position. Hence, the delay time of the MPCs of the DMC can be expressed by

$$\tau_{DMC} = \tau_{DMC,geo} + \tau_{DMC,add}, \quad (3.46)$$

where $\tau_{DMC,geo}$ is the delay from the geometry, and $\tau_{DMC,add}$ is a random additional delay that is added to each MPC of the DMC; $\tau_{DMC,add}$ is determined so that the DMC obeys an exponentially decaying PDP, as shown in Fig. 3.34(b). The base delay of the DMC clusters is determined by the centroid of the SC cluster. The peak power of the MPCs for the DMC can fluctuate and be higher or lower than the power of the corresponding SC. The MPCs for the DMC in delay domain are obtained by generating Poisson-distributed random delay taps in the range starting at the base delay and ending at the delay time where the power of the DMC has decayed to a sufficiently low level. The MPCs for the DMC are generated densely in the spatial domain and hence also in the delay domain; however, the PDP of the DMC will be sampled according to the delay resolution defined by the system bandwidth. Finally, DMC for the local cluster with a uniform angular distribution and a low power decay factor is added. In this way it is possible to model also the propagation paths that are not regarded as SC clusters, i.e., the MPCs that have random angular distribution and clusters that are weak, for instance, due to long delays.

3.6.2 COST 2100 Multilink MIMO Channel Model

Accurate modeling of radio wave propagation in multilink scenarios is crucial for system-level simulations and assessment of novel radio communication systems. In fact, many of the envisioned wireless systems base their operation on utilizing agile communications between multiple nodes in the network. Examples of such systems are macro diversity, relay schemes, and indoor localization systems. This section describes the extension of the COST 2100 channel model to a multilink scenario as first proposed by [PHL+10a].

3.6.2.1 Common Clusters

When extending GBSCMs to cover multilink scenarios, two aspects must be considered:

- the singe-link behavior should remain the same,
- the correlation between links, i.e., the interlink correlation, should be represented in a realistic way.

As described in Sect. 3.5, it was found that the interlink correlation can be seen as resulting from the interaction of common clusters, as they simultaneously influence the spatial properties of multiple radio links. Figure 3.35 shows an example of a multi-BS scenario, where links between the MS and BS1 and BS2 have one uncommon cluster per link and share one common cluster that contributes to both links. The correlation between both links therefore depends on the proportion of power that propagates through the common clusters; the larger this proportion, the stronger the inter-link correlation, as shown in Sect. 3.5 (see also [PTH+11]). The concept of the common cluster is interestingly highly compatible with the GBSCM approach in general, since it introduces interlink correlation while maintaining the intralink correlation properties.

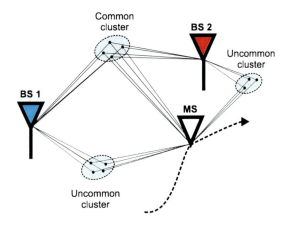

Fig. 3.35 Example of a common cluster in a scenario with one MS and two BSs. The correlation between different links is related to the amount of power that propagates through the common cluster [PTH+11] (©2010 IEEE, reproduced with permission)

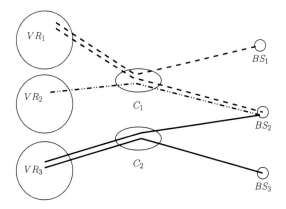

Fig. 3.36 An example of common clusters (C_x) and their link connections with 3 static BSs and 3 VRs. The *lines* describe different link connections from each VR. When an MS is located inside a VR, it obtains link connections determined by the VR through a certain common cluster

3.6.2.2 Modeling Principles

The structure of the COST 2100 channel model implicitly supports multilink scenarios simply by dropping multiple BSs or MSs into the simulation environment. However, this requires to extend the concept of VR in order to consider common clusters. In particular, the VRs and corresponding clusters need to be generated in a coordinated way among different links, so that some links share the same VRs and clusters. To this end, the following new notions are introduced regarding the VRs and clusters.

1. **BS-common cluster:** when considering multi-BS scenarios, the common clusters connect m BSs and a single MS when the MS is inside a VR. We call this common cluster (or VR) as an m BS-common cluster (or VR). Figure 3.36 illustrates an exemplary structure of BSs, VRs, and clusters, showing that cluster C_1 is 2 BS-common for BS1 and BS2. A *common cluster ratio*, $P_{m,\mathrm{BS}}$, is defined to classify attributes of BS-commonness for VRs and indicates the probability to observe m BS-common VRs. The common cluster ratio has the following prop-

erties:

$$P_{m,\text{BS}} = L_m \Big/ \sum_{m'=1}^{M} L_{m'}, \qquad \sum_{m'}^{M} P_{m',\text{BS}} = 1, \qquad (3.47)$$

where M is the number of BSs under consideration, and L_m is the number of clusters/VRs common among m BSs, $1 \le m \le M$.

2. **VR group:** analogous to BS-common clusters, multiple MSs can see the same cluster, even though they are separated by a large distance. Such cluster is, e.g., due to very large walls or buildings. The COST 2100 channel model takes these clusters into account by introducing the concept of *VR group*; when allocating a cluster to a VR, we allow n VRs to correspond to the same cluster, as shown in Fig. 3.36. Those VRs are called an n-VR group. The average number of VRs in a VR group, N_{VR}, was introduced to characterize n. A VR belonging to an n-VR group can naturally support an m BS-common cluster. Hence, such VR is called n-VR m-BS common.

Having extended the concept of VR, some care must also be taken to maintain the model consistency with the single-link channel properties. For example, the VR density is maintained by fixing the total number of links between VRs and BSs,

$$MQ = \sum_{m=1}^{M} (mL_m), \qquad (3.48)$$

where Q is the number of VRs determined in generating the first link. Another important aspect is the cluster spread parameterization. When clusters are shared by multiple BSs and MSs, there are multiple ways to shape the cluster ellipsoids in order to meet the required cluster spreads of at least one links, whereas meeting the required spreads for all links simultaneously might be very complex. The BS and VR used to shape clusters, known as reference BS and VR, are defined as follows:

- the *reference BS* for a cluster has the largest number of VRs connected through the cluster,
- the *reference VR* is one of the VRs connected to the reference BS through the cluster and is also connected to as many BSs.

The reference BS and VR are defined for every cluster separately. Finally, the same cluster selection factor, K_{sel}, and the number of VRs, Q, are applied to all links. This reduces the complexity of the multilink channel model though it makes properties of different links somewhat homogeneous. At present, no measurements enable to infirm these simplifications.

3.6.2.3 VR Assignment Table

The cluster link commonness is numerically characterized by introducing a VR assignment table. Table 3.12 provides a simple example corresponding to Fig. 3.36. The VR assignment table is generated in three steps. First, an empty VR assignment

Table 3.12 A table for the assignment of clusters to BS and VR in the case of the scenario presented in Fig. 3.36

	VR 1	VR 2	VR 3
BS 1	C_1	–	–
BS 2	C_1	C_1	C_2
BS 3	–	–	C_2

table is created. The table must be large enough to specify the link connections for all the VRs in the environment. Hence, its minimum size must be $M \times MQ$. Secondly, the randomly distributed VRs are sequentially labeled along the columns, with the total number satisfying both the VR density constraint for each BS, Q, and the BS-common cluster ratio $P_{m,\text{BS}}$. The labeled VRs are then sequentially assigned to be m-BS common by randomly connecting the VR with m BSs. The assignment stops when all the m-BS common cluster ratio is reached, $1 \leq m \leq M$. Finally, the clusters are sequentially assigned to n (Poisson-distributed) VRs that are randomly selected from the table to form its VR group. The cluster assignment is achieved when all VRs in the table are grouped. These steps are independent processes, and the final VR assignment table will provide the link connections for all clusters. Note that the VR assignment table is scalable for multiple links, by introducing additional columns representing new BSs, but should also follow the principles of the COST 2100 channel model, e.g., each VR in the table should be strictly related to only one cluster.

3.6.3 Parameterization of the COST 2100 Multilink MIMO Channel Model

This subsection summarizes parameters of the COST 2100 channel model. Many parameters are backward compatible with the COST 273 [Cor06] and the WINNER II [KMH+07] channel model, while there are new model parameters associated with the extension of the channel model with respect to polarization, diffuse multipath, and multilink features. The parameters are classified into external and stochastic parameters. The former gives description of environment under consideration and is therefore deterministic and fixed during the simulation, while the latter defines the random nature of radio channels and is described by probability density functions. Since the COST 2100 channel model is composed of a single generic structure, the same set of external and stochastic parameters is defined for various environments, and different values of parameters are used for each environment. The range of applicability of parameter values are determined by settings of the measurement, e.g., the center frequency and bandwidths.

3.6.3.1 Parameterization Campaigns

The parameterization was performed in a number of propagation environments [QOHD09a, PHL+10b, ZTW+10]. They cover LOS and NLOS indoor hall and corridor measured at 5.3 GHz [PHL+10b], NLOS office scenarios at 3.6 GHz band [QOHD09a], and LOS and NLOS outdoor suburban peer-to-peer scenario at 285 MHz [ZTW+10]. Though the propagation measurements were performed independently by several research groups using different equipment, the parameterization methodology was consistent in all the campaigns. Still, not all the parameters are given because of the limitation of measurement equipment.

3.6.3.2 External Parameters

The external parameters consist of system- and environment-related parameters. Particular values of those parameters can actually be determined by users of the channel model, while the range of the applicability of the parameters must be carefully examined. For example, simulating 5 GHz channels by parameters obtained from 2 GHz channels loses accuracy to some extent. In order to provide insights on the applicability of the parameters for users' simulations, settings of external parameters in propagation measurements are summarized in Table 3.13. As a reference, parameters from the COST 273 large urban macrocell scenario [Cor06] and the WINNER II channel model C4 (Outdoor-to-indoor macrocell) scenario [KMH+07] are also shown. The empty columns of the tables mean that the parameters are not available in the measurements and models.

1. **System-related parameters:** The parameters specify formats of channel outputs.

    - *Center frequency*, f_c, of a simulation chain and measurements in the RF domain.
    - *Bandwidth*, B, of a simulation chain and measurements.
    - *Channel snapshot rate*, R_{ss}, over the movement of MSs; a channel response is calculated once in R_{ss} seconds.
    - *Number of channel snapshots*, N_{ss}, defines how many samples are obtained in the simulation.
    - *Position of BS*, $\mathbf{r}_{BS}$, is usually an origin of coordinate system of simulations and measurements on the horizontal plane. The position is fixed in simulations and measurements.
    - *Position of MS*, $\mathbf{r}_{MS}$, evolves as it moves. In the simulation, an initial position of MS needs to be given in the Cartesian coordinate system where the ground is represented by the x–y plane.
    - *Velocity of MS*, $\mathbf{v}_{MS}$, is also defined in the same Cartesian coordinate system as $\mathbf{r}_{MS}$.

2. **Environment-related parameters:** These parameters are needed to calculate pathloss. As in the COST 273 model, the COST 231 Walfish–Ikegami model is

Table 3.13 External parameters of the COST 2100 channel model

Parameter	Unit	Values					Channel models	
		Indoor			Outdoor		COST 273	WINNER II C4
		LOS [PHL+10b]	NLOS [PHL+10b]	NLOS [QOHD09a]	LOS [ZTW+10]	NLOS [ZTW+10]	macrocell [Cor06]	O21 [KMH+07]
f_c	GHz	5.3	5.3	3.6	0.285	0.285	0.9 to 2.0	2.0 to 6.0
B	MHz	60	60	200	20	20		100
R_{ss}	s	3.9×10^{-2}	3.9×10^{-2}		1.4×10^{-3}	1.4×10^{-3}		
N_{ss}		1500	300	16	332	113		
$\mathbf{v}_{MS}$	m/s	(−2.6, 0, 0)	(0.1, 0, 0)		(−0.2, 0.9, 0)	(1.0, 0, 0)		0 to 1.4
h_{BS}[1]	m	2.0	2.0	1.6	1.8	1.8	50	25
h_{MS}	m	1.4	1.4	1.5	2.1	2.1	1.5	1.5
$\mathbf{r}_{MS}$[2]	m	(5, −5)	(24, 11)		(267, −222)	(78, 314)	Random drop in the cell	
$h_{rooftop}$	m				15	15	15	
w_r	m				7	7	25	
w_b	m				250 to 300	250 to 300	50	
ϕ_{road}	°				30 to 90	30 to 90	45	
n_{floor}		0	0	0				
r_{cell}	m	50	50	50	1000	1000	1000	500
$\mathbf{r}_{BS}$	m	(0, 0, h_{BS}) in all cases						

[1] Height above a floor in indoor scenarios
[2] Initial position of an MS along a measurement route. The z coordinate is h_{MS}

used for macrocells, and the COST259 pathloss models are used for microcells and indoor picocells. For details of the pathloss models, readers are directed to Sect. 4.4 of [De99] for the COST231 and Sect. 3.2.4.2 of [Cor01] for the COST259.[2]

- *BS height*, h_{BS}, and *MS height*, h_{MS}.
- *Average rooftop height*, $h_{rooftop}$, of a building in an outdoor scenario.
- *Road width*, w_r, *distance between buildings*, w_b, and *Road orientation with respect to LOS*, ϕ_{road}, in an outdoor scenario.
- *Number of floors between BS and MS*, n_{floor}, in an indoor scenario.
- *Cell radii*, r_{cell}, in a simulation environment.

3.6.3.3 Stochastic Parameters

The number of stochastic parameters is sensibly larger than the number of external parameters. A definition and value of each parameter is summarized in the following.

1. **VR, LOS, and cluster power parameters**

    - *Number of local clusters*, N_C, is active only on the MS side both in indoor and outdoor environments.
    - *Number of additional clusters*, $N_{C,add}$, follows the Poisson distribution with a constant value, $N_{C,add} = Pois(\lambda_{N_{C,add}}) + N_{const}$.
    - *VR radii*, R_C, determines the mean cluster lifetime.
    - *TR width*, L_C, specifies the area where gradual increase and decrease of cluster power takes place.
    - *Cluster decay factor*, k_τ, determines the slope of the exponential decay of cluster power against excess delay time.
    - *Cluster cut-off delay*, τ_B, specifies a delay beyond which cluster power becomes constant.
    - *Cluster selection factor*, K_{sel}, is a ratio of the number of single bounce cluster relative to the twin cluster.
    - *LOS cut-off distance*, d_{co}, is a distance beyond which the LOS can never be seen.
    - *LOS VR radii*, R_L, determines the mean lifetime of LOS.
    - *LOS TR width*, L_L, specifies the area where gradual increase and decrease of the LOS power takes place.
    - *LOS power factor*, K_{LOS}, is a ratio of LOS power relative to the sum of cluster power and is a log-normal random variable with mean $\mu_{K_{LOS}}$ and standard deviation $\sigma_{K_{LOS}}$.
    - *Cluster shadow fading*, σ_S, follows the log-normal distribution.

---

[2]The WINNER II channel model provides pathloss model for variety of environments. They are summarized in Table 4 of [KMH+07].

2. **Cluster location parameters (values are summarized in Table 3.15)**

    - *Cluster link delay*, $\tau_{C,\text{link}}$, is an extra delay of multipaths for twin clusters; this parameter follows the exponential distribution as

    $$f(\tau) = \begin{cases} 0, & \tau < \tau_{C,\text{link},\min}, \\ \dfrac{1}{\mu_{\tau_{C,\text{link}}}} \exp\{-(\dfrac{\tau - \tau_{C,\text{link},\min}}{\mu_{\tau_{C,\text{link}}}})\} & \text{otherwise,} \end{cases} \quad (3.49)$$

    where $\mu_{\tau_{C,\text{link}}}$ is a mean, and $\tau_{C,\text{link},\min}$ is a minimum cluster link delay.
    - *Cluster distance*, d_C, is a distance between BS (MS) and a cluster center. This parameter follows the exponential distribution with its minimum $d_{C,\min}$ and mean μ_{d_C}. The values for the BS and MS sides are separately defined.
    - *Cluster angle*, Ψ_C, is used to define directions of clusters seen from the BS and MS. This parameter is a normal random variable represented by its mean μ_{Ψ_C} and standard deviation σ_{Ψ_C} and is defined separately for the BS and MS sides and for azimuth and elevation angles.

3. **Cluster spread parameters**

    - *Number of multipaths in a cluster*, N_{MPC}, is a constant value for simplicity.
    - *Cluster spreads* in azimuth, elevation (or coelevation), and delay follow the log-normal distribution with a base of 10, mean m, and standard deviation S. The log-normal distribution is used to assure the spread values to be always nonnegative, and hence the standard deviation is given in decibels. A realization of the spread is given by $m \cdot 10^{\frac{S\xi}{10}}$ where ξ is a real Gaussian random variable. It must be noted that the mean and standard deviation of spread values reported in almost all measurement results were derived in a linear scale, and hence they do *not* fit into use with the log-normal distribution. A conversion of the measured mean and standard deviation is therefore needed as

    $$m \text{ [deg] or [ns]} = \frac{m'}{\sqrt{1 + S'^2/m'^2}}, \quad (3.50)$$

    $$S \text{ [dB]} = \sqrt{\ln\left(1 + \frac{S'^2}{m'^2}\right)} \times 10 \log_{10} e, \quad (3.51)$$

    where m' and S' are mean and standard deviation of the spread derived in the linear domain.
    - *Cluster spread cross-correlation*, ρ, characterizes correlation between cluster spread parameters and shadowing.[3] Like the COST 273 model, correlated random processes can be calculated by the Cholesky factorization [MK99]. In Table 3.16, correlation is shown, revealing moderate correlation between cluster delay and angular spreads both in outdoor and indoor scenarios.

---

[3]The WINNER channel model [KMH+07] provides similar parameters but they describe cross-correlation of *global* spreads and hence are different from parameters here. The same goes to the autocorrelation distances.

- *Autocorrelation distances* were introduced in the COST 273 channel model to simulate the variation of cluster spreads and shadowing against large, typically tens of wavelengths, movement of the MS. In the COST 2100 channel model, this variation is implicitly simulated by the geometrical relationship of the MS and multipath locations as the MS moves inside a VR. Furthermore, experimental results in [ZTW+10] showed that the autocorrelation distance was almost the same as the VR size, which is a strong indication that a large variation of the cluster spread occurs when a cluster disappears. We can therefore see that the effect of the autocorrelation distances is well modeled by the VR.
- *Rice factor of additional clusters*, K_{MPC}, is set to 0 dB in the COST 2100 channel model. Therefore, multipaths inside the cluster have Rayleigh-distributed amplitude and random phase. It was found in [PHL+10b] that the LOS power factor, K_{LOS}, influences the fading statistics much more than K_{MPC}.

4. **Polarization parameters**

   - *CPR* of a cluster is a log-normal random variable with its mean and standard deviation denoted as μ_{CPR} and σ_{CPR}.
   - *XPD* is also a log-normal random variable. This parameter is modeled in the same way as the CPR.

5. **Multilink parameters**

   - *m-BS common cluster ratio*, $p_{m,BS}$, shows how probable m-BS common clusters appear. The value is available only for a two-BS case in the measurement and is a constant value in a simulation run.
   - *Number of VRs in a VR group*, n, is a Poisson-distributed random variable with mean N_{VR}; in the propagation modeling, the parameter was derived by the average number of appearance of the same scattering object along a large movement of MSs.

6. **Diffuse multipath parameters**

   - *Radius of the DMC cluster*, r_{DMC}, determines the angular and delay spreads of the DMC clusters.
   - *Additional delay of the MPC*, $\tau_{DMC,add}$, gives a tail of the DMC clusters on the delay domain.

It should be noted that multilink parameters are available only for indoor picocell scenarios. Therefore, parameters of outdoor scenarios remain important missing parts of the channel model. Furthermore, the diffuse multipath parameters have not been derived from channel measurements yet. Values for the external parameters are summarized for different scenarios in Tables 3.14 to Tables 3.18.[4]

---

[4]For [PHL+10b], parameters from BS1 are shown in the list, and parameters from LOS routes 1 and NLOS route 4 are shown from [ZTW+10].

Table 3.14 Stochastic parameters of the COST 2100 channel model: VR, LOS, and cluster power parameters

Parameter	Unit	Values					Channel models	
		Indoor			Outdoor		COST 273	WINNER II C4
		LOS [PHL+10b]	NLOS [PHL+10b]	NLOS [QOHD09a]	LOS [ZTW+10]	NLOS [ZTW+10]	macrocell [Cor06]	O21 [KMH+07]
$\lambda_{N_{C,\text{add}}}$		0.8	0.7	1.7	4.0	4.0	1.2	12.0
N_{const}		3.7	4.3	3.0	6.3	5.2	100	
R_C	m	2.8	3.8		1.4	1.9	20	
L_C	m	0.3	0.4		16.9	3.2	1	
k_τ	dB/µs	54.2	5.9	>12[1]			10	
τ_B	µs	1	1				1	
K_{sel}		0.7	0.0		0.4	0.3	500	
d_{co}	m	100					30	
R_L	m	10					20	
L_L	m	1						
μ_K	dB	0.4	−9.7		−31.5	−27.0	$(26 - EPL)/6^2$	
σ_K	dB	2.8	7.4		8.2	8.7	6	
σ_S	dB	4.9	5.4	5.1	0.4	0.5	6	4

[1] Specifically, normalized gain by LOS power is given in dB by $-12 - 0.06\tau$ [ns]

[2] EPL stands for excess path loss and is defined by $EPL(d) = L - 20\log_{10}(\frac{4\pi d}{\lambda})$ [dB], where d and λ is a Tx-Rx distance and the center frequency of channels [Cor01], p. 191

3 Radio Channel Modeling for 4G Networks

Table 3.15 Stochastic parameters of the COST 2100 channel model: cluster location parameters

Parameter	Unit	Values Indoor	
		LOS [PHL$^+$10b]	NLOS [PHL$^+$10b]
$\mu_{\tau_{C,link}}$	ns	44.1	74.8
$\tau_{C,link,min}$	ns	27.3	14.0
$\mu_{d_{C,BS}}$	m	4.2	8.5
$d_{C,BS,min}$	m	6.7	8.0
$\mu_{d_{C,MS}}$	m	8.4	6.7
$d_{C,MS,min}$	m	1.5	1.1
$\mu_{\psi_{C,AZ,BS}}$	°	−17.1	28.4
$\sigma_{\psi_{C,AZ,BS}}$	°	47.0	30.6
$\mu_{\psi_{C,AZ,MS}}$	°	−7.6	−30.2
$\sigma_{\psi_{C,AZ,MS}}$	°	38.8	52.0
$\mu_{\psi_{C,EL,BS}}$	°	1.2	−4.8
$\sigma_{\psi_{C,EL,BS}}$	°	23.5	6.4
$\mu_{\psi_{C,EL,MS}}$	°	20.3	−0.1
$\sigma_{\psi_{C,EL,MS}}$	°	38.2	10.2

3.6.4 Implementation and Validation of the COST 2100 Multilink MIMO Channel Model

The COST 2100 multilink channel model was implemented in Mathworks MATLAB [LCO09] and by Jiang et al. in C++ with IT++ plugin [JLZT10]. The implementation loads channel model parameters and provides channel responses such as double-directional impulse response and transfer functions. They are easily integrated into a radio system simulation chain. Using the implementations, the COST 2100 channel model was validated by comparing the channel model outputs and channel measurements [HPL$^+$10, ZTE10]. Model parameters of the COST 2100 model were derived by dynamic double-directional channel measurements with which the channel realizations from the COST 2100 model were compared. The tested scenarios were an indoor hall environment with a LOS named "CS, BS1" in [PHL$^+$10b], and an outdoor suburban environment with a LOS denoted as "Route1" in [ZTE10]. The measurement consists of a fixed transmit BS location and an MS route stretching for 28 m in indoor and 303 m in outdoor scenarios. Channel model parameters of those scenarios are given in [PHS$^+$10, ZTE10] and also in Sect. 3.6.3. A set of simplifications was made for the comparison. First, only vertical copolarized path weights were considered. Second, local clusters were inactivated in the channel simulation in the indoor scenario because it was not possible to detect them in the measurements. Finally, only specular propagation paths were taken into account for the comparison.

Table 3.16 Stochastic parameters of the COST 2100 channel model: cluster spread parameters

Parameter	Unit	Values Indoor LOS [PHL+10b]	Indoor NLOS [PHL+10b]	Indoor NLOS [QOHD09a]	Outdoor LOS [ZTW+10]	Outdoor NLOS [ZTW+10]	Channel models COST 273 macrocell [Cor06]	WINNER II C4 O2I [KMH+07]
$\mu_{N_{MPC}}$		4	4	10	4	4	20	20
$m_{\psi_{BS}}$	°	0.55	2.8		17.1	19.8	6.5	8
$S_{\psi_{BS}}$	dB	3.1	3.6		2.0	2.0	0.34	
$m_{\theta_{BS}}$	°	1.4	2.6				3.2	3
$S_{\theta_{BS}}$	dB	3.4	3.8				3	
$m_{\psi_{MS}}$	°	2.8	6.0	8.7	17.8	21.3	35	5
$S_{\psi_{MS}}$	dB	3.6	1.9	1.1	2.2	1.8	0	
$m_{\theta_{MS}}$	°	3.2	1.3	7.5			$\mathcal{U}(0, 45)$	3
$S_{\theta_{MS}}$	dB	2.3	2.5	0.84			0	
m_τ	ns	0.81	4.2	3.7	135	247	400	
S_τ	dB	3.3	1.4	1.8	2.5	2.5	3	
$\rho_{\psi_{BS}\theta_{BS}}$		0.6	0.6		0.3	0.4	0.0	
$\rho_{\psi_{BS}\psi_{MS}}$		0.6	-0.2					
$\rho_{\psi_{BS}\theta_{MS}}$		0.6	0.3					
$\rho_{\psi_{BS}\tau}$		0.6	-0.2		0.6	0.6	0.5	
$\rho_{\psi_{BS}\sigma_S}$		0.1	0.0		0.0	0.0	-0.6	
$\rho_{\theta_{BS}\psi_{MS}}$		0.3	-0.2					
$\rho_{\theta_{BS}\theta_{MS}}$		0.7	0.0					
$\rho_{\theta_{BS}\tau}$		0.6	-0.2					
$\rho_{\theta_{BS}\sigma_S}$		-0.1	0.3					
$\rho_{\psi_{MS}\theta_{MS}}$		0.5	0.2					
$\rho_{\psi_{MS}\tau}$		0.6	0.8		0.8	0.9	0.0	
$\rho_{\psi_{MS}\sigma_S}$		0.0	0.1		0.1	0.0	0.0	
$\rho_{\theta_{MS}\tau}$		0.5	0.5					
$\rho_{\theta_{MS}\sigma_S}$		0.0	0.0					
$\rho_{\tau\sigma_S}$		-0.1	0.2		0.0	-0.2	-0.6	

Table 3.17 Stochastic parameters of the COST 2100 channel model: polarization parameters

Parameter	Unit	Values				
		Indoor			Channel models	
		LOS [PHL+10b]	NLOS [PHL+10b]	NLOS [QOHD09a]	COST 273 macrocell [Cor06]	WINNER II C4 O2I [KMH+07]
$m_{\mu_{CPR}}$	dB	0.9	−5.3	4.2	0	
$S_{\mu_{CPR}}$	dB			8.2		
$m_{\sigma_{CPR}}$	dB	6.6	11.9	$1.3 + \mu_{CPR}$	$-\infty$	
$S_{\sigma_{CPR}}$	dB			3.3		
$m_{\mu_{XPD_V}}$	dB	15.6	9.6	13.5	6	9
$S_{\mu_{XPD_V}}$	dB			7.8		
$m_{\sigma_{XPD_V}}$	dB	10.4	9.5	7.8	2	11
$S_{\sigma_{XPD_V}}$	dB			$1.3 + \mu_{XPD_V}$		
$m_{\mu_{XPD_H}}$	dB	14.9	11.4	3.9	6	9
$S_{\mu_{XPD_H}}$	dB			5.5		
$m_{\sigma_{XPD_H}}$	dB	11.8	8.7	$1.3 + \mu_{XPD_H}$	2	11
$S_{\sigma_{XPD_H}}$	dB			3.3		

Table 3.18 Stochastic parameters of the COST 2100 channel model: multi-link parameters

Parameter	Unit	Values	
		Indoor	
		LOS [PHL+10b]	NLOS [PHL+10b]
$p_{2,BS}$		0.02	0.56
N_{VR}		2.0	1.4

In [HPL+10], delay and angular spreads are used as comparison metrics. Figure 3.37 shows Cumulative Density Function (CDF) of the spreads from measurements and three runs of channel simulation outputs. The delay spread revealed close agreement between the measurement and channel simulation outputs. On the other hand, the channel simulation underestimated the measured angular spread as can be noticeably seen in the elevation spread on the MS side. The underestimation can be explained by several reasons. First, it is due to lack of explicit mechanism in the COST 2100 channel model to relate cluster angles to power information. The COST 2100 model determines cluster power by their delay only as defined in Eq. (3.40). The angular dependency of the power is, on the other hand, considered only in determining cluster angles and not in cluster power. Thus, the COST 2100 model has less capability to maintain the measured global angular spread compared to, for example, the WINNER II channel model [KMH+07]. Secondly, oversimplification of channel distributions could result in such a discrepancy in comparison. Channel characteristics in angular and delay domains follow multimodal distribution in many cases, while the COST channel model uses single-modal distributions for simplic-

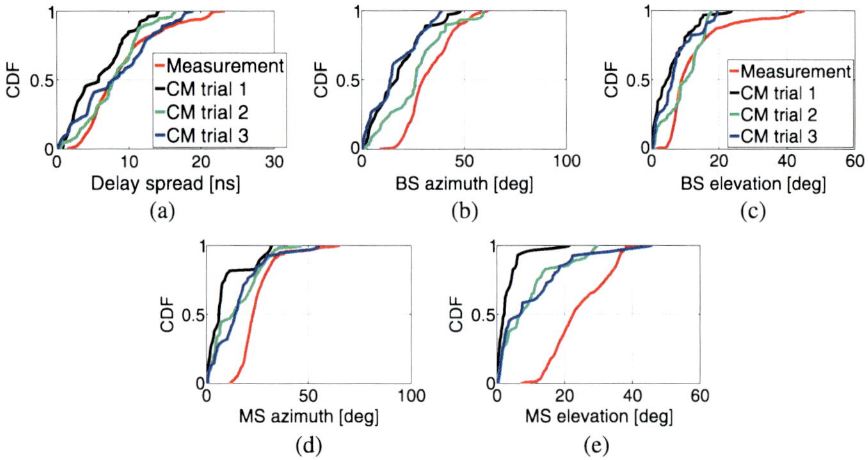

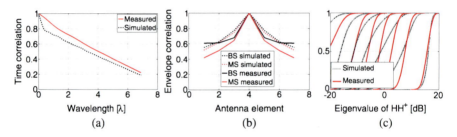

Fig. 3.37 Comparison of delay, azimuth, and elevation spreads between measurements and the COST 2100 model in an indoor hall scenario at 5.3 GHz frequency: (**a**) delay, (**b**) BS azimuth, (**c**) BS elevation, (**d**) MS azimuth, and (**e**) MS elevation [HPL+10]

Fig. 3.38 Comparison between outputs of the COST 2100 model and channel measurements in an outdoor suburban scenario at 285 MHz frequency: (**a**) time correlation, (**b**) antenna correlation, and (**c**) magnitude of singular values [ZTE10]

ity. The large elevation angular spread at the MS is attributed to a strong reflection from the ceiling as well as a LOS, but that property was not modeled in the COST model properly since a single-modal normal distribution is assumed for the cluster elevation angle. Finally, there is a difference of nature in the channel simulations and measurements. Simulated channels are generated from a stochastic channel model, while measurements are purely deterministic. Given the limited number of cluster samples available both in channel measurements and simulations, it is not sensible to expect perfect agreement between them.

The comparison in [ZTE10] for the outdoor suburban scenario was made in terms of delay spread, time and antenna correlation, and singular value distribution. The time correlation shows how the channel changes as the MS is moving. The time correlation was calculated from channel responses by assuming omni-directional antennas both on the BS and the MS sides. The envelope correlation over the displacement of the MS is shown in Fig. 3.38(a). Since the channel was in a LOS condition, a linear decay of the amplitude correlation was observed both in channel model outputs and in measurements. Figure 3.38(b) shows the antenna correlation both at the BS and MS, revealing an agreement between simulation and measurements. The antenna correlation property is an indicator of the available diversity and hence the multipath richness in the environment. A uniform circular dipole array was considered at both ends of the link. A reference antenna element was randomly selected in the circular dipole array and was named as an antenna element "4". The envelope correlation of signals was then analyzed between reference and other antenna elements and was averaged over seven different selections of the reference antenna element. Finally, Fig. 3.38(c) illustrates a comparison of singular value distributions. The singular values were derived from 7×7 MIMO channel matrix which assumed antenna arrays used in the measurements. All singular values were evaluated with 20 dB SNR. The results showed that the strongest singular value distribution matched well with the measurement, while the weaker singular values were more prone to deviation.

Overall, the comparison revealed an acceptable level of agreement in terms of the channel spreads, correlation, and eigenvalue structures. The results indicate practicality of the COST 2100 channel model for realistic MIMO channel simulations. As a final remark, it must be noted that the validation so far has been limited to single-link properties and leaves multilink characteristics as an important future task.

References

[3GP09] 3GPP TR25.996 V9.0.0. Spatial channel model for MIMO simulations. Technical Report, Third Generation Partnership Project (3GPP), December 2009.

[ALO07] H. Asplund, K. Larsson, and P. Okvist. How typical is the "Typical Urban" channel model? Technical Report TD07356, Duisburg, Germany, September 2007.

[BCE+10] M. Barbiroli, C. Carciofi, V. Degli-Esposti, F. Fuschini, P. Grazioso, D. Guiducci, D. Robalo, and F. J. Velez. Characterization of WiMAX propagation in microcellular and picocellular environments. In *Proc. 4th European Conference on Antennas and Propagation (EuCAP 2010)*, Barcelona, Spain, April 2010. [Also available as TD(10)11007].

[BCG+09] M. Barbiroli, C. Carciofi, D. Guiducci, D. Robalo, F. J. Velez, P. Sebastiano, F. Varela, and F. Cercas. Analysis of WiMAX propagation measurements in outdoor and indoor environments. Technical Report TD(09)957, Vienna, Austria, February 2009.

[BDHZ10] R. I. C. Bultitude, G. S. Dahman, R. H. M. Hafez, and H. Zhu. Double directional radio propagation measurements and radio channel modelling pertinent to mobile MIMO communications in microcells. Technical Report TD(10)11019. Aalborg, Denmark, June 2010.

[BLdC08] J. W. De Bleser, E. Van Lil, and A. Van de Capelle. Applying ray tracing to finite multilayered structures. Technical Report TD(08)628, Lille, France, October 2008.

[BLdC09] J. W. De Bleser, E. Van Lil, and A. Van de Capelle. An analysis of the low-frequency limit of the boundary integral equations. Technical Report TD(09)846, Valencia, Spain, May 2009.

[Blu03] R. S. Blum. MIMO capacity with interference. *IEEE J. Select. Areas Commun.*, 21(5):793–801, 2003. doi:10.1109/JSAC.2003.810345.

[Bul08] R. Bultitude. Methods for estimating and modelling consistency intervals and the detection and simulation of changes on mobile radio channels. Technical Report TD(08)424, Wroclaw, Poland, September 2008.

[CBOP10] N. Czink, B. Bandemer, C. Oestges, and A. Paulraj. An analytic multi-user MIMO channel model including subspace alignment. Technical Report TD-10-11028, Aalborg, Denmark, June 2010.

[CBVV$^+$] N. Czink, B. Bandemer, G. Vazquez-Vilar, L. Jalloul, C. Oestges, and A. Paulraj. Spatial separation of multi-user MIMO channels. In *Proc. IEEE Pers Indoor Mobile Radio Comm. Conf. (PIMRC)*, 2009.

[CBYZ08] N. Czink, E. Bonek, J. Ylitalo, and T. Zemen. Measurement-based time-variant channel modelling using clusters. Technical Report TD(08)435, Wroclaw, Poland, September 2008.

[CCC10] Q. H. Chu, J. M. Conrat, and J. C. Cousin. Path loss characterization for LTE-advanced relaying propagation channel. Technical Report TD(10)12019, Bologna, Italy, November 2010.

[CDNL09] J.-M. Conrat, H. Dekov, A. Nasr, and M. Lienard. Analysis of the space–time channel behavior in outdoor-to-indoor environment, February 2009. [Also available as TD(08)627].

[CKZ$^+$10] N. Czink, F. Kaltenberger, Y. Zhouz, L. Bernado, T. Zemen, and X. Yin. Low-complexity geometry-based modeling of diffuse scattering. In *10th COST 2100 MCM Meeting*, Athens, Greece, February 2010. COST 2100 TD(10)10026.

[CL08] Y. Corre and Y. Lostanlen. Methods to extrapolate 3D antenna radiation pattern ensuring reliable radio channel predictions. Technical Report TD(08)426, Wroclaw, Poland, February 2008.

[CO08a] N. Czink and C. Oestges. The COST 273 MIMO channel model—a short tutorial. Technical Report TD(08)405, Wroclaw, Poland, February 2008.

[CO08b] N. Czink and C. Oestges. The COST 273 MIMO channel model: three kinds of clusters. In *Proc. IEEE 10th Int. Sym. ISSSTA '08*, pages 282–286, Bologna, Italy, August 2008.

[Cor01] L. Correia. *Wireless Flexible Personalized Communications*. Wiley, New York, 2001.

[Cor06] L. M. Correia. *Mobile Broadband Multimedia Networks*. Academic Press, San Diego, 2006.

[CP05] J.-M. Conrat and P. Pajusco. Clusterization of the propagation channel in urban macrocells at 2 GHz. In *The European Conference on Wireless Technology, 2005*, pages 39–42, 2005. [Also available as TD(07)307].

[CTK10] K. L. Chee, S. A. Torrico, and T. Kürner. Foliage attenuation over mixed terrains in rural areas for broadband wireless access at 3.5 GHz. Technical Report TD(10)12027, Bologna, Italy, November 2010.

[Czi07] N. Czink. *The random-cluster model: a stochastic MIMO channel model for broadband wireless communication systems of the 3rd generation and beyond*. PhD thesis, Vienna University of Technology, Vienna, Austria, 2007.

[DC07a] A. Dunand and J.-M. Conrat. Dual-polarized spatio-temporal characterization in urban macrocells at 2 GHz. In *Proc. VTC 2007 Fall—IEEE 66th Vehicular Technology Conf.*, Baltimore, MD, USA, 2007. [Also available as TD(08)406].

[DC07b] A. Dunand and J.-M. Conrat. Polarization behaviour in urban macrocell environments at 2.2 GHz. In *Proc. European Conference on Antennas and Propagation (EuCap 2007)*, Edinburgh, UK, 2007. [Also available as TD(08)406].

[DDGW03] D. Didascalou, M. Doettling, N. Geng, and W. Wiesbeck. An approach to include stochastic rough surface scattering into deterministic ray-optical wave propagation modeling. *IEEE Trans. Antennas Propagat.*, 51(7):27–37, 2003. [Also available as TD(01)007].

[De99] E. Damosso and L. M. Correia (ed.). *Digital Mobile Radio Towards Future Generation Systems*. European Commission, Brussels, 1999.

3 Radio Channel Modeling for 4G Networks   141

[DE01]   V. Degli-Esposti. A diffuse scattering model for urban propagation prediction. *IEEE Trans. Antennas Propagat.*, 49(7):1111–1113, 2001.
[DEFVF08]   V. Degli-Esposti, F. Fuschini, and E. M. Vitucci. A fast model for distributed scattering from buildings. Technical Report TD(08)668, Lille, France, October 2008.
[DEFVF07]   V. Degli-Esposti, F. Fuschini, E. M. Vitucci, and G. Falciasecca. Measurement and modelling of scattering from buildings. *IEEE Trans. Antennas Propagat.*, 55(1):143–153, 2007.
[DEG+04]   V. Degli-Esposti, D. Guiducci, A. De Marsi, P. Azzi, and F. Fuschini. An advanced field prediction model including diffuse scattering. *IEEE Trans. Antennas Propagat.*, 14:1717–1728, 2004.
[dlFL+10]   G. de la Roche, P. Flipo, Z. Lai, G. Villemaud, J. Zhang, and J.-M. Gorce. Combined model for outdoor to indoor radio propagation prediction. Technical Report TD(10)10045, Athens, Greece, February 2010.
[dlRJRG07]   G. de la Roche, K. Jaffres-Runser, and J.-M. Gorce. On predicting in-building WiFi coverage with a fast discrete approach. *International Journal of Mobile Network Design and Innovation*, 2:3–12, 2007.
[DPP+07]   R. Delahaye, Y. Pousset, A.-M. Poussard, P. Combeau, and R. Vauzelle. From a deterministic transmission channel modelling to an efficiency simulation of radio links quality of ad hoc networks. Technical Report TD-07-357, Duisburg, Germany, September 2007.
[ETM07]   G. Eriksson, F. Tufvesson, and A. F. Molisch. Characteristics of MIMO peer-to-peer propagation channels at 300 MHz. Technical Report TD-07-376, Duisburg, Germany, September 2007.
[FMC08]   R. Fraile, J. F. Monserrat, and N. Cardona. Analysis and validation of a shadowing simulation model suited for dynamic and heterogencous wireless networks. Technical Report TD-08-542, Trondheim, Norway, June 2008.
[FWE10]   K. Fors, K. Wiklundh, and G. Eriksson. A stochastic channel model for simulation of mobile ad hoc networks. Technical Report TD-10-11084, Aalborg, Denmark, June 2010.
[GCC+10]   M. Gan, N. Czink, P. Castiglione, C. Oestges, F. Tufvesson, and T. Zemen. Modeling time-variant fast fading statistics of mobile peer-to-peer radio channels. Technical Report TD-10-12003, Bologna, Italy, November 2010.
[Geo06]   T. T. Georgiou. Distances between power spectral densities. arXiv:math/0607026v2. International Telecommunication Union, Radiocommunication Sector (ITU-R), July 2006.
[GJRR07]   J. M. Gorce, K. Jaffres-Runser, and G. De La Roche. Deterministic approach for fast simulations of indoor radio wave propagation. *IEEE Trans. Antennas Propagat.*, 55(3):938–948, 2007.
[GRC07]   A. P. Garcia, L. Rubio, and N. Cardona. Frequency impact on the MIMO capacity between 2 and 12 GHz in an office environment. Technical Report TD(07)216, Lisbon, Portugal, February 2007.
[GT07a]   M. Ghoraishi and J. Takada. Azimuth-power-spectrum prediction by site-specific stochastic channel modeling in a microcell line-of-sight scenario. Technical Report TD(07)389, Duisburg, Germany, September 2007.
[GT07b]   M. Ghoraishi and J. Takada. Stochastic channel model for dense urban line-of-sight street microcell. Technical Report TD(07)241, Lisbon, Portugal, February 2007.
[Gud91]   M. Gudmundson. Correlation model for shadow fading in mobile radio systems. *Electronics Letters*, 27(23):2145–2146, 1991. doi:10.1049/el:19911328.
[GvL96]   G. Golub and C. van Loan. *Matrix Computations*, 3rd edition. Johns Hopkins University Press, London, 1996.
[HCOB05]   M. Herdin, N. Czink, H. Ozcelik, and E. Bonek. Correlation matrix distance, a meaningful measure for evaluation of non-stationary MIMO channels. In *IEEE VTC Spring 2005*, vol. 1, pages 136–140, 2005.

[HPL+10] K. Haneda, J. Poutanen, L. Liu, C. Oestges, F. Tufvesson, and P. Vainikainen. Comparison of delay and angular spreads between channel measurements and the COST 2100 channel model. In *Proc. Loughborough Ant. Prop. Conf. (LAPC2010)*, Loughborough, UK, November 2010. [Also available as TD(10)11072].

[HTK+] T. Hult, F. Tufvesson, V.-M. Kolmonen, J. Poutanen, and K. Haneda. Analytical dual-link MIMO channel model using correlated correlation matrices. In *Proc. European Conf. on Antennas and Propagation (EUCAP 2010)*.

[HTK+10] T. Holt, F. Tufvesson, V.-M. Kolmonen, J. Poutanen, and K. Haneda. Canalytical multi-link MIMO channel model using rotated correlation matrices. Technical Report TD-10-12061, Bologna, Italy, November 2010.

[HWW03] R. Hoppe, G. Woelfle, and P. Wertz. Advanced ray-optical wave propagation modelling for urban and indoor scenarios. *European Transactions on Telecommunications (ETT)*, 14:61–69, 2003.

[ITU08] ITU-R M.2135. Guidelines for evaluation of radio interface technologies for IMT-advanced. Technical Report, International Telecommunication Union, Radiocommunication Sector (ITU-R), 2008.

[Jal10] N. Jalden. *Analysis and modelling of joint channel properties from multisite, multi-antenna radio measurements*. PhD thesis, KTH, Stockholm, February 2010.

[JKM08] T. Jämsä, P. Kyösti, and J. Meinilä. Approximation of WINNER channel models. Technical Report TD(08)505, Trondheim, Norway, June 2008.

[JLZT10] W. Jiang, L. Liu, M. Zhu, and F. Tufvesson. Implementation of the COST 2100 multi-link MIMO channel model in C++/IT++. Technical Report TD(10)12092, Bologna, Italy, November 2010.

[KBZ09] F. Kaltenberger, L. Bernadó, and T. Zemen. On the characterization of measured multi-user MIMO channels. In *Workshop on Smart Antennas (WSA 2009)*, Berlin, Germany, February 2009. [Also available as TD-08-640].

[KFW96] T. Kürner, R. Fauss, and A. Wäsch. A hybrid propagation modelling approach for DCS1800 macro cells. In *Proc. VTC 1996—IEEE 46th Vehicular Technology Conf.*, Atlanta, Georgia, USA, May 1996.

[KH07] M. Kwakkernaat and M. Herben. Analysis of clustered multipath estimates in physically nonstationary radio channels. Technical Report TD(07)324, Duisburg, Germany, September 2007.

[KH08] M. Kwakkernaat and M. Herben. Modelling angular dispersion in ray-based propagation prediction models. Technical Report TD(08)519, Trondheim, Norway, June 2008.

[KH09] M. Kwakkernaat and M. Herben. Modelling of angular dispersion due to rough surface scattering for implementation in ray-tracing-based propagation prediction tools. Technical Report TD(09)713, Braunschweig, Germany, February 2009.

[KHH+10] V.-M. Kolmonen, K. Haneda, T. Hult, J. Poutanen, F. Tufvesson, and P. Vainikainen. Measurement-based evaluation of interlink correlation for indoor multi-user MIMO channels. *IEEE Antennas Wireless Propagat. Lett.*, 9(4):311–314, 2010.

[KIS+08] K. Kitao, T. Imai, K. Saito, Y. Okano, and S. Miura. Estimation accuracy of ray-tracing for spatio-temporal dynamic channel properties. Technical Report TD(10)11046, Aalborg, Denmark, June 2008.

[KJ07] P. Kyosti and T. Jamsa. Complexity comparison of MIMO channel modelling methods. In *Proc. 4th Int. Symp. Wireless Communication Systems ISWCS 2007*, pages 219–223, 2007.

[KJBN08] P. Kyösti, T. Jämsä, A. Byman, and M. Narandžić. Computational complexity of drop based radio channel simulation. Technical Report TD-08-533, Trondheim, Norway, June 2008.

[KLT09] M. Käske, M. Landmann, and R. Thomä. Modelling and Synthesis of Dense Multipath Propagation Components in the Angular Domain. In *7th COST 2100 MCM Meeting*, Braunschweig, Germany, February 2009. [COST 2100 TD(09)762].

[KMH+07] P. Kyösti, J. Meinilä, L. Hentilä, X. Zhao, T. Jämsä, C. Schneider, M. Narandžić, M. Milojević, A. Hong, J. Ylitalo, V.-M. Holappa, M. Alatossava, R. Bultitude, Y. de Jong, and T. Rautiainen. IST-4-027756 WINNER II Deliverable 1.1.2. v.1.2, WINNER II Channel Models. Technical Report, IST-WINNERII, September 2007.

[KMT+09] Y. Konishi, L. Materum, J. Takada, I. Ida, and Y. Oishi. Polarization characteristics of MIMO system: measurement, modeling and statistical validation. In *Proc. PIMRC 2009—IEEE 20th Int. Symp. on Pers., Indoor and Mobile Radio Commun.*, Tokyo, Japan, September 2009. [Also available as TD(09)737].

[KT07] W. Kotterman and R. Thomä. Testing for Kronecker-separability. Technical Report TD(07)217, Lisbon, Portugal, February 2007.

[Kwa08] M. Kwakkernaat. *Angular dispersion of radio waves in mobile channels: measurement based analysis and modelling*. PhD thesis, Eindhoven University of Technology, Eindhoven, The Netherlands, December 2008.

[KZU07] F. Kaltenberger, T. Zemen, and C. W. Uberhuber. Low-complexity geometry-based MIMO channel simulation. *EURASIP Journal in Advances in Signal Processing*, 2007, 2007. Article ID 95281. doi:10.1155/2007/95281.

[LCO09] L. Liu, N. Czink, and C. Oestges. Implementing the COST 273 MIMO channel model. In *Proc. NEWCOM++—ACoRN Joint Workshop*, Barcelona, Spain, March 2009.

[LDMGP08] M. Lienard, P. Degauque, and J. M. Molina-Garcia-Pardo. Wideband large and small scale fading in tunnels. Technical Report TD(08)608, Lille, France, October 2008.

[LOC08] L. Liu, C. Oestges, and N. Czink. The COST 273 channel model implementation. Technical Report TD(08)638, Lille, France, October 2008.

[LP09] Y. Lustmann and Porrat. Indoor channel spectral statistics, K-factor and reverberation distance example. Technical Report TD(09)959, Vienna, Austria, February 2009.

[LwDBTdC09] E. Van Lil, J.w. De Bleser, D. Trappeniers, and A. Van de Capelle. On the effects of large (moving) objects like wind turbines on the accuracy and reliability of radio frequency sensors. Technical Report, Delft, The Netherlands, June 2009. [Also available in extended form as TD(08)443].

[MAH+06] A. F. Molisch, H. Asplund, R. Heddergott, M. Steinbauer, and T. Zwick. The COST 259 directional channel model—part i: overview and methodology. *IEEE Trans. Wireless Commun.*, 5(12):3421–3433, 2006.

[Man07] O. Mantel. Effect of database simplification on ray-tracing predictions for a dense urban GSM cell. Technical Report TD(07)323, Duisburg, Germany, September 2007.

[MBvDK09] O. Mantel, A. Bokiye, R. van Dommele, and M. Kwakkernaat. Measurement-based verification of delay and angular spread ray-tracing predictions for use in urban mobile network planning. Technical Report TD(09)914, Vienna, Austria, May 2009.

[MCBH10] R. Moghrani, J. M. Conrat, X. Begaud, and B. Huyart. Performance evaluation of a 3D ray tracing model in urban environment. In *Antennas and Propagation Society International Symposium (IEEE APSURSI 2010)*, pages 1–4, Toronto, Ontario, July 2010. [Also available as TD(10)10064].

[MGPLD+09] J.-M. Molina-Garcia-Pardo, M. Liénard, P. Degauque, C. Garcia-Pardo, and L. Juan-Llácer. MIMO channel capacity with polarization diversity in arched tunnels. *IEEE Antennas Wireless Propagat. Lett.*, 8:1186–1189, 2009. [Also available as TD(09)802].

[MGPRJL08] J.-M. Molina-Garcia-Pardo, J.-V. Rodriguez, and L. Juan-Llácer. Polarized indoor MIMO channel measurements at 2.45 GHz. *IEEE Trans. Antennas Propagat.*, 56(12):3818–3828, 2008. [Also available as TD(08)605].

[MK99] P. Mogensen and K. Klingenbrunn. Modelling cross-correlated shadowing in network simulations. In *Proc. VTC 1999 Fall—IEEE 50th Vehicular Technology Conf.*, pages 1407–1411, Amsterdam, The Netherlands, September 1999.

[MK07] M. Neuland and T. Kuerner. Analysis of the influence of different resolutions on radio propagation models at 900 and 1800 MHz. In *Proc. 2nd European Conference on Antennas and Propagation (EuCAP 2007)*, 2007. [Also available as TD(07)366].

[MO10] F. Mani and C. Oestges. Evaluation of diffuse scattering contribution for delay spread and crosspolarization ratio prediction in an indoor scenario. In *Proc. 4th European Conference on Antennas and Propagation (EuCAP 2010)*, Barcelona, Spain, April 2010. [Also available as TD(09)812].

[MO11] F. Mani and C. Oestges. Ray-tracing evaluation of diffuse scattering in an outdoor scenario. In *Proc. 5th European Conference on Antennas and Propagation (EuCAP 2011)*, Roma, Italy, April 2011. [Also available as TD(10)12018].

[MPC10] N. Moraitis, A. Panagopoulos, and P. Constantinou. Attenuation measurements and interference study for in-cabin wireless networks. Technical Report TD(10)10096, Athens, Greece, February 2010.

[MQO10] F. Mani, F. Quitin, and C. Oestges. Evaluation of diffuse scattering contribution in office scenario. Technical Report TD(10)10001, Athens, Greece, May 2010.

[NKK+] M. Narandžić, W. Kotterman, M. Käske, C. Schneider, G. Sommerkorn, A. Hong, and R. S. Thomä. On a characterization of large-scale parameters for distributed (multi-link) MIMO—the impact of power level differences. In *Proc. of European Conference on Antennas and Propagation (EuCAP 2010)*.

[NST08] M. Narandžić, C. Schneider, and R. Thomä. WINNER wideband MIMO system-level channel mode: comparison with other reference models. Technical Report TD(08)457, Wroclaw, Poland, February 2008.

[OC07] C. Oestges and B. Clerckx. *MIMO Wireless Communications*. Academic Press, San Diego, 2007.

[OCB+] C. Oestges, N. Czink, B. Bandemer, P. Castiglione, F. Kaltenberger, and A. Paulraj. Experimental characterization and modeling of outdoor-to-indoor and indoor-to-indoor distributed channels. *IEEE Trans. Veh. Technol.*, 59(5), 2253–2265, 2010

[OCB+10] C. Oestges, N. Czink, B. Bandemer, P. Castiglione, F. Kaltenberger, and A. I. Paulraj. Experimental characterization and modeling of outdoor-to-indoor and indoor-to-indoor distributed channels. *IEEE Trans. Veh. Technol.*, 59(5):2253–2265, 2010. [Also available as TD-09-722].

[OCC10] C. Oestges, P. Castiglione, and N. Czink. Empirical modeling of nomadic peer-to-peer networks in office environment. Technical Report TD-10-12002, Bologna, Italy, November 2010.

[Par01] J. D. Parsons. *The Mobile Radio Propagation Channel*, 2nd edition. Wiley, London, UK, 2001.

[PCRH02] C. Prettie, D. Cheung, L. Rusch, and M. Ho. Spatial correlation of UWB signals in a home environment. In *Conference on Ultra Wideband Systems and Technologies*, pages 65–69, May 2002.

[PHK+09] J. Poutanen, K. Haneda, V.-M. Kolmonen, J. Salmi, and P. Vainikainen. Analysis of correlated shadow fading in dual-link indoor radio wave propagation. Technical Report TD-09-910, Wien, Austria, September 2009.

[PHKU09] D. Porrat, A. Hayar, E. Kaminsky, and M. Uziel. Spatial dynamics in indoor channels. Technical Report TD-09-827, Valencia, Spain, May 2009.

[PHL+10a] J. Poutanen, K. Haneda, L. Liu, C. Oestges, F. Tufvesson, and P. Vainikainen. Inputs for the COST 2100 multi-link channel model. Technical Report TD(10)10067, Athens, Greece, February 2010.

[PHL+10b] J. Poutanen, K. Haneda, L. Liu, C. Oestges, F. Tufvesson, and P. Vainikainen. Parameterization of the COST 2100 MIMO channel model in indoor scenarios. Technical Report TD(10)10066, Athens, Greece, February 2010.

[PHS+09] J. Poutanen, K. Haneda, J. Salmi, V.-M. Kolmonen, F. Tufvesson, and P. Vainikainen. Analysis of radio wave scattering processes for indoor MIMO channel models. Technical Report TD-09-839, Valencia, Spain, May 2009.

[PHS+10] J. Poutanen, K. Haneda, J. Salmi, V.-M. Kolmonen, T. Hult, F. Tufvesson, and P. Vainikainen. Significance of common scatterers in multi-link radio wave propagation. In *Proc. 4th European Conf. on Antennas and Propagation (EuCAP 2010)*, Barcelona, Spain, April 2010. [Also available as TD(10)10069].

[PJJK] S. Priebe, M. Jacob, C. Jansen, and T. Kürner. Non-specular scattering modeling for THz propagation simulations. In *5th European Conference on Antennas and Propagation (EuCAP 2011)*.

[PJK] S. Priebe, M. Jacob, and T. Kürner. Polarization investigation of rough surface scattering for THz propagation modeling. In *5th European Conference on Antennas and Propagation (EuCAP 2011)*.

[PQD+10] A. Panahandeh, F. Quitin, J.-M. Dricot, F. Horlin, C. Oestges, and P. De Doncker. Multi-polarized channel statistics for outdoor-to-indoor and indoor-to-indoor channels. In *Proc. VTC 2010 Spring—IEEE 71st Vehicular Technology Conf.*, Taipei, Taiwan, May 2010. [Also available as TD(09)815].

[PSH+11] J. Poutanen, J. Salmi, K. Haneda, V.-M. Kolmonen, and P. Vainikainen. Angular and shadowing characteristics of dense multipath components in indoor radio channels. *IEEE Trans. Antennas Propagat.*, 59(1):245–253, 2011. doi:10.1109/TAP.2010.2090474. [Also available as TD(09)911].

[PTH+] J. Poutanen, F. Tufvesson, K. Haneda, V.-M. Kolmonen, and P. Vainikainen. Multi-link MIMO channel modeling using geometry-based approach. Technical Report.

[PTH+10a] J. Poutanen, F. Tufvesson, K. Haneda, L. Liu, C. Oestges, and P. Vainikainen. Adding dense multipath components to the COST 2100 MIMO channel model. In *Proc. Int. Symp. Antennas and Propagation (ISAP 2010)*, Macao, China, November 2010. [Also available as TD(10)11033].

[PTH+10b] J. Poutanen, F. Tufvesson, K. Haneda, L. Liu, C. Oestges, and P. Vainikainen. Adding dense multipath components to the COST 2100 MIMO channel model. In *11th COST 2100 MCM Meeting*, Aalborg, Denmark, June 2010. COST 2100 TD(10)11033.

[PTH+11] J. Poutanen, F. Tufvesson, K. Haneda, V.-M. Kolmonen, and P. Vainikainen. Multi-link MIMO channel modeling using geometry-based approach. *IEEE Trans. Ant. Prop.*, 2011. doi:10.1109/TAP.2011.2122296 [Also available as TD(10)11034].

[QBR+10] F. Quitin, F. Bellens, S. Van Roy, C. Oestges, F. Horlin, and P. De Doncker. Extracting specular-diffuse clusters from MIMO channel measurements. In *12th COST 2100 MCM Meeting*, Bologna, Italy, November 2010. COST 2100 TD(10)12065.

[QOHD] F. Quitin, C. Oestges, F. Horlin, and P. De Doncker. A polarized clustered channel model for indoor multiantenna systems at 3.6 GHz. *IEEE Trans. Veh. Technol.*, 59(8):3685–3693, 2010.

[QOHD09a] F. Quitin, C. Oestges, F. Horlin, and P. De Doncker. Small-scale variations of cross polar discrimination in Ricean fading channels. *Elect. Lett.*, 45(4):213–214, 2009. [Also available as TD(08)603].

[QOHD09b] F. Quitin, C. Oestges, F. Horlin, and P. De Doncker. Clustered channel characterization for indoor polarized MIMO systems. In *Proc. PIMRC 2009—IEEE 20th Int. Symp. on Pers., Indoor and Mobile Radio Commun.*, Tokyo, Japan, September 2009. [Also available as TD(09)818].

[QOHD09c] F. Quitin, C. Oestges, F. Horlin, and P. De Doncker. Polarimetric measurements for spatial wideband MIMO channels. In *Proc. VTC 2009 Spring—IEEE 69th Vehicular Technology Conf.*, Barcelona, Spain, April 2009.

[QOHD10a] F. Quitin, C. Oestges, F. Horlin, and P. De Doncker. Polarization measurements and modeling in indoor NLOS environments. *IEEE Trans. Wireless Commun.*, 9(1):21–25, 2010. [Also available as TD(08)602].

[QOHD10b] F. Quitin, C. Oestges, F. Horlin, and P. De Doncker. A spatio-temporal channel model for modeling the diffuse multipath component in indoor environments. In *Proc. 4th European Conference on Antennas and Propagation 2010 (EuCAP 2010)*, Barcelona, Spain, April 2010. p1847194. [Also available as COST 2100 TD(10)10003].

[Ric05] A. Richter. *On the Estimation of Radio Channel Parameters: Models and Algorithms (RIMAX)*. PhD thesis, TU-Ilmenau, Ilmenau, Germany, 2005. http://www.db-thueringen.de/servlets/DerivateServlet/Derivate-7407/ilm1-200500111.pdf.

[RO07] O. Renaudin and C. Oestges. Influence of measurement bandwidth on wideband mobile radio channel parameters. Technical Report TD(07)365, Duisburg, Germany, September 2007.

[SA08] K. Shoshan and O. Amrani. Polarized-MIMO capacity analysis in street-canyon using 3D ray-tracing model. In *Proc. IEEE 25th Convention of Electrical and Electronical Engineers in Israel, 2008*, Eilat, Israel, 2008. [Also available as TD(09)833].

[SA09] K. Shoshan and O. Amrani. Polarization diversity analysis in rural scenarios using 3D method-of-images model. In *Proc. European Conference on Antennas and Propagation (EuCap 2009)*, Berlin, Germany, 2009. [Also available as TD(09)833].

[Sai10] K. Saito. Time-variant MIMO channel modeling based on measurements in an urban environment. Technical Report TD(10)10050, Athens, Greece, February 2010.

[Sal07] S. Salous. Simultaneous measurements in multiple frequency bands. Technical Report TD(07)259, Lisbon, Portugal, February 2007.

[Sib01] A. Sibille. Efficient generation of spatially and frequency correlated random values for cognitive radio network simulators. *IEEE Trans. Veh. Technol.*, 59(3):1121–1128, 2001. [Also available as TD(09)949].

[SKI+11] K. Saito, T. Kitao, T. Imai, Y. Okano, and S. Miura. The modeling methods of time-correlated MIMO channels using the particle filter. Technical Report TD(10)11086, Aalborg, Denmark, June 2011.

[SPF10] G. Steinbock, T. Pedersen, and B. H. Fleury. Model for path loss of in room reverberant channels. In *Proc. Globecom 2010—IEEE Global Telecommunications Conf.*, 2010.

[SPH+10] J. Salmi, J. Poutanen, K. Haneda, A. Richter, V.-M. Kolmonen, P. Vainikainen, and A. F. Molisch. Incorporating diffuse scattering in geometry-based stochastic MIMO channel models. In *Proc. 4th European Conference on Antennas and Propagation 2010 (EuCAP 2010)*, Barcelona, Spain, April 2010. C21P1-3. [Also available as COST 2100 TD(10)11047].

[SRK09] J. Salmi, A. Richter, and V. Koivunen. Detection and tracking of MIMO propagation path parameters using state-space approach. *IEEE Trans. Signal Processing*, 57(4):1538–1550, 2009.

[ST10] C. Schneider and R. Thomä. Analysis of local quasi-stationarity regions in an urban macrocell scenario. Technical Report TD(10)10081, Athens, Greece, February 2010. To be presented at VTC Spring 2010.

[Ste01] M. Steinbauer. *The radio propagation channel, a non-directional, directional, and double-directional point-of-view*. PhD thesis, Technische Universität Wien, November 2001.

[TL]	S. Torrico and R. Lang. Millimeter wave scattering and attenuation prediction from a volume with random located lossy-dielectric discs and cylinders: forward scattering approximation. Technical Report.
[TL08]	S. Torrico and R. Lang. Diffraction and scattering effects of trees and houses on path loss in a vegetated residential environment. Technical Report TD(08)507, Trondheim, Norway, June 2008.
[VDEMO10]	E. Vitucci, V. Degli-Esposti, F. Mani, and C. Oestges. A study on polarimetric properties of scattering from building walls. In *Proc. VTC 2010 Fall—IEEE 72nd Vehicular Technology Conf.*, Ottawa, Canada, September 2010. [Also available as TD(10)11020].
[VKDEV08]	E. Vitucci, V.-M. Kolmonen, V. Degli-Esposti, and P. Vainikainen. Analysis of radio propagation in co- and cross-polarization in urban environment. In *Proc. 10th International Symposium on Spread Spectrum Techniques and Applications (ISSSTA 2008)*, Bologna, Italy, August 2008. [Also available as TD(08)666].
[VKDEV09]	E. Vitucci, V.-M. Kolmonen, V. Degli-Esposti, and P. Vainikainen. Analysis of X-pol propagation in microcellular environment. In *Proc. 11th International Conference on Electromagnetics in Advanced Applications (ICEAA 09)*, Torino, Italy, September 2009. [Also available as TD(08)666].
[VQAR07]	P. Vieira, M. P. Queluz, and A. A. Rodrigues. Clustering of scatterers in mobile radio channels—an approach for macro-cell environment over irregular terrain. Technical Report TD-07-219, Lisbon, Portugal, February 2007.
[VRZ08]	A. Valcarce, G. De La Roche, and J. Zhang. On the use of a lower frequency in finite-difference simulations for urban radio coverage. In *69th IEEE Vehicular Technology Conference (VTC Spring 2008)*, pages 270–274, Barcelona, Spain, April 2008. [Also available as TD(08)408].
[ZK10]	R. Zentner and A. Katalinic. Dynamics of multipath variations in urban environment. In *Proc. 3rd European Wireless Technology Conference, EuWiT 2010*, Paris, France, September 2010. [Also available as TD(10)12071].
[ZTE10]	M. Zhu, F. Tufvesson, and G. Eriksson. Validation of 300 MHz MIMO measurements in suburban environments for the COST 2100 MIMO channel model. Technical Report TD(10)12048, Bologna, Italy, November 2010.
[ZTM10]	M. Zhu, F. Tufvesson, and J. Medbo. Correlation properties of large scale parameters from multi-site macro cell measurements at 2.66 GHz. Technical Report TD-10-12049, Bologna, Italy, November 2010.
[ZTW[+]10]	M. Zhu, F. Tufvesson, S. Wyne, G. Eriksson, and A. Molisch. Parametrization of 300 MHz MIMO measurements in suburban environments for the COST 2100 MIMO channel model. Technical Report TD(10)11071, Aalborg, Denmark, June 2010.

Chapter 4
Assessment and Modelling of Terminal Antenna Systems

**Chapter Editor Buon Kiong Lau, Chapter Editor Alain Sibille,
Vanja Plicanic, Ruiyuan Tian, and Tim Brown**

The pervasive use and insatiable demand for high-performance wireless devices globally are driving the development of new communication technologies and standards. In this chapter, we present advancements in the area of terminal antenna systems and describe how they play a major role in enabling strict performance requirements to be met in future systems. As opposed to conventional systems that do not require explicit considerations of the terminal antenna beyond the satisfaction of some predefined design criteria in impedance bandwidth, radiation pattern and efficiency, multiple antenna systems in upcoming and future wireless networks must deliver significantly higher performance goals. To accomplish this, they must be able to function reliably when presented with different operating conditions, which include:

- Propagation environments, e.g., indoor, urban macrocell/microcell, technical/industrial/natural environments for sensor networks, etc.
- User influence, especially with the hand and head, as well as other near field disturbances.
- Time variability conditions, e.g., user movement and fixed wireless application versus high-speed train link.

From the point of view of delivering high data rates over extensive coverage areas, the Multiple-Input Multiple-Output (MIMO) principle has unique advantages. Most importantly, instead of suffering from the strong multipath character of the propagation in typical terminal usage scenarios, it exploits the multipath propagation to achieve data rates which grow linearly with the number of antennas, at no additional expense in frequency spectrum. However, despite extensive research in multiple antenna systems over the past decade, quite little in proportion has been

---

B.K. Lau (✉)
Dept. of Electrical and Information Technology, LTH, Lund University, Lund, Sweden

A. Sibille (✉)
Département Communications & Électronique, TELECOM ParisTech, Paris, France

done to assess the performance of realistic multiple antenna terminals which takes into account the various imperfections and perturbations the users will experience in true usage conditions.

In this context, Chap. 4 addresses several important issues regarding the performance and the assessment of antennas on terminals. Section 4.2 first examines the impact of practical constraints imposed on multiple antenna systems for terminals, namely the mutual coupling among closely spaced terminal antennas, the dynamics of the terminal device's MIMO eigenmodes and user interaction and modelling. Then, Sect. 4.3 expands on the subject of performance characterisation and measurements by focusing on several promising techniques which can greatly simplify efforts to characterise terminal antenna systems. For example, the composite channel approach, with or without user interaction, makes it possible for each propagation channel of interest to be measured only once, so that the channel characteristics can be combined with antenna characteristics through postprocessing. Apart from sheer convenience, the approach ensures perfect repeatability of the same channel conditions for testing antenna systems. Section 4.4 subsequently investigates the use of different techniques to improve the performance of single and multiple antennas. In particular, novel antenna structures (e.g., dielectric resonator antenna and Intelligent Quadrifilar Helix Antenna (IQHA)) and compensation techniques such as dielectric loading, decoupling stub, and neutralisation line have been found to reduce or circumvent the problem of mutual coupling, despite the compact size of the multiple antenna systems. Finally, Sect. 4.5 proposes to extend the commonly used statistical approach in channel modelling to also include antennas and user interaction. This original approach is prompted by the need for both simple and representative models of the port-to-port channel, which are able to cover the variety of terminals/propagation combinations in a parameterised manner. It is based either on advanced and generic models of the antenna electromagnetic behaviour or on the extraction of statistical models from databases of simulated or measured terminals combined with local propagation models.

4.1 Practical Considerations for Compact Terminal Antenna Systems

As an important feature in the Fourth Generation (4G) and the Wireless Local Area Network (WLAN) communication systems, multiple antenna technologies are expected to leverage high data rates, good coverage and high-quality link connections. The potential that can be harvested is conditioned by the state of the overall communication channel, which extends beyond antenna design and common antenna parameters to comprise near field user interaction and propagation environment. Hence, in comparison to single antenna systems, the practical implementation of multiple antenna systems needs to consider not only the constraints with terminal form and size, and available antenna volume and location, but also the extended user terminal utilisation in arbitrary propagation environment. Within the scope of

COST2100 some of the considerations have been highlighted and investigated. This section summarises the resulting contributions by dedicating Sect. 4.1.1 to a discussion on the mutual coupling as a consequence of the terminal's compactness, and its influence on antenna characteristics and overall diversity and MIMO performance. This is followed by Sect. 4.1.2 where the interaction between antenna characteristics and dynamic propagation channel is investigated. Finally, Sect. 4.1.3 summarises the extensive studies performed on assessment and modelling of the user interaction.

4.1.1 Compactness of Terminal Antenna Systems

The diversity and MIMO gains enabled by co-band multiple antenna systems depend on the extraction of all available degrees of freedom through high and equal efficiencies and low correlation at the antenna ports [PL09, PZL10, TPLY10a]. The lack of adequate spacing between the antennas in the terminals induces mutual coupling between the antenna ports, causing negative effects on the antenna port efficiencies and a significant impact on the correlation. Moreover the demand for broad frequency coverage with a compact antenna design can further degrade antenna efficiencies.

4.1.1.1 Mutual Coupling in Co-band Multiple Antenna Systems

Due to the high demand on compactness for mobility and handy usage, the constraints on the volume and the placement of the antennas are especially pronounced for small terminals such as mobile phones and mobile broadband dongles. The receive diversity performance of two multi-band antennas (WCDMA850/1800/2100) is studied in [AYL07]. One antenna is Planar Inverted F-Antenna (PIFA)-based, and the other monopole-based. The antennas are implemented on a common ground plane with the size of 100×40 mm^2, which is typical for a candybar-shape mobile phone. The simulation results reveal high mutual coupling at the low-frequency band (850 MHz), which is in part due to the short distance (0.24λ) between the antenna feeds. Correspondingly, an envelope correlation of 0.8 is obtained. At higher-frequency bands (1800/2100 MHz) with an adequate antenna feed spacing of equal to or more than 0.5λ, both the mutual coupling and the correlation are very low (i.e., correlation < 0.1). The diversity performance, assuming that the total efficiency performances of the antennas at all frequencies are balanced, differs on average by 3 dB between the low- and high-frequency bands. In this case, the mutual coupling decreases the correlation and the diversity performance. However, previous studies on dipoles have shown that high mutual coupling does not necessarily indicate high correlation for all cases of small antenna spacing. In some cases, the mutual coupling reshapes the radiation patterns of the antennas, which enables lower correlation in the dipole array [WJ04a].

Fig. 4.1 Dual multiband antenna system for mobile phone [PLY08] (©2008 IEEE, reproduced with permission)

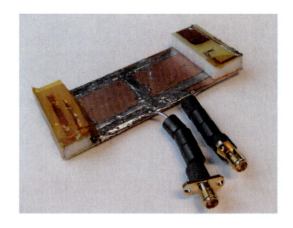

A multiple antenna system based on the one presented in [AYL07] (see Fig. 4.1) is investigated in [PLY08] and [SLDY09]. The diversity and the 2×2 MIMO capacity performance is shown to be degraded due to the imbalance of the measured antenna efficiencies despite the low mutual coupling and low measured correlation at the high frequencies.

The effect of the mutual coupling on the diversity and the MIMO performance is so far discussed for the case of Three-Dimensional (3D) uniform Angular Power Spectrum (APS) propagation. In [PLPV08], more realistic scenarios are considered by modelling mutual coupling effects on the characteristics of the MIMO sub-channels for a Line-Of-Sight (LOS) and an Non-Line-Of-Sight (NLOS) indoor propagation environments. The MIMO channel is in this contribution modelled using a 3D ray-tracing technique with assumption of no coupling between the antennas at the Receiver (Rx) and Transmitter (Tx) ends. The case with the mutual coupling present is obtained by combining the channel transfer matrix **H** with the coupling matrices of two thin wire dipoles, simulated at 5 GHz with an antenna spacing of 0.1λ at both the Rx and Tx ends. As expected, mutual coupling is shown to affect the MIMO channel characteristics: for the LOS case, the predominant signal path is influenced, whereas for the case of NLOS, all sub-channels are affected. Mutual coupling influences antenna characteristics, which, together with the propagation channel, determine the number of available orthogonal sub-channels and thus the MIMO performance. The investigation on the impact of the mutual coupling on the MIMO performance in simulated and measured propagation channels in [MBJ08] shows a case where mutual coupling has a negative impact on the antenna efficiencies but not on the correlation for a small distance between the antennas. This is illustrated in Fig. 4.2 with correlation results for two Rx half-wavelength dipoles at 5.25 GHz with antenna separations of approximately 0.25λ (1.5 cm) and 0.5λ (3 cm). Both separation distances give the same correlation, which is validated by comparing the synthetic combinations of an antenna coupling model and ray-estimated propagation channels against direct propagation channel measurements with real antennas. This is in agreement with results presented in [WJ04a]. Despite the same correlation performance, the MIMO capacity performance is lower

4 Assessment and Modelling of Terminal Antenna Systems

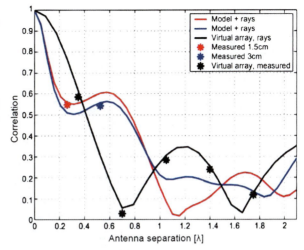

Fig. 4.2 Simulated and measured correlation between two half-wavelength dipoles for different separation distances [MBJ08]. The *solid curves* represent correlation obtained from combining an antenna coupling model and a ray traced propagation model. The *dots* represent measured correlation in a realistic indoor environment

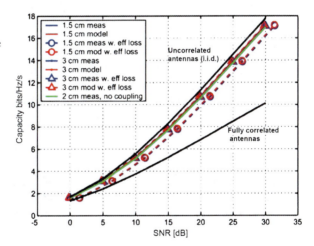

Fig. 4.3 Simulated and measured average Shannon MIMO capacity performance for 2 × 2 antenna set-up [MBJ08]

for the separation of 0.25λ, as shown in Fig. 4.3. Hence, the mutual coupling as a consequence of bringing the antennas closer appears to cause 1.5 dB of efficiency degradation.

4.1.1.2 Impedance Matching for Control of Mutual Coupling

Recently, the research on mutual coupling has been expanded to consider coupled and uncoupled matching networks at the antenna ports. These networks have been proved to enable a degree of control of mutual coupling and thus the radiation characteristics of the antenna arrays [WJ04a, WJ04b, AL06]. The results in [RD07] illustrate the positive effects on correlation performance at 900 MHz between several pairs of thin dipole elements in a nine-element Uniform Circular Array (UCA)

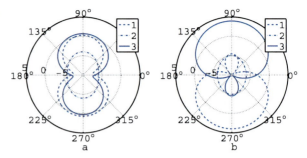

Fig. 4.4 Antenna patterns for the three-dipole ULA under uniform APS with (**a**) balanced and (**b**) unbalanced matching for maximised receive power [TL08b] (©2008 IEEE, reproduced with permission)

when loading the antenna elements with 73 Ω, as compared to the open-circuited case. Two-Dimensional (2D) Laplacian APS is assumed. The matching network reshapes the dipole radiation patterns from omnidirectional in the case of open-circuit to directive with no significant side-lobes, thus enabling lower element correlation. Furthermore, loading the antenna elements at frequencies 1600–1800 MHz attenuates the correlation oscillation observed for open-circuit case. Reshaping the radiation patterns using an impedance matching network for compact Uniform Linear Arrays (ULA), as studied in [LA07b, LA07a], has proved to be very valuable for the one- and two-signal scenarios of Direction of Arrival (DoA) estimation. In particular, the angle diversity enabled by conjugate matching at the ULA ports facilitates a dramatic improvement in the accuracy of DoA estimates for small antenna spacings, as compared to other matching conditions such as the characteristic impedance match or the self-impedance match, which cannot offer angle diversity to the same extent.

In [TL08b], a simulation study is performed on received power and MIMO capacity performance optimisation by means of uncoupled balanced and unbalanced matching networks. Three configurations of receive half-wavelength dipoles, two ULAs (of two and three antenna elements) and one Uniform Triangular Array (UTA)s, with adjacent element spacing of 0.1λ are studied. Propagation channels with 2D uniform and 2D Laplacian APS are applied. In the balanced case, the antenna ports are terminated with the same load impedances, while in the unbalanced case the termination at each of the ports is arbitrary. For all three antenna configurations, the received power in all the propagation environments relative to that of single dipole with conjugate matching is improved with the uncoupled matching, both balanced and unbalanced. Two of the three antenna configurations, the two-element ULA and the three-element UTA with the same coupling characteristics for all the ports, experience better adaptation to the propagation environment with unbalanced matching. The unbalanced matching is shown to be particularly useful for the third antenna configuration comprising a three-element ULA in which the centre element is subjected to the strongest mutual coupling. This is because the unbalanced matching can partially decouple the outer two elements as compared to the balanced matching case, where the lack of degree of freedom causes the optimisation of the centre element to limit the effectiveness in optimising the whole array (see Fig. 4.4). In this context, the unbalanced case penalises the centre antenna element in order to improve the overall capacity performance.

Fig. 4.5 Realistic and reference antenna configurations for evaluation of eigen coherence time [BHW07] (©2007 EurAAP, reproduced with permission)

4.1.2 Effect of Terminal Antenna System on Dynamics of MIMO Eigenmodes

Antenna radiation characteristics (i.e., angular and polarisation patterns and efficiencies) at the Rx and Tx ends interact with the propagation channel to determine the overall communication channel performance. Due to the terminal mobility, the propagation channels in realistic communication scenarios are rarely static. To assure sufficiently strong and orthogonal sub-channels, the changes in the spatial propagation scenarios need to be tracked and preferably adapted to some extent. In [BHW07], two new figures of merits are proposed to gain sufficient insight into the dynamics of the communication channel eigenmodes: eigen coherence time and eigen coherence bandwidth. Eigen coherence time is defined as the delay time in which the auto covariance of the eigenvectors in the communication channel goes below the value of 0.7. The coherence bandwidth is the corresponding measure in the frequency domain. Through practical evaluations of 4×4 MIMO systems under urban propagation channel conditions, the eigenmode dynamics for three antenna configurations at the mobile Rx end of a communication system are evaluated at 2 GHz (see Fig. 4.5). Two of the Rx terminals (a laptop and a Personal Digital Assistant (PDA)) are examples of realistic user equipment with antenna designs appropriate for the available space. The third Rx terminal, which also serves as a reference, comprises four ± 45 deg oriented dipoles in two orthogonal planes. The results from the measurements reveal that walking conditions give shorter coherence times and higher dynamics in the eigenmodes, in comparison to the stationary conditions (see Fig. 4.6). In particular, the reference antennas stand out with the shortest coherence time. The high dynamics in the communication channel are enabled through unobstructed omnidirectional behaviour of the dipole antennas, when compared to the laptop and PDA. The later suffer from less polarisation and angular diversity due to the terminal design and user impact during the measurements. Thus, the antenna system with more omnidirectional patterns experience more channel variations and consequently gives a shorter coherence time. Hence, by "limiting" the beamwidth or directionality of the antennas, longer coherence times can be achieved, which is beneficial from the channel feedback point of view. However, as highlighted in [Bro07], the channel feedback or eigenstate tracking is only improved at the expense of limiting the scattering and increasing the branch power ratio at the receiving ports.

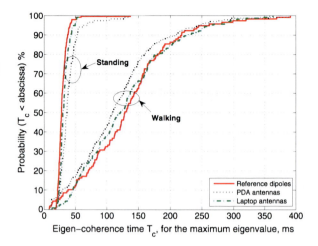

Fig. 4.6 Eigen coherence time for walking and stationary scenarios [BHW07] (©2007 EurAAP, reproduced with permission)

Thus, limiting the channel variations can only be advantageous where the scattering richness of the channel, as well as the efficiencies and correlation properties of the antennas, can together assure orthogonality in the communication channel eigenmodes. The study in [Bro07] is based on an indoor 2×2 MIMO system measurements at 2.4 GHz with omnidirectional dipoles and directional feed horn antennas used at the Rx side.

4.1.3 User as a Part of Communication Channel

In addition to the dynamics in the propagation channels, mobile terminals such as mobile phones, laptops and broadband dongles experience dynamics in the antenna characteristics due to the unavoidable and variable near field interaction with users. Therefore, terminal design has to consider the user interaction to ensure that the overall communication channel can yield good system performance.

4.1.3.1 User in the Near Field of Antenna

The user induces frequency detuning when he or she is close enough to the antenna to become a part of the radiating structure. In [WGB10], this effect is illustrated for one of four MIMO slot antennas in a smart phone size prototype. However, efficiency loss is the dominant consequence of user interaction, due to the significant absorption of energy by the user body. Extensive studies are carried out to gain understanding on the effect of the user on the antenna performance, especially for the case of user terminals of mobile phone size, since they are utilised in a large variety of user scenarios. A systematic investigation of absorption and mismatch losses for a number of antenna configurations and user cases by means of Finite Difference in Time Domain (FDTD) simulation is reported in [PFK+09a, Pel08, PFK+10, PFKP09b]. In [PFK+09a] and [Pel08], the

Fig. 4.7 Example of (**a**) soft grip, (**b**) firm hand grip and (**c**) another firm hand grip with SAM user scenarios [PFK+10] (©2010 IEEE, reproduced with permission)

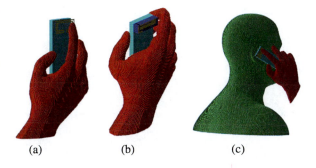

absorption and the mismatch losses are studied for a coarse model of a mobile terminal with the size of $20 \times 50 \times 110$ mm^3. Two antenna cases are presented: a top-placed single PIFA and an external quarter-wavelength monopole, both resonating at 1800 MHz. Hand models with simplified homogeneous and Visual Human Project (VHP) [Ack98] homogeneous and heterogeneous characteristics are used, and a total of six user hand-held scenarios are considered. The human torso models from SPEAG [SS10] and VHP are also studied in combination with the hand models. The results reveal the hand to be more dominant in the energy absorption than the phantom torso, causing the highest absorption losses (8–11 dB) for different levels of dielectric homogeneity. The highest loss is observed when the hand is located closest to the antenna. Regardless of the hand model, the PIFA experiences on average more absorption and mismatch losses than the monopole, due to its location and its larger radiating structure. The study on the location of the PIFA along the ground plane in the simulation model confirms the insight that a closer distance between the antenna and the human tissue will induce a higher level of energy absorption [Pel08].

In [PFK+10], the study is extended to more realistic PIFA and monopole based antenna designs with dual band coverage (GSM900 and DCS1800) suitable for internal implementation in mobile terminals. Beside the free space used as a reference scenario, five user scenarios are also evaluated. All five user scenarios are derived from a rigorous and extensive study of the hand grip positions for mobile users documented in [PFK+09a]. Soft and firm hand grips of the simulated handset model and firm hand grip with Specific Anthropomorphic Mannequin (SAM) are shown in Fig. 4.7.

Absorption and mismatch loss results for the different antenna and user configurations in Fig. 4.8 support previous conclusions on hand grip position, especially for the firm grip case, being more influential on the antenna performance than the head or the torso. Furthermore, at the lower-frequency band, the losses are higher than at higher frequency band. Another major part of the study comprises investigations on influence of the finger location along the sides of the simulated terminal model, the influence of the location of the index finger and the influence of the distance variation between the terminal and user palm. A variation of up to 1 dB is observed when the location of the fingers is changed along the whole length of the terminal. The change of location of the index finger across the antenna region induces a variation

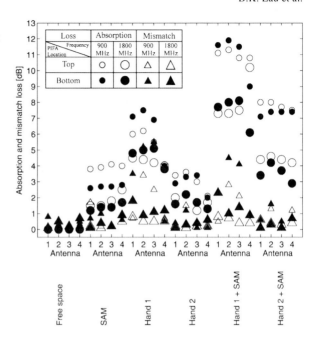

Fig. 4.8 Absorption and mismatch losses for different antenna and user scenarios [PFK+10] (©2010 IEEE, reproduced with permission)

in losses of up to 3 dB. As expected, the larger the distance between the terminal and the palm, the smaller is the absorption loss. Paper [PFKP09b] complements the described study with absorption and mismatch results for overlapped and interlaced hand grip styles, which comprise two hands with fingers either overlapped or interlaced. Again, only user tissue close to the region of the antenna structure gives a significant contribution to the total amount of absorption loss. A study of the near field of the antenna in the proximity of a user in [BCP08a] further substantiates these findings.

4.1.3.2 Impact of User Presence on MIMO Performance

The absorption and mismatch losses induced by the presence of a user alter the angular and polarisation radiation characteristics of an antenna and decrease its efficiency. For multiple antenna systems, the alteration of the polarisation and angular radiation properties directly influences their ability to extract a sufficient number of uncorrelated multipaths in a propagation channel. In Fig. 4.9, an example of the distortion of the radiation pattern is shown for a dual antenna system for a terminal model of size 100×40 mm^2 at 881.5 MHz (see Fig. 4.1). The top antenna in the simulated and measured terminal model is a PIFA-based antenna for coverage of the Rx frequencies of the WCDMA850/1800/2100 bands. The monopole-based antenna at the bottom is also a multiband antenna covering all the frequencies in the mentioned bands. The radiation patterns of both antennas are distorted when the terminal is held in the hand. The distortion enables a level of orthogonality in the two antenna patterns, thus decreasing correlation for the dual antenna system in the

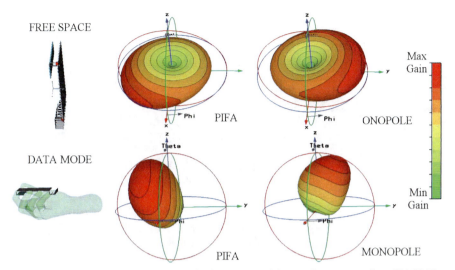

Fig. 4.9 Simulated 3D radiation patterns for free space and data mode user scenario at 881.5 MHz [PLY08] (©2008 IEEE, reproduced with permission)

case of uniform propagation [PLDY09]. At 881.5 MHz, the antennas in the terminal are closely spaced, since the distance between the feed points is 0.24λ, thus half of what is conventionally recommended to avoid significant mutual coupling. For a terminal model of size 111×59 mm^2 at 776 MHz, decreased correlation was also observed from measurements performed for realistic hand-held user scenarios in an indoor propagation environment, relative to the no user case [YCNP10].

For antennas with low mutual coupling, the correlation in a *uniform* propagation channel, which is typically low to begin with, may not necessarily be affected significantly by the altered radiation patterns. However, in an arbitrary realistic propagation channel, which is generally non-uniform, the altered polarisation and angular radiation properties may have an influence on the overall performance if the antenna characteristics do not match those of the propagation environment. The effects of different angular and polarisation properties on MIMO capacity in real environments are illustrated for the free space case in [PLA11]. Three simple antenna topologies in a smart phone size prototype were investigated at the Rx side in a 2×2 MIMO measurement campaign in a urban macrocell environment. The results show differences in capacity performances, due to different angular and polarisation properties of the antennas, that are up to 15% between the best and worst performing topologies at 20-m sections of a 2-km route. If the average capacity performance over 1.1-m sections (corresponding to the removal of small scale fading) is considered, the peak differences along the route are up to 19% for the noise-limited scenario and 28% for the interference-limited scenario. Hence, altered radiation patterns due to user influence may further contribute to this difference.

An extended measurement study on the previously mentioned multiple antenna system at 881.5 MHz is documented in [SLDY09]. The Mean Effective Gain (MEG) performances for the two antennas in the terminal and for a reference case of a single

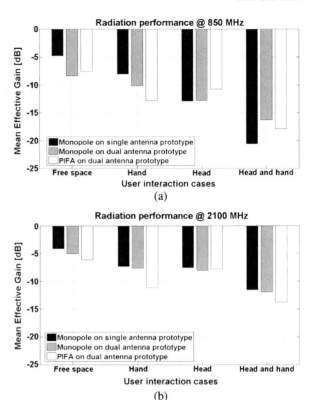

Fig. 4.10 Measured radiation performance of the antennas in a single and a dual terminal for different interaction scenarios and two frequencies (**a**) 850 MHz and (**b**) 2100 MHz in uniform 3D APS [SLDY09] (©2009 IEEE, reproduced with permission)

Table 4.1 Averaged envelope correlation for two frequencies and different user interaction scenarios in uniform 3D APS [SLDY09]

Frequency band [MHz]	Average envelope correlation			
	FREE SPACE	HAND	HEAD	HEAD&HAND
850	0.51	0.33	0.012	0.39
2100	0.006	0.007	0.004	0.05

antenna in the same size terminal are showing different effects of the user presence for the low (< 1000 MHz) and high (> 1000 MHz) frequency bands, see Fig. 4.10. The more significant effect on the antenna performance at low frequency is due to the user interacting with the antenna and the ground plane. At the low frequencies, the ground plane of the terminal is very small compared to the wavelength, and thus it becomes part of the radiating structure, with the antenna acting as the excitation point [VOKK02]. Hence, radiation properties are more easily and significantly affected by the user in comparison to antenna systems with less radiating surface in close proximity to the user.

The influence of user on the MEG and correlation as seen in Fig. 4.10 and Table 4.1 are directly translated to capacity performance for the studied Single-Input

Fig. 4.11 Ergodic Shannon capacity performance for a single and a dual terminal for different interaction scenarios and two frequencies (**a**) 850 MHz and (**b**) 2100 MHz in uniform 3D APS [SLDY09] (©2009 IEEE, reproduced with permission)

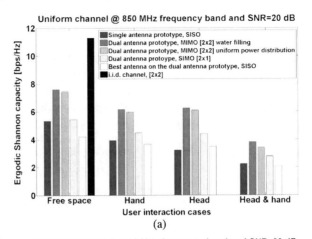

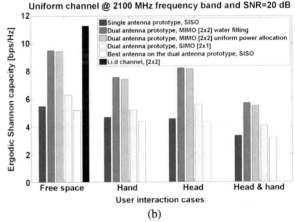

Multiple-Output (SIMO) and MIMO setups. The SIMO and MIMO capacities are lower at the low-frequency band (see Fig. 4.11). The user influence in the three interaction scenarios has different impact on the capacity depending on the alternation of the correlation, MEG and gain balance between the antenna branches in the dual antenna system. As an example, at the frequency of 850 MHz, despite the decrease in correlation by almost half, the SIMO and MIMO capacities for the user hand interaction scenario show a decrease of 20% as compared to free space. This is due to significant individual decrease in MEGs (20% and 40%) and the increase in the difference between the MEGs of the two antenna branches. At 2100 MHz, the user head only case shows the best performance, due to the lowest change in MEGs (40% and 20%) and a good balance between the MEGs of the antenna ports. The correlation at the high-frequency band is low for all the user interaction cases. It should be noted that the performance evaluation is carried out under the assumption of uniform 3D APS channel.

The effect of user on capacity performance in more realistic indoor environments is presented in [HDT10, NYB+10]. The studies conclude that, *on average*, the capacity decrease that results from the user being in the near field of the antenna is mainly due to degraded Signal-to-Noise-Ratio (SNR) performance. In [HDT10], switching to the best antenna pair in a terminal mock-up with four PIFAs, based on the strongest SNR, gives very similar performance to switching based on the highest capacity.

Like the propagation channel, the user interaction is rarely static, which can further impact on the overall communication channel performance. The extensive studies in [Bro07] and [BEP09] deal with the effects of user on (i) the antenna characteristics and thus the eigenmode dynamics and (ii) the interference in a dynamic propagation channel. The study comprises 4×4 MIMO measurements in an indoor Semi-Line-Of-Sight (SLOS) scenario with two configurations of four-printed-monopole antenna systems on their respective ground planes of the size of a mobile phone. A number of user interactions are investigated for the talk and data mode user interaction cases, while the users are moving down a corridor. It is found that a significant impact on the eigenmode dynamics is caused by random changes in the phases of the radiation properties at the antenna ports, in combination with random gain differences at the ports [Bro07]. The observed randomness caused by the user and the dynamics in the propagation channel highlights a formidable challenge in designing effective multiple antenna systems, suggesting that it may be worth putting more effort into the design of smart signal processing techniques, rather than focusing on antenna design alone.

The effects of the variations in user interactions are also identified in the interference study [BEP09]. The Signal-to-Interference Ratio (SIR) is defined as the signal power when transmitting with the highest diversity order in a channel between Base Station (BS) and main mobile terminal, relative to the channel between another BS and an interfering mobile terminal. There is a variation of 10 dB between the worst and the best SIR for different users in the talk mode. Due to more consistent usage behaviour, the difference between the best and worst SIR for the data mode is observed to be only 5 dB. The SIR performance is determined by the interaction of the antenna characteristic with the Angle of Arrival (AOA) of the signal and gain imbalance between the antenna ports, which are both affected by the user. The study shows that the higher the gain imbalance is between the Rx antennas at the terminal, the harder it is to mitigate the interference from the interfering mobile terminal. Moreover, the SIR and thus the overall capacity performance of a terminal are not only determined by the terminal's interaction with the user, but also by the user interaction that the interfering terminal experiences. A smaller gain imbalance in the antennas of the interfering mobile terminal leads to a higher interference level, and vice versa.

4.1.3.3 Modelling of User Influence

The influence of user on antenna characteristics, and thus the performance of the overall communication channel, has just been presented. However, it is not yet obvious as to how one should account for user influence in the design of multiple antenna systems. The variations in the utilisation of mobile terminals, especially mobile phones, point to the importance of modelling user influence. Today, for single antenna performance of mobile phones, the requirements are mainly based on voice application, thus the required design and testing scenario only involves the SAM head [Cellular Telecommunications & Internet Association [CTIA09]]. Since the influence of the hand is significant, as confirmed in previous discussions, the head only case is not sufficient. Moreover, the focus on packet-oriented communications in current and upcoming systems points towards the need for more design and testing scenarios. To enable these user scenarios, a realistic hand phantom is required. The CTIA has initiated the development of such a hand phantom, with focus on reproducibility, anatomical accuracy, true human-like impact on mobile terminals Over-The-Air (OTA) performance, cost and overall practical implementation [Mol08]. In [Mol08], the rationale behind the development of the hand phantom is explained, comprising hand dimension studies, RF dielectric properties, mobile terminal types, user cases and grips. The final results from the completed development work are presented in [Mol09]. Four phantom grips are defined: mono-block (bar/slider phone), fold (clamshell/jack-knife phone) and PDA (phones with widths of 56–72 mm) to be used in the talk mode with the SAM head, and the narrow data grip for the data mode case where browsing, navigation and texting are predominant.

The hand phantom has homogeneous properties as listed in [Mol09]. A comparison between homogeneous and heterogeneous hands from VHP, performed through simulations in [PFK+09a], identifies the homogeneous hand as up to 3 dB more energy absorbent. A difference was also observed from a study on the near field of the homogeneous and heterogeneous head tissues [BCP08b]. A scaling of the dielectric properties may enable a homogeneous composition that mimics the heterogeneous one, with respect to the amount of losses [PFK+09a]. However, due to the lack of complexity in simulations and good repeatability in measurements, the homogeneous hand is chosen as a good representation of the human hand. Furthermore, variations of the dielectric properties of up to 50% in a homogeneous hand are shown not to alter the absorption and mismatch losses by more than 1 dB for the firm and soft hand grip cases, with and without presence of the SAM [PFK+09a] head.

Data-intensive communications with applications such as gaming are indicating an additional need for a generic two-hand grip position for the design and testing of mobile terminals. The CTIA phantom hand is designed to only represent right-hand case, but it can be extended to the left-hand case if the two-hand case is to be developed. In [Mol09], it is suggested to use mirror image versions of the right-hand grips. Thus, considering dual antennas that are located at each end of the mobile terminal, the narrow data grip modes for left and right hand may be sufficient to represent an average two-hand grip scenario. However, there is always concern regarding increased cost and time consumption that the extended number of user cases may involve, which will be taken into account when deciding on extending the requirements for OTA performance testing.

4.2 Performance Characterisation and Measurement Techniques

This section is dedicated to the discussion of performance characterisation and measurement techniques of terminal antenna systems. In the first half of the section, the focus is on MIMO systems. In Sect. 4.2.1, the performance of different reference dipole array configurations are evaluated using measured propagation channels, based on a composite channel approach that conveniently combines antenna radiation patterns and measured propagation channel characteristics through post-processing. Due to the significance of user effects for the performance of user terminals, the composite channel approach has also been extended to include the user as part of the antenna to form the so-called *super-antenna*. In some practical situations, only the magnitude of the antenna radiation patterns is available for multiple antenna systems. The performance impact of phaseless antenna patterns on MIMO systems, together with several remedies, is summarised in Sect. 4.2.2. The discussion then continues into Ultra-WideBand (UWB) antennas in Sect. 4.2.3, where methods of modelling mutual coupling in compact UWB arrays are investigated, with the explicit purpose of including their effect into statistical descriptions of the antenna and propagation channel models. See also Sect. 4.4.

The second half of this section discusses some advanced measurement techniques. Section 4.2.4 considers the use of optical fibre for RF measurements to mitigate the negative influence of conventional cables. In Sect. 4.2.5, a spheroidal coupler is used for antenna measurements, which provides a compact, low-cost Total Radiated Power (TRP) measurement system with high sensitivity and speed. Section 4.2.6 introduces a 3-D radiation pattern measurement system for 60-GHz antennas.

4.2.1 Reference and Super-MIMO Antennas

In MIMO systems where multiple antennas are used at both the Tx and Rx ends of a wireless channel, the channel capacity can in principle increase linearly with the number of antennas under ideal rich scattering environments [Win87, JW04]. In theory, ideal multiple antenna elements are often assumed to be arranged in a co-located or co-linear fashion. This aspect is addressed in [CHV09], where the performance of the theoretical reference MIMO antennas is evaluated in realistic propagation environments, using a MEasurement Based Antenna Testbed (MEBAT) [Suv06]. Using a composite channel approach [SVS+06], the far-field radiation patterns of the MIMO antennas are combined with the propagation channel estimated from double-directional polarimetric measurement campaigns. However, users can have strong impacts on the interaction between the antennas and the propagation channel. In [HMDM07, HMM08, HMM+10], the user is considered as an integral part of the antenna, forming a super-antenna concept. The composite channel approach is further extended in order to include user effects by combining the channel with the radiation patterns of the super-antenna.

4 Assessment and Modelling of Terminal Antenna Systems

Table 4.2 List of reference MIMO antenna arrays. $\hat{x}$-, $\hat{y}$-, and $\hat{y}$-Pol denote different polarisations. The inter-element separation in ULA is $d = \lambda/2$

Config.	Rx dipole array	Tx patch ULA
1. 3×3	Co-located electric tripole	$\hat{z}$-Pol
2. 3×3	Co-located electric tripole	$\hat{y}$-Pol
3. 3×3	Co-located magnetic tripole	$\hat{z}$-Pol
4. 3×3	Co-located magnetic tripole	$\hat{y}$-Pol
5. 6×6	$\hat{x}$- and $\hat{y}$-Pol ULA	$\hat{y}$-Pol
6. 6×6	$\hat{y}$- and $\hat{z}$-Pol ULA	$\hat{y}$- and $\hat{z}$-Pol

4.2.1.1 Reference MIMO Antennas in Experimental Environments

Several reference MIMO antennas are investigated in [CHV09] as given in Table 4.2. The radiation patterns are obtained from theoretical calculations. In order to gain understanding into the limitation of MIMO systems as caused by the propagation channel, these reference MIMO antennas are evaluated in realistic channels using MEBAT. Two propagation environments are considered, featuring an outdoor-to-indoor scenario and an indoor scenario, both of which are measured at 5.3 GHz in the TKK campus area, Finland.

Results show that whereas the channel capacity increases with an increase in the number of antennas, the *spatial multiplexing efficiency* decreases. The spatial multiplexing efficiency is a measure of the effectiveness of the channel in supporting parallel sub-channels. It is defined in [CHV09] as

$$\eta_{\text{SM}} \propto \frac{(\prod_{k=1}^{K} \lambda_k)^{1/K}}{\frac{1}{K}\sum_{k=1}^{K} \lambda_k}, \tag{4.1}$$

where λ_k denotes the kth eigenvalue of the channel covariance matrix. A saturation effect of the capacity with the number of antennas is observed. This is because the rank of the channel matrix does not increase indefinitely when the environment only supports a finite number of eigen-channels. Among all configurations listed in Table 4.2, Configs. 5 and 6 achieve higher channel capacity with however worse spatial multiplexing efficiency comparing to the 3×3 systems. In terms of the total received power, Config. 5 performs the best because of the horizontal polarisation match between the Tx and Rx antenna elements. Comparing with Config. 5, Config. 6 achieves better capacity and spatial multiplexing efficiency due to the use of cross-polarised antennas.

4.2.1.2 Super-antenna

The impact of a realistic user upper-body phantom, including the arm and hand, and the terminal position inside the hand is found to be significant on both antenna and MIMO performance [YTK+07, PFPK09]. In order to establish the correctness of the composite channel approach involving users, a comprehensive study consisting of a wider range of antenna systems, propagation and user scenarios is carried out

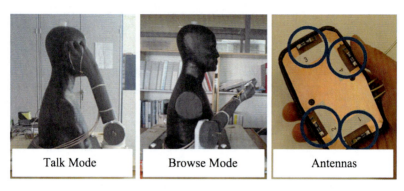

Fig. 4.12 The super-antenna: antenna mock-up plus user phantom used in [HMM+10] (©2010 IEEE, reproduced with permission)

[HMDM07, HMM08, HMM+10]. The composite channel method is employed to synthetically combine double-directional measurements of the propagation channel when the antenna and the user are absent, with the super-antenna radiation patterns when the user is present. The propagation channel is fully described by the double-directional Multi-Path Components (MPCs), whereas the user (including its head, hand and torso) together with the actual mobile terminal (antennas with casing) is treated as a super-antenna characterised by the measured far-field radiation patterns. In order to investigate how the composite channel method can appropriately account for the presence of the user in realistic scenarios, the underlying model is evaluated experimentally by comparing the obtained results with direct measurements in the same propagation environment.

Super-antenna Characterisation Both the antenna without the presence of a user and the super-antenna (inherently with the presence of a user) are characterised by their respective frequency dependent far-field radiation patterns $\mathbf{G}(\Omega, f)$ [HMM+10]. The test handset is a PDA-type mock-up equipped with four PIFA elements covering 2.5–2.7 GHz. The user phantom consists of a part of liquid filled upper body phantom (torso and head) and a part of solid hand/lower arm. Two different user operations are investigated, i.e., the talk mode and the browse mode (see Fig. 4.12).

The measurement shows that the total loss increases with the presence of the user phantom compared to a free space handset. The increase lies between 1.4 dB and 4.8 dB, depending on the user operation and the antenna positions. The "hand influence" accounts on average for a loss of about 2.2 dB, whereas it is about 2.6 dB for the "head influence" and 1.4 dB for the "body influence". The radiation patterns also show how the phantom head and body shadow the handset antennas, which accounts for the increase in antenna correlation from the median value between 0.2–0.3 to between 0.4–0.5 with the user phantom present.

Composite Channel Characterisation The propagation channel is expressed in terms of double-directional MPCs, which is independent of the Tx and Rx super-antennas. Each component is characterised by its complex gain (α), excess path delay (τ), DoA (Ω_r) and Direction of Departure (DoD) (Ω_t). The composite channel matrix $\mathbf{H}(f)$ is obtained by combining the double-directional MPCs with the field patterns of the super-antennas $\mathbf{G}_t$ and $\mathbf{G}_r$, written as

$$\mathbf{H}(f) = \sum_{l=1}^{L} \alpha_l \mathbf{G}_r^T(\Omega_{r,l}, f) \mathbf{P}_l \mathbf{G}_t(\Omega_{t,l}, f) e^{-j2\pi f \tau_l}, \tag{4.2}$$

where $\mathbf{P}$ denotes the polarimetric transfer matrix.

Two channel measurement campaigns were performed at an Ericsson office building in Kista, Sweden. The first considered a stationary outdoor-to-indoor microcell scenario [HMDM07], whereas the second considered a stationary indoor scenario [HMM08]. In either campaign, a virtual array channel sounding measurement is performed first to enable a double-directional characterisation of the propagation channel using high-resolution parameter estimators. Following that, direct measurements were performed with the super-antenna, which includes the user phantom at the mobile terminal.

The performance of the composite channel is first evaluated by comparing the average received powers with the direct measurements. The difference is found to be within 1 dB in most cases. Good match is also obtained for the eigenvalues of the channel covariance matrix, with the agreement in median values of within 1 dB even for the weakest eigenvalue. Following that, the user impact on MIMO performance is evaluated using the composite channel approach. Due to the increase of both antenna mismatch and correlation, the performance of four-antenna diversity performance with Maximum Ratio Combining (MRC) is found to be slightly reduced by 0.7 dB at the 1% outage probability level in the talk mode. The channel capacity is likewise reduced when the user is present, due to the decrease in available SNR. It is also found that the choice of the placement of the antennas is crucial when implementing a mobile handset with multiple antennas.

4.2.2 Performance Estimation Using Phaseless Radiation Patterns

Evaluating the performance of MIMO systems relies on the accurate measurement of complex radiation patterns for all antenna elements. However, the phase information is usually not available in active measurements of mobile terminals [KI08].

The effects of using phaseless patterns on MIMO performance are evaluated in [KSIV06, TL08a, TL09a, TL09b]. It is shown to result in unacceptable underestimation of MIMO performance. The study of [KSIV06] is based on the evaluation of five antenna prototypes, and it also shows that adding random phases to phaseless patterns can reduce the performance estimation errors. In [TL08a, TL09a, TL09b], the impact of phaseless patterns on capacity performance is investigated for the whole continuum of small antenna spacings in simulated dipole arrays, taking into account the use of different matching networks, and the impacts of different propagation scenarios. Phase synthesis methods for improving the estimation of capacity performance based on magnitude patterns are also proposed.

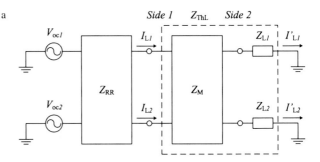

Fig. 4.13 Block diagram of a 2 × 2 MIMO receive subsystem employing an impedance matching network [TL09b] (©2009 EurAAP, reproduced with permission)

The system model is established with a focus on the Rx subsystem as shown in Fig. 4.13, where $\mathbf{Z}_{RR}$ is the antenna impedance matrix, $\mathbf{Z}_L$ denotes the load impedance, and $\mathbf{Z}_M$ represents the matching network as

$$\mathbf{Z}_M = \begin{bmatrix} \mathbf{Z}_{11} & \mathbf{Z}_{12} \\ \mathbf{Z}_{21} & \mathbf{Z}_{22} \end{bmatrix}, \quad (4.3)$$

which can be of any kind, e.g., [DBR04, WJ04a, WJ04b, LAKM06]. The transfer function between the open-circuit voltages and the output voltages is obtained as

$$\mathbf{v}'_L = \mathbf{Z}_L(\mathbf{Z}_{22} + \mathbf{Z}_L)^{-1}\mathbf{Z}_{21}(\mathbf{Z}_{RR} + \mathbf{Z}_{ThL})^{-1}\mathbf{v}_{oc}, \quad (4.4)$$

where $\mathbf{Z}_{ThL}$ denotes the Thevenin equivalent load impedance as seen by the antenna ports. A path-based channel model is used to represent the propagation channel of the complete MIMO system, which is defined by a trans-impedance matrix relating the Rx antenna output voltages $\mathbf{v}'_L$ to the Tx antenna source currents. The following cases of pattern types are evaluated:

- *Complex patterns*: with both magnitude and phase.
- *Magnitude patterns*: the antenna patterns are phaseless.
- *Random-phase patterns*: random phases $\phi \in (-\pi, \pi]$ are added to the magnitude patterns.
- *Simple phase-synthesis patterns*: phases are synthesised according to the inter-element separation using the steering vector $\mathbf{a}$.
- *Correlation-based phase-synthesis patterns*:

$$\mathbf{a}' = \mathbf{R} \odot \mathbf{a}, \quad (4.5)$$

where the matrix $\mathbf{R}$ is obtained from the correlation of the available magnitude patterns.

With uniform APS and small antenna spacings, random-phase patterns provide good performance estimation for antenna systems in which the applied matching network facilitates low-phase correlation. This is the case of the antenna systems with sophisticated matching networks [TL09b]. However, this approach overestimates capacity for simple matching networks and for APS with limited angular spread. This is because that, under these conditions, the phase correlation of the actual complex patterns is not so low as that of random phases. On the other hand, the simple phase-synthesis method exhibits underestimation in small antenna spacings, as it does not take into account either coupling or coupling compensation effects, which tend to reduce phase correlation. The improved correlation based phase-synthesis method adds a degree of randomness, in order to account for the aforesaid coupling and compensation effects. Results indicate that significantly more accurate estimation can be achieved by the improved approach of all investigated parameters. Nevertheless, this approach should be validated in the future using antenna patterns measured from several more different multiple antenna terminal prototypes.

4.2.3 Mutual Coupling in Compact UWB Array

As discussed above, it is important to consider the mutual coupling among antenna elements in compact arrays. The coupling phenomenon can be modelled using network theory analysis [WJ04a, WJ04b, LAKM06]. In [DS09], the modelling of mutual coupling is studied for compact UWB arrays. The method approximates coupling by means of frequency-dependent narrowband dipoles. For evaluation of modelling accuracy, this method is applied to two realistic UWB arrays, one with two UWB bicones [DS08] and one with two UWB dual feed microstrip monopoles [GBD06]. The model should be accurate if the current distribution is similar to that of a dipole, which is however not always the case for the more complex UWB antennas. The results show that the model is accurate when it is used to approximate the magnitude of mutual coupling. However, extra delays should be considered carefully in order to estimate the phase. To improve the modelling, the mutual coupling is further approximated by means of far-field radiation characteristics. In particular, the far-field antenna transfer function $\mathbf{H}^7(f, \hat{\mathbf{r}})$ [RBS03] in the end-fire direction of the array $\hat{\mathbf{d}}$ is considered:

$$S_{21}(f) = \frac{e^{-jkd}}{d}\left(\frac{-j\lambda}{4\pi}\right)\mathbf{H}_1^T(f, -\hat{\mathbf{d}})\mathbf{H}_2^T(f, \hat{\mathbf{d}}). \quad (4.6)$$

Good approximation can be obtained in both magnitude and phase. This approach of modelling the mutual coupling can be completed with statistical descriptions of the antenna and propagation channel models [Sib09b]. In particular, the radiation and impedance properties of the isolated antenna elements are first modelled statistically. Following that, the separation distance and relative orientation of the antenna elements are likewise modelled statistically. Mutual coupling effects can then be included using the proposed model.

4.2.4 RF Measurements over Optical Fibre

After discussing the performance characterisation of different antenna systems, the discussion is followed by some advanced measurement techniques. RF measurements of antennas, either in anechoic chambers for radiation and impedance measurements or in measurement campaigns for channel sounding, usually connect the

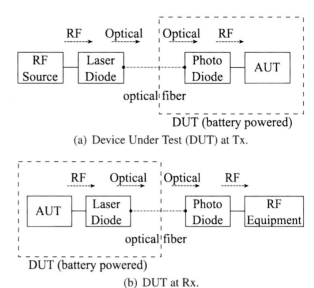

Fig. 4.14 Block diagram of RF over optical fibre measurement systems

antennas with different RF blocks using coaxial cables. However, when the DUT is small in terms of the wavelength of operation, the entire device becomes the radiating structure. This effect greatly increases the undesired influence of the cable, which can cause severe distortions in the measurement results [KPOE01].

In [IOV99, IKV04], the use of balun chokes is demonstrated to circumvent undesired currents flowing on the surface of the cable connecting the DUT. However, the performance tends to be of limited bandwidth, and it does not change the fact that the cable is physically present as a scatterer, which is not the case when operating the mobile terminal without cables. Alternatively, [YP09, YPC10b, YPC10a, YNP10, KH07, THN+08] consider the use of optical fibres for RF measurements. Two set-ups are discussed in [YP09]. One solution is for short-range measurements, where a relatively low Tx power is needed. This would typically be the case for radiation pattern measurements in an anechoic chamber. The other solution is for longer-range measurements, where a higher Tx power is needed, e.g., in channel sounding measurements.

In the short-range solution, the output of an optical detector, i.e., a photo diode (PD), can be used directly without additional amplification due to relatively low power requirements. Figure 4.14(a) shows the block diagram of such a system where the DUT is at the Tx side. The Laser Diode (LD), which directly modulates an RF signal into an optical signal, is connected to the Photo Diode (PD) via optical fibre. The PD detects the optical signal and feeds the RF signal to the antenna.

In longer-range applications with higher Tx power requirements, the DUT is used at the Rx side instead. The proposed solution features an Rx unit of size $40 \times 40 \times 10$ mm^3. The LD used is of the Distributed Feedback (DFB) type with driving circuits, consisting of a current source, implemented with an operational amplifier, with negative feedback from the photo-monitor of the LD for power stabilisation. The total power consumption of the device (drawn from the battery at 3.7 V) is

Table 4.3 Performance of the RF over optical fibre measurement system [YPC10b]

Parameter	Symbol [Unit]	776 MHz	2300 MHz
System Gain	G [dB]	10.1	4.8
System Noise Figure	NF [dB]	9.3	10.9
Intrinsic Link NF	NF_{intr} [dB]	29.8	33.8
Intrinsic Link Gain	G_{intr} [dB]	−21	−23
System Dynamic Range	IMF_3 [dB]	39.02	37.16
Current Consumption	I_q [mA]	115	
Battery Life	[min]	30	
Supply Voltage	$V_{battery}$ [V]	3.7	
Optical Power	OP [dBm]	3	

110 mA, which allows the use of a 25 × 20 × 5 mm³ Lithium-ion battery of 130-mAh capacity. It also features an SP4T switch allowing measurement of up to four antennas per phone mock-up with a single optical unit.

Various parameters of such a system are further characterised in [YPC10b] and summarised in Table 4.3. In particular, the intrinsic link gain is defined as the transducer power gain of an optical link without any amplifier, including the basic blocks shown in Fig. 4.14(b). It can be calculated theoretically as [CABP06]

$$G_{intr} = s_l^2 r_d^2, \qquad (4.7)$$

where s_l [W/A] denotes the slope efficiency of the LD, and r_d [A/W] denotes the responsivity of the PD. The complete RF link includes three stages of gain blocks, achieving the total system gain of 5–10 dB over the operating frequency range.

The measurement system is further improved in [YPC10a] by adding a second laser which enables the use of two parallel optical channels. For the MIMO OTA measurement system, the filters used in [YP09] can be also removed. This not only improves the overall noise figure (6–13 dB, depending on the frequency), but also enables its usage from about 500 MHz up to 5.5 GHz with a single gain block. The overall dynamic range is improved to 60 dB.

This measurement tool was verified by performing radiation pattern measurements of antennas in an anechoic chamber. The results show good match between simulations without the connecting RF cable and the RF optical based measurements.

4.2.5 Antenna Measurement Using a Spheroidal Coupler

The most common method of measuring the radiated power of a DUT is by scanning it with a spherical positioner in an anechoic chamber. However, this method is very time consuming. In [TTK+09], a novel TRP measurement method using a spheroidal coupler is developed, where the DUT and the Rx antenna are located around the two focus points of the spheroid. A sketch of the spheroidal coupler

Fig. 4.15 Sketch of the spheroidal coupler

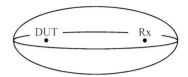

Table 4.4 Comparison of measured TRP at 2.41 GHz [TTK+09]

	Sleeve	Dipole	CTIA Std.
Radiation efficiency	83.9%	74.1%	
	−0.76 dB	−1.30 dB	
TRP (dBm)	9.14	8.82	8.86

is given in Fig. 4.15. In order to measure the TRP in this multiple-reflection environment, a displacement method is applied, which measures the transmitted power from the DUT by changing its position and the position of the Rx antenna along the long axis of the spheroid. Alternatively, it can be replaced with a matching tuner.

In Table 4.4, the measured TRP using the spheroidal coupler is shown to be comparable to that, of an anechoic chamber measurement using spherical scanning based on the CTIA standard [CTIA05]. Furthermore, in [TTK+10], an estimation method of reflection coefficients of DUT by phase rotation technique is proposed in TRP measurement using the spheroidal coupler, which improves accuracy and reliability of the displacement method. Round robin tests of the TRP are performed and compared with conventional methods. Furthermore, the Total Radiated Sensitivity (TRS) measurement method using the spheroidal coupler is also described, where actual Universal Mobile Telecommunications System (UMTS) terminals are measured to show sufficient sensitivity characteristics.

To conclude, the use of spheroidal coupler eliminates the need for the conventional anechoic chamber and spherical positioner, which is considered promising for future OTA measurements.

4.2.6 Measurement Techniques for 60-GHz Antennas

The 60-GHz band is being considered as one option to meet the increasing demand for very high-data-rate wireless communication systems [Smu02]. However, it is challenging to characterise antennas at mm wavelengths. In this context, [RKI+09] presents a quasi-full 3-D on-wafer radiation pattern measurement system for 60-GHz antennas.

A planar omnidirectional antenna of size 4.5×4.3 mm^2 fed by a coplanar waveguide is measured as a test antenna. The measured 3-D radiation pattern shows good agreement with simulations. The gap between two repeated sets of measurement data is less than 0.2 dB, which indicates good repeatability of the method. In order to show the importance of the probe-fed approach, a similar antenna fed by a V-connector is measured with the same measurement system. The use of the V-connector is shown to not only create additional resonances but also makes the omnidirectional beam narrower, which increases the gain of the beam and creates strong side-lobes. This is due to the fact that the whole structure radiates, including the original radiating element and the connector. Further details can be found in [RKI+09].

4.3 Performance Improvement Through Dedicated Single/Multiple Antenna Design

Designing compact antennas for MIMO brings challenges with regard to the close proximity of the antenna elements that often do not have high isolation between them, which can degrade efficiency. Furthermore, user interaction with the multiple antennas will have variable impact on the isolation, and this issue should be given careful attention. Several methods are proposed here that show evidence of reducing the mutual coupling effects using a range of methods, while using polarisation to gain better independent and isolated channels is shown as another means to achieve better performance. Another important factor to note with MIMO antenna design is the use of fewer Radio Frequency (RF) transceivers for cost and complexity considerations, where switched parasitics and also antenna selection is proposed.

4.3.1 Mitigating Influence of the User with Planar Antennas

Considering the interaction of the user, PIFA antenna configurations are investigated at Global System for Mobile Communications (GSM) and UMTS bands, studying the benefits of dielectric loading on both impedance mismatch loss and absorption loss. First of all, by using multiple layers of high-permittivity materials, impedance mismatch loss can be reduced at the expense of a narrower bandwidth. Secondly, it is shown how the use of narrowband antennas are more robust against mutual coupling, and finally a novel decoupling technique can be applied to reduce the coupling for antennas in general.

High-relative-permittivity materials with low losses can nowadays be used as antenna substrates, allowing antennas of smaller physical dimensions to be integrated into handsets [ZHZ97]. A further addition is the use of a "superstrate" or extra layer of substrate that can achieve further antenna size reduction [HZZ95]. However, in handsets using PIFAs, it is even more difficult to sacrifice antenna volume by making room for a superstrate. Figure 4.16 shows the impedance mismatch loss (i.e., power loss due to different input impedances of the antenna and the radio transceiver). In this instance, several combinations of substrate dielectric, ε_r^{sub} and superstrate

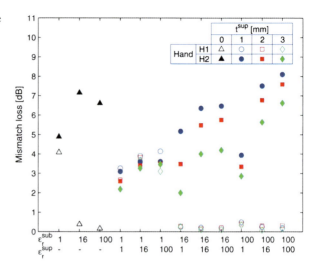

Fig. 4.16 Comparison of the impact of user handling, dielectric constant and dimensions of substrate and superstrate materials on mismatch loss in a PIFA antenna [PFKP09a]

dielectric ε_r^{sup} of 1, 16 and 100 are compared with different combinations of thickness. The comparisons are plotted by unfilled shapes for a user handling, H1, where the PIFA is not covered and another handling, H2, where the PIFA is covered by a finger. For the case of handling H1, the use of different substrates and superstrates has no real impact on the antenna, though when considering user interaction in the case of H2, a high superstrate thickness of 3 mm and differing dielectric constant can reduce the mismatch loss by over 3 dB compared to a case where there is no superstrate. It was also found that there is negligible difference made to the absorption loss of the user when different substrates and superstrates are used, thus only the impedance matching is affected.

Another important factor regarding user handling is the isolation between multiple antennas. Clearly, as shown in Fig. 4.17, a clearance gap of up to 5 mm within the PIFA antenna, which creates a narrow bandwidth, is necessary to increase the isolation to an acceptable level of above 10 dB. This is the case where the SAM head model is included and applies to several user handlings. Furthermore, UMTS bands I, II and V are compared for repeatability. Further comparisons of isolation have been recorded in [PP10a] with respect to the isolation in UMTS bands I, II and V, where band V in particular exhibits isolations as much as 6 dB less on average than the other two bands.

Another novel method proposed in [PP10b] describes the Transceiver Separation Mode (TSM). For example, a MIMO terminal may have three antennas, where antennas 1 and 2 are receiver antennas, and antenna 3 is a transmit antenna. If the transmit frequency is placed into a separate narrowband carrier to that of the two receiver antennas, then the rule will hold that $s_{31} < s_{21}$ and $s_{32} < s_{21}$ (i.e., having higher isolation from antenna 3 compared to isolation between antennas 1 and 2) whatever the physical distances involved on a handset. This enables isolation be-

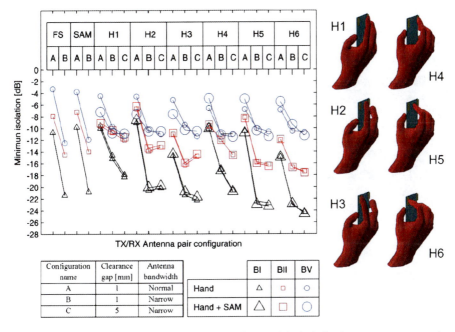

Fig. 4.17 Comparison between different user handlings and the isolation between antenna terminals between bandwidths and user handling [PPK10] (©2010 EurAAP, reproduced with permission)

Fig. 4.18 Comparison of different proposed PIFA and IFA antenna configurations [PKP09]

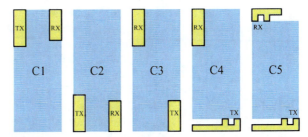

tween the transmit and receive antennas to be better than 10 dB in the majority of cases, which is much improved over when TSM is not applied.

Finally, it has been found how the use of different antenna configurations can be applied to gain improvements in reducing the mutual coupling [PKP09]. In particular, the configurations shown in Fig. 4.18 are studied. It can be seen from the results in Fig. 4.19 that up to 3 dB improvement in isolation can be gained by using the IFA options over the PIFA options, which is consistent for all user handlings in all UMTS bands. This will have benefit to the antennas' efficiency though not necessarily the correlation between the terminals.

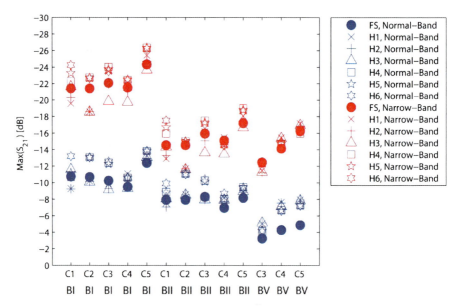

Fig. 4.19 Comparison of isolation improvement from different antenna configurations [PKP09]

4.3.2 Compact MIMO Antennas with Less RF Transceivers than Antennas

Designing MIMO antennas at the terminal is complex (and potentially costly) when more than one transceiver is required. Therefore, a number of techniques have been considered at the mobile or access point, such as switched parasitics, antenna selection and also other mutual coupling mitigating techniques, to help overcome the problem of including several transceivers with MIMO antennas for a mobile terminal.

One proposed method for switched parasitics is the Electronically Steerable Passive Array Radiator (ESPAR) antennas [KKP07]. The principle is that when the switched parasitics are either open or shorted by a PIN diode switch, they will then steer the antenna pattern transmitted from a single-source antenna. In the example used here, the transmitter is equipped with a single RF front-end, capable of mapping Binary Phase Shift Keying (BPSK) modulated symbols onto orthogonal basis functions on the wavevector domain, while at the receiver there is either a two-element ULA or an omni-directional antenna. In order to calculate the Shannon capacity, a new channel is defined whereby the receiver will have a defined coefficient, $\mathbf{A}_{Rx}(\theta_{Rx})$, for the angle at which it receives the transmitted bit streams. The transmitter will have a defined pattern for a given binary state at which it will transmit, $\mathbf{B}_{Tx}\mathbf{x}_{Tx}$, where $\mathbf{x}_{Tx}$ represents the bit transmission,

$$\mathbf{H}_{ESPAR} = \mathbf{A}_{Rx}(\theta_{Rx})\mathbf{H}_{Channel}\mathbf{B}_{Tx}\mathbf{x}_{Tx}. \qquad (4.8)$$

In the example shown in Fig. 4.20, a three-element ESPAR antenna is taken where it will have the capability to steer in two directions due to the switched para-

Fig. 4.20 Diagram of three-element ESPAR antenna

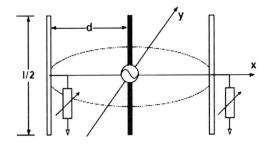

sitics and thus can compare its resulting 2 × 2 Beam Steering MIMO (BS-MIMO) capacity with traditional MIMO scenario taken from [KP07]. Comparisons are made for cases where there is full channel state information (providing feedback) at the transmitter and where there is not. In all cases, capacity is comparable and within 1 bit/s/Hz, as shown in Fig. 4.21, while at the same time the number of transceivers has been halved in this case. Earlier work has also shown comparable diversity gains with the traditional Alamouti scheme [KP07].

Another way in which the number of RF transceivers can be reduced is by using antenna selection. An example of how this can be implemented at the access point is using two Quadrifilar Helix Antennas (QHAs), such as those illustrated in Fig. 4.22, where each one consists of four MIMO antenna branches compacted into one [BS07]. If two QHAs are at the Tx while one Quadrifilar Helix Antenna (QHA) is at the Rx, then the channel consists of eight 1 × 4 column vectors, $\mathbf{h}_n$ to represent the eight transmit antennas. The antenna selection will find the four strongest transmit antennas in terms of their magnitude $|\mathbf{h}_n|$:

$$H_{\text{Sel}} = \max_4\left([\mathbf{h}_1 \quad \mathbf{h}_2 \quad \mathbf{h}_3 \quad \mathbf{h}_4 \quad \mathbf{h}_5 \quad \mathbf{h}_6 \quad \mathbf{h}_7 \quad \mathbf{h}_8]\right). \tag{4.9}$$

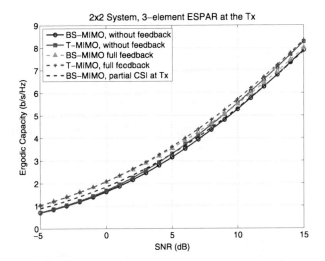

Fig. 4.21 Comparison of 2 × 2 MIMO capacity achieved with three-element BS-MIMO capacity resulting from use of an ESPAR antenna

Fig. 4.22 Illustration of the QHA

Fig. 4.23 Comparison of the improvement made in capacity when using antenna selection over a single QHA

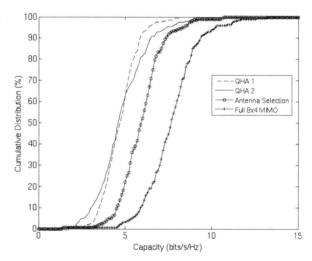

Therefore branches which are suppressed by user interaction or other near field effects can be ignored, and thus the best branches can be selected [BK08]. It is also assumed that all the eight branches of the two separated QHAs antennas have low correlation, or otherwise there would be a chance that all four best branches are contained within one QHA and the other becomes redundant. Results in Fig. 4.23 indicate that there is comparable improvement in the capacity compared to one single QHA at the access point as branches with a low branch power ratio or in an instantaneous deep fade become redundant and only the four highest branches are taken into account.

4.3.3 Use of Polarisation to Achieve Compact Multiport Antennas

Given that spatial separation in compact mobile terminal antennas is limited, polarisation and angular pattern diversity has received much attention as an alternative in recent days [BSE05]. The results presented here consider how polarisation is exploited both at the Tx and Rx as shown in Figs. 4.24 and 4.25, where there are

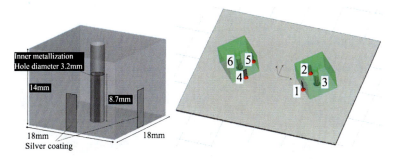

Fig. 4.24 Illustration of the multiport dielectric resonator antenna array [TPLY10b] (©2010 IEEE, reproduced with permission)

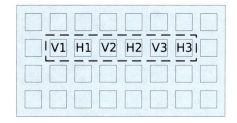

Fig. 4.25 Use of dual polarisations at the access point from a patch array [TPLY10b] (©2010 IEEE, reproduced with permission)

antennas with dual polarisation capability at both ends. This brings about the opportunity to exploit polarisation from both the transmitter and receiver, which can assist in creating more compact solutions to the mobile terminal antenna [ILYT08]. Comparisons of capacity with respect to conventional half wavelength spaced monopoles are shown in Fig. 4.26 [TPL[+]09, TPLY10b], which indicate there is similar capacity in the NLOS scenario. In LOS, this is not the case, as would be expected since there is not enough scattering to de-polarise the channels and make them independent. However, there is clearly an opportunity to exploit polarisation in such a scenario.

Even though the six-port dielectric resonator antenna is relatively compact in comparison with the reference monopole array, it still partly relies on spatial diversity to achieve good performance. In a well-cited paper published in 2001 [AMd01], it is claimed that up to six degrees of freedom can be obtained from electromagnetic polarisation states of *co-located* electric and magnetic dipoles. However, no experiment is performed to substantiate the claim. In [TL10], six-port antennas are designed to prove the theoretical prediction. In particular, the receive array is fully contained in a cube with sides of 0.24λ, so that the electric and magnetic dipoles are as co-located as possible, without causing unreasonably high mutual coupling. The measured antenna patterns are combined with a simulated channel that can be used in a way to superimpose the channel onto the antenna pattern, in order to obtain the overall MIMO channel and the Shannon capacity. It is confirmed that the compact receive array can offer the predicted six degrees of freedom, with the pattern correlation being less than 0.32 between any two antenna ports.

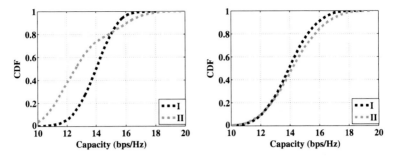

Fig. 4.26 Illustration comparing the capacity of Cases I and II with Case I as the reference case [TPLY10b] (©2010 IEEE, reproduced with permission). The *left* and *right plots* show the LOS and NLOS scenarios, respectively

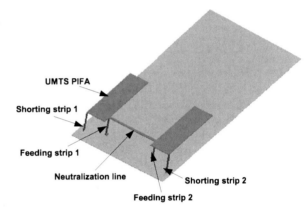

Fig. 4.27 Diagram illustrating the application of a neutralisation line to two PIFA antennas

4.3.4 Mitigating Mutual Coupling Effects

As has been seen earlier in Sect. 4.3.1, the isolation between two antennas will be low in cases of close proximity. In this subsection, a number of new methods that can alleviate the problem of low isolation are proposed. One example is to use a neutralisation line between two PIFAs [Lux07] shown in Fig. 4.27. When the line is inserted, the isolation improves from less than 10 dB to 20 dB, while also allowing a good impedance match to the PIFA antennas. The benefit of this isolation improvement is that the efficiency increases by around 1 dB, as seen in simulations and reverberation chamber measurements. Similar results to that in Fig. 4.27 have been found to be possible when using other decoupling mechanisms such as [LLT10] that increases the relative permittivity in the dielectric loading of two PIFAs to as high as 20. However, in [LLT10], the decoupling is achieved by reducing the excitation of the ground plane shared by the two PIFAs. Also in [TLBA10], the use of parasitics to reduce coupling between two dipoles is considered, where it undertakes further analysis to show that bandwidth can be enhanced when changing suitably the lengths of the dipoles and the parasitic element for a given separation distance.

4 Assessment and Modelling of Terminal Antenna Systems

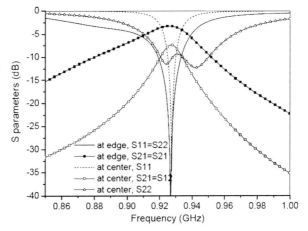

Fig. 4.28 Comparison of the S-parameters when two PIFAs are configured at two ends of the ground plane or alternatively with one PIFA in the centre of the ground plane

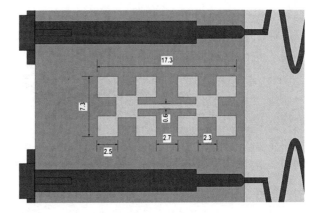

Fig. 4.29 Example case of sensor node antenna with loaded DGS

For monopole and PIFA antennas configurations, analysis has shown that for a small ground plane, particularly where the wavelength is comparably large at frequencies below 1 GHz, the positioning of antennas is critical [LTL10]. Figure 4.28 compares the S-parameters of two antenna configurations. The first configuration features two antennas, a monopole and PIFA, located at the opposite (short) edges of a 100 mm × 40 mm ground plane, which has a low isolation, i.e., high value of s_{21}. Surprisingly, when the PIFA antenna is moved into the centre of the ground plane, i.e., the antennas are closer, then the isolation is improved. This phenomenon is due to a lower current density that is present in the centre of the ground plane, thus reducing mutual impedance. It must be noted, however, that this improvement is achieved at the price of a smaller bandwidth.

Printed arrays such as those used in sensor networks shown in Fig. 4.29 are based on sinusoidal reduced size monopoles, which can be brought closer than the expected half wavelength distance apart using a Defected Ground Structure (DGS) [KC10]. This will operate reliably for the narrowband characteristics of many sensor networks in the Industrial, Scientific and Medical (ISM) band at 2.4 GHz. Clearly

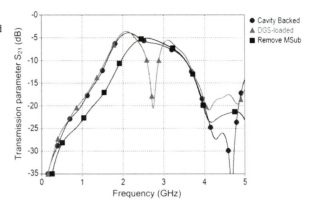

Fig. 4.30 Comparison between non-DGS loaded and DGS loaded antenna structures to improve isolation

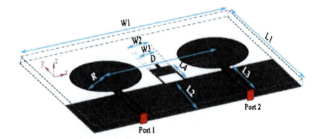

Fig. 4.31 Illustration of UWB array with stub to reduce mutual coupling

the mutual coupling suppression is illustrated in Fig. 4.30 that shows improved isolation compared to no DGS or a cavity backed layout (i.e., ground planes printed on both sides between the antennas).

For wideband antennas and indeed UWB antennas, the reduction of coupling over the desired bandwidth is a great challenge. A recent novel design in Fig. 4.31 shows the insertion of a stub with critical widths of substrate $W_1 = 68$ mm, length of substrate $L_1 = 40$ mm, length of ground plane $L_2 = 11.5$ mm, length of feed line $L_3 = 12$ mm, radius of each radiating element $R = 12$ mm, and distance between two conventional circular printed UWB disc monopoles [CPKP04] $D = 34$ mm. Using this configuration [NDT09], there is improvement in isolation while the reflection coefficient is kept to below an acceptable value of -10 dB, as shown in Fig. 4.32.

4.4 Statistical Antenna-Channel Modelling

Traditionally, antennas are designed, simulated and measured as isolated as possible form any electromagnetic perturbation, i.e., in a rather ideal case which is not realised when the antenna is used in practice. A certain level of disturbances on the antenna characteristics will occur when the true ground plane and the casing are taken into account, or when objects are present in the vicinity of the antenna. While some

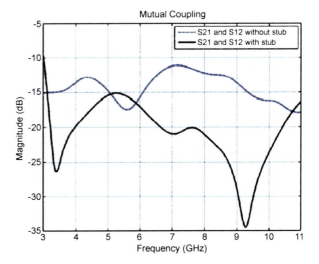

Fig. 4.32 Comparison of mutual coupling in UWB array with and without the stub

of these disturbances can be properly computed using deterministic techniques, others are highly variable in nature and do not lend themselves easily to deterministic modelling. Most generally, these effects are just ignored or described very roughly by assuming an ad hoc gain margin. However, the growing complexity of modern wireless communication systems, the greater variability of application contexts and the greater needs for enhanced performance plead for a better approach to account for the terminal in the performance evaluation of physical layer schemes and of radio access networks in general. Thus, advanced approaches intended to achieve a good trade-off between complexity and accuracy of the terminal description are needed. Statistical methods provide a suitable framework for this purpose, since they intend to condense this complexity into a small number of statistical quantities, such as moments of distributions functions. In this section we address such issues, by first concentrating on the statistical modelling of the antennas themselves and then by including the radio channel variability in order to combine both antenna and channel variabilities into a single statistical description of an antenna in its local propagation environment.

4.4.1 Statistical Antenna Modelling

In this part, the elaboration of statistical methods accounting for the variabilities of antenna characteristics is addressed. Early works in another context have been carried out a long time ago [Shi74]. More contemporarily, there seems to be a trend to incorporate statistical methods into electromagnetism in order to account for uncertainties [sta08]. However, providing representative and simple statistical models of antennas in their usage context is all but a simple problem. Several general issues have been discussed in [Sib08b], such as the relevance of categorising antennas

Fig. 4.33 Schematic view of the input–output relation between the statistics of input (parameters defining the terminal) and output (antenna characteristics) random variables for an antenna system

through major antenna characteristics. Another question is the choice of the maximum distance within which disturbing objects may be considered to be part of the antenna. Last but not least, the quality criteria for statistical antenna modelling were also addressed.

Generally speaking, a statistical model can be seen as an input–output relation (Fig. 4.33): the input includes the set of stochastic variables which constitutes the input stochastic space and the multivariate law of probability for these variables. This law may factorise if the various variables are independent, but most often it will not be the case. For instance, if the size of a casing is allowed to vary strongly, the length and width will not be entirely independent. In this case, a suitable choice or transformation of the input variables may be useful, such as keeping the length but using the aspect ratio rather than the width. The output is the set of antenna characteristics of interest, together with their probability density function (PDF). In the general case again, these characteristics will not be independent.

The input–output relation embodies the complex electromagnetic phenomena involved in the change of output characteristics with a modification of the input variables. This relation may be (highly) nonlinear if the input parameters vary strongly. In the work [Sib08a], a small signal analysis of the input–output relation is carried out, whereby the author expresses the output variables as a Taylor expansion of the vector of input variables. To the first order, any output variable is Gaussian distributed if the input variables are Gaussian distributed and independent. A deviation from Gaussianity is observed for large input parameters variations or when the output variable is e.g. close to an extremum, which requires a 2nd-order expansion and leads to asymmetric distributions. The computation of the expansion coefficients may require a large database of antennas for sufficient accuracy. Through a suitable choice of the input variables vectors, it is possible to minimise the size of this database or alternatively to increase the accuracy.

4.4.1.1 Modelling Using Mode Expansion

The pair of companion papers [RD08] and [DR08] address the statistical analysis of antenna properties operated on intermediate quantities (e.g., antenna mode amplitudes) which contain all the specificities of antenna radiation in a compact way,

Fig. 4.34 Planar UWB antenna (Dual-Fed Microstrip Monopole (DFMM), from [DR08])

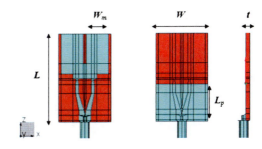

Table 4.5 Randomly varied geometrical parameters for the DFMM following a uniform law (from [DR08])

	DFMM	
	Original Design	range
W	20 mm	[20; 28] mm
L	33 mm	[33; 43] mm
t	1.524 mm	[1,5; 1,6] mm
W_m	9.1 mm	[9.1; 12.1] mm
L_p	14 mm	[14; 18] mm
ε_r	2.33	[2.1; 2.6]

rather than on the output quantities of interest themselves. The method exploits parametric modelling based on Spherical Mode Expansion Method (SMEM) combined with the Singularity Expansion Method (SEM) [Bau76]. These techniques allow a very high compression rate of the amount of data necessary to express accurately the antenna radiation over a large frequency range, after proper truncation of the unnecessary modes. Owing to this high compression, a statistical analysis of a database of antennas may be performed on the intermediate parameters, which are the poles (natural frequency and damping factor) and the modal residues. In [DR08], this approach has been tested on a class of three kinds of UWB planar antennas, of different designs but having common major features. Several geometrical parameters of each design have been randomly changed according to relevant rules (Fig. 4.34) and the statistics of the most significant intermediate parameters have been determined. They were generally found to be Gaussian distributed (Fig. 4.35).

Further statistical analysis of the dependencies between the various intermediate parameters may be necessary to obtain a complete statistical description. In particular, as regards the angular radiation patterns, the statistics of the phase difference between the spherical modes was found to be nonuniform and again Gaussian distributed. As a result of this analysis, it appears that generating an antenna statistical set from a convened multivariate distribution on these intermediate parameters appears both feasible and efficient. However, further work is still needed to confirm the validity of the method and extend it to a variety of antennas, particularly in the directional case.

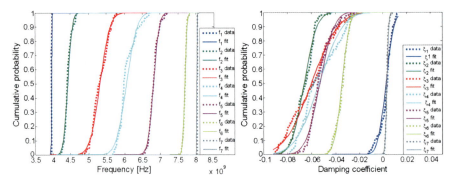

Fig. 4.35 Statistics of the intermediate parameters for a class of random UWB antennas (from [DR08])

4.4.2 Joint Antenna and Channel Statistical Modelling

While in the previous paragraph the analysis bore on the antennas as such, a series of works have intended to describe the effective antenna system behaviour in its local environment. From a utility perspective indeed, a terminal antenna is always operating in a local propagation environment, and both fully determine the characteristics and performance of the radio link. For that reason, these works intend to provide an effective terminal antenna performance from the radioelectric point of view, which implies the consideration of propagation characteristics in addition to antenna characteristics.

In the work [Sib09c] and more completely in [Sib09a], the author addresses this problem in the context of cognitive radio simulators, which need to take into account a broad frequency spectrum. The Effective Gain (EG) combines the angular and polarisation characteristics of the local propagation and the antennas through the following definition:

$$EG = \sum_n |A_{nH}|^2 G_{rH,n} + \sum_n |A_{nV}|^2 G_{rV,n}$$
$$\text{subject to the normalisation } \sum_n |A_{nH}|^2 + \sum_n |A_{nV}|^2 = 1, \quad (4.10)$$

where A_{nH} and A_{nV} are the amplitudes of path n in H and V polarisations incident on the terminal, and $G_{rH,n}$ and $G_{rV,n}$ are the antenna realised power gains in the incoming directions. This equation expresses EG as a simple sum of received powers, and it implies that interference terms between multipaths are neglected in order to express long-term averaging of the inter-path interferences.

It is possible to statistically analyse and model the global antenna-channel behaviour of the radio terminal in its environment, by incorporating both the stochastic character of the terminal and of the channel. In [Sib09a], this has been done for a few examples of terminals and channels. Personal computers (PC) with an UWB antenna have been investigated through electromagnetic simulations on one hand,

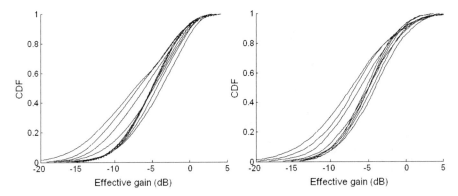

Fig. 4.36 True distribution (*left*) and lognormal model (*right*) of the effective gain statistics computed for an indoor NLOS channel (from [Sib09a])

and commercial dual frequency handsets have been measured on the other. Radio channels have been described using a subset of the WINNER project channel model [Wi007], providing the angular power spectrum and the cross-polarisation discrimination and allowing the generation suitable stochastic realisations of the channel. Such characteristics are related to the type of local environment (such as indoor, urban LOS or NLOS, etc.), according to the WINNER channel model or other standardised models. Combining the local propagation characteristics and the antenna characteristics by post-processing in this way bears strong similarity with the "composite channel approach" discussed in Sect. 4.2.

As a result of the stochastic character of both the terminal and the local propagation, EG is a random variable, the statistical distribution of which is obtained after combining the terminal randomness and the propagation randomness. Since both are affected by strong variations especially when there are obstructions, the lognormal distribution may be anticipated as a good candidate for the EG statistics. This is shown in Fig. 4.36, where the CDF of the EG from $\approx$ 1.5 GHz to 6 GHz every 500 MHz is shown for a simulated PC. The comparison with a perfect lognormal distribution is indeed quite acceptable.

Another example is shown in Fig. 4.37 for the handset. This graph presents the MEG together with the spread of the EG. The camel-like structure of the MEG highlights the dual band performance, with peak values in the order of -10 dBi. This small value has three main origins: (i) the imperfect matching between the antenna radiation lobes/polarisation and the channel wave angles/polarisation; (ii) the masking effect by the head which may hide the antenna from powerful paths; and (iii) the head and hand absorption and antenna detuning which reduces the total antenna efficiency. In addition, the spread of 5 dB shows that large deviations of EG around its mean can occur, due to the various sources of randomness. The Root Mean Square Error (RMSE) between the true distribution and the lognormal model is also shown, being smaller than 1 dB in most cases. Again, using the antenna patterns of the handset in the presence of the user is done in the spirit of the "super-antenna" concept discussed earlier in Sect. 4.2.

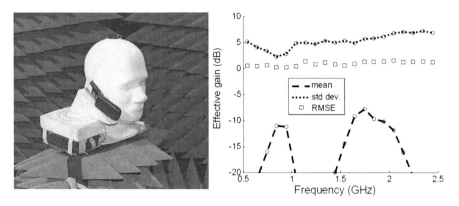

Fig. 4.37 Handset on a phantom head (*left*) and corresponding EG vs. frequency computed for an urban LOS channel (from [Sib09a])

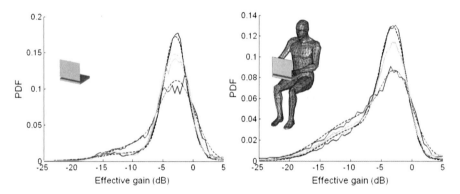

Fig. 4.38 PDF of the EG for a laptop without (*left*) or with (*right*) the presence of a human body, for three different channels. The full line is the true PDF and the dashed is the mixture model (from [SGMP09])

This approach was also applied to an electromagnetically simulated laptop PC [SGMP09], without or with the presence of a human body. The various masking effects produced by parts of the laptop itself or by the body result in more or less thick tails in the CDF of the EG, which may prevent a proper fitting by a lognormal distribution. In such a case a mixture of distributions is more appropriate, e.g. a mixture of Gaussian distributions applied to the EG expressed in dB, which is a generalisation of the lognormal. Through this slight complication enhancement, the RMSE between the true distribution can again fall well below 1 dB, which can be considered very sufficient for many needs. This can be seen in Fig. 4.38, computed for three types of channels (indoor NLOS, urban LOS and highway LOS).

These first results have started to be extended to the case of MIMO systems in the work [Sib10]. Again, the goal was to generate statistical output characteristics from statistical input variables. The complexity is greater than for single antenna systems, since the specificities of the antennas at both the receiver and transmitter sides affect the MIMO link performance. In order to very much simplify the situation, the author chose to use the Kronecker model from the very beginning, which allowed him to factor the transmit and receive side antenna system description in the MIMO channel matrix. Through this simplification, it is possible to describe statistically the multiple antennas terminal and the local propagation only, assuming full decorrelation and antenna ideality at the base station side. In [Sib10], it was in particular shown that the terminal variability may result in strongly non-Gaussian entries of the channel matrix, requiring care in the stochastic generation of the channel matrices.

4.5 Conclusions and Perspectives

The results and discussions of the previous sections highlight the very lively character of antenna research for terminals. Indeed, far from being restricted to isolated antennas in an ideal usage context, the antennas designed for wireless communications terminals must now cope with several key challenges:

- Compactness and integration into highly space-limited terminal devices.
- Coupling and efficiency loss in multiple antennas terminals that are required for MIMO-based standards.
- Major influence of the casing and electronics on the electrical and radiation characteristics.
- Variable and detrimental influence of surrounding electromagnetic disturbances, such as the presence of a human body.

In the present chapter, these challenges have been tackled through a variety of scientific and technical approaches. Indeed, a major step forward in wireless communications will require a convergence of advances in antenna design, antenna characterisation, antenna modelling and antenna processing, and hence a better understanding of the physics underlying multiple antenna systems operating in realistic usage conditions is essential. Most of the cases considered here explicitly or implicitly address hand-held wireless communications terminals. However, the distinction between terminals and network wireless nodes such as access points or base stations may eventually disappear in the future. Such nodes will also most often be equipped with multiple antennas and will operate in far-from-ideal environmental conditions. On the other end, an increasing number of wireless-connected small objects with many operational constraints are being dispersed into our surroundings. The performance optimisation of such wireless devices will require dedicated efforts such as the topics addressed in this chapter. Thus, there are still many exciting research opportunities ahead on compact antenna systems in a radio channel context for the years to come.

References

[Ack98] M. J. Ackerman. The visible human project. *Proc. IEEE*, 86(3):504–511, 1998.

[AL06] J. B. Andersen and B. K. Lau. On closely coupled dipoles in a random field. *IEEE Antennas Wireless Propagat. Lett.*, 5(1):73–75, 2006.

[AMd01] M. R. Andrews, P. O. Mitra, and R. deCarvalho. Tripling the capacity of wireless communications using electromagnetic polarization. *Nature*, 409:316–318, 2001.

[AYL07] J. Avendal, Z. Ying, and B. K. Lau. Multiband diversity antenna performance study for mobile phones. In *2007 International Workshop on Antenna Technology: Small and Smart Antennas Metamaterials and Applications*, pages 193–196, 2007. [Also available as TD-07-220].

[Bau76] C. E. Baum. *The Singularity Expansion Method*. Springer, Berlin, 1976.

[BCP08a] I. B. Bonev, M. Christensen, and G. F. Pedersen. Hearing aids compliance investigation of mobile terminals. Technical Report TD-08-550, Trondheim, Norway, June 2008.

[BCP08b] I. B. Bonev, M. Christensen, and G. F. Pedersen. Impact of the ear and head shapes on the near fields radiated by mobile phones. Technical Report TD-08-647, Lille, France, October 2008.

[BEP09] T. W. C. Brown, P. C. F. Eggers, and G. Pedersen. Analysis of user impact on interference between two 4×4 MIMO links at 5 GHz. In *2nd COST2100 Workshop, Multiple Antenna Systems on Small Terminals (Small and Smart)*, Valencia, Spain, May 2009.

[BHW07] M. A. Beach, M. Hunukumbure, and M. W. Webb. Dynamics of spatial eigen modes in measured MIMO channels with different antenna modules. In *Antennas and Propagation (EuCAP 2007)*, 2007.

[BK08] T. W. C. Brown and U. H. Khan. Analysis of using MIMO antenna selection with compact IQHAs. Technical Report TD-08-619, October 2008.

[Bro07] T. Brown. Analysis of the antenna impact on eigenspace tracking. Technical Report TD-07-311, Duisburg, Germany, September 2007.

[BS07] T. W. C. Brown and S. R. Saunders. The intelligent quadrifilar helix: a compact antenna for IEEE 802.11n. In *Proc. EUCAP 2007—European Conf. Antennas Propagat.*, November 2007.

[BSE05] T. W. C. Brown, S. R. Saunders, and B. G. Evans. Analysis of mobile terminal diversity antennas. *IEE Proc. Micro. Antennas Propagat.*, 152:1–6, 2005.

[CABP06] C. Cox, E. Ackerman, G. Betts, and J. Prince. Limits on the performance of RF-over-fiber links and their impact on device design. *IEEE Trans. Micro. Theory*, 54(2):906–920, 2006.

[CHV09] M. S. Cortes, K. Haneda, and P. Vainikainen. Performance study of reference MIMO antenna configurations using experimental propagation data. Technical Report TD-09-979, Vienna, Austria, September 2009.

[CPKP04] S. H. Choi, J. K. Park, S. K. Kim, and J. Y. Park. A new ultrawideband antenna for UWB applications. *Microwave and Optical Technology Lett.*, 40(5):399–401, 2004.

[CTIA05] CTIA. Test plan for mobile station over the air performance: method of measurement for radiated RF power and receiver performance, Revision 2.1. Technical Report, 2005.

[CTIA09] CTIA. Test plan for mobile station over the air performance: method of measurement for radiated RF power and receiver performance, Revision 3.0. Technical Report, 2009.

[DBR04] S. Dossche, S. Blanch, and J. Romeu. Optimum antenna matching to minimise signal correlation on a two-port antenna diversity system. *Elect. Lett.*, 40(19):1164–1165, 2004.

[DR08] R. D'Errico and C. Roblin. Statistical analysis of UWB antenna radiation and scattering: applications and results. Technical Report TD-08-656, Lille, 6–8 October 2008.

[DS08] R. D'Errico and A. Sibille. Single and multiple scattering in UWB bicone arrays. *Int. J. Antennas Propagat.*, 2008.

[DS09] R. D'Errico and A. Sibille. Mutual coupling in UWB compact arrays. In *2nd COST2100 Workshop, Multiple Antenna Systems on Small Terminals*, Valencia, Spain, May 2009.

[GBD06] H. Ghannoum, S. Bories, and R. D'Errico. Small-size UWB planar antenna and its behaviour in WBAN/WPAN applications. In *The Institution of Engineering and Technology Seminar on Ultra Wideband Systems, Technologies and Applications*, pages 221–225, 2006.

[HDT10] F. Harrysson, A. Derneryd, and F. Tufvesson. Evaluation of user hand and body impact on multiple antenna handset performance. In *Proc. IEEE International Symposium on Antennas and Propagation*, Toronto, Canada, July 2010. [Also available as TD-10-12035].

[HMDM07] F. Harrysson, J. Medbo, A. Derneryd, and A. Molisch. Performance of a MIMO terminal including a user phantom in a stationary micro-cell scenario with comparison between a ray-based method and direct measurements. Technical Report TD-07-379, Duisburg, Germany, September 2007.

[HMM08] F. Harrysson, J. Medbo, and A. Molisch. Indoor performance of a MIMO handset including user influence by comparing a composite channel method with direct measurements. Technical Report TD-08-661, Lille, France, October 2008.

[HMM+10] F. Harrysson, J. Medbo, A. Molisch, A. Johansson, and F. Tufvesson. Efficient experimental evaluation of a MIMO handset with user influence. *IEEE Trans. Wireless Commun.*, 9(2):853–863, 2010.

[HZZ95] Y. Hwang, Y. P. Zhang, and G. X. Zheng. Planar inverted F antenna loaded with high permittivity material. *Elect. Lett.*, 31(20):1710–1712, 1995.

[IKV04] C. Icheln, J. Krogerus, and P. Vainikainen. Use of balun chokes in small-antenna radiation measurements. *IEEE Trans. Instrum. Meas.*, 53(2):498–506, 2004.

[ILYT08] K. Ishimiya, J. Långbacka, Z. Ying, and J. I. Takada. A compact MIMO DRA antenna. In *Proc. IWAT 2008—Int. Workshop Antenna Technol.*, pages 286–289, March 2008.

[IOV99] C. Icheln, J. Ollikainen, and P. Vainikainen. Reducing the influence of feed cables on small antenna measurements. *Elect. Lett.*, 35(15):1212–1214, 1999.

[JW04] M. Jensen and J. Wallace. A review of antennas and propagation for MIMO wireless communications. *IEEE Trans. Antennas Propagat.*, 52(11):2810–2824, 2004.

[KC10] C. Kakoyiannis and P. Constantinou. Wireless-sensor-targeted printed arrays with embedded mutual coupling mitigation. Technical Report TD-10-10093, February 2010.

[KH07] S. Kurokawa and M. Hirose. A new balun for antenna measurement using photonic sensor. In *The Second European Conference on Antennas and Propagation (EuCAP 2007)*, November 2007.

[KI08] J. Krogerus and C. Icheln. Considerations on anechoic chamber test methods for performance evaluation of multi-antenna mobile terminals. Technical Report TD-08-470, Wroclaw, Poland, February 2008.

[KKP07] A. Kalis, A. G. Kanatas, and C. Papadias. An ESPAR antenna for beamspace-MIMO systems using PSK modulation schemes. In *Proc. ICC 2007—IEEE Int. Conf. Commun.*, June 2007.

[KP07] A. G. Kanatas and C. Papadias. An Alamouti transmit diversity scheme using a single active antenna element. Technical Report TD-07-039, February 2007.

[KPOE01] W. A. T. Kotterman, G. F. Pedersen, K. Olesen, and P. Eggers. Cable-less measurement set-up for wireless handheld terminals. In *12th IEEE International Symposium on Personal, Indoor and Mobile Radio Communications, vol. 1*, pages B-112–B-1161, September 2001.

[KSIV06] J. Krogerus, P. Suvikunnas, C. Icheln, and P. Vainikainen. Evaluation of diversity and MIMO performance of antennas from phaseless radiation patterns. In *First European Conference on Antennas and Propagation, (EuCAP 2006)*, Nice, France, November 2006.

[LA07a] B. K. Lau and J. B. Andersen. Direction-of-arrival estimation for closely coupled arrays with impedance matching. In *IEEE International Conference on Information, Communications and Signal Processing (ICICS'07)*, Singapore, December 2007.

[LA07b]   B. K. Lau and J. B. Andersen. Impact of impedance matching on direction-of-arrival estimation of compact antenna arrays. Technical Report TD-07-377, Duisburg, Germany, September 2007.

[LAKM06]   B. K. Lau, J. B. Andersen, G. Kristensson, and A. F. Molisch. Impact of matching network on bandwidth of compact antenna arrays. *IEEE Trans. Antennas Propagat.*, 54(11):3225–3238, 2006.

[LLT10]   H. Li, B. K. Lau, Z. Tan, and Y. Ying. Isolation enhancement of compact MIMO antennas with current localization. Technical Report TD-10-12076, November 2010.

[LTL10]   H. Li, Y. Tan, Z. Lau, and B. K. Ying. Antenna design tradeoff of multiple antenna terminals with ground plane excitation. Technical Report TD-10-11080, June 2010.

[Lux07]   C. Luxey. Multi-antenna systems for UMTS cellular phones: diversity performance in different propagation environments. Technical Report TD-07-301, September 2007.

[MBJ08]   J. Medbo, J. E. Berg, and M. Jovic. Validation of antenna coupling and channel modeling in a real propagation environment. Technical Report TD-08-624, Lille, France, February 2008.

[Mol08]   P. Moller. CTIA hand phantom development status. Technical Report TD-08-523, Trondheim, Norway, June 2008.

[Mol09]   P. Moller. CTIA hand phantom update homogeneous hand phantoms in four grips will allow hand held OTA testing of modern cell phones. Technical Report TD-09-803, Valencia, Spain, May 2009.

[NDT09]   A. I. Najam, Y. Duroc, and S. Tedjini. UWB MIMO antenna with reduced mutual coupling. In *COST2100 Multiple Antenna Systems on Small Terminals (Small and Smart)*, May 2009.

[NYB+10]   J. O. Nielsen, B. Yanakiev, I. B. Bonev, M. Christensen, and G. F. Pedersen. MIMO channel capacity for handsets in data mode operation. Technical Report TD-10-12028, Bologna, Italy, November 2010.

[Pel08]   M. Pelosi. Total efficiency dynamics of handheld devices influenced by human hand. In *Student Paper, 2008 Annual IEEE Conference*, pages 1–4, 15–26 February 2008. [Also available as TD-07-319]. doi:10.1109/AISPC.2008.4460565.

[PFK+09a] M. Pelosi, O. Franek, M. B. Knudsen, M. Christensen, and G. Pedersen. A grip study for talk and data modes in mobile phones. *IEEE Trans. Antennas Propagat.*, 57(4):856–865, 2009. [Also available as TD-07-036,TD-08-431,TD-07-320]. doi:10.1109/TAP.2009.2014590.

[PFK+10]   M. Pelosi, O. Franek, M. B. Knudsen, G. Pedersen, and J. B. Andersen. Antenna proximity effects for talk and data modes in mobile phones. *IEEE Antennas Propagat. Magazine*, 52(3):15–27, 2010. [Also available as TD-08-525]. doi:10.1109/MAP.2010.5586570.

[PFKP09a] M. Pelosi, O. Franek, M. B. Knudsen, and G. F. Pedersen. Efficiency improvement of PIFA antennas in close proximity with the user's body with buffer dielectric loading. Technical Report TD-09-721, February 2009.

[PFKP09b] M. Pelosi, O. Franek, M. B. Knudsen, and G. F. Pedersen. Hand phantoms for browsing stance in mobile phones. In *Antennas and Propagation Society International Symposium, 2009. APSURSI '09. IEEE*, pages 1–4, 1–5 June 2009. [Also available as TD-08-611]. doi:10.1109/APS.2009.5171882.

[PFPK09]   M. Pelosi, O. Franek, G. Pedersen, and M. Knudsen. User's impact on PIFA antennas in mobile phones. In *IEEE 69th Vehicular Technology Conference, VTC Spring*, April 2009. doi:10.1109/VETECS.2009.5073899.

[PKP09]   M. Pelosi, M. B. Knudsen, and G. F. Pedersen. A novel decoupling technique: isolation potential of narrow-band radiators influenced by the user's body proximity. Technical Report TD-09-909, September 2009.

[PL09]   V. Plicanic and B. K. Lau. Impact of spacing and gain imbalance between dipoles on HSPA throughput performance. *Elect. Lett.*, 45(21):1063–1065, 2009.

[PLA11] V. Plicanic, B. K. Lau, and H. Asplund. Performance of handheld MIMO terminals in noise- and interference-limited urban scenarios. In *Proc. EuCAP 2011: The 5th European Conference on Antennas and Propagation*, Rome, Italy, April 2011. [Also available as TD-10-12050].

[PLDY09] V. Plicanic, B. K. Lau, A. Derneryd, and Z. Ying. Actual diversity performance of a multiband diversity antenna with hand and head effects. *IEEE Trans. Antennas Propagat.*, 57(5):1547–1556, 2009.

[PLPV08] C. Pereira, F. Lepennec, Y. Pousset, and R. Vauzelle. Impact of mutual coupling on MIMO channel modelling. Technical Report TD-08-413, Wroclaw, Poland, February 2008.

[PLY08] V. Plicanic, B. K. Lau, and Z. Ying. Performance of a multiband diversity antenna with hand effects. In *2008 International Workshop on Antenna Technology: Small Antennas and Novel Metamaterials*, pages 534–537, 2008. [Also available as TD-08-428].

[PP10a] M. B. Pelosi, M. Knudsen, and G. F. Pedersen. Multiple antenna systems with high isolation. Technical Report TD-10-12032, November 2010.

[PP10b] M. B. Pelosi, M. Knudsen, and G. F. Pedersen. Narrowband MIMO antennas with transceiver separation. Technical Report TD-10-12033, November 2010.

[PPK10] M. Pelosi, G. F. Pedersen, and M. B. Knudsen. A novel paradigm for high isolation in multiple antenna systems with user's influence. In *Proc. Europ. Conf. Antennas Propagat (EuCAP)*, April 2010.

[PZL10] V. Plicanic, M. Zhu, and B. K. Lau. Diversity mechanisms and MIMO throughput performance of a compact six-port dielectric resonator antenna array. In *Proc. IEEE International Workshop on Antenna Technology (IWAT2010)*, Lisbon, Portugal, March 2010.

[RBS03] C. Roblin, S. Bories, and A. Sibille. Characterization tools of antennas in the time domain. In *Int. Workshop on Ultra Wideband Systems IWUWBS*, Oulu, Finland, June 2003.

[RD07] H. Rogier and W. Dullaert. Spatial correlation in uniform circular arrays subject to mutual coupling. Technical Report TD-07-332, Duisburg, Germany, September 2007.

[RD08] C. Roblin and R. D'Errico. Statistical analysis of UWB antenna radiation and scattering: theory and modelling. Technical Report TD-08-657, Lille, 6–8 October 2008.

[RKI+09] S. Ranvier, M. Kyro, C. Icheln, C. Luxey, R. Staraj, and P. Vainikainen. Compact 3-D on-wafer radiation pattern measurement system for 60 GHz antennas. *Microwave and Optical Technology Lett.*, 51(2):319–324, 2009. [Also available as TD(08)669].

[SGMP09] A. Sibille, J. Guterman, A. Moreira, and C. Peixeiro. Performance evaluation of 2.4 GHz laptop antennas using a joint antenna-channel statistical model. Technical Report TD-09-930, Vienna, Austria, 28–30 September 2009.

[Shi74] Y. Shifrin. *Statistical Antenna Theory*. The Golem Press, Boulder, Colorado, 1974.

[Sib08a] A. Sibille. A small signal analysis of statistical antenna modelling. Technical Report TD-08-635, Lille, France, 6–8 October 2008.

[Sib08b] A. Sibille. Statistical antenna modelling. Technical Report TD-08-524, Trondheim, Norway, 4–6 June 2008.

[Sib09a] A. Sibille. A frequency wise statistical model of the effective gain incorporating terminals and local propagation variability. Technical Report TD-09-735, Braunschweig, Germany, 16–18 February 2009.

[Sib09b] A. Sibille. A statistical approach to account for terminal variability in radio access simulators. Technical Report TD-09-733, Braunschweig, Germany, February 2009.

[Sib09c] A. Sibille. A statistical approach to account for terminal variability in radio access simulators. Technical Report TD-09-733, Braunschweig, Germany, 16–18 February 2009.

[Sib10] A. Sibille. A first step towards statistical modeling of MIMO terminals accounting for local propagation. Technical Report TD-10-10054, Athens, Greece, 3–5 February 2010.

[SLDY09] V. P. Samuelsson, B. K. Lau, A. Derneryd, and Z. Ying. Channel capacity performance of multi-band dual antenna in proximity of a user. In *2009 IEEE International Workshop on Antenna Technology*, pages 1–4, 2009. [Also available as TD-08-667].

[Smu02] P. Smulders. Exploiting the 60 GHz band for local wireless multimedia access: prospects and future directions. *IEEE Commun. Mag.*, 40(1):140–147, 2002.

[SS10] Schmid and Partner Engineering AG (SPEAG). http://www.speag.com, 2010.

[sta08] Session BKF: Stochastic modeling and uncertainty management in electromagnetics. In *29th URSI General Assembly and Scientific Symposium*, Chicago, USA, 10–16 August 2008.

[Suv06] P. Suvikunnas. *Methods and criteria for performance analysis of multiantenna systems in mobile communications*. PhD thesis, Helsinki University of Technology, Espoo, Finland, August 2006.

[SVS+06] P. Suvikunnas, J. Villanen, K. Sulonen, C. Icheln, J. Ollikainen, and P. Vainikainen. Evaluation of the performance of multiantenna terminals using a new approach. *IEEE Trans. Instrum. Meas.*, 55(5):1804–1813, 2006.

[THN+08] H. Tanaka, M. Hirose, M. Nagatoshi, S. Kurokawa, and H. Morishita. Photonic balun for antenna measurements free of coaxial cables. In *International Workshop on Antenna Technology (iWAT 2008)*, pages 255–258, March 2008.

[TL08a] R. Tian and B. K. Lau. On prediction of MIMO capacity performance with antenna magnitude patterns. Technical Report TD-08-651, Lille, France, October 2008.

[TL08b] R. Tian and B. K. Lau. Uncoupled antenna matching for performance optimization in compact MIMO systems using unbalanced load impedance. In *Proc. IEEE Vehic. Technology Conf. Spring*, pages 299–303, Singapore, May 2008. [Also available as TD(08)438].

[TL09a] R. Tian and B. K. Lau. Correlation-based phase synthesis approach for MIMO capacity prediction using antenna magnitude patterns. Technical Report TD-09-727, Braunschweig, Germany, February 2009.

[TL09b] R. Tian and B. K. Lau. Simple and improved approach of estimating MIMO capacity from antenna magnitude patterns. In *Third European Conference on Antennas and Propagation (EuCAP 2009)*, Berlin, Germany, March 2009.

[TL10] R. Tian and B. K. Lau. Six-port antennas for experimental verification six degrees-of-freedom in wireless channels. Technical Report TD-10-12054, November 2010.

[TLBA10] Y. Tan, B. K. Lau, and J. Bach Andersen. On bandwidth enhancement of compact multiple antennas with parasitic decoupling. Technical Report TD-10-12085, November 2010.

[TPL+09] R. Tian, V. Plicanic, B. K. Lau, J. Långbacka, and Z. Ying. MIMO performance of diversity-rich compact six-port dielectric resonator antenna arrays in measured indoor environments at 2.65 GHz. In *COST2100 Multiple Antenna Systems on Small Terminals (Small and Smart)*, May 2009.

[TPLY10a] R. Tian, V. Plicanic, B. K. Lau, and Z. Ying. A compact six-port dielectric resonator antenna array: MIMO channel measurements and performance analysis. *IEEE Trans. Antennas Propagat.*, 58(4):1369–1379, 2010.

[TPLY10b] R. Tian, V. Plicanic, B. K. Lau, and Z. Ying. A compact six-port dielectric resonator antenna array: MIMO channel measurements and performance analysis. *IEEE Trans. Antennas Propagat.*, 58(4):1369–1379, 2010.

[TTK+09] J. I. Takada, T. Teshirogi, T. Kawamura, A. Yamamoto, and T. Sakuma. Total radiated power measurement for antenna integrated radios using a spheroidal coupler. Technical Report TD-09-954, Vienna, Austria, September 2009.

[TTK+10] J. I. Takada, T. Teshirogi, T. Kawamura, A. Yamamoto, T. Sakuma, and Y. Nago. OTA measurements of small radio terminals by a spheroidal coupler. Technical Report TD-10-12060, Bologna, Italy, November 2010.

[VOKK02] P. Vainikainen, J. Ollikainen, O. Kivekas, and K. Kelander. Resonator-based analysis of the combination of mobile handset antenna and chassis. *IEEE Trans. Antennas Propagat.*, 50(10):1433–1444, 2002.

[WGB10]	M. Webb, D. Gibbins, and M. Beach. Slot antenna performance and signal quality in a smartphone prototype. Technical Report TD-10-12025, Bologna, Italy, November 2010.
[Wi007]	Winner channel models. In *IST-4-027756 WINNER II D1.1.2*, September 2007.
[Win87]	J. Winters. On the capacity of radio communication systems with diversity in a rayleigh fading environment. *IEEE J. Select. Areas Commun.*, 5(5):871–878, 1987.
[WJ04a]	J. W. Wallace and M. A. Jensen. Mutual coupling in MIMO wireless systems: a rigorous network theory analysis. *IEEE Trans. Wireless Commun.*, 3(4):1317–1325, 2004.
[WJ04b]	J. Wallace and M. Jensen. Termination-dependent diversity performance of coupled antennas: network theory analysis. *IEEE Trans. Antennas Propagat.*, 52(1):98–105, 2004.
[YCNP10]	B. Yanakiev, M. Christensen, O. J. Nielsen, and G. F. Pedersen. Simulated and measured small terminal antenna correlation. Technical Report TD-10-11050, Aalborg, Denmark, June 2010.
[YNP10]	B. Yanakiev, J. Nielsen, and G. F. Pedersen. On small antenna measurements in a realistic MIMO scenario. In *Fourth European Conference on Antennas and Propagation (EuCAP 2010)*, April 2010.
[YP09]	B. Yanakiev and G. F. Pedersen. Using optical links for radio propagation measurements with small mobile devices. Technical Report TD-09-968, Vienna, Austria, September 2009.
[YPC10a]	B. Yanakiev, G. F. Pedersen, and M. Christensen. MIMO OTA optical measurement device. Technical Report TD-10-12068, Bologna, Italy, November 2010.
[YPC10b]	B. Yanakiev, G. F. Pedersen, and M. Christensen. Using optical links for radio propagation measurements, Part 2. Technical Report TD-10-065, Athens, Greece, February 2010.
[YTK+07]	A. Yamamoto, H. Toshiteru, O. Koichi, K. Olesen, J. Nielsen, N. Zheng, and G. F. Pedersen. Comparison of phantoms in a browsing position by a NLOS outdoor MIMO propagation test. In *ISAP 2007*, Niigata, Japan, August 2007.
[ZHZ97]	Y. P. Zhang, Y. Hwang, and G. X. Zheng. A gain-enhanced probe-fed microstrip patch antenna of very high permittivity. *Microwave and Optical Technology Lett.*, 15:89–91, 1997.

Chapter 5
"OTA" Test Methods for Multiantenna Terminals

Chapter Editor Gert F. Pedersen, Chapter Editor Mauro Pelosi, Jan Welinder, Tommi Jamsa, Atsushi Yamamoto, Miia Nurkkala, Soon L. Ling, Werner Schroeder, and Tim Brown

This chapter focuses on Over-The-Air (OTA) test methods for multiantenna systems. The chapter is organized into six sections: the first section presents a brief introduction to OTA; the second section describes some proposed OTA measurement setups; third and fourth sections address evaluations and comparison of OTA setups; the fifth section briefly summarizes the main aspects about regulation issues of Ultra-WideBand (UWB) technology; the second section describes antenna design and characterization; the third section discusses UWB channel modeling and measurements; real environment references for OTA measurements are discussed in the fifth section; finally, the last section presents the conclusions.

5.1 Introduction to OTA

5.1.1 Recent Trends in Mobile Communications

The development of mobile communication systems is undergoing a rapid evolution, leading to a significant increase in the use of small antennas for handheld devices [KLM+07]. The ever increasing data traffic needs have fostered significant progress, opening new scenarios and possibilities and paving the way for the future advent of Fourth Generation (4G). Unfortunately, conventional urban and indoor propagation environments create hostile conditions for the transmission of wireless signals. However, the performance of mobile terminals subject to strong multipath fading can be improved using diversity techniques. By combining in some optimum or practical way the signals from more than one antenna, the overall system performance can be improved [SI08]. Though the exploitation of time, frequency, and code diversity is now mature, the potential of spatial diversity and multiplexing with the use of multiple antennas needs yet to be uncovered. The strength of

G.F. Pedersen (✉) · M. Pelosi (✉)
Aalborg University, Aalborg, Denmark

Multiple-Input Multiple-Output (MIMO) is that it offers an increased data throughput/capacity without the need for increasing the frequency bandwidth, as the scarcity of spectrum is becoming more and more challenging. In fact, MIMO systems use multiple antennas at both ends of the communication link, turning multipath propagation into a more beneficial environment.

5.1.2 Potential and Challenges in Handset MIMO

The most recent developments in wireless standards clearly indicate that multiple antennas will be utilized also in the mobile terminal side. In fact multiple antennas in both terminals and base stations compose a MIMO system which targets to enhance the total system performance. Typical performance improvements are higher spectral efficiency, better Quality-of-Service (QoS) and link quality even in harsh fading scenarios, improved cell coverage and performance at the cell edges [KKN$^+$08]. 3GPP-Long Term Evolution (LTE) is one of the most promising upcoming air interfaces for cellular mobile telephony, and its development is an important step towards 4G [PLA10]. LTE requires mandatory support for MIMO in downlink, where a minimum of two antennas will be in place at the mobile station, while so far the uplink will take advantage of multi-user MIMO [Gag09]. The Swedish carrier TeliaSonera has recently deployed the first commercial LTE network. Though in the area of central Stockholm unprecedented downlink throughputs around 80 Mbps were achieved, the distance of the terminal to the base station was of crucial importance in determining the gap between theoretical and practical MIMO performance. In fact propagation and antennas are often overlooked in MIMO system design [KKN$^+$08]. In multiantenna communication systems antennas have significant impact on total system performance. Both antenna placement and design has to be considered carefully, especially in mobile terminals covering several frequency bands. The current trend in the form factor of mobile terminals demands for smaller and smaller devices. Before the advent of handset MIMO, there was already a problem in the coexistence of multiband antennas in a reduced volume. In fact fundamental theoretical limitations link in a tradeoff antenna size, efficiency and bandwidth. When multiple antennas on the same ground plane are radiating in the presence of each other mutual coupling phenomena emerge. This causes an unwanted transfer of electromagnetic energy from one antenna to another, causing a loss in the antenna system efficiency. Mutual coupling causes also an increase in the envelope correlation coefficient with a detrimental effect on both diversity and MIMO gain. Beside the challenges posed by the propagation environment itself, it is also important to take into account the interaction of the MIMO enabled handset with the close by objects. It is in fact very important to design the antenna system paying attention to the interaction with the human user. The user can be seen as a lossy material load from the antenna system point of view, causing frequency detuning, absorption, coupling and radiation pattern deterioration. Beside the well-known influence of the head of the user, the presence of the hand has direct consequences in determining a successful handset MIMO performance. In fact the gain imbalance among different branches may strongly reduce the benefits of MIMO. This shows that often too optimistic predictions such as high antenna efficiency, favorable channel conditions and reduced user interaction can hinder the theoretical performance [Gag08]. It is therefore obvious that both proper design guidelines and test procedures are necessary to deploy reliable and efficient mobile communication systems.

5.1.3 Single Antenna OTA

Despite digital cellular systems having been in existence for nearly 20 years, it is only in very recent times that realistic radiated performance tests for Single-Input Single-Output (SISO) terminals have been developed [Rum09]. In fact this was motivated by a compromise in the mobile communication performance between vendor offerings and operator desires. This is achieved by OTA measurements, which determine the Over-The-Air performance of mobile terminals. Though it can be said that the term OTA methods has a fairly wide range of application, it may embrace most non-conducted tests. In fact different kinds of tests exist [SFP09]:

1. Test of compliance with regulatory limits (Specific Absorption Rate (SAR), Hearing Aid Compatibility (HAC), out of band emissions).
2. Conformance and interoperability tests based on the specifications of the applicable standard (Radio Frequency (RF) test, physical layer procedures and coexistence).
3. Performance tests as specified in the applicable standard and additional performance tests beyond the minimum specifications as usually required by mobile operators before approving User Equipment (UE) for a given network (throughput, coverage).

Device manufacturers have the challenging task of jointly optimizing spurious emission limits, SAR values, and OTA performance [Gag08]. As optimizing one end will deteriorate another one finding an acceptable optimum is a very tricky task. Good OTA performance requires that the antenna pattern both in the transmitting and receiving direction need to be smooth and to avoid nulls. In order to compensate for a weak transmission, both transmitted power and receiving sensitivity are crucial. Most of SISO antenna parameters and figures of merit can be derived from radiation pattern measurement. Because of both limited size and power resources, highly efficient systems are needed. In fact the network operators have high interest in making sure that reliable devices are used [Gag08]. Standardized figures of merit were also needed to replace expensive and time-consuming field tests, allowing different test labs to use a common measurement framework. The test requirements are typically relaxed by the test system uncertainty, avoiding the possibility of failing a good terminal on one hand but reducing the probability of failing a bad one on the other hand [Rum10b]. It is highly desirable therefore if the test is to have any practical use, that the test system uncertainty is kept as low as possible. It is important to note that OTA performance was a retrospective requirement which would apply to existing designs so that it was decided to base the performance requirements on a measurement campaign conducted using existing terminals [Rum10b]. OTA measurements are performed in fully anechoic chambers in a controlled environment [Gag08]. In an anechoic chamber a direct line of sight exist between the Device Under Test (DUT) and the probe antenna with no reflections from the walls [SI08]. Because of the rapid development of mobile communication, the use of small antennas has highly increased [KLM$^+$07], making more important to have fast and flexible systems to test the antenna performance [SI08]. A compromise between testing time accuracy and cost can be realized by multi-probe systems [KLM$^+$07]. The development of methods for the evaluation and comparison of the true performance of mobile terminal antennas in a mobile network has gained clearly less attention in the industry compared to the developing traditional [KKN$^+$08] radiation pattern measurement systems [KLM$^+$07]. In part, this tendency has been recently strengthened by the developed standards both in Europe and North America that define the Total Radiated Power (TRP) and Total Isotropic Sensitivity (TIS). The European pre-standard on the OTA performance test method was prepared by the sub-working group of COST 273 and then adopted in the 3rd Generation Partnership Project (3GPP) technical specifications. In North America, a working group formed mainly by wireless carriers and mobile device manufacturers has prepared the OTA Test Plan within the Cellular Telecommunications & Internet Association (CTIA) Certification Program [KLM$^+$07]. During research and product development, the measurements are typically carried out by conducted testing. In the final product development and certification testing, the figure of merit is the TRP rather than antenna efficiency. The TRP is an indicator of how much power the DUT actually radiates, when both mismatch and losses in the antenna are taken into account and it is defined as the integral of the Equivalent Isotropic Radiated Power (EIRP) in different directions over the full spherical surface enclosing the DUT [KLM$^+$07]. By using a phantom the impact of body of the user can also be taken into account. Currently, the antenna performance testing of mobile terminals only specifies the inclusion of a Specific Anthropomorphic Mannequin (SAM) phantom head in the measurement, while the adoption of a phantom hand is in progress [KLM$^+$07]. The corresponding figure of merit in the receive mode is the TIS, defined as the minimum received power required to achieve a certain specified Bit Error Rate (BER) [KLM$^+$07]. The TIS is obtained by integrating the spherical pattern [KLM$^+$07]. The TRP and TIS statistics gathered from the SISO OTA process were used to get a better understanding of the spread of performance across a wide variety of commercial UEs [Rum10b]. From this data, performance requirements were set being a compromise between operator goals and vendor capabilities, with the end result of setting the performance requirements so that the majority of existing terminals would pass. Over time the standardized test method resulted in a general improvement in the mobile terminals' performance [Rum10b]. As TRP and TIS do not take into account the multipath propagation environment, they are not the best possible figures of merit [JKW$^+$10]. The Mean Effective Gain (MEG) is a more appropriate figure of merit, as includes the effect of the gain pattern of the antenna and the realistic signal distribution of incoming multipath signals at the mobile terminal [KLM+07]. Though MEG is a practical figure of performance for single-element mobile terminal antennas, it is not adequate for the evaluation of multiple antenna systems. Driven by the development and utilization of diversity and MIMO techniques in particular, interest in methods for testing the true performance of mobile terminal antennas is of great research interest at the moment. As the current test method is not suitable for evaluating mobile terminals that would include a diversity or MIMO antenna system, it is of utmost importance to develop an applicable way to test the OTA performance of multiantenna mobile terminals [KLM+07].

5.1.4 Multiantenna OTA Features and Challenges

In MIMO systems antennas will have an even more significant impact on system performance. It is still common to overlook the propagation and antennas in the design of MIMO algorithms and systems [KKN+08]. Most of the conventional simulations, performance tests and conformance tests neglect the antenna effects or take them into account only in limited extent. Therefore they will not give the complete picture of the radio performance [KKN+08]. Recently is strongly emerging a need to perform proper MIMO OTA tests. A candidate MIMO OTA test method requires several characteristics. It should include realistic radio channel spatial and temporal effects and take into account the effect of the human user in a controlled environment [KMV08]. As future needs related to new mobile terminals' form factors, channel models, and operating modes may arise, the test should be open to longer term scenarios. In order to ensure good test repeatability, particular care should be devoted to the choice of realistic figures of merit, allowing also a fast and low-cost test implementation [KMV08]. Spatial and temporal characteristics of the radio channel have significant effect on the MIMO system performance. The development of an effective MIMO OTA test procedure would be valuable for all parties in the field, e.g. the users of the mobile devices, device manufacturers, network operators, and both antenna and test equipment manufacturers [KMV08]. Testing of antenna performance of mobile terminals is widely based on radiation pattern measurements [KI08]. The most common test method makes use of spherical radiation pattern measurement. For single antennas, amplitude-only radiation pattern measurement provides in many cases sufficient information for characterizing antenna gain, efficiency and the [KKN+08] far-field radiation pattern. For multiantenna configurations, the performance evaluation involves parameters like envelope correlation coefficient, diversity gain, and effective array gain. To evaluate all necessary parameters, complex radiation patterns of all the elements in the antenna configuration are typically needed. However, in active measurements of mobile terminals, the phase information is usually not directly available [KI08]. In the SISO OTA case every terminal on the market was immediately available for measurement providing a strong foundation for the definition of the performance requirements [Rum10b].

Concerning the MIMO OTA case, currently there are almost no commercial UE on the market which have implemented MIMO. Moreover, the performance of a MIMO terminal in a complex radio environment is hugely more complex than the SISO equivalent being a scalar figure integrating the TRP and TIS in a sphere around the UE [Rum10b]. Though in SISO OTA it has been possible to define simple figures of merit, in MIMO OTA it is hard to condense in a unique figure of merit the manifold applications of MIMO. In MIMO OTA it is not allowed to separate the measurement of basic quantities like TIS from the complex data processing and error control process used during MIMO transmissions [Gag08]. Measuring each antenna individually does not mirror the later performance of the device in the network, as antenna, propagation and signal processing aspects are combined [Gag08]. UE vendors would not have agreed on SISO OTA performance requirements until they had measured their devices using the standard procedure [Rum10b]. It can therefore safely be assumed that for the MIMO case it is even less likely that agreement on performance requirements will be achievable until actual commercial devices have been developed and tested in a standard environment [Rum10b]. In fact many issues concur in the definition of a proper MIMO OTA candidate test method. It is important to specify the channel models/propagation scenarios to be used, comparing direct test route measurements of MIMO terminals and Spatial Fading Emulator (SFE) laboratory testing and actual network performance [KMV08]. Because of the inherent complexity of any MIMO system, MIMO OTA figures of merit are hard to define [KJW+10b]. Important antenna performance parameters for multi-antenna configurations include parameters like envelope correlation coefficient, diversity gain, effective array gain, capacity and eigenvalue dispersion, while system level ones are throughput, capacity, BER, and sensitivity [KMV08]. The efficacy of a testing method can be considered as its ability to differentiate between good and bad performance [Rum10a]. The studies on MIMO OTA testing indicate that there are numerous MIMO-specific factors which will increase the test uncertainty [Szi10]. This may lead to a situation where the probability of failing undesirable products drops [Rum09]. The difficulty in deciding what to measure is further complicated by the growing gap between the open-loop performance requirements traceable to link level simulations used for conformance testing and the far more complex closed-loop operating conditions which will prevail in real networks [Rum09]. In fact some of the factors that will influence real-life performance are not only related to the antenna design but also to the ability of the entire transceiver chain to properly respond to the dynamics of the radio environment including narrowband scheduling, precoding and adaption to the ideal MIMO configuration [Rum09]. The ultimate goal of MIMO OTA is to develop tests that enable measurement of UE MIMO performance which can tell the difference between acceptable and unacceptable performance [Rum10b]. A test which has a too high uncertainty leading to relaxed test requirements would not be acceptable since the probability of failing a bad design would be too low. Equally, any test whose figure of merit is not sensitive to actual OTA performance is of little or low value, regardless of how accurate it may be [Rum10b].

5.1.5 Multiantenna OTA Standardization Efforts

In the European COST 2100 Action on "Pervasive Mobile & Ambient Wireless Communications" the sub-working group Compact Antennas Systems for Terminals (CAST) addressed the testing of the OTA performance on MIMO devices in great detail [Gag09]. Since COST 2100 has a rather large participation by people from academia, also theoretical background is covered well. Similar to COST 273 which prepared the OTA test plan now implemented under CTIA, COST 2100 has jointly worked with CTIA and 3GPP towards a test plan for OTA measurements on MIMO devices. The realization of a MIMO OTA method is not trivial. Seeking for the best solution, beside high-end methods simplified and cost, effective methods may complement them in the process of commercial certification and approval of a device by a network operator [SF09]. In fact it is a challenging question what degree of sophistication and detail in the emulation of channel models is needed for purposes of commercial device testing as it is more important to be statistically meaningful than exact for a specific scenario [FS09a]. The following sections will discuss in detail the proposed MIMO OTA measurement setups with laboratory evaluations and round robin tests, providing also comparisons with real environment references and recommendation on candidate MIMO OTA setup to the relevant standardization bodies.

5.2 Proposed OTA Measurement Setups

5.2.1 Introduction

There are a number of different proposals for measurements setups that have different properties, capabilities and price.

5.2.1.1 Circular Antenna Array

The basic idea is to emulate a Two-Dimensional (2D) propagation model as described e.g. in the Winner documents. Each antenna transmits a wave that represents a reflected wave in the multipath environment. Several antennas that could be combined with sophisticated signal processing can generate clusters as described in the clustered delay line models. The signal processing is done at RF frequency using attenuators, phase shifters, etc., or in the base band using channel emulators.

5.2.1.2 Single Spatial Cluster Array

The complexity but also the possibilities are reduced if the array includes a few antennas in a limited sector.

5.2.1.3 Reverbertion Chambers

COST 2100 has no intention to study reverb chambers, but they are anyway a possible measurement setup. They generate a Three-Dimensional (3D) isotropic environment with no possibilities to emulate clustered wave propagation models, at least in their basic configuration.

5.2.1.4 Decomposition

The fact that there exist a large number of tests and test equipment developed for conducted test gives the idea that as little as possible should be done over the air. OTA test should be limited to antenna characterization and then combined with conducted tests.

5.2.1.5 Simulation

Simulation is actually not proposed as a full test but instead as a useful tool for basic investigations of system concepts and test setups (e.g. [FKS10]).

5.2.2 The Multi-Path Simulator (MPS) and the Spatial Fading Emulator (SFE)

The spatial fading emulator presented by Tokyo Institute of Technology and Panasonic has been further developed in parallel at SP and Sony Ericsson. Various aspects and measurements have been presented in the COST 2100 Transmit Diversity (TD)s [Wel09, WFBM09, Tak09, YTH[+]10, HYS[+]10, Hal10] (see Fig. 5.1 for SFE test setup).

Both consist of a number of antennas placed in a circular array around the tested user equipment and hardware that operates at the carrier frequency. Each antenna generates a ray that can be controlled in several manners. To fully cover the requirements in a propagation model like e.g. the clustered delay line models in Winner, it is necessary to control the following parameters for a number of rays:

1. Direction
2. Amplitude
3. Doppler shift
4. Polarization
5. Delay

Both systems solve 1–4 above at the carrier frequency using reasonably cheap components that are flexibly controlled from computers. The signal path includes e.g. attenuators and phase shifters and thus can be used with any type of cellular system within the frequency range without modifications.

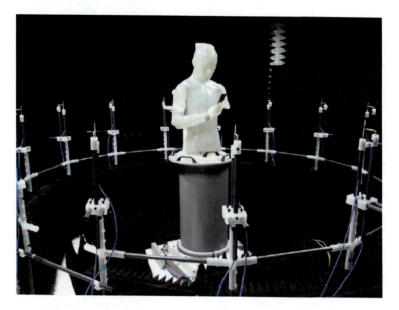

Fig. 5.1 The SFE test setup

5.2.2.1 The Antenna Array

The Multi-Path Simulator (MPS) and the SFE have arrived at different numbers of antennas in their designs. The difference is significant since the MPS uses 16 independent antennas mounted in pairs with one vertical and one horizontal broadband dipole. The SFE on the other hand has 62 dipole antennas in 31 pairs.

It is shown for the MPS that eight antennas having the same polarization are enough to create a good Rayleigh distribution (see Fig. 5.2). SFE emulates the angular power spectrum quite accurately. A more detailed analysis on the antenna array has been presented for the SFE discussing measurement accuracy. The adequate number of antennas has been also discussed in other papers, but the final answer does not seem to be in place yet. However, it is clear that a large number of antennas allows a more accurate emulation of propagation models but increases cost and also internal scattering and coupling. The dual polarizations are used to control the polarization in the test area. The MPS antennas have by themselves a polarization discrimination of -30 dB. This number cannot be achieved in the test area because of scattering in cables and other equipment. The experimental systems presented in the TDs are not designed to minimize cross polarization and are limited to -14 dB.

5.2.2.2 Signal Processing

A Doppler shift is added to each of the 16 channels using a phase shifter that runs through 360° phase change continuously at a certain rate. If e.g. the period is set

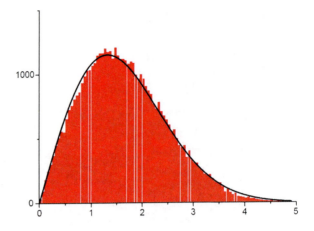

Fig. 5.2 Field strength probability density from a simulation using 8 vertical antennas (not normalized). Rayleigh distribution is also shown in the figure. The results are verified in the experimental system

to 5 ms a Doppler shift of 200 Hz is generated. By selecting the individual rates carefully any Doppler spectrum can be generated with realistic angles of incidence. Individual attenuators create the angular power spectrum. Handling delay is done differently in the MPS and SFE. The SFE does not include any provision for generating delay at the carrier frequency. Measurements including delay are done using a channel emulator. It is discussed to introduce delay in the various signal paths in the MPS, and several technologies are proposed but not implemented in the presentations.

5.2.2.3 Measurements

Both the MPS and the SFE have been used for measurement that has been presented. Most work has so far been done with the SFE. Examples of characteristics that have been measured are:

- BER for Global System for Mobile Communications (GSM) handset, ongoing work on throughput for High-Speed Downlink Packet Access (HSDPA) (see Fig. 5.3)
- Characteristics of passive antennas, e.g. MEG and correlation
- MIMO capacity for passive antennas
- Influence of delay for IEEE 802.11n Wireless Local Area Network (WLAN)

5.2.2.4 Proposed Performance Test Method

A test method is briefly outlined for the MPS. It is recognized that the antenna ring is strictly 2D as well as most propagation models, while the real world is slightly more complicated. It is currently not clear if emulating the real environment is important or if simplified test cases will give an equal result. Simplifications are anyway necessary. Two tests are proposed:

Fig. 5.3 GSM BER as a function of the Doppler spectrum. Measured with the MPS

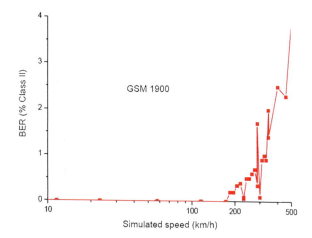

- Static test in Line-Of-Sight (LOS) measuring maximum performance. MIMO is achieved using polarization diversity.
- Mobile test in Non-Line-Of-Sight (NLOS) emulating walking and driving.

5.2.3 Elektrobit [KKN+08]

The MIMO OTA test setup is composed of a base station emulator, a multidimensional fading emulator, an anechoic chamber, a number of OTA chamber antennas and a DUT. Figure 5.4 depicts an example of the OTA concept. The purpose of the figure is not to restrict the implementation, but rather to clarify the general idea of the MIMO OTA concept. For simplicity, uplink cabling is not drawn here.

The DUT is located at the center of the anechoic chamber. The idea of locating DUT into the center provides a possibility to create a radio channel environment where the signal can arrive from various possible directions simultaneously to the DUT.

The directional information in the channel model is linked to the antenna constellation in the chamber by allocating the paths to the most geometrically suitable antennas. Based on the transmitting antenna and OTA antenna configurations, OTA models are created by taking in to account the geometrical (Angle Of Arrival (AOA) and Angle Of Departure (AOD)) information. The amplitudes are weighted so that the different channel models being routed to OTA antennas create the desired channel effects. The signal at the terminal antenna will be such that Power-Delay Profile (PDP), fast fading and spatial effects are as close as possible the same as in the original standard channel model. Base Station (BS) antenna characteristics, AOD, PDP, Doppler, and fading are included in the channel model in the channel emulator. For DUT side, the antenna effect is taken into account via real radio transmission in the OTA chamber. AOA is created by the channel emulator and the probe antennas.

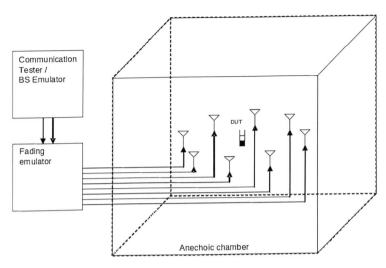

Fig. 5.4 Test set-up of OTA performance measurement

The desired channel model can be created in a limited area, called quiet zone. The size of the quiet zone depends on the number of probe antennas. One possibility to reduce the number of required probe antennas is to use single spatial cluster model. In single cluster model, the AOA of each cluster is moved to the same angle. Obviously, one AOA requires lower number of antennas than multiple AOAs especially when angular spread is narrow, e.g., 35 degrees.

5.2.4 Agilent [Kon09]

The assumption of the two-stage MIMO OTA method is that the measured far-field antenna pattern of multiple antennas can fully capture the mutual coupling of the multiple antenna arrays and their influence. Thus, to do the two-stage MIMO OTA test, the antenna patterns of the antenna array needs to be measured accurately in the first stage.

Stage 1: Test multiple antennas system in traditional chamber. For the MIMO antenna pattern measurement, most of the test configurations are the same as that in Annex A of [3GP]. The DUT is placed against a SAM phantom, and the characteristics of the SAM phantom are specified in Annex A.1 of [3GP]. The chamber is equipped with a positioner, which makes it possible to perform full 3D far-zone pattern measurements for both Transmitting and Receiving radiated performance. The measurement antenna should be able to measure two orthogonal polarizations (typically linear theta (q) and phi (f) polarizations as shown in Fig. 5.5).

Stage 2: Combine the antenna patterns measured in stage 1 into the MIMO channel model, emulate the MIMO channel model with the measured antenna patterns

Fig. 5.5 The coordinate system used in the measurements

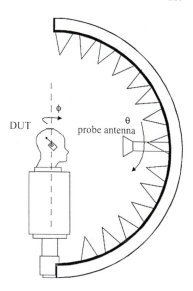

incorporated in the channel emulator and perform the OTA test in conducted approach. There are two different approaches to combine the antenna patterns with MIMO channel model.

Apply antenna patterns to ray-based channel models. The ray-based method used here is the geometry-based stochastic channel model, which is also called double directional channel model. The time-variant impulse responses are calculated from the antenna pattern and the parameters of rays. These parameters are, e.g., angle of departure, angle of arrival, delay, polarization. The method is the same as in SCM/SCME, WINNER and IMT-Advanced.

The full method is illustrated in Fig. 5.6.

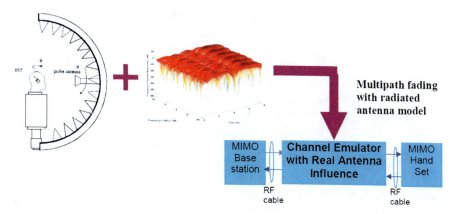

Fig. 5.6 Two-stage method

5.2.5 Cost-Effective OTA Measurements Using Channel Decomposition

Gagern [Gag09] discusses the way to reduce cost when testing MIMO performance over the air. The main possibility to optimize OTA testing is to decompose the signal chain. The specific property of the OTA test is that it includes the antenna performance. The author argues that features handled by smart software and error correction are normally tested in a conducted set-up and have not to be duplicated in OTA. A full end-to-end test is not achieved. The key function of the OTA test is to evaluate the role of the antenna. The antenna must be characterized regarding its MIMO performance. This will include correlation and antenna pattern.

5.2.5.1 Simulated Scenarios

There is a wide variation in the real propagation scenarios. To test all aspects or even a representative subset is prohibitively time consuming and thus not possible. The author instead wants to define the goal of the OTA performance test to find out how the MIMO antennas couple to each other and how that influences the throughput. Using many antennas in a circular array, it is obviously possible to generate many scenarios, but cost and complexity increase, and also so does the complexity of the system calibration. To represent a single cluster only requires a single antenna. Using a second antenna with variable position will have a minimum configuration that can be used to estimate antenna correlation and pattern. The terminal is supposed to be turned to cover all relevant angles of arrivals. The paper does not give a detailed description how this set-up should be used to evaluate the antenna characteristics. The authors of [TBG10] present in detail the theoretical background for a new OTA method that quantifies uniquely the sensitivity of MIMO wireless DUTs. This method assumes a static channel model concerning the coupling of the electromagnetic field between the base station and the UE. The influence of the design of a multiple antenna system for a particular UE on its radiated sensitivity is captured, being a complement of the verification tests in conducted mode which include dynamic channel models. By adding a second test antenna with its angular positioner, the two-channel method can be easily implemented in already standardized SISO OTA test systems (Fig. 5.7). Thanks to the extension of the SISO TIS to the MIMO case, it is possible to define the MIMO effective isotropic sensitivity. Based on this quantity, it is also possible to derive proper integrated figures of merit that will be useful in characterizing the MIMO antenna system performance.

5.2.5.2 Throughput Estimation Using Fast Channel Quality Feedback

Throughput is the most discussed Figure Of Merit (FOM). To obtain sufficiently small measurement uncertainty, it is necessary to transmit an often large number of bits or packets. This might be time consuming, and the [Gag09] proposes a faster

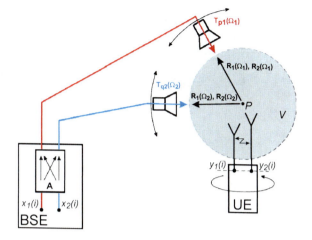

Fig. 5.7 Two-channel method. Static model of DL transmission in MIMO 2×2 mode (incoming signals are linearly polarized and have single AoAs)

way. To use Channel Quality Indicator (CQI), Rank Indicator (RI) and Precoding Matrix Indicator (PMI) would be a faster way since these are verified in conducted tests.

5.3 Lab Evaluations of OTA Setups

This section presents evaluation results of all the types of methodologies of OTA testing proposed in COST 2100 activities. Sections 5.3.1–5.3.4 describe MIMO OTA measurements for cellular radios. Section 5.3.5 presents an OTA testing for Global Positioning System (GPS) receiver.

5.3.1 Two-Stage Method

The two-stage method is an OTA method demonstrated by Agilent Technologies, Inc., USA [WKJ+10]. Figure 5.8 illustrates the flow chart of two-stage MIMO OTA test with DUT. The first stage of two-stage method is to measure the antenna patterns of the antenna array of DUT. In the second step, the influence of the antenna is introduced into the channel emulator to test how it impacts the DUTs receiver. The test platform includes the DUT, the baseband signal generator and channel emulator, the RF signal generator and the Personal Computer (PC). The DUT is a MIMO Mobile WiMAX Universal Serial Bus (USB) dongle with 2 antennas, which is connected to a PC through the USB interface. Figure 5.9 shows the Frame Error Rate (FER) and channel capacity as a function of antenna orientation using the two-dipole antenna with 0.225 wavelength spacing and two-dipole antenna with 0.45 wavelength spacing. From Fig. 5.9 it can be clearly seen that when the channel capacity is high, the measured FER will be low and when the channel capacity is low, the measured FER

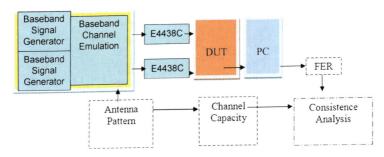

Fig. 5.8 Flow chart of two-stage MIMO OTA test with DUT by Agilent Technologies, Inc.

is high. From this we see that the measured FER using two-stage approaches agrees well with channel capacity analysis. These results validate the performance of the two-stage method.

5.3.2 Reverberation Chamber

The OTA testing using a reverberation chamber is proposed by Bluetest AB, Sweden, NTT DOCOMO, Inc., Japan [KONO09, FOKO10], EMITE Ing [M. 10] and investigated by Motorola, Inc., USA [MV09]. Figure 5.10 shows the testing results of HSDPA using three different OTA methods, including the reverberation chamber method performed by NTT DOCOMO, Inc. [KONO09]. It is confirmed from Fig. 5.10 that equivalent results can be obtained from the 3D uniform distribution in the reverberation chamber, the single cluster model with angular spread of 70 degrees in the anechoic chamber, and the uniform distribution with a single tap in the spatial fading emulator. Based on the results, it is concluded that any type of MIMO OTA testing methodology can be employed for the minimum MIMO OTA testing to differentiate the MIMO antenna performance.

Figure 5.11 shows three different types of Antenna Under Test (AUT) used in a collaboration between Bluetest AB and NTT DOCOMO, Inc. [FOKO10]. Figure 5.12 shows the measured correlations and throughputs using the reverberation chamber (RC) and spatial fading emulator (SFE) from the OTA measurements using WLAN IEEE 802.11n equipment. It is clear from Fig. 5.12 that all three channel models are able to distinguish the low-correlation case from the high-correlation case. It can be concluded that in all three channels the throughput is clearly reduced when the correlation increases, and the high correlation can be measured irrespective of which methodology that is used.

Figure 5.13 shows a reverberation chamber in call set-up presented by Motorola, Inc., USA [MV09]. The Wideband Code Division Multiple Assess (WCDMA) UE is placed into a phone call, whilst the number of absorber panels is adjusted from 0 to 5 per surface. The WCDMA call box then reports Error Vector Magnitude (EVM) of the uplink signal. Figure 5.14 shows the resultant EVM as a function of

Fig. 5.9 FER and channel capacity with regard to antenna orientation using the two-dipole with 0.225 wavelength spacing and two-dipole with 0.45 wavelength spacing

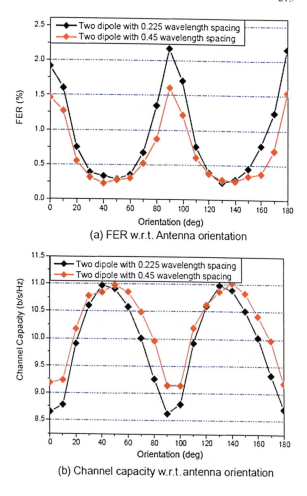

(a) FER w.r.t. Antenna orientation

(b) Channel capacity w.r.t. antenna orientation

the number of panels per surface. It is interesting to note that for EVM values of roughly 30 percents (peak), the call box is unable to report the value, even though the UE remained in a call. This occurred with two absorber panels on three surfaces, and three panels on the remaining three surfaces, thus the nomenclature of 2 and 1/2 panels. With fewer panels, say two on all six surfaces, the WCDMA call would drop and/or the call box was unable to report any EVM (error code reported). It is thus concluded that a WCDMA phone call and resultant sufficiently low EVM is only possible with chamber loading in excess of about three panels per surface.

5.3.3 Analog Fading Emulator Method

The RF-controlled spatial fading emulator [HYS+10] has been proposed by Panasonic Corporation, Japan. Figure 5.15 illustrates four handset arrays tested. Each

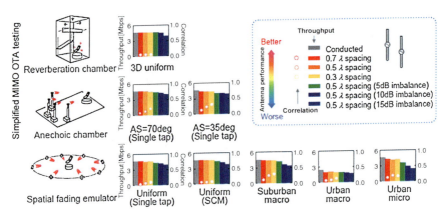

Fig. 5.10 Test results of dependency of MIMO OTA throughput performance on spatial channel models by NTT DOCOMO, Inc.

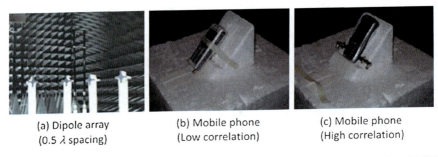

Fig. 5.11 Three AUTs for WLAN IEEE 802.11n in the 2.45-GHz band by Bluetest AB and NTT DOCOMO, Inc.

handset has two monopole antennas and a ground plane with a size of 90 mm by 45 mm, and is operated at 2.4-GHz band. The MIMO OTA tests with and without delay waves, which have a uniform-power distribution in the horizontal plane, were performed by using a WLAN IEEE 802.11n system and two digital fading simulators. Figure 5.16 shows mean values of the measured MIMO OTA throughputs of four handsets with and without delay waves using the 2-by-2 MIMO WLAN system. Doppler frequency was set at 6 Hz. The radio frequency was 2.412 GHz. It is found from Fig. 5.16 that the handsets B and C exhibited good MIMO throughput characteristics, whilst the throughput of the handset D was worst among all the handsets. Moreover, it is observed that the throughputs decreased as the number of the delay waves increased, regardless of handset type. From a comparison between throughputs with and without delay waves in the cases of uniform-power distribution, the delay wave can degrade throughput.

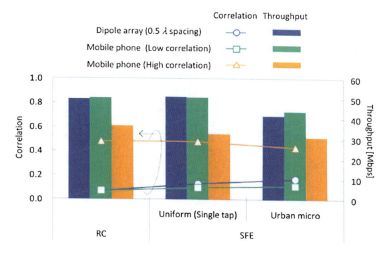

Fig. 5.12 Measurement results from the OTA measurements using WLAN IEEE 802.11n equipment in the 2.45-GHz band by Bluetest AB, and NTT DOCOMO, Inc.

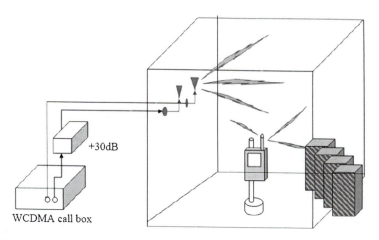

Fig. 5.13 Reverberation chamber in-call set-up by Motorola, Inc.

5.3.4 Digital Fading Emulator Method

Digital fading emulator method is based on active fading emulation using commercially available fading emulator equipment, which can reproduce standard radio channel profiles and models. Test signal from communication tester is fed through a fading channel emulator to an anechoic chamber. A probe antenna array is utilized in the test system, which makes possible to reproduce realistic spatial responses of real radio channels. A spherical array can be considered to be the most flexible geometry for the test antenna array. It provides possibility to emulate multiple realistic directions of arrival distributions during the testing [KNM09] (see Fig. 5.17).

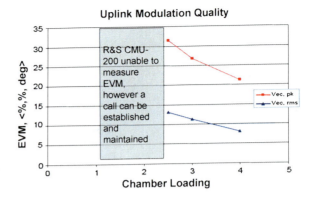

Fig. 5.14 EVM vs. the number of loading panels for reverberation chamber

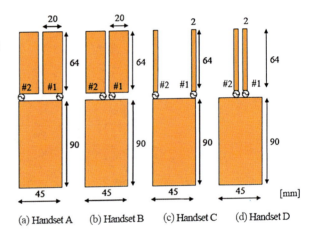

Fig. 5.15 Handset arrays with two monopole elements for MIMO OTA measurement by Panasonic

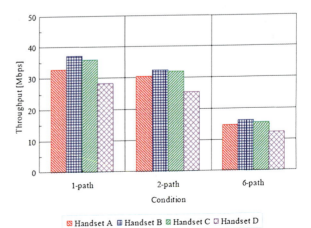

Fig. 5.16 Measured throughput using WLAN IEEE802.11n by the spatial fading emulator

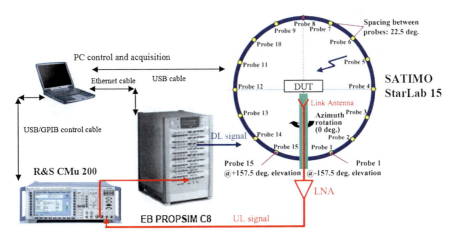

Fig. 5.17 Example of OTA test set-up using digital fading emulator [SF09]

The advantages of this approach are that any geometry-based propagation model and existing standardized radio channel models used in conducted performance testing can be utilized and changing of the propagation scenario is easy. The test system also enables testing of radiated performance of terminals without altering the structure of the terminal. Test labs do not need to access HardWare (HW) or SoftWare (SW) of Equipment Under Test (EUT). Greatest drawback is the cost of the implementation, if compared with the other types of fading emulation techniques. An important objective thus is to find a minimum set-up for achieving a sufficient accuracy and reliability on OTA MIMO performance results in practice [KNM09, SF09].

5.3.4.1 Verifications of OTA Setups

Okano et al. studied impact of amount of probe antennas used in a spatial channel emulator considering the radiation characteristics of AUTs. Antenna correlation, that has a substantial impact on MIMO performance, was taken as the criterion, employing Power Angular Spectrum (PAS) of uniform and cluster model. Five set-ups were considered; four, six, eight, twelve and sixteen probes with the fixed spaced linear alignment of 90, 60, 45, 30 and 22.5 degrees, respectively. AUTs were a vertical dipole array representing mobile terminals with nearly omnidirectional radiation patters and a horizontal dipole array representing terminals with non-omnidirectional radiation patterns. Element spacing in array was within the range of 0.1 to 2λ. The characteristics under these conditions were seen to represent performance of multiantennas mounted on small mobile phones or laptops.

Figure 5.18 shows a simulated Root-Mean-Square (RMS) error, which is a measure of the differences between antenna correlations of a uniform or Laplacian distribution and that of probe models, corresponding to the element spacing of AUTs. The low RMS error indicates that the degree of accuracy in emulating an intended

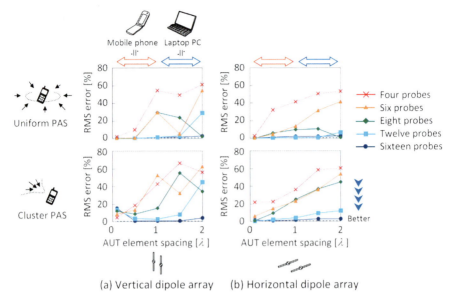

Fig. 5.18 Simulated RMS error for antenna correlation corresponding to element spacing of AUTs [OKI09]

PAS for an antenna correlation is sufficient. Commonly observed feature in both uniform and cluster PAS for both AUTs, it was found that the RMS error decreases according to an increase in the number of probe antennas. Based on measured and simulated results, Okano et al. concluded that for a spatial channel emulator that has a limited number of probe antennas, fixed-spaced linear alignment may not be best option for the MIMO OTA testing in terms of the accuracy in emulating an intended PAS for an antenna correlation [OKI09].

Since the number of OTA antennas set upper bounds for the performance and accuracy of the OTA test set-up, it has been a matter of great interest during the work of COST 2100. The maximum physical size of DUT antenna array is constrained by spacing of OTA antennas in azimuth plane. Nuutinen et al. reported in [NKF+09] spatial correlation results that support well conclusions of Okano et al. Figure 5.19 depicts the spatial correlation as a function of the DUT antenna separation.

Ideally, the spatial correlation should be a Bessel function of first kind, but, as seen, the three antennas in a quadrant (i.e. eight OTA antennas) support accurately the DUTs with antenna separation of wavelength. If the number of OTA antennas is increased to five antennas in a quadrant (i.e. 16 OTA antennas), DUTs with antenna separation around two wavelengths can be tested [NKF+09].

Capabilities of digital fading emulator method to create an appropriate propagation environment to an anechoic chamber has been investigated in [NKF+09] and in [KNB09]. In the latter document testing was done using an 8-probe setup, previous study used a 3-probe setup. Radio channel models used as a reference were 3GPP Spatial Channel Model (SCM), 3GPP Spatial Channel Model Extended (SCME) and TGn, and the evaluated propagation characteristics were amplitude distribution

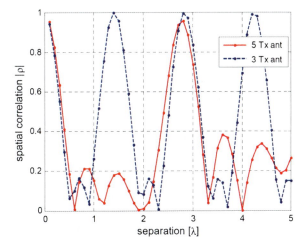

Fig. 5.19 Spatial correlation as function of the DUT antenna separation [NKF+09]

of the fading coefficient, power delay profile, Doppler spectrum and spatial correlation function. In [KNB09] OTA measurements for power delay profiles were performed by using a Propsound™ channel sounder. Measurement data for SCM amplitude distribution was obtained as an offshoot of the throughput measurement by using a RACE 4G platform from Elektrobit. The amplitude distribution was found to deviate from Rayleigh with SCME model but has good agreement with the theory with TGn model. Measured power delay profiles were found to overlap closely to the original profiles of SCME Urban macro and micro. Measured power delay profiles of SCME Urban macro- and micro-models are illustrated in Figs. 5.20(a) and (b).

Simulations presented in [KNB09] show that Doppler spectra of OTA models match well to the corresponding spectra of reference models. Also, the simulated spatial correlation in the OTA case coincides well with the theoretical Bessel function. Two different OTA antenna configurations were simulated; 8 OTA antennas with 45° spacing and 16 antennas with 22.5° spacing. In the OTA simulation a single cluster with Laplacian shaped power azimuth spectrum was modeled. Again it was observed in the study that with higher number of OTA antennas and more dense spacing, we can obtain accurate spatial correlation for a larger test volume. Spatial correlation was also experimentally studied in [SF09] using an 8-probe linearly spaced test set-up. Test frequency was 2050 MHz. DUT was three dipoles spaced less than half wave lengths away from each others, resulting DUT size of 18 cm. Spatial correlation was found to match theoretical values when angular spread was 10°, at 35° divergence was found.

5.3.4.2 Calibration Procedures

Common idea in calibration procedures proposed for digital fading emulator method is to compensate positioning differences of antenna placements and other measurement set-up non-idealities like cable lengths. The calibration process is started by measuring amplitude and phase response of OTA antennas $1 - n$ with Vector Network Analyzer (VNA). The result values are stored into memory. Next step is to

Fig. 5.20 Measured power delay profiles of SCME scenarios. Urban macro in picture (**a**) and Urban micro in picture (**b**). Original SCM power delay profiles are plotted using circles [KNB09]

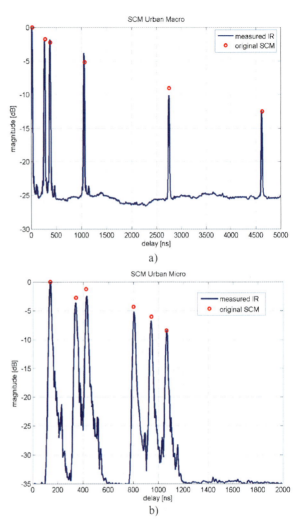

adjust fading emulator's output settings so that each probe has the same amplitude and phase in the center of the test volume [KLN09a, NKF$^+$09, SF09, KLN09b]. Set-up for calibration is depicted in Fig. 5.21. Kallankari et al. [KLN09b] introduce a calibration method where the goal is to produce a known signal level to an anechoic chamber. This is necessary if inter-laboratory correlation is desired.

5.3.4.3 Measurements Results of Commercial DUTs

Kallankari et al. reported HSDPA OTA measurement results, measured by using a digital fading emulator method [KLN09a]. Set-up of eight dual polarized probe antennas was used with SCM urban micro-fading profile. EUT was placed on a turning table, and testing was done in free space and in a browsing hand phantom. Field

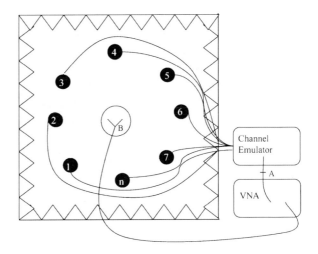

Fig. 5.21 Calibration setup [KLN09b]

strength in the chamber was altered by a programmable step attenuator. HSDPA throughput was measured in a fixed reference channel. Channel coding was H-set 3 (Quaternary Phase Shift Keying (QPSK)), enabling a maximum downlink bit rate of 1601 kbps. Throughput was found to be strongly dependent on EUT orientation, and therefore EUT was rotated in horizontal plane using angle step of 30°. EUT used were from four different mobile phone vendors. The following steps were repeated with six different devices. Throughput was measured in each twelve angles. Signal level in the chamber attenuated using 3-dB steps. Average of rotation throughput was calculated for each attenuation level. Average throughput results are presented in Fig. 5.22.

Increasing mobile speed and measuring DUT in a hand phantom was found to decrease throughput. Effect of the location of EUT in respect to the calibration center was studied. EUT was displaced on a line in horizontal plane in range of 15 cm with 1-cm steps. Results indicated that the field in chamber was uniform enough to allow differences in antenna phase center placement due to different sizes and form factors of DUTs [KLN09a].

Throughput results of Linksys IEEE 802.11n device with 2×2 MIMO were reported in [NKMF09]. The device was operated in the 2.4-GHz band using 20-MHz channel bandwidth. Another matching device from the same family was used as the test equipment making up the other side of the link outside the chamber to ensure compatibility with all modes of MIMO operation of the DUT. The "downlink" signal from the tester to the DUT was routed through the channel emulator then through eight 35-dB gain amplifiers to a circular array of eight vertically polarized broadband measurement antennas. DUT was placed on an azimuth positioning turntable in the center of the test volume. A laptop was located nearby to serve as the client side for performing throughput measurements between the tester and DUT. Experiments were performed using three different types of channel models. The first one was the modified 3GPP SCM Urban Micro model, where the delays were squeezed to zero while keeping the original angular behavior. Next, a modified

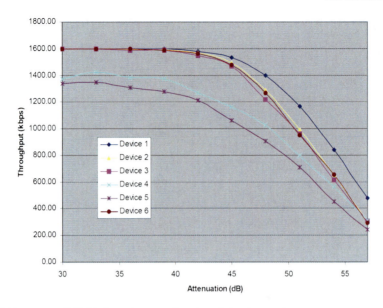

Fig. 5.22 Average HSDPA throughput as a function of signal strength

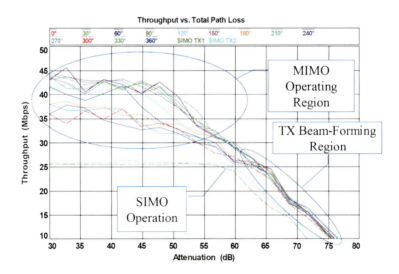

Fig. 5.23 Throughput vs. attenuation azimuth cut for low-correlation TGn-C model illustrating different regions of operation and comparing to Single-Input Multiple-Output (SIMO) behavior for same DUT

TGn-C model, consisting of the delay profile from TGn-C, but the angular behavior from the SCM Urban Micro model was developed, followed by another modified TGn-C model having low correlation (10 wavelength separation) on the client antennas. Figure 5.23 depicts the throughput vs. attenuation results as a function of

5 "OTA" Test Methods for Multiantenna Terminals 223

Fig. 5.24 Principle of A-GPS by Rohde and Schwarz

rotating the DUT on the turntable in 30-degree increments using the modified TGn-C low-correlation model.

From Fig. 5.23 it can be seen that there are at least two modes of MIMO operation in this device. The curves at the low attenuation end are clustered to 35–40 Mbps and to 40–45 Mbps, indicating some level of fall-back between at least two different modulation and coding schemes (MCS) within the 2×2 MIMO specification of 802.11n. As the attenuation increases, the MIMO performance converges to performance that is independent of the DUT orientation due to the symmetry of the sleeve dipole elements. The same "waterfall" curve of the lower 802.11 data rates seen in the SIMO test cases is evident but with a 3-dB shift indicating the likely presence of transmit beam forming in the test system.

Conclusions from the measured data were following. Digital fading emulator method is capable of testing MIMO performance in the laboratory conditions. Throughput is highly dependent on the channel model, thus in order to have realistic results, the channel model should be appropriately selected. MIMO performance is dependent on antenna orientation, and therefore angular information plays an extremely important role when assessing the MIMO performance [NKMF09].

5.3.5 OTA Measurement for a GPS Receiver

The OTA test system for Assisted GPS (A-GPS) has been proposed by Rohde and Schwarz, Germany [vG08]. Figure 5.24 shows the principle of A-GPS. The A-GPS uses not only the satellite information but also assistance information from the cellular network in order to speed up the time for a position fix, and also to allow transmission of the mobile's position back to the network. Figure 5.25 depicts the experimental configuration of the emulator. As shown in Fig. 5.25, a typical test system for OTA tests on A-GPS devices is similar to the OTA systems already in use. The key components are a fully anechoic chamber with a positioner to turn the UE,

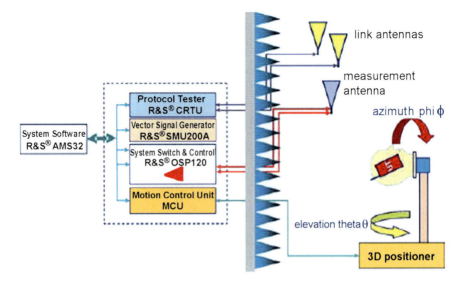

Fig. 5.25 Components of A-GPS test system TS8991 by Rohde and Schwarz

possibly a positioning device for the measurement antenna, and the base station simulator allowing to establish a link in the required technology and to do the required measurements. In addition, a signal generator has to be provided for simulating the signals of eight GPS satellites including their movements over the sky.

5.3.6 MIMO OTA Measurements of LTE Devices According to the Two-Channel Method

In [TAK+10] the two-channel method was used in evaluating the radiated sensitivity performance of a commercial LTE modem that was operating in spatial diversity MIMO 2×2 mode, while also the SISO mode was used for reference purposes. By performing two different TIS measurement campaigns it was possible to study the LTE modem with either internally integrated antennas or with external antennas arranged for maximum spatial diversity. It was found that the UE arrangement having external antennas had better SISO and MIMO sensitivities by 3.5 dB and 7 dB, respectively, than the internal antennas one, while the two-channel method was able to quantitatively evaluate UE spatially integrated sensitivity for both SISO and MIMO cases. The hardware system presented allowed also the evaluation of several MIMO OTA parameters (Fig. 5.26).

5.4 Lab Comparison (Round Robin)

The development of MIMO OTA measurement methodology in COST 2100 has resulted in several candidate methodologies, each with its own unique concept,

5 "OTA" Test Methods for Multiantenna Terminals

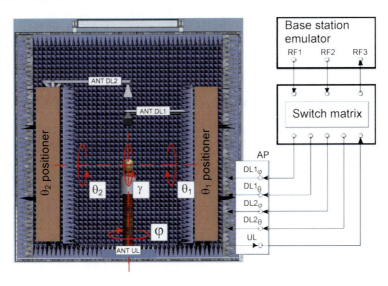

Fig. 5.26 Block diagram of MIMO OTA test system supporting two-channel method

Table 5.1 Table (MIMO OTA candidate measurement methodologies)

Company	Candidate Methodology	Category
Nokia	Multi-probe SFE anechoic chamber	Anechoic
SATIMO	Stargate SFE anechoic chamber	Anechoic
Agilent	Two-Stage: Stage 1: anechoic chamber, Stage 2: conductive	Multi-Stage
Bluetest	Generic reverberation chamber	Reverberation
EMITE	Multi-Slot reverberation chamber	Reverberation
Wiesbaden	Reduced-probe SFE anechoic chamber	Simplified Anechoic
R&S	2-Probe anechoic chamber	Simplified Anechoic
NTT	Multi-probe SFE anechoic chamber	Anechoic
ETSL	Multi-probe SFE anechoic chamber	Anechoic
Panasonic	RF-controlled SFE in anechoic chamber	Anechoic

measurement configurations, measurement technique and capability. These test methods can be categorized into four main categories: Anechoic, Reverberation, Simplified Anechoic and Multi-Stage method, as given in Table 5.1.

The simulation and/or analytical comparison approach will be too complex to prove that each method can effectively be used to measure the OTA performance of Rx diversity and MIMO devices. Hence, an High-Speed Packet Access (HSPA) round robin measurement campaign was performed, where each proponent of methodology would conduct measurements using the same DUTs and a common test plan. The proponents from RAN4 and CTIA were also involved in the round robin campaign, which lasted for approximately five months.

Fig. 5.27 Mapping between HSPA DUT category and F-RMC

HSPA DUT category	F-RMC
Category 1	H-Set 1
Category 2	H-Set 1
Category 3	H-Set 2
Category 4	H-Set 2
Category 5	H-Set 3
Category 6	H-Set 3
Category 7	H-Set 3
Category 8	H-Set 3
Category 9	H-Set 6
Category 10	H-Set 6
Category 11	H-Set 4
Category 12	H-Set 5
Category 13	H-Set 6
Category 14	H-Set 6
Category 15	H-Set 6, H-Set 9
Category 16	H-Set 6, H-Set 9
Category 17	H-Set 6, H-Set 9
Category 18	H-Set 6, H-Set 9
Category 19	H-Set 6, H-Set 9, H-Set 11
Category 20	H-Set 6, H-Set 9, H-Set 11

MIMO ⟹ (Categories 15–20)

5.4.1 HSPA Round Robin Measurement Campaign

The HSPA round robin measurement campaign is aimed to provide practical evaluation and comparison of the candidate methodologies proposed in Table 5.1. The measurement campaign will also give valuable insights into whether the test methods can be used to assess the OTA performance of MIMO devices.

5.4.2 Test Plan

In order to run the measurement campaign effectively, a test plan was created. The test plan established a common measurement foundation for all test methods, i.e. the OTA throughput as the figure of merit was measured, the same DUTs were used, a common sub-set of channel models was defined, a common set of NodeB emulator parameters was used, and the common test procedure was also adopted. For anechoic chamber-based methods, the SCME urban micro-cell and macro-cell model were used. The urban micro-cell model is mainly characterized by the relatively small PDP spread (i.e. 0.3 µs) and small AOD and AOA (i.e. 5° and 35°, respectively), as compared to urban macro-cell model (i.e. 0.8 µs PDP spread, with 8° and 60° of AOD and AOA, respectively). For the simplified anechoic chamber-based methods, the SCME channel models were used, where only a few clusters (1–2 clusters) were being emulated. For the reverberation chamber-based methods, the uniform channel models based on predefined PDP were used. The test plan also specified the NodeB emulator parameters. The parameters are set according to the DUT category. The NodeB emulator generates the realistic NodeB RF signal based on the fixed reference measurement channel (F-RMC). The mapping of DUT category to F-RMC is given in Fig. 5.27. F-RMC contains the physical channel parameters, modulation and coding parameters, transport block size, HARQ parameters, etc. In the measurement campaign, F-RMC H-Set3 and H-Set6 were used. A set of reference DUTs consists of two DELL Laptops (model E6400 and E4300), two

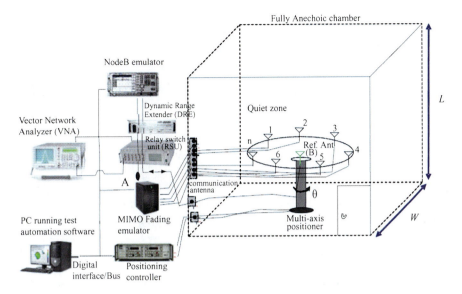

Fig. 5.28 Calibration setup of anechoic-chamber-based methods

Nokia HSPA USB dongles (model CS-15) and two Huawei HSPA MIMO USB dongles (model K4505) were provided by Vodafone during the measurement campaign. These off-the-shell reference DUTs were used as if they were provided to the commercial test lab/"test house". A reference host laptop (powered by battery) was used to test the USB dongle-type DUTs connected to a specific USB port. To reduce the impact of the laptop on the results, the laptop lid is kept closed; the display, hard disk and system standby is kept on during the measurement. The WLAN and Bluetooth radios are also switched off during the measurement. To ensure comparable results among the test methods, the host laptop orientation with respect to the turn-table is also specified. For testing of laptop type DUTs, the same power management settings and laptop orientation are used, except that the laptop lid is kept open during the measurement. When the laptop is placed on the turn-table, the effect of turn-table on radiation pattern of laptop antennas should be captured. The test plan also provided the calibration procedure for each category of test methods due to the similarity of the test configurations and test system components in each category. The anechoic-chamber-based methods are calibrated using the setup illustrated in Fig. 5.28. Firstly, using the known characteristics of the reference antenna, the pathloss from the input of the channel emulator (Point A) to the DUT position (Point B) can be calculated. The same pathloss between each antenna probe and DUT position is kept by adjusting the channel emulator output power (i.e. apply a pathloss compensation factor) recursively. For fading propagation channels, the average channel output power is measured at Point B. The average channel output power is the sum of signal powers via a number of calibrated probe antennas and averaged over 30 seconds. In order to verify that each antenna probe has been properly calibrated, the DUT sensitivity measurement is performed for each antenna

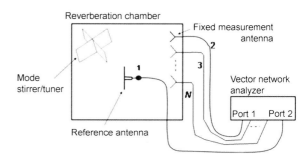

Fig. 5.29 Calibration setup of reverberation-chamber-based methods

probe using the static propagation channel model. The result of this check can be used to verify that the probe antennas have been properly calibrated. The pathloss compensation is considered successful if the path average results are within ± 1 dB. For reverberation-chamber-based test methods, the calibration is performed using the setup illustrated in Fig. 5.29. The multiport vector network analyzer VNA is used to measure the S-parameters. From the S-parameter values obtained, the average chamber transfer function and power correction factor can be then calculated [PSF[+]10]. The average chamber transfer function is proportional to the radiated power of the DUT or all the measurement antennas. The power correction factor is the amount of power loss due to the reflection of measurement antenna port. During the calibration, several essential parameters are also recorded: Chamber Q-factor or average power received from a lossless antenna, propagation channel-related parameters such as RMS delay, coherence bandwidth and Doppler spread. For Multi-Stage method, calibration is performed to measure the RF impedance of the DUT in the first stage. In the second-stage, when an RF cable is connected to the antenna ports of the DUT, the impedance is matched between the RF cable and the DUTs antenna ports by adjusting the channel emulator output level. Finally, the test plan also outlined the test procedure for MIMO OTA throughput. For F-RMC, the transmitted TBS is a fixed value. The average channel power is varied over a sufficiently large dynamic range from the mean (50%) throughput point. The average channel power step size is set to 2 dB.

5.4.3 Measurement Results

The results from the HSPA round robin measurement campaign have been presented in COST 2100 TDs [WJK10b, JWK10a, WJK10a, NS10, SHVVH[+]10, KNL10, PSF[+]10, FSJR10, JWK10b]. In [WJK10b] the authors compared the anechoic-chamber-based test methods with the two-stage test method and demonstrated that both methods can efficiently perform the MIMO OTA test on the reference DUTs. When the antenna pattern of DUT can be accurately obtained, the two-stage and anechoic-chamber-based test methods exhibit consistent OTA throughput results (obtained from two DUTs: E6400, K4505), as illustrated in Fig. 5.30. The channel model selection and its parameters (e.g. angular spread, DUT speed) were also shown to influence the OTA throughput results.

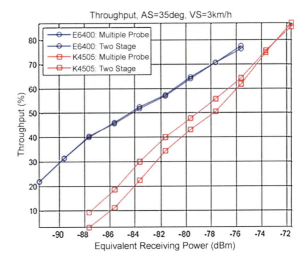

Fig. 5.30 HSPA throughput measurement results comparison between two-stage method and multiple probe antenna method (AS = 35 deg, V = 3 km/h)

When the receiver diversity mode is considered, the antenna radiation performance is mainly determined by its gain, while the influence of power imbalance and the correlation coefficient is small. However, when the spatial multiplexing mode is considered, the antenna radiation performance is more sensitive to the power imbalance and the correlation coefficient beside the antenna gain. The uniform channel models are less sensitive to the antenna radiation performance than both the multiple- and single-cluster-based models [WJK10a]. Further measurement results were reported for anechoic-chamber-based test methods [NS10, KNL10]. For certain combination of channel models, NodeB parameters and DUTs, the results from SATIMO and Nokia test methods were shown to be closely matched with each other, as illustrated in Fig. 5.31. However, some discrepancy was observed from the measurement results, which could be attributed to difference in the NodeB emulators. In [KNL10] a diversity gain of 6 dB was reported when measured via the single-polarized antenna probes for Nokia's test method. However the result was not general, as diversity gain of 6 dB was obtained with gain imbalance of 1.5 dB. It is still unclear what the realistic diversity gain will be in real networks and with different channel models. Different device orientations were also shown to impact the diversity gain of DUT, as illustrated in Fig. 5.32. However, the study showed that the need for device orientations may diminish by choosing the dual-polarized channel models, but further studies are still needed. In [FSJR10], for the Wiesbaden method, all measurements happen to results in the same relative ranking of DUT performance. The azimuthal plane results are not sufficient to characterize the performance of all DUTs. In fact its significance depends on the particular 3D pattern of the DUT. Moreover, when indoor and outdoor-to-indoor scenarios are considered, a considerable contribution to the PAS is found at non-zero elevation, so that predictions based on 2D measurements may not be representative [FSJR10]. Both polarizations are required to characterize a DUT, and moreover antenna systems in electrically small devices must necessarily exploit polarization selectivity. The

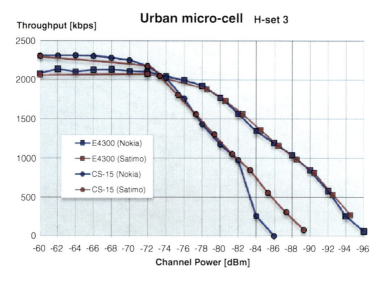

Fig. 5.31 HSPA (DUTs: E4300, CS-15) OTA throughput results comparison between Nokia and SATIMO method (SCME Urban micro-cell model, F-MRC = H-Set6)

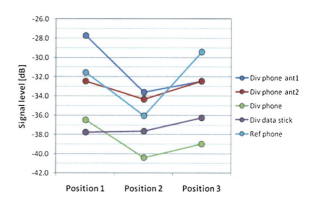

Fig. 5.32 Impact of different DUT orientations (position 1, 2, 3) to DUT diversity gain

emulation of a fading channel may not be required to differentiate DUTs with respect to performance. In fact measurements using the simplified SCM model and the measurement in which the two downlink paths were only decorrelated by using a small Doppler shift show the same trend. Statistical metrics are believed to be of high interest, as they give additional insight into the relative performance of the DUT [FSJR10]. In [PSF+10] the authors found that the reverberation chamber produced both repeatable and reproducible results, as illustrated in Fig. 5.33. The results showed two variants of reverberation-chamber-based methods: Bluetest method (designated as lab2) and modified Bluetest method (designated as lab1). The modified Bluetest method incorporates the channel emulator in order to emulate SCME channel models. A good performance ranking was observed in testing the two DUTs (CS-15 and E4300) using the test methods. The measurement re-

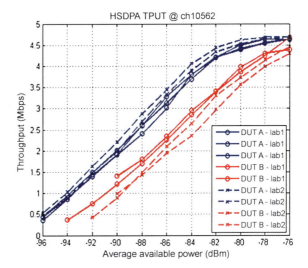

Fig. 5.33 HSPA OTA throughput results for reverberation-chamber test methods, showing repeatability across different labs (DUT A-E4300, DUT B-CS-15)

sults for spatial correlation, (antenna) gain imbalance and channel capacity FOMs were reported in [JWK10a]. Assuming certain AWGN level and SNR value for the DUTs, the channel capacity can be calculated from the measured spatial correlation results. Applying large sample averaging effect, the obtained channel capacity results shown in Fig. 5.34 were comparable between the anechoic-chamber-based test methods and two-stage test methods, albeit some differences due to the possible chamber imperfections. The obtained spatial correlation and gain imbalance results showed that DUT1's antenna outperforms DUT2's antenna. These results were consistent across both test methods. In [SHVVH[+]10], LTE MIMO OTA throughput performance was reported. It was observed that higher decreasing slopes with decreasing received power are observed as higher throughputs are required from the DUT. Such slope is larger than the slopes previously found for HSPA DUTs. The obtained performance also depends on both channel bandwidth and fading channel model.

5.5 Real Environment References and Comparison to OTA

5.5.1 General Aspects

Real environment references are vital to establish relevance of any FOM defined for MIMO OTA tests as well as the suitability of the underlying measurement methodology. This subject has been addressed by in-the-field measurements [YTH[+]09, SKI[+]09] as well as by theoretical arguments and numerical simulation based on existing Geometry-Based Stochastic Channel Models (GSCMs).

The question which kind of environment to choose as reference may find different answers. The most general descriptions of physical propagation environments we have at our hands are GSCMs as e.g. the SCM, SCME, WINNER and

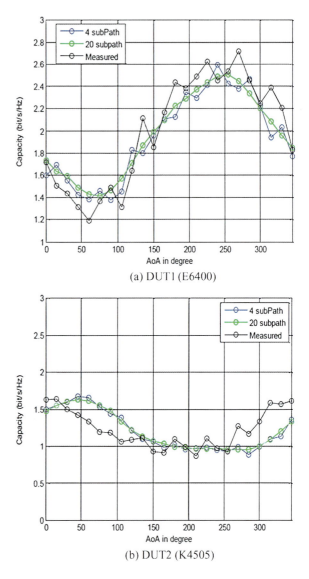

Fig. 5.34 Channel capacity results comparison between two-stage method (4-subpath, 20 subpath) and multiple probe antenna method (measured) (single cluster, AS = 35 deg)

IMT-Advanced models. These models include the interplay of various spatial and temporal aspects of multipath propagation and a considerable number of statistical parameters. In any specific instance of a propagation channel as realized in a statistical simulation model or in a measurement on the other hand, generality of the model is lost. Even if a large number of channel realizations is taken into account, an exhaustive comparison between real propagation environments and OTA test methodologies appears virtually impossible, because any metric considered for comparison, e.g. throughput or channel capacity, in addition depends on the characteristics of the selected UE antenna system and properties of the communication

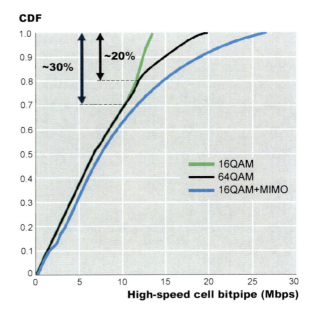

Fig. 5.35 Cumulative Density Function (CDF) of throughput for different UE categories in an HSPA network

standard. Researchers have chosen different approaches to deal with this situation. These also reflect differences in focus and assumed application context for OTA measurement.

A number of investigations focuses on the fidelity in reproducing in detail all relevant physical aspects of real propagation scenarios within an OTA test method. They are concerned with the joint emulation of all spatial and temporal aspects of real multipath propagation. This approach answers the needs of basic research and of measurements accompanying device development: coverage of a wide range of typical scenarios. Cell throughput, in the end, depends on device performance in these scenarios.

Starting point of several other studies is the question which kinds of propagation scenarios are best suited to discriminate quickly and reliably between good and bad antenna (and UE) designs. The background idea is that aspects which are less critical for a specific standard or not genuinely related to the spatial properties of the UE antenna system might be addressed separately, e.g. by conducted tests.

With view on UE certification for instance, consideration of "MIMO favourable" propagation conditions is advocated in [Man09]. The argument is made specifically for investigation of MIMO performance in Spatial Multiplexing (SM) mode. It is based on traffic data from a real HSPA network. Three different UE categories are considered whereof two are of interest in the present context: single-antenna UEs that use 16QAM, and dual-antenna UEs that use 16QAM and MIMO in Spatial Multiplexing (SM) mode. Figure 5.35 shows the CDF of throughput per UE category. Comparing the 16QAM data with the 16QAM + MIMO (SM) data reveals that a benefit from SM is only available for the 20%–30% or so best channel realizations. Users with median or typical channel conditions experience almost no advantage from having a MIMO capable UE. The conclusion is made that emulation of a sta-

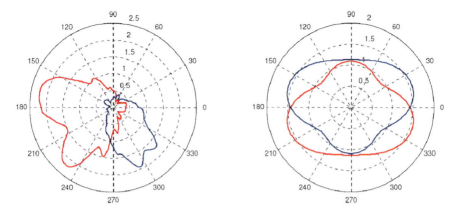

Fig. 5.36 Normalized dual-antenna patterns considered for comparison of 2D spatial against uniform propagation scenario: a measured pattern (*left*) and the canonical dual-dipole arrangement with $\frac{\lambda}{2}$ spacing [Ree10]

tistically varying propagation scenario in which MIMO-favorable conditions occur only with small probability is likewise not apt to discriminate between UE with good or bad performance in SM mode. Differences would just disappear in the average performance.

At least for evaluation of UE performance in SM mode it is therefore worthwhile to have a closer look at possible MIMO favorable environments.

The isotropic Rayleigh scattering scenario with uniform PAS might at first glance be attractive as an average over any conceivable real world multipath propagation scenario and the prototype of a rich scattering environment. However, this scenario is never realized in reality (except as a statistical average in a reverberation chamber). Describing a MIMO OTA measurement procedure conceptually as a mathematical operator whose argument is a propagation scenario and whose result is some FOM for the DUT, a uniform PAS amounts in fact to averaging over the arguments instead as over the results of the operator. This potential pitfall is illustrated in an enlightening way in [Ree10]. The study compares the SCME urban micro model and an SCME single cluster model with 35° Angular Spread (AS) against a 2D scenario with uniform PAS. Two substantially different UE antenna systems are employed in the comparison. Their magnitude patterns are reproduced in Fig. 5.36. Channel capacity is used as metric for the comparison. As an example, the observed channel capacities under conditions of the single cluster model with 35° AS are shown in Fig. 5.37. Capacity is displayed as a function of cluster AOA in the azimuthal plane. Obviously, not only is there a pronounced difference in the variation of capacity with AOA but also in the average capacities (see insets in Fig. 5.37). For the 2D uniform model on the other hand, not only remains any dependency on AOA hidden, but also are the average capacities for the two antenna systems according to [Ree10] virtually identical (5.4 and 5.5 b/s/Hz, respectively). These results indicate that a propagation scenario with isotropic or 2D uniform PAS is not suitable to discriminate between spatial properties of an UE antenna system.

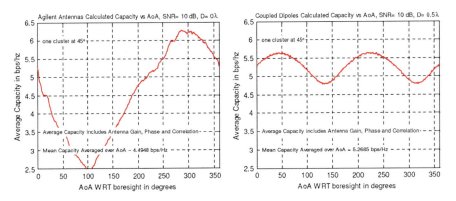

Fig. 5.37 Capacity over azimuthal AOA for the SCME single cluster model with 35° AS applied to the antennas after Fig. 5.36

The spatial properties of an UE antenna system are of course only one aspect. Antenna efficiency and receiver performance may in many cases dominate the picture. In this context, although not directly related to a physical propagation scenario, some more traditional performance metrics which are obtained under static channel conditions like Total Radiated Sensitivity (TRS) may be considered "real-world references" because a bulk of measured data (also for SISO UEs) is available for comparison. A generalization of TRS to MIMO terminals operated in transmit diversity mode (as in the 3GPP LTE standard) is proposed in [FS09b]. It is built on the fact that all relevant MIMO standards support an Space-Frequency Block-Code (SFBC) or an Space–Time Block-Code (STBC) downlink transmission mode, the Alamouti scheme being its most prominent implementation in case of two transmit antennas. In this case two blocks of signals s_1 and s_2 are once combined to a vector (s_1, s_2) and mapped to two transmit antennas in this order in a first transmission and rearranged to the vector $(-s_2^*, s_1^*)$ for a second transmission. The two signal vectors are orthogonal. Assuming perfect channel estimation and only Additive White Gaussian Noise (AWGN) an optimum M-antenna receiver would approximate the signal blocks s_1 and s_2 by

$$\tilde{s}_1 = \sum_{m=0}^{M-1} \left(|h_{m1}|^2 + |h_{m2}|^2 \right) s_1 + h_{m1}^* n_{m1} + h_{m2} n_{m2}^*, \tag{5.1}$$

$$\tilde{s}_2 = \sum_{m=0}^{M-1} \left(|h_{m1}|^2 + |h_{m2}|^2 \right) s_2 - h_{m1} n_{m2}^* + h_{m2}^* n_{m1}, \tag{5.2}$$

where h_{mi} denotes channel coefficient, and n_{mi} represents noise. Augmenting the conventional SISO TRS measurement procedure in that the two transmit signals are mapped to the two orthogonal polarizations of a dual-polarized probe thus obviously yields a test case in which the received signals are weighted with the norm of the channel matrix and hence, by sampling over the whole sphere, with the total efficiency of the UE antenna system. Since the metric is applicable to UE receivers

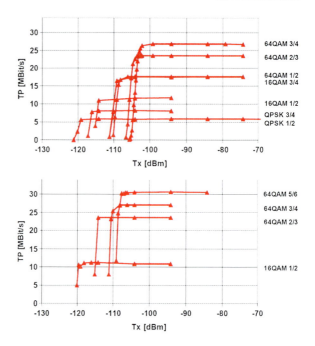

Fig. 5.38 Throughput over nominal Tx power for two different WiMAX devices (fixed MCS) [Gag10]

with single or multiple antennas, and in addition a straightforward extension of the conventional SISO TRS definition, it allows, among others, for direct comparison between MIMO and SISO UEs.

Static channels should in fact not be excluded from considerations as highlighted also in an experimental investigation [Gag10]. Although performed in a conducted measurement set-up and not yet OTA, it is a valuable background for discussion and selection of MIMO OTA methodology because it provides a vivid picture of the kind of results which can be obtained. The study reports throughput measurements for two different WiMAX devices. Figure 5.38 displays throughput over nominal transmit power of the base station emulator for the two WiMAX devices operating in Matrix B mode (SM) for several of the supported Modulation and Coding Schemes (MCSs). It is observed that the throughput break-off is steep in all cases and extends over a few decibels in Tx power only. The break-off power levels are obviously good indicators of receiver performance. If a time varying (faded) channel model had been applied, the base station would of course have adjusted MCS in accordance with reported channel quality, and smoothed throughput vs. power level relations would have been observed. Similar results may be expected for OTA measurement. The decision whether to prefer static channel conditions or time-varying channel models may be made in accordance with the objective of the measurement.

Gagern [Gag10] also illustrates that further speed up of measurements may be possible by evaluating channel indicators instead of e.g. Block Error Rate (BLER). For the WiMAX DUTs at least, a very good coincidence between applied transmit power level and reported Received-Signal Strength Indicator (RSSI) levels was ob-

Fig. 5.39 Measured Carrier to Interference plus Noise Ratio (CINR) for the two Worldwide Interoperability for Microwave Access (WiMAX) devices after Fig. 5.38

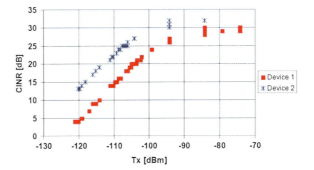

served. Furthermore the difference between the CINR values (reported from the UE and refined with support from the test system), which is given in Fig. 5.39, corresponds well to the differences between the break-off levels for the two devices after Fig. 5.38. Evaluation of these quantities instead of BLER, which requires measurement over a sufficient number of frames, is therefore proposed in [Gag10] in order to reduce test time.

5.5.2 Emulation of a GSCMs in an Anechoic Chamber

Emulating a Geometry-Based Stochastic Channel Model (GSCM) in an anechoic chamber is the most general approach to OTA measurement. In a GSCM multipath propagation scenarios are generated by superposition of a large number of individual rays with per ray selection of AOD, AOA, time delay, Doppler shift and polarization based on statistical parameters. The latter have been derived from measurements, and it is assumed that main physical characteristics of real propagation scenarios are preserved. Eligible GSCMs are for instance SCM, SCME, WINNER and IMT-Advanced models. Correlation-based channel models are not suitable in this context. Within each model, a number of environments, e.g. urban, rural or indoor environments, may be selected.

Key model parameters for the temporal (frequency) characteristics of a GSCM model are:

- Delay Spread (DS), relevant for time dispersion (inter-symbol interference) and frequency selectivity of the channel (frequency/multipath diversity),
- Doppler Spread, relevant for time selectivity of the channel and frequency dispersion (inter-carrier interference),

and for the spatial and polarimetric characteristics,

- Angular Spread (AS), relevant for the spatial selectivity of the channel (spatial diversity and multiplex, beam-forming),
- Cross-Polarization Ratio (XPR), relevant for polarization diversity and polarization multiplex.

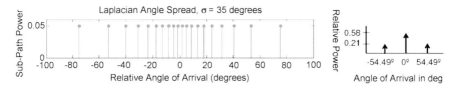

Fig. 5.40 Mapping of subpaths to probes for analysis of single cluster model reproduction made in [Ree09]

By emulating a GSCM it is possible, in principle, to account also for the interplay of all the above phenomena, e.g. correlation between spatial and temporal characteristics (such as between Delay Spread (DS) and AS or between angular profile and sign of Doppler shift).

In view of the complexity of these models, investigations in connection with MIMO OTA methods have so far focused mostly on what is new and critical when moving from SISO to MIMO UE testing, i.e. on the spatial aspects. The PAS in particular, is considered a critical aspect when approximating a GSCM in an anechoic chamber. The SCM for instance specifies six clusters with 20 subpaths per cluster, thus allowing for a total of 120 paths with distinct AOA. For OTA measurement, these must be mapped to a realistic, i.e. much smaller number of antennas (probes) in the anechoic chamber. The question arises whether the relevant properties of the channel model can be preserved in such mapping. The angular power spectrum, described e.g. in the SCME by tabulated angular offsets which follow a Laplacian azimuthal distribution with prescribed angular spread, is a key property in this context. At least in the case of a roughly omnidirectional characteristic of the UE antenna system, where spatial selectivity is mainly due to different propagation delays for different antenna elements (as e.g. in case of the canonical example of two coupled dipoles at half-wavelength separation), AS basically determines signal correlation.[1]

To investigate how well a multi-probe set-up can reproduce the spatial characteristics of a GSCM, correlation between received signals is selected in [KNR+09] as the metric for comparison. A UE antenna system with half-wavelength spacing is assumed in the simulation. Firstly, comparison is made between a single cluster with 35° angular spread modeled after the SCME urban macro model (20 subpaths) and a reduced version of the model in which all subpaths are mapped to three probes (Fig. 5.40).

It is found in [KNR+09] that the magnitude of the correlation coefficient as a function of mean AOA is well reproduced for the assumed dual antenna arrangement. So as to reproduce also the fading statistics, individual components must, however, be pre-faded before mapping to the small number of probes. Moreover, to preserve the wideband characteristics of the channel, the probe powers must include the proper delay dispersion so that each path of the channel realization contains

---

[1] The picture changes somewhat for UE antenna systems more directive patterns and/or polarization selectivity.

Fig. 5.41 Comparison between SCME model and an 8-probe approximation in terms of correlation observed with two half-wavelength space antennas [Ree09]

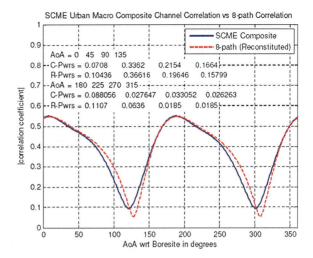

the correct narrow angle spread, i.e. the pathwise PAS. To this end, the composite channel model is decomposed in [KNR+09] into its constituent paths and the appropriate delay applied before superposition by mapping to probes. Doing so for an arrangement of eight probes, located at 45° separation along a circle centered about the UE in the azimuthal plane, is found to reproduce also the SCME Urban Macro model with sufficient accuracy. The comparison between the full SCME model and the 8-path approximation is shown in Fig. 5.41.

In summary, a very interesting approach has been developed in [Ree09] to compare MIMO OTA measurement setups against existing GSCMs. Application of the same approach to a larger set of more realistic UE antenna patterns, in particular also patterns with a more rapid angular variation, more directive behavior and polarization selectivity, may lead to conclusive answers about the number probes required for trustworthy emulation of a GSCM.

Investigations as to the fidelity of reconstructing Power-Delay Profile (PDP) and XPR as specified by a GSM in a multi-probe OTA measurement set-up are reported in [KNH09]. The focus of the investigation here is on verification of the implementation of setup which employs two 8-channel fading emulators and a circular arrangement of eight dual-polarized probes in an anechoic chamber. The reconstruction of the PDP is verified for vertical polarization using a channel sounder and a dipole antenna in the DUT position. As an example, the measured impulse response is given in comparison to the desired PDP for the SCM urban macro scenario in Fig. 5.42. The results confirm that the temporal characteristics of the channel can be reproduced to very good accuracy.

For XPR verification, the dipole antenna is [KNH09] in sequence mounted in three orthogonal orientations. This set-up allows for direct comparison against target values only for selected values of the AOA of a cluster, i.e. for cases where it coincides with the broadside orientation of a horizontal dipole. In these cases at least it is observed that the measured ratio follows the target value.

Fig. 5.42 Reproduction of SCME urban macro PDP with an 8-probe arrangement: measured impulse response [KNH09]

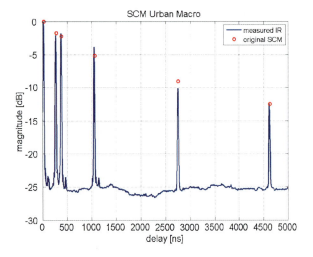

Fig. 5.43 Observed throughput in a 2 × 2 MIMO system in an OTA measurement system: comparison of full SCME Urban Micro model and simplified models without DS (noDS), without AS (noAS) and neither (noAS/DS)

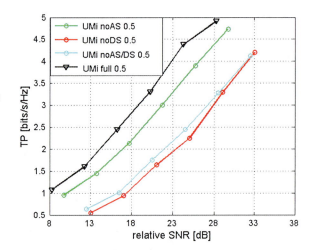

An investigation which goes beyond analysis of a reconstructed propagation channel itself only but includes performance results for a 10 MHz OFDM 2 × 2 adaptive MIMO system is reported in [NKB09]. Also the approach is different: in the sense of negative testing, simplifications are introduced at will into the propagation model to confirm their relevance for performance investigations. Starting point is the SCME Urban Micro TDL model. Results are compared against the case that (i) delay spread, (ii) angular spread and (iii) both parameters are set to zero. Observed throughput over Signal-to-Noise Ratio (SNR) is shown in Fig. 5.43, the original model and the three simplified models. The results confirm that a frequency flat channel (in the absence of delay spread) or a propagation scenario with no angular spread are lacking diversity or expose increased correlation, respectively, and hence differ from the original channel model.

5 "OTA" Test Methods for Multiantenna Terminals 241

With view on OTA testing, discrimination between good and bad antenna (and UE) designs is also in the context of GSCM emulation an important aspect. Clearly, if a propagation scenario would tend to produce baseline results for the communication system under study, weighted perhaps only by overall UE sensitivity, its emulation would amount to a waste of time.

This aspect is addressed in [KJW10a] by comparing several scenarios from a simplified WINNER II channel model, the SCME sub-urban macro, urban macro and urban micro models and single clusters with 15° and 68° angular spread using a large set of different antennas. The latter includes, amongst dual dipole arrangement with different spacings, also some realistic patterns, e.g. from commercial WLAN equipment, and one example in which the influence of a SAM phantom is taken into account. The metrics for comparison are (i) median channel capacity and (ii) the range of capacity variation during rotation of the UE about the vertical axis.

The extensive results given in [KJW10a] are interesting in several respects. Despite the wide range of propagation scenarios included in the study, the median capacity shows only moderate deviation among models if applied to the same antenna (at the order of $\pm 10\%$). Differences in median capacity between any pair of antennas subject to the same propagation scenario are typically of comparable magnitude but in some cases, particularly for SCM models, also much smaller. If these simulation results can be confirmed by measurements, it will have to be concluded that this type of measurements may not always be suitable to discriminate between good and bad antenna designs as far as their spatial characteristics are concerned.[2] After all, throughput changes of the order of 10% may correspond to SNR differences down to the order of 1 dB, which also appears critical with view on reproducibility.

The second metric considered in [KJW10a], i.e. the range of the capacity variation during UE rotation shows moderate dependency on antenna characteristics within one and the same propagation scenario. But the variation is found to be consistently larger for the WINNER II scenarios, whereas only negligible variation is observed for some of the SCME scenarios. The study links this observation to the differences in angular spread assumed in the models. A detailed investigation of the range of capacity variation as a function of angular spread in a single-cluster model confirms this. Figure 5.44 displays the difference observed in median capacity when comparing selected pairs of antenna systems. It is observed that the differences between the antenna pairs compared in this figure tend to smear out for too large angular spread. These results seem to be in agreement with the statement (see Sect. 5.5.1) that propagation scenarios which approach a uniform PAS result in little differentiation between DUTs.

5.5.3 Channel Measurements and OTA Comparison

In the aim of consistently attempting to validate the OTA test method, there are repeated comparisons made on the same handset by comparing OTA against real

---

[2]Nevertheless, they may still be sensitive to antenna efficiency and receiver performance.

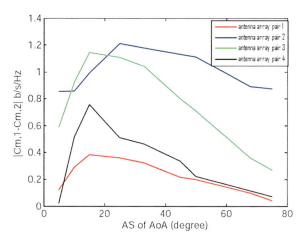

Fig. 5.44 Differences in median capacity for selected pairs of UE antenna systems as functions of AS

test measurements. There are two important aspects to this validation. The first is that clear and consistent angles of arrival models are assumed that are used in the OTA, where repeated propagation measurements and resulting stochastic models can give valid outputs. However, further detail is required in several channel models in terms of the angle of arrival models adopted and the accuracy with which they are modeled. Measurements carried out using cylindrical patch array antennas has allowed higher resolution predictions using the SAGE algorithm.

Comparisons were made in an urban scenario to compare real measurements with that of OTA shown in Fig. 5.45. Comparisons repeatedly show that for differing handsets with comparable user interaction, the predicted average capacity according to OTA is within 10% difference. The chosen environment was a street canyon with an operating frequency of 2.4 GHz that would have experienced typical variation in azimuth angle of arrival as recorded in the other measurements. The base station antennas in this case are suitably separated apart so as to allow sufficient decorrelation.

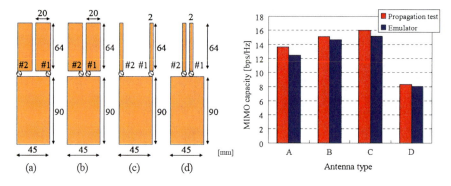

Fig. 5.45 Comparison of capacity results for OTA and full measurement campaign tests with the same handsets

5.6 Conclusion/Final Remarks on MIMO OTA Set-up

This chapter has presented the latest advancements on the standardizations efforts of COST 2100 SWG 2.2. This joint work has been unanimously recognized one of the main success of the Action COST 2100, bringing Academia and Industry together in an unprecedented synergy, with partners well beyond the European boundaries. As the mass deployment of LTE networks is very close, it has been of primary importance to build a common framework to be used as enabler of the challenging upcoming 4G systems. The first section has briefly introduced the topic, starting first with the inherent challenges in handset MIMO, moving later on to the description of SISO and MIMO OTA characteristics. Many competitive MIMO OTA set-ups were examined and compared in Sect. 5.2, with different levels of cost/complexity trade-offs. In Sect. 5.3 several lab evaluations of MIMO OTA setups were investigated, building the foundations of the round robin measurement campaign of Sect. 5.4, while Sect. 5.5 described the comparison between real environment references and MIMO OTA set-ups. Here we report a technical summary of the obtained findings, which will however need to be integrated with upcoming LTE round robin measurements. Due to its impact in boosting the development of innovative solutions, the MIMO OTA work will naturally continue in the COST Action IC1004 that will soon start in 2011.

5.6.1 Summary of Findings

The HSPA round robin measurement campaign has provided valuable insights and more complete understanding of the test methodologies. From the results some findings can be obtained, and brief overview of each of these findings is given below:

5.6.1.1 Finding One: Some Results Indicated that DUT Cannot Reach the Expected OTA Throughput

From the HSPA measurement campaign some test methods (e.g. the reverberation-chamber test methods) produced OTA throughput values that were significantly less than the theoretical throughput calculated from transmitted TBS, UE category and MCS level for a DUT. This finding is more obvious for channel models with larger delay spread. Therefore, one plausible reason is the large delay spread from the channel's PDP. The large delay spread could result in severe ISI and degrade the BER performance, thus degrading the DUTs throughput. It is also well known that UMTS RAKE receiver can suffer from irreducible BER due to excessive multipath delay. For LTE MIMO system, the excessive delay spread could still cause performance degradation if the RMS delay is larger than the cyclic prefix length of the OFDM symbol. Advanced receivers such as MMSE equalizer receiver can mitigate the problem. Hence, the selection of channel models and detailed channel parameterization are needed for comparison. The SCME channel models produced less throughput degradation. Another plausible reason is related to antenna radiation performance of DUT. It was found that under receiver diversity mode, DUT with poorer diversity gain will generally result in lower OTA throughput, and the influence of power imbalance and spatial correlation is small. It was also postulated that the opposite effect will be observed under spatial multiplexing mode.

5.6.1.2 Finding Two: Anechoic-Chamber-Based Methods Show Good Repeatability and Reproducibility Across Different Labs

Measurement results also showed that anechoic-chamber-based test methods produced consistent results across different labs (Nokia, SATIMO, NTT, ETSL). Nokia's results also highlighted the influence of NodeB emulator on the throughput. It was recommended that detailed and thorough parameter settings should be performed to overcome the influence. The reverberation-chamber-based test methods were shown to produce good repeatability also but with relatively larger deviation. However, only two labs were compared.

5.6.1.3 Finding Three: Anechoic-Chamber-Based and Two-Stage Test Methods Show Similar Results

The measurement results from both anechoic-chamber-based and two-stage test methods were found to be closely matched on H-Set6 F-RMC under SCME Umi channel models, under the assumption that the antenna pattern of DUT can be measured accurately. In addition, various stages of computation and signal combining have to be controlled to achieve comparable results.

5.6.1.4 Finding Four: Reverberation and Anechoic Methods Produce Different Results

Conversely, it was found that the measurement results from the reverberation-chamber test methods are significantly different from the two-stage and simplified anechoic-chamber methods. This is based on the simplified SCME channel models (1–2 clusters). This means that the DUT behavior could be significantly different between the reverberation-chamber-based and two-stage/simplified test methods. However, the comparison did not consider 3D channel propagation effect in the anechoic chamber. The mean throughput point was found to be consistent across both categories of test methods.

References

[3GP] 3GPP TS 34.114: User equipment (UE)/mobile station (MS) over the air (OTA) antenna performance conformance testing.

[FKS10] Y. Feng, A. Krewsky, and W. L. Schroeder. Towards a simulation tool for comparison of MIMO OTA measurement methods and metrics. In *COST 2100 TD(10)10058*, February 2010.

[FOKO10] M. Franzén, C. Orlenius, D. Kurita, and Y. Okano. Comparison of MIMO OTA performance characteristics as measured in reverberation chamber and spatial fading emulator. In *COST 2100 TD(10)10084*, February 2010.

[FS09a] Y. Feng and W. Schroeder. Discussion of low-effort MIMO OTA testing approaches & suggested applicable FOM. In *COST 2100 TD(09)951*, September 2009.

[FS09b] Y. Feng and W. Schroeder. Extending the definition of TIS and TRP for application to MIMO OTA testing. In *COST 2100 TD(09)866*, May 2009.

[FSJR10] Y. Feng, W. Schroeder, J. Jonas, and M. Rumney. Results from the COST 2100 SWG 2.2 MIMO OTA round robin measurement campaign. In *COST 2100 TD(10)12086*, November 2010.

[Gag08] C. V. Gagern. New wireless technologies and OTA measurements. In *COST 2100 TD(08)441*, February 2008.

[Gag09] C. V. Gagern. Cost-effective over-the-air performance measurements on MIMO devices. In *COST 2100 TD(09)804*, May 2009.

[Gag10] C. Gagern. Measurement of channel parameters vs. throughput. In *COST 2100 TD(10)10062*, February 2010.

[Hal10] P. Hallbjörner. Test zone disturbance level in three multipath simulators. In *COST 2100 TD(10)12012*, November 2010.

[HYS+10] T. Hayashi, A. Yamamoto, T. Sakata, K. Ogawa, K. Sakaguchi, and J. Takada. Effect of delay waves on MIMO OTA throughput evaluation of a handset MIMO array using a spatial fading emulator. In *COST 2100 TD(10)10018*, February 2010.

[JKW+10] Y. Jing, H. Kong, Z. Wen, S. Duffy, and M. Rumney. Considerations on TIS definition for MIMO OTA test. In *COST 2100 TD(10)11075*, June 2010.

[JWK10a] Y. Jing, Z. Wen, and H. Kong. MIMO OTA round robin test report: capacity, correlation and power imbalance measurement results for multiple probe antenna based method and two-stage method, November 2010.

[JWK10b] Y. Jing, Z. Wen, and H. Kong. Statistical property analysis and verification of multi-probe MIMO OTA test method, November 2010.

[KI08] J. Krogerus and C. Icheln. Considerations on anechoic chamber test methods for performance evaluation of multi-antenna mobile terminals. In *COST 2100 TD(08)470*, February 2008.

[KJW10a] H. Kong, Y. Jing, and Z. Wen. MIMO OTA channel models comparison, analysis and recommendations. In *COST 2100 TD(10)10048*, February 2010.

[KJW+10b] H. Kong, Y. Jing, Z. Wen, S. Duffy, and M. Rumney. Considerations on criteria for good and bad MIMO antenna design. In *COST 2100 TD(10)11076*, June 2010.

[KKN+08] P. Kyösti, J. Kolu, J. P. Nuutinen, M. Falck, and P. Mäkikyrö. OTA testing for multiantenna terminals. In *COST 2100 TD(08)670*, October 2008.

[KLM+07] J. Krogerus, T. Laitinen, M. Mustonen, P. Suvikunnas, J. Villanen, C. Icheln, and P. Vainikainen. Antenna performance testing of mobile terminals—current state and future prospects. In *COST 2100 TD(07)383*, September 2007.

[KLN09a] J. Kallankari, S. Laukkanen, and M. Nurkkala. OTA HSDPA throughput measurements in a fading channel. In *COST 2100 TD(09)836*, May 2009.

[KLN09b] J. Kallankari, S. Laukkanen, and M. Nurkkala. Test plan for OTA throughput comparison measurements in a fading channel environment. In *COST 2100 TD(09)964*, September 2009.

[KMV08] J. Krogerus, P. Mäkikyrö, and P. Vainikainen. Towards an applicable OTA test method for multi-antenna terminals. In *COST 2100 TD(08)671*, October 2008.

[KNB09]	P. Kyösti, J. Nuutinen, and A. Byman. Verification of MIMO OTA set-up via simulations and measurements. *COST 2100 TD(09)990*, September 2009.
[KNH09]	P. Kyösti, J. Nuutinen, and P. Heino. Reconstruction and measurement of spatial channel model for OTA. In *COST 2100 TD(09)860*, May 2009.
[KNL10]	J. Kallankari, M. Nurkkala, and S. Laukkanen. HSDPA OTA throughput test results. In *COST 2100 TD(10)12079*, November 2010.
[KNM09]	J. Krogerus, M. Nurkkala, and P. Mäkikyrö. Discussion on some topical issues related to the spatial fading emulation based OTA test method for multi-antenna terminals. In *COST 2100 TD(09)780*, February 2009.
[KNR+09]	P. Kyösti, J. Nuutinen, D. Reed, T. Jämsä, and R. Borsato. Requirements for channel models for OTA multi-antenna terminal testing. In *COST 2100 TD(09)859*, May 2009.
[Kon09]	H. Kong. Two-stage MIMO OTA method. In *COST 2100 TD(09)924*, September 2009.
[KONO09]	D. Kurita, Y. Okano, S. Nakamatsu, and T. Okada. Experimental comparison of MIMO OTA testing methodologies. In *COST 2100 TD(09)932*, September 2009.
[M. 10]	M. Á. Garcia-Fernández and D.A. Sanchez-Hernandez. Recent advances in mode-stirred reverberation chamber emulation for 4G wireless terminals testing. In *COST 2100 TD(10)12004*, November 2010.
[Man09]	L. Manholm. Discussion on channel conditions for MIMO OTA measurements. In *COST 2100 TD(09)978*, September 2009.
[MV09]	P. Moller and L. Vannatta. Investigation of a 3 meter reverberation chamber for its suitability in WCDMA OTA measurements. In *COST 2100 TD(09)821*, May 2009.
[NKB09]	J. Nuutinen, P. Kyösti, and A. Byman. Effect of channel model simplification on throughput in MIMO OTA. In *COST 2100 TD(09)971*, September 2009.
[NKF+09]	J. Nuutinen, P. Kyösti, M. Falck, P. Heino, H. Lehtinen, and T. Jääskö. Experimental investigations of OTA system. In *COST 2100 TD(09)753*, February 2009.
[NKMF09]	J. Nuutinen, P. Kyösti, J. Malm, and M. Foegelle. Experimental investigations of MIMO performance of IEEE802.11n device in MIMO OTA test system. In *COST 2100 TD(09)972*, September 2009.
[NS10]	M. Nurkkala and A. Scannavini. Laboratory correlation of 3GPP MIMO OTA measurement campaign. November 2010.
[OKI09]	Y. Okano, K. Kitao, and T. Imai. Impact of number of probe antennas for MIMO OTA spatial channel emulator. In *COST 2100 TD(09)929*, September 2009.
[PLA10]	V. Plicanic, B. K. Lau, and H. Asplund. Free space performance comparison of MIMO terminal antennas in noise- and interference-limited urban environments. In *COST 2100 TD(10)12050*, November 2010.
[PSF+10]	C. L. Patanén, A. Skaarbratt, M. Franzén, J. Asberg, and C. Orlenius. OTA round robin measurement campaign: experiences from measurements in reverberation chamber. In *COST 2100 TD1012082*, November 2010.
[Ree09]	D. Reed. Experiments with correlation. In *COST 2100 TD(09)856*, May 2009.
[Ree10]	D. Reed. Measuring device antennas with spatial and uniform channels. In *COST 2100 TD(10)10092*, February 2010.
[Rum09]	M. Rumney. Efficacy criteria of MIMO OTA test. In *COST 2100 TD(09)925*, September 2009.
[Rum10a]	M. Rumney. Figures of merit for MIMO OTA testing. In *COST 2100 TD(10)11063*, June 2010.
[Rum10b]	M. Rumney. Selecting figures of merit and developing performance requirements for MIMO OTA. In *COST 2100 TD(10)10028*, February 2010.
[SF09]	W. Schroeder and Y. Feng. Proposal for a built-in MIMO test function for E-UTRA user equipment and its application. In *COST 2100 TD(09)853*, May 2009.
[SFP09]	W. Schroeder, Y. Feng, and M. Pesavento. Discussion of some options and aspects in over-the-air (OTA) testing of multiple input—multiple output (MIMO) user equipment (UE). In *COST 2100 TD(09)740*, February 2009.

[SHVVH+10] J. D. Sánchez-Heredia, J. F. Valenzuela-Valdés, J. P. Hidalgo, A. Torrecilla, S. Lobato, and D. A. Sánchez-Hernández. Evaluation of MIMO OTA parameters for LTE using a mode-stirred reverberation chamber. In *COST 2100 TD(10)12073*, November 2010.

[SI08] A. Scannavini and P. O. Iversen. Radiated performance testing of diversity and MIMO enabled terminals. In *COST 2100 TD(08)618*, October 2008.

[SKI+09] K. Saito, K. Kitao, T. Imai, Y. Okano, and S. Miura. 2 GHz channel measurement in the urban environment and the investigation of channel dynamic properties. In *COST 2100 TD(09)998*, September 2009.

[Szi10] I. Szini. LTE antenna system figure of merit techniques and trade-offs probing antennas on LTE devices. In *COST 2100 TD(10)11026*, June 2010.

[Tak09] J. Takada. Handset MIMO antenna testing using a RF-controlled spatial fading emulator. In *COST 2100 TD(09)742*, February 2009.

[TAK+10] A. Tankielun, J. A. Antón, R. Koller, E. Böhler, and C. V. Gagern. Two-channel method for evaluation of MIMO OTA performance of wireless devices. In *COST 2100 TD(10)12046*, November 2010.

[TBG10] A. Tankielun, E. Böhler, and C. V. Gagern. Two-channel method for evaluation of MIMO OTA performance of wireless devices. In *COST 2100 TD(10)12046*, November 2010.

[vG08] C. von Gagern. OTA performance measurements for A-GPS devices. In *COST 2100 TD(08)612*, October 2008.

[Wel09] J. Welinder. Multi path simulator. In *COST 2100 TD(09)704*, February 2009.

[WFBM09] J. Welinder, L. Fast, T. Bolin, and L. Manholm. Towards a low cost over the air performance test method. In *COST 2100 TD(09)807*, May 2009.

[WJK10a] Z. Wen, Y. Jing, and H. Kong. Device performance under different channel models using two-stage method. November 2010.

[WJK10b] Z. Wen, Y. Jing, and H. Kong. MIMO OTA round robin test report: throughput measurement results for multiple probe antenna based method and two-stage method. In *COST 2100 TD(10)12055*, November 2010.

[WKJ+10] Z. Wen, H. Kong, Y. Jing, S. Duffy, and M. Rumney. Test MIMO antenna effects on a MIMO handset using two-stage OTA method. In *COST 2100 TD(10)10049*, February 2010.

[YTH+09] A. Yamamoto, S. Tsutomu, T. Hayashi, K. Ogawa, J. O. Nielsen, G. F. Pedersen, J. Takada, and K. Sakaguchi. MIMO performance evaluation in a street microcell using a spatial fading emulator in comparison with a radio propagation test. In *COST 2100 TD(09)912*, September 2009.

[YTH+10] A. Yamamoto, S. Tsutomu, T. Hayashi, K. Ogawa, K. Sakaguchi, and J. Takada. Procedure of designing the structural parameters of a spatial fading emulator. In *COST 2100 TD(10)10016*, February 2010.

Chapter 6
RF Aspects in Ultra-WideBand (UWB) Technology

Chapter Editor Grzegorz Adamiuk, Jens Timmermann, Christophe Roblin, Wouter Dullaert, Philipp Gentner, Klaus Witrisal, Thomas Fügen, Ole Hirsch, and Guowei Shen

The chapter focuses on radio frequency aspects in UWB and is organized into four sections: the first section briefly summarizes the main aspects about regulation issues of UWB technology; the second section describes antenna design and characterization; the third section discusses UWB channel modeling and measurements; and the last section describes localization and radar imagining with UWB technology.

6.1 UWB Technology and Regulatory Issues

The utilization of UWB technology is restricted by regulations. There are several regulations worldwide that take into consideration the respective national frequency plans. UWB regulations describe the following items:

- Application (indoor, outdoor, portable, fixed installed)
- Frequency range
- Maximal Power Spectral Density (PSD) in the sense of Equivalent Isotropic Radiated Power (EIRP)
- Mitigation techniques to reduce possible interference

Up to now, there exist only UWB regulations in the United States, Europe, Japan, Korea, Singapore and China. The first regulation was released by the FCC for the United States (Feb. 2002). It allocates a technical usable band between 3.1 and 10.6 GHz for indoor applications according to Table 6.1 [FCC02]. Since March 2006, also Europe released a regulation [ECC08] for indoor applications. It allocates the two bands [4.2, 4.8] GHz and [6, 8.5] GHz, whereas the first band can only be used with additional mitigation techniques (else, the maximal PSD is −70 dBm/MHz instead of −41.3 dBm/MHz). Table 6.2 summarizes the maximal PSD versus frequency for the European regulation.

G. Adamiuk (✉)
Karlsruhe Institute of Technology, Karlsruhe, Germany

Table 6.1 FCC regulation: Limit of the PSD for indoor applications

Frequency range in GHz	PSD in dBm/MHz
0–0.96	−41.3
0.96–1.61	−75.3
1.61–1.99	−53.3
1.99–3.1	−51.3
3.1–10.6	−41.3
>10.6	−51.3

Table 6.2 ECC regulation: Limit of the PSD for indoor applications (*—with additional mitigation techniques)

Frequency range in GHz	PSD in dBm/MHz
0–1.6	−90.0
1.6–2.7	−85.0
2.7–3.4	−70.0
3.4–3.8	−80.0
3.8–4.2	−70.0
4.2–4.8	−70.0/−41.3*
4.8–6.0	−70
6.0–8.5	−41.3
8.5–10.6	−65
>10.6	−85

Table 6.3 Usable frequency ranges for UWB communication in different nations

Nation	1st frequency band in GHz	2nd frequency band in GHz
USA	[3.1 10.6]	–
Europa	[4.2 4.8]	[6.0 8.5]
Japan	[3.4 4.8]	[7.25 10.25]
Korea	[3.1 4.8]	[7.2 10.2]
Singapore	[4.2 4.8]	[6.0 9.0]
China	[4.2 4.8]	[6.0 9.0]

Finally Table 6.3 lists the technical usable frequency ranges for all regulations worldwide. The allocated maximal PSD is always −41.3 dBm/MHz.

Due to the large bandwidth, the main applications of UWB are high data rate transmission (> 100 Mbit/s) and applications that exploit the ultrafine time resolution (localization and imaging). The integration of the PSD between 3.1 and 10.6 GHz leads to a maximal transmit power of 0.56 mW for the FCC mask. Since the transmit power is hence very low, UWB transmission is limited to short-range applications. UWB transmission can be performed by two different techniques:

- Pulse-based transmission using pulses with a bandwidth of several GHz (called impulse radio)
- Subdivision of the bandwidth into several broadband channels and transmission of the signals by Orthogonal Frequency Division Multiplexing (OFDM)

Classical pulse shapes for impulse radio transmission are for example the Gaussian Monocycle and its derivatives. They can be generated by low-cost devices, and the pulses are directly fed into the transmit antenna in the base band. There is no necessity of mixers which leads to a reduced complexity compared to conventional narrowband transceivers. However, a disadvantage of impulse radio is that classical pulse shapes do not exploit the spectral mask very well which leads to a loss of signal-to-noise ratio.

In contrast, transmission by OFDM can exploit the given spectral mask very well. A drawback of this technique is however the necessity of an increased signal processing.

6.2 Advances in UWB Antenna Research

6.2.1 UWB Antenna Characterization

6.2.1.1 Compact Model of UWB Antenna Radiation Pattern

Let us consider the radiation pattern of a UWB antenna, $\mathbf{F}(f, \phi, \theta)$. The main difference between standard narrowband radiation patterns and UWB radiation patterns is that it can no longer be considered constant over frequency: the dependency on f explicitly needs to be taken into account. This makes UWB radiation pattern data very large and difficult to manipulate and optimize. The angular dependency can be easily modeled using vector spherical waves. The frequency dependency can either be modeled using the Singularity Expansion Method (SEM) (Singularity Expansion Method), [Rob07] or Slepian modes, [Dul08, DR09]. Here the method based on Slepian modes will be presented. For more information on the SEM, the reader is referred to [Rob06]. First the theoretical background is presented. Second the model is verified using both simulated and measured antenna data.

Phase Modes We will present the model for a fixed planar cut $\theta = \theta_0$ of a sampled radiation pattern, $\mathbf{F}(f, \phi, \theta_0)$. The remaining angular dependency ϕ will be expanded into phase modes. The frequency dependency in Slepian modes. This model can be easily expanded to 3D by replacing the phase mode expansion with a related spherical mode expansion [DR10].

To obtain the phase modes, each polarization of the radiation pattern is expanded into complex exponentials:

$$F(f, \phi) = \sum_{m=-\infty}^{\infty} F_m^{\phi}(f) \exp(jm\phi). \tag{6.1}$$

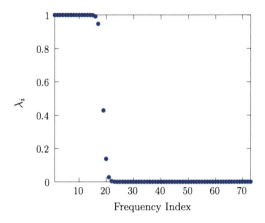

Fig. 6.1 λ_i for $N = 73$ and $c = 0.1254$

Rogier [Rog06] shows that the amount of phase modes needed to accurately describe a radiation pattern is limited by the dimensions of the antenna. We can therefore limit the summation in (6.1) to $M = \lceil k_0 d \rceil$, with k_0 the wave number at the highest considered frequency and d the largest dimension of the antenna.

Slepian Modes We now model the frequency dependency of the coefficients $F_m^\phi(f)$, by expanding them into discrete prolate spheroidal sequences (DPSSs):

$$F_m^\phi(f) = \sum_{k=0}^{N} F_{m,k}^{\phi,f} \psi_{k,c}(f), \qquad (6.2)$$

where N is the total number of frequency samples, $\psi_{k,c}(f)$ is the discrete prolate spheroidal sequence of order k, with bandwidth parameter $0 < c < 0.5$. The DPSSs are the discrete equivalent of the prolate spheroidal wave functions (PSWFs), both discovered by Slepian in [Sle78] and [SP61] respectively. The DPSSs are the time-limited sequences, in a predefined interval $[-t_0, t_0]$, with the most energy in BW, the bandwidth of interest. This makes them ideally suited to serve as a basis for modeling frequency dependencies and allows us to truncate the series (6.2) after $K \ll N$.

Making use of some properties of the DPSSs, a rule of thumb for K has been presented in [DAR10]. For proof of these properties and a more detailed overview of these sequences, the interested reader is directed to [Sle78]. The DPSSs are the eigenvectors of the sinc-matrix:

$$B(N, M)_{k,l} = \frac{\sin(2\pi c(k-l))}{\pi(k-l)} \qquad (6.3)$$

with distinct eigenvalues λ_n: $1 > \lambda_0 > \lambda_1 > \cdots > \lambda_n > 0$. It can be proven that these λ_i are almost 1 for $i \leq 2cN$ and almost 0 for $i > 2cN$, where c is the bandwidth parameter defined before. In Fig. 6.1 these λ_i are plotted for $N = 73$ and $c = 0.1254$.

The λ_i also represent the fraction of the energy of the ith-order DPSS that falls in $|f| < BW$. Since the DPSSs form a complete orthonormal basis, Parceval's theorem

Fig. 6.2 Original Simulated Radiation Pattern [DR10] (©2010 IEEE, reproduced with permission)

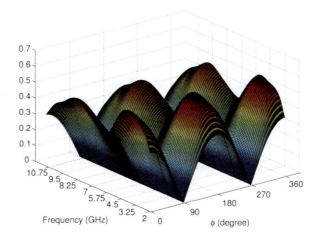

shows that all BW-bandlimited functions can be described by the first $2cN$ Slepian modes. In Fig. 6.1 we see that there is a small transition region for the eigenvalues, where they are not close to 0 nor close to 1. We therefore propose $K = 2cN + 10$ as a rule of thumb. Accuracies around 90% can be obtained using this rule of thumb.

Verification We will verify the model by applying it to the θ-polarization of the radiation pattern of a planar monopole antenna. The design details of the antenna can be found in [DR10]. The simulated radiation pattern has been sampled each degree in the azimuthal plane and every 125 MHz between 2 GHz and 11 GHz, resulting in 26280 samples. For the model, a c value of 0.1254 was used. This results in a Slepian truncation boundary of $K = \lceil 2cN \rceil + 10 = 28$. The largest dimension of the antenna is 46 mm, resulting in a phase mode truncation boundary of $M = \lceil k_0 d \rceil = 11$.

The original simulated radiation pattern and the reconstructed radiation pattern have been plotted side by side in Figs. 6.2 and 6.3 respectively. It can be visually verified that they are in good agreement. The amplitude of the model coefficients is nearly zero at the truncation boundaries, as can be seen in Fig. 6.4. This means that almost all the energy of the original radiation pattern is maintained in the model. Finally in Fig. 6.5 the relative error between a polarization of the original simulated radiation pattern $F(f, \phi, \theta_0)$ and its reconstruction $\hat{F}(f, \phi, \theta_0)$, defined as

$$Error(f) = \frac{\sum_\phi |F(f, \phi, \theta_0) - \hat{F}(f, \phi, \theta_0)|}{\sum_\phi |F(f, \phi, \theta_0)|}, \quad (6.4)$$

is shown. Apart from one spike to 14%, the relative error stays well below 10% over the entire frequency range.

We will also verify the model on the measured θ-polarization of this radiation pattern, using the same model coefficients as before. The radiation pattern has been sampled 51 times in the same frequency range, and every 3 degrees in the azimuthal plane. The original data is shown in Fig. 6.6 and is very noisy, especially at higher

Fig. 6.3 Reconstructed Simulated Radiation Pattern

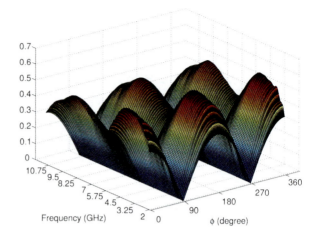

Fig. 6.4 Coefficients of the Simulated Radiation Pattern [DR10] (©2010 IEEE, reproduced with permission)

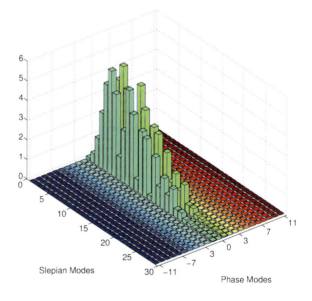

frequencies. On the reconstructed radiation pattern, Fig. 6.7, it can be seen that the model managed to remove a large portion of the noise, while still capturing the useful antenna data. This is possible because the noise contribution is spread equally over all modes, while the useful data is clustered in the lower ones. Truncating at a certain M and K, effectively filters out most of the noise (cf. Fig. 6.8).

6.2.1.2 Statistical Analysis of UWB Antennas and Scatterers

Multidimensional models of the Far Field (FF) radiation characteristics of multi-band, wideband or UWB antenna—i.e. frequency (or time) and angular dependent—

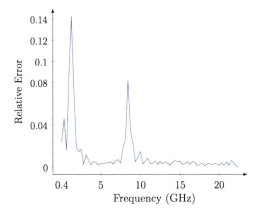

Fig. 6.5 Relative Error of the Simulated Radiation Pattern [DR10] (©2010 IEEE, reproduced with permission)

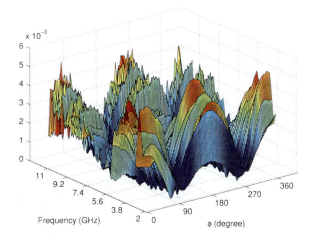

Fig. 6.6 Original Measured Radiation Pattern [DR10] (©2010 IEEE, reproduced with permission)

or Multiple Antenna System (MAS), can be particularly useful in the context of end-user applications. Here, the main objective is an efficient representation (both parsimonious and accurate) of small to moderate size UWB antennas and/or terminals used in radiocommunication systems, for example in simulators at the radio link level or in ray-tracing tools. The topic is actually manifold: 1. Modeling of all FF radioelectric properties of single UWB (or multiband) antennas (belonging to various antenna "classes") with an extremely "light" database and at a low computational cost [RD08]. 2. Modeling of MAS with either a given array geometry for various radiators or variable geometries for a particular radiator [DR08]. 3. Modeling of small or moderate-size antennas or terminals perturbed by their close environment (human body, close scatterers, integration/casing, etc.). The latter is addressed in Sect. 4.5. The formers will be summarized in the sequel. The variability of the situations for all theses topics is such that the combinatory is unreachable, suggesting that—in the manner of the analysis and the modeling of the propagation channel—a statistical approach seems particularly appropriate.

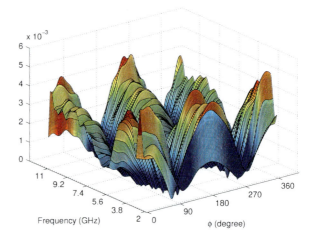

Fig. 6.7 Reconstructed Measured Radiation Pattern [DR10] (©2010 IEEE, reproduced with permission)

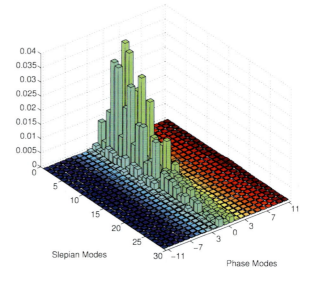

Fig. 6.8 Coefficients of the Measured Radiation Pattern [DR10] (©2010 IEEE, reproduced with permission)

Statistical Models of UWB Antennas Based on a Parametric Modeling An indirect approach is used here to reduce the complexity of the statistical analysis, which is applied to the parameters extracted from a parametric model based on a high-order reduction technique, the SEM [Bau71, Bau73] and the Spherical Mode Expansion Method (SMEM) (Spherical Mode Expansion Method) [Han88]. This is an alternative to the direct approach in which the statistical analysis is directly performed on "primary" radioelectric quantities (efficiency, directional gain or UWB antenna transfer function, etc.). To build up the stochastic population of antennas, several generic designs are considered. A "generic design" is the union of a geometry represented by a set of design parameters and of the space these parameters are allowed to span. The Monte Carlo method is used for each generic design: N_A "antenna realizations" are generated from random sampling of the design parameters

6 RF Aspects in UWB Technology

(assuming an a priori probabilistic model). These antennas are then electromagnetically simulated, and some of their properties verified: for example, realizations which are not sufficiently matched (S_{11} above an a priori threshold) or present an insufficient realized gain (G_r below a threshold) are considered outliers and discarded (and replaced by a new shot if required).

The transmit Antenna Transfer Function (ATF) used here is defined as [RBS03]

$$\mathcal{H}(s, \hat{\mathbf{r}}) = \frac{re^{\gamma r}}{a_1(s)} \sqrt{\frac{4\pi}{\eta_0}} \cdot \mathbf{E}^{\infty}(s, \mathbf{r}), \quad (6.5)$$

where s is the complex frequency, $\gamma = s/c$ ($= jk$ in the harmonic case), $\mathbf{r}$ is the radial vector, $\hat{\mathbf{r}}$ the unit radial vector ("$\hat{\mathbf{r}} = (\theta, \varphi)$" in the functions argument), a_1 is the incident partial wave, $\mathbf{E}^{\infty}$ the electric FF, and η_0 the free space impedance.

Applying the SEM and the SMEM to the FF, it has been shown in [Rob06] that, by linearity, the ATF—and the Antenna Impulse Response (AIR), given by inverse Fourier Transform—can be expanded as

$$\mathcal{H}(s, \hat{\mathbf{r}}) \approx \widetilde{\mathcal{H}}_{N,P}(s, \hat{\mathbf{r}}) = \sum_{p=1}^{P}\left[\sum_{n=1}^{N}\sum_{m=-n}^{n}\sum_{u=1}^{2} R_{nmp}^{(u)} \hat{\boldsymbol{\psi}}_{nm}^{(u)}(\hat{\mathbf{r}})\right] \cdot (s-s_p)^{-1}, \quad (6.6)$$

$$h(t, \hat{\mathbf{r}}) \approx \widetilde{h}_{N,P}(t, \hat{\mathbf{r}}) = \sum_{p=1}^{P}\left[\sum_{n=1}^{N}\sum_{m=-n}^{n}\sum_{u=1}^{2} R_{nmp}^{(u)} \hat{\boldsymbol{\psi}}_{nm}^{(u)}(\hat{\mathbf{r}})\right] \cdot e^{s_p t}, \quad (6.7)$$

where $\{s_p\}$ is the poles set, the $\{\hat{\boldsymbol{\psi}}_{nm}^{(u)}\}$ are the vector spherical wave functions with $u = \{1, 2\} = \{TE, TM\}$, $\{R_{nmp}^{(u)}\}$ is the set of *modal residues* (for the pole p and the TM or TE (n, m) mode) [Rob06], the sums have been truncated to spherical mode order N, and P poles are assumed to appear in complex conjugate pairs $s_p = \sigma_p \pm j\omega_p$. The modal residues verify the general relation $R_{nmp^*}^{(u)} = (-1)^m [R_{n,-m,p}^{(u)}]^*$ where $s_{p^*} = s_p^*$ [Rob08], so that the total number of complex parameters N_T of the truncated representation is $N_T = P/2 + PN(N+2) \approx PN(N+2)$. Note that N_T is considerably smaller than the initial dataset $\mathcal{H}(f_n, \theta_q, \varphi_m)|_{n=1,\ldots,N_f, q=1,\ldots,N_\theta, m=1,\ldots,N_\varphi}$.

The modeling procedure is as follows: 1. The ATF $\mathcal{H}(f, \hat{\mathbf{r}})$ is computed in the Frequency Domain (FD) from Electromagnetic (EM) simulations of the FF. 2. The AIR $h(t, \hat{\mathbf{r}})$ is computed from $\mathcal{H}$ by inverse Fourier Transform, after, if required, appropriate processing (windowing, etc.). 3. The SEM is applied to h to extract the first P dominant poles s_p and the residues $\mathbf{R}_p(\hat{\mathbf{r}})$, with the Generalized Matrix-Pencil (GMP) algorithm [SPKR00, HS90, SP95]. 4. The modal residues are computed with the SME of the preceding residues, with a truncation to order N. 5. The modeled ATF is reconstructed following "backward" the previous procedure, and the Total Mean Squared Error (TMSE) is assessed. The "free" parameters P and N are chosen according to a predefined TMSE threshold during the preliminary step of the parametric modeling of the initial design—generally optimized under usual design constraints (matching, bandwidth, etc.).

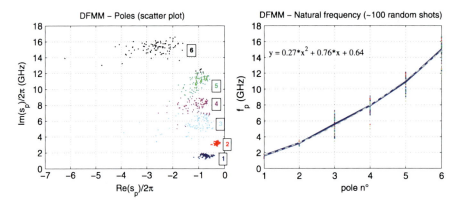

Fig. 6.9 DFMM statistics: (*left*) first poles scatter plot; (*right*) empirical means and dispersion of f_p and quadratic fit (SEM for $P = 12$) [RD09] (©2009 EurAAP, reproduced with permission)

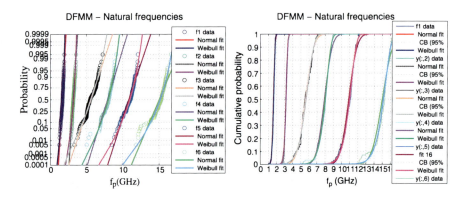

Fig. 6.10 DFMM statistics (SEM for $P = 12$): probability plots (*left*) and CDFs (*right*) of natural frequencies f_p [RD09] (©2009 EurAAP, reproduced with permission)

For a considered generic design (e.g. Fig. 4.34 [GDB06, DGRS06, Gha06, D'E08]), a representative statistical set is generated from EM simulations (WIPL® here), each realization is submitted to the above-mentioned process and represented by its parametric model dataset $\{s_p, R_{nmp}^{(u)}\}$. For example, a model with $P = 12$ poles and a truncation to order $N = 5$ of the SMEM—corresponding to a convenient trade-off between order reduction and accuracy (TMSE)—gives a data compression rate of 99.67% (compression ratio of 300), considering the initial simulated dataset of $129{,}600 \, (= 200f \times 18\theta \times 36\varphi)$ complex parameters.

Figure 6.9 (left) is a scatter plot of the $P/2 \, (= 6)$ first poles with positive natural frequencies $f_p = |\Im m(s_p)/2\pi|$. The empirical distribution of the latter are analyzed and fitted with several models (Fig. 6.10). Their empirical means are given Table 6.4. Figure 6.9 (right) shows the f_p empirical means and dispersion. To further reduce the number of model parameters, a quadratic fit as a function of the pole

Table 6.4 Natural frequencies empirical mean (DFMM "class") [RD09]

p	1	2	3	4	5	6
$\langle f_p \rangle$ (GHz)	1.628	3.176	5.591	7.915	10.897	14.996

index is also proposed Eq. (6.8):

$$\langle f_p \rangle = a(p-1)^2 + b(p-1) + \langle f_1 \rangle. \tag{6.8}$$

To fit the empirical distribution, several models are tested, in particular Normal, Weibull and Nakagami. The probability plots and CDFs are presented Fig. 6.10. The parameters of the models are evaluated with the Maximum Likelihood Estimation (MLE) method and the Akaike Information Criterion (AIC) is used for selecting the best model among the chosen set. For the f_p, Weibull is generally the winner, although the normal fit performs also well most of the time. Furthermore, the f_p are significantly correlated so that second-order statistics are required: the covariance matrix **C**, computed from the data, complete the model. In practice, implementing this model involves generating $P/2$ correlated normal variables (for each pole pair) Y_p. This can be done from a normalized (unit mean and variance), uncorrelated, Gaussian vector **X**, introducing the Cholesky decomposition of the covariance matrix **C**, and the means vector $\mathbf{M_{fp}}$ (given in Table 6.4 or from Eq. (6.8)), as follows: $\mathbf{Y} = \mathbf{X} \cdot \mathbf{chol}(\mathbf{C}) + \mathbf{M_{fp}}$. Figure 6.11 gives an example of random samples generated from this model (to be compared to Fig. 6.10). The statistics of the damping factor and first modal residues are given Figs. 6.11 and 6.12.

Statistical Models of Scattering in UWB MAS A comparable approach has been used in [DR08] to model efficiently UWB MAS. The field radiated by any radiator of the MAS has been shown [DS08, DS07] to be composed of three terms: the field radiated by the *isolated* element, the field re-radiated by all the other *loaded* elements—represented e.g. in the **Z** formalism by the $Z_{k\ell}|_{\ell \neq k}$ called *coupling*—and the difference between the field radiated by the isolated element and in the presence of the others (open-circuited), called *scattering*. This last term is modeled by

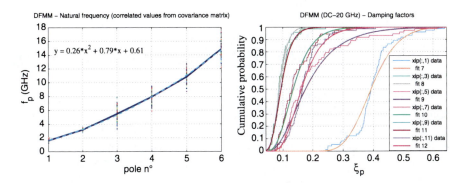

Fig. 6.11 DFMM statistics: *(left)* randomly generated samples for f_p (from correlated model); *(right)* damping factors ξ_p. Inverse Gaussian fit model [RD09] (©2009 EurAAP, reproduced with permission)

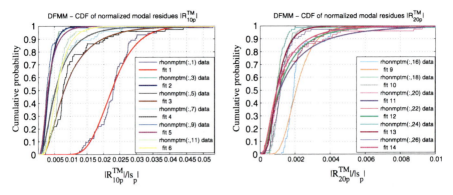

Fig. 6.12 DFMM statistics: normalized modal residues (*left*) $|\hat{R}_{10p}^{TM}| = |R_{10p}^{TM}|/|s_p|$, (*right*) $|\hat{R}_{20p}^{TM}|$. Inverse Gaussian fit model

means of a Scattering Coefficient *SC*—evaluated in the FF approximation—which in turn is represented by a parametric model based on the SEM and SMEM expansions. Without any particular assumption, the radiated field in the presence of another *open-circuited* radiator can be written as

$$\mathbf{E}(f, \mathbf{r}, d) = \mathbf{E_{isol}}(f, \mathbf{r}) \cdot \left[e^{j(\pi f d/c)\sin\theta\cos\varphi} + SC(f, \hat{\mathbf{r}}, d) \cdot e^{-j(\pi f d/c)\sin\theta\cos\varphi} \right], \quad (6.9)$$

where both radiators are symmetrically placed at $d/2$ from the origin, in their common antenna plane. Now, the FF approximation is a first simplification: SC (Scattering Coefficient) is assessed thanks to the computation of the Radar Cross Section (RCS) with an e.m. solver. This gives, restricting to the main polarization, $SC(f, \hat{\mathbf{r}}, d) \approx \frac{1[m]}{d} e^{-j2\pi f d/c} \frac{E_s(f,\hat{\mathbf{r}})}{E_{pw}(f,\hat{\mathbf{r}}_i)}$. This approximation gives excellent results, for example, in the cases of two bicones 50 mm apart and three bicones in a circular array of 50-mm radius [DS08, D'E08, DS07]. The SEM—applicable to the scattered field—is applied to *SC*, the residues being expanded on the spherical vector wave functions, as previously done:

$$SC(s, \hat{\mathbf{r}}) \approx \sum_{p=1}^{P} \frac{R_p(\hat{\mathbf{r}})}{s - s_p} \approx \sum_{p=1}^{P} \sum_{n=1}^{N} \sum_{m=-n}^{n} \sum_{u=1}^{2} R_{nmp}^{(u)} \hat{\boldsymbol{\psi}}_{nm}^{(u)}(\hat{\mathbf{r}}) \cdot (s - s_p)^{-1}. \quad (6.10)$$

If the SMEM is limited to conical cuts (i.e. $\theta = \vartheta$), the modal residues R_{nmp} reduce to the Fourier coefficient of $R_p(\theta = \vartheta, \varphi)$, which must be single valued and 2π-periodic in φ. As a consequence, we can express the pth residue as follows—where the $C_{\vartheta mp} = \int_0^{2\pi} R_{\vartheta p}(\varphi) e^{-jm\varphi} d\varphi$ are nothing else than the Fourier coefficients of $R_{\vartheta p}(\varphi)$:

$$R_p(\theta = \vartheta, \varphi) = R_{\vartheta p}(\varphi) = \lim_{M_\varphi \to \infty} \left[\sum_{m=-M_\varphi}^{M_\varphi} C_{\vartheta mp} e^{-jm\varphi} \right]. \quad (6.11)$$

Fig. 6.13 CDF of f_p, data ($\cdots$), model (—)

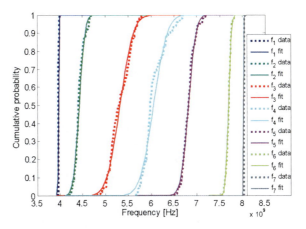

Table 6.5 Poles first two moments

p	DFMM			
	f_p		ξ_p	
	μ_{f_p} [GHz]	σ_{f_p} [MHz]	μ_{ξ_p}	σ_{ξ_p}
1	3.97	7.99	0.0024	0.0006
2	4.42	103.02	−0.0066	0.0084
3	5.32	230.64	−0.0601	0.0175
4	6.05	234.55	−0.0539	0.0156
5	6.81	135.32	−0.0531	0.0093
6	7.69	52.09	−0.0348	0.0052
7	8.039	2.98	0.0032	0.0014

As for single antennas, a statistical analysis of the same population (Fig. 4.34) is performed (over a 4–8 GHz band) on the Scattering Coefficient parametric model dataset $\{s_p, C_{\vartheta mp}\}$, truncated to the first dominant poles ($P = 12$) and modes ($M_\varphi = 3$), giving a good compression/accuracy trade-off. The natural frequencies f_p and damping factors ξ_p are found to be normally distributed ($\mathcal{N}(\mu_{f_p}, \sigma_{f_p})$, $\mathcal{N}(\mu_{\xi_p}, \sigma_{\xi_p})$) (see Table 6.5 and Fig. 6.13). It has also been found that the modulus of the $C_{\vartheta mp}$ follows a Weibull distribution: $f_x(x|a,b) = ba^{-b}x^{b-1}e^{-(\frac{x}{a})^b}$.

The phase of $C_{\vartheta mp}$ is the most sensitive parameter. In particular the phase of the fundamental coefficient is basically determined by the inter-antenna distance and p. To correctly model the way the different modes overlap (in the TD), it is the differential modal phases (defined as $\Delta_{\vartheta mp} = \angle C_{\vartheta mp} - \angle C_{\vartheta 0p}$) which are investigated. They are found to be normally distributed ($\mathcal{N}(\mu_{\Delta_{\vartheta mp}}, \sigma_{\Delta_{\vartheta mp}})$) [DR08, D'E08].

6.2.1.3 Quantities Characterizing UWB Antennas

Since antennas change their radiation properties over the frequency, the characterization of them over an ultrawide frequency range requires new specific quantities and representations. Quality measures for the efficiency of a particular more or less dispersive UWB antenna under test (AUT) can be derived directly from the complex antenna transfer function in the frequency domain $H(f,\theta,\psi)$. The relationship of the quantity with the common effective gain of the antenna is

$$G(f,\theta,\psi) = \frac{4\pi}{c^2} f^2 |H(f,\theta,\psi)|^2. \tag{6.12}$$

To avoid a distortion of a UWB signal during radiation, it is required that the transfer function possesses constant group delay, i.e. linear phase response. A more intuitive quantity for the characterization of distorting properties of an antenna is the impulse response $h(t,\theta,\psi)$. It is created by a transformation of $H(f,\theta,\psi)$ into the time domain with the inverse Fourier transform [WAS09]. For the evaluation of the antenna properties in the time domain, an envelope of the impulse response $|h(t,\theta,\psi)|$ is considered. It describes the distribution of the energy over the time and is hence a direct quantity describing dispersiveness of the radiator.

The parameters describing a quality of $h(t,\theta,\psi)$ are [WA07]:

- Peak value $p(\theta,\psi)$ is the maximum of the envelope of $|h(t,\theta,\psi)|$ over the time in a specified direction. It is proportional to the maximal amplitude of the radiated electric field strength:

$$p(\theta,\psi) = \max_t |h(t,\theta,\psi)|. \tag{6.13}$$

- Full Width at Half Maximum (FWHM) value $\tau_{FWHM}(\theta,\psi)$ is defined as the length of $|h(t,\theta,\psi)|$ at half maximum value $p(\theta,\psi)$. It describes the temporal distortion of the radiated signal:

$$\tau_{FWHM} = \tau_1|_{|h(\tau_1)|=p/2} - \tau_2|_{|h(\tau_2)|=p/2,\,\tau 1 > \tau 2}. \tag{6.14}$$

- Ringing τ_r is the time until the envelope has fallen from the peak value to the certain level r (usually $r = 0.22$). It is an undesired effect which is usually caused by resonances or multiple reflections in the antenna. It results in oscillations of the radiated pulse after the main peak:

$$\tau_r = \tau_1|_{|h(\tau_1)|=r\cdot p} - \tau_2|_{|h(\tau_2)|=p,\,\tau 1 > \tau 2}. \tag{6.15}$$

From a UWB antenna it is expected that its impulse response $|h(t,\theta,\psi)|$ has a high peak value $p(\theta,\psi)$, which assures a high amplitude of the radiated pulse. The τ_{FWHM} and τ_r should be as short as possible. This avoids an undesired spread of the signal over the time (crucial in Impulse Radio applications), which may result e.g. in lower resolution in radar/localization or inter-symbol interference in communication systems.

For UWB antennas, a one single value characterizing an amplitude for the given direction in the frequency domain is advantageous. A commonly used quantity is e.g. the mean gain G_m. It is an arithmetical mean of the gain over the given frequency range in the specified direction:

$$G_m(\theta,\psi) = \frac{1}{f_2 - f_1} \int_{f_1}^{f_2} G(f,\theta,\psi)\,df. \tag{6.16}$$

6.2.2 Design of Advanced UWB Antennas

6.2.2.1 Dual-polarized UWB Antennas

Resonant antennas radiate generally a relatively pure polarization due to clearly defined field or current distribution at the resonance frequency. However this kind of antennas can be hardly applied in the UWB technology due to their narrowband behavior. During a broadband radiation, the current distribution changes generally over the frequency, which leads to higher cross-polarization level. However some broadband antennas, if designed properly, are able to radiate purely polarized wave. One of these is the tapered slot antenna (also called Vivaldi Antenna). In order to design a dual-polarized version of this radiator, the crossing of two elements is needed [ASZW08, AZW08]. Such solution results in an acceptable HF-performance; however the manufacturing is complicated, and the devices are sensitive to the mechanical vibrations. For that reason, another solution is introduced.

For the creation of pure linear polarization in the far field of the antenna, the antenna array theory is applied. In order to radiate a single linear polarization, a second identical radiator is placed symmetrically to the original one. In such a configuration, the E-field vectors radiated from the respective elements are mirrored as well. This causes different influence of the array arrangement on the co- and cross-polarized components of the electric field. In order to suppress the cross-polarized components and enhance the co-polarization, the orientation of the second e-field vector must be inverted, which is realized by a differential feeding of both antennas. The radiation from the differentially fed mirrored elements causes in an ideal case a complete annihilation of the cross-polarization in the certain plane. At the same time the co-polarization interferes constructively.

In order to create a dual-polarized version of the antenna, a second pair of the radiators must be placed orthogonal to the first one. An example of such configuration is shown in Fig. 6.14 [AWZ09a]. It is fed by the four microstrip lines, which feed the four monopole radiators oriented symmetrically to each other w.r.t. the center of the structure. On the bottom side of the antenna a circle is cut out from the ground plane. The antenna is fully symmetrical considering the center point of the antenna as the point of symmetry. Such an arrangement assures the same radiation properties for both polarizations in the equivalent planes and similarity of the radiation patterns in E- and H-planes.

In order to radiate the single polarization the antenna has to be fed at two ports. Accordingly to Fig. 6.14, the pairs of ports 1.1–1.2 and 2.1–2.2 have to be fed for horizontal and vertical polarization, respectively.

The schematic electric field distribution in the antenna fed at the ports 1.1 and 1.2 with the differential signals is shown in Fig. 6.16. On the left side the cross section through the antenna is introduced. As can be noticed, the orientations of the electric field vectors at the feeding points are opposite, which yields the phase shift between signals of 180 degrees. The upper side introduces the plane with the monopole radiators and the bottom side the ground plane with the circle. As can be concluded from the scheme, the electric fields coming from the ports interfere in the middle

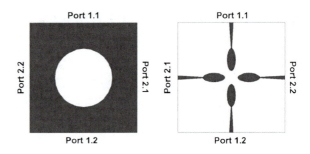

Fig. 6.14 Schematic layout of dual-polarized UWB antenna with port indication [AWZ09a] (©2009 IEEE, reproduced with permission)

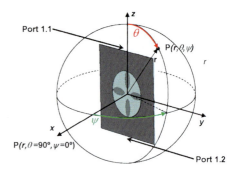

Fig. 6.15 Dual-polarized UWB antenna in a coordinate system

of the structure constructively. The signals are merged with each other, and since the radiation conditions are fulfilled, the wave is emitted into the top and bottom of the antenna surface. In the center of the Fig. 6.16 a schematic electric field distribution in the opening in the ground plane is presented. A symmetrical arrangement of the field lines can be observed. The distribution of the respective vectors in co- and cross-polarization is shown in the right part of Fig. 6.16. It can be clearly observed that the co-polarized components are in-phase and can be radiated constructively, whereas the cross-polarized ones are out-of-phase and annihilate each other. This results in very high polarization purity of the radiated signal, which is achieved by the differential feeding of the mirrored elements.

The additional advantage of the configuration is the same position of the phase center of radiation for both polarizations, which is in the middle of the structure. Due to the application of two radiators, the position of the phase center over frequency is very stable. The benefit is a lower distortion of the radiated pulse [APP+07]. Furthermore it can be observed that the orientation of the electric field produced by one pair of the ports, is orthogonal to the microstrip lines at the ports for the orthogonal polarization. Such oriented field is not able to propagate in the orthogonal lines, which implies a very good decoupling of the ports for orthogonal polarization.

The measured mean gain for single polarization in the E- and H-plane for co- and X-polarization is shown in Fig. 6.17. The placement of the antenna in the respective coordinate system is presented in Fig. 6.15. The measurement is performed with differential power divider, of which outputs are connected via microwave cables with the input ports of the antenna. It can be observed that for the co-polarization, the antenna radiates with two beams oriented oppositely to each other. The radiation

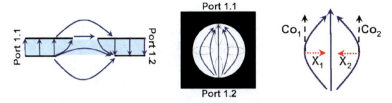

Fig. 6.16 Schematic electric field distribution in the antenna: *left*—cross section, *middle*—bottom view, *right*—distribution of the vectors in co- and cross-polarized components [AWZ09a] (©2009 IEEE, reproduced with permission)

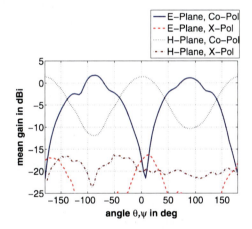

Fig. 6.17 Measured mean gain of dual-polarized UWB antenna in E- and H-plane for co- and cross-polarization

pattern is similar for both planes, which is due to similar aperture dimensions in the respective directions. The cross-polarized components are very weak in comparison to the co-polarized. For the main beam direction the mean polarization decoupling is well above 20 dB, which is mainly sufficient for polarization diversity.

The measured impulse response of the antenna in H-plane for co-polarization is shown in Fig. 6.18. Accordingly to Fig. 6.17, two main beams can be observed in the directions of 0 and 180 degrees. The impulse response is very short, and its delay is nearly constant over the angle. It confirms an applicability of the antenna in pulse-based UWB systems. The large delay of the signal is caused by the implemented differential power divider and connecting cables. Their influence on the delay and amplitude of the measured quantities is not calibrated out from the data set.

The introduced concept presents an excellent possibility for the realization of UWB dual-linearly-polarized antennas with high polarization purity, stable phase center of radiation over the frequency, high decoupling between the ports and similar radiation properties for both polarizations. It can be successfully applied in enhancement of the system performance in UWB radar/imaging, localization and communication, especially UWB-MIMO communication systems.

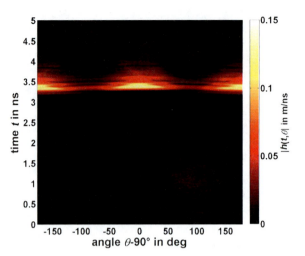

Fig. 6.18 Measured Impulse response of the dual-polarized UWB antenna in the H-plane for co-polarization

6.2.2.2 UWB-Antenna with Reconfigurable Radiation Pattern

In certain applications such as MIMO, localization or radar the antennas with reconfigurable radiation pattern are desired. Such property can be applied especially in monopulse radar technique, where the angular position of the target can be determined by evaluation of the amplitude differences of the signals received in different modes.

The antennas with reconfigurable radiation pattern can be realized by the application of a multimode feed. Different guided modes in the feed cause different electric field distributions in the antenna and thus change of its radiation pattern. In order to be applied in UWB, such feed should exhibit not only broadband behavior, but also a linear phase response when used in pulse-based systems. A solution has been presented in [AWZ09b].

In order to take an advantage of a multi-mode feed and enable a creation of different radiation patterns, two tapered slot antennas are used. Each antenna is connected to the single slot at the output of the feed presented in [AWZ09b]. At the input of the antenna two modes are created: CPW (Coplanar Waveguide) and CSL (Coupled Slot Line). They differ in the relative orientation of the electric fields in the feeding slot lines, which is schematically presented in Fig. 6.19. In the figure also a schematic creation of the radiation pattern in both modes is presented. In the CSL mode one single beam is created, which results from superposition of the fields radiated from the single slots. On the other hand, in the CPW mode two beams are created. The minimum of the radiation in this mode is oriented in the direction of the main beam in the CSL mode.

A prototype of the antenna is presented in detail in [AWZ09b]. Each radiation mode of the antenna is created by an excitation of a separate port. Since the multi-mode characteristics of the presented antenna can be observed clearly only in the E-plane, in the following the measurement results in the E-plane of the antenna are presented. In Fig. 6.20 the measured mean gain over the angle is shown. The result

Fig. 6.19 Schematic distribution of the electric field distribution in the dual-mode Vivaldi antenna (*left*—CSL mode, *right*—CPW mode) [AWZ09b] (©2009 IEEE, reproduced with permission)

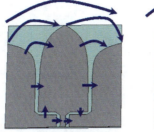
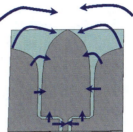

Fig. 6.20 Measured mean gain of the pattern diversity antenna in the E-plane for CSL (sum beam) and CPW mode (difference beam)

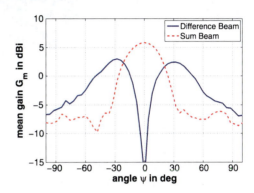

confirms the theoretical assumption that the antenna radiates in CPW mode with two beams with minimum radiation at 0 degrees. The maxima of both beams point the direction of approximately ±45 degrees. On the other hand, the measured characteristics during excitation of the CSL mode shows a single beam. Its maximum is oriented at 0 degrees and is relatively symmetric w.r.t. this value. The maximal value of the mean gain in CSL mode is around 3 dB higher than in the CPW mode. This is due to the distribution of the feeding power into two separate beams in CPW mode.

In Fig. 6.21 the measured impulse response of the complete antenna is presented. It can be noticed that the impulse response is concentrated in time, which implies small pulse distorting properties during radiation. The amplitude distribution is analog to the gain measurements. The minimum for the CPW mode is located at 0 degrees and is symmetric to this value. The ringing of the impulse response in CPW mode is very weak. In the CSL mode the maximum of the impulse response amplitude occurs at 0 degrees.

6.2.3 Analysis of UWB Antenna Arrays in the Time Domain

Time domain (TD) analysis of UWB antenna arrays is an important factor in beamforming and beamsteering research for modern ultrawideband applications. This topic becomes more and more interesting [RK05] for future wireless ad hoc

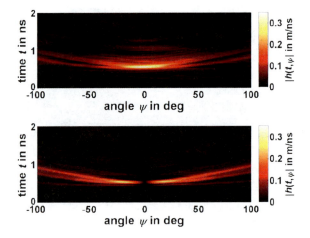

Fig. 6.21 Measured impulse responses of pattern diversity antenna in the E-plane (*top*—CSL mode, *bottom*—CPW mode) [AWZ09b] (©2009 IEEE, reproduced with permission)

networks in indoor environments for power-efficient short-range connectivity at medium to high data rates suitable for home entertainment [HCR08]. Examples are transmission of uncompressed high-definition television data to a display or delivery of music streams to loudspeakers.[1] In the literature, such as Sörgel et al. [WS05], UWB antenna arrays are well presented. However, the antennas used in [WS05] were directional planar horn antennas which cannot be embedded in the devices mentioned above due to their inappropriate form factor. Furthermore, due to the unique advantages of UWB, to resolve the distance from one node to another, location-aware coding can stabilize or even enhance the achievable data rate.

For beamforming, two methods of connecting the antennas are explained and evaluated in TD in this section. Both methods are sketched in Fig. 6.22 and marked with (a) and (b) respectively. Small monopole antennas according to [JCC+05] are manufactured and driven with a pulser prototype PCB. This pulser creates a Gaussian pulse with a pulse width of 480 ps from a baseband rectangular signal [BKM+06].

On the top of Fig. 6.23 the measurement result of a single antenna element with attached ground plane is presented. The measurement was carried out with a digital sampling oscilloscope connected to a reference horn antenna (gain = 11 dBi), placed at a distance of 0.3 m from the antenna element. The antenna element was rotated 360° in azimuth (with a step size of 10°), while the oscilloscope stored the received voltage waveform at every step. For comparison, on the bottom of Fig. 6.23 the result of a four antenna element array is shown. For both measurements, the pulser circuitry is used; in the array measurement a 4–1 Wilkinson combiner was placed between the pulser and the four antennas (see Fig. 6.22a).

---

[1]This work was performed as part of the project "Smart Data Grain" which is embedded into the program "Forschung, Innovation, Technologie—Informationstechnologie" (FIT-IT) of the "Bundesministerium für Verkehr, Innovation und Technologie" (BMVIT). This program is funded by the "Österreichische Forschungsförderungsgesellschaft" (FFG).

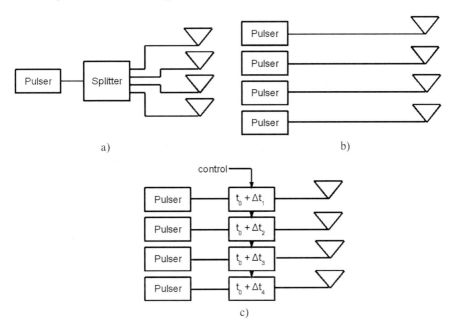

Fig. 6.22 Methods for UWB beamforming (**a** and **b**) and proposed block diagram for UWB beamforming (**c**)

The single element measurement results show that the transmitted pulse is radiated almost uniformly over the azimuth angle. At broadside of the antenna element (azimuth = 180°) a slight rise is observed. The result in this TD-representation implicates that the captured pulse shape is almost invariant regarding the azimuth angle. In case of the antenna array in endfire direction, the expected pulse minima due to beamforming were measured (at azimuth 90° and 270°). The measurements were carried out in an office scenario.

Another idea to realize ultrawideband beamforming is to drive each antenna element with one pulser circuitry. The feasibility to drive each antenna individually is demonstrated by the measurement plotted in Fig. 6.24. Again the broadband horn antenna is used as a reference and is placed about 60 cm apart from the array [GGH+10]. Due to the higher transmit power (4 times higher), the measurement distance was increased. The measurement results are plotted in Fig. 6.24. The angle of tilt of the main beam is approximately 12° in azimuth. This can be explained that the pulser connected to the antenna, which has the longest distance towards the reference antenna, has less gain than the other pulsers. Undesired pulse maxima besides the main lobe appear in the TD measurement results of an UWB array if the element distance increases [GHBM10]. Gain measurements in an anechoic chamber [GGH+10] delivered almost identical results as the TD measurements, which confirms that wideband beamforming and beamsteering can be achieved with the proposed antennas.

Fig. 6.23 TD antenna measurement with a single antenna element (*top*), and four linear arranged wideband monopoles (interelement distance 2 cm), *bottom*

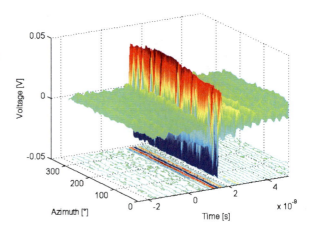

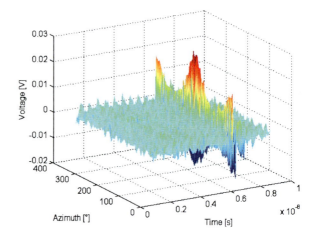

6.3 UWB Signal Propagation and Channel

6.3.1 Channel Measurements, Analysis and Characterization

6.3.1.1 Measurements in Different Scenarios and Channel Properties

Motivated by different application scenarios and by different research questions, UWB channel measurements have been performed in a variety of environments. The COST 2100 action has received contributions on measurements in (office, residential and industrial) indoor environments, in cars, aircraft, a road tunnel, anechoic chambers and for Body Area Network (BAN) scenarios. This section seeks to summarize the key features investigated.

Table 6.6 gives an overview of the measurement campaigns. It lists the approximate dimensions of the environments, the measured frequency band, the channel

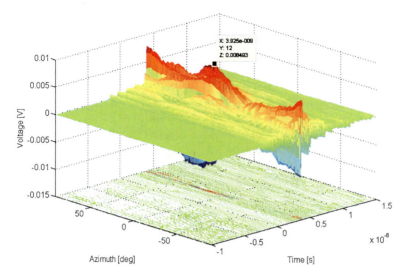

Fig. 6.24 Azimuth and time versus received voltage of the measured pulse at a distance of 60 cm with an interelement distance of 5 cm ($\lambda/2$) [GGH+10] (©2010 IEEE, reproduced with permission)

Table 6.6 Overview of UWB channel measurement campaigns

Environm.	Dims. [m]	F. Band [GHz]	Sounder	PL/Epx. [dB]/[–]	RDS [ns]	K [–]	Ref.
Domestic	10×7	3.5–10.5	MLBS	–/2.3	7	–	[CPB07] & [CMUB09]
Office	11×28 13×31	3.5–10.5	MLBS-MIMO	–/–	14 19	–	[CVT09] & [TCAB09]
Office	9×7	3.1–10.6	VNA	–/–	–	–	[HTTV09]
Various	–	4.25–4.75	802.15.4a	–/1–5.6	10–70	–	[GBG+09]
In-Car	pass. cabin	3–8	VNA	40–50/–	5–10	1–2	[SJP+08] & [SGS+09]
In-Car		67–70	VNA	>56/–	1	>3	[SJK10]
Aircraft cabin	7×5	3–8	VNA	65/2.5	15	–	[JLS+09] & [APC+10]
Tunnel	$d = 50{-}500$	2.8–5	VNA	85/0.6	–	0.1–10	[MGPLND08]
UHF-RFID	$40 \times 15 \times 7$	0.5–1.5	VNA	–/–	10–50	1–10	[AAMW10]

sounding technology (see Sect. 2.1.4) and—where applicable—some typical channel parameters that have been reported. The channel parameters are the pathloss (PL), the pathloss exponent (Exp.), the RMS delay spread (RDS) and the Ricean K-factor (in linear scale).

Indoor Scenarios Indoor scenarios have been studied by Cepeda et al. from TRL Toshiba in Bristol UK and University of Bristol, with a focus on frequency-dependent pathloss, channel dynamics, diversity and channel capacity [CPB07, CTB08, CMUB09, CVT09, TCAB09]. The time domain sounding equipment used is discussed in Sect. 2.1.4. References [CPB07, CMUB09] consider a domestic indoor environment, while [CVT09, TCAB09] study an office scenario. In [CPB07, CTB08], measurements in an anechoic chamber are discussed as well, to verify observations from the domestic indoor channels.

The frequency-dependent path loss describes systematic variations as a function of frequency. It is often modeled using an exponential law, for instance of the form $PL(f) \propto f^\delta$, where δ is the pathloss exponent. The measurements presented in [CPB07] (for a domestic environment) show surprisingly large variations of δ. For the complete set of measurements, δ varies over -1.32 ± 1.1, which is the mean $\pm$ one-sigma range. The minimum and maximum values are as large as -5.2 and 5.8. These variations apparently cannot be explained by deterministic propagation mechanisms as frequency-dependent antenna characteristics or Friis' equation. It can only originate from small-scale effects "due to a combination of diffraction, scatter, frequency selective reflection, and multipath interference" [CPB07].

To shed light on these mechanisms, the authors have conducted measurements in artificial multipath environments inside an anechoic chamber [CTB08]. In the first experiment, a metal cylinder has been inserted, and indeed, rapid and large power variations of δ have been reproduced when the diameter of the cylinder was greater than 90% of the wavelength. This explains that the effect is more pronounced at higher frequencies. The second experiment with three metallic plates instead of the cylinder shows that with more multipath components the effect becomes even more pronounced. Final conclusions are left open. But it is apparent that small-scale effects can dominate over deterministic mechanisms when studying the frequency-dependent pathloss in UWB channels.

The same residential indoor environment has been investigated in [CMUB09] with the goal to derive *distance* dependent Path Loss (PL) models. In addition to static measurements, dynamic channel data has been acquired. Both data sets show only small deviations, and thus the summary here is restricted to the largest measurement set for a dynamic measurement throughout the ground floor of the apartment. The various data samples can be classified in Line-Of-Sight (LOS) at very short range below 1.6 m, LOS at farther range up to 3.3 m, and Non-Line-Of-Sight (NLOS) measurements above that range. The paper discusses that a multi-slope PL model better fits the data than a single-slope PL model. The multi-slope model employs a separate PL exponent for each data class. The Chow test and t-tests are used to prove statistical significance of the additional model complexity invoked. (See [CMUB09] and references therein for further details.) Finally Ordinary Least Squares (OLS) regression is used to fit the data, see Fig. 6.25.

The two papers [CVT09, TCAB09], from the same group, analyze diversity and channel capacity respectively, for measured UWB channels in two spacious and mostly open office environments of dimensions in the order of 10×30 m. A 2×4 Multiple-Input Multiple-Output (MIMO) version of the channel sounder has been used. Results are discussed in Sect. 6.3.3.

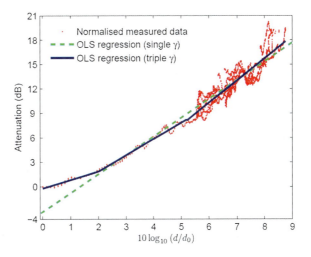

Fig. 6.25 Multi-segment pathloss model for dynamic indoor channel measurements [CMUB09] (©2009 VDE, reproduced with permission)

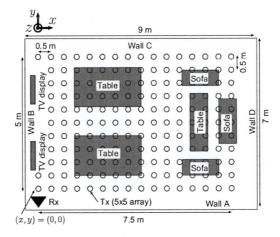

Fig. 6.26 Floor plan of the office environment measured using a room wide spatial scanner [HTTV09] (©2009 IEEE, reproduced with permission)

Another remarkable indoor measurement campaign has been presented by Haneda et al. in [HTTV09]. It is remarkable in the sense that a complete 7.5×7 m office room has been measured over the full 3.1–10.6 GHz band, using a room-wide spatial scanner and a Vector Network Analyzer (VNA). The measurement area was divided in 168 positions, in each of which data was collected on a 5×5 grid with a 2.5 cm spacing. All but one positions were in LOS conditions. Figure 6.26 shows the measurement scenario. The measured data have been used to identify and track Multi-Path Components (MPCs) (cf. Sect. 6.3.1.2 below).

The measurement campaign reported in [GBG+09] compares a wide variety of scenarios, ranging from domestic indoor environments, to industrial, office and even some outdoor measurements. The data has been acquired using an experimental demonstration system for the standardized IEEE 802.15.4a air interface, as discussed in Sect. 2.1.4.

In-car Scenarios Moritz Schack et al. from Technische Universität Braunschweig presented measurements of the UWB channel inside several different cars [SJP+08, SGS+09]. Data was acquired for the 3–8 GHz band using a VNA and different antenna placements. A large difference is seen between measurements in an empty and an occupied car. In the Volkswagen Golf studied in [SJP+08], the pathloss coefficient increases from 0.75 to almost 2, while the RMS delay spread is reduced from values around 10 ns to 5 ns and the maximum excess delay from up to 80 ns down to maximum values around 40 ns. These effects can be attributed to attenuation of MPCs in presence of passengers. A larger car by Audi, equipped with heavy leather seats, shows much lower RMS delay spreads anyway and also less influence of passengers, which is again explained by higher attenuation of MPCs [SGS+09]. The study also extends to a convertible in open and closed configurations. Finally, the extracted data has been used to verify the possibility of establishing a multiband OFDM link at 200 Mbit/s. It has been concluded that pathloss is too high in many antenna configurations to get a satisfactory SNR in the link budget. This holds for the trunk mounted antenna in all cases except for the Golf, and at the rear seats for the dashboard antenna in particular when the car is occupied.

Aircraft Cabin UWB may be an attractive technology for replacing cabling in in-flight entertainment systems. Large data rates are needed for distributing multimedia content to individual passengers on demand. Researchers from Technische Universität Braunschweig also conducted measurements in a mock-up of a wide-bodied aircraft [JLS+09]. Again a VNA was used to scan the 3–8 GHz band, and the influence of passengers was studied. The PL is greater in general for the occupied cabin. For LOS channels, the difference increases with distance and frequency, from about 0 to 3 dB. In NLOS channels, the difference is constant for all frequencies. It changes with distance between about 3 and 5 dB. The difference between LOS and NLOS is typically very small, up to only ±2 dB.

A distance-dependent model has been derived for the RMS delay spread, based on a quadratic function and an additive Gaussian random component. The RMS delay spread varies from only 5 or 10 ns at close range (less 1 m) up to above 20 ns at distances of around 8 m. While there is little difference between LOS and NLOS conditions, the influence of passengers is significant. It reduces the RMS delay spread by between 3 and 8 ns. This behavior agrees with in-car UWB channels, as the presence of passengers attenuates reflected MPCs.

In [APC+10], the theory of "room electromagnetics"—an analogy to the well-established principles of room acoustics—has been used to characterize the diffuse scattering in an aircraft cabin. The enclosed environment is considered as a lossy cavity, with a given volume and so-called "absorption area". These two parameters define the reverberation time in an environment, essentially the slope of the exponentially decaying part of the typical, dense impulse response in indoor scenarios. Having this parameter available, delay spread and also the pathloss of the environment can be specified. The theoretical results are compared with channel measurements, showing remarkably close resemblance.

Tunnel Environments In tunnel environments, UWB can be used as a wireless backbone technology for train-to-train or train-to-track communications, and it is envisioned that simultaneous distance measurements may be employed to optimize the link [SHER06]. In [MGPLND08] VNA measurements in a road tunnel of 3.3-km length are presented for the 2.8–5 GHz band and for ranges from 50 to 500 m. RF-to-fiber convertors have been used to avoid excessive attenuation in RF cables. The analysis shows some rather untypical results due to the tunnel environment. For instance, the PL exponent has been found to be as low as 0.57 due to waveguiding effects. Furthermore the Ricean K-factor *increases* with distance, a result that normally indicates that the LOS becomes more prominent compared with other MPCs.

UHF-RFID Channel in Industrial Environments The technical document [AAMW10] describes a measurement campaign performed to characterize the UHF band around 900 MHz used for a certain type of RFID systems. The study is included in this section, as a measurement system has been designed, which supports an ultrawide bandwidth of around 1 GHz. This allows a detailed study of the multipath propagation in such scenarios. Ranging based on the propagation delay has been motivating this work, where it is needed to identify the direct line-of-sight component from the received signal. The measurement setup mimics an RFID setup as it arranges the antennas like the readers in a typical gate setup, and it copies the directional characteristics of the reader antennas. A rather high K-factor and low RMS delay spread are needed to support ranging with narrowband systems [AMW09, AAMW10]. The measurements clearly show that such favorable conditions are only found within the main beam of the reader antennas. Objects in the LOS path will deteriorate the situation, as well as metal backplanes behind the gate, which are used to better concentrate the RF field within the gate.

6.3.1.2 Analysis Techniques

We turn our discussion towards analysis techniques for UWB channel data. UWB channel sounding allows for a fine time resolution of the acquired channel impulse responses such that—in theory—individual MPCs can be resolved. Several attempts have been made to automatically decompose channel measurement data into MPCs, which is not only interesting for channel modeling but also for sensing and positioning applications in indoor environments [ZST05, MSW10]. In the literature, applications of the Space-Alternating Generalized Expectation-maximization (SAGE) [HT03, FTH+99] and CLEAN [CWS02] algorithms can be found. As expected, both algorithms can resolve large numbers of MPCs, but their application is not straightforward in the UWB regime. On the one hand, large fractions of the CIR energy are contained in stochastic components, which cannot be assigned clearly to scattering objects in the environment [KP03]. Deterministic MPCs (also called specular reflections), on the other hand, may suffer from distortions [Mol09].

This section reviews two contributions to this area of research that have been presented to the COST 2100 action. One is a modification to the SAGE approach that makes the algorithms significantly more suitable for UWB while hardly increasing its complexity [HWM+10]. The second paper [SKA+08] resembles the two mentioned algorithms as well. It is new in the sense that it allows the analysis of data that extends over the stationarity region of the channel. We start our discussion with the second work.

Successive Detection and Cancellation of Scatterers In the work by Santos et al. from Lund University [SKA+08], scatterers[2] are detected on the basis of measurements taken along a "long" measurement track, where stationarity of scattering processes cannot be assumed any longer. The analysis is performed in several steps. First, a high-resolution peak search is performed upon the frequency domain data. This is basically an iterative maximum likelihood estimation to find the highest peak of a channel impulse response and its successive cancellation from the data. Amplitude and delay of all peaks (above a defined threshold) are stored for each point along the measurement track. Next, the 2D coordinates of hypothesized candidate scatterers are used to assign the detected MPCs to possible scatterers. This yields for each candidate location a vector of MPC amplitudes across antenna array locations. On this basis, the birth and death of scatterers are determined, which is of fundamental importance to account for the non-stationarity of the data. A sliding window is shifted across the MPC amplitude vectors ensuring that only scatterers are detected which exist over a sufficiently long path. Or, in other words, to avoid detection of a single strong peak. The window length corresponds to the range where stationarity is expected. Finally the "strength" of each scatterer is calculated by summing the MPC amplitudes between a scatterer's birth and death. The MPCs for the strongest scatterer found in this way are canceled from the MPC data (i.e. from the channel data), and the scatterer detection is repeated in an iterative fashion.

The essential steps of this algorithm are illustrated in Fig. 6.27(a)–(d) for an outdoor UWB measurement campaign at a gas station [SKA+08]. All estimated Channel Impulse Response (CIR) peaks from the first step are shown in Fig. 6.27(a) along with the peak locations of two hypothesized scatterers, denominated as "A" and "B". In Fig. 6.27(b), the peak amplitudes are plotted across array positions for these two hypothesized scatterers. A sliding window is shifted over these data, whose output defines birth and death of identified scatterers. Hypothesis "A" apparently does not correspond to an existing scatterer as the averaged magnitude of the MPCs never crosses the threshold. Hypothesis "B" does correspond to an existing scatterer that persists over a significant part of the measurement track. The final result after completion of the iterative scatterer identification process is depicted in Figs. 6.27(c) and (d), in the delay domain and in a 2D map respectively. Scatterers map very well to physical objects in the propagation scenario.

The analysis performed by Haneda et al. on the room-wide channel scans (see [HTTV09] and above in this section) resembles the idea discussed here. Measurements taken over a large geographic area are used to estimate parameters of MPCs and to identify clusters of scatterers. The key principle is the tracking of MPCs across stationarity boundaries.

---

[2]A scatterer is a feature in space that can be identified as the source location of a received MPC.

6 RF Aspects in UWB Technology

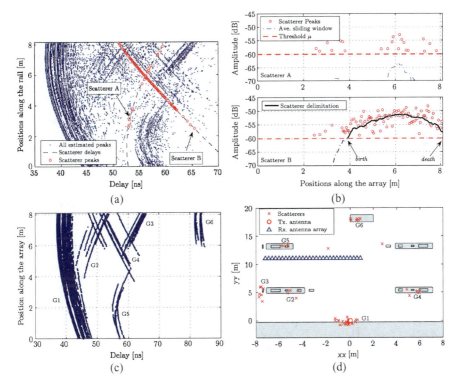

Fig. 6.27 Illustration of the successive scatterer detection algorithm presented in [SKA+08] (©2008 IEEE, reproduced with permission). (**a**) All peaks estimated from the channel transfer functions across receiver positions and delays of two hypothesized scatterers. (**b**) Peak amplitudes as functions of array position for the two hypotheses leading to the identification of "birth" and "death" of MPCs. (**c**) Plot of all detected scatterers and their visibility regions in the delay domain. (**d**) Detected scatterers indicated in a map of the propagation scenario

SAGE for UWB The SAGE (Space-Alternating Generalized Expectation-maximization) algorithm [FH94] is an extension to the Electromagnetic (EM) algorithm for parameter estimation. An essential difference is that only a subset of the components of the parameter vector is updated in each iteration of the SAGE procedure while the other elements are kept fixed. This leads to faster convergence and lower complexity compared with the EM algorithm.

The application of the SAGE algorithm to the joint estimation of delay, angle of arrival, Doppler frequency and complex amplitudes of MPCs in mobile radio channels has been described in [FTH+99]. An antenna array of M elements has been assumed at the receiver. The algorithm is based on a signal model that describes the contribution of the lth MPC to the M received signals as

$$\mathbf{s}(t, \boldsymbol{\theta}_l) = \mathbf{c}(\phi_l)\alpha_l \exp(j2\pi \nu_l t)u(t - \tau_l), \qquad (6.17)$$

where $\boldsymbol{\theta}_l = [\phi_l, \alpha_l, \nu_l, \tau_l]$ is the parameter vector for the lth MPC, comprising its arrival angle ϕ_l, complex amplitude α_l, Doppler frequency ν_l and delay τ_l. The vector $\mathbf{c}(\phi)$ is the steering vector of the array, whose elements are defined as $c_m(\phi) = f_m(\phi) \exp\{j2\pi \lambda^{-1} \langle \mathbf{e}(\phi), \mathbf{r}_m \rangle\}$, where λ denotes wavelength, $\mathbf{e}(\phi)$ is a unit vector pointing in direction ϕ, $\mathbf{r}_m$ is the location of antenna element m, and $f_m(\phi)$ is the antenna gain pattern. Finally, $u(t)$ is the sounding signal used. The signal model (6.17) illustrates that a phased-array-type beamformer is used in [FTH+99], which is a simplification as pointed out in the paper. It is assumed that the inverse of the bandwidth of $u(t)$ is much greater than the ratio of the array dimension to the speed of light. However, this assumption is easily violated in UWB channels. A bandwidth of 1 GHz corresponds to an inverse of 1 ns, where an EM wave travels a distance of just 30 cm at speed of light. This is clearly the magnitude order of the array dimensions for UWB frequencies; hence the delay varies notably between antenna elements, and a more complex delay-and-sum beamformer would be needed in the SAGE algorithm.

The Transmit Diversity (TD) [HWM+10] proposes a small twist to the SAGE procedure, which makes it remarkably well suited for UWB channels, despite using the original phased-array beamformer. After detection of an MPC, a small time window is defined around its delay, which is excluded from the search space for other MPCs. This avoids multiple detection of MPCs due to the delay offsets. Figure 6.28 illustrates the achieved performance difference for a bandwidth of 4 GHz centered at 5.5 GHz and a 10×10 array spaced by 2 cm. An LOS environment has been considered.[3] Figure 6.28(a) shows the output of the original SAGE. Estimated MPCs are plotted by shaded bullets on top of a delay-azimuth scatter plot. Only very few MPCs are actually found, but MPCs are found multiple times if the delay differences among antenna signals are neglected. The problem is largely avoided by the time gating extension, as shown in Fig. 6.28(b). Multiple detection of MPCs is strongly suppressed, and many additional MPCs are found, corresponding to additional important propagation paths.

The paper also investigates the bandwidth dependence of the MPC extraction and compares data reconstructed from these components with the original CIRs. An example for the latter investigation is shown in Fig. 6.29. In particular, at relatively small bandwidths below 1 GHz, the resemblance is remarkable, which illustrates the good performance of the proposed algorithm.

6.3.2 Channel Modeling

Multiple UWB channel models for a variety of environments and application scenarios have been proposed within the COST action. The contributions can be divided in statistical and deterministic ones and are summarized in the following subsections.

---

[3] The data analyzed in this example has been collected by J. Kunisch and J. Pamp within the "whyless" project [KP02].

Fig. 6.28 Detection of MPCs from 10 × 10-array data at 4-GHz bandwidth. (**a**) Result for the original SAGE algorithm [FTH+99] showing multiple detection of MPCs due to the large signal bandwidth. (**b**) Time gating clearly improves the performance while avoiding a significant complexity gain

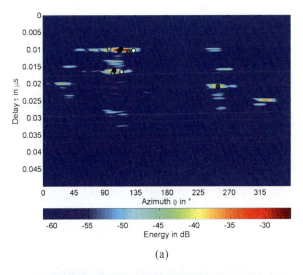

(a)

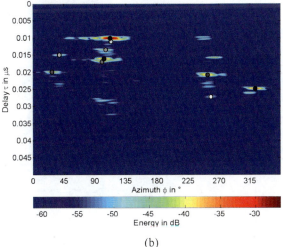

(b)

6.3.2.1 Statistical Channel Modeling

UWB Channel Modeling in Aircraft Cabines A popular stochastic UWB channel model is the Saleh–Valenzuela (S–V) model [SV87]. It captures the most important characteristics of the UWB channel and is well established for the simulation of the UWB propagation channel in indoor environments. With the intention to adapt the model to empty and occupied aircraft cabins, the authors of [CJK09] propose a systematic and efficient best fit algorithm to derive the main parameters of the S–V model from measurement data. The measurements were carried out at the front section upper deck of a double-decker large wide-bodied aircraft cabin. The extracted parameters are given in Table 6.7 and are the cluster arrival rate, the cluster decay factor, the ray decay factor and the mean Rice factor. Figure 6.30 shows an example

Fig. 6.29 Reconstruction of measurement data based on the extracted MPCs at a bandwidth of 500 MHz. (**a**) Original CIRs as functions of antenna displacement. (**b**) Reconstructed CIRs

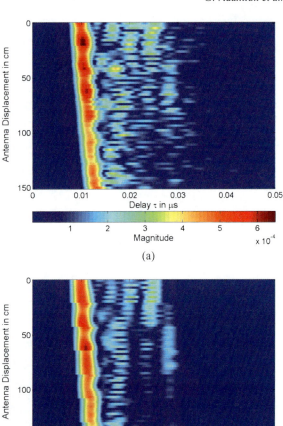

of a measured power delay profile and a regenerated power delay profile with the parameterized S–V model for NLOS in an empty aircraft cabin. A good agreement is obtained.

Channel Characteristics for Statistical Channel Modeling A number of channel characteristics have been extracted from several different UWB measurement campaigns. These are the distant and frequency-dependent path loss under both line-of-sight as well as non-line-of-sight condition for outdoor [GBG+09] (office, residential and industrial) indoor [CPB07, CMUB09, GBG+09], aircraft [JLS+09], in-car [SJP+08, SGS+09], anechoic chambers [CTB08] and tunnel environments [MGPLND08]. In addition, the fading statistics, the RMS delay spread as well as the excess delay are given in most contributions. The channel parameters are use-

6 RF Aspects in UWB Technology

Table 6.7 Summary of channel parameters for S–V model [CJK09]

Parameter	Empty cabin		Occupied cabin	
	LOS	NLOS	LOS	NLOS
Cluster arrival rate [1/ns]	0.23	0.22	0.20	0.23
Cluster decay factor [ns]	4.82	6	3.97	4.71
Ray decay factor [ns]	0.68	0.48	0.87	0.57
Mean Rice factor [dB]	−7.56	−5.62	−7.20	−4.90

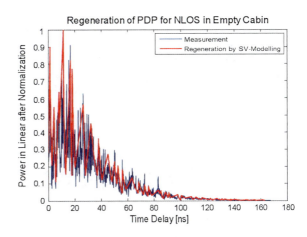

Fig. 6.30 Regenerated power delay profile for NLOS in empty aircraft cabin [CJK09] (©2009, reproduced with permission)

ful for the parameterization and verification of statistical channel models and are summarized in Sect. 6.3.1.

6.3.2.2 Deterministic Channel Modeling

Statistical channel models are much helpful in the early design phase of UWB systems and applications. However, adequate parameterization requires a large number of measurements. Especially, if a system has to be tested in a specific environment, deterministic channel models are superior as they allow for a site-specific channel prognoses.

Despite the advances in computer technology, full wave solutions are still too computationally intensive for most scenarios. Therefore, the wave propagation modeling is usually based on geometrical optics (i.e. ray-optical methods) and the uniform geometrical theory of diffraction (UTD). As the determination of the multi-paths within the deterministic model is based on ray concepts, it is referred as ray tracing.

However, comparisons with measurements have recently shown that the measured UWB channel impulse response is much more dense than the simulated one and that the received power and delay spread predicted by ray tracing are typically too low. Different possibilities to improve the quality of ray tracing have therefore been proposed within the COST action.

Hybrid Deterministic Statistical Model The motivation behind the modeling approach proposed by Nasr et al. in [NC07] is to improve the model accuracy without increasing the complexity of the environment model or the propagation models. For this purpose, a hybrid model is proposed that is based on a 2.5D ray tracing tool in combination with a field strength and an rms delay spread correction term. The correction terms reflect the statistical differences obtained from UWB indoor measurements from 2 GHz to 6 GHz. It is shown that the error in field strength prediction reduces to a fraction of a dB and in delay spread to around 1 ns in the worst case compared to the measurements. The hybrid model is therefore appropriate for a quick estimation of the coverage of an UWB device, the rms delay spread as well as the study of coexistence and interference issues. However, it does not allow for the evaluation of directional channel characteristics.

Calibrated Ray Tracing Model In contrast to the hybrid deterministic statistical modeling approach, Jemai et al. [JKS08] focus on the improvement of the propagation environment. A calibration method is proposed that retrieves the optimal material parameters of the environment objects. The calibration is performed on few measured power delay profiles by simulated annealing, whereby the cost function is the root mean square error between the measured and the predicted power of multipath components [JK07].

In Fig. 6.31 the measured PDP is compared with the PDP predicted by the uncalibrated and calibrated ray-tracing model. Though the calibrated ray-tracing model does consider neither the complete environment details, nor the diffuse scattering due to rough surfaces and interactions with small objects, it provides an improved accuracy in terms of path loss, delay spread and maximum excess delay is obtained.

Hybrid Ray Tracing/FDTD Model Another approach to improve the prediction quality is to focus on the propagation models. A requirement for using GO and UTD is that the wavelength should be small compared to the dimensions of the environment objects (high-frequency approximation). However, in indoor scenarios, the encountered objects may be smaller than the wavelength. A possible solution to this problem is to combine ray tracing with FDTD, as proposed by Janson et al. in [JSS+09].

By comparison of simulation and measurements in a complex laboratory scenario it is shown that the combined ray-tracing/FDTD model provides higher precision for propagation paths with small excess delay. However, the overall improvement of the simulation quality is limited and, in comparison to the potential improvement by considering diffuse scattering in the ray-tracing simulation, almost negligible.

Deterministic-Stochastic Channel Modeling With the intention to model dense multipath components within a ray-tracing model, Janson et al. propose in [JFZW09, JPZW10] a new deterministic-stochastic channel model. As conventional

Fig. 6.31 Power delay profiles from the standard model before calibration (*top*) and the calibrated model (*bottom*) [JKS08]

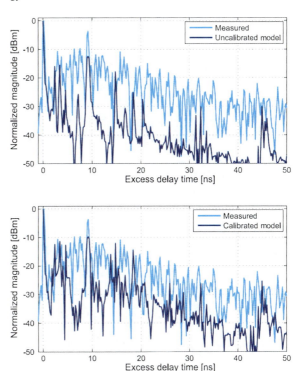

ray-tracing model is based on the GO and the UTD for the calculation of reflection and diffraction but is complemented with a scattering model that uses randomly distributed point scatterers placed on the surface of large objects like walls.

In order to trace pure reflection paths, the image theory is used. Diffracted paths are determined by the Fermat principle. For the scattering, a number of scatterers are placed randomly on the wall surface within a certain radius around the reflection points (cf. Fig. 6.32). These scatterers represent small structures on the surface not reflected in the scenario data, as well as interactions with inhomogeneities inside the wall and with objects behind the wall. Once the scatterers are generated, the wave propagation calculation is done in a deterministic way. Thus the direction information is preserved in the prediction.

In [JFZW09] the simulated angle-dependent impulse response in front of a plaster, brick, wood and concrete wall is compared to UWB measurements from 3.1 GHz to 10.6 GHz. Verification results in a rich scattering indoor environment are presented in [JPZW10]. As an example, the measured and simulated delay spread, receive power and 4×4 MIMO capacity for the scenario given in Fig. 6.33 are presented in Fig. 6.34. Compared to conventional ray tracing, the hybrid model shows a much higher accuracy. As the result is angle resolved, it is appropriate for the simulation of MIMO-UWB systems.

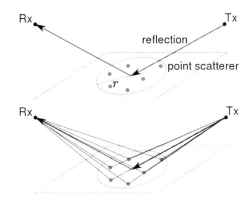

Fig. 6.32 Modeling of scattering points [JPZW10] (©EurAAP, reproduced with permission)

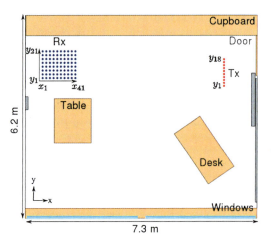

Fig. 6.33 Measurement scenario [JPZW10] (©EurAAP, reproduced with permission)

6.3.3 Impact of the Channel on UWB Systems

Interest in UWB systems originates from the fact that the channel's impact on the system is very different to conventional wideband or even narrowband systems. Due to the extremely large bandwidth, the multipath channel is observed with far more detail. In a narrowband system, we essentially see the interference between the carrier waves, which leads to amplitude fading and can be described simply but accurately by a Rayleigh or Ricean probability distribution that derives from Gaussian statistics. In conventional wideband systems, signal distortions become relevant. The channel is frequency selective, and Inter-Symbol Interference (ISI) is a typical effect seen from a system's viewpoint. An appropriate model of the channel can still resort to Gaussian statistics, arguing that a large number of actual propagation paths are superimposed within the resolvable time bins of a sampled channel response.

The ultrawide bandwidth of UWB systems implies a much finer time resolution. Therefore, the central limit theorem cannot be used any longer to justify Gaussian channel statistics [Mol09]. This makes UWB channels much harder to describe, and it raises the question what effects of the channels are relevant to a system? The

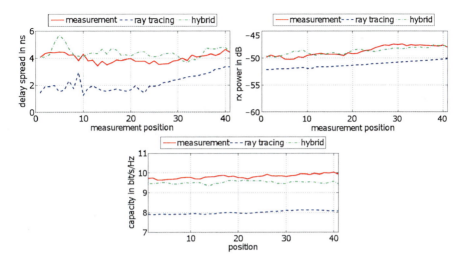

Fig. 6.34 Comparison of measured and simulated channel parameters along the line $y = y_1$ [JPZW10] (©EurAAP, reproduced with permission)

answer highly depends on the system type and application. A UWB localization system may require a different model than a simple low-data-rate BAN employing noncoherent receivers. As far as fading is concerned, individual MPCs interferes much less in UWB systems, hence the received energy fluctuates much less than in case of (Rayleigh) fading narrowband systems.

Noncoherent UWB Systems The implication of noncoherent receiver processing has been studied in [Wit07, WP08]. In such receivers, the signal energy is collected by squaring the received UWB pulses and integrating over some time interval. The integration has the effect that again a large number of multipath components becomes superimposed, when studying the receiver output. Therefore statistical methods can be used to accurately model the behavior of the multipath channel.

The focus in [Wit07, WP08] has been on the so-called Received Pulse AutoCorrelation Function (RP-ACF) and Received Pulse CrossCorrelation Function (RP-CCF), because these functions relate the response of the UWB channel (denoted as $g(t)$) to the output samples of energy detection or autocorrelation receivers [WLJ+09]. The RP-ACF is defined as

$$I_{[a,b]}(\tau) = \int_a^b g(t) g^*(t+\tau) \, dt \qquad (6.18)$$

for complex equivalent lowpass pulses, where $[a,b]$ is the integration interval. Based on a mathematical model of the received pulse $g(t) = w(t) * h(t)$—the convolution of a pulse $w(t)$ and the CIR $h(t)$ that is modeled as a sum of discrete Dirac pulses—the first and second-order moments have been derived from the RP-ACF and the RP-CCF [Wit07, WP08]. The moments can be expressed in terms of the average power delay profile of the channel and a second function that relates to the

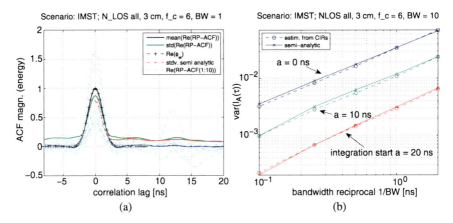

Fig. 6.35 (a) Statistics of the RP-ACF as a function of the correlation lag τ for measured channel impulse responses. (b) The variance of the RP-ACF at a lag of $\tau = 10$ ns for several signal bandwidths and integration intervals (from [Wit07])

variance of the power delay profiles. The moments also depend on the pulse shape $w(t)$ and therefore on the system bandwidth (which is defined by the time extent of $w(t)$). Figure 6.35(a) illustrates ten individual realizations of the (real part of the) RP-ACF and the mean and standard deviation of an ensemble of realizations. The sample means are compared with "semi"-analytical curves that are based on the mean and variance of the power delay profiles derived from the same measurement data.

Equation (6.18) represents the channel's impact on noncoherent UWB systems in various forms. For a lag of $\tau = 0$, for instance, it corresponds to the received energy in the pulse $g(t)$. Mean and variance of the received energy describe the signal fading, and the equations presented in [Wit07, WP08] demonstrate how the fading is related to system and channel parameters as bandwidth or delay spread. The result clearly shows that the fading reduces when the system bandwidth is increased.

For a lag greater than zero, (6.18) expresses inter-pulse interference experienced by an autocorrelation receiver when the pulse pair is spaced closer than the maximum excess delay of the channel. Again the impact of system and channel parameters can be evaluated. Figure 6.35(b) shows how the variance of the RP-ACF increases as a function of a reduced system bandwidth. This is because multipath components interfere more when the pulse duration increases.

Channel Capacity and Diversity The analysis in [CBB07] shows that *variations* of the channel capacity decrease similarly with bandwidth as the variations of the signal energy. Indeed, the latter can be used as an explanation for the former. Due to the ultrawide bandwidth of the channel, a single channel transfer function samples the complete distribution of the fading statistics. The overall capacity (and energy) of the channel relate(s) to the statistical mean, thus it will not change when a single snapshot already contains information on the full probability distribution of the channel. The analysis in [CBB07] is based on a linearization of the capacity equation that is only valid at low channel SNR. The capacity decreases with frequency, according to this work, and it is higher in LOS scenarios.

Interestingly, the conclusions in [TCAB09] seem to contradict the results in [CBB07]. They report increasing capacity with frequency and a general advantage in NLOS scenarios. The key difference is that MIMO channels have been investigated in [TCAB09], where *spatial* correlation has a fundamental impact on the channel capacity. Less correlation means higher capacity, and this is found at higher carrier frequencies and in NLOS environments, where no dominant LOS path is present to induce correlation. The latter work is based on channel measurement data and does not provide analytical approximations for the channel capacity.

Diversity was studied in [CVT09] for a static setting of the antennas, while low mobility inside two office environments affects the channels. The goal is to quantify the diversity induced by these channel variations. Its relation to bandwidth and center frequency is studied by processing the measurement data accordingly. Interestingly, the results are quite different for the two environments. The scenario with larger delay spread shows less diversity because a smaller fraction of multipath components is affected by time variations. But the absolute numbers of affected MPCs is similar in both cases.

6.4 UWB Localization and UWB Radar Imaging

6.4.1 UWB Localization

Precise and robust localization is required in many sensor network applications [GZG+05, THSZ07], e.g. indoor navigation and surveillance, tracking of persons or objects [RRSS08], or in safety and healthcare applications [AAM09]. Both passive localization of objects and active localization, e.g. of persons carrying a transmitter, is possible.

UWB localization attracts significant research interest, especially for localization in short-range cluttered environments. UWB technology possesses a great potential for accurate positioning in such environments thanks to its extremely large bandwidth [GZG+05, THSZ07, GP09, SGG08, SZHT10]. It allows separation of multipath components and achieves robust results. The incorporated low frequencies can penetrate non-metallic obstacles.

To date a large number of different localization approaches exploiting UWB technology have been developed. They can be divided into methods based on received signal strength intensity (RSSI), angle of arrival (AOA), time of arrival (TOA) and time difference of arrival (TDOA) [SZHT10, GZT08, SZT08]. Using UWB systems, the ranged-based schemes, TOA and TDOA, provide the best performance due to the available excellent time resolution. Two-step positioning is the common technique in most range-based localization systems. It includes a range (-difference) estimation step and a location estimation step [GP09, SGG08]. The goal of the range

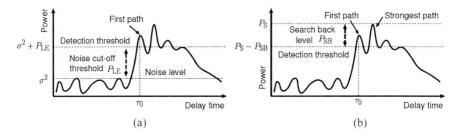

Fig. 6.36 Two threshold-based first path detection methods: (**a**) leading edge method and (**b**) search back method. The figure has been taken from [HiTiT+09] (©2009 EurAAP, reproduced with permission)

estimation is determination of distances between the target and several nodes of the localization system. From the multitude of range estimates the coordinates of the target are estimated in the location estimation step.

To achieve a high signal-to-noise ratio, correlator or matched filter (MF) receivers are used for UWB ranging (TOA, TDOA estimation) [Gez08]. However, even by using this equipment, range estimation is affected by noise, multipath components, Non-Line-Of-Sight (NLOS) situations and changes in propagation speed caused by obstacles.

Within COST 2100 localization problems have been discussed several times. The ranging accuracy achievable with the Europcom demonstrator has been analyzed based on point-to-point range measurements in [DWL08]. Here, as in other cases, the major challenge is the presence of strong biases in NLOS and dense multipath scenarios. Generally, the first path (direct path) is not always the strongest one, making the range estimation challenging. Different algorithms attempt to solve this problem and to achieve precise range estimations from received multipath signals. The threshold-based range estimation has attracted considerable interest due to its simplicity. The main advantage of the threshold-based estimator is that it has the potential for simple hardware implementation. An adaptive threshold could be introduced depending on the operating situation (e.g. SNR).

Other TOA estimation methods are closely related to channel parameter estimation. A customary approach is the application of a maximum likelihood (ML) estimator to find the most likely direct path channel delay. The computational complexity of these methods limits their implementation. Performance characterization of threshold-based estimators is of considerable importance. For two threshold-based ranging methods, this is done in [HiTiT+09]. The authors compare the performance of leading edge detection and search back methods. The basic principles of both methods are shown in Fig. 6.36. In [HiTiT+09] it is shown that the leading edge detection outperforms the search back method. Further improvement of range estimation can be achieved by incorporation of the antenna radiation pattern. The influence of this pattern on the TOA estimation was investigated in [DLG+09]. Here a virtual antenna array was used to produce different radiation patterns. The authors report a reduction of range variance after incorporation of the known array pattern in the estimation procedure. Range or range-difference estimation is followed by

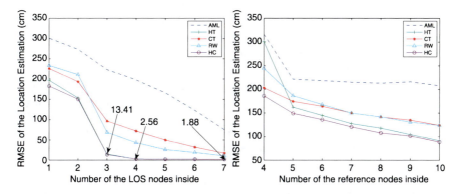

Fig. 6.37 Performance comparison between the method explained in [SZHT10] and some other methods. The number of the NLOS nodes is a random variable [SZHT10] (©2010 Guowei Shen et al., reproduced with permission)

location estimation. The algorithms, developed for this purpose, aim at high localization precision [GZT08]. Uncertain NLOS identification of course influences the location estimation.

A localization approach for NLOS situations is described in [SZHT09]. Here, the authors propose a novel NLOS identification and mitigation algorithm. This method handles the NLOS problem in an UWB sensor network applying a hypothesis test. It assumes an over-determined system exploiting more range estimates than the minimum number necessary for unambiguous localization. It iteratively determines the NLOS nodes by comparing the mean square error of the range estimates with the variance of the estimated LOS ranges. Identified NLOS nodes are excluded from the location estimation. Additionally, situations where only three or less LOS nodes are available (in a 2D case) are discussed in [SZHT09]. In this case, statistics of the geometrical arrangement of observing nodes are taken into account. The performance of the proposed method has been compared with other selected methods by means of computer simulation in a 2D area. Results are displayed in Fig. 6.37.

6.4.2 UWB Radar Imaging

Introduction Radar-imaging identifies shape and spatial distribution of reflecting objects within an area under investigation [HSZT10]. For this purpose, classical radar principles are applied: EM waves are transmitted by a Radar device and subsequently scattered and reflected at the surface of objects and barriers. A part of these scattered waves is received by the device shortly afterwards. From the delay between transmission and reception of a signal the total propagation path-length can be deduced, but the received signal does not directly provide the direction to the detected object. To achieve the directivity, which is necessary in imaging applications, conventional radar systems use antennas with a narrow beam. Antenna dimensions of the order of several meters are necessary in this case. These antennas are moved continuously to scan the observation area. Application areas of these systems are e.g. air traffic monitoring and bank detection in river navigation.

Ultrawideband (UWB) radar imaging describes a number of methods that make use of the specific properties of UWB radar signals: large bandwidth and low transmission power. These systems are suitable both for indoor and outdoor short-range applications, typically at distances of less than 20 meters. UWB radar devices can be manufactured as small, inexpensive units. To preserve this advantage for the complete system, antennas connected to the UWB chip should be small as well. But since the beam pattern of small antennas extends over a large angular area, directivity must be achieved by other technical means.

Fundamentals Before discussing different UWB imaging methods, we will clarify some fundamental terms of radar imaging.

1. Domains. In communications, signals are described in time or frequency domain. For imaging purposes, the spatial domain (unit m) is of special importance. It forms a Fourier pair with the wave number domain (unit m^{-1}). In a medium with propagation speed c, the quantities time t, frequency f, distance s and wave number k are linked as follows:

$$\mathbf{s} = c \cdot t \cdot \mathbf{e}, \qquad \mathbf{k} = \frac{2\pi f}{c}\mathbf{e} = \frac{2\pi}{\lambda}\mathbf{e}. \tag{6.19}$$

The propagation vector $\mathbf{e}$ (unit vector) points in the direction of wave propagation. λ is the wavelength.

2. Aperture. This is the opening of the antenna system. Because typical UWB antennas have small dimensions, a sufficiently long aperture must be synthesized by movement of at least one antenna. The path of this antenna is referred to as "synthetic aperture".

3. Resolution. Range resolution ρ_z describes the ability to resolve two objects that are close together along the direction of wave propagation $\mathbf{e}$. Cross range resolution ρ_x gives an estimate for the resolution power perpendicular to the direction of. Approximations for both quantities are:

$$\rho_x = \frac{\lambda R}{2A}, \qquad \rho_z = \frac{c}{2B}, \tag{6.20}$$

where R is the range, A is the aperture, and B is the bandwidth of the transmitted signal. For exact calculations, the actual pulse shape has to be taken into account. One can easily see that the large bandwidth of UWB systems leads to a good range resolution ρ_z, while ρ_x depends on the ratio of wavelength and aperture and on the range. The further the object is away, the worse becomes cross range resolution.

4. Clutter. Besides signals scattered at the objects, a variety of other signals is received at the same time. Especially in indoor environments, signals reflected at walls, at the floor, etc. create a signal background which is called "clutter". It can cause image distortions.

The Imaging Process UWB imaging is executed in the following steps: 1. data acquisition, 2. data pre-processing, 3. generation of a focused image, 4. image post-processing.

Fig. 6.38 Principle of range migration

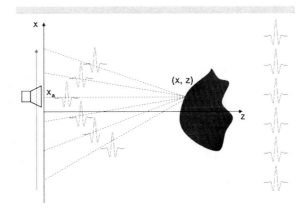

1. Data are acquired during the antenna movement. The synthetic aperture can be a straight line, a circle or even a random walk.
2. This point involves de-convolution of received data with a standard signal, distance calibration and echo detection.
3. The received data contain contributions from all scatterers within the observed area. A migration algorithm mathematically back-propagates the waves in a way that a focused image of the objects is created. See the next paragraph for details.
4. Image processing methods, feature extraction methods or classification algorithms can be deployed to extract information from the radar image.

Focusing Principle The principle function of a focusing algorithm is illustrated in Fig. 6.38 at the example of range migration. The antenna aperture is identical with the x-axis in this case. The distance between antenna position $(x_a, 0)$ and object point (x, z) gives raise to a delay τ between transmission and reception of a signal (Eq. (6.21), right-hand side). This delay changes in the course of antenna movement and causes the echo arrival time to "migrate" from its original time. The signal arrival times form a curve through the receiver data set s_r, which is unique for each object point (a hyperbola in case of this acquisition geometry). The migration algorithm first computes the delay τ for each combination of object point and antenna position and then integrates all signal contributions that originate from the same object point:

$$p(x,z,t) = \int_A s_r(x_a, t-\tau)\,dx_a \quad \text{with } \tau = \frac{1}{2c}\sqrt{(x-x_a)^2 + z^2}. \quad (6.21)$$

The result is a focused image of the distribution of scattering objects.

Distributed UWB Transceivers Besides the above-mentioned synthetic aperture radar (SAR) imaging, methods that apply a distributed network of UWB transceivers are of increasing importance [THSZ07, JTM+08]. An example is shown in Fig. 6.39. Here the Tx is moved on an arbitrary path. Waves scattered at the objects are received by the two Rx antennas. From these data an image is computed that shows the dominant objects [THSZ07].

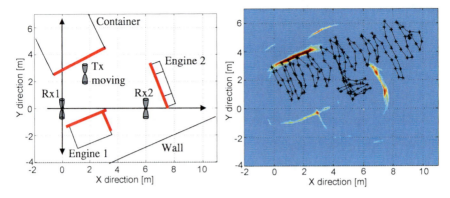

Fig. 6.39 UWB imaging in industrial environment. A network was used that consists of a mobile node and 2 fixed nodes

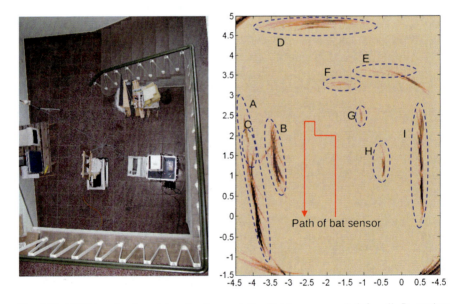

Fig. 6.40 UWB imaging in a stairwell using a rotating Bat-type sensor. *A*: left wall, *B*: monitor and cover, *C*: shadow of cover, *D*: front wall, *E*, *F*: handrail, *G*: front side of box, *H*: front side of UWB sounder, *I*: right wall. The image is an overlay of 15 measurements along the *red path*. The bat sensor can be seen in the *middle* of the *left figure*

"Bat" Sensor Another imaging method uses a rotating sensor, consisting of one Tx in the middle and two Rx-antennas at the ends of a bar (Bat-type sensor) [Li09] and Fig. 6.40. The sensor can be moved to different places and acquires data during a full rotation of the bar. Images are computed from data collected at each individual position. The final result is a combination of these images.

Inclusion of additional quantities and features increases the quality and the information content of UWB images. The application of dual polarized antennas in UWB imaging has been described in [LJACH09]. This system shows improved detection capabilities and can distinguish between different orientations of metal bars.

References

[AAM09] A. Adalan, H. Arthaber, and C. Mecklenbräuker. On the potential of IEEE 802.15.4a for use in car safety and healthcare applications. In *COST 2100 TD(09)865*, Valencia, Spain, 18–19 May 2009.

[AAMW10] D. Arnitz, G. Adamiuk, U. Muehlmann, and K. Witrisal. UWB channel sounding for ranging and positioning in passive UHF RFID. Aalborg, Denmark, June 2010. [TD(10)11085].

[AMW09] D. Arnitz, U. Muehlmann, and K. Witrisal. Multi-frequency continuous-wave radar approach to ranging in passive UHF RFID. *IEEE Trans. Microwave Theory Tech.*, 57(5):1398–1405, 2009.

[APC+10] J. B. Andersen, G. F. Pedersen, K. L. Chee, M. Jacob, and T. Kürner. Room electromagnetics applied to an aircraft cabin with passengers. Aalborg, Denmark, June 2010. [TD(10)11001].

[APP+07] G. Adamiuk, E. Pancera, M. Porebska, C. Sturm, J. Timmermann, and W. Wiesbeck. Antennas for UWB-systems. In *COST 2100 TD(07)358*, Duisburg, Germany, 10–12 September 2007.

[ASZW08] G. Adamiuk, C. Sturm, T. Zwick, and W. Wiesbeck. Dual polarized traveling wave antenna for ultra wideband radar application. In *International Radar Symposium, IRS*, pages 1–4, Wroclaw, Poland, May 2008. [Also available as TD(08)449].

[AWZ09a] G. Adamiuk, W. Wiesbeck, and T. Zwick. Differential feeding as a concept for the realization of broadband dual-polarized antennas with very high polarization purity. In *Antennas and Propagation Society International Symposium, 2009. APSURSI '09. IEEE*, pages 1–4, June 2009. [Also available as TD(09)850].

[AWZ09b] G. Adamiuk, W. Wiesbeck, and T. Zwick. Multi-mode antenna feed for ultra wideband technology. In *Radio and Wireless Symposium, 2009. RWS '09. IEEE*, pages 578–581, January 2009. [Also available as TD(08)662].

[AZW08] G. Adamiuk, T. Zwick, and W. Wiesbeck. Dual-orthogonal polarized Vivaldi antenna for ultra wideband applications. In *17th International Conference on Microwaves, Radar and Wireless Communications, MIKON*, pages 1–4, May 2008. [Also available as TD(08)449].

[Bau71] C. E. Baum. On the singularity expansion method for the solution of electromagnetic interaction problems. *Interaction Notes*, Note 88, December 1971.

[Bau73] C. E. Baum. Singularity expansion of electromagnetic fields and potentials radiated from antennas or scattered from objects in free space. *Sensor and Simulation Notes*, Note 179, May 1973.

[BKM+06] M. G. D. Benedetto, T. Kaiser, A. F. Molisch, I. Oppermann, C. Politano, and D. Porcino. *UWB Communication Systems: A Comprehensive Overview*. Hindawi Publishing Corporation, Cairo, Egypt, 2006.

[CBB07] A. Czylwik, O. Bredtmann, and S. Bieder. Simulations and experimental results on the capacity of ultra-wideband radio channels, pages 316–321, Singapore, September 2007. [Also available as TD(07)390].

[CJK09] K. L. Chee, M. Jacob, and T. Kürner. A systematic approach for UWB channel modeling in aircraft cabins. In *IEEE Vehicular Technology Conference VTC*, September 2009. [Also available as TD(09)940].

[CMUB09] R. Cepeda, J. McGeehan, M. Umana, and M. Beach. Static and dynamic measurement of the UWB distance dependent path loss. In *European Wireless Conference, 2009. EW 2009*, pages 6–10, May 2009. [Also available as TD(09)714].

[CPB07] R. Cepeda, S. C. J. Parker, and M. Beach. The measurement of frequency dependent path loss in residential LOS environments using time domain UWB channel sounding, Singapore, September 2007. [Also available as TD(07)306].

[CTB08] R. Cepeda, W. Thompson, and M. Beach. On the mathematical modelling and spatial distribution of UWB frequency dependency. In *2008 IET Seminar on Wideband and Ultrawideband Systems and Technologies: Evaluating Current Research and Development*, pages 1–5, November 2008. [Also available as TD(08)456].

[CVT09] R. Cepeda, C. Vithanage, and W. Thompson. Analysis of diversity from dynamic channel measurements. Vienna, Austria, September 2009. [TD(09)921].

[CWS02] R. J.-M. Cramer, M. Z. Win, and R. A. Scholtz. Evaluation of an ultra-wide-band propagation channel. *IEEE Trans. Antennas Propagat.*, 50(5):561–570, 2002.

[DAR10] W. Dullaert, G. Adamiuk, and H. Rogier. Compression of measured 2D UWB antenna transfer functions. *Electronics Letters*, 46(8):552, 2010.

[D'E08] R. D'Errico. *Analysis and modeling of multiple antennas in ultra wideband*. PhD thesis, Ecole Doctorale STITS, Université Paris Sud 11 and Università degli Studi di Bologna, Orsay, France, December 2008. [Ch. 2 and Ch. 4].

[DGRS06] R. D'Errico, H. Ghannoum, C. Roblin, and A. Sibille. Small semi directional antenna for UWB terminal applications. In *EuCAP*, Nice, France, November 2006.

[DLG+09] M. Dashti, T. Laitinen, M. Ghoraishi, K. Haneda, J. i. Takada, and P. Vainikainen. Influence of antenna radiation pattern on accuracy of ToA estimation. In *COST 2100 TD(09)942*, Vienna, Austria, September 2009.

[DR08] R. D'Errico and C. Roblin. Statistical analysis of UWB antenna radiation and scattering: applications and results. In *COST 2100*, Lille, France, October 2008. [TD(08)656].

[DR09] W. Dullart and H. Rogier. Compression of measured 2D UWB transfer functions. In *COST 2100 TD(09)907*, Vienna, Austria, September 2009.

[DR10] W. Dullaert and H. Rogier. Novel compact model for the radiation pattern of UWB antennas using vector spherical and Slepian decomposition. *IEEE Trans. Antennas Propagat.*, 58(2):287–299, 2010.

[DS07] R. D'Errico and A. Sibille. Scattering vs. coupling in UWB multiple antennas. In *COST 2100*, Lisbon, Portugal, February 2007. [TD(07)234].

[DS08] R. D'Errico and A. Sibille. Single and multiple scattering in UWB bicone arrays. *International Journal of Antennas and Propagation*, 2008. Guest Editors: James Becker, Dejan Filipovic, Hans Schantz, and Seong-Youp Suh.

[Dul08] W. Dullaert. Compact 3D radiation pattern model for UWB antennas. In *COST 2100 TD(08)518*, Trondheim, Norway, June 2008.

[DWL08] V. Dizdarevic, K. Witrisal, and R. Lobnik. Distance measurement tests using the emergency UWB radio positioning and communications EUROPCOM demonstrator. In *COST 2100 TD(08)462*, Wroclaw, Poland, February 2008.

[ECC08] Electronic Communications Committee ECC. Decision of 1 December 2006 amended 31 October 2008 on supplementary regulatory provisions to decision ECC/DEC/(06)04 for UWB devices using mitigation techniques, 31 October 2008.

[FCC02] Federal Communications Commission FCC. Revision of part 15 of the commission's rule regarding ultra-wideband transmission systems, February 2002.

[FH94] J. A. Fessler and A. O. Hero. Space-alternating generalized expectation-maximization algorithm. *IEEE Trans. Signal Processing*, 42(10):2664–2677, 1994.

[FTH+99] B. H. Fleury, M. Tschudin, R. Heddergott, D. Dahlhaus, and K. I. Pedersen. Channel parameter estimation in mobile radio environments using the SAGE algorithm. *IEEE J. Select. Areas Commun.*, 17(3):434–450, 1999.

[GBG+09] T. Gigl, T. Buchgraber, B. Geiger, A. Adalan, J. Preishuber-Pfluegl, and K. Witrisal. Pathloss and delay spread analysis of multipath intensive environments using IEEE 802.15.4a UWB signals. Vienna, Austria, September 2009. [TD(09)965].

[GDB06]	H. Ghannoum, R. D'Errico, and S. Bories. Small-size UWB planar antenna and its behaviour in WBAN/WPAN applications. In *IEE Seminar on Ultra Wideband Systems, Technologies and Applications*, London, UK, April 2006.
[Gez08]	S. Gezici. A survey on wireless position estimation. *Wirel. Pers. Commun.*, 44(3):263–282, 2008.
[GGH+10]	P. K. Gentner, W. Gartner, G. Hilton, M. A. Beach, and C. F. Mecklenbräuker. Towards a hardware implementation of ultra wideband beamforming. In *2010 International ITG Workshop on Smart Antennas (WSA)*, pages 408–413, December 2010. [Also available as TD(10)10034].
[Gha06]	H. Ghannoum. *Etude conjointe antenne/canal pour les communications ultra large bande en présence du corps humain*. PhD thesis, Ecole Doctorale d'Informatique, Télécommunications et Electronique de Paris, Paris, France, December 2006. [Ch. 3].
[GHBM10]	P. K. Gentner, G. Hilton, M. A. Beach, and C. F. Mecklenbräuker. Near and farfield analysis of ultra wideband impulse radio beamforming in the time domain. In *ICUWB*, 2010.
[GP09]	S. Gezici and H. V. Poor. Position estimation via ultra-wide-band signals. *Proceedings of the IEEE*, 97(2):386–403, 2009.
[GZG+05]	S. Gezici, T. Zhi, G. B. Giannakis, H. Kobayashi, A. F. Molisch, H. V. Poor, and Z. Sahinoglu. Localization via ultra-wideband radios: a look at positioning aspects for future sensor networks. *IEEE Signal Processing Mag.*, 22(4):70–84, 2005.
[GZT08]	S. Guowei, R. Zetik, and R. S. Thomä. Performance comparison of TOA and TDOA based location estimation algorithms in LOS environment. In *5th Workshop on Positioning, Navigation and Communication, 2008. WPNC 2008*, pages 71–78, March 2008.
[Han88]	J. E. Hansen. *Spherical Near-Field Antenna Measurements*. IEE Electromagnetic Waves Series, vol. 26. Peter Peregrinus, London, UK, 1988.
[HCR08]	H. Hashemi, T. Chu, and J. Roderick. Integrated true-time-delay-based ultra-wideband array processing. *IEEE Commun. Mag.*, 46(9):162–172, 2008.
[HiTiT+09]	K. Haneda, K.-i. Takizawa, J.-i. Takada, M. Dashti, and P. Vainikainen. Performance evaluation of threshold-based UWB ranging methods. In *Proc. European Conference on Antennas and Propagation (EuCAP)*, pages 3673–3677, March 2009.
[HS90]	Y. Hua and T. K. Sarkar. Matrix pencil method for estimating parameters of exponentially damped/undamped sinusoids in noise. *IEEE Trans. on Acoustics, Speech, and Signal Processing*, 38(5):814–824, 1990.
[HSZT10]	O. Hirsch, G. Shen, R. Zetik, and R. S. Thomä. UWB radar imaging. In *COST 2100 TD(10)10097*, Athens, Greece, February 2010.
[HT03]	K. Haneda and J. I. Takada. An application of SAGE algorithm for UWB propagation channel estimation, pages 483–487, November 2003.
[HTTV09]	K. Haneda, J. I. Takada, K. I. Takizawa, and P. Vainikainen. Ultrawideband spatio-temporal area propagation measurements and modeling. In *Proc. IEEE Interf. Conf. on UWB*, pages 326–331, Vancouver, Canada, September 2009. [Also available as TD(09)955].
[HWM+10]	K. Hausmair, K. Witrisal, P. Meissner, C. Steiner, and G. Kail. SAGE algorithm for UWB channel parameter estimation. Athens, Greece, February 2010. [TD(10)10074].
[JCC+05]	J. Jung, W. Choi, J. Choi, P. Miskovsky, C. Ibars, J. Mateu, and M. Navarro. A small wideband microstrip-fed monopole antenna. *IEEE Microwave Wireless Compon. Lett.*, 15(10):703–705, 2005.
[JFZW09]	M. Janson, T. Fügen, T. Zwick, and W. Wiesbeck. Directional channel model for ultra-wideband indoor applications. In *Proc. ICUWB 2009—IEEE International Conference on Ultra-Wideband*, pages 235–239, Vancouver, Canada, September 2009. [Also available as TD(09)908].

[JK07] J. Jemai and T. Kürner. Calibration of indoor channel models. In *Proc. ITG/VDE, Osnabrueck*, pages 31–36, Germany, May 2007.

[JKS08] J. Jemai, T. Kürner, and I. Schmidt. UWB channel: from statistical aspects to calibration-based deterministic modeling. In *Proc. German Microwave Conference (GeMiC)*, Hamburg-Harburg, Germany, 2008.

[JLS+09] M. Jacob, C. K. Lien, I. Schmidt, J. Schuur, W. Fischer, M. Schirrmacher, and T. Kürner. Influence of passengers on the UWB propagation channel within a large wide-bodied aircraft. In *3rd European Conference on Antennas and Propagation, 2009. EuCAP 2009*, pages 882–886, March 2009. [Also available as TD(09)748].

[JPZW10] M. Janson, J. Pontes, T. Zwick, and W. Wiesbeck. Directional hybrid channel model for ultra-wideband MIMO systems. In *Proc. of the European Conference on Antennas and Propagation EuCAP, Barcelona*, Barcelona, Spain, April 2010. [Also available as TD(10)11021].

[JSS+09] M. Janson, R. Salman, T. Schultze, I. Willms, T. Zwick, and W. Wiesbeck. Hybrid ray tracing/FDTD UWB-model for object recognition. *Frequenz, Journal of RF-Engineering and Telecommunications*, 63:220–271, 2009. [Also available as TD(07)337].

[JTM+08] L. Jofre, A. Toda, J. Montana, P. Carrascosa, J. Romeu, S. Blanch, and A. Cardama. UWB short-range bifocusing tomographic imaging. *IEEE Transactions on Instrumentation and Measurement*, 57(11):2414–2420, 2008.

[KP02] J. Kunisch and J. Pamp. Measurement results and modeling aspects for the UWB radio channel, pages 19–23, Baltimore, MD, May 2002. Measurements available under http://www.imst.de/de/funk_wel_dow.php.

[KP03] J. Kunisch and J. Pamp. An ultra-wideband space-variant multipath indoor radio channel model. Reston, VA, November 2003.

[Li09] Y. Li. *Abbildung von Objekten mit UWB-Radarsensoren ("Fledermaus"-Sensor)*. Master's thesis, Technische Universität Ilmenau, 2009.

[LJACH09] X. Li, M. Janson, G. Adamiuk, T. Zwick, and C. Heine. A 2D ultra-wideband indoor imaging system with dual-orthogonal polarized antenna array. In *COST 2100 TD(09)9101*, Vienna, Austria, September 2009.

[MGPLND08] J.-M. Molina-Garcia-Pardo, M. Lienard, A. Nasr, and P. Degauque. Wideband analysis of large scale and small scale fading in tunnels. In *8th International Conference on ITS Telecommunications, 2008. ITST 2008*, pages 270–273, October 2008. [Also available as TD(08)608].

[Mol09] A. F. Molisch. Ultra-wide-band propagation channels. *Proceedings of the IEEE*, 97(2):353–371, 2009.

[MSW10] P. Meissner, C. Steiner, and K. Witrisal. UWB positioning with virtual anchors and floor plan information. Dresden, Germany, March 2010.

[NC07] K. M. Nasr and J. Cosmas. A hybrid channel model for ultra wideband (UWB) applications. Duisburg, Germany, September 2007. [TD(07)314].

[RBS03] C. Roblin, S. Bories, and A. Sibille. Characterization tools of antennas in the time domain. In *IWUWBS*, Oulu, Finland, June 2003.

[RD08] C. Roblin and R. D'Errico. Statistical analysis of UWB antenna radiation and scattering: theory and modelling. In *COST 2100*, Lille, France, October 2008. [TD(08)657].

[RD09] C. Roblin and R. D'Ericco. Statistical analysis of a parametric model of a "population" of UWB antennas. In *3rd European Conference on Antennas and Propagation, EuCAP*, March 2009.

[RK05] S. Ries and T. Kaiser. Highlights of UWB impulse beamforming. In *EUSIPCO 13*, September 2005.

[Rob06] C. Roblin. Ultra compressed parametric modelling of UWB antenna measurements. In *2006 First European Conference on Antennas and Propagation*, pages 1–8. IEEE Press, New York, 2006.

[Rob07]   C. Roblin. Analysis of the parameters of a model-based parsimonious representation of UWB antenna radiation characteristics. In *COST 2100 TD(07)375*, Duisburg, Germany, September 2007.

[Rob08]   C. Roblin. Ultra compressed parametric modeling for symmetric or pseudo-symmetric UWB antennas. In *ICUWB*, Hannover, Germany, September 2008.

[Rog06]   H. Rogier. Spatial correlation in uniform circular arrays based on a spherical-waves model for mutual coupling. *AEU—International Journal of Electronics and Communications*, 60(7):521–532, 2006.

[RRSS08]   M. Rydström, L. Reggiani, E. G. Ström, and A. Svensson. An algorithm for locating point scatterers using a wide-band wireless network. In *COST 2100 TD(08)420*, Wroclaw, Poland, February 2008.

[SGG08]   Z. Sahinoglu, S. Gezici, and I. Guvenc. *Ultra-wideband Positioning Systems: Theoretical Limits, Ranging Algorithms, and Protocols*. Cambridge University Press, New York, US, 2008.

[SGS+09]   M. Schack, R. Geise, I. Schmidt, R. Piesiewicz, and T. Kürner. UWB channel measurements inside different car types. In *3rd European Conference on Antennas and Propagation, 2009. EuCAP 2009*, pages 640–644, March 2009. [Also available as TD(09)761].

[SHER06]   H. Saghir, M. Heddebaut, F. Elbahhar, and J. M. Rouvaen. Evaluation of a tunnel ground to train UWB communication, pages 1–5, September 2006.

[SJK10]   M. Schack, M. Jacob, and T. Kürner. Comparison of in-car UWB and 60 GHz channel measurements. In *2010 Proceedings of the Fourth European Conference on Antennas and Propagation (EuCAP)*, April, pages 1–5, 2010. [Also available as TD(10)11037].

[SJP+08]   M. Schack, J. Jemai, R. Piesiewicz, R. Geise, I. Schmidt, and T. Kürner. Measurements and analysis of an in-car UWB channel. In *Vehicular Technology Conference, 2008. VTC Spring 2008. IEEE*, pages 459–463, May 2008. [Also available as TD(08)455].

[SKA+08]   T. Santos, J. Karedal, P. Almers, F. Tufvesson, and A. F. Molisch. Scatterer detection by successive cancellation for UWB—method and experimental verification. In *Vehicular Technology Conference VTC Spring*, pages 445–449, May 2008. [Also available as TD(08)411].

[Sle78]   D. Slepian. Prolate spheroidal wave functions, Fourier analysis, and uncertainty. V. The discrete case. *Bell Systems Technology Journal*, 57:1371–1430, 1978.

[SP61]   D. Slepian and H. O. Pollak. Prolate spheroidal wave functions, Fourier analysis, and uncertainty. I. *Bell Systems Technology Journal*, 40(1):43–64, 1961.

[SP95]   T. K. Sarkar and O. Pereira. Using the matrix pencil method to estimate the parameters of a sum of complex exponentials. *IEEE Antennas Propagat. Mag.*, 37(1):48–55, 1995.

[SPKR00]   T. K. Sarkar, S. Park, J. Koh, and M. S. Rao. Application of the matrix pencil method for estimating the SEM poles of source-free transient responses from multiple look directions. *IEEE Trans. Antennas Propagat.*, 48(4):612–618, 2000.

[SV87]   A. Saleh and R. Valenzuela. A statistical model for indoor multipath propagation. *IEEE J. Select. Areas Commun.*, 5:128–137, 1987.

[SZHT09]   G. Shen, R. Zetik, O. Hirsch, and R. S. Thomä. Range based localization under NLOS conditions within UWB sensor networks. In *COST 2100 TD(09)708*, Braunschweig, Germany, February 2009.

[SZHT10]   G. Shen, R. Zetik, O. Hirsch, and R. S. Thomä. Range-based localization for UWB sensor networks in realistic environments. *EURASIP J. Wirel. Commun. Netw.*, 2010:1, 2010.

[SZT08]   G. Shen, R. Zetik, and R. S. Thomä. Performance evaluation of range-based location estimation algorithms under LOS situation. In *Proc. of the German Microwave Conference (GeMiC'08)*, Hamburg, Germany, 2008.

[TCAB09] W. Thompson, R. Cepeda, S. Armour, and M. Beach. Frequency dependency of capacity in MIMO UWB LOS and NLOS environments. Vienna, Austria, September 2009. [TD(09)977].

[THSZ07] R. S. Thoma, O. Hirsch, J. Sachs, and R. Zetik. UWB sensor networks for position location and imaging of objects and environments. In *The Second European Conference on Antennas and Propagation, 2007. EuCAP 2007*, pages 1–9, November 2007.

[WA07] W. Wiesbeck and G. Adamiuk. Antennas for UWB-systems. In *2nd International ITG Conference on Antennas, 2007. INICA '07*, pages 67–71, March 2007.

[WAS09] W. Wiesbeck, G. Adamiuk, and C. Sturm. Basic properties and design principles of UWB antennas. *Proceedings of the IEEE*, 97(2):372–385, 2009.

[Wit07] K. Witrisal. UWB channel characterization for system studies. In *COST 2100*, Duisburg, Germany, September 2007. [TD(07)378].

[WLJ+09] K. Witrisal, G. Leus, G. J. M. Janssen, M. Pausini, F. Troesch, T. Zasowski, and J. Romme. Noncoherent ultra-wideband systems. *IEEE Signal Processing Mag.*, 26(4):48–66, 2009.

[WP08] K. Witrisal and M. Pausini. Statistical analysis of UWB channel correlation functions. *IEEE Trans. Veh. Technol.*, 57(3):1359–1373, 2008.

[WS05] W. Wiesbeck, W. Soergel, and C. Sturm. Impulse responses of linear UWB antenna arrays and the application to beam steering. In *ICUWB*, September 2005.

[ZST05] R. Zetik, J. Sachs, and R. Thomae. Imaging of propagation environment by UWB channel sounding, January 2005. [TD(05)058].

Part II
Transmission Techniques and Signal Processing

Chapter 7
MIMO and Next Generation Systems

Chapter Editor Alister Burr, Ioan Burciu, Pat Chambers, Tomaz Javornik, Kimmo Kansanen, Joan Olmos, Christian Pietsch, Jan Sykora, Werner Teich, and Guillaume Villemaud

For the past decade or more, Multiple-Input Multiple-Output (MIMO) systems have been the subject of very intensive research. For example, it was one of the most important topics in the previous COST Wireless Action, COST 231. However in the past few years, in the time-frame of COST 2100 these techniques have begun to be implemented in practice. In particular they have appeared in the standards for next generation systems such as 3GPP-Long Term Evolution (LTE), 3GPP-LTE Advanced (LTE-Adv) and Worldwide Interoperability for Microwave Access (WiMAX), as well as the latest versions of WiFi. In COST 2100 MIMO has remained an extremely important subject of our research. In addition to extending the science of MIMO systems to develop new approaches, it has also tended to focus on the techniques such as precoding and MUltiuser Multiple-Input Multiple-Output (MU-MIMO) which will appear in these new systems. Moreover there has also been work on the practical implementation of MIMO techniques for next generation systems, and especially on practical terminal architectures. It has also been very important to provide means of modeling and testing the physical layer of these next generation systems.

This chapter therefore brings together the MIMO systems used in next generation systems with other work on the implementation and simulation of these systems. It also describes advances in MIMO techniques in a number of areas. The first section is divided into two sub-sections dealing first with simulators and testbeds which are used in system-level simulators to evaluate overall system capacity, as discussed in later chapters of this book. Secondly the development of terminals for next generation MIMO systems is considered, especially considering the additional Radio Frequency (RF) hardware required for MIMO. Section 7.2 then discusses especially precoding techniques used in many of the recent standards to implement MIMO. In particular precoding allows the implementation of closed-loop or adaptive MIMO. There is also in next generation systems much increased attention on MU-MIMO

A. Burr (✉)
University of York, York, UK

and on multi-terminal MIMO in general, including so-called "networkMIMO" approaches, which appear in LTE as Coordinated Multiple Point (CoMP): this is covered in Sect. 7.3. Various advanced MIMO transmission and detection approaches are covered in Sects. 7.4 to 7.6, including some interesting work on MIMO techniques involving Continuous Phase Modulation (CPM), giving advantages in terms of Peak-to-Average Power Ratio (PAPR).

7.1 Next Generation Systems: Architecture and Simulation

7.1.1 Simulators and Testbeds

The evaluation procedures of Next Generation Systems include simulators and testbeds to ensure that the specifications meet the desired performance level and to research on future improvements leading to system upgrades. From the user's point of view, the system performance is measured by several Quality-of-Service (QoS) parameters like throughput, latency, packet loss, etc. In addition to these parameters, operators are also interested in the system spectral efficiency in bit/s/Hz/cell, since a high spectral efficiency translates into more users per cell for a given bandwidth.

The simulation of a mobile communications system is a complex task, and it is usually split into a Link Level Simulator and a System Level Simulator. The Link Level Simulator basically characterizes the link in terms of the error rate of the received code blocks (Block Error Rate (BLER)) versus the Signal-to-Interference-plus-Noise Ratio (SINR) for all the possible transport formats, physical layer configurations and channel environments, while the System Level Simulator models a cell deployment with several tiers and a lot of users spread around a given scenario. Propagation losses, fading, intracell and intercell interference, frequency-reuse patterns, as well as traffic models and scheduling aspects are addressed in this type of simulator.

The mapping from Link-to-System (L2S) simulation must capture all the link properties while allowing for fast processing of system level simulations. Usually L2S mapping takes the form of a set of Look-Up-Table (LUT) that summarize the relevant link behavior. With the introduction of wideband packet mode mobile communication systems, for low mobility or pedestrian environments, the channel can be considered almost constant during one Transmission Time Interval (TTI), and the link can be properly characterized by the BLER curves under Additive White Gaussian Noise (AWGN) conditions. Since each subcarrier experiments a different frequency selective fading, the coded bits are exposed to different SINR conditions, and this introduces the need to predict the BLER performance of codes in multistate channels.

7.1.1.1 Effective Signal-to-Noise Ratio (SNR)

Given an experimental BLER measured in a multistate channel with a specific Modulation and Coding Scheme (MCS), the Effective Signal-to-Noise Ratio (ESNR) of

7 MIMO and Next Generation Systems

that channel is defined as the SNR that would produce the same BLER, with the same MCS, in AWGN conditions.

For a given multistate channel with N different SINR measurements $\{\gamma_1, \gamma_2, \ldots, \gamma_N\}$, the ESNR can be computed as the value γ_{ef} that accomplishes the following equation [BAS+05]:

$$I\left(\frac{\gamma_{ef}}{\alpha_1}\right) = \frac{1}{N} \sum_{k=1}^{N} I\left(\frac{\gamma_k}{\alpha_2}\right), \tag{7.1}$$

where the function $I(\cdot)$ is a sigmoidal like curve that can take several forms. The constants α_1 and α_2 are a refinement of the model to adjust the ESNR for different MCS and must be obtained by using a Link Level Simulator. The best results in the ESNR estimation are obtained by setting $I(\cdot)$ to be the Mutual Information (MI) between transmitted and received modulation symbols in AWGN conditions. This is the so-called Mutual Information Effective Signal to Noise Mapping (MIESM) link abstraction model.

The MIESM training methodology is as follows: The first stage is to obtain the reference BLER curves under AWGN channel. The second stage is to generate a big number of snapshots of the wideband multipath channel with known set of measurements, $\{\gamma_1^i, \gamma_2^i, \ldots, \gamma_N^i\}$, where γ_k^i denotes the SINR for the subcarrier k and the snapshot i. Notice that if linear MIMO equalization is involved, the MIMO post-processing SINR must be used in order to take into account the noise increase due to MIMO spatial channel equalization. For each snapshot of the multipath channel, a big number of codeblocks are simulated and a BLER measurement for the snapshot i is obtained. The channel transfer function remains constant during the simulation of a given channel snapshot. Finally, α_1 and α_2 are obtained as the parameters that minimize, given the whole set of snapshots, the mean square error between the measured BLER and the reference BLER.

7.1.1.2 A LTE Link Level Simulator

The LTE specifications define 15 Channel Quality Indicator (CQI) indexes that are used by the User Equipment (UE) to signal to the Evolved NodeB (eNB) the preferred modulation and code rate for the next TTI. By means of an Evolved UMTS Terrestrial Radio Access (E-UTRA) Down-Link (DL) LTE Link Level Simulator developed within COST 2100 action [OSR+09b, OSR+09a, ORGLMS10], a MIESM link abstraction model was trained for all the possible 15 CQIs specified by the 3rd Generation Partnership Project (3GPP). Figure 7.1 summarizes the behavior of the obtained MIESM model. The solid curves in Fig. 7.1 are the reference BLER curves for each CQI under AWGN channel, and the scattered dots are the experimental BLER values of each of the simulated channel snapshots.

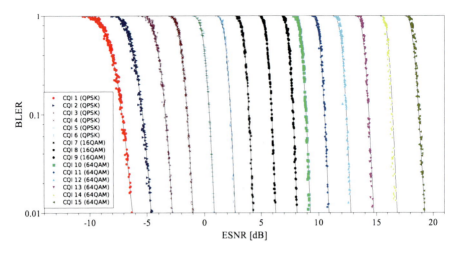

Fig. 7.1 MIESM estimated BLER vs. ESNR for LTE DL

7.1.1.3 MIMO Testbeds

MIMO testbeds have been built by several research teams in order to assess the throughput that can be obtained with MIMO schemes under field conditions. In [RGJ+07] a MIMO-High-Speed Packet Access (HSPA) testbed implemented by Ericcson Research in 2007 is described. The testbed was based on Ericcson R'99 commercial Wideband Code Division Multiple Access (WCDMA) products, enhanced with industrial PCs and Digital Signal Processor (DSP) boards, to generate the MIMO-HSPA physical channels. A 2×2 MIMO single codeword scheme ("vertical encoding") with Minimum Mean Squared Error (MMSE) equalization was implemented instead of the dual codeword scheme (Double Transmission Antenna Array (DTxAA)) standardized by 3GPP for HSPA Evolved (HSPA+), although the conclusions are valid for future standard compliant products as long as a linear MMSE receiver is used, since the performance gains of the DTxAA scheme are only realized when Successive Interference Cancellation (SIC) MIMO processing is applied at the receiving side.

The testbed used commercial dual polarized ($+45°/-45°$) antennas at the Base Station (BS) and a magnetic loop and an electrical dipole at the UE to achieve a dual polarization antenna set-up. DL throughput measurements were recorded along a drive path in Kista (Sweden), with a maximum car speed of 30 km/h and for three different configurations: Single-Input Single-Output (SISO) (HSPA with single antenna receiver), Single-Input Multiple-Output (SIMO) (HSPA with two receiving antennas in diversity mode) and 2×2 MIMO HSPA spatial multiplexing. At every TTI the UE reported the preferred transport format (single- or dual-stream) to the BS using CQI feedback.

Figure 7.2 shows the Cumulative Density Function (CDF) of the measured layer 1 throughput with the three configurations. The scenario is single-cell, single-user. The throughput is higher in sector one because it is an area where Line-Of-Sight

7 MIMO and Next Generation Systems

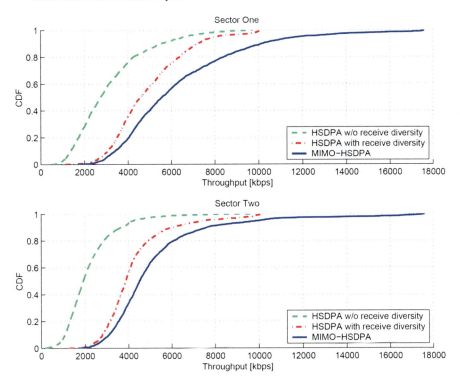

Fig. 7.2 CDF of measured layer 1 DL throughput [RGJ+07] (©2007 IEEE, reproduced with permission)

(LOS) propagation conditions were often found. The conclusions show that when dual-polarized antennas are used, dual-stream transmission is preferred in more that 70% of the measured locations, and that MIMO introduced a mean throughput gain of 115% over SISO and a 25% gain over SIMO.

In [MROZ09] a 4×4 MIMO-Orthogonal Frequency Division Multiplexing (OFDM) testbed at 5.6 GHz and 2.4 GHz is described. Emphasis is put on Time Division Duplex (TDD) in order to use the estimated Up-Link (UL) channel to apply optimal precoding in the DL in real-time. This approach simplifies the processing at the UE, since no channel estimation and feedback is required. Notice also that the RF transmit and receive branches need to be calibrated to guarantee channel reciprocity between UL and DL. A dual polarized 4-port antenna, with low correlation among antenna ports, suitable for MIMO 4×4 schemes is also described. The testbed implements 256 OFDM subcarriers and a transmitted power of 10 dBm per antenna port. The MIMO receiver is based on Zero Forcing (ZF) or MMSE processing, and for a maximum distance of 14 m between UE and BS achieves a Bit Error Rate (BER) of 10^{-5} with Quaternary Phase Shift Keying (QPSK) and 10^{-2} with 16 Quadrature Amplitude Modulation (QAM).

The OpenAirInterface LTE testbed implements LTE release 8 with two transmit antenna ports, 5-MHz bandwidth and TDD configuration. It supports transmission modes 1 (single-antenna transmission), 2 (Alamouti transmit diversity) and 6 (single-layer precoding). Additionally to throughput measurements, raw channel estimates are stored for further post processing. The testbed hardware is described in Sect. 2.1.1. In [KGL+10] the Eurecom OpenAirInterface testbed has been used to perform measurements of an LTE system in rural areas at 850 MHz using a 3-sector, dual RF high-power eNB and one UE with two receive and one transmit antenna.

7.1.2 Terminal Architectures

The solutions proposed for next generation radio transmission lead to an increasing complexity of terminals in terms of cost and power consumption. Particularly from the RF front-ends point of view: OFDM imposes high PAPR constraints and good linearity [vNP00], MIMO and discontinuous bandwidths conduct to the multiplication of RF chains, and scalable bandwidth imposes wider bandwidth characteristics of RF components [Kai05]. Over the past decade, a great deal of effort and literature has been devoted in proposing solutions that can reduce the complexity of the next generation of radiofrequency terminals. This section is focusing on the proposed single front-end architectures dedicated to multiantenna and/or multiband radiofrequency terminals. Among these proposed architectures we mention three examples that we consider revealing: one dedicated to the dual-band reception, one dedicated to the multiantenna reception of a dual-band signal, and finally one dedicated to the MIMO transmission.

Single Front-End Architecture for Biband Receivers The majority of efforts made in order to propose novel designs of single front-end receivers capable of simultaneously processing two independent frequency bands rely on the use of a multiplexing technique that allows the mutualization of certain elements of the reception chain. Meanwhile, due to the high constraints imposed to the mutualized electronic blocks of the processing chain, most of these multiplexing techniques do not represent an alternative to the front-end stack-up. By taking into account these assessments, a novel single front-end architecture was proposed in [BVV09b]. In order to multiplex the two useful frequency bands, this architecture is relying on the double orthogonal frequency translation technique, previously used only for image band rejection. As shown in Fig. 7.3, the two useful frequency bands are first separately filtered and amplified. The two RF signals thus obtained are then added. The implementation of the double orthogonal frequency translation (double IQ) technique is characterized by the choice of the frequency of the first local oscillator. This choice is done in such a manner that each of the two useful RF bands is situated in the image frequency band of the other. Each of the four signals issued from this double IQ is the result of the spectral overlapping of two baseband components, each of them corresponding to one of the useful RF bands. In the digital domain, the demultiplexing stage consists in two separate series of basic operation using the four baseband signals. From a theoretical point of view, the consequences of each of

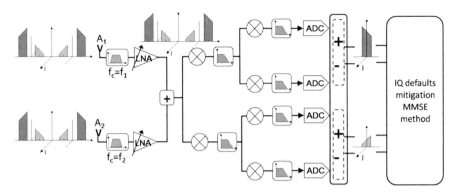

Fig. 7.3 Single front-end dual band RF receiver

the two series of basic operation is the reconstruction of one of the two useful baseband components, as well as the complete rejection of the other. However, because of the presence of gain and phase mismatches between the two orthogonal paths of the electronic blocks that realize the frequency translation, the rejection of the complementary component is not perfect at the output of the demultiplexing stage. The quality of each of the useful baseband component can be therefore deteriorated by the presence of the other useful component, called complementary component.

In order to mitigate the influence of the complementary component on the biband reception quality provided by the double IQ front-end, several technical solutions have been taken into account. In order to limit the complexity increase of the analog part of the receiver, low-complexity digital signal processing methods have been proposed [BVVG09, CKM04]. These methods realize a supplementary rejection of the baseband complementary component. Their primary goal is the estimation of the complex weight characterizing the presence of the complementary component in each of the demultiplexer's outputs based on the MMSE technique. Once the estimation step is finished, the second demultiplexer exit is multiplied with the estimated complex weight, and the product is subtracted from the baseband received signal. The validation of the theoretical results is realized by using the ADS software provided by Agilent technologies in order to simulate a simultaneous Universal Mobile Telecommunications System (UMTS) and 802.11g transmission. The results of several series of simulation show that the RF receiver integrating the double IQ front-end along with the digital signal processing detailed above has similar performance to those of a front-end stack-up receiver. Moreover, these results are confirmed by different series of measurements of a prototype of the receiver integrating the single front-end architecture.

In order to evaluate the performance-power consumption trade-off offered by the integration of the double IQ architecture in a biband receiver, a study estimates the power consumption of this type of receiver in comparison with the one of the front-end stack-up receiver [BVV09a]. The results of this analysis show that, for an RF receiver capable of simultaneously treating a UMTS signal as well as an 802.11g signal, the implementation of a single double IQ front-end leads to a 20% power consumption gain. Therefore, by taking into account its good performance-power consumption trade-off, the double IQ front-end architecture represents a good candidate for the integration in RF receivers dedicated to the next generation discontinuous spectrum transmissions, such as the ones of LTE-Adv incoming standard [Par08].

Single Front-End Architecture for Multiantenna Biband RF Receiver Among the proposed radio access techniques for next generation communication standards, multiantenna transmission and discontinuous spectrum allocation have a direct influence on the front-end part of the receiver. Each additional antenna or each additional band induces an important complexity increase of the receiver's analog part. In order to moderate this inherent complexity augmentation of the RF receiver compliant with a multiantenna and biband transmission, a novel single front-end architecture was proposed. It relies on a two-step method in order to be able to realize a multiantenna reception processing of a biband signal.

The first step consists in multiplying each biband antenna contribution with one of the orthogonal sequences of a Walsh code. As a result of this operation, when analyzing each antenna contributions, we observe that each useful frequency band is spread while being centered on the same central frequency. Therefore, the output signal of the addition block that follows this coding step is a biband RF signal. Each of its two frequency bands represents a spectral overlapping of the coded contributions corresponding to the different antenna reception of one of the useful frequency bands. This biband signal is then translated in the baseband domain by using a double IQ structure. In the digital domain, a two-step demultiplexer is being implemented. Each of the four baseband signals used by this demultiplexer is composed of a spectral overlapping of components corresponding to the different contributions issued from the multiantenna reception of a biband signal. The first step of the demultiplexing technique separately reconstructs the baseband components of each of the two useful frequency bands that compose the RF signal at the output of the adder. The second step is using digital filters followed by subsampling operations in order to process each of these two baseband signals. The validation of the theoretical results concerning the performance of this type of single front-end receiver has been realized through system level simulations.

MIMO Transmission Using a Single RF Front-End In [KKP08], it has been shown that it is possible to achieve spatial multiplexing gains using wireless transmitters that are equipped with a single RF front-end and an array of parasitic antenna elements that are spaced as closely as $\lambda/16$ from each other. In the proposed transmission technique, information symbols are mapped onto the radiation patterns of parasitic antenna arrays, by modulating orthogonal functions directly in the wave-vector domain of the antennas; this results in diverse symbol streams being simultaneously transmitted towards different angles of departure. In rich scattering environments, these symbol streams experience multipath fading in a manner similar to conventional MIMO transmission. Using a traditional MIMO receiver, it is then feasible to retrieve the transmitted information at a rate that is equal to that of comparable conventional MIMO systems. Overall, unlike conventional wisdom, spatial multiplexing is indeed possible with wireless transmitters that are equipped with a single RF front end and have critical size and power constraints. Therefore, this newly shown capability can help MIMO technology make its way into small wireless devices, thus enabling and benefiting a large variety of future wireless networking paradigms and applications.

To conclude, one of the main reasons to the increase of cost and consumption of terminals is the use of a transceiver analog part architecture composed of a front-end stack-up, each of the processing chains being dedicated to an antenna contribution or to a useful frequency band. In order to limit this, the terminal architectures presented in this section are based on different multiplexing or transmission techniques that allow the use of a single front-end complying to all the constraints imposed by next generation radio communications processing. Preliminary power consumption studies estimate that this type of architecture offers a real alternative to the front-end stack-up in terms of the performance-power consumption trade-off.

7.2 Adaptive MIMO and Precoding

The term "precoding" is used for a wide variety of purposes: here it is used for the preprocessing used in MIMO transmitters to optimize link capacity. It is usually a linear process consisting in the multiplication of a vector of data symbols by a precoding matrix to generate a vector of signals to be transmitted via the multiple antennas. However nonlinear forms of precoding are also possible, including Tomlinson–Harashima Precoding (THP) and Dirty Paper Coding (DPC). Precoding can also be used to control transmission of symbols in the time domain as well as the frequency domain: we refer to this as *spatio-temporal precoding*. In this way any transmission scheme, including spatial multiplexing and Space–Time Block-Code (STBC), can also be treated as a form of precoding.

Precoding can be used in adaptive MIMO systems, where the transmitter has Channel State Information (CSI) available. The optimum form of precoding if perfect CSI is available employs the transmit side singular vector matrix as a precoding matrix and uses Shannon's "water-filling" principle based on the singular values to allocate power between the data symbols. However in practice it is not usually feasible to provide CSI to this degree of precision at the transmitter, and a more commonly used alternative is to select at the receiver the optimum precoding matrix from a *codebook* of precoding matrices and transmit the index of this matrix to the transmitter.

Precoding in this form is widely used in emerging wireless standards, including WiMAX, LTE and LTE-Adv, and hence is of great interest both for the implementation of these standards and for future evolution of wireless systems. However in this section the emphasis is on generic versions of precoding, rather than specifically those used in the standards, although these will be used as examples in some cases.

In this section we will first introduce the general spatio-temporal precoding framework, which will then be used to classify other forms of precoding and adaptive MIMO techniques, including selection-based schemes. Finally we will consider the use of adaptive MIMO systems on some measured channels.

7.2.1 Spatio-Temporal Precoding

[Bur08b] introduces a generalized framework for precoding based on tensor notation. Spatial precoding conventionally uses a precoding matrix $\mathbf{F} \in \mathbb{C}^{N_t \times M}$ such that the transmitted $N_t \times 1$ signal vector may be written $\mathbf{s} = \mathbf{Fd}$, where $\mathbf{d}$ is an $M \times 1$ vector of data symbols, assuming that there are N_t transmit antennas and M data symbols are transmitted per time slot. Spatio-temporal precoding generalizes this to a precoding tensor $\mathcal{F} \in \mathbb{C}^{N_t \times T \times M}$, where T is the number of channel uses (i.e. code symbols transmitted) per time slot. We can identify the dimensions of this tensor with the domains of the transmitted signal: the first dimension is the spatial domain, corresponding to the antenna used; the second is the time domain, corresponding to the code symbol, and the third corresponds to the data symbol. Hence the element $f_{i,t,m}$ defines the weight of the mth data symbol transmitted on the ith antenna in the tth code symbol period. The transmitted signal for the time slot is then represented by an $N_t \times T$ matrix given by $\mathbf{S} = \mathcal{F} \otimes_3 \mathbf{d}$ where $\otimes_3$ denotes the *mode 3 product* of the tensor with the vector: that is, the implicit summation of the product runs along the third dimension of the tensor. This can also be written $s_{i,t} = \sum_{m=1}^{M} f_{i,t,m} d_m$.

Burr [Bur08b] shows that this generalized framework can be used to describe space–time block codes and conventional spatial multiplexing schemes as well as spatial precoding. MIMO-OFDM can also be described in this way: in this case the temporal element of the precoding is provided by a Fourier matrix. Similarly MIMO-MultiCarrier CDMA (MC-CDMA), where a different temporal spreading matrix is used. The frequency-selective MIMO channel can also be treated using a tensor model: in this case the tensor is fourth order (i.e. it is a four-dimensional array). The framework also encompasses linear dispersion codes [HH02]. MIMO-OFDM systems however are more likely to be treated on a per-sub-carrier basis.

This form of precoding can be used in non-adaptive MIMO systems, where no CSI is available at the transmitter. As mentioned, it can be applied to STBC, and Sect. 7.6 describes how it can be used to devise new full diversity, full rate space–time block codes. It can also be used where full CSI is available: however in this case spatial-only precoding is more likely to be used.

It can also be used where only partial CSI is available at the transmitter. For example [YBTL08] describes a combined spreading and precoding scheme for OFDM systems where partial CSI, in the form of the transmit correlation matrix (rather than the actual channel matrix), is available per sub-carrier. Simple OFDM, frequency domain spreading in the form of MC-CDMA, and joint spatial and frequency domain spreading using MultiCarrier Cyclic Antenna Frequency Spreading (MC-CAFS) [YTL05] are compared. A spatial precoder is then separately applied per OFDM sub-carrier, formed from the transmit-end eigenvectors of the transmit correlation matrix, and weighted by the water-filling power allocation. Hence the approach is less general than that described above but can still be regarded as a special case of spatio-temporal precoding. The results show that the precoding provides an advantage only when spreading, especially MC-CAFS joint spatial and frequency spreading, is also present to provide diversity.

Reference [CB08] considers spatial-only precoding based on a precoder codebook, as in LTE, assuming that perfect CSI is available at the receiver to select a precoder, and discusses the criteria used for the selection. Note that while standards define the codebook, they do not usually define the selection criteria, since this allows manufacturers to optimize the performance of their receivers. Also it is essential to take into account the detection method used in the receiver. This paper compares precoder selection for an Maximum Likelihood (ML), compared with an MMSE detector, deriving a criterion based on expected BER rather than post-detection SINR or Symbol Error Rate (SER). The system allows a variable number of streams from 1 to N_t to be transmitted, varying the modulation order to maintain a constant link throughput. This provides full diversity, fully exploiting both transmit end (through the potential of antenna selection) and receive end diversity (through ML detection). The results show superior coding gain using ML detection with the BER criterion, compared with MMSE with either criterion. Different codebooks are also considered, from codebooks allowing only antenna selection or spatial multiplexing with a fixed number of streams to a codebook based on the LTE standard. Restricting the number of streams to more than one reduces the diversity order; otherwise increasing codebook size increases the coding gain; for example, for two antennas, the coding gain with the full LTE codebook is around 0.5 dB better than for a codebook allowing only antenna selection of simple spatial multiplexing, and is 3 dB better than for MMSE detection.

7.2.2 Adaptive Schemes and Measured Channels

As we have seen, because of the equivalence of spatial precoding and other MIMO transmission schemes, there is no sharp distinction between precoding and other schemes where the transmission is adapted to the characteristics of the channel. Such schemes can be regarded as adaptive precoding schemes, where each scheme is represented by a different precoding tensor.

For example [YCC08] describes an approach involving scheme selection according to the channel. The system chooses between transmit diversity (in the form of Space-Frequency Coding (SFC)), spatial multiplexing and beamforming using the largest eigenmode of the transmit correlation matrix. In all cases OFDM is used, and in the case of spatial multiplexing Per Antenna Rate Control (PARC) is used, in which separate coding, modulation and OFDM multiplexing is applied to each antenna, using Exponential Effective Signal-to-Noise Mapping (EESM) [CC06] as a metric to select the modulation and coding. Note that the selection of the scheme is applied to the whole multiplex, not per sub-carrier. The system is able to adapt to the channel correlation conditions: on a relatively uncorrelated channel it will select transmit diversity at low $\frac{E_b}{N_0}$ and spatial multiplexing at high $\frac{E_b}{N_0}$; for a strongly correlated channel, the choice will be between beamforming and spatial multiplexing.

The context of COST 2100 means that channel measurements are readily available, which allows schemes to be evaluated using measured channels, rather than channel models, so that the performance on realistic channels can be evaluated.

Thus several papers consider the effect of the requirement for feedback in precoded systems. Reference [WB08] considers the sensitivity of precoding schemes to errors in the CSI due to errors in channel estimation errors or transmission of channel parameters. Three MIMO schemes are considered: zero-forcing precoding assuming full feedback of transmit and receive beamforming matrices; codebook-based precoding as used in the IEEE 802.16e standard; and a one bit feedback scheme. Two channel models are used: the 3GPP Spatial Channel Model (3GPP-SCM) and the IEEE 802.11n model. The effect of errors in the estimation of three quantities is determined: the Ricean K-factor, the receive end azimuth spread, and the amplitude of one channel tap, including also the effect of transmission errors. The results show that the zero forcing precoding, which relies on accurate estimates of the beamforming matrices to retain orthogonality, is much more sensitive both to errors in the channel estimation and to transmission errors than either of the other two schemes. In fact it shows that the codebook-based precoder is remarkably insensitive to channel errors, both those caused by estimation error and those due to transmission errors.

Colman [Col10] considers the effect of feedback errors in Orthogonal Diversity-Multiplexing Precoding (ODMP) on measured channels, showing that ODMP is more robust to errors and delay in channel state information, resulting in much less variation in BER, and hence reduced outage, even though the occurrence of very low error rates is also reduced. The scheme uses a set of spatial precoding matrices, allowing a trade-off between diversity and multiplexing: the increased diversity also has the effect of increasing robustness to channel state errors. This could also be modeled using the tensor-based spatio-temporal precoding approach described above.

Reference [HZNW10] also discusses the effect of limited feedback in measured channels, this time for the case of a precoder used at a MIMO Amplify and Forward (AF) relay. It shows that losses decrease with the size of the codebook, and hence the number of feedback bits required. 16-bit feedback results in a loss of around 0.3 bits/s/Hz compared to no quantization; 4 bits in a further 0.3 bits/s/Hz loss.

An adaptive MIMO system is capable of providing two types of gain: *diversity gain*, which is measured by the rate of decrease of BER with SNR at a constant link throughput, and *multiplexing gain*, measured by the increase in link throughput with SNR at a constant BER. Specifically the diversity gain is given by $d = -\lim_{SNR \to \infty} [\frac{\partial \log(BER)}{\partial \log(SNR)} | R]$ while multiplexing gain $r = \lim_{SNR \to \infty} [\frac{\partial R(SNR)}{\partial \log(SNR)} | BER]$. Zheng and Tse [ZT03] showed that a trade-off is possible between these quantities and gave an asymptotic bound at high SNR on the diversity and multiplexing gain for a Rayleigh independent identically distributed (iid) channel. [WC07] considers this diversity-multiplexing trade-off for more practical conditions: at finite SNR and using measured rather than iid channels. It shows that the bound of

[ZT03] is very optimistic compared with these more practical situations, especially for large numbers of transmit and receive antennas. (Note that multiplexing gain is also considered in the context of multiuser MIMO systems in Sect. 7.4.)

7.3 Multiple Terminal MIMO

The focus of this section is on (i) multiple-user MIMO systems, (ii) multiple-user diversity (iii) multiple-BS MIMO and (iv) multiple virtual user MIMO. The various analyses are supported by a mixture of channel measurement and channel modeling-based approaches.

7.3.1 The Finite Scattering Channel Model and the Media Access Control (MAC) Multiuser MIMO Capacity

In Single User (SU) MIMO systems, the effect of correlated channels is to reduce capacity by increasing the spread of channel matrix eigenvalues. In [KHHP10], the effect of interlink channel correlation for a two-user Multiuser (MU) MIMO system was examined using measurements and was seen to reduce capacity. For a more precise understanding of this effect, a finite scattering channel model is introduced [Bur09, Bur03]. Given a BS that is equipped with N antennas and that receives signals from a series of K users, which are each equipped with M_k antennas, there are a total of K channel matrices. These channel matrices are denoted individually as $\mathbf{H}_k$ with dimension: $N \times M_k$ and may be decomposed as $\mathbf{H}_k = \mathbf{\Psi}_{R,k} \Xi_k \mathbf{\Psi}_{T,k}$, where

$$\mathbf{\Psi}_{R,k} = \left[\exp\left(2i\pi n \frac{l_R}{\lambda} \sin(\phi_{R,ks}) \right), \; n = 1, \ldots, N, \; s = 1, \ldots, S \right], \quad (7.2)$$

$$\mathbf{\Psi}_{T,k} = \left[\exp\left(2i\pi m \frac{l_T}{\lambda} \sin(\phi_{T,ks}) \right), \; m = 1, \ldots, M_k, \; s = 1, \ldots, S \right], \quad (7.3)$$

$$\Xi_k = diag([\xi_{ks}, \; s = 1, \ldots, S]), \quad (7.4)$$

l_R and l_T are the transmit and receive element spacing, respectively, and λ is the wavelength with $\{\frac{l_R}{\lambda}, \frac{l_T}{\lambda}\} = \{\frac{1}{2}, \frac{1}{2}\}$ throughout. From a given kth user, and as the result of the sth scatterer, there is a multi-path signal component that has an Azimuth of Departure (AoD), $\phi_{T,ks}$, and arrives at the BS at an Azimuth of Arrival (AoA), $\phi_{R,ks}$, with a corresponding path gain of ξ_{ks}. In Eq. (7.2), $\phi_{T,ks}$ is defined as $\phi_{T,ks} \sim \mathcal{U}(0, 2\pi) \; \forall k, s$ with $\sim \mathcal{U}(\mu, \phi)$ denoting the uniform random distribution of mean μ and angular spread ϕ. However in Eq. (7.3), $\phi_{R,ks}$ is defined as $\phi_{R,ks} \sim \mathcal{U}(\theta_k, \phi_{spr}) \; \forall k, s$, thus $\phi_{R,ks}$ is a uniform iid angle with an angular spread of ϕ_{spr} about a mean angle θ_k. θ_k is in turn defined as $\theta_k \sim \mathcal{U}(0, 2\pi) \; \forall k$. In summary, the user is free to move to any position over an angular range of $-\pi$ to π

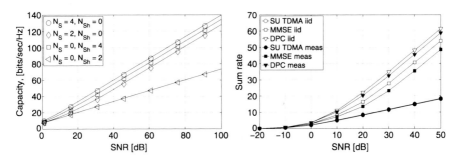

Fig. 7.4 *Left*—Capacity calculations for a two-user MU MIMO MAC system for the cases of different configurations of scatterers, i.e. distinct (N_S) and shared (N_{Sh}), using the finite scattering model. *Right*—Sum capacity for DPC, MMSE and Time-Division Multiple Access (TDMA)

around the BS, but the signals from this user are focused into a limited angular spread of ϕ_{spr} at the BS. The elements of $\boldsymbol{\Psi}_{R,k}$ and $\boldsymbol{\Psi}_{T,k}$ have unit magnitude, and $\mathbf{H}_k$ is normalized such that $\mathscr{E}\{\mathbf{H}_k\} = M_k$. As a result, ξ_k, from Eq. (7.4), is defined as [Bur09] $\xi_k \sim \mathscr{CN}(0, \frac{1}{2NS})$ for $k = 1, \ldots, K$, where $\mathscr{CN} \sim (\mu, \sigma^2)$ is the complex normal distribution of mean μ and variance σ^2. By defining the composite channel matrix as $\widetilde{\mathbf{H}} = [\mathbf{H}_1 \mathbf{H}_2 \cdots \mathbf{H}_K]$, then, in the absence of waterfilling and given an AWGN variance σ_n^2, an appropriate metric for system performance, i.e. the MAC capacity, $\widetilde{C}_{MAC}$, may be written as

$$\widetilde{C}_{MAC} = \mathscr{E}_{\mathbf{H}}\left[\log_2\left\{\det\left(\mathbf{I} + \frac{1}{\sigma_n^2}\widetilde{\mathbf{H}}\widetilde{\mathbf{H}}^H\right)\right\}\right]. \quad (7.5)$$

On the left of Fig. 7.4, the capacity of a two-user MU MIMO MAC system is considered for various configurations of distinct scatterers, N_S, and shared scatterers, N_{Sh}. It is clear from these plots that for a fixed number of scatterers, shared scatterers have the effect of reducing $\widetilde{C}_{MAC}$ by reducing the rank of $\widetilde{\mathbf{H}}$. Also, the number of distinct AoAs, set by $KN_S + N_{Sh}$, determines the capacity [Bur09], and thus the configuration $N_S = 0$, $N_{Sh} = 2$ exhibits the lowest capacity in Fig. 7.4.

7.3.2 Multiuser MIMO Broadcast Channel (BC) Capacity and Transmission Techniques

In the MU MIMO BC, the BS is the transmitter, and thus each channel matrix $\mathbf{H}_k$ is of dimension $M_k \times N$. The capacity of the MIMO BC, C_{BC}, is commonly cited as follows, under the assumption of unity noise variance [GKH+07, KKGK09]:

$$C_{BC}(\mathbf{H}_1, \ldots, \mathbf{H}_K) = \max_{\Sigma_k \geq 0, \sum_{k=1}^K \text{Tr}(\Sigma_k) \leq P} \sum_{k=1}^K \log_2 \frac{\det(\mathbf{I} + \mathbf{H}_k(\sum_{j=1}^K \Sigma_j)\mathbf{H}_k^H)}{\det(\mathbf{I} + \mathbf{H}_k(\sum_{j \neq k}^K \Sigma_j)\mathbf{H}_k^H)}. \quad (7.6)$$

In contrast to (7.5), this expression incorporates waterfilling. In effect, a maximization must be performed over all downlink covariance matrices $\boldsymbol{\Sigma}_1, \ldots, \boldsymbol{\Sigma}_K$.

This is challenging from a algorithm design point of view since such a maximization is not a convex function. This problem is overcome by first applying waterfilling to C_{MAC}, which remarkably yields a maximization over a convex function, and then mapping the appropriately generated covariance matrices using a MAC–BC transformation algorithm [JRV+05, see Appendix 1].

The only known technique that achieves C_{BC} is DPC. DPC is difficult to implement; thus linear downlink precoders have been implemented on the BC. In [GKH+07, KKGK09], the MMSE precoder is considered for the case where $M_k = 1$, thus yielding a composite channel matrix, $\hat{\mathbf{H}} = [\mathbf{h}_1^T \ \mathbf{h}_2^T \cdots \mathbf{h}_K^T]$. Then for each channel $\mathbf{h}_k$, a transmit weight $\mathbf{w}_k$, is designed based on the regularized Moore–Penrose pseudoinverse of $\hat{\mathbf{H}}$, and thus,

$$\mathbf{W} = \hat{\mathbf{H}}^H \left(\hat{\mathbf{H}} \hat{\mathbf{H}}^H + \frac{N \sigma_n^2}{P} \mathbf{I} \right)^{-1}. \qquad (7.7)$$

$\mathbf{w}_k$ is the kth column of $\mathbf{W}$ normalized such that $\|\mathbf{w}_k\|_2^2 = 1$. $\frac{N\sigma_n^2}{P}$ is a regularization factor [PHS05] and seeks to ensure reasonable SNRs at the receivers whenever $\hat{\mathbf{H}}$ approaches singularity, but it is a trade-off against levels of Co-Channel Interference (CCI). The rate of the MMSE precoder, $\mathscr{R}_{MMSE}$, is [PHS05, KKGK09]

$$\mathscr{R}_{MMSE} = \sum_{k=1}^{K} \log_2 \left[1 + \frac{|\mathbf{h}_k \mathbf{w}_k|^2}{\sum_{j \neq k} |\mathbf{h}_k \mathbf{w}_j|^2 + \frac{K\sigma_n^2}{P}} \right]. \qquad (7.8)$$

On the right of Fig. 7.4, the ergodic rates (capacities) of DPC, MMSE precoding and a SU TDMA MIMO scheme are plotted for the case of a measurement scheme detailed in Sect. 2.1.1 and also for Rayleigh iid channels (randomly generated). In the SU TDMA MIMO scheme, each user is scheduled fairly in time, and waterfilling is implemented for the purposes of reasonable comparison. In each case the average rate is taken over 4 users. At high SNR, i.e. > 20 dB, the DPC outperforms all other schemes, but the MMSE precoding provides a reasonable trade-off between complexity and performance since it has a comparable capacity to DPC. It appears that the Rayleigh iid channel is a somewhat idealistic scenario since it outperforms the measured channels.

Another linear precoding technique is that of leakage-based precoders [STS07b, STS07a, TSS05]. In common with the MMSE precoder, a certain amount of CCI must be tolerated. For a given ith user out of the total K users, the composite channel matrix in this case is $\check{\mathbf{H}} = [\mathbf{H}_1^T \cdots \mathbf{H}_{i-1}^T, \mathbf{H}_{i+1}^T \cdots \mathbf{H}_K^T]^T$.

This excludes the presence of the ith user. Setting $\mathbf{A} = \mathbf{H}_i^H \mathbf{H}_i$ and $\mathbf{B} = \check{\mathbf{H}}_i^H \check{\mathbf{H}}_i$ allows the leakage-based precoder of the ith user, $\mathbf{w}_{i(R)}$, to be written as

$$\mathbf{w}_{i(R)} = \mathscr{P}_R \left\{ \left(M_i \sigma_{n(i)}^2 \mathbf{I} + \mathbf{B} \right)^{-1} \mathbf{A} \right\}. \qquad (7.9)$$

$\mathscr{P}(\cdot)$ extracts the dominant eigenvector, and $(\cdot)_R$ specifies the right eigenvector. The power gain of the ith user is G_i, and the sum of the CCI from the ith user to others and the AWGN variance at the ith user is L_i. They are defined respectively as

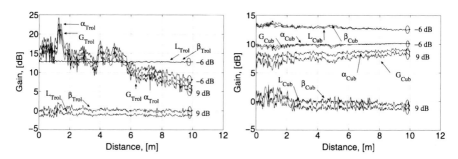

Fig. 7.5 *Left*—Comparison of the pairs $\{G_i, \alpha\}$ and $\{L_i, \beta\}$ for SNRs of -6 dB and 9 dB for the trolley-based user ('Trol'). *Right*—Comparison of the pairs $\{G_i, \alpha\}$ and $\{L_i, \beta\}$ for SNRs of -6 dB and 9 dB for the cubicle-based user ('Cub'). Note: at -6 dB, the terms within either pairs, $\{G_i, \alpha\}$ or $\{L_i, \beta\}$, are almost indistinguishable

$G_i = \mathbf{w}_{i(R)}^H \mathbf{A} \mathbf{w}_{i(R)}$ and $L_i = \mathbf{w}_{i(R)}^H (M_i \sigma^2 \mathbf{I} + \mathbf{B}) \mathbf{w}_{i(R)}$. Another viewpoint for calculating $\mathbf{w}_{i(R)}$ comes from the generalized eigenvalue problem, which may be stated as

$$\beta \mathbf{A} \mathbf{w}_{i(R)} = \alpha (M_i \sigma_i^2 \mathbf{I} + \mathbf{B}) \mathbf{w}_{i(R)}. \tag{7.10}$$

Remarkably, the two scalar factors, namely α and β, that arise from this may be calculated efficiently by applying the QZ decomposition [MS72] to $\mathbf{A}$ and $(M_i \sigma^2 \mathbf{I} + \mathbf{B})$. Thus, calculation of α and β is possible given knowledge of all users' channels and the AWGN at the receiver but does not require implementation of $\mathbf{w}_{i(R)}$. However, α and β may be expressed in terms of $\mathbf{w}_{i(R)}$ and the difference between G_i and α, as well as between L_i and β, is, in either case, simply the substitution of one of the right eigenvectors for that of the appropriate left one [Ste75]. The pairs $\{G_i, \alpha\}$ and $\{L_i, \beta\}$ are plotted in Fig. 7.5 [CO09] for the case of two SNRs of -6 dB and 9 dB. The results are based on the outdoor-to-indoor (O2I) distributed measurement campaign outlined in [CBV+08, p. 28], where trolley-based user ('Trol') is moved along a 10-m route while a cubicle-based user ('Cub') remains stationary within the confines of a cubicle. For either pair $\{G_i, \alpha\}$ and $\{L_i, \beta\}$, a reasonable and highly correlated match is observed, and thus α and β are proposed in [CO09] as propagation-motivated metrics for leakage-based precoder performance. They are also arguably good candidates for the development of a scheduling algorithm for this particular precoder.

Finally in relation to leakage based precoders, the study in [CO10] investigates the robustness of the precoder in relation to mismatches as the precoder is applied at the user and BS ends. Using the same measurement campaign, i.e. [CBV+08, p. 28], the precoders are computed, and pairs of precoder that satisfy a certain inner product are selected. These inner products, denoted ξ, are less than one, and thus either precoder in a given pair is mismatched. Along with the aforementioned CCI, denoted σ_{CCI}^2, various other terms are defined that quantify additional CCI, denoted $\sigma_{CCI'}^2$, and receive signal offset, denoted σ_{OFF}^2. Unlike σ_{CCI}^2, the terms $\sigma_{CCI'}^2$ and σ_{OFF}^2 arise purely due to the precoder mismatch. In order to characterize the performance degradation due to precoder mismatch, the terms $\sigma_{CCI}^2 + \sigma_{OFF}^2$ and $\sigma_{CCI}^2 + \sigma_{CCI'}^2$ are cal-

7 MIMO and Next Generation Systems

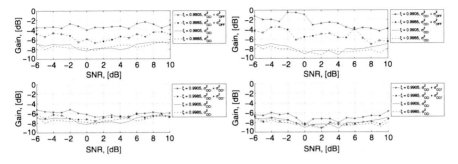

Fig. 7.6 *Both upper plots*: calculation of σ^2_{CCI} and $\sigma^2_{CCI} + \sigma^2_{OFF}$ for precoder mismatches of $\xi = \{0.9985 \pm 0.0005, 0.9905 \pm 0.0005\}$. *Both lower plots*: calculation of σ^2_{CCI} and $\sigma^2_{CCI} + \sigma^2_{CCI'}$ for precoder mismatches of $\xi = \{0.9985 \pm 0.0005, 0.9905 \pm 0.0005\}$. *Left-hand plots* pertain to the cubicle measurements and *right-hand plots* pertain to trolley measurements. The SNR varied from -6 dB to 10 dB

culated for precoder mismatches of: $\xi = \{0.9985 \pm 0.0005, 0.9905 \pm 0.0005\}$ in Fig. 7.6 [CO10].

Clearly, the term σ^2_{OFF} is significantly more detrimental to system performance than $\sigma^2_{CCI'}$. It is noted that there exists an SU beamforming precoder, which is analogous to this leakage-based precoder and where a term similar to σ^2_{OFF} occurs when there is precoder mismatch. Efforts to palliate this term are described in [ASDG06] and thus could be beneficial if they were applied to leakage-based precoder.

7.3.3 MU Diversity

In MU diversity, the link capacity can be improved when multiple users schedule their channel access such that each user transmits during a favorable channel state by using a Proportional Fair Scheduling (PFS) algorithm. Considering a configuration where $\{M_k, N\} = \{1, 1\}$, i.e. single antenna systems, each user k experiences a propagation channel, d_k, which is the product of a short-term fading term, f_k, and a path loss, p_k, both of which are random iid variables across all users. It was shown in [VTL02] that, for the case of the rayleigh iid channels, the system capacity, in respect of the number users, K, scales approximately according to the function $\ln(\ln(K))$ when the PFS algorithm is implemented. This implies a somewhat ideal capacity behavior as K increases, and thus a finite scatterer model, introduced in [PH06], has been used to evaluate the system behavior of PFS algorithm scheduled MU diversity systems. In [KM08], the capacity of the kth user has been evaluated using

$$R_{PFS,k} = \int_0^\infty \log_2(1 + xSNR)\, dF_{p,f_{K:K}}(x), \qquad (7.11)$$

where $dF_{p,f_{K:K}}$ denotes the joint distribution of p_k and f_k for the user with the best value of p_k. For a noise spectral density, N_0, $SNR = 1/N_0$ and furthermore

Fig. 7.7 Capacity with increasing user number, K, for a PFS algorithm scheduled MU diversity system for channels with $S = \{5, 10, 15\}$ scatterers and for Rayleigh fading. $(E_b/N_0)_{sys} = -5$ dB [KM08] (©2008 IEEE, reproduced with permission)

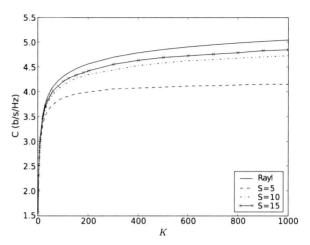

$(\frac{E_b}{N_0})_{sys} = \frac{SNR}{R_{PFS,k}}$. The computed capacities for various values of S, namely $S = \{5, 10, 15\}$, are compared with the Rayleigh iid case in Fig. 7.7 for a fixed $(\frac{E_b}{N_0})_{sys}$ of -5 dB.

Clearly when $S = 5$, the capacity saturates with respect to K relatively earlier than for the cases of $S = 10$ or $S = 15$. However, the relative difference in capacity for the cases of $S = 10$ and $S = 15$ indicates a behavior of diminishing returns.

7.3.4 Multiple BS MIMO

Due to the normally elevated positions of BSs, their angular spread can be severely limited, which in turn leads to channel correlation and hence capacity degradation. In order to palliate this effect, the downlink transmit antennas are distributed over several BSs leading to a co-operative multiple BS MIMO system. Given the obvious complexity of such a scheme, the macro-cellular measurements and analysis in [LMF10b, LMF10a] seek to investigate whether its potential benefits would justify its deployment.

The system configuration is $\{N_1, N_2, N_3, M\} = \{1, 1, 1, 4\}$, where each BS has one $45°$ linearly polarized antenna and the single user has four antennas comprising two loop and two dipole antennas, thus forming the 3×4 MIMO channel matrix. The Tx power is 35 dBm, the center frequency is 2.66 GHz, and the bandwidth is 20 MHz, thus leading to 432 frequency bin evaluations of the channel matrix. The user is driven by van along two routes that encompass LOS, obstructed LOS and highly shadowed scenarios. Each measurement point is normalized with respect to the individual SIMO link that has the strongest power and a moving average is performed over ten wavelengths, which removes the effect of fast fading from the analysis.

Considering the capacity analysis at an SNR $= 0$ dB in Fig. 7.8, a hierarchy of capacity may be observed: the 1×4 link with interference $\leq$ best single 1×4 link

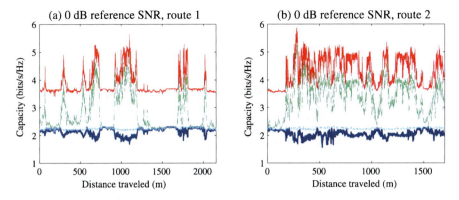

Fig. 7.8 The capacity of the best 1×4 link with interference (*lower most plot* in either case) $\leq$ best single 1×4 link without interference $\leq$ co-operative 4×3 link with equal power allocation $\leq$ co-operative 4×3 link with waterfilling [LMF10a] (©2010 IEEE, reproduced with permission)

without interference $\leq$ co-operative 4×3 link with equal power allocation $\leq$ co-operative 4×3 link with waterfilling. Applying waterfilling to the 4×3 link yields at least a two-fold capacity gain with respect to either of the 1×4 links. Equal power allocation in the 4×3 link proves to be beneficial, but more so in the case of route 2, and is at its most beneficial when the 1×4 link capacity with interference is at its lowest levels. In [LMF10b, LMF10a], this analysis was repeated for an SNR of 20 dB and, as would be expected, the relative capacity gain due to waterfilling was found to be less. It should be noted finally here that a very recent study [GLG10] has also examined the mutual information and spectral efficiency of an algorithm known as CoMP, which can co-ordinate multiple BSs to transmit to a given user should the user be at point in space where there signal would be weak, i.e. at the edge of cells. The study argues that the performance enhancement of multiple BS MIMO using CoMP is significant and fully justifiable from a complexity point of view.

7.3.5 Multiple Virtual-User MIMO

Due to the size restrictions of mobile users, the inherent channel correlations, due to the limited antenna spacing, can restrict capacity. A possible solution is to distribute the construction of the MIMO system across its constituent users. In effect, the users cooperate with one and other to form virtual arrays and hence reduce channel correlation. In order to examine whether the potential benefits of this scheme could justify its deployment, macro-cellular measurements and analysis are performed in [YWB09]. The approach taken is to consider the following configuration: $\{N, M_1, M_2\} = \{4, 4, 4\}$, i.e. a BS with two $\pm 45°$ dual polarized UMTS antennas (effectively a total of four BS antennas) and two users, each equipped with four inverted-F antennas. A Medav channel sounder was

Table 7.1 Comparative analysis of virtual MIMO and conventional MU MIMO

%	Standing pairs (190 total)			Walking pairs (36 total)		
	Improves both	Improves one	Degrades both	Improves both	Improves one	Degrades both
Capacity (40 dB) ↑	55	30	14	44	44	11
K-factor ↓	4	85	12	8	92	0
Correlation ↓	41	58	1	69	31	0

used at a center frequency of 2 GHz operating over a bandwidth of 20 MHz, which in turn recorded measurements in 128 frequency bins. Each measurement lasted 6 s, which, after some averaging to increase SNR values, allowed for a total of 1024 time snapshots in each bin. Measurements were made in various places in the city of Bristol in the United Kingdom (UK) and furthermore, in any given case, the two user terminals were either moving, i.e. 'walking', or motionless, i.e. 'standing'. 20 standing measurements lead to 190 possible constituent pairs and similarly 9 walking measurements lead to 36 possible constituent pairs.

The measurements can then be analyzed in two ways. Firstly, the BS communicates in the conventional MU MIMO manner, and the two resultant 4×4 arrays that have been formed can be analyzed. Alternatively, a virtual 4×4 array can be formed where both users activate only the two antennas at the extreme end of each other's arrays, which is a scheme that seeks to minimize channel correlations. These three 4×4 arrays are then analyzed comparatively using the matrices of: (i) path loss normalized capacity [YWB09, see Eq. (1)], (ii) the K factor as evaluated using the estimator-based approach in [TAG03] and (iii) the correlations between the constituent SISO channels in the channel matrix [YWB09, see Eq. (3)]. Some of the main findings of the analysis are given in Table 7.1. In this table, arrows indicate what is meant by improvement and the columns entitled, 'Improves one' indicate an improvement of one user and simultaneous degradation of the other.

Clearly the virtual MIMO capacity is better than that of the two constituent systems approximately 50% of the time, however, simultaneous degradation of these systems also occurs, though less so. In contrast, the improvement with respect to one user (simultaneous degradation of the other) occurs quite a significant amount of times particularly in the case of the walking pairs. K-factor improvement occurs mainly to the detriment of one user and the benefit of the other. Rarely are both users' K-factors simultaneously improved. In contrast, whenever the virtual MIMO is implemented, there is a reduction in correlation with respect to both constituent users for a considerable amount of the time. Thus the benefit of virtual MIMO is thought to be more due to correlation reduction rather than K-factor reduction.

7.4 MIMO Detection

7.4.1 An Adaptive Zero Forcing Maximum Likelihood Soft-Input Soft-Output MIMO (AZFML-SfISfO-MIMO) Detector

The ML detection involves an exhaustive search of all possible transmit signal vectors, which in the case of MIMO approach results in the exponential increase of the ML algorithm computational complexity regarding to the number of transmit antennas and the size of the constellation. Furthermore, applying the concept of joint detection considering also time correlation properties of the received signal, either inserted by communication system design in the form of forward error correction codes or introduced by multipath propagation environment, leads the prohibitive computational complexity. A fundamental idea to decrease the complexity of joint detection is to split the complex detection process into a number of concatenated processes and introduce an iterative (turbo) approach by passing soft information from one process to another. Each concatenated process calculates the output soft values based on the received soft information and its internal information. An Adaptive Zero Forcing Maximum Likelihood Soft Input Soft Output MIMO (AZFML-SfISfO-MIMO) detector presented in [JKJ09] is an example of the reduced complexity MIMO detector.

An iterative algorithm for joint detection of the block encoded MIMO signals is implemented in the AZFML-SfISfO-MIMO detector. The detector consists of two key signal processing blocks namely the MIMO soft-input soft-output (SfISfO) detector and the Forward Error Correction (FEC) soft-input soft-output detector. A simplified algorithm based on Chase–Pynidiah [Pyn98] proposal is applied as a FEC SfISfO detector, while a new algorithm for SfISfO-MIMO detection is proposed [JKJ09]. While the classical SfISfO detector is based on ML approach with exhaustive search, the proposed algorithm computational complexity is reduced by estimating soft information from a subset of the most probable symbol vectors, instead of the whole set of symbol vectors. The most probable set of symbol vectors is estimated by applying the Adaptive Zero Forcing Maximum Likelihood (AZFML) algorithm [LHLF00, HG03, VH05], which applies ZF detection as a first guess for the transmitted symbol. The hypersphere within which the ML solution exists and consequently the area of local search is defined by the center of the hypersphere determined as the received symbol vector and hypersphere radius calculated as a Euclidean distance between received symbol vector and ZF initial estimate. The size of the local search area is adapted to the channel conditions. That means that, in a good channel, the local search area is small while, in very bad channel conditions, the local search area is extended to a set containing all symbol vectors, which leads to the ML solution.

The proposed algorithm was tested in a MIMO system either with Hamming block codes or turbo product codes specified in the IEEE 802.16 standard for single carrier broadband wireless access, and with 16-QAM and 64-QAM modulation schemes. The relation between algorithm complexity, i.e. the choice of hypersphere radius, and the system performance is illustrated in Fig. 7.9 for MIMO system with

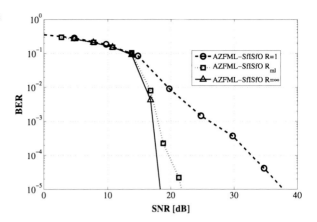

Fig. 7.9 The influence of the choice of the radius R on the system performance of the iterative MIMO system applying an AZFML-SfISfO detector to 16-QAM signal

three transmit and three receive antennas and Hamming (64, 57) encoder and following variation of the iterative MIMO detection: (i) ML-MIMO detector (AZFML-SfISfO R_∞), (ii) ZFML-SfISfO detector with search limited by a polygon which includes sphere with radius $R = 1$ for each transmit antenna (AZFML-SISO $R = 1$) and (iii) AZFML-SfISfO detector with search limited to a polygon around the sphere which includes ML solution, i.e. the radius of sphere is adapted to the channel conditions. If the radius is fixed and equal for each transit antenna, either limited to four neighboring symbols at each transmit antenna or assuming a fixed normalized radius equal to 1, the system property is poor comparing to the ML solution. A significant improvement of system performance is observed if the radius is limited so to include the ML solution. This basically means that the radius is different for each transmit antenna. At high SNR a difference from ML solution is observed, mainly due to inaccurate calculation of the soft information.

These results lead to the conclusion that the algorithm is an excellent candidate to supplement the ML algorithm. The algorithm may be implemented in software or in combing software and hardware implementations.

7.4.2 Receive Subarray Formation in MIMO Systems

Linear antenna combining at the radio frequency can be used to reduce the number of required receiver chains in multiantenna receivers. By combining several antenna signals into one fewer RF chains are required, while keeping the receiver processing complexity low. In addition, the complexity can be further reduced by constraining the linear combiners only perform phase correction of the received signals before combining, instead of also including amplification.

Hardware may impose different constraints on the structure of the linear combining matrix **A**, which left multiplies the MIMO channel matrix. By constraining **A** to have orthogonal columns, every antenna is guaranteed to connect to one combiner only. By constraining the number of nonzero elements in **A**, the number of multipliers can be controlled. The dimensionality of **A** defines the number of inputs and outputs to the combiner. A Frobenius norm-based capacity bound can be used in each of these cases to find a good combining matrix. In [TK07], these different cases are evaluated in Rayleigh fading channels with performance close to exhaustive search solutions. In measured channels, the evaluations reveal that restricting the linear operations to phase shifts brings little loss compared to the more complex methods.

7.4.3 MIMO OFDM BICM Decoding Under Imperfect Channel State Information

Pilot symbols are often used for channel state estimation in practical systems that utilize multiple antennas and multiple carriers for transmission. Since the pilot symbol overhead should in general be minimized, even coherent detectors must in practice use an estimate of the channel state in the detection. The detection algorithm should then be developed to account for channel estimation error instead of simply using the channel estimate as if it was accurate channel state information (mismatched detection).

An improved strategy is to use a different decoding metric that was used in the case of space–time decoding of MIMO channels in [TB05]. The metric of [TB05] can be derived as the average of the likelihood that would be used if the channel is perfectly known, over all realizations of the channel uncertainty which mitigates the impact of channel estimation error on the decoding performance and provides a robust design. The averaging of this metric is performed in the Bayesian framework provided a posteriori pdf of the perfect channel conditioned on its estimate that characterizes the channel estimation process and matches well the channel knowledge available at the receiver. Based on that metric, a decoding rule for Bit Interleaved Coded Modulation (BICM) Multibeam Orthogonal Frequency Division Multiplexing (MB-OFDM) and MIMO-OFDM can be formulated [SPD07].

The performance of the alternate metric and a comparison to the mismatched metric is given in Fig. 7.10.

7.5 MIMO CPM

7.5.1 STC CPM

The Space–Time Coded (Space–Time Coded (STC)) nonlinear modulations, particularly constant envelope CPM class, represent an alternative to linear STC modulation schemes. Multi-channel CPM modulations in MIMO channel have a range of very useful properties and provide a number of very attractive advantages to linear modulations. They are inherently resistant to nonlinear (e.g. C-class amplifier)

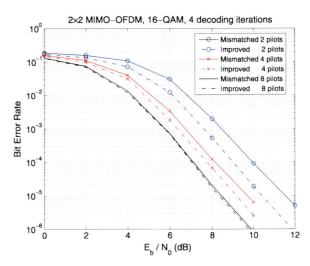

Fig. 7.10 BER performance of the proposed and mismatched decoders in the case of a 2A-2 MIMO-OFDM Rayleigh fading channel with $M = 100$ subcarriers for training sequence length $N \in 2, 4, 8$

distortion at the transmitter. As a consequence of the modulation nonlinearity, the dimensionality of the waveform per transmit antenna is higher than one. This opens more dimensions of the channel and also allows one to decouple a particular waveform shape from its second-order properties. Most of those properties are particular to the transmitter side. On the other hand, the receiver side is believed to require more complicated processing (compared to the linear modulation) due to a higher dimensionality of the waveform. Some performance aspects of MC-CPM, like information capacity, code design for selected channels and decoder performance, were investigated [ZF03a, ZF03b, CL05, Syk01, Syk04].

The ith transmitter of STC CPM produces the signal $s_i(t) = \exp(j\phi(t, \mathbf{q}_i))$, where $\mathbf{q}_i = [\ldots, q_{i,n}, \ldots]^T$ is a vector of STC real-valued discrete finite alphabet channel symbols, and $\phi(t, \mathbf{q}_i) = \sum_n q_{i,n} \beta(t - nT_S)$ is an instant phase of the signal. A phase function $\beta(.)$ is continuous and controls the trajectory of the phase. It determines a particular type of CPM. Its first derivative is called a frequency pulse $\mu(t) = \partial \beta(t)/\partial t$. Popular choices of the frequency pulse are rectangular (Continuous Phase Frequency Shift Keying (CPFSK)), raised cosine or convolution of rectangular and Gaussian (Gaussian Minimum Shift Keying (GMSK)). The modulation index (a multiplicative scaling constant at instant phase) is inherently contained in a proper scaling of channel symbols and phase function.

7.5.2 Multiplexing Properties of Two-Component Information Waveform Manifold Phase Discriminator

A completely new approach for receiver processing using specific waveform properties of multi-channel CPM was proposed in [Syk07b, SS08, Syk05a, Syk05b]. It builds on a fact that multi-channel CPM useful signal spans a curved space (the manifold) when passed through the MIMO channel. Particularly it has a form of a cylinder on cylinder for two-component CPM signal. It is called Information Waveform Manifold. At the receiver, we use a three-step phase discriminator: (1) nonlinear preprocessor consisting of nonlinear projector on the manifold, (2) optional processing on curved space and (3) decoding isomorphism. The advantage of this approach is two-fold. First, it reduces the dimensionality of the processing. It does not depend on the particular CPM modulation type, and it is typically much lower than the dimensionality of the traditional Euclidean signal space expansion. Second, the useful signal lies on known curved space. This opens a wide range of nonlinear processing possibilities not possible on traditional Euclidean signal space. One of the most attractive possibilities is the direct separability of the CPM components not requiring temporal processing.

The nonlinear projection operator $z(t) = T[x(t)]$ on received signal $x(t) = Xe^{j\psi}$ is the actual operation that reduces the dimensionality of the problem. A particular form of sample space manifold projector for two-component CPM signal can be based on constrained ML criterion $z = \arg\max_{\tilde{a}\in\psi} p(x|a)$ where $\Psi = \{A_1 e^{j\alpha_1} + A_2 e^{j\alpha_2}\}_{\alpha_1,\alpha_2}$ is the useful received signal. Received CPM components at one antenna have instant phases α_i and amplitudes A_i. The solution (with negligible information loss [Syk05a]) has a form of parametric limiter $z = \chi(X)e^{j\psi}$ with AM/AM conversion $Z = \chi(X) = \{A_1 + A_2, X > A_1 + A_2; |A_1 - A_2|, X < |A_1 - A_2|; X,$ elsewhere$\}$. The phase discriminator has generally a two-fold ambiguity. One ambiguity-free region is $\alpha_1 < \alpha_2 < \alpha_1 + \pi$, $\alpha_1 \in [0, 2\pi)$. The discriminator solution is then

$$\hat{\alpha}_1 = \psi - \arccos((A_1^2 - A_2^2 + Z^2)/(2A_1 Z)), \quad (7.12)$$

$$\hat{\alpha}_2 = \psi + \arccos((A_2^2 - A_1^2 + Z^2)/(2A_2 Z)). \quad (7.13)$$

This manifold-based phase discriminator has multiplexing (component phase separation) capabilities. The information carrying phases $\phi_i(t, \mathbf{q}_i)$ are separated at receiver (at the level of composite phases $\alpha_i = \phi_i + \eta_i$), thus allowing information multiplexing of the individual data $\mathbf{q}_i$ streams. It is very important to stress that the whole multiplexing processing of the discriminator is purely done in value domain without any time-domain processing. Therefore it is easily applicable with arbitrary outer space–time coding performed on phases ϕ_i.

The multiplexing performance is evaluated by modified capacity region. The bounding conditions for these rates are given by the capacity region

$$R_1 < C_1 = I(\alpha_1; \hat{\alpha}_1|\alpha_2), \quad (7.14)$$

$$R_2 < C_2 = I(\alpha_2; \hat{\alpha}_2|\alpha_1), \quad (7.15)$$

$$R_1 + R_2 < C_{12} = I(\alpha; \hat{\alpha}), \quad (7.16)$$

where $\alpha = [\alpha_1, \alpha_2]^T$ and $\hat{\alpha} = [\hat{\alpha}_1, \hat{\alpha}_2]^T$. This capacity region is somewhat similar to that of classical multiple access channel [CT91]. It has however one subtle but important difference. The conditional mutual information in first-order bounds has the output variable corresponding only to one of the output, unlike as it is in multiple access case where both of them are considered. We define the multiplexing separation ratio as $\kappa = C_{12}/(C_1 + C_2)$. Clearly, any value $\kappa < 1$ causes the capacity region

Fig. 7.11 Capacity region. (**a**) $A_1 = 1$, $A_2 = 1$ *solid line*, (**b**) $A_1 = 3$, $A_2 = 1$ *dashed line*, (**c**) $A_1 = 10$, $A_2 = 1$ *dash-dotted line*

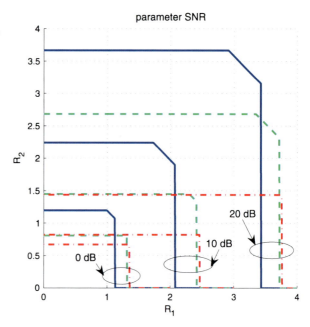

to have upper right corner cut-out by C_{12} condition. This corresponds to imperfect channel separation. Values $\kappa > 1$ indicate separation with nonzero "safety" margin. The coefficient κ describes a quality of channel separation; the lower κ, the worse separation. A numerical evaluation of capacity region and $\tilde{\kappa}$ are shown in Fig. 7.11.

7.5.3 Burst Alamouti STC MSK with Linear Subspace Projector

A reduction of the dimensionality of the STC CPM can be also done by a linear subspace projection. We derive a closed-form [SH08] solution of the optimal linear waveform subspace projector for burst Alamouti space–time coded Minimum Shift Keying (MSK) signal (see also a numerical optimization in [Syk07c]). We utilize a special structure of the code that allows us to get a simple analytical solution for the optimal projection basis. The projection basis is required to be orthogonal and Nyquist in order to facilitate the receiver metric computation (Laurent expansion [Lau86] including its various tilted forms does not satisfy this). In a particular case of MSK, the waveform space basis is $N_s = 2$-dimensional (complex), and the signal space points are $\mathbf{s} \in \mathscr{A}_s = \{\mathbf{s}^{(i)}\}_{i=1}^{8}$. We assume a block-wise burst Alamouti code build from the segments of the CPM modulated signal having a total length K (even).

The signal of each receive antenna passes through a linear dimensionality reducing projector. In a special case of waveform-only projector, it was shown in [Syk05b] and [Syk07c] that the system with the projector can be modeled by equivalent projected transmitted constellation points $\mathbf{u} = \mathbf{P}\mathbf{s}$ where $\mathbf{P}$ is the common waveform space projection $N_s \times N_s$ matrix with rank $N_P < N_s$, where N_P is the dimensionality of the projection subspace. Due to a specific structure of the block burst Alamouti STC-CPM codeword, the Gram matrix of codeword differences $\mathbf{R_u}$ has a diagonal form, and the determinant is $\det(\mathbf{R})_u = (\sum_{n=1}^{K} \|\mathbf{P}\Delta\mathbf{s}_n\|^2)^r$. An optimal projector is defined in terms of the rank and the determinant criterion. Clearly, any nonzero projector applied on arbitrary orthogonal design code preserves the rank. The determinant criterion should now provide the optimal projector $\hat{\mathbf{P}}$ and its associated basis $\hat{\mathbf{B}}$. We constrain the optimization only to the most vulnerable codeword pair with difference $\Delta\mathbf{s}$

$$\hat{\mathbf{P}} = \arg \max_{\mathbf{P}} \min_{\Delta\mathbf{s}} \sum_{n=1}^{K} \|\mathbf{P}\Delta\mathbf{s}_n\|^2. \qquad (7.17)$$

The constraint optimization can be solved by the Lagrange multiplier method. The basis vectors $\mathbf{b}$ are get as a solution of $\sum_{n=1}^{K} \mathbf{Q}_n \mathbf{b} = \lambda \mathbf{b}$ where $\mathbf{Q}_n = \Delta\mathbf{s}_n \Delta\mathbf{s}_n^H$. Since $\alpha_n = \Delta\mathbf{s}_n^H \mathbf{b}$ are scalars, we clearly see that the optimal basis is a linear combination of the vectors $\Delta\mathbf{s}_n$ with weights $\mathbf{b}^H \Delta\mathbf{s}_n / \lambda$, where λ is such that the basis has an unit energy. This means that the basis vector lies in the expansion space of $\{\Delta\mathbf{s}_n\}_n$.

7.5.4 Multi-channel Irregular CPM

Multi-channel Irregular CPM utilizes multidimensional nature of the CPM waveforms (signal space constellation) and allows us to develop a technique for waveform-coded signal construction for MIMO channel [SA09a, SA09b]. The waveform domain coding (we can view this also as a multi-D constellation shaping) is shown to be an approach of designing the waveforms with certain constraints (e.g. constant envelope, continuous signal) and also having desired performance parameters (e.g. the diversity and the coding gain). It could be utilized independently of the spatial and/or temporal domains. We take a simple binary full response irregular two-channel CPFSK and optimize its waveforms for the determinant and full-rank criterion. The result of the optimization leads to a spatially irregular modulation that can be elegantly implemented as Spatially Mirrored Tilted Phase CPFSK. The resulting scheme is shown to have a full diversity rank. It is achieved by the waveform dimensionality expansion within an approximately constant bandwidth w.r.t. the reference MSK scalar modulation.

We allow the signals on individual antennas to have a general modulation index κ, an arbitrary continuous phase function $\beta(.)$ (generally a nonlinear function of data) and a symbol alphabet size M_c. The complex envelope signal on the ith antenna is $s_i(t) = \exp(j 2\pi \kappa_i \sum_n \beta_i(c_{ni}, t - nT_S))$, where T_S is the symbol period, and $c_{ni} = c_{ni}(d_n)$ the coded symbol depending on the data symbol d_n. We restrict ourselves to binary alphabet ($M_c = 2$, $c_{ni} \in \{1, 2\}$) and CPM with linear full-response phase function, i.e. CPFSK. We define a proto-phase function $\beta_0(t)$ which is the phase function of traditional MSK. The irregular phase functions are

$\beta_i(c_{ni}, t) = \{\alpha_{1i}\beta_0(t)$ for $c_{ni} = 1; \alpha_{2i}\beta_0(t)$ for $c_{ni} = 2\}$, where $\alpha_{ki} \in [-1, 1]$ are constants controlling the instant frequency for individual symbols and antennas.

A waveform-code-based design utilizes additional multi-D waveform domain (on top of temporal and spatial) to construct the code. We can create a tailor-made, matched or optimized waveforms suited for the outer $q_{i,n,k}(d_n)$ code. The optimization target for the irregular CPM waveforms is the full-rank and the maximum determinant of the codeword difference Gram matrix of inner products. A semi-numerical optimization gave the resulting optimal values $\boldsymbol{\alpha}_1 = [+1, -1]^T$, $\boldsymbol{\alpha}_2 = [\tilde{\alpha}_1^1, \tilde{\alpha}_2^1]^T$ where $\tilde{\alpha}_1^1 = \hat{\alpha}_1^1 \approx -0.26$, $\tilde{\alpha}_2^1 = \hat{\alpha}_2^1 \approx 0.26$. Observing the results, we can draw an important and rather surprising observation. The optimal values do not correspond to the two-antenna MSK with mutually swapped (i.e. mutually orthogonal) symbols, the resulting waveforms do have (as was conjectured) increased dimensionality $N_w = 4$ and are nicely mirror-symmetrical.

The optimization leads to CPM with phase tree on Tx1 and Tx2 with tilted and mutually mirrored trajectories. This immediately leads to an elegant implementation with Spatially Mirrored Tilted Phase CPFSK

$$s_1(t) = s_0(t)e^{j2\pi \tilde{f} t}, \qquad s_2(t) = s_1^*(t) = s_0^*(t)e^{-j2\pi \tilde{f} t}. \qquad (7.18)$$

The signal $s_0(t) = \exp(j2\pi \kappa_e \sum_n \tilde{c}_n \beta_0(t - nT_S))$ is an ordinary CPFSK signal with effective modulation index κ_e and symbols $\tilde{c}_n \in \{\pm 1\}$. The effective modulation index is $\kappa_e = 1/3$, and the frequency tilt is $\tilde{f} = 1/(12T_S)$.

7.5.5 Factor Graph Sum-Product Algorithm for Limiter Phase Discriminator Receiver

A topic closely related to the MIMO CPM receiver structures is the iterative joint detection and synchronization using Sum-Product algorithm in a Factor Graph. In [Syk07a], we develop the factor node marginalization rules involving canonical distributions with finite number of parameters. The core is in the efficient representation of the densities passed over the edges between Factor Nodes and Variable Nodes. These messages must cope with the circular nature of the modulo 2π operations of the signal phase processing. On the other side, the message update rules must be simple enough. This would cause a problem if we strictly apply all modulo operations on all random variables in the factor graph. This would lead to non-Gaussian densities which are difficult to cope with. We show the solution based on modulo component mean approximation. This keeps the circular nature of the phase modulo operations by applying them on variable component means. The Gaussian tails are however allowed to extend out of the $[0, 2\pi)$ range. It will keep the update rules simple since all the densities keep to be Gaussian ones.

7 MIMO and Next Generation Systems

7.6 Transmit Diversity and Space–Time Block Coding

The use of multiple antennas at the transmitter and at the receiver was a major technical breakthrough in improving the reliability of a digital transmission and enabling higher data rates. Employing two or more transmit antennas together with transmit diversity schemes, such as the Alamouti scheme [Ala98], leads to an increased link reliability in a fading environment. This is especially important when there is no other source of diversity available. For multiple antennas at the transmitter and at the receiver, STBCs on the other hand, do not only improve the link quality, but also lead to an increased data rate. When constructing STBCs, three conflicting goals have to be considered: maximizing the data rate, maximizing the error performance and minimizing the receiver complexity.

In Sect. 7.6.1 three different design methods for STBCs are presented. An improved differential STBC is introduced in Sect. 7.6.2. Finally, Sect. 7.6.3 discusses various system aspects of STBCs.

7.6.1 Design Methods for Space–Time Block Codes

7.6.1.1 Full-Diversity Full-Rate Tensor-based STBCs

A wide range of MIMO transmission schemes, including both spatial multiplexing and STBCs, can be modeled using tensor-based precoding [Bur08a]. This framework is exploited in [LB10] to design STBCs which provide full rate and full diversity as defined in [TSC98]. STBCs based on orthogonal designs [TJC99] provide full diversity for any signaling constellation but exist only for specific numbers of transmit antennas and transmission rates. In contrast to that, the conditions derived for tensor-based STBCs depend on the signaling constellations employed. Results are given for Binary Phase Shift Keying (BPSK), QPSK, and 16QAM.

Thus an STBC can be described in terms of a third-order precoding tensor. A parallel factor decomposition [Har70] is used to obtain conditions on the properties of the component matrices in order to provide full diversity. To accomplish this, the rank criterion is applied. Based on these results, a design procedure for full-rank STBCs is established. For three transmit antennas, the following two transmit matrices can be found:

$$T_{3,1} = \begin{bmatrix} 4s_1 - 2s_3 & 2s_2 - 2s_3 & s_1 + s_2 - s_3 \\ 2s_2 - 2s_3 & 4s_1 - 2s_3 & s_1 - s_2 + 3s_3 \\ s_1 + s_2 - s_3 & s_1 - s_2 + 3s_3 & 3s_1 + s_2 - s_3 \end{bmatrix},$$

$$T_{3,2} = \begin{bmatrix} 3s_1 - s_2 & 4s_1 - 3s_2 & 4s_1 - 3s_2 + s_3 \\ s_1 + s_2 - 2s_3 & -2s_1 + s_2 - 2s_3 & -2s_1 + 3s_2 - 5s_3 \\ 2s_2 - s_3 & -s_1 + 2s_2 - 3s_3 & -3s_1 + 2s_2 - 2s_3 \end{bmatrix},$$

where s_i are the data symbols transmitted in each time slot. Figure 7.12 shows the BER of these two schemes operating with both BPSK and 16QAM. With both modulation schemes the same diversity order is achieved, while 16QAM has a larger offset due to the larger size of the constellation.

Fig. 7.12 BER versus E_b/N_0 for two tensor-based STBCs. A block flat fading channel is assumed

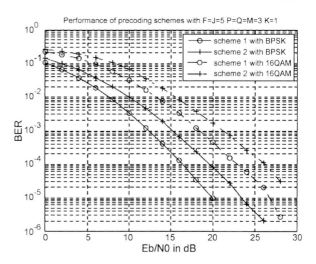

7.6.1.2 Half-Symbol-Decodable Orthogonal STBCs

Another transmit diversity scheme supporting full rate and achieving full diversity is proposed in [ZGN09]. It is based on sending two real orthogonal STBCs with Offset Quadrature Amplitude Modulation (OQAM) on the I- and Q-channels, respectively. By an appropriate joint design of these orthogonal STBCs and the OQAM pulse shape, an excellent coding gain can be achieved. The I- and Q-channel symbols can be decoupled and separately decoded symbol-by-symbol. Due to this half-symbol decoding feature, the decoding complexity is much lower than for conventional QAM. Figure 7.13 shows that the BER for the appropriately chosen orthogonal STBC with offset QPSK with shaping parameter $\beta = 0.25$ shows the same performance as the best-performing quasi-orthogonal STBC with joint decoding of two QPSK symbols suggested by Sharma and Papadias [SP03]. In contrast to conventional orthogonal STBC, the results for the full-rate, full-diversity orthogonal STBC with offset QPSK can be extended to an arbitrary number of transmit antennas. Coding matrices are given for up to eight transmit antennas.

7.6.1.3 Coherent STBCs and Grassmannian Packings

In [PL07], a relationship between orthogonal STBCs [Ala98, TJC99, LS03] and packings on the Grassmann manifold is established, see [EAS98] for a concise introduction on Grassmann manifolds. A general criterion was derived for packings on the Grassmann manifold that yield coherent space–time constellations with the same code properties that make orthogonal STBCs so favorable. The following two theorems summarize the main results.

Theorem 7.1 *Let* $\mathbf{U}^+$ *be* $n_t \times n_t/2$ *and satisfy* $\mathbf{U}^{+H}\mathbf{U}^+ = \mathbf{I}$. *That means* $\mathbf{U}^+$ *represents a subspace or, equivalently, a point on the Grassmann manifold. Further,*

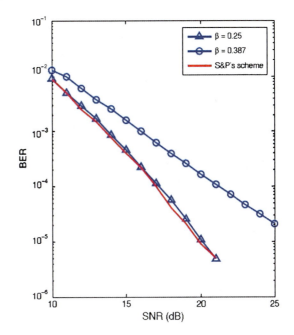

Fig. 7.13 BER performance comparison at 2 bit/s/Hz between the full-rate orthogonal STBC with offset QPSK and two shaping parameters $\beta = 0.25$ and 0.387 and Sharma–Papadias quasi-orthogonal STBC with joint decoding of two QPSK symbols [SP03] for four transmit antennas

let $\{\mathbf{U}_i^+\}$ be a set of points on the Grassmann manifold with all principal angles between any two subspaces from this set being the same, i.e. the singular values of $\mathbf{U}_i^{+H}\mathbf{U}_j^+$ are equal for arbitrary i and j. Then, the set of all

$$\mathbf{X}_{i,k,l} = a_k \mathbf{I} + b_l \begin{pmatrix} \mathbf{U}_i^+ & \mathbf{U}_i^- \end{pmatrix} \begin{pmatrix} j\mathbf{I} & \mathbf{0} \\ \mathbf{0} & -j\mathbf{I} \end{pmatrix} \begin{pmatrix} \mathbf{U}_i^{+H} \\ \mathbf{U}_i^{-H} \end{pmatrix} \quad (7.19)$$

constitutes a space–time constellation for coherent transmissions, where $\mathbf{U}^-$ is the orthogonal complement of $\mathbf{U}^+$, and where a_k and b_l are both chosen from discrete alphabets with values $a_k \in \mathbb{R}$ and $b_l \in \mathbb{R}^+$. Any such constellation inherits the same code properties known from orthogonal STBC, possibly, however, requiring a more complex receiver. The known orthogonal STBCs are special cases of the space–time constellations defined in (7.19).

Theorem 7.2 *A set of dispersion matrices $\{\mathbf{C}_i\}$ that denotes an orthogonal STBC is linked with a packing on the Grassmann manifold $\{\mathbf{U}_i^+\}$, $i \in \{2, \ldots, n\}$, where the principal angles are $\boldsymbol{\Phi} = \pi/4\,\mathbf{I}$ for any pair of distinct subspaces. The connection between the dispersion matrices and the subspaces is*

$$\mathbf{C}_i = \begin{pmatrix} \mathbf{U}_i^+ & \mathbf{U}_i^- \end{pmatrix} \begin{pmatrix} j\mathbf{I} & \mathbf{0} \\ \mathbf{0} & -j\mathbf{I} \end{pmatrix} \begin{pmatrix} \mathbf{U}_i^{+H} \\ \mathbf{U}_i^{-H} \end{pmatrix} = j(2\mathbf{U}_i^+\mathbf{U}_i^{+H} - \mathbf{I}) \quad \forall i \in \{2, \ldots, n\},$$

where $\mathbf{U}_i^-$ is the orthogonal complement of $\mathbf{U}_i^+$. Additionally, $\mathbf{C}_1$ is chosen to be the identity matrix.

Table 7.2 Throughput and diversity order of various differential MIMO systems. N (M) denotes the number of transmit (receive) antennas

Differential SM system	Throughput (symbol per channel use)	Diversity order
Square DOSM	up to $\min\{N, M\}$	M
Rectangular DOSM	$\min\{N, M\}$	M
Differential quasi-orthogonal STBCs [ZJ04]	1	$N \cdot M$
Differential BLAST [SBdL07]	$\min\{N, M\}$	1
Differential SM [JNK+06]	$\min\{N - (N-1)N/(2T), M\}$	up to M
Differential SM [CS06]	G ($G \leq \min\{N, M\}$)	$N \cdot M / G^2$

Previously, Grassmannian packing were predominantly used for non-coherent space–time constellations [ZT02]. Having identified these new properties of coherent STBCs, it establishes a new relationship between coherent and non-coherent space–time constellations. A suitable example is the packing that is derived from the popular rate 3/4 orthogonal STBC [HH02, TJC99], which yields an extension of an already existing non-coherent space–time constellation.

7.6.2 Differential Space–Time Block Codes

Another MIMO transmission scheme which does not require channel knowledge at the receiver is presented in [DBSdL08]. Deng et al. propose a differential Spatial Multiplexing (SM) scheme based on complex orthogonal designs (Differential Orthogonal Spatial Multiplexing (DOSM)). The decision in the receiver is based on two consecutively received code words. Thus, the receiver does not require to know channel fading coefficients, channel power, signal power or noise power to decode the data symbols. This is an advantage when channel estimation becomes impracticable or even impossible as in highly mobile environments or when multiple transmit and receive antennas are deployed. Unlike standard differential STBCs, the throughput is not limited to one data symbol per channel use, but increases linearly with the number of antennas. Two schemes are proposed: square DOSM based on complex square orthogonal designs and rectangular DOSM based on complex rectangular designs. Both schemes achieve full receive diversity. The maximum transmission rate of square DOSM with N transmit antennas is $N(N+1)/2$ symbols per channel use. The rectangular DOSM can achieve full data rate. A constellation rotation is used to enhance the transmission rate of square DOSM. Table 7.2 compares throughput and diversity order of various differential MIMO systems.

Based on a Pairwise Error Probability (PEP) upper bound for the Rayleigh fading channel, a set of design criteria for the constellation rotation angle is found. Simulation results for MIMO systems with $N = M = 2$ and $N = M = 3$ antennas show that both DOSM schemes outperform full-diversity differential STBCs [ZJ04] and differential SM systems [CS06] in quasi-static Rayleigh fading channels. DOSM even has a comparable performance to coherent SM, while at the same time greatly reducing the receiver complexity and bandwidth requirements for channel estimation.

7.6.3 System Aspects for Space–Time Block Codes

Combining transmit diversity techniques with Single Carrier Frequency Division Multiple Access (SC-FDMA) modulation is not straightforward. Using standard STBC or Space-Frequency Block-Code (SFBC) techniques, suppose either restrictions on the frame duration or result in a significantly increased PAPR of the SC-FDMA signal. For a system with two transmit antennas, a solution is given in [CCMS07]. In [CCMS08] the case of four transmit antennas is addressed. A new quasi-orthogonal SFBC is proposed which is compatible with SC-FDMA and keeps the flexibility and robustness of classical quasi-orthogonal SFBCs. Compared to quasi-orthogonal STBCs, there is a small performance degradation which can be tolerated due to the relaxation of the strict framing constraints imposed by quasi-orthogonal STBCs.

Cyclic Delay Diversity (CDD) is a transmit diversity scheme that converts space selective fading into frequency selective fading. It is typically used where other diversity schemes such as STBCs or beamforming are not available. In [JCHP09] it is shown that the combination of CDD on the DL with antenna selection on the UL not only improves the DL performance but also improves the UL performance in a MIMO TDD system. Based on BS received UL SNR distributions, a theory is developed to prove this phenomenon. Thus an antenna selection criterion is derived which maximizes the BS received UL SNR, even so the subscriber station does not have knowledge of the full MIMO channel matrix. Measurement-based evaluations support the theoretical derivations. Larger improvements are seen from UL transmit diversity based on antenna selection in measured channels that have significant frequency selectivity when DL CDD is applied.

Space–time block codes can also be applied in a distributed mode in relaying systems. Dynamic Decode and Forward (DDF) protocols [AGS05] provide a substantial performance improvement, while at the same time limiting the use of the relay. A DDF protocol is composed of two phases. In the first phase the relay and the destination listen to the source. It ends when the relay correctly decodes the code word. This allows the relay to generate symbols for a distributed STBC which are transmitted from the relay in the second phase together with the symbols from the source. The disadvantage of this scheme is that the symbols of the first phase are not protected by the relay-destination link. This limits the diversity order. In [PGO10] a patched distributed STBC for the DDF relaying protocol is proposed. In this patched distributed STBC the relay transmits in the second phase a combination of symbols from the first and the second phase and not only from the second phase. After linear combination of the received symbols from the two phases at the destination, an STBC is built. In this way the diversity of the code word is enhanced, or in other words, the length of phase two to achieve full diversity can be reduced. Three examples of patched distributed STBCs are presented, the patched Alamouti, the patched golden code and the patched silver code. These schemes are compared based on the outage probability. Cooperative systems of this sort are considered in more detail in the next chapter.

References

[AGS05] K. Azarian, H. E. Gamal, and P. Schniter. On the achievable diversity-multiplexing tradeoff in half-duplex cooperative channels. *IEEE Trans. Inform. Theory*, 51(12):4152–4172, 2005.

[Ala98] S. M. Alamouti. A simple transmit diversity technique for wireless communications. *IEEE J. Select. Areas Commun.*, 16:1451–1458, 1998.

[ASDG06] A. Abdel-Samad, T. Davidson, and A. B. Gershman. Robust transmit eigen beamforming based on imperfect channel state information. *IEEE Trans. Signal Processing*, 54(5):1596–1609, 2006.

[BAS+05] K. Brueninghaus, D. Astdlyt, T. Silzert, S. Visuri, A. Alexiou, S. Karger, and G. Seraji. Link performance models for system level simulations of broadband radio access systems. In *Proc. PIMRC 2005—IEEE 16th Int. Symp. on Pers., Indoor and Mobile Radio Commun.*, pages 2306–2311, 2005.

[Bur03] A. Burr. Capacity bounds and estimates for the finite scatterers MIMO wireless channel. *IEEE J. Select. Areas Commun.*, 21(5):812–818, 2003.

[Bur08a] A. G. Burr. Tensor-based linear precoding models for frequency domain MIMO systems. In *1st COST2100 Workshop MIMO and Cooperative Communications*, Trondheim, Norway, June 2008.

[Bur08b] A. G. Burr. Tensor-based linear precoding models for frequency domain MIMO systems. In *Proc. 1st COST2100 Workshop on MIMO and Cooperative Communications*, Trondheim, Norway, June 2008.

[Bur09] A. Burr. Multiplexing gain of multiuser MIMO on the finite scattering channel. Technical Report TD-09-871, Valencia, Spain, May 2009.

[BVV09a] I. Burciu, J. Verdier, and G. Villemaud. Low power multistandard simultaneous reception architecture. In *European Wireless Technology Conference*, September 2009.

[BVV09b] I. Burciu, G. Villemaud, and J. Verdier. Multiband simultaneous reception frontend with adaptive mismatches correction algorithm. In *IEEE Personal Indoor and Mobile Radio Communications Symposium*, September 2009.

[BVVG09] I. Burciu, G. Villemaud, J. Verdier, and M. Gautier. A 802.11g and UMTS simultaneous reception front-end architecture using a double IQ structure. In *IEEE Vehicular Technology Conference*, April 2009.

[CB08] D. Chen and A. Burr. Adaptive stream mapping in MIMO-OFDM with linear precoding. Technical Report TD-08-643, Lille, France, October 2008.

[CBV+08] N. Czink, B. Bandemer, G. Vazquez, A. Paulraj, and L. Jalloul. July 2008 measurement campaign: measurement documentation. Technical Report, 2008.

[CC06] A. Camargo and A. Czylwik. PER prediction for bit-loaded BICM-OFDM with hard-decision Viterbi decoding. In *Proc. 11th Int. OFDM Workshop (InOWo)*, Hamburg, Germany, August 2006.

[CCMS07] C. Ciochina, D. Castelain, D. Mottier, and H. Sari. A novel space-frequency coding scheme for single-carrier modulation. In *Proc. PIMRC—IEEE 18th Int. Symp. on Pers., Indoor and Mobile Radio Commun.*, Athens, Greece, September 2007.

[CCMS08] C. Ciochina, D. Castelain, D. Mottier, and H. Sari. A novel quasi-orthogonal space-frequency block code for single-carrier Frequency Division Multiple Access (FDMA). Technical Report TD-08-468, Wroclaw, Poland, February 2008.

[CKM04]   E. Cetin, I. Kale, and R. C. S. Morling. Adaptive self-calibrating image rejection receiver. In *IEEE International Conference on Communication*, vol. 5, page 2731, June 2004.
[CL05]   C.-C. Cheng and C.-C. Lu. Space–time code design for CPFSK modulation over frequency-nonselective fading channels. *IEEE Trans. Commun.*, 53(9):1477–1489, 2005.
[CO09]   P. Chambers and C. Oestges. Experimental and heuristic examination of a multi-user MIMO precoder. Technical Report TD-09-926, Vienna, Austria, September 2009.
[CO10]   P. Chambers and C. Oestges. Mis-match and variation of leakage-based precoders for multi-user MIMO systems in measured channels. Technical Report TD-12-006, Bologna, Italy, November 2010.
[Col10]   G. Colman. Low rate feedback precoding for robust MIMO communications. Technical Report TD-11-027, Aalborg, Denmark, June 2010.
[CS06]   S.-K. Cheung and R. Shober. Differential spatial multiplexing. *IEEE Trans. Wireless Commun.*, 5(8):2127–2135, 2006.
[CT91]   T. M. Cover and J. A. Thomas. *Elements of Information Theory*. Wiley, New York, 1991.
[DBSdL08] Y. Deng, A. G. Burr, L. Song, and R. C. de Lamare. Differential spatial multiplexing from orthogonal designs. Technical Report TD-08-644, Lille, France, October 2008.
[EAS98]   A. Edelman, T. A Arias, and S. T. Smith. The geometry of algorithms with orthogonality constraints. *SIAM Journal on Matrix Analysis and Applications*, 20(2):303–353, 1998.
[GKH$^+$07] D. Gesbert, M. Kountouris, R. W. Heath, C. B. Chae, and T. Sälzer. From single user to multiuser communications: Shifting the MIMO paradigm. *IEEE Signal Processing Mag.*, 24(5):36–46, 2007.
[GLG10]   V. Garcia, N. Lebedev, and J. M. Gorce. Selection of stations for multi cellular processing for MIMO rayleigh channels. Technical Report TD-12-074, Bologna, Italy, November 2010.
[Har70]   R. A. Harshman. Foundations of the PARAFAC procedure: Model and conditions for an "explanator" multi-mode factor analysis. *UCLA Working Papers in Phonetics*, 16:1–84, 1970.
[HG03]   L. He and H. Ge. Reduced complexity maximum likelihood detection for V-BLAST systems. In *Proceedings of the IEEE Military Communications Conference (MILCOM)*, pages 1386–1391, IEEE, Boston, USA, 2003.
[HH02]   B. Hassibi and B. M. Hochwald. High-rate codes that are linear in space and time. *IEEE Trans. Inform. Theory*, 48(7):1804–1824, 2002.
[HZNW10] C. Huang, J. Zhang, X. Nie, and Z. Wu. Capacity evaluation with limited feedback for amplify-and-forward MIMO relay channel in urban environment. In *Proc. 5th International ICST Conference on Communications and Networking in China, Chinacom 2010*, Beijing, China, August 2010.
[JCHP09] L. M. A. Jalloul, N. Czink, B. M. Hochwald, and A. Paulraj. Why downlink cyclic delay diversity helps uplink transmit diversity. Technical Report TD-09-715, Braunschweig, Germany, February 2009.
[JKJ09]   I. Jelovcan, G. Kandus, and T. Javornik. An adaptive zero forcing maximum likelihood soft input soft output MIMO detector. *IEICE Trans. Commun.*, (2):507–515, 2009. [Also available as TD(07)253 and TD(07)353].
[JNK$^+$06] Y.-H. Jung, S. H. Nam, Y. Kim, J. Chung, and V. Tarokh. Differential spatial multiplexing for two and three transmit antennas. In *Proc. ICC—IEEE Int. Conf. Commun.*, Istanbul, Turkey, June 2006.
[JRV$^+$05] N. Jindal, W. Rhee, S. Vishwanath, S. A. Jafar, and A. Goldsmith. Sum power iterative waterfilling for multi-antenna Gaussian broadcast channels. *IEEE Trans. Inform. Theory*, 51(4):1570–1580, 2005.

[Kai05] T. Kaiser. *Smart Antennas: State of the Art*. Hindawi Publishing Corporation, New York, 2005.

[KGL+10] F. Kaltenberger, R. Ghaffar, I. Latif, R. Knopp, D. Nussbaum, H. Callewaert, and G. Scot. Comparison of LTE transmission modes in rural areas at 800 MHz. In *COST 2100 12th MCM*, Bologna, Italy, 2010. [TD(10)12080].

[KHHP10] V. M. Kolmonen, K. Haneda, T. Hult, and J. Poutanen. Measurement-based evaluation of interlink correlation for indoor multi-user MIMO channels. Technical Report TD-10-10070, Athens, Greece, February 2010.

[KKGK09] F. Kaltenberger, M. Kountouris, D. Gesbert, and R. Knopp. On the trade-off between feedback and capacity in measured MU-MIMO channels. *IEEE Trans. Wireless Commun.*, 8(9):4866–4875, 2009. [Also available as TD-09-407].

[KKP08] A. Kalis, A. Kanatas, and C. Papadias. A novel approach to MIMO transmission using a single RF front end. *IEEE J. Commun.*, 26:972–980, 2008.

[KM08] K. Kansanen and R. R. Müller. Multiuser diversity in channels with limited scatterers. In *Proc. PIMRC 2008—IEEE 19th Int. Symp. on Pers., Indoor and Mobile Radio Commun.*, 2008. [Also available as TD-08-529].

[Lau86] P. A. Laurent. Exact and approximate construction of digital phase modulations by superposition of amplitude modulated pulses (AMP). *IEEE Trans. Commun.*, COM-34(2):150–160, 1986.

[LB10] D. Liu and A. Burr. Full diversity full rate tensor-based space time codes. Technical Report TD-10-10038, Athen, Greece, February 2010.

[LHLF00] X. Li, H. C. Huang, A. Lozano, and G. I. Foschini. Reduced-complexity detection algorithms for systems using multi-element arrays, pages 1072–1076, San Fransisco, USA, November 2000. IEEE.

[LMF10a] B. K. Lau, J. Medbo, and J. Furuskog. Downlink cooperative MIMO in urban macrocell environments. In *IEEE Int. Symp. Antennas Propagat. (APSI 2010)*, Toronto, Canada, July 2010.

[LMF10b] B. K. Lau, J. Medbo, and J. Furuskog. Single-user capacity performance of downlink cooperative MIMO in urban macrocell vehicular routes. Technical Report TD-10-10063, Athens, Greece, February 2010.

[LS03] E. G. Larsson, and P. Stoica. *Space–Time Block Coding for Wireless Communications*, 1st edition. Cambridge University Press, Cambridge, 2003.

[MROZ09] D. Martini, M. Rädeker, and C. Oikonomopoulos-Zachos. Realization of a realtime 4×4 MIMO-OFDM testbed. In *COST 2100 7th MCM*, Braunschweig, Germany, 2009. [TD(09)723].

[MS72] C. B. Moler, and G. W. Stewart. An algorithm for the generalised matrix eigenvalue problems. Technical Report No. 3, The University of Michigan, 1972.

[ORGLMS10] J. Olmos, S. Ruiz, M. García-Lozano, and D. Martín-Sacristán. Link abstraction models based on mutual information for LTE downlink. In *COST 2100 11th MCM*, Aalborg, Denmark, 2010. [TD(10)11052].

[OSR+09a] J. Olmos, A. Serra, S. Ruiz, M. García-Lozano, and D. Gonzalez. Exponential effective sir metric for LTE downlink. In *Proc. PIMRC 2009—IEEE 20th Int. Symp. on Pers., Indoor and Mobile Radio Commun.*, 2009.

[OSR+09b] J. Olmos, A. Serra, S. Ruiz, M. García-Lozano, and D. Gonzalez. Link level simulator for LTE downlink. In *COST 2100 7th MCM*, Braunschweig, Germany, 2009. [TD(09)779].

[Par08] S. Parkval. LTE-Advanced—Evolving LTE towards IMT-Advanced. In *IEEE Vehicular Technology Conference*, September 2008.

[PGO10] M. Plainchault, N. Gresset, and G. R.-B. Othman. Patched distributed space–time block codes. Technical Report TD-10-10027, Athens, Greece, February 2010.

[PH06] M. Pätzold and B. O. Hogstad. Classes of sum-of-sinusoids rayleigh fading channel simulators and their stationary and ergodic properties—part I. *WSEAS Transactions on Mathematics*, 5(2):222–230, 2006.

[PHS05]	C. B. Peel, B. M. Hochwald, and A. L. Swindlehurst. A vector-perturbation technique for near-capacity multi-antenna multi-user communication—part 1. Channel inversion and regularization. *IEEE Trans. Commun.*, 53(1):195–202, 2005.
[PL07]	C. Pietsch and J. Lindner. On orthogonal space–time block codes and packings on the Grassmann manifold. Technical Report TD-07-204, Lisbon, Portugal, February 2007.
[Pyn98]	R. Pyndiah. Near-optimum decoding of product codes: block turbo codes. *IEEE Trans. Commun.*, 46(8):1003–1010, 1998.
[RGJ+07]	M. Riback, S. Grant, G. Jöngren, T. Tynderfeldt, D. Cairns, and T. Fulghum. MIMO-HSPA testbed performance measurements. In *Proc. 18th IEEE Int. Symp. (PIMRC)*, September 2007.
[SA09a]	J. Sykora, and K. Anis. Space-waveform coded CPFSK with full rank optimized waveforms, pages 1–6, Braunschweig, Germany, February 2009. [TD-09-770].
[SA09b]	J. Sykora and K. Anis. Spatially irregular space-waveform coded CPFSK with full rank optimized waveforms, pages 1–5, Tokyo, Japan, September 2009.
[SBdL07]	L.-Y. Song, A. G. Burr, and R. C. d. Lamare. Differential bell-labs layered space–time architectures. In *Proc. ICC—IEEE Int. Conf. Commun.*, Glasgow, Scotland, June 2007.
[SH08]	J. Sykora and M. Hekrdla. Determinant maximizing and rank preserving waveform subspace linear projector for burst Alamouti STC MSK modulation, pages 1–7, Lille, France, October 2008. [TD-08-655].
[SP03]	N. Sharman, and C. B. Papadias. Improved quasi-orthogonal codes through constellation rotation. *IEEE Trans. Wireless Commun.*, 51(3):332–335, 2003.
[SPD07]	S. Sadough, P. Piantanida, and P. Duhamel. MIMO-OFDM optimal decoding and achievable information rates under imperfect channel estimation. In *Proc. SPAWC.—Sig. Proc. Advances in Wireless Commun.*, pages 1–5, Helsinki, Finland, June 2007. [Also available as TD(07)035].
[SS08]	J. Sykora, and R. Schober. Multiplexing precoding scheme for STC-CPM with parametric phase discriminator IWM receiver, pages 1–6, Cannes, France, September 2008.
[Ste75]	G. W. Stewart. Gershgorin theory for the generalised eigenvalue problem $\mathbf{Ax} = \lambda \mathbf{Bx}$. *Mathematics of Computation*, 29(130):600–606, 1975.
[STS07a]	M. Sadek, A. Tarighat, and A. H. Sayed. Active antenna selection in multi-user MIMO communications. *IEEE Trans. Signal Processing*, 55(4):1498–1510, 2007.
[STS07b]	M. Sadek, A. Tarighat, and A. H. Sayed. A leakage-based precoding scheme for downlink multi-user MIMO channels. *IEEE Trans. Wireless Commun.*, 6(5):1711–1721, 2007.
[Syk01]	J. Sykora. Constant envelope space–time modulation trellis code design for Rayleigh flat fading channel. In *Proc. IEEE Global Telecommun. Conf. (GlobeCom)*, pages 1113–1117, San Antonio, TX, USA, November 2001.
[Syk04]	J. Sykora. Symmetric capacity of nonlinearly modulated finite alphabet signals in MIMO random channel with waveform and memory constraints. In *Proc. IEEE Global Telecommun. Conf. (GlobeCom)*, pages 1–6, Dallas, USA, December 2004.
[Syk05a]	J. Sykora. Information waveform manifold based preprocessing for nonlinear multichannel modulation in MIMO channel. In *Proc. IEEE Global Telecommun. Conf. (GlobeCom)*, pages 1–6, St. Louis, USA, November 2005.
[Syk05b]	J. Sykora. Linear subspace projection and information waveform manifold based preprocessing for nonlinear multichannel modulation in MIMO channel, pages 1–6, Cape Town, South Africa, May 2005. Invited paper.
[Syk07a]	J. Sykora. Factor graph sum-product algorithm update rules in sampled phase space of limiter phase discriminator receiver, pages 1–7, Duisburg, Germany, September 2007. [TD-07-362].

[Syk07b]   J. Sykora. Multiplexing properties of 2-component information waveform manifold phase discriminator for multichannel CPM modulation in MIMO system, pages 1–6, Duisburg, Germany, September 2007. [TD-07-361].

[Syk07c]   J. Sykora. Receiver constellation waveform subspace preprocessing for burst Alamouti block STC CPM modulation. In *Proc. IEEE Wireless Commun. Network. Conf. (WCNC)*, pages 1–5, Hong Kong, March 2007.

[TAG03]   A. Tepedelenlioğlu, A. Abdi, and G. B. Giannakis. The Ricean k factor: estimation and performance analysis. *IEEE Trans. Wireless Commun.*, 4, July 2003.

[TB05]   G. Taricco and E. Biglieri. Space–time decoding with imperfect channel estimation. *IEEE Trans. Wireless Commun.*, 4(4):1874–1888, 2005.

[TJC99]   V. Tarokh, H. Jafarkhani, and A. R. Calderbank. Space–time block codes from orthogonal designs. *IEEE Trans. Inform. Theory*, 45:1456–1467, 1999.

[TK07]   P. Theofilakos and A. G. Kanatas. Capacity performance of adaptive receive antenna subarray formation of MIMO systems. Technical Report TD-07-038, Lisbon, Portugal, February 2007.

[TSC98]   V. Tarokh, N. Seshadri, and A. R. Calderbank. Space–time codes for high data rate wireless communication: performance criteria and code construction. *IEEE Trans. Inform. Theory*, 44:744–765, 1998.

[TSS05]   A. Tarighat, M. Sadek, and A. Sayed. A multi-user beam-forming scheme for downlink MIMO channels based on maximizing signal-to-leakage ratios. In *IEEE Trans. Acoust., Speech, Signal Processing*, vol. 3, March 2005.

[VH05]   H. Vikalo and B. Hassibi. On the sphere-decoding algorithm I. expected complexity. *IEEE Trans. Signal Processing*, 53(8):2806–2818, 2005.

[vNP00]   R. van Nee and R. Prasad, editors. *OFDM for Wireless Multimedia Communications*. Artech House, Norwood, 2000.

[VTL02]   P. Viswanath, D. Tse, and R. Laroia. Opportunistic beamforming using dumb antennas. *IEEE Trans. Inform. Theory*, 48(6), 1277–1294, 2002.

[WB08]   M. Webb, and M. Beach. Sensitivity of closed-loop MIMO to imperfect feedback. Technical Report TD-08-537, Trondheim, Norway, June 2008.

[WC07]   T. J. Willink, and G. W. K. Colman. Diversity-multiplexing trade-off for mobile urban environments. *Electronics Letters*, 43(4):231–232, 2007.

[YBTL08]   D. Y. Yacoub, B. Baumgartner, W. G. Teich, and J. Lindner. Precoding and spreading for coded MIMO-OFDM. In *Proc. International ITG Workshop on Smart Antennas*, February 2008.

[YCC08]   D. Yao, A. Camargo, and A. Czylwik. Adaptive MIMO transmission scheme for spatially correlated broadband BICM-OFDM systems. In *Proc. IEEE 68th Vehicular Technology Conference*, September 2008.

[YTL05]   D. Y. Yacoub, W. G. Teich, and J. Lindner. MC-cyclic antenna frequency spread: a novel space-frequency spreading for MIMO-OFDM. In *8th International Symposium on Comm. Theory and Applications (ISCTA 05)*, Ambleside, Lake District, UK, September 2005.

[YWB09]   M. Yu, M. Webb, and M. Beach. Statistics, metrics, propagation characteristics for virtual MIMO performance in a measured outdoor cell. Technical Report TD-09-960, Vienna, Austria, September 2009.

[ZF03a]   X. Zhang, and M. P. Fitz. Space–time code design with continuous phase modulation. *IEEE J. Select. Areas Commun.*, 21(5):783–792, 2003.

[ZF03b]   X. Zhang, and M. P. Fitz. Symmetric information rate for continuous phase channel and BLAST architecture with CPM MIMO system. In *Proc. IEEE Internat. Conf. on Commun. (ICC)*, 2003.

[ZGN09]   K. Zhong, Y. L. Guan, and B. C. Ng. Half-symbol-decodable OSTBC system achieving full rate full diversity with offset QAM and pulse shaping. Technical Report TD-09-906, Vienna, Austria, September 2009.

[ZJ04]   Y. Zhu and H. Jafarkhani. Differential modulation based on quasi-orthogonal codes. *IEEE Trans. Wireless Commun.*, 5(1):531–536, 2004.

[ZT02]　　　L. Zheng and D. N. C. Tse. Communication on the Grassmann manifold: a geometric approach to the noncoherent multiple-antenna channel. *IEEE Trans. Inform. Theory*, 48(2):359–383, 2002.

[ZT03]　　　L. Zheng and D. N. C. Tse. Diversity and multiplexing: a fundamental tradeoff in multiple-antenna channels. *IEEE Trans. Inform. Theory*, 49(5):1073–1096, 2003.

Chapter 8
Cooperative and Distributed Systems

Chapter Editor Jan Sykora, Vasile Bota, and Tomaz Javornik

Cooperative and distributed communication systems represent an import shift of the classical historic role of the physical layer functionality. Traditionally, the physical layer was viewed in a point-to-point communication perspective. All communication tasks related to the information transfer in a more complicated multiple-source and multiple-node network were delegated to upper layers. Cooperative and distributed systems can be viewed as the *physical layer coding and processing algorithms aware and actively utilizing the knowledge of the network structure/topology*. These algorithms do not depend or rely on the orthogonal spectrum sharing technique (e.g., CDMA, OFDMA). Even the simplest scenario (e.g., the bidirectional relay communication) shows a substantial throughput improvement.

The chapter is organized into three sections according to the particular system scenario. The first focuses on the single-source and multiple-node or relay scenario. The second focuses on scenario with multiple sources and multiple interim nodes or relays. The last one concentrates mainly on the distributed signal processing and coding.

8.1 Virtual MIMO and Cooperative Diversity and Relaying

The research activities performed in this area were focused mainly in two major directions:

- the study of performances provided by cooperative algorithms, based either on virtual MIMO or on cooperative relaying, which proposed various such algorithms and analyzed the improvements brought by them, compared to noncooperative transmissions, in different environments, and studied some aspects regarding their implementation;

J. Sykora (✉)
Czech Technical University in Prague, Prague, Czech Republic

- the study of relay-assignment algorithms that dealt with algorithms used to build up the cooperation clusters, the activation/deactivation of cooperation, and with the impact of a given assignment upon the performance of a cooperation-enhanced environment.

Therefore, this section is subdivided into two subsections dedicated to each of the two directions mentioned above.

8.1.1 Virtual MIMO and Cooperative Algorithms

Some potential models of the relaying and mesh topologies, strategies applicable to such topologies, and their potential advantages are reviewed in [Bur07]. The use of relays is expected to extend the range of cells and/or to increase their throughput at relatively low cost, while mesh networks could improve capacity and/or diversity using multiple routes through the network. Cooperative diversity or cooperative relaying can similarly improve capacity and/or diversity by exploiting MIMO principles. The models considered for the mesh networks are network coding and the wireless relay channel.

Network coding is regarded as a mean that allows multi-hop transmission through a mesh network, and also information flows to be shared between multiple routes simultaneously, which is more than a coding technique to be applied at intermediate nodes.

The wireless relay channel is designed specifically to model the simultaneous interaction of multiple wireless communication nodes. It operates on a peer-to-peer level and is intended for the case where the relay nodes, other than the source and destination node, cooperate with all other nodes to improve their communication. Within this model, all nodes access a shared channel representing the common wireless medium, and all transmissions from each node are received by every other node, albeit at different power levels.

The vector of transmitted and received signals of such a model is related by the channel square matrix $\mathbf{H}$ (whose elements are referred to as the gains between each pair of nodes). Since a wireless node cannot transmit and receive at the same time, a submatrix of $\mathbf{H}$, relating a signal vector over the receiving nodes to that over the transmitting nodes, should be invoked for any given transmission.

The paper then briefly summarizes the operational principles of two main *relaying strategies* in such scenarios, namely the *decode-and-forward* and *amplify-and-forward*, with its variant *compress-and-forward*. It also presents the main requirements imposed to the channels by the relaying strategies.

The *amplify-and-forward* strategy, as a particular case of the *compress-and-forward* strategy, is considered to be the strategy used in the proposed *virtual antenna array* concept. The main conclusion is that even if the relay may not be able to decode the message without error, it still may provide useful information which could improve the decoding at destination.

The special case where there is no direct source-destination link is also considered. Here the author introduces an extension of the wireless relay channel, called the *wireless relay network*, which divides the relays into groups (supernodes) and supports multiple "hops" and proposes a cooperation scheme that does not require communications within a group. This approach requires at most $2h - 1$ phases, for an h-hop connection, plus a further phase for each of the ending nodes.

The paper then discusses the issue of capacity upper bound of the proposed cooperation scheme by using the cut-set bound based on Multiple-Input Multiple-Output (MIMO) principles. The proposed approach determines the capacity upper bound of a route as the minimum MIMO mutual information of any cut on that route. Since the source-destination connection could be accomplished by mare routes in the network, the overall capacity is that of the route with maximum capacity. This upper bound is shown to be unachievable since it assumes error-free links between all the other nodes of the partitions involved. This approach could then be used to upper bound the capacity of the wireless relay network by using the corresponding to the MIMO links between the supernodes.

The question of deriving tight bounds for decode-and-forward relay networks using specific codes is addressed in [VK08] by applying the combining technique between the mutual information of each link, over nonfrequency selective fast fading channels and binary input, from the source via the relays to the destination. It is shown that the signals forwarded by the relays can still improve the throughput of the cooperative system, even if they are not capable of decoding the whole received sequence correctly, especially in the lower Signal-to-Noise Ratio (SNR) region, by finding and applying suited coding schemes.

The transmission between a source s and a destination d via one relay r is split into a serial concatenation of two independent links that are defined by their mutual information I_{sr} and I_{rd}. A lower bound on the entire mutual information is obtained by concatenating two Binary Erasure Channel (BEC) leading to $I_{serlow}(I_{sr}, I_{rd}) = I_{sr} \cdot I_{rd}$, which is then generalized to N-link transmissions. An upper bound of such a transmission is obtained by the serial concatenation of N Binary Symmetric Channel (BSC). At the destination, the signals carrying the same message received from the relays and the source are combined and processed by a common decoder, which is equivalent to a parallel concatenation of N independent $s - r - d$ links. This way the entire mutual information at the destination's decoder input is lower bounded by assuming N BSC and upper bounded by concatenating N BEC.

The lower and upper bounds are analyzed, across Additive White Gaussian Noise (AWGN) and Rayleigh channels, both for hard and soft decision at the relays. The transfer function of the relays is determined numerically. The results presented in the paper for the two types of channels, see Fig. 8.1, illustrate that the information combining bounds tightly predict the true behavior even for specific codes and decoders. Moreover, it shows that the results obtained using the min-cut max-flow theorem are rather pessimistic.

The increase of the spectral efficiency in cooperative transmissions that use relaying is discussed in [SA09]. The cooperation technique employed is the signal superposition, i.e., the transmission of local and relayed data by the relay node us-

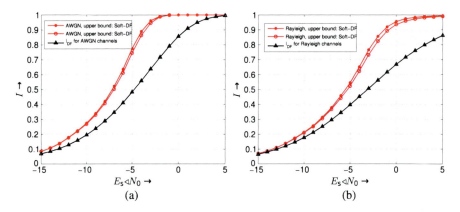

Fig. 8.1 (**a**) Lower and upper bounds of mutual information for a simple relay network with nonrecursive convolutional code $R_c = 1/2$ and constraint length $L_c = 3$ for the AWGN channel. (**b**) Lower and upper bounds of mutual information for a simple relay network with nonrecursive convolutional code $R_c = 1/2$ and constraint length $L_c = 3$ for the Rayleigh channel

ing an encoding scheme in which it transmits the Euclidean superposition of the local bits and the relayed bits.

A new decoding approach at destination which relies on two iterations is proposed: in the first one, each transmitted information is evaluated from two consecutive received sequences (first as local and then as relayed information), while in the second, decoding is performed assuming that the previous and next information sequences were evaluated correctly. The paper defines the metric employed in decoding and shows that by increasing the number of iterations performed by the decoder only small additional coding gain is obtained.

The effect of the percentage of power that is allocated to the relayed data upon the overall Bit Error Rate (BER) at destination is analyzed, and the results show that the proposed method provides smaller BER, or a coding gain of about 1.5 dB for the particular Forward Error Correction (FEC) code used in simulations, for a greater percentage of power dedicated to the relayed data, compared to the reference approach.

The spectral efficiency provided by the repetition coding algorithm within the relay channel, where the two-phase cooperation is distributed in time, is analyzed in [VPB09]. In order to decrease the additional resources required by the relaying phase, the paper considers the use of different Quadrature Amplitude Modulation (QAM) modulations in the two cooperation phases while still ensuring a BER that is below the target value BER_t after the joint decoding at destination.

For flexible resource allocation, the spectral efficiency is expressed as the average number of payload bits correctly decoded on a QAM symbol, n_{usQ}, which is depending on the numbers of bits/QAM symbol n_{BM} and n_{RM} used in the two co-operation phases, the coding rate R_c and the global PER_g provided. The spectral efficiency of the cooperative algorithm is compared to the one of the direct link, which employs n_d bits/QAM symbol and coding rate R_d and ensures PER_d, see

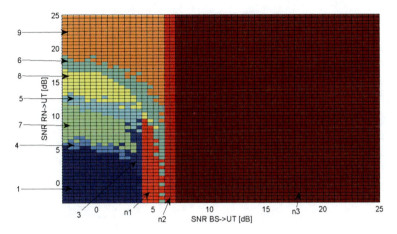

Fig. 8.2 Configurations with the greatest n_{usQ} vs. $\text{SNR}_{\text{BS-UT}}$ and $\text{SNR}_{\text{RN-UT}}$ with $\text{SNR}_{\text{BS-RN4}} \in (7; 10]$ dB

(8.1); this comparison might be employed to decide whether the cooperative transmission (left-hand side) should be employed instead of the direct link (right-hand side):

$$\frac{n_{BM} \cdot R_c}{1 + n_{BM}/n_{RM}}(1 - \text{PER}_g) > n_d \cdot R_d(1 - \text{PER}_d). \quad (8.1)$$

Since the BER provided by turbocodes cannot be derived analytically, the Packet Error Rate (PER) provided by the repetition coding cooperative algorithm and by the direct transmission, both employing the same set of 4-, 16-, and 64-QAM modulations, and the same set of convolutional codes ($R = 1/2, 2/3, 3/4$) was established by computer simulations in terms of the SNR of the Base Station (BS)-User Terminal (UT), BS-Relay Node (RN), and RN-UT channels. Then the configuration (modulation, code, and cooperation or not) that provides the highest spectral efficiencies and a global $\text{BER} < \text{BER}_t = 10^{-6}$ was selected for each tridimensional SNR domain and stored as a set of bidimensional tables, i.e., n_{usQ} vs. $\text{SNR}_{\text{BS-UT}}$ and $\text{SNR}_{\text{RN-UT}}$, as shown in Fig. 8.2.

Each table corresponds to a domain of the SNR on BS-UT channel. The value of $n_{RM}, = 0$ or $\neq 0$, can be used for an adaptive employment of cooperative and noncooperative transmissions, besides the adaptive use of modulation and coding in terms of the effective SNR values of the component channels.

The outage probability and the energy and spectral efficiencies of cooperative transmissions with dual hop relaying and diversity combining at the receiver in log-normal correlated fading channels are analyzed in [SSPK10], considering two access schemes of cooperating devices, namely a Time-Division Multiple Access (TDMA) or Frequency Division Multiple Access (FDMA) one (scheme A) and an Space Division Multiple Access (SDMA) scheme that requires smart antennas (scheme B).

The paper gives integral expressions of the outage probability provided by the adaptive employment of the cooperative or direct transmissions vs. a normalized

Fig. 8.3 Outage probability versus rate-normalized threshold of the cooperative diversity system using MRC and SC under both multiple access protocols, compared to the reference direct link channel

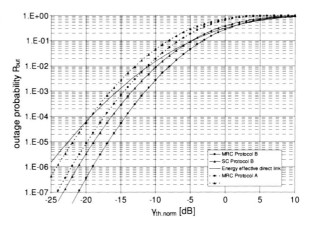

SNR for two combining methods used at receiver, i.e., Selection Combining (SC) and Maximum Ratio Combining (MRC) and for the two access schemes. The simulation results, see Fig. 8.3, show that for access scheme B, the cooperative transmissions, using MRC or SC at the receiver, outperform the direct transmission in terms of outage probability. But for access scheme B, the cooperative-MRC transmission should be used adaptively with the direct transmission, while the cooperative-SC provides greater outage probabilities in almost all SNR ranges considered.

The paper also analyzes the effects of the source-destination distance, also taking into account the correlation between the log-normal channels involved, using the gain of the rate-normalized SNR threshold defined for a reference outage probability $G_{\gamma\text{th}}$ as a metric. This metric is also employed for the indirect evaluation of the spectral efficiency. The energy efficiency of the cooperative systems discussed is evaluated by the "energy-effective direct link" reference-system, which measures the cooperation efficiency in terms of total transmission energy.

Finally, the paper analyzes the effects of the correlation factors between the channels included in the cooperative transmission upon $G_{\gamma\text{th}}$ having the variance of the log-normal fading as parameter. The results of the analysis show, in the case of SDMA scheme, that the cooperative approach employing MRC outperforms the one with SC and the direct transmission, that the performances decrease linearly with the increase of the correlation factors between the constituent channels, and that performances of all transmission schemes decrease with the increase of the log-normal fading's variance, as shown in Fig. 8.4.

The potential capacity benefits brought by using indoor relays to the outdoor-to-indoor transmissions are analyzed in [OCBP09], using experimental data recorded on stationary channels at 2.45 GHz. By contrast to the referenced papers, the authors also address the issues of *power normalization* when estimating the capacity gain provided by relaying, analyze the impact of the *relay transmit power*, and investigate the potential benefits of using *multiple relays*, cooperating in transmission mode and in reception mode.

The cooperation schemes considered are the *Broadcast/Multiple Access (BMA)*, where the source (BS) transmits in both cooperation phases, the *Broadcast/Point-to-*

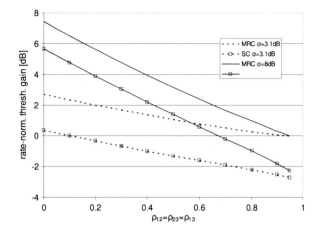

Fig. 8.4 Impact of (equal) correlation on the SNR threshold gain of MRC and SC for various log-normal variances

Point *(BP)*, where the source transmits only in the broadcast phase, the *Strict Two-Hop (PP)* scheme, where no direct link is available, and the *No relay* one, which is equivalent to the direct source-destination transmission.

The *power normalization* is analyzed for two different constraints: *Network Power Constraint (NPC)*, i.e., by setting the overall available power irrespective of the number of BS and relays, and *BS Power Constraint (BSPC)*, where only the transmit power at the BS is constrained to P, and each relay brings its additional power αP to the system. The capacities of the cooperative schemes are evaluated by computing the bit rates achievable, assuming Gaussian random codebooks and using the channels' parameters obtained by measurements.

The comparison between the cooperation schemes, performed under the NPC constraint, shows that BMA does not always provide higher rates than BP, meaning that the additional signal transmitted by the source during relaying might be too "weak" to increase the overall performance. The comparison performed under the BSPC constraint shows that the PP scheme, involving *multiple relays*, is preferable over both BMA and BP, due to the additional power brought in by relays.

The authors also conclude that relay cooperation in reception mode increases the attainable rates significantly when the transmissions from the BS to some relays are much weaker than the ones to the other relays.

The study performed in the previous paper is extended in [SBM+10] in a scenario that considers the same real-channel measurements and the same decode-and-forward relaying schemes, i.e., BMA, BP, PP, and No relay, for reference. The paper studies the maximization of the transmission rates by optimizing the durations of the two cooperation-phases and evaluates the spectral efficiencies provided by means of their ergodic transmission and outage rates. Moreover, the paper analyzes the impact of instantaneous versus average channel knowledge at the Mobile Station (MS) and the relays upon the transmission rates ensured.

The authors derive the optimum values of the durations of the two cooperation phases as an optimization problem which maximizes the capacity (transmission rate) of the cooperative transmission for each of the algorithms. But, because the capacity is strongly depending on the channel knowledge at the transmitting nodes, the authors analyze the impact of three types of channel knowledge upon the optimum durations of cooperation phases. The durations are computed for perfect channel knowledge (inst), where the channels coefficients are updated at every frame, average channel knowledge (avg), which assumes stationary channels and no channel knowledge (50), for which the durations of the two phases are set to be equal. The results presented indicate that optimizing the phase duration, based on any kind of channel knowledge, brings significant increases of the spectral efficiency provided by each algorithm. They also show that the *avg* channel knowledge performs relatively close to *inst* channel knowledge, in spite of the fact that the computation of *avg* is based on an upper bound instead of the true expected rate. In terms of outage probability, the use of *avg* channel knowledge for optimization is slightly less robust than the use of *inst* channel knowledge.

Finally, regarding the issue of relay-assignment, the paper concludes that the best performance was obtained with the relays that had the best channels from the MS, assuming that the relays-BS channels have good qualities.

The more general issue of the tradeoff between diversity gain and relaying gain in Amplify and Forward (AF) or Decode and Forward (DF) relay-assisted transmissions is analyzed in [HRKW08] based on outdoor-to-indoor channel measurements. In order to analyze jointly and independently the impact of signal levels, such as receiver's SNR and transmitted power, the benefits in channel capacity and BER that could be brought by such schemes, the paper considers two factors, the relay gain and diversity order, and presents preliminary results on the analysis of the tradeoff between relay gain and diversity order improvement. The main aspects targeted by the paper are: (1) difference of the tradeoff characteristics between AF and DF relaying; (2) variation of the tradeoff characteristics in different propagation conditions; and (3) identification of Rx locations where the relay transmission improves channel capacity relative to the direct transmission.

After defining the relay gain and diversity order improvements, the authors study the case of equal transmitted powers at source and relay nodes. The results revealed a clear difference of tradeoff between AF and DF relaying and showed that the DF cooperation scheme provides higher channel capacity than AF, due to higher gain improvement, and that relaying provides greater channel capacity than direct transmission for normalized mean Tx power below a certain threshold. The effect of propagation environment on the tradeoff was also investigated, and a criterion of whether introduction of relaying would be effective over the direct transmission was derived.

Then, the paper discusses the effect of power adaptation as a mean to reach the tradeoff between relay gain and diversity order and shows that, as opposed to MIMO systems, cooperative diversity benefits of the relay gain as an additional degree of freedom in optimizing the end-to-end system performance. The maximum relay gain is accomplished when the power of source-relay and combined source-destination and relay-destination links is close with each other, but the diversity order is not improved. As the power source increases further, the diversity order begins to improve while the relay gain decreases. The maximum improvement of the diversity order is achieved when the power of the source-destination and relay-destination links is close with each other. Finally, if all power is allocated to the

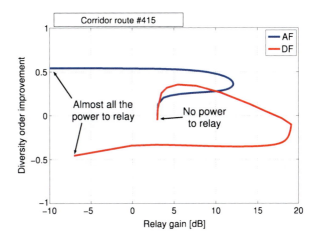

Fig. 8.5 Gain-diversity tradeoff of AF and DF schemes due to the power adaptation

source, the tradeoff curve converges to a point with 3-dB relay gain and no diversity order improvement, as shown in Fig. 8.5 for one particular case.

The authors conclude that the AF can achieve better diversity order than DF, while DF produces larger relay gain than AF, but the effect of relay gain upon the end-to-end throughput is more significant. Therefore, obtaining the maximum relay gain is the primary interest of practical systems, and most likely, the operation range of practical relay systems over the tradeoff curve is between the point achieving the maximum relay gain and that with the maximum diversity order improvement. Moreover, the effect of the relay gain upon capacity is depending on the received SNR. Even small relay gains would increase capacity for a low SNR direct transmission, while for high SNR of the direct link, the relaying approach would not bring any benefits.

The more efficient use of the available frequency band is the topic dealt with in [BTHP09], which proposes a cooperative DF scheme that allows a secondary (unlicensed) user to transmit together with primary (licensed) user with no mutual interference, by using a cognitive radio approach. The cooperative approach proposed is combined with a spectrum sharing scheme, which uses multiple antennas and block space–time coding at the secondary transmitter to cancel out the mutual interferences between the two transmission systems involved. The paper shows that the likelihood of secondary spectrum for the secondary user increases with an increase of its transmit power; thus it exhibits the property to trade-off transmit power with spectrum access opportunity.

The performance of primary and secondary systems were quantified in terms of outage probability, and the authors demonstrate that the primary user can improve its Quality-of-Service (QoS) (target rate) by taking assistance from the secondary user, while simultaneously providing spectrum access to secondary system. Figure 8.6 presents the outage probabilities of the primary and secondary systems vs. the transmitted power of the secondary one for different values of the primary transmitted power. The outage probability of the primary direct link, p_d, is also presented as reference.

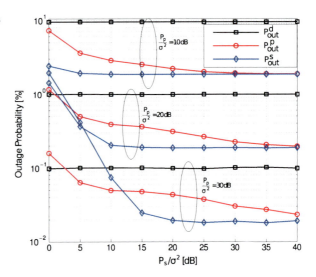

Fig. 8.6 Outage probabilities vs. P_s for different values of P_p

The paper also contains an analysis of the conditionalities between the SNR at both receivers (or transmitted powers of the two transmitters and the distances between the equipments involved in the cooperation scheme for a given propagation model) for which both the primary and secondary systems have outage probabilities smaller than the one of the primary system's direct link.

The data rates that could be achieved by the two cooperating systems are also derived in the paper, taking into account the impact of the outage probabilities upon these data rates.

Another important fact that affects significantly the relay-assisted transmissions is the synchronization between the nodes involved. The impact of noisy carrier phase synchronization upon cooperative transmissions that employ AF relaying is analyzed in [BW07]. The authors consider N_R AF single-antenna relays that assist the communication between N_{SD} source-destination pairs, which uses a channel-flattening technique like Orthogonal Frequency Division Multiplexing (OFDM) and take place in a quasi-static or slow fading environment. The channels are assumed to be uncorrelated, frequency-flat Rayleigh fading. The communication follows a two-hop relay traffic pattern, i.e., each transmission cycle includes two channel uses: one for the first hop transmission from the sources to all relays and one for the second hop transmission from the relays to the destinations. The direct link is not taken into account.

Using a scheme based on the estimation of phase differences, which provides a global phase reference at the relays and which allows the relays to achieve a spatial multiplexing (or diversity gain), the authors characterize the inaccuracy of the phase synchronization caused by the additive noise signal and relay's Local Oscillator (LO) phase noise. Then, using the individual phase offsets provided by the phase estimation error, the relay gain factors are calculated so that all source/destination links would be orthogonalized. The relay gain factors are calculated using a multiuser zero-forcing gain allocation method which takes into account the individual

LO phase offsets. The paper analyzes the influences upon the performance of the instability factor of the relay's LO (α) and of the time interval between phase synchronization and channel estimation t, on one hand, and of the length of the synchronization sequence, on another hand.

The performances are evaluated by means of the average sum rate for different lengths of the synchronization sequence and for different values of the product αt. In order to point out the effects of phase noise and of additive noise, the paper analyzes separately the average sum rate in the presence of one of the two noise signals. The authors conclude that the relay phase noise only has a limited impact on the performance, which is more affected by the phase uncertainty during the time between phase estimation and channel measurements at the relays, t, and by the phase estimation inaccuracy.

The problem of Delay-Tolerant Cooperative Diversity Routing (DT-CDR) schemes for Mobile Ad Hoc Networks (MANETs) is addressed in [RYYS10] where the authors propose such a scheme in which the network nodes cooperate to enhance the robustness of routing through cooperative space–time transmission. The delay-tolerant cooperative diversity is provided using a cross-layer approach, which involves a new Shift-Orthogonal Space–Time Block Code (OSTBC) and a simpler Media Access Control (MAC) protocol for MANETs. The OSTBC allows a low-complexity symbol-wise decoding at the PHY layer and is delay-tolerant, not requiring any longer a strict synchronism between the cooperating nodes, and therefore, the needed MAC protocol is simpler than the ones required by conventional approaches.

The above features are achieved by: (1) *a loose path formation*, obtained by only setting the maximal size M of the relay cluster, and so the cluster could have a random size; (2) *cooperative diversity*, due to which more than one packet copy could be received at the destination; this also provides robustness, since it is very unlikely that all relays in a cluster fail during the same packet transmission; and (3) *delay-tolerant cooperative Space–Time Block-Code (STBC)*, which does not require the synchronization algorithm specific to Synchronous Cooperative Diversity Routing (S-CDR) MANETs. This simplifies the required MAC protocol and decreases the processing and power consumption in all nodes.

The paper presents the construction of a delay-tolerant orthogonal space–time block code and provides a proof of its orthogonality. Moreover, it shows that in practical systems the delay-tolerance is ensured by a Cyclic Prefix (CP), which allows the periodic correlation to be achieved by the aperiodic correlation. The insertion of the cyclic prefix, longer than the maximum timing difference between adjacent clusters, slightly decreases the OSTBC's coding rate if the CP's duration is much smaller than the duration of the transmitted packet.

Using this DT-OSTBC, the authors develop a delay-tolerant PHY protocol, which is used to enable the routing algorithm.

The proposed routing approach is evaluated by comparing its end-to-end BER to the ones provided by the classical Dynamic Range (DR) and S-CDR approaches, in the same topology and wireless environment. Simulation results show that the proposed routing algorithm provides smaller BER for the same percentage of node failures; moreover, it maintains the source-destination connection for a greater percentage of node failure, thus avoiding the necessity of rerouting required by the DR MANET. Finally, the authors also claim that the proposed delay-tolerant OSTBC performs symbol-wise decoding with very small computational complexity, which is an important factor for relay nodes with low power and limited computational capability.

8.1.2 Relay Assignment Algorithms

The performances provided by four Relay-Assignment Algorithm (RAA) for the Up-Link (UL) connection of a Relay-Enhanced Cell were studied in a comparative manner in [BBP08]. The RAA analyzed are three centralized algorithms, i.e., the Optimal Relay Assignment (ORA), the Sequential Relay Assignment (SRA), a modified version of SRA, called Sequential Unique Relay Assignment (SURA) proposed by the authors, and a Semi-Distributed Relay Assignment (SDRA). The scenarios used for evaluation had different densities of nodes, randomly positioned and randomly moving, and different ratios between the number of sources N_S and number of relays N_R.

The performance metrics employed were:

1. The average capacity C_{avg} and the minimum capacity C_{min} of cooperative links built up by these algorithms, assuming that cooperation is used only if the capacity of the cooperative link is greater than the one of the direct link. The capacity was evaluated for the ideal case, i.e., the relay assignment was performed at every coherence time interval and for the case where the assignment is updated at constant intervals, to see the effect of the signaling duration. The effect of cooperation was also expressed in a relative manner by the capacity improvement factors, i.e., $C_{min_{Imp}} = (C_{xcoop} - C_{xdir})/C_{xdir}$.
2. The average durations of relay-assignments provided by the four algorithms.

The results obtained show that the assignment duration strongly depends on the density of nodes in the cell, becoming shorter as the density increases. Another expected result is that the relay assignment process occurs more often, as the users' speeds increase, decreasing the average duration of an assignment.

The capacity performances showed that cooperation would bring significant improvements for nodes that are situated far from the base station, with a small capacity provided by the direct link. The capacity improvement factors are shown in Table 8.1 for two N_S/N_R ratios across AWGN channels.

The ORA and SURA algorithms have the best performances, while SDRA performs poorly. The SRA algorithm has good performances decreased by the fact that it assigns a relay to multiple sources. The paper also shows that the capacity improvement factors are significantly greater across Rayleigh-faded channels, where even the SDRA brings some improvements. An approximate evaluation of the computational complexity of the four Relay Assignment (RA) algorithms is presented as well.

Table 8.1 Capacity improvements for $N_S < N_R$ and $N_S > N_R$; AWGN channels; ideal case

	$N_S = 40, N_R = 60$		$N_S = 60, N_R = 40$	
	$C_{min_{Imp}}$	$C_{avg_{Imp}}$	$C_{min_{Imp}}$	$C_{avg_{Imp}}$
ORA	18.3%	4.5%	2.9%	2.2%
SRA	11.6%	2.4%	6.9%	−0.4%
SURA	16.5%	4.0%	15.4%	2.3%
SDRA	−1.6%	−37.6%	−11.1%	−34.8%

The effects of the relay position upon the performances of Space–Time Coded (STC) cooperative wireless systems, over Block Fading Channel (BFC), are studied in [ZCT09] in connection with three strategies of power allocation at the relay node R. The authors employ Pragmatic Space–Time Code (P-STC), which consists in the use of convolutional encoders and Viterbi decoders over multiple transmitting and receiving antennas, and describe a feasible method to search for good cooperative codes, which generates Cooperative Overlay Pragmatic Space–Time Code (COP-STC).

The power allocation strategies considered in the paper are: (1) *Uniform power allocation*: the source and relays transmit with equal powers; (2) *Ideal power control*: the power among source and relays are balanced so that the average received power at the destination is the same; and (3) *Optimal power allocation*: the power among source and relays are balanced so that the outage or the error probability is minimized at destination.

The performance metric used is Frame Error Rate (FER), which is analyzed in terms of the SNR, relay location, the variability degree of the BFC and power allocation strategy. The scenario considered consists in a source that transmits to a destination assisted by one potential relay.

At first, the paper studies the spatial distribution of the cooperation probability depending on the relay's position, for the selected propagation model and two of the power allocation strategies, and concludes that, for uniform power allocation, the cooperation probability is large when the relay is placed around the source and for channels with higher variability, while for the power control, the regions with high probability of cooperation move toward the destination. The study of the FER performances at destination evaluates the cooperation regions, i.e., the regions where the cooperative FER is smaller than the direct link's FER, for different SNR, BFC variabilities, and the three power allocation strategies.

The authors conclude that, for uniform power allocation, the cooperation region is large and centered in a point between source and destination, and its contour enlarges, while the minimum FER decreases for an increased BFC variability.

For the case of ideal power control, the minimum FER is greater, compared to the previous case, but at the same time the FER is less sensitive to the relay's position, while for optimal power allocation, the results show that the achieved FER is smaller, while the cooperation region is larger. Regarding the relay allocation strategy, the results indicate that the choice of relay's position depends on a proper balancing between the FER at destination and the transmitted power at the relay node.

Finally, the authors summarize their results by the following considerations: (1) the relay should be located between source and destination, preferably closer to the destination than to the source; (2) by using ideal power control the benefits of cooperation on the FER are less sensitive to the relay's position, although best performance is obtained with the optimal power allocation; and (3) with ideal power control and optimal power allocation, the power level transmitted by the relay is lower near the destination.

Paper [CSNM10] proposes a partner selection algorithm, which allows cooperation as long as the energy required for data transmission of the partner message is below a given (application-dependent) threshold. This algorithm aims at minimizing the maximum energy consumed by devices during a two-phase cooperation algorithm, with a constraint on the energy loss (or gain) for the single user. The energy consumption is evaluated, imposing that the outage probability stays below a target value p, for single (noncooperating) users, for cooperating pairs with User Fairness (UF) constraint, i.e., the transmit energy for these users is chosen such that the outage probability at the BS for both users is lower than or equal to p, and for a Multiple-Input Single-Output (MISO) system with two antennas, which does not use cooperation.

The paper defines a user energy gain, i.e., the ratio between the energy required by direct transmission and that in case of cooperation and decomposes it into a product of three factors linked to causal factors, namely the penalty factor paid for the rate increase from time slot splitting gain, the micro-diversity gain from the exploitation of redundancy in the partner link, and the macro-diversity gain from asymmetric links with different statistics. To apply the UF constraint, the user energy gain is compared to a UF threshold, and three possible domains for the threshold values are discussed.

The analysis performed shows that the largest minimum energy gain that could be guaranteed is upper bounded, and this upper bound is obtained for symmetric up-link channels, which means that the system can guarantee the largest UF only if no macro-diversity, i.e., no cooperation, is available. The energy gain of a 2×1 MISO noncooperative system, where users employ the Alamouti scheme, is also computed, as an alternative to the cooperative transmission. The analysis shows that the energy gain of this system is only due to the micro-diversity gain with a penalty factor $1/2$ due to the power splitting between the two transmit antennas.

Two relay-assignments approaches that observe the energy constraints stated above are proposed and analyzed for this case:

1. *Optimal pairing for the min-max energy consumption with UF constraint.* This approach determines the optimal pairing, considering candidate pairing sets ξ and the corresponding single-user sets S_ξ, by determining the maximum energy consumed by a user in the network as the maximum between the energy of a pair and the energy of an unpaired user.
2. *Worst-Link-First Coding-Gain (WLF-CG)-based algorithm with UF constraint.* This two-phase algorithm is a modified version of the "worst-link-first" algo-

rithm, known in literature, which reduces the complexity, and gives better performance than its "parent" algorithm using second-order statistics.

Network Lifetime (NLT) performances of the partner selection algorithms proposed are evaluated in a simulated network topology defined in the paper. The results show that for small UF constraint values, the AF cooperation is at least as efficient as the MISO solution, achieving larger NLT gains when the user-fairness constraint becomes less severe and cooperation is on average extremely beneficial for the network. The performance (i.e., macro-diversity) can be further improved by increasing the density of the users. Nevertheless, the difference between the optimal and the WLF-CG pairing algorithms is high.

As for the average percentage of cooperating users given a certain UF constraint, the results show that the more severe the UF constraint becomes, the less likely it is for the users to cooperate (especially for small density of users). Hence, for large UF constraints, an efficient multi-antenna resource allocation must avoid detrimental macro-diversity effects in order to exploit the benefits of the micro-diversity gain.

The partner selection in cooperative ad hoc and wireless sensor networks, in particular for Indoor-to-Outdoor (I2O) or Indoor-to-Indoor (I2I) applications, is important to increase the lifetime of the network. The partner selection in such networks is significantly affected by the fading statistics, and paper [CSNZ09] provides an analytical study to evaluate the effects of the knowledge of fading distribution to a greater degree (K-factors of the Ricean distribution) than the knowledge of only the second-order statistics (such as signal strength or path-loss), on the performances, in terms of required energy, and on the optimality of the grouping algorithms.

Using measurements of I2O channels, where the Rayleigh fading model does not apply, the authors derived a distributed multilink channel model in order to simulate practical scenarios, where the proposed analytical evaluation is tested. Finally, the authors introduce a novel partner selection strategy that exploits the distributed knowledge of the effective coding gains provided by the fading statistics of the wireless links. The optimality criterion used is the minimization of a network goodput metric, e.g., the maximum outage probability or the maximum energy consumption.

After providing simplified models for the I2I and I2O channels by using Ricean distribution, but considering only the static shadowing of the pathloss, the authors derive the coding gain of the AF relaying on these channels and analyze its variation, concluding that coding gains on Ricean faded channels are greater than the ones on Rayleigh faded channels, and proposing the effective coding gain as a key metric to evaluate the optimal node pairing. To analyze the node pairing, the authors compute first the energy consumptions of single nodes and of cooperating nodes and then present and analyze two partner selection strategies, i.e., the "Optimal for the min-max energy consumption" (Opt) and the WLF-CG, which were analyzed also in [CSNM10] and are briefly summarized above.

The performances provided by the two cooperation strategies are evaluated by the ratio between the maximum energy consumptions of the cooperative (paired according to the studied algorithm) and noncooperative transmissions vs. the number of users in the scenario. For comparison, the study includes the Worst-Link-First Path Loss (WLF-PL) algorithm, which assumes Rayleigh fading, and the random pairing. The simulations showed that the proposed WLF-CG algorithm almost always reaches the *Opt* approach and outperforms the WLF-PL, and the relative gain decreases with the increase with the number of users involved. Both greedy algorithms outperform remarkably the random pairing, which requires even more energy than the direct transmission in some scenarios.

Finally, the paper studies the effects of the errors that occur at the estimation of the K-factors, the higher-order statistics used, upon the energy performance of WLF-CG, and concludes that it is robust for small networks, whereas, for large ones, it is more sensitive to the K-factor estimation errors.

The issue of relay-selection is also the topic of paper [Bur09] which proposes multirelay selection algorithm for fixed DF relaying in a clustered network. It considers a system model composed of the source and the N selected relays which transmit, with the same power using Binary Phase Shift Keying (BPSK), in $N+1$ different time slots in order to avoid interference at destination; the destination combines the multiple signal copies using MRC. The model considers symmetrical slow-fading Rayleigh channels.

The criterion used to select the candidate relays is obtained by imposing the condition that the BER performance is not limited by the S-R channel; this condition is derived for the particular case of BPSK, without FEC, in terms of the BER vs. SNR performance of each link. Then, imposing a target global BER, the paper derives the SNR threshold and the corresponding S-R limit distance, which could be used to adaptively select the forwarding relays. This approach aims at reducing the effect of error propagation, by choosing the nodes which are close enough to the source as candidate relays.

The paper also proposes an adaptive MAC protocol to select the relays, using the criterion previously derived, based on IEEE 802.11 MAC that uses Carrier Sense Multiple Access/Collision Avoidance (CSMA/CA). The proposed protocol is divided into the control plane, which is responsible for selecting the candidate relay nodes, and the data plane, which completes the data exchange.

The performance of this relay selection algorithm is compared to the performance provided by the best-relay selection algorithm and by the direct S-D transmission. The simulation results, see Fig. 8.7, show that both relay selection algorithms achieve full order diversity and that the proposed adaptive relay selection algorithm requires a smaller SNR than the best-relay selection algorithm to ensure the same BER.

The study also shows that the average number of relay nodes employed by the proposed algorithm is smaller than one within an SNR domain of great interest and hence increases power efficiency.

The performances of a DF cooperative diversity scheme in the context of Impulse Radio Ultra-WideBand (IR-UWB), where nodes are affected by Multi-Access Interference (MAI), are studied in paper [CC10]. The authors suggest that the ad hoc configuration gives the shape of the MAI distribution which then is accurately represented by symmetric α-stable distributions. Due to the stability property, it is shown that α-stable approach is an attractive MAI modeling solution in interfering cooperative transmission applications.

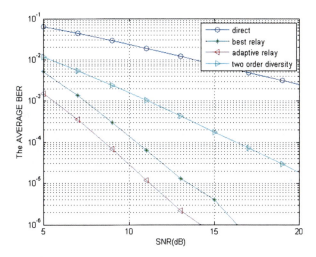

Fig. 8.7 The BER performance comparison of different schemes

The paper presents analytical models for the direct link, for the DF Dual-Hop (DF-DH) link, and for MAI. Then it derives an α-stable interference modeling for cooperative transmissions which is employed to characterize the MAI contribution at the receiver's output by two α-stable distribution parameters. Using this approach, the authors derive a formula to compute the bit error probability in terms of MAI.

A second issue dealt with in the paper is the relay selection considering the MAI. After defining the cooperation selective DF protocol employed, the authors briefly recall the principles of conventional relay selection algorithms and extend the selection criterion, to consider MAI besides the Gaussian noise, proposing an Interference-Aware Relay Selection (IARS) algorithm. In order to validate their MAI modeling, the authors consider a wireless scenario characterized by a high density of nodes and a Poisson distribution of the interfering links in which the receiver performs either MRC or Equal Gain Combining (EGC) combining. The MAI modeling proposed is validated by the comparison between computer simulated and analytically computed BER performances of both DF cooperative and noncooperative transmissions. The paper then compares the simulated and computed BER performances of DF and noncooperative transmissions for different levels of MAI. Both comparisons indicate a very good match between the simulated and analytically computed performances and point out the effects of strong MAI that are present in the studied environments.

The BER provided by the DF transmissions that use relays selected using IARS are compared, by computer simulations, to the ones provided by DF algorithms provided by a conventional relay selection algorithm. The simulation results show that the IARS algorithm provides smaller BER in the medium Signal-to-Interference-plus-Noise Ratio (SINR) domain, above a crossover value, where the MAI is greater than the AWGN. For low SINR values, the conventional relay-selection algorithm outperforms the IARS because in this region the AWGN dominates the system degradation. The study also concludes that the optimum relay position in an MAI-dominated environment is closer to destination.

8.2 Wireless Network Coding

Multinode and multiterminal wireless communication scenarios are currently under intensive investigation in the research community. Generally, these can be seen as similar to the Network Coding (NC) paradigm [YLCZ06]. The NC operates with a discrete (typical binary) alphabet over loss-less discrete channels. It is in fact an operation on data rather than on the channel codewords. The NC has a great potential in substantially increasing the throughput of complicated communication networks. An extension of these principles into the wireless (signal space) domain is however nontrivial. Some attempts have been carried out using a simple concatenation of the NC and a single-link physical layer modulation and coding technique. This has number of drawbacks and only a limited optimality. An optimal solution is a direct signal space domain code synthesis.

The Wireless Network Coding (WNC) term is used to denote a modulation and coding technique *aware* and *utilizing* the knowledge of the *network structure*. It is a major conceptual step from traditional understanding of the physical layer technique (modulation, coding) as a tool for a point to point link connection where all routing and topology related management is done by upper layers. Various WNC strategies are investigated. One of them is the Hierarchical Decode and Forward strategy.

8.2.1 Hierarchical Decode and Forward Strategy

In [SB09] (also [SB10a]), authors present a relaying strategy for the 2-Way Relay Channel (2WRC) based on hierarchical exclusive code relay processing. The communication of the two sources over a shared relay is divided into two phases, Multiple Access Channel (MACh) and Broadcast Channel (BC) phase. Hierarchical relay processing handles hierarchical data symbols which uniquely represent both source data only in a combined way, not distinguishing individual sources. The original individual data from source A and B can be reconstructed only by providing side information on the complementary data at the final destination. An advantage of the hierarchical relay processing is its higher achievable rate region, extending outside the rate region of the classical MACh channel that would correspond to individual data stream decoding at the relay.

A direct synthesis of the proper hierarchical exclusive constellation code suitable for such processing appears to be too complex. A *layered* hierarchical exclusive code design is proposed. It combines a hierarchical inner constellation symbol mapping layer and an outer ordinary single-user-link capacity approaching code (e.g., Low Density Parity Check Code (LDPC) or turbo code). It offers a tractable low-complexity solution. Basic provided theorems show that the layered scheme forms a *hierarchical exclusive code*. The alphabet limited *achievable rate region* is analyzed for the hierarchical MACh stage of the relaying, and it is compared to the alphabet limited and unconstrained cut-set bound capacity limits. Also, the impact of the channel *parameterization* on the above-stated capacities and the impact of the amount of the *side information* available at the destination are analyzed.

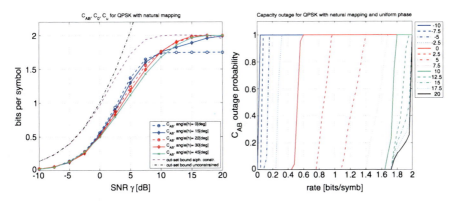

Fig. 8.8 A comparison of capacities C_{AB}, C_0, C_u and capacity outage for QPSK alphabet and channel parameterization h

The Exclusive Code is in fact a 2-source uniquely invertible signal space codebook $\mathbf{v} \in \mathcal{C}_R$: $\mathbf{v} = \mathcal{X}_\mathbf{v}(\mathbf{d}_A, \mathbf{d}_B)$. The exclusive law must hold at arbitrary processing stage $\mathbf{v}(\mathbf{d}_{AB}(\mathbf{d}_A, \mathbf{d}_B)) \neq \mathbf{v}(\mathbf{d}_{AB}(\mathbf{d}'_A, \mathbf{d}_B))\ \forall \mathbf{d}_A \neq \mathbf{d}'_A$. The invertibility implies conditions on a codebook cardinality given by $\max(|\mathcal{C}_A|, |\mathcal{C}_B|) \leq |\mathcal{C}_R| \leq |\mathcal{C}_A||\mathcal{C}_B|$. The lower bound is called minimal mapping. The upper bound corresponds to a classical MACh relaying. Anything between is an extended mapping. These mappings imply different levels of required complementary side information at the destination ranging from perfect (minimal mapping) to no side information (classical MACh).

The layered hierarchical codebook design theorem [SB09] ([SB10a]) states that for any outer linear capacity achieving code, inner exclusive alphabet, and minimal mapping $c_{AB} = \mathcal{X}_c(c_A, c_B)$, the achievable MACh rate region is rectangular, and it is constrained by the hierarchical alphabet and channel parameterization $R_A = R_B = R_{AB} \leq I(c_{AB}; x)$. Throughput rate region is calculated for given hierarchical alphabet class $\mathcal{A}_u(c_{AB})$ by $C_{AB} = I(c_{AB}; x) = H[x] - H[x|\mathcal{A}_u(c_{AB})]$. The $\mathcal{A}_u(c_{AB})$ is generally a *set* or a multisymbol *class*. All channel parameterization influences are inherently contained inside. Numerical results for the Quaternary Phase Shift Keying (QPSK) component alphabet are in Fig. 8.8.

In [SB10c] (also [SB10b]), authors concentrate on the BC phase of the two-source relay communication. Now, the sources and the destinations are allowed to be generally disjoint in order to model an imperfect complementary side information link. See the system model in Fig. 8.9. The MACh phase uses layered hierarchical exclusive code with the minimal mapping $d_{AB} = \mathcal{X}_d(d_A, d_B)$ and $c_{AB} = \mathcal{X}_c(c_A, c_B)$. The same outer code $\mathbf{b}_{AB} = \mathcal{C}(\mathbf{d}_{AB})$ is used for the 2nd-stage BC relay transmission. The destination node has available received signal from both MACh and BC phases, $z_A = s(c_B) + \xi_A$ and $y_A = v(b_{AB}) + w_A$. The first one carries the complementary side-information $\mathbf{c}_B$, and the second one the hierarchical codewords $\mathbf{b}_{AB} \equiv \mathbf{c}_{AB}$.

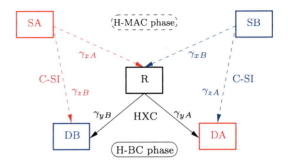

Fig. 8.9 2-Source relay network with partial complementary side information

It is shown that the destination observation provides a symbol-wise soft decoding metric for decoding the source data

$$\mu(b_A) = p(y_A, z_A|b_A) = \sum_{c_B} p(y_A|\mathcal{X}_c(b_A, c_B))p(z_A|c_B)p(c_B). \quad (8.2)$$

This metric is used by a destination decoder. It also allows to evaluate the constellation constrained BC capacity region. This region is shown to be again a rectangular one with maximum rates given by $C_{HBC} = I(b_A; y_A, z_A)$. A numerical evaluation of the hierarchical BC capacity and also the corresponding complementary side-information link (between the complementary source and the destination) for the QPSK alphabet are shown in Fig. 8.10.

Numerical results demonstrate an expected fact that the BC stage of hierarchical code with minimal exclusive mapping is upper bounded by complementary side-information link capacity. This simply comes from the fact that the minimal mapping requires reliable (full) complementary side-information. The second observation is rather surprising. There is an *approximately* linear law between hierarchical BC capacity and complementary side-information link capacity $C_{CSI}(\gamma_z)$ for given

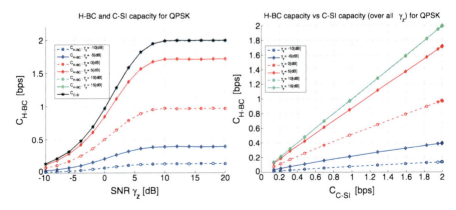

Fig. 8.10 QPSK alphabet constrained capacity of H-BC stage and capacity of the complementary side-information link

alphabet size $|\mathcal{A}|$:

$$C_{\text{HBC}}(\gamma_y, \gamma_z) = \frac{C_{\text{HBC}}^{\text{perfCSI}}(\gamma_y) C_{\text{CSI}}(\gamma_z)}{\lg |\mathcal{A}|}, \qquad (8.3)$$

which holds for finite values of γ_y. It is a product of two simple to evaluate single-link capacities $C_{\text{HBC}}^{\text{perfCSI}}(\gamma_y) = C_{\text{HBC}}(\gamma_y, \gamma_z \to \infty)$ and $C_{\text{CSI}}(\gamma_z)$. The linear law holds very well for number of alphabets. It can be conveniently used by upper layers for the resource management control.

One of the problems of the hierarchical decode and forward strategy at the relay is the need of full MACh and BC phase decoding and reencoding of the hierarchical symbols. While the usage of the hierarchical symbols saves on the required rates w.r.t. classical sharing protocols, it does not save on the latency introduced by the codeword decoding and reencoding. An approach for reducing the latency of the relay processing is presented in [PS10]. From the perspective of the hierarchical decode and forward strategy, it would essentially be sufficient if the final destination had access to the relay observation or at least to the relay decoding metric. Provided this, the final destination could do all the joint relay and destination processing at one place. This would reduce all the relay introduced latency. However, the destination does not have a direct access to the relay decoding metric. The only possibility is to pass the decoding metric (or any equivalent form) in some form between the relay and the destination instead of the reencoded data. The relay metric however needs a proper source coding.

A solution of [PS10] reduces the amount of the information transfer by sending per-symbol survival path patterns of the Viterbi relay decoder instead of the pure decoding metric. This can be viewed as a form of metric source coding. The bounds on the entropy contained in the surviving patterns are analyzed.

8.2.2 Wireless Network Coding in Parametric Channel

One of the major problems of the hierarchical code design is the impact of a channel parameterization (e.g., a phase rotation). The channel parameterization strongly influences how the signal space codewords fulfill the exclusive law. Some channel parameters values (e.g., phase rotations) can have a disastrous effect on the code exclusivity. Then it directly reduces the achievable constellation and channel parameter constrained rates.

There are several options how to solve the problem. Some of them are based on adaptive change of the relay demodulation maps and/or adaptive predistortion (pre-rotation) at the source depending on the actual channel parameterization [KAPT09]. In [Syk09, US09, USH10], authors investigate an alternative approach. The hierarchical code components are designed in such a way that the relay decision regions are invariant of the MAC channel parameterization, namely the relative phase rotations of the received signals. It is called a parametric hierarchical exclusive code—the codewords fulfilling the exclusive law span a subspace defined by the channel parameterization.

Fig. 8.11 Demonstration of extended and generalized (full) parametric design

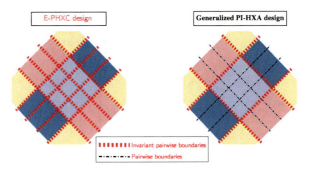

The paper [Syk09] concentrates on a codebook design criterion for a wireless 2-way relay channel which has a channel parameter-invariant processing (pairwise decision regions) at the relay. The codewords $\mathbf{s}^i \in \mathbb{R}^2$ are pairwise parameter invariant if

$$\langle \mathbf{s}^{i_A} - \mathbf{s}^{j_A}; \mathbf{s}^{i_B} + \mathbf{s}^{j_B} \rangle = 0 \quad \text{and} \quad \langle \mathbf{s}^{i_B} - \mathbf{s}^{j_B}; \mathbf{s}^{i_B} + \mathbf{s}^{j_B} \rangle = 0, \tag{8.4}$$

where $\langle \cdot ; \cdot \rangle$ is the inner product of differences for any index pairs forming the relay decision regions $(i_A, i_B) \neq (j_A, j_B)$ drawn from the permissible sets satisfying the exclusive law.

The criterion of [Syk09] solves only pairwise invariance. In [US09], authors have used it to develop an extended parametric design criterion. The extended criterion requires all pairwise boundaries to be invariant. This includes even the boundaries between the region pairs that fall under the same hierarchical symbol. This is a requirement that is sufficient but definitely not necessary. In fact, there is a large number of such pairwise boundaries that have a common hierarchical symbol. But, on the other side, the criterion allows a straightforward design verification procedure. It is also proved that the extended parametric design is not possible for any two identical binary alphabets.

A parametric design which makes *only relevant* hierarchical relay decision boundaries parametric invariant is treated in [USH10] (see Fig. 8.11). The design is strongly motivated by geometrical interpretations and also relies heavily on a 2-mode relay mapping. The received equivalent signal u is obtained by rescaling of the true channel response ($u' = h_A s_A + h_B s_B$) by $1/h_A$. The only purpose of this rescaling is to obtain a simplified expression of the useful signal, which is (after rescaling) parameterized only by a single complex channel parameter $h = h_B/h_A$. Alternatively, it can be rescaled it by $1/h_B$; hence two alternative models (modes) of the useful signal are obtained:

$$M_1: \quad u_{M_1} = s_A + h s_B, \qquad M_2: \quad u_{M_2} = s_A/h + s_B. \tag{8.5}$$

As a consequence, by selecting proper mode, it can be always assured that $0 \leq |h| \leq 1$. This simplifies the parametric design.

A geometric interpretation of the parametric design is in Fig. 8.12. Lines in the figure show trajectory of the corresponding hierarchical constellation points as a

Fig. 8.12 Parametric design with hierarchical constellation point trajectories

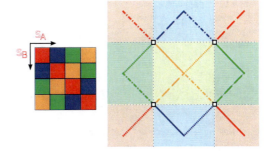

function of the channel parameterization. Assuming constrained values of the relative channel gain h, it must be assured that the trajectory stays completely in one color corresponding to the correct hierarchical codeword which the sources A and B pair map to.

The minimal distance between the hierarchical regions of two different colors can be used as a simplified, but easier to evaluate, performance metric. Namely in a parametric channel, this can provide more straightforward parametric design rules and also a particular impact of given channel parameter is easier to understand. The general analysis of the Euclidean distance of the hierarchical symbols (received at the relay) was introduced in [US10], followed by the derivation of the hierarchical distance bounds under various alphabet cardinality, dimensionality, and indexing conditions. The results of the Euclidean distance analysis have lead to a proposal of simple alphabet construction algorithm (for arbitrary alphabet cardinality), which provides the parameter invariant min-distance performance. Many important conclusions are observed from the analysis. For binary alphabet, the hierarchical min-distance is always equal to its upper bound. The upper bound $d^2_{\min,\mathrm{UB}} = \min\{d^2_{\mathrm{row}}, d^2_{\mathrm{col}}\}$ is given by row and column minima since hierarchical component symbol pairs which have one index in common ($i = i'$ or $j = j'$) cannot be clustered under the same hierarchical symbol for minimal cardinality code. Hence, the binary alphabets are highly resistant to the channel parameterization. For a higher alphabet cardinality (QPSK, 8-PSK, and 16-QAM), an increasing number of overlapping min-distance reducing influences quickly emerge and significantly affect the resulting values of the minimum distance.

An extension of this work for *nonlinear* modulation formats with multidimensional symbols is presented in [HS10]. It was inspired by the previous observations of the impact of the symbol cardinality and dimensionality on the parametric invariance of the design. It uses XOR map and the simplest nonbinary multidimensional FSK modulation with the optimized modulation index. It is found that such minimal modulation index that leads to the same performance as invariant orthogonal modulation, surprisingly with lower modulation index (bandwidth) than orthogonal case. The design is verified by means of the parametric minimal distance evaluation and the error rates in the hierarchical MACh stage with Rayleigh–Rice fading.

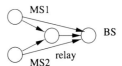

Fig. 8.13 Cooperative relay-based transmission with combined channel and network coding

8.2.3 Discrete Network Coding Combined with Channel Coding

A traditional discrete NC [YLCZ06] can help significantly increase the throughput over the relay network with multiple paths between the source and the final destination. *Unlike WNC*, it however requires *full individual decoding* of all incoming MACh phase channel symbols at the relay. The individually decoded symbols are combined by XOR binary discrete operation to produce the network coded source data symbol. Then it is passed to the 2nd stage of the wireless transmission. This procedure increases the throughput of the second stage of the network. The technique can be combined with a number of channel coding strategies including iterative message passing decoding algorithms.

The paper [SPB08] compares a performance of the NC combined with channel coding in a selection of the transmission schemes for cooperative up-link connection. The cooperation is achieved by sharing a common relay node (Fig. 8.13) which performs network coding and forwards the NC symbols to the base station. The base station also receives direct links from the mobile station. A number of scenarios is defined where some of the links can be considered as error-free and some are introducing errors under various signal-to-noise ratios. The signal observation at the destination is a function of both direct encoded signals and relayed encoded signals and can be viewed as the signal with distributed channel coding. Various channel coding options are considered.

The results show that the algorithms incorporating network coding perform better in terms of the error rate than the algorithms using purely distributed channel coding. The gain is particularly high in a situation where the direct signal from one mobile station and the signal from relay are having higher signal-to-noise ratios than the link from the other mobile station. The proposed schemes improve the efficiency of the time-frequency resources due to the fact that the relay is shared by two mobile stations.

The scenario of [SPB08] is further examined in [PSVS09], where low complexity cooperation algorithms are developed. The complexity reducing idea uses the linearity property of the turbo code used as a channel code. The soft-decoding messages (log-likelihood ratios) of the destination decoder are obtained from direct links and from the network coded relay link. The network coded relay log-likelihood ratio can be obtained as a combination of the two direct link log-likelihood ratios:

$$L_{x_1 \oplus x_2} = \ln \frac{1 + e^{L_{x_1} + L_{x_2}}}{e^{L_{x_1}} + e^{L_{x_2}}} \approx \text{sign}(L_{x_1})\text{sign}(L_{x_2}) \min(|L_{x_1}|, |L_{x_2}|). \quad (8.6)$$

This operation can be also inversed, and the decoder of mobile station 1 is fed by the soft-message from both direct and relay link. This enables to use only one turbo decoder with the input being a soft demapper of direct and relay likelihoods.

Fig. 8.14 2-way relay communication using 3-stage scenario with network coding

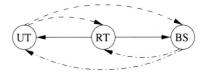

The principles of network coding applied over channel coding are further extended into a 2-way communication scenario in [SBBS09]. The system uses three stages (Fig. 8.14). In the first stage, the user terminal sends its encoded block, and both the relay node and the base station decode and store the received block. In the second stage, the base station sends its encoded block, and both the relay node and the user terminal decode and store the received block. In the third stage, the relay node combines both received blocks and forms the network coded block which is sent to both the user terminal and the base station. The network coded block is encoded by the same mother turbo code as it is used by both the user terminal and the base station. The integrity of all blocks is checked (e.g., by Cyclic Redundancy Check (CRC)).

In the first scenario, the user terminal (and similarly the base station) first decodes the data received by the direct path and waits for additional information from the relay node. If the direct link received block is error-free, then it is output disregarding of the third stage. On the other side, if the block received from the relay is error-free, then it is combined with its own node source data, and using the network coding rules, the other node data are decoded.

In the second scenario, a joint soft information combining decoding using both stage received signals is employed. The direct and the relay signals enter the component soft output decoders which exchange their extrinsic information. The extrinsic information is properly deinterleaved and also incorporates the knowledge of the terminal own source data. The performance of the above-stated scenarios is compared in terms of their spectral efficiency and error rates.

A generalization of the relay-based cooperative communications using network coding with multiple sources and multiple relays is investigated in [SBBP09]. The scenario assumes architectures with the number of available relays smaller than the number of sources. Again, the cooperation strategy is based on XOR-type network coding. First, the graph representation of the cooperative cellular network, called cooperation graph, is introduced. Using the cooperation graph, the network decoding performance is analyzed for some cooperation architectures.

The cooperation graph is a bipartite graph with user and relay nodes. The graph defines various cooperation clusters and the links describe "who cooperates with whom." The extended form of the graph reflects multistage operation and additionally also the base station nodes. The network coded cooperation strategy employs the network coding on relay-base station links. The motivation for using the network coding stems from the fact that one relay node can serve more than one user terminal using the same physical resources.

All links in a network are modeled as block erasure channels. Provided that the knowledge of the erasure probability is available, the network decoder performance can be evaluated. In the worst case, the network decoder will not be able to recover

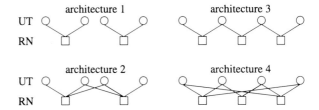

Fig. 8.15 Cooperation graphs of the network architectures

the data. This situation is called an outage state of the network decoder. The paper [SBBP09] analyses the probability of the outage and shows that this probability can be used as an overall network performance indicator. Numerical results are provided for number of network architectures (Fig. 8.15).

The problem of a bidirectional two-step relaying with an NC transmission over the multiple antenna BC stage is considered in [EYCG10]. Two scenarios are investigated. The first is based on a max-min antenna selection where only one antenna is selected on the relay to transmit the NC BC message. In the second strategy, the relay transmit beamforming vectors are optimized. The destination nodes have the complementary side information about their own data, and therefore the beamforming vectors can be optimized for the best point-to-point communication.

8.2.4 Amplify and Forward Strategy

An AF strategy is an alternative to the Decode and Forward strategies (including all the flavors like, separate decoding with NC or Hierarchical Decode and Forward). In [UK07], a two-way relay communication with multiple antennas and AF relay operation is considered. During the first stage, the source nodes S1 and S2 simultaneously transmit to the relay equipped with multiple antennas. The relay receives a superposition of the signals which is retransmitted in the second stage. The second stage has a form of broadcast transmission. Both terminal nodes receive this superposition and can detect the desired signal from the other node by subtracting their own known signal. The channel state information of relay to destination and source to relay links is required at the destination receiver in order to properly design its receive filter. A performance criterion adopted in [UK07] is the network sum-rate.

The overall (source-relay-destination) equivalent channel model embodies all channel matrices $y_i = A_i x_i + B_i w_i$ where $i \in \{1, 2\}$ distinguishes the direction of the communication, y_i is the observation, x_i is the source signal, w_i is the Gaussian noise, and A_i, B_i are equivalent channel and noise scaling matrices. The sum-rate is $C = C_1 + C_2$ where

$$C_i = \frac{1}{2} \log_2 \left(\det \left(I + \frac{A_i R_x A_i^H}{B_i R_w B_i^H} \right) \right), \qquad (8.7)$$

and R_x, R_w are transmit and noise vector covariance matrices. The power assignment is optimized, and a closed-form solution for the optimal power split between

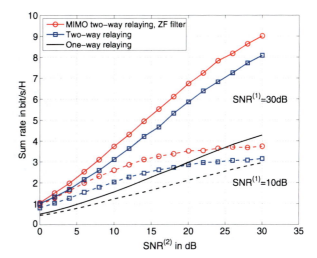

Fig. 8.16 Sum-rate of different AF protocols; S1-Relay (SNR1), S2-Relay (SNR2)

source nodes is derived in [UK07]. Under this optimal power assignment and a zero-forcing transceiver filter at the relay, the sum-rate performance is evaluated and compared to the traditional two-way and one-way relaying (Fig. 8.16).

A specific special-case scenario of *analog* transmission in a two-hop Gaussian cascade network is considered in [KK10]. The scenario is motivated by the applications in sensor networks, where the analog mutually correlated information is collected by the nodes, and each node can also serve as a relay.

The system is formally defined by two correlated analog Gaussian sources which are jointly source and channel encoded. The performance criterion is a quadratic distortion. In the first hop, the transmission is uncoded. The relay performs an optimal minimum mean-squared error decoding which uses the other source (present at the relay) as side information. In the second hop, Karhunen–Loeve transform is fist applied to decorrelated the sources. Then, the dimension reduction mapping is applied. The final destination then reconstructs both sources.

First, the performance limits are derived in [KK10]. The performance is measured by average required transmission power for given minimum mean square distortion. There is also suggested an analog joint source-channel encoder (mapper) for the second hop. After Karhunen–Loeve decorrelation, a two-dimensional mapping scheme (Fig. 8.17) is suggested, and its performance against the performance bounds is compared.

8.3 Distributed Coding and Processing

This section encompasses two topics connected to the cooperative and distributed systems, namely the Multi Cell Processing (MCP) and distributed code design for distributed network message detection.

MCP is a form of inter-cell cooperation also known as macro-diversity in a wireless communication system. In down-link multiple base stations coordinate

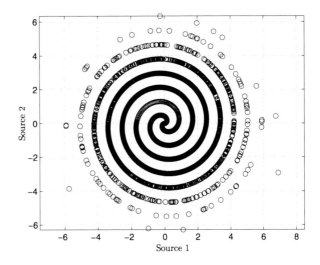

Fig. 8.17 Two-dimensional codebook mapping [KK10] (©2010 IEEE, reproduced with permission)

their transmissions to provide user terminal possibility to combine transmitted signal [JTS+08]. It is expected that the MCP method will be an important method in emerging communication systems based on 3GPP-Long Term Evolution (LTE) or IEEE 802.16m standard to decrease interference and achieve high capacity gains.

The 4G technologies among others offer the possibility for full spectral reuse in each cell (reuse 1), which may lead to very poor SINR especially at the cell boundaries. A promising method to reduce the SINR at cell boundaries is also MCP [GLG09]. It applies the Alamouti MISO technique as a method to improve the uniform capacity of a wireless networks. The main goal of the uniform capacity approach is to achieve constant and SINR independent capacity at any location at given cell. One- and two-dimensional scenarios are studied in [GLG09] to analyze the impact of the MCP concept on sum cell capacity. In both cases the uniform distribution of the users are assumed.

The optimal MCP transmission method is shown to be efficient to reduce the total resource consumption and thus improving the uniform capacity compared to other classical methods. Moreover, it is more resistant to shadowing phenomenon and reduces the impact of the worst SINR users on uniform capacity. The simulation results for two-dimensional scenario are depicted in Fig. 8.18. The optimal MCP performs slightly below the Fractional Frequency Reuse (FFR), when no or little shadowing is present, due to the lack of diversity. In the practical scenarios where shadowing standard deviation is beyond 4 dB, the MCP outperforms FFR4 and frequency reuse 1.

In many emerging wireless networks, such as wireless sensor networks, the message of interest is distributed across many nodes or even across complete network. In such a scenario the basic problem is how to simply and efficiently encode a distributed network message of the length k information symbols into a distributed network codeword of the length n encoded symbols, such that any $(1 + \varepsilon)k$ encoded symbols are sufficient for network message recovery with high probability. The problem can be solved using a novel distributed approach for the design of

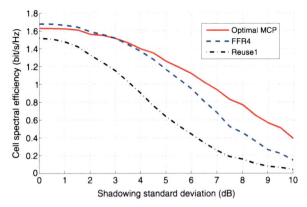

Fig. 8.18 Uniform capacity for 2D simulation versus shadowing standard deviation level (dB)

LDPC codes, such as the Low Density Generator Matrix (LDGM) and Irregular Repeat Accumulate (IRA) codes [GFW03, JKKM00, YRL04].

The proposed distributed encoding process is based on a packet centric approach for the distributed code design, where the distributed encoding is performed using random walks over the network graph. The network codeword consists of k information packets and $m = n - k$ parity packets, where n is the length of the codeword. The information packet corresponds to the k nodes of a random graph. The parity packets are formed in the encoding process, which consists of three steps: (i) initialization, (ii) random sampling, and (iii) dispersion step. In the initialization step parity packets are created at m nodes with initial value equal to the information packet of the node. In the sampling phase, parity packets perform random walks over the network and after a number of predefined random hops (called mixing time), the parity packet incorporates information packet of the current node by using simple bitwise XOR. The above procedure continuous until all parity packets incorporate a predefined number of information packets. In the dispersion phase the parity packets make yet another series of mixing time number of hops, finishing their walks in random network nodes. The decoding is performed using iterative belief propagation algorithm, and any randomly collected $(1 + \varepsilon)k$ network codeword packets should suffice for the reconstruction of the k information packets.

The two algorithms to obtain uniform node sampling are tested in [SVS09], which are Normal Random Walk (NRW) and Metropolis-Hastings (MH) [DK07], and obtained results are compared to centralized algorithm. The results are depicted in Fig. 8.19. The simulations show that lower mixing times τ yield higher PER, because the collected packets have shorter random walks and show higher degree of correlation. The advantage of MH over NRW algorithm is evident, as MH algorithm provides more uniform sampling of the information packets. NRW reaches its error floor quickly with the increase of mixing time, due to the fact that NRW favors nodes with higher degrees, causing undesirable correlations among encoded packets. It can be noted that the error floor of the distributed version of the code is higher than the error floor of the centralized version, a consequence of the fact that degrees of some information packets are not met due to the constraints on the duration of the encoding procedure.

Fig. 8.19 PER of rate $R = 1/2$, length $n = 2000$, distributed (10, 10) LDGM code

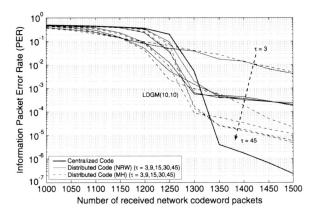

References

[BBP08]   M. A. Badiu, V. Bota, and Z. Polgar. Performance comparison of relay-nodes assignment algorithms for cooperative networks. Technical Report TD-08-653, Lille, France, October 2008.

[BTHP09] V. A. Bohara, S. H. Ting, Y. Han, and A. Pandharipande. Orthogonal cognitive spectrum access with cooperative space time coding. Technical Report TD-09-905, Vienna, Austria, September 2009.

[Bur07]   A. Burr. Wireless relay networks. Technical Report TD-07-231, Lisbon, Portugal, February 2007.

[Bur09]   A. Burr. Adaptive relay selection scheme for decode-and-forward cooperative clustered networks. Technical Report TD-09-986, Vienna, Austria, September 2009.

[BW07]   S. Berger and A. Wittneben. Impact of noisy carrier phase synchronization on linear amplify-and-forward relaying. Technical Report TD-07-351, Duisburg, Germany, September 2007.

[CC10]   J. Chen and L. Clavier. Alpha-stable interference modelling and relay selection for regenerative cooperative IR-UWB systems. Technical Report TD-10-10100, Athens, Greece, February 2010.

[CSNM10] P. Castiglione, S. Savazzi, M. Nicoli, and G. Matz. Partner selection: Efficiency vs fairness in Ricean environments. Technical Report TD-10-10020, Athens, Greece, February 2010.

[CSNZ09] P. Castiglione, S. Savazzi, M. Nicoli, and T. Zemen. Impact of fading statistics on partner selection in indoor-to-outdoor cooperative networks. Technical Report TD-09-902, Vienna, Austria, September 2009.

[DK07]   S. Datta and H. Kargupta. Uniform data sampling from a peer-to-peer network. In *ICDCS '07: Proceedings of the 27th International Conference on Distributed Computing Systems*, page 50. IEEE Comput. Soc., Washington, DC, USA, 2007.

[EYCG10] M. Eslamifar, C. Yuen, W. H. Chin, and Y. L. Guan. Multi-antenna relaying with network coding for 2-step bi-directional communication, pages 1–6, Aalborg, Denmark, June 2010. [TD-10-11038].

[GFW03]   J. Garcia-Frias and Z. Wei. Approaching Shannon performance by iterative decoding of linear codes with low-density generator matrix. *IEEE Communications Letters*, 7(6):266–268, 2003. doi:10.1109/LCOMM.2003.813816.

[GLG09]   V. Garcia, N. Lebedev, and J.-M. Gorce. Multi-cell processing for uniform capacity improvement in full spectral reuse system. In *COGnitive Systems with Interactive Sensors*, Paris, France, November 2009. [Also available as TD(09)989].

[HRKW08] K. Haneda, T. Riihonen, V.-M. Kolmonen, and R. Wichman. Measurement based analysis of gain-diversity tradeoff in relay transmission. Technical Report TD-08-549, Trondheim, Norway, June 2008.

[HS10] M. Hekrdla and J. Sykora. Channel parameter invariant network coded FSK modulation for hierarchical decode and forward strategy in wireless 2-way relay channel, pages 1–8, Aalborg, Denmark, June 2010. [TD-10-11087].

[JKKM00] H. Jin, A. Khandekar, A. Kh, and J. R. McEliece. Irregular repeat accumulate codes. In *2nd Int. Symp. Turbo Codes and Related Topics*, pages 1–8, Brest, France, September 2000. IEEE Comput. Soc., Los Alamos, 2000.

[JTS+08] S. Jing, N. C. D. Tse, B. J. Soriaga, J. Hou, E. J. Smee, and R. Padovani. Multicell downlink capacity with coordinated processing. *EURASIP Journal on Wireless Communications and Networking*, 2008(5):1–19, 2008.

[KAPT09] T. Koike-Akino, P. Popovski, and V. Tarokh. Optimized constellations for two-way wireless relaying with physical network coding. *IEEE J. Sel. Areas Commun.*, 27(5):773–787, 2009.

[KK10] A. N. Kim and Kimmo Kansanen. Analogue transmission over a two-hop Gaussian cascade network. *IEEE Commun. Lett.*, 14(2):175–177, 2010.

[OCBP09] C. Oestges, N. Czink, B. Bandemer, and A. Paulraj. Spectral efficiency of outdoor-to-indoor relay schemes in measured radio channels. Technical Report TD-09-813, Valencia, Spain, May 2009.

[PS10] P. Prochazka and J. Sykora. Per-symbol representation of sufficient statistics for FSM decoder metric used in BC phase of 2WRC with HDF strategy, pages 1–8, Athens, Greece, February 2010. [TD-10-10087].

[PSVS09] Z. Polgar, M. Stef, M. Varga, and A. De Sabata. Low complexity separate network and channel coding algorithm for multiple access relay channel, pages 1–10, Braunschweig, Germany, February 2009. [TD-09-731].

[RYYS10] T. P. Ren, Y. L. Guan, C. Yuen, and R. J. Shen. Delay-tolerant cooperative-diversity routing for asynchronous MANET. Technical Report TD-10-11039, Aalborg, Denmark, June 2010.

[SA09] A. J. Salomon and O. Amrani. Improved signal superposition coding for cooperative diversity. Technical Report TD-09-703, Braunschweig, Germany, February 2009.

[SB09] J. Sykora and A. Burr. Hierarchical exclusive codebook design using exclusive alphabet and its capacity regions for HDF strategy in parametric wireless 2-WRC, pages 1–9, Vienna, Austria, September 2009. [TD-09-933].

[SB10a] J. Sykora and A. Burr. Hierarchical alphabet and parametric channel constrained capacity regions for HDF strategy in parametric wireless 2-WRC. In *Proc. IEEE Wireless Commun. Network. Conf. (WCNC)*, pages 1–6, Sydney, Australia, April 2010.

[SB10b] J. Sykora and A. Burr. Network coded modulation with partial side-information and hierarchical decode and forward relay sharing in multi-source wireless network, pages 1–7, Lucca, Italy, April 2010.

[SB10c] J. Sykora and A. Burr. Physical network coding with partial side-information and hierarchical decode and forward relay sharing in multi-source wireless network, pages 1–8, Athens, Greece, February 2010. [TD-10-10053].

[SBBP09] M. P. Stef, L. Boita, A. Botos, and Z. Polgar. Network-coded cooperation analysis for multiple source—multiple relay architectures, pages 1–9, Valencia, Spain, May 2009. [TD-09-852].

[SBBS09] M. Stef, V. Bota, A. Botos, and D. Stoiciu. Combined network and distributed FEC coding cooperation schemes for the two-way relay channel, pages 1–10, Braunschweig, Germany, February 2009. [TD-09-732].

[SBM+10] F. Sánchez, B. Bandemer, G. Matz, C. Oestges, and F. Kaltenberger. Performance of transmission-time optimized relaying schemes in real-world channels. Technical Report TD-10-10030, Athens, Greece, February 2010.

[SPB08] M. P. Stef, Z. A. Polgar, and V. Bota. Performance analysis of some distributed network and channel coding schemes in cooperative transmissions, pages 1–13, Lille, France, October 2008. [TD-08-654].

[SSPK10] V. K. Sakarellos, D. Skraparlis, A. D. Panagopoulos, and J. D. Kanellopoulos. Cooperative diversity performance in correlated lognormal channels. Technical Report TD-10-10083, Athens, Greece, February 2010.

[SVS09] C. Stefanovic, D. Vukobratovic, and V. Stankovic. On distributed LDGM and LDPC code design for networked systems, pages 208–212, Piscataway, NJ, USA, 2009. [Also available as TD(09)863]. doi:10.1109/ITW.2009.5351420.

[Syk09] J. Sykora. Design criteria for parametric hierarchical exclusive constellation space code for wireless 2-way relay channel, pages 1–6, Valencia, Spain, May 2009. [TD-09-855].

[UK07] T. Unger and A. Klein. On the performance of relay stations with multiple antennas in the two-way relay channel, pages 1–11, Lisbon, Portugal, February 2007. [TD-07-206].

[US09] T. Uricar and J. Sykora. Extended design criteria for hierarchical exclusive code with pairwise parameter-invariant boundaries for wireless 2-way relay channel, pages 1–8, Vienna, Austria, September 2009. [TD-09-952].

[US10] T. Uricar and J. Sykora. Hierarchical exclusive alphabet in parametric 2-WRC—Euclidean distance analysis and alphabet construction algorithm, pages 1–9, Aalborg, Denmark, June 2010. [TD-10-11051].

[USH10] T. Uricar, J. Sykora, and M. Hekrdla. Example design of multi-dimensional parameter-invariant hierarchical exclusive alphabet for layered HXC design in 2-WRC, pages 1–8, Athens, Greece, February 2010. [TD-10-10088].

[VK08] S. Vorköper and V. Kühn. Information combining for relay networks. Technical Report TD-08-513, Trondheim, Norway, June 2008.

[VPB09] M. Varga, Z. Polgar, and V. Bota. Efficiency of repetition coding in relay-enhanced transmissions. Technical Report TD-09-983, Vienna, Austria, September 2009.

[YLCZ06] R. W. Yeung, S.-Y. R. Li, N. Cai, and Z. Zhang. *Network Coding Theory*. Now Publishers, Hanover, 2006.

[YRL04] M. Yang, W. Ryan, and Y. Li. Design of efficiently encodable moderate-length high-rate irregular LDPC codes. *IEEE Transactions on Communications*, 52(4), 2004.

[ZCT09] L. Zuari, A. Conti, and V. Tralli. Effects of relay position and power allocation in space–time coded cooperative wireless systems. Technical Report TD-09-878, Valencia, Spain, May 2009.

Chapter 9
Advanced Coding, Modulation and Signal Processing

Chapter Editor Laurent Clavier, Dejan Vukobratovic, Matthias Wetz, Werner Teich, Andreas Czylwik, and Kimmo Kansanen

This chapter deals with advanced coding, physical layer and signal processing issues. The evolution of technologies make the specific activities on these subjects in the COST 2100 action disparate. One reason is that the amount of work available in the community is huge and it can appear that the real challenge for future networks is not in algorithms or new coding schemes. However this point of view can be mitigated because new networks induce new constraints that modify the way to optimize traditional techniques. A second reason is that many works are connected to specific applications. Several chapters of this book are devoted to applications or to specific technologies (Mobile Ad Hoc Network (MANET), Fourth Generation (4G), Digital Video Broadcasting (DVB), mobile to mobile communications, body communication, Ultra-WideBand (UWB), Multiple-Input Multiple-Output (MIMO), cooperation, etc.). Some contributions in coding and signal processing have been left to these sections.

However some works remain and several contributions justify this chapter, which can then be seen as a noncomprehensive selection of topics about coding, modulation and signal processing aspects. The different topics that have attracted contributions in the COST 2100 action are channel coding, iterative techniques, multi-carrier and adaptative schemes, channel estimation and synchronization. Finally, several contributions linked to interference have been addressed. All those topics are gathered in this section which can appear as a mixture of contributions with little cohesion. We have however tried to extract the main ideas in order to give an overview of what topics appeared to be important in this action. In Sect. 9.1, we consider advanced coding techniques and, more specifically, the design of physical layer Forward Error Correction (FEC) schemes based on Low Density Parity Check Code (LDPC) codes. The *coding over bits* idea is then replaced by *coding over packets* and packet-based application layer FEC techniques are presented. Section 9.2 is dedicated to another essential topic in coding: iterative techniques. Iterative demodulation and decoding, turbo equalization are discussed, and some

---

L. Clavier (✉)
University of Lille, Lille, France

more general tools are presented for application and design of iterative schemes: factor graphs, Sum-Product Algorithm (SPrA), and Extrinsic Information Transfer (EXIT) charts. A lot of consideration has also been put on multicarrier and adaptative schemes. They are discussed in Sect. 9.3. Section 9.4 discusses channel estimation and synchronization. Some solutions to face network evolutions are discussed. Especially, ways to reduce the needed signalization in Frequency-Division Duplex (FDD) context, considerations on complexity/performance trade-offs, channel estimation in high mobility and frame synchronization using the constraints imposed by coding are proposed. This chapter ends on the important topic of network interference. A modeling approach based on α-stable distribution is proposed when Impulse Radio Ultra-WideBand (IR-UWB) is concerned, and the consequences on receiver design and multihop transmissions are discussed. Finally, solutions for interference limited Orthogonal Frequency Division Multiplexing (OFDM) systems are proposed.

9.1 Forward Error Correction (FEC) Coding

Modern coding theory emerged over the last two decades leading to exciting developments of coding schemes that approach the Shannon capacity over a number of standard channel models [RU07]. The solution for capacity-approaching performance is found as a conjunction of carefully designed codes defined on sparse graphs and the iterative decoding procedure that operates by exchanging messages over code-defining sparse graphs. Due to sparsity of code-defining graphs, capacity-approaching code ensembles with linear encoding and decoding complexity are found within a number of different ensembles of turbo and LDPC codes [BGT93, MN96].

Although it is fair to say that the capacity-approaching code design problem for the most important channel models is well understood and a number of capacity-approaching schemes have been proposed, a number of open issues regarding the optimal code design still remain. In particular, as soon as the focus is shifted from standard channel models to real-world channel models inspired by specific wireless communication systems, the design of capacity-approaching codes may require insights that are substantially different from the mainstream theory. The design of physical layer FEC schemes based on LDPC codes for specific wireless channels is the topic that attracted significant attention within the COST 2100 action and is discussed in the first of the following subsections.

In the second subsection, the problem of capacity-approaching code design is shifted from the physical layer FEC and "coding over bits," toward the application layer FEC and "coding over packets." Linear time encoding and decoding algorithms for sparse-graph codes provide opportunity for efficient application layer software implementations of packet-based FEC codecs. The interest in this field was dramatically increased by introduction of rateless (or digital fountain) coding concepts [BLMR98]. The strong interest for packet-based application layer FEC techniques within COST 2100 action is reflected in the second subsection covering selected topics in application layer LDPC and rateless code design.

9.1.1 Physical Layer Coding for Wireless Channels

9.1.1.1 LDPC Codes for Non-Gaussian Channels

Recent advances in coding theory based on sparse-graph codes and the iterative decoding have been mainly accomplished under standard channel model assumptions, such as additive white Gaussian noise (AWGN), binary symmetric channel (BSC), and binary erasure channel (BEC). Although they are adapted to many real-world environments, in some channels, the ambient noise found through experimental measurements is shown to be non-Gaussian and to exhibit impulsive nature. Examples include urban and indoor radio channels, power line channels, digital subscriber lines, etc. Impulsive noise usually appears as random bursts of relatively large instantaneous power, often exceeding the power of the useful signal, leading to significant reduction in system capacity. Impulsive noise models such as stable distributions have already been used to model non-Gaussian noise, theoretically justified by the generalized central limit theorem.

In [MGCG10], the asymptotic performance of LDPC codes over the Additive Independent samples Symmetric α Stable Noise (AISαSN) channel is presented and compared to the channel capacity. The AISαSN channel model considered is a memoryless channel with a binary input and symmetric output. The input X of the channel is a BPSK modulated signal, and the noise W is impulsive and modeled by an SαS random variable (RV) so that the channel output is $Y = X + W$. An RV W is SαS distributed if the Fourier transform of its density function (characteristic function) is of the form $\Phi(t) = \mathbb{E}(e^{tW}) = \exp^{-|\gamma t|^{\alpha}}$, where α ($0 < \alpha \leq 2$) is a shape parameter known as the characteristic exponent. It measures how heavy-tailed the distribution is; the smaller the value of α, the heavier is the tail (and hence the more impulsive is the behavior). The scale parameter $\gamma > 0$ called dispersion measures the spread of the RV.

The results of [MGCG10] demonstrate that the performance of LDPC codes over AISαSN channel can be close to the Shannon limit for different values of α, i.e., for different degrees of the noise impulsiveness. However, the optimal receiver remains difficult to implement. Indeed, the calculation of the log-likelihood ratio at the decoder input is based on the noise probability density function. However, this function has not an analytical expression for most of the α-stable distributions. Several promising approaches to overcome this difficulty are proposed and evaluated.

9.1.1.2 LDPC Codes for DVB-SH Applications

Digital Video Broadcasting-Satellite to Handheld (DVB-SH) is a transmission system standard designed to deliver multimedia and data services to small handheld devices, such as mobile phones and personal digital assistants (PDAs). Using S-band frequencies, a hybrid satellite/terrestrial architecture is employed with (i) a satellite to achieve wide coverage region and (ii) subordinate terrestrial gap fillers to cover the areas with inadequate satellite signal reception. In order to compensate

Fig. 9.1 BER performance of DVB-S2 LDPC codes [SJM+09] (©2009 IEEE, reproduced with permission)

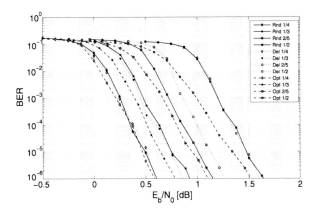

for time-varying channel conditions, the 3GPP2 turbo codes with a specific frame size of 12282 bits have been selected as FEC solution and combined with a flexible channel interleaver.

Handheld devices are characterized by limited processing capabilities and tight memory constraints, so an alternative FEC scheme based on rate-compatible irregular repeat-accumulate codes (RC-IRA) are considered for DVB-SH applications. The work presented in [SJM+09] analyzes three padding and puncturing schemes for DVB-S2 LDPC codes, in order to obtain coding rates equivalent to DVB-SH specification. The performances of punctured DVB-S2 LDPC codes are compared against the 3GPP2 turbo codes for random, deliberate, and optimal puncturing scheme. Simulation results for bit and frame error rate (BER/FER) show that the proposed coding schemes exhibit 0.5 dB to 0.75 dB penalty loss compared to 3GPP2 turbo codes [SJM+09] for random and deliberate puncturing, while the optimal puncturing provides 0.2 dB better performance (Fig. 9.1). It is important to note that the main advantage of IRA codes as compared to turbo codes is in hardware implementation, where faster decoding and higher data throughput can be achieved by the use of parallel decoding architectures.

9.1.1.3 LDPC Codes for OFDM Based Systems

The spectral efficiency provided by the coded Quadrature Amplitude Modulation (QAM) constellations is an important parameter in wireless communication systems. Adaptive modulation schemes employed in radio transmission require a set of coded modulation techniques whose spectral efficiency changes in small steps over the range of received SNR values. A possible way to obtain flexible coded modulation scheme is to map noncoded bits, besides the coded ones, onto QAM symbols. This approach results in an increase in the coding rate of the applied coded modulation. However, by appropriate bit mapping and soft decision of the noncoded bits, the error-rate of the noncoded bits can be made closer to the coded ones. This result in the overall error-rate comparable to the one of the coded bits and increased spectral efficiency for the same SNR value.

9 Advanced Coding, Modulation and Signal Processing

The work presented in [PVBS07] investigates Bit Error Rate (BER) and spectral efficiency performances of QAM constellations coded with array-based LDPC codes, where noncoded bits are mapped besides the coded ones in an Orthogonal Frequency-Division Multiple Access (OFDMA) scheme. Procedures applied for mapping and demapping the noncoded bits are presented and their error-probability performance are discussed. An influence of noncoded bits mapping upon the rates, BER, and spectral efficiency provided by different configurations adaptively employing the proposed coded modulations within the OFDM-A transmission scheme are detailed.

9.1.2 Application Layer Coding for Content Distribution and Collection

9.1.2.1 AL-FEC Codes for IPDC Services in DVB-H

DVB-Handheld (DVB-H) is an IP-based wireless broadcast technology in which users are offered, among other services, the Internet Protocol DataCast (IPDC) service for downloading files to their mobile terminals. In [NVB09], different Application Layer FEC (AL-FEC) solutions providing simple and efficient file delivery are explored. Performance comparisons of various possible AL-FEC schemes in a simulated DVB-H environment are provided, and the benefits and drawbacks of different solutions are discussed.

The DVB-H system setting assumes the file, divided into k information packets of length L bits, is available at the AL of the IPDC server. The rate $R = k/n$ AL-FEC code encodes the file into a codeword of length n encoded packets. Encoded packets are placed in one IP packet each and transmitted to service end-users over a real-world emulated packet erasure channel. The realistic ingredient of this study is usage of traces of IP datagram losses obtained from a DVB-H testbed at the University of Turku. The traces are fed as a training sequence for DVB-H packet erasure channel model introduced in [PP06]. At each receiver, the encoded packets are decoded in parallel with the reception process, and as soon as decoder is finished, the decoder starts receiving the next codeword.

The average numbers of encoded packets needed for successful recovery of the transmitted codeword, $n' = (1 + f)\Delta k$, where f is the reception overhead factor, are compared for different AL-FEC codes. Example results are presented in Fig. 9.2. Channel model and physical layer parameters are: TU6 DVB-H channel model with a constant Doppler frequency of 10 Hz, 8k OFDM mode and guard interval 1/4, and QPSK modulation encoded with a rate 1/2 convolutional code.

9.1.2.2 Rateless Codes for Multicast Streaming Services in DVB-H/NGH

Rateless (or fountain) codes were introduced as an efficient and universal FEC solution for data multicast over lossy packet networks. They have recently been pro-

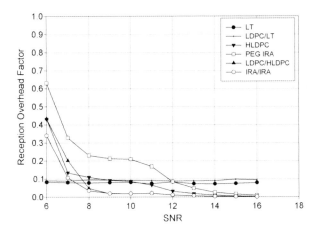

Fig. 9.2 Average reception overhead factor for various AL-FEC schemes [NVB09] (©2009 IEEE, reproduced with permission)

posed for large-scale multimedia content delivery in practical multimedia distribution systems. However, standard fountain codes, such as LT or Raptor codes, are not designed to meet unequal error protection (UEP) requirements typical in real-time scalable video multicast applications.

The work presented in [VSS+09] proposes an application of recently introduced Expanding Window Fountain (EWF) codes as a flexible and efficient solution for real-time scalable video multicast. EWF codes are a novel class of UEP fountain codes based on the idea of "windowing" the source block to be transmitted. The layered source data is organized into nested data sets called windows, the ith window accommodating the first i importance layers. Rateless coding is performed over different windows instead of the whole source block, thereby creating dependencies in the code graph where inner windows are progressively better protected than the outer ones (Fig. 9.3). EWF encoding applies probabilistic window selection for each encoded symbol, which enables their density evolution analysis and make their performance unaffected by bursty nature of the underlying channel.

The design flexibility and UEP performance make EWF codes ideally suited for multimedia streaming scenario as EWF codes offer a number of design parameters to be "tuned" at the server side to meet the different reception conditions of heterogeneous receivers. Performance analysis of H.264 Scalable Video Coding (SVC) multicast to heterogeneous receiver classes confirms the efficiency of the proposed EWF-based solution [VSS+09]. The study on AL-FEC UEP rateless solutions is extended with investigation on applicability of EWF codes and Receiver Controlled Parallel Fountain (RCPF) codes for video streaming applications in DVB-H and forthcoming DVB-NGH networks [NGB10, NV10].

In [BNM09], a multicast scenario in IEEE 802.11 wireless network is investigated. Due to the unreliable nature of the multicast MAC service, a reliable transport is realized at the application layer. A straightforward approach for the transport protocol that uses an Automatic Repeat reQuest (ARQ) method, controlled with feedback from receivers, is compared with the approach based on fountain codes. The ARQ-based approach offers low latencies and an inherent congestion control. However, ACK implosion and feedback information limit the achievable throughput

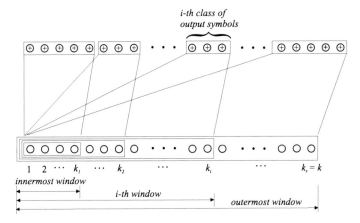

Fig. 9.3 Expanding Window Fountain (EWF) codes [SVD+09] (©2009 IEEE, reproduced with permission)

and its scalability with respect to the number of receivers. The fountain code-based approach delivers excellent results with respect to its throughput performance and scales to a large number of receivers requiring less feedback information from the receivers. Both approaches are compared analytically and by simulation with respect to their channel efficiency and their throughput performance. The superior performance of the fountain code-based approach, especially with respect to its scalability, is demonstrated.

9.1.2.3 Distributed Rateless Codes for Data Gathering in WSN

Appealing properties of rateless codes can be efficiently explored in distributed scenarios such as Wireless Sensor Networks (WSNs). An example of such application, called Rateless Packet scheme, is presented in [VSC+09]. Rateless Packets scheme is applied on WSNs described as a connected graph of N network nodes, each node initially storing a single equal-length information packet. The goal of the scheme is to create and uniformly disperse over the network a predefined number of rateless packets in a fully decentralized manner where ideally, each rateless packet shares the same properties as if it was created by centralized rateless encoder.

The Rateless Packets scheme proceeds in three phases. In the first, initialization phase, each node initializes b copies of its own information packet and assigns to each packet independently and randomly a degree d from the degree distribution $\Omega(d)$. Each initialized rateless packet should uniformly select and encode into its content remaining $d - 1$ information packets. This is done during the second, encoding phase, where rateless packets are let to randomly walk over the network and encode a new information packet into their content after each number of hops selected as the random walk mixing time. Once they collect their degree, controlled by the header data, rateless packets are stored in a random network node during the

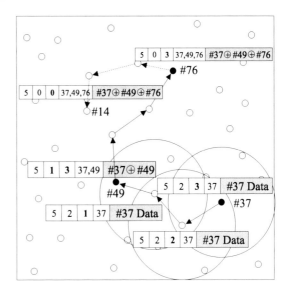

Fig. 9.4 Rateless Packet Scheme Example [VSC+09] (©2009 IEEE, reproduced with permission)

third, dispersion phase. An encoding example of a single rateless packet is given in Fig. 9.4. Finally, the network contains bN uniformly dispersed rateless packets out of which only slightly more than N is sufficient for recovery of information packets from all network nodes.

An application of this scheme for constrained data gathering in nonaccessible WSNs is discussed in [SVN+10]. The possible improvement over this scheme is to use distributed versions of random linear codes (RLC), as presented in [SVCS10]. The RLC-based scheme has benefits of exploiting broadcasting and may result in reduced power expenditure for distributed encoding process, in particular for small-to-medium sized WSN.

9.2 Iterative Techniques

Iterative techniques have a long history in wireless communications. In the early 1960s, Gallager described the iterative decoding of low-density parity-check codes [Gal62]. After the presentation of turbo codes [BGT93], iterative methods became very popular and the "turbo principle" [Hag97] was recognized as a general method in communication systems.

One of the possible applications of the turbo principle is iterative demodulation and decoding. In this case the demapper and the decoder of the channel code repeatedly exchange soft information to reduce the number of bit errors. We will discuss this topic with the aid of an example in Sect. 9.2.2.

Another popular method is turbo equalization [DPD+95], where Inter-Symbol Interference (ISI), introduced by a linearly distorting channel, is iteratively removed in the receiver. Turbo equalization can be extended by including channel estimation in the iterative process [OT04]. An application related to turbo equalization is the removal of interference introduced by clipping in an OFDM system. Section 9.2.3 will take a closer look at this application.

When looking at iterative methods, there are some convenient tools to visualize the iterative structure of the employed algorithms. One of them is the use of factor graphs and the related SPrA. A factor graph is a graphical model which can be used in a wide variety of signal processing applications including iterative receivers. The SPrA is used to efficiently compute marginal distributions in the factor graph. More details on this topic and its application to Continuous Phase Modulation (CPM) can be found in Sect. 9.2.4. Another tool is the EXIT chart. It was introduced by ten Brink [tB00] and is a powerful tool to analyze the behavior of iterative systems. It considers the exchange of mutual information between two components. For this, each component is simulated separately, using artificial a priori information. Then the characteristics of both components are plotted into one plot, considering the fact that in general the output of one component is the input of the other. Based on this graphical representation, the iterative behavior of the total system can be predicted, without running time consuming Monte Carlo simulations.

9.2.1 Reliability Information in Iterative Receivers

Usually it is advantageous to consider soft information instead of hard decisions in iterative receivers. For this soft information, a measure of reliability has to be used. Where binary variables are involved, it is useful to use the Log-Likelihood Ratio (LLR). If we consider the transmission of binary data c_j, the a posteriori LLR, conditioned on the receive vector $\mathbf{y}$, can be defined as

$$L_j = \ln \frac{P(c_j = 0|\mathbf{y})}{P(c_j = 1|\mathbf{y})}. \tag{9.1}$$

Using Bayes' rule, this can be expressed as

$$L_j = \ln \sum_{\mathbf{a}^{(i)} \in S_j^0} p(\mathbf{y}|\mathbf{x}=\mathbf{a}^{(i)}) P(\mathbf{x}=\mathbf{a}^{(i)}) - \ln \sum_{\mathbf{a}^{(i)} \in S_j^1} p(\mathbf{y}|\mathbf{x}=\mathbf{a}^{(i)}) P(\mathbf{x}=\mathbf{a}^{(i)}). \tag{9.2}$$

The set S_j^1 contains all hypothetical transmit vectors $\mathbf{x} = \mathbf{a}^{(i)}$ where $c_j = 1$, and S_j^0 contains the transmit vectors where $c_j = 0$. $P(\mathbf{x} = \mathbf{a}^{(i)})$ is the a priori probability that vector $\mathbf{x} = \mathbf{a}^{(i)}$ was transmitted. This probability can be calculated using extrinsic feedback from previous iterations. In the first iteration this feedback information is not available, and thus the a priori probability for all symbols is assumed to be equal.

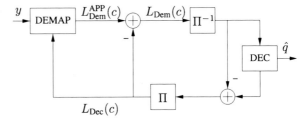

Fig. 9.5 Principle of iterative demodulation and decoding for BICM

9.2.2 Iterative Demodulation and Decoding

In this section we will have a closer look at iterative demodulation and decoding of Bit Interleaved Coded Modulation (BICM). The basic principle is shown in Fig. 9.5. The demapper (DEMAP) calculates LLRs according to Eq. (9.2). After deinterleaving, the information is passed forward to the decoder (DEC). The decoder may use the BCJR algorithm to calculate improved LLRs which are interleaved and fed back to the demapper. To avoid positive feedback, only extrinsic information, which was gained from other bits, is passed between the components. This means that the information already available at the input is subtracted from the output of each component.

In such a system, the performance gain of the iterative demodulation and decoding is depending on the constellation and on the bit mapping and channel code. It is known that "anti-Gray" mapping may improve performance. However, the simulated performance of such schemes with turbo coding has been disappointing, being poorer than Gray mapping with no feedback. [Bur07] indicates the reason for this: although anti-Gray mapping can improve the mutual information for perfect feedback, it also has a loss for no feedback. This means that a larger Signal-to-Noise Ratio (SNR) is required for the convergence of the iterative process to start. Since the performance is mainly limited by the start of convergence and not the final level of convergence, Gray mapping may outperform anti-Gray for a turbo-coded system.

The metric used in the demapper is a crucial point. It is based on the Probability Density Function (PDF) of the received symbols, which depends on the channel, the noise at the receiver, and the transmit signal. The feedback information has also to be processed adequately in this device. In [SD07] the authors state that the performance of an iteratively detected MIMO system can be improved by using an adapted decoding metric that considers the imperfectness of the channel estimate obtained from the statistics of the estimation errors.

One example of iterative demodulation and decoding was studied in [WDTL08], where a transmission scheme is used that is a combination of OFDM and M-ary Frequency Shift Keying (MFSK) (see Sect. 9.3). It was shown that the iterative receiver can improve the power efficiency by more than 1 dB compared to a noniterative receiver for OFDM-4FSK. This is also true for fast time variant frequency selective channels as shown in Fig. 9.6. The receive symbols are detected noncoherently, and therefore exact Channel State Information (CSI) is not needed. The dashed curves show the corresponding result for a two-path channel with equal path amplitudes

Fig. 9.6 BER curves for noncoherently detected OFDM-4FSK with iterative detection for a two-path channel with a Doppler spread of DS = 0.135. Comparison of conventional modulation (*dashed lines*) with extended mapping and precoding. Channel code: convolutional code with code rate r_c and memory m [WHTL09] (©IEEE, reproduced with permission)

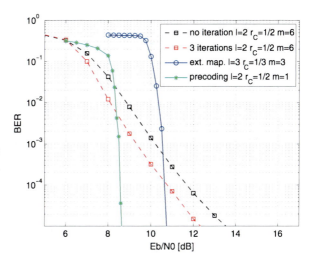

and a Doppler spread of $DS = 2vf_c/(c\Delta f) = 0.135$. Here v denotes the receiver velocity, c the propagation velocity, and f_c the carrier frequency. DS is normalized to the OFDM subcarrier spacing Δf. To optimize the performance of the iterative receiver, the channel code and the bit mapping have to be adapted to each other. One possibility is to use "extended mapping" where the number of bits per symbol is increased above $l = \log_2 M$, leading to an increased data rate and an ambiguous mapping. The increased data rate can be used by a stronger channel code with lower rate to compensate the increased error rate caused by the ambiguous mapping. A second possibility is to adapt the channel code by using precoding in the form of an additional inner recursive convolutional code with rate one, which is increasing the mutual dependency among the code bits. Both methods have been discussed in [WHTL09], and the results are also shown in Fig. 9.6. It can be seen that both methods lead to a steeper waterfall region and a reduced BER for high SNR. A method to further increase the power efficiency of noncoherent OFDM-MFSK is to jointly detect multiple symbols and exploit the correlation of the channel transfer function in frequency direction [WLTL10]. By Multiple Symbol Detection (MSD) the loss in power efficiency due to noncoherent detection can be partly avoided. In addition, the iterative receiver can profit from the larger number of jointly detected bits in MSD, increasing the gain during the iterations.

9.2.3 Iterative Interference Cancellation

9.2.3.1 Code Division Multiplexing

Generally, for linear modulation schemes, bandwidth efficiency is improved by increasing the size of the complex-valued transmit symbol alphabet. However, this leads to a decreased power efficiency. Alternatively, the modulation can be kept simple, e.g., Binary Phase Shift Keying (BPSK), and the data rate is increased by multiplexing the data stream on several subchannels. The subchannels can be formed by spreading codes or also by different interleavers in the case of Interleave Division Multiplexing (IDM) [PLWL06]. Increasing the number of subchannels (codes) beyond a threshold given by the dimension of the signal space inevitably leads to interference between the subchannels. Therefore all these schemes have in common that they require powerful iterative detection algorithm. Vanhaverbeke et al. [VMS02] proposed a multiple access scheme based on sets of orthogonal spreading codes. The sets are formed by orthogonal Walsh–Hadamard codes. Different sets are separated by complex scrambling sequences. This allows us to control the interference to some extent. Teich and Kaim [TKL05] present a description of m-Orthogonal Code Division Multiplexing (m-OCDM) in the general framework of vector-valued transmission [Lin99]. Interference is mitigated by a powerful iterative vector equalization algorithm (soft-decision iterative interference cancellation) [TS96]. For a spreading length of $N = 128$, the bandwidth efficiency can be increased up to three bit/s/Hz without loss in power efficiency. The convergence of the iteration process is investigated by a variance transfer chart analysis [AGR98] in the limit of large code length N. Results for small code length, i.e., $N = 1$, are also given. In this case the signal space is two-dimensional, and general QAM symbol alphabets are recovered. In this limit, m-OCDM can be linked to the recently introduced method of superposition mapping [WH10].

9.2.3.2 Turbo Equalization

The idea of turbo equalization is to use the redundancy of the channel code to improve the equalization process. This is achieved by iteratively exchanging soft information between the equalizer and the channel decoder. This leads to a structure similar to Fig. 9.5, where the demapper block now includes the equalizer. The LLRs provided by the channel decoder are now used to improve the calculation of the estimated interference, which shall be removed from the receive signal. The equalizer itself can also be realized as an iterative device. In [PFSTL10] the authors use the same iterative vector equalization algorithm as in the previous subsection on code division multiplexing to remove the ISI in an 3GPP-Long Term Evolution (LTE) uplink scenario. The modulation scheme adopted for the LTE uplink is Single Carrier Frequency Division Multiple Access (SC-FDMA). The advantage of SC-FDMA is the reduced Peak-to-Average Power Ratio (PAPR) at the transmitter compared to an OFDMA transmission. However, the Fourier spreading of SC-FDMA leads to ISI when transmitting over a multipath channel, and therefore a more complex equalizer is required. It was shown that the performance can be improved by up to 5 dB by using soft-decision interference cancellation compared to an Minimum Mean Squared Error (MMSE) block linear equalizer in the medium SNR range. By using a turbo equalizer, where the output information of the channel decoder is fed back to the vector equalizer, the performance can be further increased.

9.2.3.3 Clipping Noise Cancellation

The received signal in many OFDM systems contains distortions due to amplitude clipping in the transmitter. Amplitude clipping can be used to reduce the high PAPR of OFDM signals which would reduce the system performance, but it destroys the orthogonality of the subcarriers and therefore leads to interference. In [DCG07] the authors examined receive algorithms which iteratively cancel clipping distortion. Turbo Decision Aided Reconstruction (Turbo-DAR) [CGD04] jointly uses the channel coding gain and a time-domain reconstruction loop to iteratively improve the signal detection. In a second algorithm [CH03], the distortion caused by clipping is iteratively estimated and subtracted from the received signal. The performance of this algorithm can be improved by using a modified soft decision receiver [DCG07].

9.2.4 Factor Graph Sum-Product Algorithm

A factor graph based SPrA is widely used in communications for iterative detection and synchronization algorithms. A pristine canonical form of the messages passed among the factor and variable nodes should be the probability density function. In systems with discrete-only variables, the message density function can be easily parameterized by a small number of parameters (e.g., the log-likelihood ratio) which are passed among the nodes. But this does not hold for systems with a mixture of discrete (data) and continuous (e.g., channel parameters) variables. The messages are complicated functions which are very difficult to be passed among the nodes in their pristine form.

The solution lies in finding a proper approximate representation by a parameterizable set of kernels; it is called a canonical message representation. The procedure of finding a suitable set of kernels with the minimal number of parameters for a given fidelity criterion is however not an easy one. In some cases, the form of the kernels can be inferred from the system scenario. Namely in linear systems with linear modulation formats, it frequently leads to mixture-Gaussian canonical kernels. In [LBB09] the authors look at the SPrA applied to joint channel estimation and decoding for multipath channels. They propose to replace the involved probability densities by standard messages where the channel estimate is modeled as such a Gaussian mixture. However, the situation in nonlinear systems (e.g., with nonlinear CPM modulation) is much more complicated.

The problem of canonical message representation for CPM-based systems and its bit error rate performance was addressed in [Syk07] and [SP08]. The proposed solution relies on modulo 2π circular phase-space messages transformed, as an approximation, into a linear infinite support. It was shown that this solution effectively reduces the dimensionality of the message support from two (real and imaginary parts of complex envelope) to a one-dimensional unwrapped phase-space with a negligible fidelity drop.

The above stated procedure of obtaining the kernels heavily depends on a particular system scenario, and it is mostly an ad hoc approach. A general systematic procedure for finding the kernels was developed in [PS09]. It exploits the fact that the messages are random functions depending on iteration number and random observation inputs of the factor graph. The Karhunen–Loève-based transform is used to obtain the eigenfunctions from the empirical message correlation function (obtained by offline simulation). These eigenfunctions serve as kernels for canonical message representation. An essential feature of such an approach is the fact that the approximation decomposition is directly tied to the fidelity criterion (mean square error) through the second-order moments of the eigenvalues. This, together with the orthonormality of the kernels, gives an easy tool for designing a message representation with minimal number of parameters for a given fidelity requirement. A variety of numerical results is shown in [PS09]. One of the very surprising qualitative results shown there is the fact that kernel functions obtained from Karhunen–Loève analysis tend to create harmonic-like (sinusoidal) shapes for kernels in various particular scenarios. This has also a very pleasing effect on the implementation efficiency of the factor node update equations.

9.3 Modulation, Adaptive Techniques, Multicarrier

Within COST 2100 most advances in the field of modulation have been obtained in the field of multicarrier modulation.

9.3.1 Modulation Techniques

Although today many mobile communication standards are based on multicarrier techniques, single-carrier transmission is still of relevance for specific applications. Most important is the uplink channel where the mobile terminal should transmit as much power as possible while consuming as little power as possible from the battery. For this application, CPM shows very good performance since low-cost and highly efficient nonlinear amplifiers can be used.

In order to minimize the required bandwidth, long-duration frequency impulses have to be used. Correspondingly, intersymbol interference at the receiver occurs which has to be equalized. The optimum detection method is maximum likelihood sequence estimation which can be implemented using the Viterbi algorithm. In order to reduce the computational complexity, in [HS09] it is proposed to use a low-dimensional approximation which is based on the orthogonal subspace projection in the constellation space. It is found that this approach is equivalent to a principal component analysis presented in [MA03]. By simulation it is shown that a receiver using the low-dimensional approximation shows almost the same BER performance as a receiver using no projection.

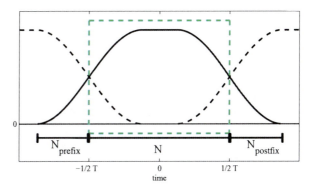

Fig. 9.7 Raised cosine window with roll-off factor $\alpha = 0.75$ for weighting the samples at the receiver input

9.3.2 Multicarrier-Related Techniques

In mobile radio communications, OFDM is the most widely used multicarrier transmission technique.

By using a large number of subcarriers, intersymbol interference can be avoided effectively. On the other hand, a large number of subcarriers is associated with a narrow bandwidth of the individual subchannels. Therefore, OFDM is more sensitive to the time variance of radio channels than single-carrier transmission techniques: The Doppler spread of fast fading channels causes Inter-Carrier Interference (ICI) which leads to a severe performance degradation. In [PDTL09], a windowing technique at the receiver is proposed to reduce the effect of fast fading channels.

It is proposed to extend the symbol duration at the transmitter using an additional cyclic pre- and postfix. At the receiver a temporal window function (see Fig. 9.7) is applied before applying the Discrete Fourier Transform (DFT).

The performance of a Worldwide Interoperability for Microwave Access (WiMAX) system using the windowing approach is shown in Fig. 9.8, which displays the bit error ratio (BER) versus the normalized velocity of the mobile station. It can be observed that windowing significantly reduces ICI. This results from the effect that the window functions lead to a faster decrease and a narrower main lobe of the corresponding frequency spectrum.

Another OFDM-based technique which is robust against fast fading is MFSK. In [WPTL08] it is proposed to use adjacent OFDM subcarriers for Frequency Shift Keying (FSK) transmission (see Fig. 9.9). This method is quite robust because non-coherent reception can be carried out.

The main drawback of OFDM-based FSK transmission is its low spectral efficiency (0.5 bit/(s × Hz) for the example of Fig. 9.9). To mitigate this, it is proposed to combine FSK transmission with Differential Phase Shift Keying (DPSK). DPSK can be carried out with respect to both: time or frequency. For fast fading channels, it is advantageous to modulate differentially in the frequency domain if the coherence bandwidth is large enough. As an alternative to the increase of bandwidth efficiency, in [WPTL08] it is also proposed to use the degrees of freedom of the phases of the FSK subcarriers to reduce the PAPR.

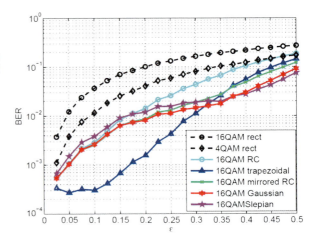

Fig. 9.8 Bit error ratio versus normalized Doppler frequency $\varepsilon = f_{\text{Doppler}}/\Delta f$ (with subcarrier spacing Δf) for $10\lg E_b/N_0 = 60$ dB and different window functions. (rect = rectangular window, RC = raised cosine window with $\alpha = 1$)

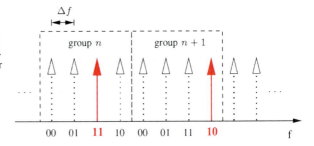

Fig. 9.9 Principle of OFDM-4FSK modulation: Two bits are assigned to each subcarrier using Gray coding. One out of each group of four subcarriers is taken for transmission, indicated by a *solid arrow* [WHTL09] (©IEEE, reproduced with permission)

Interleaved Frequency Division Multiple Access (IFDMA) is a method to solve the multiple access problem between different users and distribute the information of individual users on the available frequency bandwidth [SBS99]. In [SDL08] two types of Block-Interleaved Frequency Division Multiple Access (B-IFDMA) methods are investigated: the joint DFT B-IFDMA, which is based on the application of one DFT over all subcarriers assigned to a given user, and the added-signal B-IFDMA, which is based on the assignment of multiple IFDMA signals to one user. Especially in the uplink, ICI is a fundamental problem, since the radio channels of different users in general exhibit different Carrier Frequency Offsets (CFOs). In [SDL08] the power of ICI due to CFOs is calculated analytically. The effect of CFOs in the downlink is much less critical since each mobile station receives just a single signal so that only one single CFO has to be compensated. The signal components for each mobile station are affected by the same CFOs.

The sensitivity of OFDMA with respect to CFOs is also discussed in [MAC07]. Here, a system is considered where adjacent subcarriers are assigned to a specific mobile user. Different receiver architectures for CFO compensation can be designed with different computational complexity.

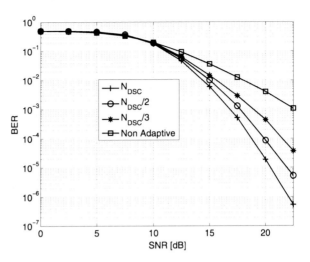

Fig. 9.10 BER for N_{DSC} individually adapted data subcarriers, $N_{DSC}/2$ groups of subcarrier pairs, and $N_{DSC}/3$ groups of subcarrier triplets. Possible modulation schemes are taken from the set {no transmission, BPSK, 4-QAM, 16-QAM, 64-QAM} [CHC09] (©2009 VDE, reproduced with permission)

9.3.3 Adaptive Techniques

Because of the finite time dispersion of multipath radio channels, they are frequency-selective. To mitigate this effect, modulation and coding schemes can be adapted depending on the quality of the subchannel (also called bit-loading). Assuming a multipath Rayleigh fading channel with log-normal shadowing, the probabilities that specific modulation formats are used are calculated in [Bot07].

For an OFDM transmission system with adaptive modulation, the use of different modulation schemes for the individual subcarriers may be summarized in a so-called Bit Allocation Table (BAT). The information of the BAT has to be synchronized between transmitter and receiver. A simple straightforward solution for this synchronization may be that the BAT is transmitted via a signaling channel so that the receiver can use the same BAT. In order to reduce the signaling overhead, adjacent subcarriers can be grouped together so that the same modulation scheme is used for each group of subcarriers.

On the other hand, grouping subcarriers causes that the adaptation of modulation schemes becomes less flexible resulting in a degradation of the bit error probability (see Fig. 9.10, [CHC07, CHC09]).

In [CHC07] the signaling overhead is analyzed in detail, and—for a WLAN-type system—it is found that the signaling overhead is 11.5% if the BAT is encoded efficiently. Source coding allows one to further reduce the signaling overhead: Huffman coding gives already some gain. But significant gain is obtained if state-dependent Huffman coding is used, which leads to an overhead of only 5.5%.

An alternative to signaling the BAT from transmitter to receiver or vice versa is to avoid signaling at all and to blindly estimate the BAT at the receiver. This so-called automatic modulation classification is discussed in detail in [HCC09]. The proposed maximum a posteriori algorithm achieves very high probabilities of a correct classification (close to 1) even for short data packets of 10 symbols at an SNR of 10 dB.

In [HCC07] a cross-layer comparison of two different design philosophies for discrete-rate adaptive transmission is carried out. A transmission via slowly time-varying flat-fading wireless channels is considered. The first method maximizes the average spectral efficiency for adaptive coded modulation (MASA scheme). The second method maximizes the average reliable throughput (ART scheme). To facilitate reliable communications, the ART scheme is extended by a retransmission option.

9.4 Channel Estimation and Synchronization

In practical communication systems, channel estimation and synchronization are inescapable. Although the subjects have been widely studied, some new constraints and the need for always better performance make relevant some more research activities. This section focuses on those two aspects. In a first part, some difficulties linked to the network evolutions are discussed, and a solution is proposed. Ways to reduce the needed signalization in FDD context and considerations on complexity/performance trade-offs are proposed. It is to be noticed that increasing the receiver complexity allows one to reduce retransmission and, consequently, energy dissipation and network interference. Finally, a solution for channel estimation updates based on Space–Time Block-Code (STBC) properties is made for medium speed mobiles. In the second part some proposals to improve frame synchronization using the constraints imposed by coding are made, and a brief word on the specific case of cooperation is given.

9.4.1 Channel Estimation

Channel estimation has been widely studied in literature. We do not address in this project general methods on this subject, but we are concerned by specific contexts. One application field is the use of repeaters in Digital Video Broadcasting services either terrestrial or more recently hand held. High isolation is required between transmit and receive antennas to reduce the impact of unwanted echoes. Physical separation is generally impractical and signal processing appears to be an effective solutions [NCBG07b, NCBG07a]. A training sequence is added to the transmitted signal and allows estimating the response of a Finite Impulse Response (FIR) filter to realize an echo cancellation on the received signal.

Some other difficulties remain due to the necessary exchanges between the receiver and the transmitter. Especially in an FDD system, it would be an important gain if the Channel Impulse Response (CIR) could be directly estimated from the incoming link before being used for transmission. This can only be done if this estimation can be accurately transformed from one frequency band to another frequency band. When using multiple antenna, this can be done using the Direction of Arrival (DoA) estimation. In [PW07], it is shown that the channel matrix can be written as

$$\mathbf{H}(f_0) = \mathbf{H}_d(f_0)\mathbf{A}^T(f_0), \tag{9.3}$$

where $\mathbf{H}_d(f_0)$ is a directional matrix, and $\mathbf{A}(f_0)$ a steering matrix. Based on (9.3), a change in carrier frequency from f_0 to $f_0 + \Delta f$ can be obtained by an elementwise multiplication on the steering matrix and a matrix multiplication on the directional matrix. Comparisons with measurements have shown that these operations allow an accurate channel estimation in another frequency band.

Another challenge is the complexity/performance trade-offs. When ISI is considered, equalizers like A Posteriori Probability (APPb) or MMSE equalizers need an estimation of the channel coefficients. A factor graph with Belief Propagation (BPr) approach can help defining a joint channel estimation and decoding. One difficulty is to handle continuous variables. A proposed solution [WS01] is to propagate quantized probability distributions, but this does not fill the complexity gap between BPr and other algorithms and let BPr unfeasible when multipath channels are considered. In [LBB09], the approximation of channel estimates is done with a mixture of Gaussian distributions. It allows one both to propagate continuous distributions and reduce complexity in the factor graph for multipath channels with performance close to the APPb equalizer and better than the MMSE.

However, complexity in the receiver has an impact on the global energy dissipation and network interference. Increasing signal processing complexity can induce a reduction in the number of retransmission and, in the sequel, reduce the energy dissipation but also network interference. In [Wsz08, WL10], the association of a linear chip level equalizer with a nonlinear symbol level equalizer for a Wideband Code Division Multiple Acess (WCDMA) system is considered. The nonlinear equalization is based on artificial neural networks. The nonlinear correction is only applied when channel or interference conditions prevent the linear equalizer from an efficient decoding. Many retransmissions can then be saved leading to energy savings.

Finally, applications in high-speed environment remain a difficult challenge. A proposed solution [KL07] is to update the channel estimate using the properties of STBC. Let $\mathbf{X}$ be a matrix representing the transmitted symbols from an STBC scheme. The transmission can be describe as

$$\mathbf{Y} = \mathbf{H}.\mathbf{X} + \mathbf{N}, \tag{9.4}$$

where $\mathbf{H}$ is the channel transfer matrix, and $\mathbf{N}$ the additive white Gaussian noise. It was proven, concerning STBC, that [TJC99]

$$\mathbf{X}.\mathbf{X}^H = A.\mathbf{I}_n, \tag{9.5}$$

where A is a constant equal to the sum of the powers of all transmitted symbols, and $\mathbf{I}_n$ is the $n \times n$ identity matrix. Using (9.4) and (9.5), we obtain:

$$\mathbf{Y}.\mathbf{X}^H = A.\mathbf{H} + \mathbf{N}.\mathbf{X}^H. \tag{9.6}$$

The multiplication of the matrix of the received signals and the matrix of the estimated transmitted signals (transpose and conjugate) gives the channel transfer matrix $\mathbf{H}$ disturbed by noise. $\mathbf{H}$ can be iteratively estimated after decoding of each block. The proposed algorithm is to first estimate the channel by calculate the mean of the estimates on T training blocks. Then, when a data block is received, the oldest training block is replaced by the new estimate using (9.6) and so on. This allows one to update the channel estimate without frequent training sequences. This algorithm showed to be very effective as long as SNR is high and speed not too large (up to about 60 km/h). At larger speed, differential STBC gets better results.

9.4.2 Synchronization

We first consider here some propositions for frame synchronization. It is usually achieved by periodic insertion of a predefined sequence (so-called synchro-sequence, synchro-pattern, or frame alignment word) at the beginning of each frame. Although very simple, this method suffers from an inherent drawback of random simulations of the synchro-sequence in the data portion of the frame [Nie73b]. The probability of these simulations decreases as the synchro-sequence length increases and also depends on the sequence structure. It was shown that for the practical applications, so-called bifix-free sequences are the best choice, since they have the maximum expected period between two simulations in random data stream [Nie73a]. A sequence is bifix-free if there is no subsequence which is both prefix and suffix of the sequence. Following this work, in [BSV05, SB07] the probability of the first simulation of the sequence on a given position in the random data stream was derived, from which other statistics, such as the expected duration of the search and the probability that the sequence is not simulated in a given portion of the stream, can be obtained. A detailed overview of different synchro-sequence selection criteria is given in [Sch80].

In case of the coded data transmission, frame synchronization can be aided using the constraints imposed by the code, resulting in higher probability of correct frame synchronization.

A first strategy is to use pilotless or blind synchronization, based on the code constraints only, which has the advantage of bandwidth reduction. It gained an increased attention [MI02, SV04, LKJV07, IHD02] with the recent introduction of the capacity-achieving codes decoded by the iterative decoding algorithms, but these algorithms typically suffer from large computational complexity.

In [MI02], the metric used to discriminate the beginning of the frame (i.e., the LDPC code codeword) among the candidate positions is the mean of the absolute value of variable node LLR, which are outputs of the sum-product algorithm after a single iteration. A simplified approach to the LDPC coded frame synchronization is to use the metric based on constraints (check nodes). In [LKJV07], a simple hard-decision detection of the beginning of the frame is employed in LDPC coded transmission. For every bit position in the time-frame of one whole codeword, the syndrome is formed using the hard values of the received bits. The output of the algorithm is the position for which the syndrome contains the largest number of satisfied constraints. In [IHD02], both hard-decision and soft-decision detection algorithms were analyzed. While the hard-decision variant is essentially the same as in [LKJV07], the soft-decision variant is based on the soft-values of the check node LLR. The output of the soft-decision algorithm is the position for which the sum of the LLR of the check nodes is the largest (see [SV04] for a related work). In [SVB08a] LDPC convolutional codes are preferred for the frame synchronization. Tested with hard or soft decision, they allow a considerably shorter acquisition time and offer a more flexible solution.

However it can be preferred, as proposed in [Rob95] and [CG04], to proceed to a two-stage frame synchronization. The first stage is the standard synchronization based on the synchro-sequence which outputs the list of the candidate positions. In the second stage, decoding is performed just for these positions. The final output is the position for which the sum of the sequence detection metric and the decoding metric is the highest. General optimal rule for locating the correct frame starting position which jointly considers synchro-sequence detection and constraints enforced on the codeword symbols is given in [HK06]. Another interesting approach that jointly considers synchro-sequence and code constraints was given in [WSBM06], where the expectation maximization technique was used to estimate the frame starting position. In [SVB08b], a simple two-stage list frame synchronizer for the LDPC coded transmission over the Additive White Gaussian Noise (AWGN) channel is proposed. It is shown that, by trading small bandwidth expenditure on the short synchro-sequence used for the list generation, frame synchronization is achieved with improvements in terms of complexity and without loss in terms of the synchronization Frame Error Rate (FER), as compared to blind frame synchronization LDPC coded schemes. On the other hand, in comparison with the frame synchronizers based on the synchro-sequence only, the proposed synchronizer provides several orders of magnitude smaller FER for the same synchro-sequence lengths.

One last specific aspect is to be mentioned, the case of cooperative communications which is discussed in the previous chapter. To implement such a technology, the different carrier frequency offsets and timing errors from the individual nodes provide a challenge. It is shown in [SZM+10] that individual carrier frequency offsets lead to a strongly time-variant compound channel (even if each individual physical channel is static). Timing offsets may lead to the misalignment of the space–time code used. Both problems are solved by combining space–time codes with OFDM transmission, together with time-variant channel estimation at the receiver. The cyclic prefix of OFDM turns the timing offset into a complex rotation, while time-variant channel estimation takes care of the channel variability. By this system design, the bit error rate performance has been improved by an order of magnitude.

9.5 Interference Modeling and Suppression

9.5.1 Modeling Ultra-wideband Interference

In networks, multiple access is the main component of interference. In the past, it has often be considered as the sum of numerous independent and identically distributed random variables. As a consequence, a Gaussian approximation was used and often gave accurate results. However, the evolution of systems (for instance, IR-UWB and ad hoc networks) significantly modify the Multi-Access Interference (MAI) distribution. Sousa shows in [Sou92] that in ad hoc networks, MAI can be modeled as an α-stable random variable. This result is also presented with lognormal shadowing and Rayleigh fast fading in [IH96]. In [PCG+06], Pinto derives an exact expression for the error performance of a narrowband communication system subject to multiple UWB interferers and AWGN. Similarly, in IR-UWB it is shown in [DR02] that the Gaussian approximation significantly underestimates the Bit Error Rate (BER) of practical Time Hopping (TH) Pulse Position Modulation (PPM) systems. Several modeling approaches have been proposed in this situation [BY09] to better represent the impulsive nature of the IR-UWB. Win et al. have proposed a mathematical framework for network interference in [WPS09]. They show that, when interferers are scattered according to a Poisson process, the aggregate interference amplitude follows a symmetric stable distribution.

We consider here a situation where the TH-PPM-UWB interferers are spatially scattered according to an infinite Poisson field and are operating asynchronously [GCR07, GCA+10, CCX+10]. Let κ_R be the random variable representing the number of pulses that interfere. The MAI random variable Z can then be written as

$$Z = \sum_{k=1}^{\kappa_R} \gamma_k \psi_k, \qquad (9.7)$$

where γ_k represents the channel attenuation (eventually including path loss, shadowing and multipath [WPS09]), and ψ_k depends on the pulse correlation function.

In this context, the MAI characteristic function can be written as

$$\varphi_Z(\omega) = \log\big(\mathbb{E}\big[e^{j\omega(\sum_{k=1}^{\kappa} \gamma_k \psi_k)}\big]\big) = \frac{\bar{N}q}{R^2}|\omega|^{\frac{4}{a}} F, \qquad (9.8)$$

where $\bar{N}$ is the mean number of interferers in the considered area (a circle of radius R), q is related to the repetition period of the pulses, and F depends on the pulse shape. The value F is independent of ω so that Z is a symmetric α-stable random variable with parameters $\alpha = 4/a$ and $\sigma = -(\bar{N}q/R^2)F$ (the two remaining parameters are zero). This result validates the α-stable assumption. The main unrealistic hypothesis is to neglect the near field, but it has a very limited impact [Sou92, GCA+10].

9.5.2 Impact on Receiver Design

Geometric SNR A difficulty arises in the use of the α-stable model when $\alpha < 2$: variance is infinite. The current notion of noise power N_0 is mathematically undefined, and the standard SNR is not defined strength. Hence, an alternative form of the noise power is necessary. One practical solution is to use a new indicator of this process strength, namely the geometric noise power S_0. This estimator is a scale parameter, and as such, it can be effectively used as an indicator of process strength

Fig. 9.11 Asymptotic performance of LDPC codes with different demapper strategies. The X axis is the geometric E_b/N_0. The Y axis gives the capacity (Shannon limit) or the code rate. We can see how close from the capacity we can expect to be for different code rates and for different LLR calculations [MGCG10] (©2010 IEEE, reproduced with permission)

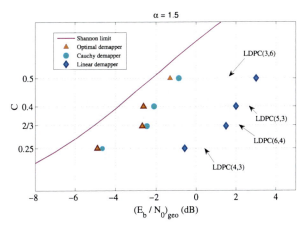

when second-order methods are inadequate. This new power [GPA06] is defined for a Random Variable (RV) X by:

$$S_0 = S_0(X) = \exp\{\mathbb{E}[\log|X|]\}. \qquad (9.9)$$

For a binary source signal taking values in $[-1, 1]$, we the define the geometric E_b/N_0 by

$$\frac{E_b}{N_0} = \frac{1}{2C_g S_0^2 R}, \qquad (9.10)$$

where $C_g = 1.78$ is the exponential of the Euler constant, and R is the code rate. The normalizing constant $2C_g$ ensures that we have the standard energy per bit ratio when $\alpha = 2$.

Log Likelihood Ratio The LLR of the input X associated with the channel output Y is given by

$$\Lambda(y) = \ln \frac{\mathbb{P}\{y|X = +1\}}{\mathbb{P}\{y|X = -1\}} = \ln \frac{f(y-1)}{f(y+1)}, \qquad (9.11)$$

where $f(\cdot)$ is the PDF of the noise. When the noise is SαS, $f(\cdot)$ has, in general, no analytical expression. However, the density evolution tool [RU07] allows evaluating the asymptotic performance of LDPC codes and the impact of different demappers. We show in Fig. 9.11 the performance of such codes with different rates. The optimal demappers use the exact PDF. This is possible because, although no explicit form is known when $\alpha = 1.5$, we can rather easily generate stable RV [NS95]. We also consider a linear demapper that would be optimal in a Gaussian noise. Finally, we consider the Cauchy demapper. Its expression is obtained based on the Cauchy distribution, which is a special case of stable distributions for $\alpha = 1$. The demapping operation is given by

$$\Lambda_{\text{Cauchy}}(y) = \ln \frac{\gamma^2 + (y+1)}{\gamma^2 + (y-1)}, \qquad (9.12)$$

where γ is the dispersion of the noise. When the received value y tends to $\pm\infty$, $\Lambda_{Cauchy}(y)$ tends to 0. It means that the confidence we put in large values is very low contrary to the linear approach which gives a large LLR for large received values. We see in Fig. 9.11 that the linear demapper results in a significant performance degradation when compared to the optimal LLR. On the contrary, the Cauchy demapper takes into account the impulsive nature of the noise and largely limits its impact. We even notice that this second demapper behaves really closely to the optimal one.

9.5.3 Multihop Transmissions

System Model It is also interesting to evaluate the performance of a multihop system. We consider, for instance, a Decode and Forward (DF) relaying scheme in the presence of MAI modeled by symmetric α-stable RV. The signal from the direct and the relay links at the receiver can be written in matrix form yielding

$$\begin{bmatrix} r_{s,d}^{(k)} \\ r_{r,d}^{(k)} \end{bmatrix} = \sqrt{E^{(k)}} \begin{bmatrix} h_{s,d}^{(k)} & 0 \\ 0 & h_{r,d}^{(k)} \end{bmatrix} \cdot \begin{bmatrix} S^{(k)} \\ \hat{S}^{(k)} \end{bmatrix} + \begin{bmatrix} n_{s,d}^{(k)} \\ n_{r,d}^{(k)} \end{bmatrix}. \tag{9.13}$$

Indexes s, d and r, d refer to the source-relay and to the relay-destination links, respectively. The received signals from user k are $r_{s,d}^{(k)}$ and $r_{r,d}^{(k)}$. $S^{(k)}$ is the source information, and $\hat{S}^{(k)}$ is the estimated source transmitted from the relay. Finally, the h are the channel attenuation, and E the transmitted signal energies, equal at the relay and the destination. The receiver output after correlation can be written as

$$Z_d = Z_1^u + Z_2^u + Z_1^{MAI} + Z_2^{MAI} + N_1 + N_2, \tag{9.14}$$

where index 1 refers to the direct link, and index 2 to the relay link. Fortunately, Gaussian random variable combination $N_1 + N_2$ still follows Gaussian distribution with variance $2\sigma_n^2$ if assuming the same noise factor at the relay and the receiver. Thus, the nature of addition of MAI contributions $Z_1^{MAI} + Z_2^{MAI}$ has to be investigated. Taking advantage of the stability property, a linear combination of α-stable noise remains α-stable whose parameters can be calculated (see, for instance, [ST94]). The MAI contribution at the receiver output can then be characterized by an α-stable distribution:

$$Z_1^{MAI} + Z_2^{MAI} \sim S_\alpha\left((\sigma_1^\alpha + \sigma_2^\alpha)^{1/\alpha}, 0, 0\right). \tag{9.15}$$

Where α and σ have been obtained in Sect. 9.5.1. With the standard linear receiver, the average bit error rate for independent and equiprobable transmitted bits is

$$P_{e,d}|_{N_1,N_2} = \int_{-\infty}^{+\infty} F_{Z_1^{MAI}+Z_2^{MAI}}(-x) \cdot f_X(x)\,dx, \tag{9.16}$$

where $x = Z_1^u + Z_2^u + N_1 + N_2$.

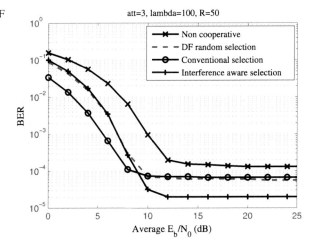

Fig. 9.12 Performance of DF relaying scheme with different strategies for the relay choice [CCX+10] (©2010 EuMA, reproduced with permission)

Interference Aware Relay Selection Issue Conventional selection criteria has been designed for environments without interference and thus does not take into account the effects of interference. We present a relay selection criteria that takes into account the MAI. The interference aware relay selection policy can be expressed as

$$k_{\text{Inter}} = \arg \max_{k \in Srelay} \min \left\{ \frac{h_{sr,j}}{\gamma_{MAI\_relay}}, \frac{h_{rd,j}}{\gamma_{MAI\_dest}} \right\}. \quad (9.17)$$

Note that this criteria does not take into account the Gaussian noise but only the MAI whose strength is quantified by the dispersions γ_{MAI_relay} and γ_{MAI_dest} at, respectively, the relay and the destination and the channel attenuations from source to relay $h_{sr,j}$ and relay to destination $h_{rd,j}$.

Figure 9.12 investigates the relay choice strategy. The comparison with the direct link considers that the global transmitted power is the same, i.e., the source uses twice more power when no relay is involved. Three possible relay nodes are randomly located in a rectangle where the X axis is in $[0.1d_{sd}, 0.9d_{sd}]$ and ordinate Y in $[0, 0.3d_{sd}]$. We show the BER performance versus the SNR for different selection schemes with adaptive DF. Firstly, it can be seen that the random relay selection is inefficient. Furthermore, we observe that the conventional relay selection criterion is more efficient at low SNRs since in this region, AWGN dominates the system degradation. However, from the crossover point, the interference aware strategy outperforms the conventional relay selection criterion. We also observed that the preferred relay position gets closer to the destination when the relay and the destination suffer from interference.

This framework can be extended to take into account both AWGN and MAI. It importantly improves the selection process, and the MAI is characterized by two parameters, highly dependent on the system design and also on the channel attenuation and the network configuration. Another important advantage to the method is that the involved parameters in the relay selections are not varying (system parameters) or only slowly varying (channel attenuation coefficient, interferers' density). Finally,

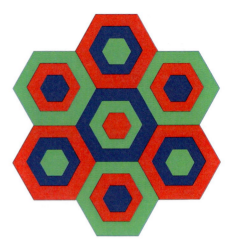

Fig. 9.13 Fractional frequency reuse to mitigate inter-cell interference. *Colors* indicate orthogonal channel groups

performance can be significantly improved, taking into account the impulsive nature of the noise, for instance, in the LLR calculation as suggested in Sect. 9.5.2.

9.5.4 Methods for Interference Limited OFDM(A)-Systems

OFDMA is the modulation and multiple access method of choice for WiMAX and LTE systems alike. In this subsection, three methods for interference-limited OFDMA systems are presented. The first one deals with multiuser detection and channel estimation in OFDM multiple access channels, concentrating on interference between users communicating with a common access point. The second one considers interference between users communicating to different access points, i.e., intercell-interference. The third one analyzes the potential of reusing unused spectral resources in OFDMA systems.

Multiuser interference can be suppressed within the cell by applying multiuser detection followed by single-user channel decoding [RTR+08]. The soft information, i.e., soft symbol estimates based on channel decoder information, can be utilized to iteratively suppress interference and detect desired signals. The method reported in [RTR+08] applies soft interference cancellation that is applicable when the signal structures in question can be estimated down to reasonable accuracy, which is the case of own cell transmissions and coherent detection with channel estimation. The channel estimator in [RTR+08] applies linear MMSE estimation for the channel, using decoder soft information on modulated symbols as additional training symbols.

Systems which can utilize frequency assignment can mitigate intercell interference by assignment optimization. In [SMR08], channels are assigned with an intelligent reuse scheme to users at different distances from cell center. Figure 9.13 illustrates the principle. Optimization strategies to allocate users to orthogonal channel groups are proposed in [SMR08] at network deployment time as a fixed allocation

9 Advanced Coding, Modulation and Signal Processing

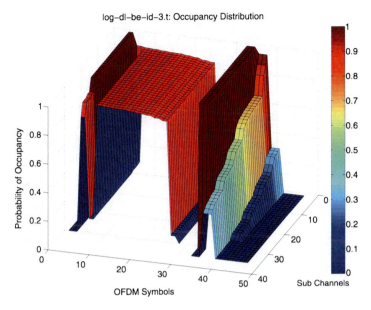

Fig. 9.14 OFDMA DL and UL subframes: three-dimension occupancy distributions. Three users with best effort FTP traffic over TCP [GPE09] (©2009 IEEE, reproduced with permission)

or at runtime as a self-configurable, cooperative allocation working in the absence of network deployment information.

Regardless of system resource allocation optimization, some system resources may be left unused from time to time, potentially to be used by secondary systems operating on the principle of cognitive radio. Both as a measure of efficiency of system efficiency and potential for secondary use opportunities, secondary reuse potential is of interest. In [GPE09], a WiMAX system is simulated, and dynamic spectrum reuse opportunities are detected. The authors estimate the occupancy distribution of the OFDMA resource elements over a consecutive sequence of the OFDMA frames. Such occupancy distribution map gives the average and instant representations of the allocated and available OFDMA resource elements from the underlying primary system. An example of the uplink and downlink subframe occupancies is given in Fig. 9.14.

References

[AGR98] P. D. Alexander, A. J. Grant, and M. C. Reed. Iterative detection in code-division multiple-access with error control coding. *European Transactions on Telecommunications (ETT)*, 9(5):419–425, 1998.

[BGT93] C. Berrou, A. Glavieux, and P. Thitimajshima. Near Shannon limit error correcting coding and decoding: turbo codes. In *IEEE ICC International Conference on Communications*, pages 1064–1070, Geneva, Switzerland, 1993.

[BLMR98] J. Byers, M. Luby, M. Mitzenmacher, and A. Rege. A digital fountain approach to reliable distribution of bulk data. In *ACM SIGCOMM 1998*, pages 56–67, Vancouver, Canada, 1998.

[BNM09] C. Blankenhorn, M. Necker, and C. Muller. Improving reliable multicast streaming in IEEE 802.11 wireless networks. Technical Report TD-09-826, Valencia, Spain, May 2009.

[Bot07] V. Bota. Joint modeling of the multipath channel and user-access for adaptive modulation OFDM schemes. Technical Report TD-07-348, Duisburg, Germany, September 2007.

[BSV05] D. Bajic, C. Stefanovic, and D. Vukobratovic. Search process and probabilistic bifix approach. In *Proc. of the International Symposium on Information Theory, ISIT 2005*, pages 19–22, Adelaide, Australia, September 2005.

[Bur07] A. Burr. Analytical derivation of performance of soft-in, soft-out demodulation. Technical Report TD-07-318, Duisburg, Germany, September 2007.

[BY09] N. C. Beaulieu and D. J. Young. Designing time-hopping ultrawide bandwidth receivers for multiuser interference environments. *Proceedings of the IEEE*, 97(2):255–284, 2009.

[CCX+10] J. Chen, L. Clavier, Y. Xi, A. Burr, N. Rolland, and P. Rolland. Alpha-stable interference modelling and relay selection for regenerative cooperative IR-UWB systems. In *European Wireless Technology Conference (EuWiT)*, September 2010. [Also available as TD(10)10100].

[CG04] T. M. Cassaro and C. N. Georghiades. Frame synchronization for coded systems over AWGN channels. *IEEE Trans. Commun.*, 52:484–489, 2004.

[CGD04] M. Colas, G. Gelle, and D. Declercq. Analysis of iterative receivers for clipped COFDM signaling based on soft Turbo-DAR. In *Proc. 1st International Symposium on Wireless Communication Systems, ISWCS'04*, pages 110–114, Mauritius, September 2004.

[CH03] H. Chen and A. M. Haimovich. An iterative method to restore the performance of clipped and filtered OFDM signals. In *Proc. ICC 2003—IEEE Int. Conf. Commun.*, vol. 5, pages 3438–3442, Anchorage, AK, May 2003.

[CHC07] Y. Chen, L. Häring, and A. Czylwik. Signaling overhead in OFDM systems with adaptive modulation. Technical Report TD-09-864, Valencia, Spain, May 2007.

[CHC09] Y. Chen, L. Häring, and A. Czylwik. Reduction of AM-induced signalling overhead in WLAN-based OFDM systems statistics. In *Proceedings of the 14th International OFDM Workshop 2009, Hamburg*, 2009.

[DCG07] R. Déjardin, M. Colas, and G. Gellé. On the iterative receivers mitigating the clipping noise in coded OFDM systems. Technical Report TD-07-046, Lisbon, Portugal, February 2007.

[DPD+95] C. Douillard, A. Picart, P. Didier, M. Jézéquel, C. Berrou, and A. Glavieux. Iterative correction of intersymbol interference: turbo-equalization. *European Transactions on Telecommunications (ETT)*, 6(5):507–512, 1995.

[DR02] G. Durisi and G. Romano. On the validity of Gaussian approximation to characterize the multiuser capacity of UWB TH-PPM. In *IEEE Conf. on Ultra Wideband Systems and Technologies*, pages 20–23, May 2002.

[Gal62] R. G. Gallager. Low-density parity-check codes. *IRE Trans. Inform. Theory*, IL-8:21–28, 1962.

[GCA+10] H. E. Ghannudi, L. Clavier, N. Azzaoui, F. Septier, and P. A. Rolland. α-stable interference modeling and Cauchy receiver for an IR-UWB ad hoc network. *IEEE Trans. Commun.*, 58(6):1748–1757, 2010. [Also available as TD(08)423]

[GCR07] H. E. Ghannudi, L. Clavier, and P. A. Rolland. Modeling multiple access interference in ad hoc networks based on IR-UWB signals up-converted to 60 GHz. In *European Conference on Wireless Technologies*, pages 106–109, Munich, Germany, October 2007.

[GPA06] J. G. Gonzalez, J. L. Paredes, and G. R. Arce. Zero-order statistics: a mathematical framework for the processing and characterization of very impulsive signals. *IEEE Trans. Signal Processing*, 54(10):3839–3851, 2006.

[GPE09] P. Grønsund, H. N. Pham, and P. E. Engelstad. Towards dynamic spectrum access in primary OFDMA systems. In *Proc. PIMRC 2009—IEEE 20th Int. Symp. on Pers., Indoor and Mobile Radio Commun.*, pages 848–852, Tokyo, Japan, September 2009. [Also available as TD(09)875].

[Hag97] J. Hagenauer. The turbo principle: tutorial introduction and state of the art. In *Proc. International Symposium on Turbo Codes*, pages 1–11, Brest, France, September 1997.

[HCC07] L. Haring, A. Czylwik, and Y. Chen. A cross-layer comparison of two design philosophies for discrete-rate adaptive transmission. Technical Report TD-07-312, Duisburg, Germany, September 2007.

[HCC09] L. Haring, A. Czylwik, and Y. Chen. Automatic modulation classification in application to wireless OFDM systems with adaptive modulation in TDD mode. Technical Report TD-08-536, Trondheim, Norway, June 2009.

[HK06] H. Huh and J. V. Krogmeier. A unified approach to optimum frame synchronization. *IEEE Trans. Wireless Commun.*, 5:3700–3711, 2006.

[HS09] M. Hekrdla and J. Sykora. CPM constellation subspace projection maximizing average minimal distance—sufficiency condition and comparison to PC analysis. Technical Report TD-09-870, Valencia, Spain, May 2009.

[IH96] J. Ilow and D. Hatzinakos. Impulsive noise modeling with stable distributions fading environments. In *8th IEEE Signal Processing Workshop on Statistical Signal and Array Processing*, pages 140–143, June 1996.

[IHD02] R. Imad, S. Houcke, and C. Douillard. Blind frame synchronization on Gaussian channel. In *Proc. of EUSIPCO 2007*, Poznan, Poland, September 2002.

[KL07] P. Kulakowski and W. Ludwin. Iterative channel estimation for space–time block codes. Technical Report TD-07-347, Duisburg, Germany, September 2007.

[LBB09] Y. Liu, L. Brunel, and J. J. Boutros. Joint channel estimation and decoding using Gaussian approximation in a factor graph over multipath channel. In *Proc. PIMRC 2009—IEEE 20th Int. Symp. on Pers., Indoor and Mobile Radio Commun.*, pages 3164–3168, September 2009. [Also available as TD(09)782].

[Lin99] J. Lindner. MC-CDMA in the context of general multiuser/multisubchannel transmission models. *European Transactions on Telecommunications (ETT)*, 10(4):351–367, 1999.

[LKJV07] D. Lee, H. Kim, C. R. Jones, and J. D. Villasenor. Pilotless frame synchronization via LDPC code constraint feedback. *Elect. Lett.*, 11:683–685, 2007.

[MA03] P. Moqvist and T. Aulin. Orthogonalization by principle components applied to CPM. *IEEE Trans. Commun.*, 51:1838–1845, 2003.

[MAC07] K. Maliatsos, A. Adamis, and P. Constantinou. Frequency synchronization, receiver architectures and polyphase filters for OFDMA systems with subband carrier allocation. Technical Report TD-07-381, Duisburg, Germany, September 2007.

[MGCG10] H. B. Maad, A. Goupil, L. Clavier, and G. Gelle. Asymptotic analysis of performance LDPC codes in non-Gaussian channel. In *Proc. IEEE Intern. Workshop on Signal Proc. Advances in Wirel. Comm. (SPAWC 2010)*, 2010. [Also available as TD(10)10082].

[MI02] W. Matsumoto and H. Imai. Blind synchronization with enhanced sum-product algorithm for low-density parity-check codes. In *Proc. WPMC—Wireless Pers. Multimedia Commun.*, vol. 3, pages 966–970, Honolulu, USA, 2002.

[MN96] D. J. C. MacKay and R. M. Neal. Near Shannon limit performance of low-density parity-check codes. *Elect. Lett.*, 32(18):1645–1646, 1996.

[NCBG07a] K. M. Nasr, J. Cosmas, M. Bard, and J. Gledhill. Channel estimation for an echo canceller in DVB-T/H single frequency networks. Technical Report TD-07-315, Duisburg, Germany, September 2007.

[NCBG07b] K. M. Nasr, J. Cosmas, M. Bard, and J. Gledhill. Echo canceller design and performance for on-channel repeaters in DVB-T/H networks. Technical Report TD-07-203, Lisbon, Portugal, February 2007.

[NGB10] K. Nybom, S. Grönroos, and J. Björkqvist. Expanding window fountain coded scalable video in broadcasting. In *IEEE International Conference on Multimedia and Expo, ICME 2010*, Singapore, July 2010. [Also available as TD(09)934].

[Nie73a] P. T. Nielsen. On the expected duration of a search for a fixed pattern in random data. *IEEE Trans. Inform. Theory*, 19:702–704, 1973.

[Nie73b] P. T. Nielsen. Some optimum and suboptimum frame synchronizers for binary data in Gaussian noise. *IEEE Trans. Commun.*, 21:770–772, 1973.

[NS95] C. Nikias and M. Shao. *Signal Processing with α-Stable Distributions and Applications*. Wiley-Interscience, New York, 1995.

[NV10] K. Nybom and D. Vukobratovic. Receiver-controlled fountain coding for layered source transmission over parallel broadcast channels. Technical Report TD-10-12078, Bologna, Italy, November 2010.

[NVB09] K. Nybom, D. Vukobratovic, and J. Bjorkqvist. Sparse-graph AL-FEC solutions for IP datacasting in DVB-H. In *IEEE International Symposium on Broadband Multimedia Systems Broadcasting, BMSB 2009*, Bilbao, Spain, May 2009. [Also available as TD(07)355].

[OT04] R. Otnes and M. Tuchler. Iterative channel estimation for turbo equalization of time-varying frequency-selective channels. *IEEE Trans. Wireless Commun.*, 3(6):1918–1923, 2004.

[PCG+06] P. Pinto, C.-C. Chong, A. Giorgetti, M. Chiani, and M. Win. Narrowband communication in a Poisson field of ultrawideband interferers. In *The IEEE 2006 International Conference on Ultra-WideBand (ICUWB)*, pages 387–392, September 2006.

[PDTL09] E. Peiker, J. Dominicus, W. G. Teich, and J. Lindner. OFDM with windowing in the receiver for fast fading channels. Technical Report TD-09-823, Valencia, Spain, May 2009.

[PFSTL10] E. Peiker-Feil, N. Schneckenburger, W. G. Teich, and J. Lindner. Improving SC-FDMA performance by time domain equalization for UTRA LTE uplink. In *Proc. 15th International OFDM-Workshop*, Hamburg, Germany, September 2010. [Also available as TD(10)12095].

[PLWL06] L. Ping, L. Liu, K. Wu, and W. K. Leung. Interleave division multiple access. *IEEE Trans. Wireless Commun.*, 5(4):938–947, 2006.

[PP06] J. Poikonen and J. Paavola. Error models for the transport stream packet channel in the DVB-H link layer. In *Proc. ICC—EEE Int. Conf. Commun.*, pages 1861–1866, Istanbul, Turkey, June 2006.

[PS09] P. Prochazka and J. Sykora. Karhunen–Loève based reduced-complexity representation of the mixed-density messages in SPA on factor graph. Technical Report TD-09-958, Vienna, Austria, September 2009.

[PVBS07] Z. A. Polgar, M. Varga, V. Bota, and D. Stoiciu. Performances of LDPC-coded QAM mapping of non-coded bits. Technical Report TD-07-359, Duisburg, Germany, September 2007.

[PW07] N. Palleit and T. Weber. Uplink-downlink-transformation of the channel impulse response in FDD-systems. Technical Report TD-07-329, Duisburg, Germany, September 2007.

[Rob95] P. Robertson. *Optimal frame synchronization for continuous and packet data transmission*. PhD thesis, VDI-Verlag, Dusseldorf, Germany, 1995.

[RTR+08] A. Richter, F. Tufvesson, P. S. Rossi, K. Haneda, J. Koivunen, V. M. Kolmonen, J. Salmi, P. Almers, P. Hammarberg, K. Polonen, P. Suvikunnas, A. Molisch, O. Edfors, V. Koivunen, P. Vainikainen, and R. Muller. Wireless LANs with high throughput in interference-limited environments—project summary and outcomes. Technical Report TD-08-432, Wroclaw, Poland, February 2008.

[RU07] T. Richardson and R. Urbanke. *Modern Coding Theory*. Cambridge University Press, Cambridge, UK, 2007.

[SB07] C. Stefanovic and D. Bajic. Bifix analysis and acquisition times of synchronization sequences. Technical Report TD-07-331, Duisburg, Germany, September 2007.

[SBS99] M. Schnell, I. De Broeck, and U. Sorger. A promising new wideband multiple-access scheme for future mobile communications systems. *European Transactions on Telecommunications (ETT)*, 10:417–427, 1999.

[Sch80] R. A. Scholtz. Frame synchronization techniques. *IEEE Trans. Commun.*, 28:1204–1212, 1980.

[SD07] S. Sadough and P. Duhamel. On optimal turbo decoding of wideband MIMO-OFDM systems under imperfect channel state information. Technical Report TD-07-035, Lisbon, Portugal, February 2007.

[SDL08] E. Simon, V. Degardin, and M. Lienard. Sensitivity of block-IFDMA systems to carrier frequency offsets. Technical Report TD-08-615, Lille, France, October 2008.

[SJM+09] M. Smolnikar, T. Javornik, M. Mohorcic, S. Papaharalabos, and P. T. Mathiopoulos. Rate-compatible punctured DVB-S2 LDPC codes for DVB-SH applications. In *Proc. IEEE International Workshop on Satellite and Space Communications (IWSSC)*, Siena, Italy, September 2009. [Also available as TD(10)10021].

[SMR08] L. Smolyar, H. Messer, and A. Reichman. Interference reduction for downlink multi-cell OFDMA systems. Technical Report COST TD-08-633, Lille, France, October 2008.

[Sou92] E. S. Sousa. Performance of a spread spectrum packet radio network in a Poisson field of interferers. *IEEE Trans. Inform. Theory*, 38(6):1743–1754, 1992.

[SP08] J. Sykora and P. Prochazka. Performance evaluation of the factor graph CPM phase discriminator decoder with canonical messages and modulo mean updates. In *Proc. APCC 2008—14th Asia-Pacific Conference on Communications*, Tokyo, Japan, Oktober 2008. [Also available as TD(08)548].

[ST94] G. Samorodnitsky and M. S. Taqqu. *Stable Non-Gaussian Random Processes: Stochastic Models with Infinite Variance*. Chapmann & Hall, London, 1994.

[SV04] J. Sun and M. C. Valenti. Optimum frame synchronization for preamble-less packet transmission of turbo-codes. In *Proc. of the 38th Asilomar Conf.*, vol. 1, pages 1126–1130, Pacific Grove, USA, 2004.

[SVB08a] C. Stefanovic, D. Vukobratovic, and D. Bajic. Frame synchronization using LDPC convolutional codes. Technical Report TD-08-445, Wroclaw, Poland, February 2008.

[SVB08b] C. Stefanovic, D. Vukobratovic, and D. Bajic. Low-complexity frame synchronization for LDPC coded transmission. Technical Report TD-08-516, Trondheim, Norway, June 2008.

[SVCS10] C. Stefanovic, D. Vukobratovic, V. Crnojevic, and V. Stankovic. A random linear coding scheme with perimeter data gathering for wireless sensor networks. Technical Report TD-10-12016, Bologna, Italy, November 2010.

[SVD+09] D. Sejdinovic, D. Vukobratovic, A. Doufexi, V. Senk, and R. Piechocki. Expanding window fountain codes for unequal error protection. *IEEE Trans. Commun.*, 57(9):2510–2516, 2009.

[SVN+10] C. Stefanovic, D. Vukobratovic, L. Niccolai, F. Chiti, V. Crnojevic, and R. Fantacci. Distributed rateless coding with constrained data gathering in wireless sensor networks. In *ACM International Conference on Wireless Communications and Mobile Computing IWCMC 2010*, Caen, France, June 2010. [Also available as TD(10)11010].

[Syk07] J. Sykora. Factor graph framework for serially concatenated coded CPM with limiter phase discriminator receiver. In *Proc. VTC 2007 Fall—IEEE 66th Vehicular Technology Conf.*, Baltimore, USA, October 2007. [Also available as TD(07)362].

[SZM+10] F. Sanchez, T. Zemen, G. Matz, F. Kaltenberger, and N. Czink. Cooperative space–time coded OFDM with timing errors and carrier frequency offsets. Technical Report TD-10-12005, Bologna, Italy, November 2010.

[tB00] S. ten Brink. Designing iterative decoding schemes with the extrinsic information transfer chart. *AEU Int. J. Electron. Commun.*, 54:389–398, 2000.

[TJC99] V. Tarokh, H. Jafarkhani, and A. R. Calderbank. Space–time block codes from orthogonal designs. *IEEE Trans. Inform. Theory*, 45(5):1456–1467, 1999.

[TKL05] W. G. Teich, P. Kaim, and J. Lindner. Bandwidth and power efficient digital transmission using sets of orthogonal spreading codes. In *Proc. 5th International Workshop on Multi-Carrier Spread-Spectrum*, Oberpfaffenhofen, Germany, September 2005. [Also available as TD(09)847].

[TS96] W. G. Teich and M. Seidl. Code division multiple access communications: multiuser detection based on a recurrent neural network. In *Proc. IEEE 4th International Symposium on Spread Spectrum Techniques and Applications—ISSSTA 1996*, Mainz, Germany, 1996.

[VMS02] F. Vanhaverbeke, M. Moeneclaey, and H. Sari. Increasing CDMA capacity using multiple orthogonal spreading sequence sets and successive interference cancellation. In *Proc. ICC 2002—IEEE Int. Conf. Commun.*, New York, USA, April 2002.

[VSC$^+$09] D. Vukobratovic, C. Stefanovic, V. Crnojevic, F. Chiti, and R. Fantacci. Rateless packet approach for data gathering in wireless sensor networks. *IEEE J. Select. Areas Commun.*, 28(7):1169–1179, 2010. doi:10.1109/JSAC.2010.100921. [Also available as TD(07)752].

[VSS$^+$09] D. Vukobratovic, V. Stankovic, D. Sejdinovic, L. Stankovic, and Z. Xiong. Scalable video multicast using expanding window fountain codes. *IEEE Trans. Multimedia*, 11:1094–1104, 2009. [Also available as TD(07)604].

[WDTL08] M. Wetz, M. A. Dangl, W. G. Teich, and J. Lindner. Iterative demapping and decoding for OFDM-MFSK. In *13th International OFDM-Workshop*, Hamburg, Germany, August 2008. [Also available as TD(08)503].

[WH10] T. Wo and P. A. Hoeher. Superposition mapping with application in bit-interleaved coded modulation. In *Proc. SCC 2010—ITG Conference on Source and Channel Coding*, Siegen, Germany, January 2010.

[WHTL09] M. Wetz, D. Huber, W. G. Teich, and J. Lindner. Robust transmission over frequency selective fast fading channels with noncoherent turbo detection. In *Proc. VTC 2009 Fall—IEEE 70th Vehicular Technology Conf.*, Anchorage, AK, USA, September 2009. [Also available as TD(09)716].

[WL10] J. Wszolek and W. Ludwin. Simulation analysis of the neural network based combined chip and symbol level equalizer. Technical Report TD-10-11094, Aalborg, Denmark, June 2010.

[WLTL10] M. Wetz, N. Lux, W. G. Teich, and J. Lindner. Noncoherent iterative multiple symbol detection for OFDM-MFSK. In *Proc. 15th International OFDM-Workshop*, Hamburg, Germany, September 2010. [Also available as TD(10)11016].

[WPS09] M. Win, P. Pinto, and L. A. Shepp. A mathematical theory of network interference and its applications. *Proceedings of the IEEE*, 97(2):205–230, 2009.

[WPTL08] M. Wetz, I. Perisa, W. G. Teich, and J. Lindner. Robust transmission over fast fading channels on the basis of OFDM-MFSK. *Wireless Personal Communications*, 47(1):113–123, 2008. [Also available as TD(07)205].

[WS01] A. P. Worthen and W. E. Stark. Unified design of iterative receivers using factor graphs. *IEEE Trans. Inform. Theory*, 47(2):843–849, 2001.

[WSBM06] H. Wymeersch, H. Steendam, H. Bruneel, and M. Moeneclaey. Code-aided frame synchronization and phase ambiguity resolution. *IEEE Trans. Signal Processing*, 54:2747–2757, 2006.

[Wsz08] J. Wszolek. Application of multilayer perceptron in combined chip and symbol level equalization for downlink WCDMA reception. Technical Report TD-08-665, Lille, France, October 2008.

Part III
Radio Network Aspects

Chapter 10
Deployment, Optimisation and Operation of Next Generation Networks

Chapter Editor Thomas Kürner, Paolo Grazioso, Andreas Eisenblätter, Guillaume de la Roche, Fernando Velez, Andreas Hecker, Matias Toril, Michal Wagrowski, Mario Garcia-Lozano, and Philipp P. Hasselbach

When discussing the deployment, optimisation and operation of mobile radio networks, three principal characteristics have to be addressed. The first flavour is related to the network coverage and is largely independent of the specific system. Hot topics currently under discussion are strategies to reduce the level of electromagnetic exposure while still providing a high level of coverage especially indoors. In order to achieve that goal, new concepts like for example femtocells are being introduced into various networks. A number of contributions within COST 2100 have been dedicated to these coverage related aspects and are discussed in Sect. 10.1.

The second flavour deals with the optimisation of the network, which depends critically on the standard. The two most widespread standards families used in mobile networks are being developed by the third generation partnership project (3GPP) and IEEE 802, which are discussed separately in two different sections. Section 10.2 is dedicated to the 3GPP networks and includes optimisation techniques mainly for UMTS and presented in COST 2100, whereas Sect. 10.3 deals with the IEEE 802.16 (WiMAX) and IEEE 802.11 (WLAN). It has to be noted that the optimisation techniques introduced in these two sections are based on classical approach using input data available in radio network planning tools and are different from the approaches described in the following Sects. 10.4 to 10.6, where optimisation makes use of real measurement values.

The approach of exploiting real measurement data in a structured way for configuration, optimisation and managing the network is the third flavour addressed in this chapter. The operators' effort in the initial configuration and optimisation of mobile radio networks consumes a considerable amount of their Operational Expenditures (OPEX). Therefore, currently developed solutions and standardisation activities in this area are focused on the automation of the network organisation process. Almost all existing optimisation software solutions for Third Generation (3G) networks are based on simulations which provide expected values of KPIs. They are by their nature inaccurate. Propagation predictions are rough, traffic forecasts vague, service

T. Kürner (✉)
Technische Universität Braunschweig, Braunschweig, Germany

mixes not known in detail, and finally, environment models become too simplified. As a result, the software optimises a network but not the one maintained by the operator—another one, truly falsified by the modelling of the simulation process. The only reasonable solution to overcome the above-mentioned problems is to use real measurement data instead of the data obtained from simulations. Real data means measured data. Hence, the concept of measurement-based network optimisation is developed. New self-organisation functionalities encompass the network-automated initial configuration, long-term optimisation, short-term tuning and reacting to failures in the network called automated troubleshooting or self-healing. These processes are assumed to give a certain autonomy to the network, decreasing thereby decisions made by humans. Within COST 2100, a specific sub-working group (SWG 3.1) has been initiated to deal with this specific issue of measurement-based optimisation yielding a couple of substantial contributions during the life-time of COST 2100. Different levels for the integration of measurement data exist and are reflected by the three following sections. Section 10.4 deals with off-line processing of measurement data applied to 3GPP standards and describes how this data can be fed back to the planning process. The main focus is on traffic and mobility related issues of second generation networks based on GSM and GPRS and points out also standard-independent issues like consistency and integrity of the collected measurement data. The next step in the evolution of measurement-based optimisation is subject to Sect. 10.5 and introduces auto-tuning concepts which are applied mainly to UMTS and HSPA. Although using a higher degree of automation, these concepts still rely on the traditional planning process. Finally, Sect. 10.6 deals with self-organising networks (SON), where the collection of measurement data in the live network and the optimisation and configuration tasks are intended to be performed by the network itself circumventing classical planning and optimisation in the final step of this evolution. The introduction of SON concepts goes hand in hand with the standardisation of 4G networks, especially LTE, and will help to handle the more dynamic and flexible systems coming with the next generation of networks supporting a wider range of services and business areas.

10.1 Coverage Aspects

Customers of today's mobile radio networks require access to an ever increasing range of advanced, bandwidth-hungry services and applications, and they expect these services to be available anywhere, anytime and irrespective of user speed. Therefore, operators have to deploy an ever increasing number of base stations to meet the customers' expectations in terms of coverage and quality of service. The number of base stations to be deployed is further increased owing to the larger variety of mobile radio and wireless communications systems in operation and to the entry of new operators in the market. On the other hand, the deployment of an increasing number of transmitting antennas causes concern among the population about possible negative effects that exposure to electromagnetic fields could have on human health. For this reason, several European countries adopted strict exposure

Table 10.1 Network scenarios considered in [Bud07]

Scenario	No. of sites	Sectors per site	Antenna height
1	1	6	40 m
2	7	3	25 m
3	1	4	30 or 60 m
4	2	4	30 or 60 m
5	5	3	30 or 60 m

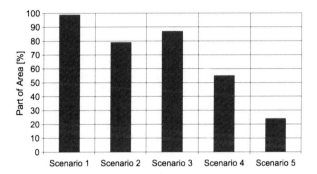

Fig. 10.1 Area with at least 90% coverage probability (384 Kb/s data service) [Bud07] (©2007 H. Buddendick, reproduced with permission)

limits and started monitoring campaigns to verify that these limits are respected. Obviously, the major concerns occur in dense urban areas, where there are a high number of base stations close to homes and working places. Two TDs presented in COST 2100 dealt with these aspects, namely [Bud07] and [BCG+08].

Paper [Bud07] provides results from a study, promoted by a major operator, of how different network topologies affect coverage and exposure. Coverage was evaluated by means of the so-called semi-deterministic propagation model described in [WWWW05]. The model is based on the concept of dominant propagation paths and can also account for indoor penetration loss in a semi-empirical way. Performance was evaluated by means of a dynamic 3G system simulator, which collected several performance indicators, including network throughput and call blocking and dropping rates. Simulations were performed for a district of the city of Bonn, Germany, that extends for about 8 km^2 and has about 40,000 inhabitants. Five different network topologies were considered, shown in Table 10.1: In scenarios 1 and 2 base stations were inside the built-up area, while in the other three scenarios base stations were placed outside this area.

Finally, four different service classes were considered: voice, video telephony, WWW browsing and a generic high-speed data transfer. As an example of coverage results, Fig. 10.1 shows the portion of area where coverage probability, evaluated with the dynamic simulator, is at least 90% for the 384 Kb/s data service. As might be expected, the covered area decreases for scenarios where the base stations are placed outside the urban area.

Figure 10.2 compares the total system throughput (magenta curve) with the 90th percentile of exposure (blue curve). This comparison shows that network topologies with base station sites outside the city yield the lowest exposure, yet at the cost of a

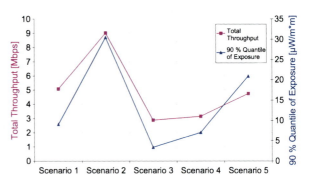

Fig. 10.2 Comparison of exposure and capacity for all scenarios [Bud07] (©2007 H. Buddendick, reproduced with permission)

large reduction of the network capacity. Scenario 1 yields low exposure values at a reasonable network capacity.

In conclusion, regarding all three aspects (exposure, coverage and capacity), the scenarios with sites within the urban area were shown to be in advantage. Using only one external site yields low exposure but insufficient capacity, while with five external sites the coverage is insufficient, in spite of an exposure comparable to that of scenarios with inner-city sites.

The authors of [BCG+08] concentrated mainly on the usage of microcells to provide coverage in an urban area, both outdoors and indoors. The studied area was a central district of the city of Bologna, Italy, of about 500 × 500 m; this area could be covered either by a single, three-sectored macrocellular, or by 28 microcells, or by a combination of the two techniques. The macrocellular site was supposed to be 30 m high, while microcells were at 3 m above street level. A three-dimensional ray-tracing tool was used for propagation, and a penetration loss of 14 dB was assumed. Coverage was considered satisfactory if at least 90% of locations, both outdoors and indoors, were covered; furthermore, it was required that the electrical field should not exceed the limit imposed by the Italian law (i.e. 6 V/m). The paper shows that using only microcells would provide poor indoor coverage, especially at higher floors. This could be overcome by increasing the microcell power, but it would cause excessive field exposure outdoors, where the limit would be exceeded in a non-negligible fraction of locations. On the other hand, using only the macrocellular site would not allow sufficient coverage at lower floors of buildings. The authors found that a combination of the two techniques would lead to excellent coverage both outdoors and indoors while respecting the exposure limits, as shown in Table 10.2.

It was shown how an optimal outdoor coverage at street level does not guarantee a sufficient coverage of indoor environment as well, particularly in high buildings with several storeys.

This topic was also addressed in [dlRZ09], which presented the main issues to be considered within the European project CWNetPlan, that at the time of writing was about to start. As a matter of fact, major operators verify that nowadays most of wireless communications take place when at least one of the terminals is indoor, as authors of [dlRZ09] point out. For this reason, indoor radio coverage optimisation is going to be a key challenge for operators in upcoming years. To fulfil indoor service requirements, operators may follow two main roads:

Table 10.2 Indoor coverage (% of locations) by floor according to [BCG+08]

Floor	Microcells only	Macrocells only	Both
0 (ground floor)	83.5	2.9	91.2
1	82.6	91.6	100
2	82.8	97.7	99.5
3	82.3	100	100
4	80.4	100	100
5	79.9	100	100

- The first consists in increasing the indoor radio coverage by modifying the outdoor network, e.g. by adding new base stations or by increasing their output power. This solution has drawbacks such as high costs and the difficulty to plan the network, as well as health issues and exposure limits that have already been recalled in [Bud07] and [BCG+08].
- The other possibility consists in adding low-power base stations directly inside the buildings. They can be
  - picocells, i.e. operator-owned and managed base stations, generally deployed in public premises such as shopping malls or airport concourses or
  - femtocells, i.e. user-owned and installed base stations, consisting of small units that need only to be plugged onto the customer's Internet connection.

The widespread availability of femtocells "off the shelf" will mean that operators will lose the complete control of the network structure. The first femtocell deployments are expected to occur in a closed access mode, i.e. only authorised users will be allowed to connect to the femtocell, all the other calls being redirected to the macrocell. This will produce interference when unauthorised users will pass across or close to the coverage area of the femtocell. A possible way to solve this problem is to endow picocells with self-configuration and self-optimisation capabilities (see also Sect. 10.6). Figure 10.3 shows an example of the benefits achievable with self-optimisation of antenna patterns, i.e. beamforming is controlled by a self-optimisation algorithm. We notice that it allows a remarkable decrease of the interference area, shown in red, and that it favours the reduction of the *ping-pong* effect.

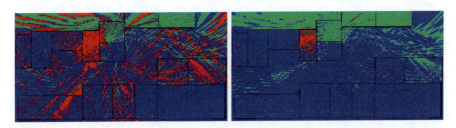

Fig. 10.3 Service areas of three open access femtocells without (*left*) and with (*right*) antenna pattern optimisation. Mutual femtocell–macrocell interference area is shown in *red*

In the discussion above we have seen the key role played by propagation prediction tools in wireless systems planning and optimisation. It is well known that models based on ray optics yield accurate field prediction in urban environment, owing to their capability of taking into account the characteristics and the topology of the urban environment.

The paper [OM08] discusses some key research challenges to be solved in order to favour the widespread adoption of ray-based tools by the operator community. Actually, current tools are more suitable for scientific work and for case studies, rather than for the massive computation effort required to simulate a whole network and to optimise base station parameters in order to improve the system performance.

The main requirements of a ray-based propagation tool, from an operator's point of view, are the following:

- improved accuracy justifying the effort in developing or adopting the propagation model and procuring an adequate topographic database;
- a computation time allowing excessive usage of the tool;
- the integration of statistical models with deterministic models including outdoor-to-indoor modelling;
- allowing wideband characterisation.

The authors show the main research challenges to be tackled to meet the above requirements.

- The integration with the existing statistical models shall be pursued by developing modules able to interface with any statistical model, without the need to adapt to any proprietary tool.
- Computation time can be significantly reduced by a simplification of the database that is generally obtained by aerial or satellite photographs, reducing the number of surfaces and of edges to be considered.
- Outdoor-to-indoor prediction accuracy can be improved by using information about the internal structures of the buildings, the position of the user terminal within the building and the angles of arrival of rays at the building external surfaces.
- Wideband predictions can be improved by refining the models and particularly also by considering diffuse scattering phenomena.

Combining the four challenges above leads to the propagation model architecture shown in Fig. 10.4, where the research challenges are indicated as four separate blocks.

10.2 3GPP Networks

The initial deployment of a network and its subsequent expansions always have and are still posing considerable challenges to a network operator. Throughout the lifetime of a network, its radio coverage and capacity are important quality metrics. These two dimensions can largely be considered in isolation for Global

10 Deployment, Optimisation and Operation of Next Generation Networks

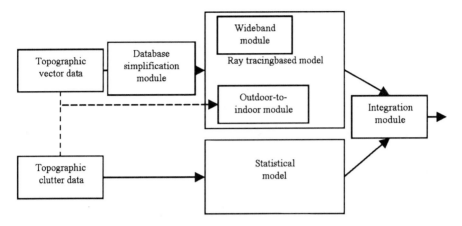

Fig. 10.4 Architecture of a propagation model combining a statistical and a ray-based component

System for Mobile Communications (GSM) networks. This is because GSM uses Time-Division Multiple Access (TDMA) and Frequency Division Multiple Access (FDMA) in such a way that interference can ideally be controlled by means of a proper frequency allocation.

The increasing service diversity over the recent years, however, has rendered the traditional capacity notion in terms of Erlang hardly effective. In addition, strongly interference-limited technologies such as Code Division Multiple Access (CDMA)-based Universal Mobile Telecommunications System (UMTS) and Orthogonal Frequency-Division Multiple Access (OFDMA)-based 3GPP-Long Term Evolution (LTE) come with a coupling of coverage and capacity that makes it futile to look at one dimension only while ignoring the other.

This section addresses the following aspects related to the deployment of 3rd Generation Partnership Project (3GPP) networks. First, examples of service requirements are reviewed, which can be used in the design of cost-effective network layouts. Second, results on network capacity optimisation by means of cell/sector configuration are presented. Third, options of increasing network capacity through the use of smart antennas or Multiple-Input Multiple-Output (MIMO) antenna systems are reported. Finally, results on the use of a cognitive radio system as an underlay to a GSM network are given.

10.2.1 Quality Requirements and Profit Maximisation

Both network technology design and network planning depend on a clear understanding of the capabilities that a network shall have. A prominent driver for recent and future network generations is service beyond speech telephony. Such services include web browsing, video streaming, e-mail, video telephony, file transfer, and remote gaming. The requirements for supporting such services in terms of the reliability of the transport, the maximum latency, and the desired bit rate have been

Table 10.3 Overview of service characteristics and requirements from [AMCV07, Table 2] including references, where the references translate as follows: [5] →[EU98], [6] → [EU03], [7] → [FCX+03], [8] → [3GP03], [9] → [AL03], [10] → [FV05], [11] → [Mon05], [12] → [Vel00], [13] → [LWN02], [14] → [ed03], [18] → [HAS96], [20] → [Kwo95], [22] → [Ant04]

Services/ applications	Reliability		Latency (link radio)	Latency (end-to-end)	
	BER	FER	Max Delay (ms)	Jitter (ms)	Max Delay (ms)
Voice	10^{-4} [7]	< 3% [7]	≤ 30 [11] 20 [5]	< 1 [13]	150 [7] 100 [20]
Web Browsing	10^{-6} [7]	< 1% [7]^3	≤ 300 [11]	N/A [13]	Few seconds [7] < 4000/page [13]
Video Steaming	10^{-4} [7] 10^{-6}–10^{-3} [6]	< 1% [7]^3	< 1000 [22] 10000 [13]	–	200 [7] < 10000 [13] 150–400 [14]
E-Mail	10^{-6} [7]	0% [8]	–	N/A [13]	5 min [12][1] 4000 [14][2]
Video Telephony	10^{-4} [14] 10^{-6} [7]	< 1% [7]^3	150 [18] 50 [5]		200 [17] < 150 [13] < 400 [13] 300 [5]
FTP	10^{-6} [7]	0% [8]	500 [14]	N/A [13]	10000 [7] < 10000 [13]
Remote Gaming	10^{-7}:10^{-6} [10]	0% [8]	–	N/A [13]	50 [10] 100 [12] < 300 [9][4] 250 [8] < 900 [9][5]

[1]Server-to-server delay, [2]user-to-server delay, [3]UMTS system, [4]RT action games, [5]RT strategy games, N/A—Not available, RT—Real Time

considered by various research teams and standardisation bodies. In an effort to create one single reference for these seven services (including speech telephony), contribution [AMCV07] compiles information into an overview that is reproduced in Tables 10.3 and 10.4.

The deployment and operation of a radio network for the mass market incurs high costs that are to be paid up by (future) revenues. Optimising the roll-out of a network according to a cost/revenue trade-off—or more generally, according to expected profit—is therefore a common goal in practice. Achieving this goal is by no means trivial in the presence of various sources of uncertainty, such as future service types, service demand, operating cost, and equipment prices. Nevertheless, the paradigm of cost-effective investments prevails roll-out in practice. Scientifically, the topic has yet received less attention than one might expect. This is possibly due to the intrinsic complexity of network planning without considering uncertainty and the various (partially related) sources of uncertainty.

The contribution [FV05] studies a cost/revenue trade-off for a UMTS network in a business city centre. The scenario entails a service mix of demanding data services, a Manhattan grid layout, micro-cellular base stations with omni-directional antennas at the crossings, and pedestrian outdoor users. The desired grade of service is defined via a metric that measures call blocking and handover failure probability and

Table 10.4 Overview of service characteristics and requirements from [AMCV07, Table 2] including references, where the references translate as follows: [5] → [EU98], [6] → [EU03], [7] → [FCX+03], [8] → [3GP03], [9] → [AL03], [10] → [FV05], [11] → [Mon05], [12] → [Vel00], [13] → [LWN02], [14] → [ed03], [18] → [HAS96], [20] → [Kwo95], [22] → [Ant04]

Services/ applications	Bit rate Rb [kb/s]	Traffic Characterisation		
		Intrinsic Time Dependency	Delivery Requirements	Burstiness
Voice	4–25 [7] 4.75–12 [6]	TB [7]	RT [7]	1 [14]
Web Browsing	< 2000 [7]	TB [7]	RT [7]	1–20 [14]
Video Steaming	32–384 [7] 24-128 [6]	TB/NTB [7]	RT [7]	1–5 [14]
E-Mail	< 1500 [12]	TB [12]	NRT [12]	1–20 [12]
Video Telephony	32–384 [7]	TB [7]	RT [7]	1–5 [7]
FTP	64–400 [12] 384 [12]	NTB [10]	RT [10]	1–50 [7]
Remote Gaming	64–1000 [14]	TB/NTB [10]	RT [10]	1–30 [10]

RT—Real Time, NRT—Non RT, TB—Time Based, NTB—Non TB

a target value of 1%. Simulations for the Down-Link (DL) only are performed using a system-level simulator based on NS-2. Several variations are analysed, which differ in the amount of capital expenditure and operational expenditure required for building and maintaining the network. In the context of a simple (and fixed) economic model, this analysis provides insights into how service demand, expected revenue per megabyte, quality requirements, cell sizes, license and equipment costs may influence operating profit.

10.2.2 Maximising Cell Capacity

Capacity maximisation in interference-limited networks is strongly related to the management of interference by the system at run-time and by the operator during network planning and optimisation. As part of the deployment of a network, cell sectorisation, the choice of antennas, and azimuth/downtilt of antennas are decided. Each of these parameters may have a strong impact on interference.

The authors of [PSN07] study means to optimise the sector configuration of a multi-sector cell in a CDMA network as to maximise the overall cell capacity. A necessary and sufficient optimality criterion for the sector configuration is known for perfectly homogeneous environments in the case of networks consisting only of one multi-sector cell. The criterion basically states that each sector shall serve the same amount of traffic. In order to be able to exploit this criterion under non-homogeneous environment conditions, the authors propose a transformation function that maps the cell power vector of a multi-sector cell operating in

Table 10.5 Computational results for coverage and capacity optimisation of realistic UMTS network scenarios [EG08]: uncovered traffic lacks pilot power or quality; other-to-own-cell interference ratio represents cell coupling, and "congested" cells have $\geq 2\%$ blocking

		Uncovered load [%]	Total TX power [W]	Avg. cell load [%]	Max cell load [%]	Other/own cell intf. ratio [%]	Blocked traffic [%]	Congested cells [#]
Berlin	initial	1.4	773.2	35.9	63.4	114.7	5.96	42
	optimised	0.2	647.9	30.1	49.5	89.2	1.87	14
Lisbon	initial	1.3	469.8	24.5	60.0	81.8	2.52	12
	optimised	0.9	422.6	22.1	51.6	65.8	0.99	6
Turin	initial	35.1	998.3	37.6	63.0	234.4	1.02	14
	optimised	35.3	854.4	32.2	57.1	208.5	0.41	6
Vienna	initial	5.2	2257.1	31.3	56.0	191.6	1.59	24
	optimised	5.3	2173.6	30.2	54.5	173.0	1.08	11

a non-homogeneous environment into a cell power vector for an "ideal" network in a homogeneous environment. Based on this transformation, the same optimality criterion and optimisation procedure as in the homogeneous case shall become applicable, see [PSN07] for details.

The author of [Pom09] proposes an approach to identifying an optimal site sectorisation in CDMA networks. He also argues that realistic multi-site scenarios can be appropriately captured using the method from [PSN07] for single multi-sector cell networks.

During network planning and optimisation, the cell capacity of UMTS networks can be improved by explicitly designing for minimal interference. The contribution of [EG08] is to introduce planning methodologies that allow to minimise interference overhead while maintaining the established network coverage. Table 10.5 shows computational results for the two public scenarios Berlin and Lisbon from the Models and Simulations for Network Planning and Control of UMTS (MOMENTUM) library [EGT05] as well as for the two scenarios Turin and Vienna from the COST 273 MObile Radio Access Reference Scenarios (MORANS) library. Furthermore, the authors present the first practicable approach to assess how well a network is configured with respect to interference avoidance. One benchmark is derived from the revised pole equation [EG07], and another one comes from homogenised versions of the scenarios. A comprehensive treatment of these topics can be found in [Gee08].

LTE cell capacity is limited by interference, like in all modern radio network systems. The main approach to mitigating inter-cell interference is soft frequency reuse: demanding users at cell edges follow a frequency reuse pattern, whereas users with low power requirements, e.g., those at the cell centre, can freely use all resources. The reuse pattern is prescribed by power masks. The authors of [BEGT09] use equation systems that describe interference coupling among cells—previously established in the context of CDMA/UMTS—as a tool for adapting power masks

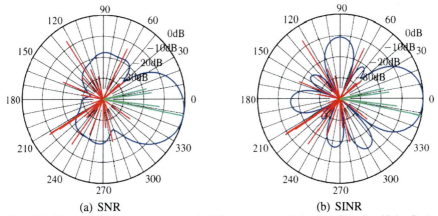

Fig. 10.5 The radiation patterns are optimised for an example link using Signal-to-Noise Ratio (SNR)- and Signal-to-Interference-plus-Noise Ratio (SINR)-based metrics. The *green* and *red lines* indicate the paths to the desired and non-desired Mobile Stations (MSs), respectively. Line lengths correspond to path gain [BHC07, Figs. 4, 5]

to traffic demand. A concise model is introduced that allows us to efficiently evaluate alternative power masks w.r.t. their impact on inter-cell interference. Based on this model, the authors propose a simple algorithm that optimises power masks for a given user demand distribution. Results from dynamic simulations suggest that adaptive power masks beat uniform, hard, and fixed static reuse patterns in terms of total cell throughput and weakest users' performances.

In [NPP10], a concept for the optimisation of resource allocation in 4G networks is introduced. This contribution contains a detailed study of related work and finally proposes a derivative-free multi-objective optimisation approach for LTE and LTE-Advanced. In order to obtain faster results, grid technologies are suggested.

10.2.3 Advanced Radio Systems

The authors of [CGV09] investigate aspects of using beamforming antenna arrays with four commercial antennas per sector in a 3-sectorised UMTS-Frequency-Division Duplex (FDD) network. They study the impact of array positioning and alignment errors and the influence of alternative feasible methods of obtaining the covariance matrix used for beamforming. The variations in the ability to support users in the DL is analysed in a wrap-around scenario with seven cells using Short-Term Dynamic (STD) system-level simulations. The key observations are that limitations in properly estimating the required DL covariance matrix have a considerable impact on the capacity of the network. Moreover, poor vertical alignment of the antennas within an array causes significant degradations in the achievable capacity. In contrast, horizontal alignment and the type of array (uniform circular or uniform triangular) do not show a strong influence on the metric. In case beamforming is performed at each base station separately, then using the SNR instead of the SINR performs better, see Fig. 10.5.

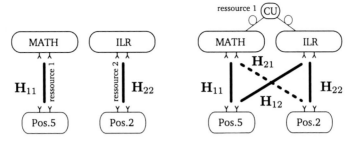

Fig. 10.6 In the traditional setup (*left*), two users are served from different sites (MATH, ILR) using distinct resources. In the cooperative approach (*right*), the same users are served from the same sites on a shared resource. The benefit comes from a joint (*central*) processing (CU)

The contribution [JJJ+09] reports on radio measurement campaigns at 5.2 and 2.53 GHz in outdoor multi-cell environments from Berlin and Dresden (Germany). The performance of cooperative multi-cell systems is under study. The authors experimentally present evidence that cooperation enhances the rank of the compound channel matrix (see Fig. 10.6). This allows us to exploit new spatial degrees of freedom. While in traditional settings, cell capacity decreases when sharing the same spectrum among cells, cooperation allows us to raise the capacity even beyond the mean capacity of isolated cells.

In [VTFDE11], a novel multi-link MIMO scheme for indoor applications, called Interleaved F-DAS (Fiber-Distributed Antenna System) MIMO, is proposed. The performance of the deployment of such a system using 2×2 MIMO applied to the LTE standard in a typical indoor office environment has been analysed. The analysis is based on simulations using ray-tracing predictions and an LTE link-level simulator. Results show that the proposed scheme, although simpler and relatively less expensive than conventional co-located MIMO schemes, achieves a similar level of performance in terms of radio coverage and system throughput. The achieved throughput depends on the signal-to-noise ratio (SNR) and the power unbalance at the two MIMO branches at the Rx, see Fig. 10.7. A key issue for the performance of F-DAS MIMO is to find the optimum deployment in order to achieve a good overlap of multiple MIMO branches at each mobile terminal with a minimum number of antennas and cables.

The contribution [GLG10] provides a statistical analysis for Coordinated Multi Point (CoMP) transmission in MIMO Rayleigh channels. Open-loop MIMO technique is used, which does not require perfect channel state information. The simulation results provide numbers for the probability of mobiles to improve their spectral efficiency by selecting several base stations.

The importance of realistic traffic models for wireless network evaluations is stressed in [Mül10], where sample evaluations of the application layer performance of CoMP transmission schemes are presented. Traffic models with different level of detail are used, and it is shown which effects can be captured by which models. The simulations presented in this contribution revealed that traffic characteristics, if not taken into account in the design of CoMP algorithms, can lead to largely reduced

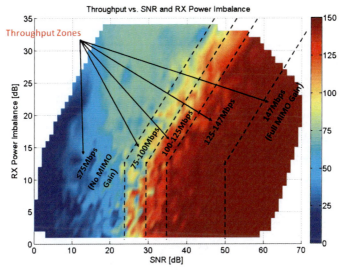

Fig. 10.7 LTE throughput as a function of SNR and Rx power unbalance for the Interleaved F-DAS MIMO configuration [VTFDE11] (©2011 EurAAP, reproduced with permission)

coordination gains or additional transmission delays with a possibly negative impact on user experience.

10.2.4 Spectrum Re-use by Cognitive Radio Systems

The authors of [CGV09] consider alternative approaches to implement a cognitive radio system as an unlicensed user in the GSM licensed frequency bands. Two proposals, an interweaved frequency use and an underlay approach, are considered [GJMS09]. The latter approach is favoured and studied in more detail. An algorithm for the selection of frequency resources by the cognitive radio system (secondary system) operating according to the underlay approach is presented, and its impact on the primary system, GSM in this case, is investigated. This investigation is based on a realistic network scenario in the city of Bologna, Italy, that is analysed in a detailed simulation environment. The preliminary results suggest that usage by a secondary system is indeed possible without causing harmful interference to GSM.

10.3 IEEE 802 Networks

10.3.1 IEEE 802.16 WiMAX

Worldwide Interoperability for Microwave Access (WiMAX) was designed by the WiMAX forum in 2001. It is a technology based on the IEEE 802.16 standard (also

Table 10.6 Coverage and interference areas for the whole region of Beira Interior

Type of antenna	Covered area	Area of interference	Non-covered area
Omnidirectional	50.8%	36.4%	12.8%
Sectorial	86.9%	0.3%	12.8%

called Broadband Wireless Access) [Zha08]. It was initially developed as a last mile connectivity solution, i.e. a solution to efficiently bring the Internet in areas where no copper is deployed. WiMAX is also a possible replacement candidate for cellular phone technologies and the future release is candidate for the Fourth Generation (4G).

10.3.1.1 Advantages of WiMAX

WiMAX covers long distances, and it uses unlicensed spectrum to provide access to a network. It can be deployed for Point to Point (PTP) or Point to Multi-Points (PTM) links [MAR+07b]. WiMAX uses OFDMA, where individual users are assigned to different subsets of sub-carriers, in order to achieve higher data rates and reduce interference. Thus, it is important, when deploying a WiMAX network, to carefully plan not only the radiated power, but also the distribution of the sub-carriers.

As explained in [MAR+07b], different received power levels correspond to different physical bit rates and modulation and coding schemes. Thus, a WiMAX receiver can automatically choose its modulation from Binary Phase Shift Keying (BPSK) until 64-state Quadrature Amplitude Modulation (QAM), displaying its ability to overcome Quality-of-Service (QoS) issues with dynamic bandwidth allocation over the distance between the base station and the mobile user. In IEEE 802.16-2004, channels of 3.5, 7 and 10 MHz are defined. This ability to ensure a good QoS is one of the best characteristics of WiMAX and makes it suitable to support multimedia and IP (Internet Protocol) communications, e.g., videoconference, voice over IP, and communication of high-resolution video/image.

In [MAR+07b], the deployment of a PTP link is presented. Such deployment is performed using relays, and it is important to verify that there is a line of sight between the transmitters for the placement of the repeaters and absence of obstructions to the first Fresnel ellipsoid. Five different scenarios were tested, covering distances from 5 km until 20 km. The field trials fit very well the Friis formula, and it is also verified that the beam width, which is small in such PTP links, can have a negative impact if it is not perfectly oriented. PTM networks have also been tested and the covered distance was up to 5 km. In such a scenario, the impact of the antenna is also very important. As represented in Table 10.6, it was verified, when covering the whole region of Beira Interior in Portugal, that using sectoral antennas improves the radio coverage and reduces interference.

In order to compare the performance of WiMAX with a more traditional approach such as Wireless Local Area Network (WLAN), an isolated village scenario

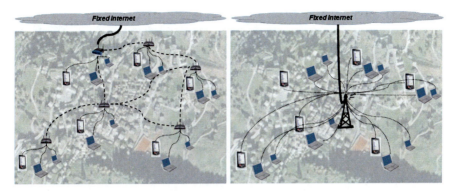

Fig. 10.8 Solutions to cover a village, using WiFi mesh network (*left*), or WiMAX (*right*) [FCFN07] (©2009 IST press, reproduced with permission)

has been considered [FCFN07]. First, as represented in Fig. 10.8(left), a mesh network of IEEE 802.11 access points, also called Mesh Access Points (MAPs), was deployed. The fixed Internet gateway was connected to a wireless Mesh Point Portal that established communication with MAPs spread over the village, each of them providing connectivity to nearby users. MAPs were equipped with two wireless interfaces: one acting as a classical 802.11g infrastructured access point, with a rate of 11 Mbps, transmission power of 5 mW and sensitivity of -95 dBm, another as an 802.11a mesh router, communicating at 54 Mbps, with a transmission power of 1 W and a sensitivity of -82 dBm, recurring to directional antennas that enable high communication ranges.

Then, as represented in Fig. 10.8(right), a single WiMAX base station was installed in the village. Mobile terminals were equipped with WiMAX interfaces. A bandwidth of 20 MHz and 2048 sub-carriers was considered. From these, 367 were guard sub-carriers. In DL, clusters of 28 sub-carriers were defined, with 4 pilot sub-carriers. This resulted in 1440 data sub-carriers for DL. For Up-Link (UL), tiles of 12 sub-carriers with 4 pilots were considered. This resulted in 1120 data sub-carriers. A transmission power of 500 mW was chosen for both the base station and mobile users.

From these two deployments it was verified that the solution based on WiMAX could cover a higher number of users, since it could support up to 555 voice sessions, whereas the mesh network of Wireless Fidelity (WiFi) access points could only serve up to 41 voice sessions. However, it was also pointed out that a WLAN-based solution is still a cheaper and less complex solution compared to WiMAX, that is why WLAN could be sufficient for small villages with low density of users.

If WiMAX seems to be an optimal approach to cover large distances, many parameters such as the positions of the emitters, the type of antennas, or the distribution of the sub-carriers have to be optimised. Hence, before deploying such network, it is helpful to perform on-site measurements or off-line simulations using planning tools.

Table 10.7 WiMAX field trial results

Distance (km)	Data rate (kbps)	C/N (dB)	Modulation	Time (s)
5.9	9774	24	8	46
14.9	9448	15	6	47
17.6	7584	15	6	49
17.9	9407	17	6	47
20.9	7016	15	6	49

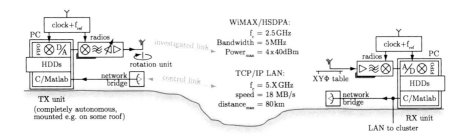

Fig. 10.9 MIMO WiMAX measurement setup

10.3.1.2 On-site Measurements

Field trials are very important to help deploy a WiMAX network. In [MAR+07a], PTP links were established using relays. First measurements were performed to check the validity of the Friis formula. But if radio measurements are useful, it is also necessary to check the real throughput. Therefore, two computers, one as a server and one as a client, were connected and used to transmit data. Different tests were performed, one of them consisted in sending a large 40 Mb file. The performance of such a file transfer is summarised in Table 10.7. In these field trials, it was verified that the measured data rate (throughput) and C/N decrease with the increase of the distance between the antennas (except in one case). The tests also demonstrated that WiMAX performs well for large distances.

As described in [VSC+09], more measurement campaigns were also performed using a spectrum analyser and a LabVIEW application especially designed to load the measurement data into the computer.

Other WiMAX field measurements are reported in [CMLR10] in the context of MIMO. Such measurements are usually time and money consuming, that is why a simple approach was proposed to study static scenarios, where only the physical layer was analysed, and a maximum of 4 antennas (at both receiver and emitter) was used. The scenario was evaluated on a block-by-block basis in real-time, with a flexible Matlab code. The measurement setup is represented in Fig. 10.9.

The advantage of such a measurement platform is that it is easily reconfigurable (by changing either the code or the position of the receiver) and many measurements can be carried out. For example, with such a system it was possible to verify that in different scenarios (alpine or urban), the turbo codes outperform the convolutional codes by about 3 dB. Moreover, an additional gain of about 1 dB was achieved when implementing LDPC channel coding. However, at low SNR, the performance of the turbo code was worse than the convolutional one.

10.3.1.3 WiMAX Planning Tools

When deploying WiMAX networks, another approach is to use planning tools. Even if field trials will always perform better than simulation, simulators are helpful to plan the pre-configuration of the network and to study dense networks. First, such tools must be based on accurate radio coverage prediction, that is why measurement campaigns are always necessary to calibrate them. In [VaC+09], the Friis model is modified to model different propagation environments by different propagation exponents α, e.g. $\alpha = 2$ for free space, $\alpha = 3$ in urban areas, and $\alpha = 4$ in shadowed urban areas. In [dlRVLP+08], a more accurate propagation is proposed, based on an FDTD model. The advantage is that such a model can lead to a higher accuracy, but very accurate scenario data must be available, containing the position of the walls and their materials. Hence, the propagation models must be carefully chosen depending on the scenarios and the requirements.

In [VaC+09], the challenges when deploying a WiMAX network are investigated and illustrated using simulations. It is explained that planning tools are needed for operators to optimise both cost and revenues:

- Cost comprises the fixed cost (e.g. spectrum licenses) plus costs proportional to the number of cells, and the number of transceivers. It usually depends on the size of the cells and on the reuse pattern.
- Revenues depend on the throughput, and the prices and are sensitive to the number of supported users.

In [VAH10], a cost/revenue function which incorporates the cost of building and maintaining the infrastructure and the effect of the available resources on revenues is proposed. In order to reach these objectives, operators will have to carefully implement interference cancellation techniques, as simple interference avoidance design presents limitations. In OFDMA, the amount of interference depends on how the sub-channels (subsets of orthogonal sub-carriers) are allocated. For example, it was verified with this tool [VaC+09] that with a reuse pattern $K = 7$, cell throughputs near the maximum are only achieved in the UL if sub-channelisation is used together with sectorisation.

As explained in [LPGSZ+08], the number of available sub-channels depends on the channel bandwidth and the permutation scheme (which indicates how the sub-channels are formed), since the sub-carrier spacing is fixed. The sub-channels may be built by using contiguous or pseudo-random distributed sub-carriers. Sub-channels using contiguous sub-carriers, e.g. Adaptive Modulation and Coding (AMC), enjoy multi-user diversity (appropriate for static and nomadic traffic), while sub-channels using distributed sub-carriers, e.g. Partial Usage of Sub-Channels (PUSC), enjoy frequency diversity (appropriate for mobile traffic).

In order to choose between the different allocation strategies, a good approach is to use system level simulators where the performance of the users is evaluated depending on the parameters of the network.

Hence, in [LPGSZ+08] the implementation of a WiMAX system level simulator is described. The simulator takes multiple Monte Carlo snapshots to observe the network behaviour over long time scales. The Monte Carlo snapshots are independent from each other since the users are randomly spread over the planning area with different requirements for throughput and QoS. The simulator takes four steps to calculate the final performance for each user, cell and snapshot:

- The network configuration such as users, sectors, services and traffic map parameters is read in.
- The path loss for each user is calculated, and the best server for each user is computed.
- The admission queue is created, according to certain admission policies. Subsequently, DL and UL are analysed separately.
- Power control is carried out, and the results of both downlink and uplink are calculated.

Different experiments have been carried out to evaluate the behaviour of the simulator and the performance of Mobile WiMAX networks.

10.3.1.4 Case Study: Deployment of WiMAX Femtocells

In [dlRVLP+08], simulations of WiMAX femtocells have been performed. Femtocells [ZdlR10] are very small base stations directly installed by the users inside their home. Since such devices have just started to be commercialised, there is no large-scale femtocell deployment at the moment. That is why planning tools are very useful in this case. Different sub-channel allocation strategies have been tested:

- Same Channel (Worst case): In this case, the same group of sub-channels from the palette of available ones is given to all the macrocells and femtocells (i.e. the same four sub-channels are taken from the 16 available ones for each cell).
- Random allocation: The sub-channels of all the macrocells and femtocells are randomly chosen from the palette of available sub-channels (i.e. four random sub-channels are taken from the 16 available ones).
- FRS 1X1X3: The palette of available sub-channels is divided into three sub-groups. Afterwards, each femtocell gets sub-channels only from its given sub-group. Neighbouring femtocells are assigned to different sub-groups to reduce interference probability (i.e. the 16 available sub-channels are divided in three sub-groups, and then each sub-group is assigned to one femtocell).
- Femtocell Optimisation: A Simulated Annealing optimisation algorithm is used to allocate the sub-channels. In this case, only the sub-channels of the femtocells are planned together.
- Femtocell and Macrocell Optimisation (Best Case): Same as previous strategy, however, not only the sub-channels of the femtocells are planned together here, but also those of the macrocells.

Table 10.8 System level simulation results with a scenario made of 63 WiMAX mobile users served by one macrocell and 30 femtocells

Method	Successful users	Total Throughput	Cost function
Same Channel	3	3168.0	2256.0
Random	46	4752.0	607.9
FRS 1X1X3	56	5913.6	143.5
Femtocell Optim	60	6336.0	22.5
Femto/Macro Optim	63	6652.8	12.5

The results are given in Table 10.8, where the number of successful users, the total throughput and a cost function (representing the amount of interference) are given. It is verified that optimisation methods provide very good results compared to the use of standard methods such as FRS 1X1X3. Moreover, it was also demonstrated that, when both the macrocells and femtocells are optimised, the performance is higher, that is why it may be useful in the future to implement Self Organization (SON) not only in the macrocells, but also in the femtocells, so that all the cells can automatically choose their optimal parameters, see also Sect. 10.6.

10.3.1.5 Best Deployment Strategies

In order to deploy a WiMAX network, both approaches have advantages and drawbacks. Field trials are more accurate, but they are time consuming and expensive. The use of planning tools suffers from a lower precision since it is based on models, but it is helpful to test dense and complex scenarios. That is why it is recommended, since all the scenarios are different, to plan the network by combining measurements and simulations. Moreover, as shown in the previous paragraph, it is verified that the use of optimisation methods performs well and is necessary in order to reach high performance.

10.3.2 IEEE 802.11 WLAN

The main goal of radio network planning is to provide widely available wireless service of high quality at a reasonable price. In IEEE 802.11 WLAN, decisions on Access Point (AP) placement and channel assignment are traditionally taken sequentially. AP placement is often modelled as a facility location problem while channel assignment is represented as an (extended) graph colouring problem. As treating these key decisions separately may lead to suboptimal designs, the authors from [EGS07a, EGS07b, EGGP08] proposed an integrated model that addresses both aspects whilst considering the Media Access Control (MAC) layer issues and maximising the throughput. The interaction between the physical and the MAC layer in IEEE 802.11e was addressed in [CSV07, CSV08, CVP09]. The achieved performance was analysed via simulation in the presence of acknowledgements and block acknowledgements, and a weighted scheduling algorithm was proposed. In [BCR09], the influence of the propagation phenomena on the performance of wireless mesh networks is analysed.

Mesh networks are emerging as a wireless networking solution to provide coverage in areas where it has previously either been prohibitively expensive or not feasible. Traditional wireless network solutions require that each Access Point (AP) has a fixed connection to a backhaul network; this means that deploying a wireless network over large geographical areas normally requires the installation of expensive cabling. Wireless Mesh Networks provide a solution to this problem by greatly reducing the number of nodes and APs which require a wired connection to a backhaul network. For Wireless Mesh Networks to achieve similar levels of performance as traditional wired networks, multiple radios are required in each device. Unfortunately, the co-location of multiple radios in a single device causes a number of interference problems; these must be solved before Wireless Mesh Networks have any chance of being deployed by network operators. In [RFMM10], the impact of non-overlapping channel interference in IEEE 802.11a-based multi-radio nodes is investigated. The primary contribution is the discovery of a channel interference effect which is present over the entire 802.11a frequency space. This interference appears if two radios are located less than 50 cm from each other while both are behaving independently without any coordination.

The authors from [STV$^+$09] considered system capacity aspects as well as economic aspects. A simple WLAN planning tool was used to optimise the position and number of APs considering the cost of the required equipment. Propagation measurements were considered, and a comparison with the Dominant Path (DP), modified Friis and COST 231 models was performed. An algorithmic approach is addressed which considers the mixture of applications and the resulting system capacity in the optimisation process.

10.3.2.1 Access Point Placement and Channel Assignment

One common but major drawback of the heuristic approaches available in the literature is their inability to provide information on how much better an alternative design might be. In [EGS07a, EGS07b], the following optimisation models were applied. For positioning APs, coverage planning models are fairly effective and can often be solved to optimality. Frequency assignment has been studied extensively, for example, in the context of GSM network planning [AHK$^+$03]. The channel assignment problem belongs to the hardest problems in wireless network design. But due to a small number of available channels in IEEE 802.11 technology, the problem is still within reach of integer programming techniques applied to the combinatorial optimisation. IEEE 802.11g networks operating in the infrastructure mode at 2.4 GHz were considered, a typical configuration in office environments.

10 Deployment, Optimisation and Operation of Next Generation Networks

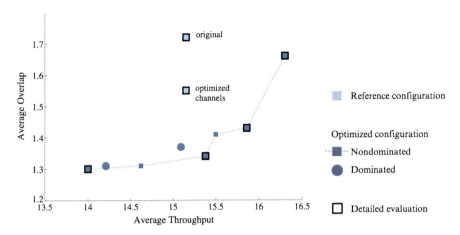

Fig. 10.10 Trade-off between throughput and overlap

The authors from [EGS07a] show how the individual models can be merged into a complete optimisation model for WLAN planning. This model allows one to optimise the trade-off between high throughput and little cell overlap. In a case study based on realistic data for an indoor office environment, optimal network designs are computed with respect to the integrated model, where the MAC contention is considered. It was demonstrated how emphasis on maximising throughput or on avoiding overlap changes the structure of the resulting solution. The different optimisation objectives and their trade-off are taken into consideration simultaneously. Computational results from [EGS07a] show that indeed the integrated approach is superior to the sequential one.

In [EGGP08], the automated planning of IEEE 802.11 was also addressed, as an extension of the work from [EGS07a]. As the main network performance criterion, the net throughput (or goodput) that the WLAN delivers to the network layer and to the user application was considered. The candidate AP locations for an office scenario are the ones from [EGGP08].

A new method for dealing with the complex WLAN planning problem was proposed. In order to reduce complexity, two simplified evaluation schemes were employed for the preferred planning solution, throughput and overlap. Using multi-criterial optimisation methods, several network configurations were generated that represent different trade-offs between overlap reduction and throughput maximisation. To pick the best configuration, a new detailed analysis was conducted by simulation. Because only a few candidate configurations are left, this is computationally feasible. The first step consists of trading-off throughput and overlap and has already been tried in computational experiments, as shown in Fig. 10.10, while the second one (detailed simulation) is work in progress.

Simulation results will enable us to decide for a preferred configuration, validate the success of simulation, and better understand the relation of our simplified measures to realistic network behaviour.

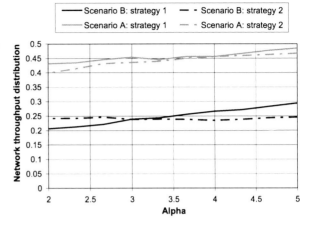

Fig. 10.11 Network throughput distribution: comparison between scenario A and scenario B for both strategies

10.3.2.2 Impact of Propagation Issues on Wireless Mesh Networks Performance

The effect of dominant propagation contributions in Wireless Mesh Network (WMN) throughput and delay has been investigated in [BCR09], considering different path loss exponent values, varying from 2 in free space to 5 in severe Non-Line-Of-Sight (NLOS) conditions. While in scenario A only the effect of traffic interference is considered, in scenario B both fading and noise are introduced. Strategy 1 from [BCR09] corresponds to no variation of the average number of neighbouring nodes, N, and a variation of the transmission power, P_t. Strategy 2 corresponds to a case where P_t is kept constant and the total number of nodes, n, varies.

In Fig. 10.11, the total network throughput of scenario A and scenario B was compared for both strategies (strategy 1: varying P_t; strategy 2: varying n and thus traffic density).

As expected, due to the presence of multipath propagation phenomena, adding a Rayleigh fading and an AWGN noise to the model, the total throughput of the network decreases by about 30–50% (this specific value depends just on the assumed transmission power and propagation exponent). Signal fluctuations and noise cause a lower carrier-to-interference-plus-noise ratio, Carrier to Interference plus Noise Ratio (CINR) value and the loss of a higher number of packets. From Fig. 10.11 it can be noted that beyond a propagation exponent 3, the first strategy provides a higher value of throughput. In fact, the throughput is strongly dependent on the number of hops to reach the destination: As the number of hops increases, the probability to lose a packet increases.

When the fall of connectivity is compensated by an increase of the transmitting power, the transmission range of each node is the same, and the number of hops to reach the destination remains constant as the propagation exponent increases. Otherwise, when the fall of connectivity is compensated by an increase of the number of nodes, the number of hops increases. Therefore, the higher the number of hops, the lower is the probability that the packet can successfully reach the destination. This

Fig. 10.12 Multi-radio mesh network experimental setup

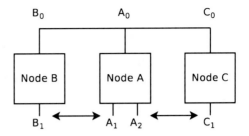

is justified by the fact that as the density of nodes increases, the number of hops increases, too. Consequently, also the packet loss probability increases. In turn, when the transmitting power is increased, the same node coverage is ensured, and so the number of hops slightly differs.

10.3.2.3 Co-channel Interference in IEEE 802.11a Multi-radio Mesh Networks

The co-channel interference results were obtained by conducting experiments in a well-planned testbed to produce reliable and reproducible results. The presented results incorporate multiple parameters including transmission power, modulation coding scheme, channel separation and physical layer effects such as adjacent channel interference, carrier sensing, retransmissions and packet distortion.

The most important factor when planning the experimental testbed was to provide reliable and reproducible results with the least number of external dependencies as possible. In order to achieve this goal, the testbed was deployed as shown in Fig. 10.12 whilst accounting for the Fresnel zone to design the radio links properly and considering the configuration and scripts proposed in [RFMM10]. The reliability was ensured by verifying the confidence intervals. In IEEE 802.11a, there is a trade-off between Adjacent Channel Interference (ACI) and the separation between co-located antennas, which is limited to 50 or 60 cm maximum because of the dimensions of the node casing. A numerical example is given in [RFMM10] for the MikroTik R52 802.11a/b/g card for the 5.2-GHz carrier with a 20-MHz bandwidth.

The results show that by increasing the channel separation between co-located radios, the level of co-channel interference decreases. All presented results take the radio parameters TxPower, MCS, channel separation and physical layer effects into account to explain the performance degradation due to carrier sensing, packet distortion and backing-off. The results obtained will be used to develop an algorithm that takes the radio parameter settings, external dependencies and some prior knowledge as an input and provides the optimal global configuration of nodes in a WMN so that the co-channel interference will be minimised.

10.3.2.4 Presentation of the Wireless Planning Tool and Its Models

The Wireless Planning Tool (WPT) enables to optimise the number of the required access points (APs) or Base Stations (BSs) to be deployed, their positions, and the total cost of the equipment by considering the choice of the characteristics for different WiFi technology suppliers. It supports the IEEE 802.11 a/b/g and WiMAX standards, and it is able to estimate the Transition Region (TR) needed by the network, according to the capacity model. Items such as obstacles, APs characteristics and network card types can be directly updated into the software. Different propagation models can also be introduced and tested. The WPT helps in the process of making a complete plan for the coverage of a given area based solely on a digital format of the floor plan, obstacles, their materials and the locations for the wireless terminals. The program then generates an output with the layout, showing the received power/capacity and the positions for the APs. The optimum location for the APs, so as to minimise their number, was achieved using two methods. One method considers the received power in each position of the coverage area using empirical propagation models. The other one allows one to choose the most probable position for each user and its capacity, according to the foreseen applications and the respective User's Simultaneous Factor (USF), via a specific algorithm. Details on the tool can be found in [STV$^+$09, SVC$^+$09], as well as a comparison with other tools for WLAN planning available in the market.

10.4 3GPP Radio Access Networks Optimisation

The installation and upgrade of a Radio Access Network (RAN) consists of a planning phase, in which rough parameter sets are determined, and an optimisation phase after the roll-out of the determined configuration, in which the planned parameters are adjusted. This section deals with measurement-based network planning, which aims to find optimal network parameter settings from measurements taken from the live network. The introduction of measurements into planning enables improvements in both phases in drawing them closer to each other.

Here, the term *optimisation* is used to refer to the improvement of the network in case a mathematical network model is available. Then, an appropriate optimisation algorithm can be used to determine the optimal network parameter settings. Optimisation and modelling for RANs have been investigated since start of their operation. Most investigations in the past were based on simple network models, on which all kinds of optimisation methods have been tested. Due to lack of knowledge about real operating networks in science, much of past research work has been focused on the refinement of models in attempt to reach for more and more realistic scenarios. The start of cooperation between researchers and network operators has caused a new impulse: Real scenarios have become available and, along with them, new challenges presented in the following.

In Sect. 10.4.1, the general framework for measurement-based planning is introduced. Section 10.4.2 presents two important examples of re-planning procedures using optimisation techniques. Section 10.4.3 states the importance of field measurements, especially regarding access technologies still under construction.

10.4.1 Processing Performance Statistics

Vendors of network elements implement counters into their equipment to monitor signalling and payload traffic that has to be handled by the node. Values for these counters from all nodes are collected and monitored in the Network Management System (NMS). These Performance Counter Statistic (PCS)s deliver quantitative information about signalling requests, speech and data traffic or successful and failed assignments. Measurements of PCS provide real information about the current state of network, which can be used to improve the accuracy of network models. Thus, optimisation can make most of the situation. In addition to the gain of reliability, PCS-based algorithms (see Sect. 10.4.2) in operation would shorten time periods of planning iterations.

PCSs look like an answer for many applications. Nevertheless, considering PCS for development of measurement-based planning algorithms, some points have to be taken into account:

- PCSs are obtained from the network infrastructure. Unlike drive tests, which are related to the experience of single users, PCSs deliver a global view of the network in operation. This makes localisation of events and tracking of single users impossible, which can only be done by the corresponding network element controlling a certain geographical area through its assigned base stations (e.g. Base Station Controller (BSCo) in GSM-EDGE Radio Access Network (GERAN) or Radio Network Controller (RNC) in UMTS). However, information about position and time of individual users is not required. To optimise the overall network performance, only aggregated measurements are required. For this task, statistics about the network usage are more than enough. Applications that require more details cannot use PCS.
- PCSs deliver aggregate information collected in time periods of minutes at best. However, the significance of statistics increases with increasing time periods. Therefore, PCS can only be used for long-term planning tasks meaning the determination of parameter sets which are valid for a couple of days or weeks, at least (for examples, see Sect. 10.4.2). In contrast, short-term aspects like power control cannot be based on PCSs; in this case, the numerous measurements of User Equipment (UE)s with time periods of a fraction of a second are more promising.
- In an operating environment, the use of PCS requires an automatical feed of PCS data from NMS to the data bases of a planning tool. However, PCSs were never specified for such usage. Accordingly, the data bases of commercial planning tools and the PCS data base of the NMS are strictly separated. The flow of information from operation to planning is difficult and, therefore, in most cases,

ineffective. A combination of the data bases of NMS and planning tool is required, but has not been intended within former development. For this reason, common planning tools are not designed for an upgrade to measurement-based planning features.
- The combination of data bases in NMS and planning tool has a major problem in the time shift of their network representations. The parameter settings in a planning tool represent a network of the next few days or weeks. In contrast, statistics in the NMS represent measurements of a network in the last few days or weeks. Therefore, a perfect match cannot exist. For instance, some base stations in the planning tool may not be implemented yet. Likewise, NMS statistics may be missing for temporarily locked base stations. Thus, some basic principles are required to merge the two data bases. The basic idea takes advantage of the fact that both representations consist of coded hierarchy trees with the network elements (from transceivers to the highest controller) as nodes. These codes can be compared in detail. Another attribute is the name of the vendor [HK09a], indicating the properties of the equipment.
- Speaking of vendors: PCSs are vendor-specific. Even though events in a 3GPP RAN are standardised, the implementation of counters in the network elements is not. Therefore, different vendors may count the same event by name, but differ in the implementation. Thus, pay attention in combining PCSs of different vendors.
- PCS deliver tera-bytes of information per day. This amount of statistical data has to be processed. At least, a small subset of available PCSs are required for a specific planning task. However, appropriate counters have to be identified in the extensively large documentations of the vendors.
- Network models might require non-measurable information. In this case, it is recommended to examine possibilities to estimate the required value out of given ones. For example, the optimisation of an Location Area (LA) plan requires the paging contribution of single base stations as input. However, paging requests are counted per LA which consist of a group of base stations, meaning that the required information per single base station is not available. In this case, the number of Mobile Terminated Call (MTC) events per base station is used to estimate the required paging value per base station, because MTC and paging have a strong correlation [KH05].

Figure 10.13 shows a basic concept concerning the corresponding enhancement of planning tools [HK09a]. The main parts are the following:

- A planning algorithm handles one specific planning task by determining a specific parameter or set of parameters to be saved in the data base of the planning tool. It defines specific input data which consists of data sets from the planning tool and estimates of traffic values for network model generation.
- The databases of the planning tool and the NMS deliver required data sets for the algorithm or estimation process.
- In order to build estimates for missing data, the estimation block requires specific input data itself, which is obtained in the form of geographically referenced measurements. To build these measurements, appropriate tools for data processing

Fig. 10.13 Framework for introduction of measurement-based planning in mobile radio network planning tools [HK09a]

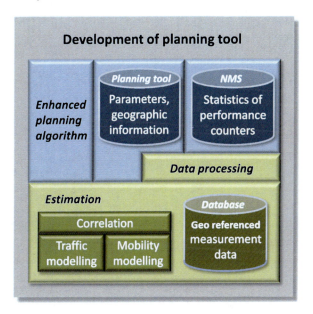

must be designed to combine information in the network planning tool (e.g. site coordinates, cell service areas) with network parameter settings in the NMS (e.g. handover neighbourhoods). Note that not all this information is readily available, but has to be calculated (e.g. cell service areas computed from cell coordinates, power settings, antenna diagrams and propagation models).

- The estimation process itself can be separated into three parts:
  - Correlation between measurable counters for estimation of non-measurable values has been investigated in [KH05, HK07c].
  - Cell traffic modelling is used for estimation of cell traffic values (e.g. statistics about call set-up in cells) based on correlation between cell planning and cell traffic changes [HK07a, HK07b]. Linear regression analysis has been investigated to determine conversion factors [HK07a].
  - Inter-cell mobility modelling is used for estimation of user mobility between cells (e.g. handover between cells) based on *Shortest Path Problem* and correlation between handover and joint coverage areas [HK09b].

10.4.2 Optimisation Algorithms

From available data, a precise RAN model can be built. In some cases, the resulting model can describe the relationship between network inputs (i.e. RAN parameters) and outputs (i.e. Key Performance Indicator (KPI)s) analytically. Over these models, classical optimisation techniques can be used to find optimal parameter settings. In the following paragraphs, several examples of network re-planning procedures inspired in the optimisation theory are given. Note that, even if all of them

Fig. 10.14 Graph of a real base station controller [TGGLRW10]

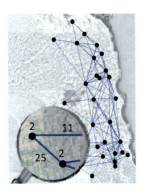

are restricted to GERAN, similar principles can be applied to other radio access technologies.

10.4.2.1 Network Re-structuring

In current cellular networks, the mere structuring of the network is one of the most time-consuming tasks. During network design, operators decide which element in the lower layers (usually base stations) is assigned to elements in higher layers (referred to as controllers or control areas). Such an assignment, which has to be updated frequently, has a strong impact on both signalling load and perceived quality of service in the network. When deciding on the cluster of elements grouped under the same controller, the aim is to minimise signalling and delays in data transmission by keeping strongly related elements (e.g. adjacent base stations) in the same controller. At the same time, controllers must be kept below their capacity limit. This clustering problem can be formulated as a *graph partitioning problem* [SKK03]. As shown in Fig. 10.14, the network is modelled as a graph, where the vertices are base stations, and the edges are adjacencies defined by the operator for handover purposes. The weight of each vertex represents the contribution of the base station to the load of the controller, while the weight of each edge represents the number of users moving between base stations. The partitioning of the graph into sub-domains reflects the assignment of base stations to controllers. The aim of partitioning is to minimise the weight of edges joining vertices in different sub-domains, while keeping the weight of sub-domains within certain limits (or, in other words, minimise the number of users moving between controllers while equalising the load of controllers). To achieve this goal, statistics of traffic (load contribution) and handovers (user movement) are gathered in the NMS to keep track of user trends.

One of such clustering problems is the re-assignment of base stations to Packet Control Unit (PCU)s in GERAN [TMWW10]. The PCU assignment should minimise the number of changes of PCU experienced by users after a cell re-selection, while also ensuring that the load of all PCUs is within certain limits. Such an assignment, performed by the operator when optimising the network, has a strong impact on the quality of packet-data services in GERAN [TWB06]. To build a good

PCU plan, the network is modelled as a graph, where vertex weight is the number of time slots devoted to data traffic per base station, and edge weight is the number of packet-data users moving between neighbouring base stations. In [TGGLRW10], an exact graph partitioning algorithm is proposed to find the optimal PCU plan from handover statistics. The method is based on an integer linear programming model of the problem, which is solved by the branch-and-cut algorithm implemented in a commercial optimisation package. In that work, it is exploited that, unlike graphs in other application domains, the size of graphs in cellular network structuring is small, making exact approaches feasible. Results show that the exact method gives a five-fold reduction of inter-PCUs cell reselections compared to the manual operator solution. Its main drawback is a large execution time (compared to heuristic methods), which is still acceptable for network re-planning purposes.

A similar problem is LA re-planning [DECD04]. An ideal LA plan should minimise the number of Location Update (LUD) requests caused by users moving between LAs, while keeping the number of mobile terminated calls per LA below certain limits so as not to exceed the paging capacity of cells in the LA. In order to find the best grouping of cells into LAs, graph partitioning algorithms can be used again. Unlike PCU planning, vertex weight now denotes the number of paging requests in the busy hour due to mobile terminated calls in the cell, while edge weight denotes the number of users in idle state moving between cells. In the absence of idle-state mobility statistics, operators use handover statistics to build the network graph. Several studies (e.g. [KH05, TLRWS09]) have shown that incoming inter-LA handovers and LUDs in a cell are correlated. However, the number of handovers is not directly linked to the load in dedicated signalling channels, namely the Standalone Dedicated Control CHannel (SDCCH) in GERAN. Therefore, it is difficult to predict the impact of changing an existing LA plan. In [TWLRJ09], a network performance model is proposed to estimate changes in the peak traffic on the SDCCH after changes in the LA plan in GERAN. Such a model can be used to quantify the number of SDCCH time slots that can be saved by improving an existing LA plan. The model is validated in [TWLRJ09] on measurements collected before and after changing the LA plan of a limited geographical area. The model is then used to compare several LA re-planning algorithms, both exact and heuristic. The comparison is applied to graphs built from measurements in a larger area covering a whole NMS. Results show that proper planning of LAs can decrease the average SDCCH traffic in a live network by 30%. Such a figure is remarkable, since LUDs only represent half of the total carried traffic in the SDCCH in existing GERAN networks. Nonetheless, the required number of SDCCH time slots can only be reduced by 10%. Thus, the main benefit of changing LAs is reducing SDCCH blocking in cells at the border of LAs (where most LUDs take place), while also reducing the total number of time slots devoted to the SDCCH.

In the above-mentioned model, the required number of SDCCH time slots per cell is computed from peak traffic estimates. For the same purpose, operators have traditionally used the Erlang B formula based on average traffic measurements. Thus, it is assumed that the network behaves as a loss system with Poisson arrivals. However, automatic retrial mechanisms and correlated arrivals in these channels cause that these assumptions do not hold. On the contrary, analysis of real data [LRTW10] shows that the Erlang B formula underestimates congestion and blocking on the SDCCH. These limitations are overcome by a queueing model with both retrials and correlated arrivals [LRTW10]. Time correlation between arrivals can be modelled by a Markov-modulated Poisson process. The resulting model, unlike more sophisticated models proposed in the literature, can easily be adjusted on a per-cell basis using statistics in the NMS. Estimation results for the main SDCCH performance indicators (i.e. traffic, congestion ratio and blocking ratio) show that, with such a model, the sum of squared errors is more than halved compared to the Erlang B formula.

10.4.2.2 Cell Traffic Sharing

In live cellular networks, the matching between the spatial distribution of traffic demand and resources becomes looser as the network evolves, causing localised congestion problems. In legacy networks, where capital expenditures must be kept to a minimum, these problems are solved by adjusting cell service areas, which can be achieved by tuning handover parameters on a per-adjacency basis [TW08]. To fix such parameters, operators have traditionally used heuristic rules in the absence of an analytical network model. These rules aim at equalising congestion problems across the network in the hope that, in doing so, call blocking is minimised, although no proof of the latter has been given. In [LRTWFN09], an analytical teletraffic model of the traffic sharing problem is presented for real time services in GERAN. From this model a closed-form expression of the optimality conditions can be obtained by solving a constrained optimisation problem, where the goal is to minimise call blocking, and the decision variables are the traffic demand allocated to each cell, constrained by spatial concentration of users. Such an analytical expression defines an optimal indicator to be balanced among cells. Showing that this indicator is not the call blocking probability, it is a clear evidence that the procedure for equalising congestion across the network, as currently applied by the operators, is not the optimal strategy. On the contrary, a comprehensive analysis in realistic scenarios shows that using the optimal sharing criterion instead of balancing blocking ratios can increase network capacity by up to 3%, even under restrictions in the cell re-sizing process. Such a figure is not negligible in terms of operator revenues, and, more important, the benefit is obtained without changing network equipment, which is a key in mature technologies such as GERAN.

10.4.3 Field Measurements

NMS statistics are used to monitor and improve network performance since they provide a global view of the network. Alternatively, field measurements can be used to evaluate performance from the user's point of view. Such measurements are the best means to launch the optimisation process in networks still under deployment, such as UMTS and High-Speed Packet Access (HSPA), in the absence of well-proven optimisation methods.

Drive-test tools are able to detect problems in the roll-out stage of the network, when the number of subscribers is still small. In [MMSMC08], a set of processing methods are described to detect common problems in High-Speed Downlink Packet Access (HSDPA) based on pilot scanning. The proposed methods use network parameters, such as neighbour lists and handover parameters, and performance indicators, such as Common Pilot Channel (CPICH), Received Signal Code Power (RSCP) and CPICH Ec/Io. To assess the methods, outdoor measurements were collected from a live HSDPA network in a small geographical area (i.e. a university campus). Measurements were taken in static positions to allow continuous measures. Thus, proper windowing and averaging of data was possible at the expense of an increased collection time. Experiments showed that a conventional drive-test tool can detect critical problems, namely poor coverage, interference coverage, pilot pollution, pilot overshooting and pilot surprise.

Drive-test tools can also be used for benchmarking purposes. In some countries, public authorities force operators to perform measurement campaigns and make the results publicly available to subscribers. A methodology for evaluating the overall QoS of packet-data services in an HSPA network is presented in [PPC09]. The methodology is based on drive surveys, where a mobile terminal is configured to repeatedly establish sessions, including Packet Data Protocol (PDPr) context activation/deactivation, File Transfer Protocol (FTP) download, web browsing and ping. Service accessibility, service retainability and service integrity are used as KPIs [25007, 25008]. The core of the method is the post-processing of data in a time-space relational database, offering enhanced drill down analysis, data mining and correlation analysis. To test the method, a measurement campaign was conducted outdoors and indoors under both static and mobile conditions. Reported statistics showed that a real downlink throughput of up to 5 Mbps and an RTT of 200 ms is achieved with HSDPA 7.2 Mbps technology in a live scenario.

10.5 Auto-tuning of RAN Parameters

Currently, both online and offline concepts of measurement-based optimisation are developed. In this sense, auto-tuning can be seen as an evolutionary step from the offline methods described in Sect. 10.4 and the SON solutions presented in Sect. 10.6. An automated dynamic optimisation of RAN procedures, which is to adapt the RRM parameters to short-term variations of the network operation conditions, is called auto-tuning. The research carried out in this topic showed the efficiency improvement of present and future technologies. Studies on the benefit of auto-tuning have increased in significance as they have been launched in the framework of the 3GPP TSG-RAN as a part of SON functionalities [3GPa, 3GPb] for Evolved UTRAN (E-UTRAN) being a part of the LTE of UMTS.

Fig. 10.15 Online optimisation loop

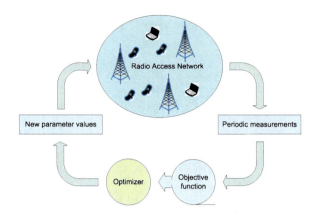

The topic of auto-tuning in UMTS appeared in 2002 and was initiated by research groups from Nokia. The propositions included common pilot power tuning aiming at load balancing while also assuring pilot coverage [VHPF02], downlink load target and the maximum transmitted power for one link [HV02] as well as the service- and bit-rate-specific E_b/N_0 requirements [HVHL02]. Recently, interesting research on auto-tuning has been done in the Eureka-Celtic GANDALF project [Gan]. It was focused on large-scale network monitoring, advanced Radio Resource Management (RRM) rules and appropriate quality of service evaluation in order to achieve automation of network management tasks in a multi-system environment. The concept of auto-tuning was extended to Joint RRM (JRRM) for multi-RAN to improve the access system cooperation.

10.5.1 Online RAN Optimisation Loop

Online measurement-based automated optimisation is a new, very promising approach arousing interest of operators. It allows one to avoid inaccuracy of simulations caused mainly by modelling of propagation and details of the environment. Instead of building a standalone simulator, the optimisation or tuning algorithm can be implemented and work in a real network online. Hence, successive optimisation engine iterations are based on real KPI measurements, not on the simulated ones [NWS+08]. The scheme of such a process is illustrated in Fig. 10.15.

Scheduling of the optimisation process is strongly related to and based on the KPI measurements. For short-term measurements used in auto-tuning, their reliability has to be taken into account [WJ08].

The implementation of complex auto-optimisation mechanisms requires a careful selection of the mathematic algorithms used and the design of efficient computation mechanisms in order to minimise the response time of the developed tools [MOMC09]. Automatic optimisation and tuning solutions were developed mainly for the UMTS/HSPA networks.

10.5.2 Input Data for Mobile Network Online Optimisation Process

The radio network online optimisation requires specific conditions to efficiently operate in a live network. Since the optimisation process is based on measurements, the Performance Indicator (PI)s and parameters that can be extracted from the system and used for the cost function construction should be identified [NWS+09]. Regardless of the purpose for which the data are collected, they are usually located initially in RNC for Third Generation (3G) networks and in Binary Symmetric Channel (BSC) for 2G and 2.5G networks. Then, every predefined period of time all of the collected data from RNC or BSC are transferred to the Operational System Support platform (OSS; another frequently used name is Network Support System, NSS). Depending on the network load, it may happen that the data transfer is interrupted, corrupted or does not start at all. Usually, the transfer attempt is retried for a number of times before it is abandoned completely. In the high network load scenario the system may also abandon data collection at all.

In the majority of cases, collected data are transferred from network elements in the form of binary files that are then parsed by vendor-supplied software to a database (typically Oracle or Sybase). However, direct access to the databases is limited because of the potential critical impact on the stability and dependency of the major parts of OSS on database as the main source of various types of information. Apart from the collected network performance data, such database usually contains mission-critical data (e.g. definition of all network elements and the entire network configuration).

The most common solution that makes the collected data available through the interface other than built into OSS applications is using vendor-supplied data exporters. Unfortunately, such access is impractical from the point of the online optimisation view due to its dependency on additional components, which are difficult to control in case of exceptions. Elimination of data exporters from the loop is therefore highly desirable. Thanks to that, the entire logic and decision making inside the optimisation loop is simplified and can be controlled by less pieces of software.

As mentioned above, accessing the central OSS database directly through the ODBC interface is seen as potentially dangerous. However, with a proper approach the risk can be eliminated. The correct approach includes database access techniques eliminating the need for nested queries, splitting the retrieved data into small data packs that prevent locking a part of the database being read from write access for periods longer than write timeouts etc. It has been tested that even obtaining very large amounts of data from the database can be handled in a manner not causing noticeable OSS CPU loads.

Robust methods for the data acquisition and processing and for efficient and safe access to the network are very important issues related to the online optimisation process. For the purpose of short-term tuning, the reliability of the measurements is also very important, especially when KPI values are based on counters.

10.5.3 Automated Tuning of Soft Handover Parameters

Dynamic optimisation of UMTS systems has been gaining a growing interest by the research community since it provides to the network more flexibility, dynamism and reactivity in terms of time-varying conditions. This is particularly important in the context of new data applications, with different services usage and UEs mobility. All of them are factors that imply that UL and DL requirements do not remain constant, and so capacity unbalance between both links is likely to appear and vary with time.

Most of the existent auto-tuning proposals fail in studying UL and DL effects jointly, and unbalance is hardly checked. In this context, we propose an Automatic Tuning System (ATS) with the objective of detecting whether one of the links has capacity problems and favour it to delay congestion control actions [GLRBALOC07]. This approach is novel and not previously addressed, since most works dealing with CDMA UL and DL unbalance consider the TDD mode duplexing.

The solution is based on a functional auto-tuning architecture that adapts RRM parameters to service mix dynamics and overcomes capacity problems. This architecture is composed of three blocks (*Learning & Memory*, *Monitoring* and *Control*) and two interfaces.

From the *Learning & Memory* stage, Soft HandOver (SHO) has been revealed as a feasible option to achieve the balancing objective. In particular, the Addition Window and the maximum number of cells that can be included in the Active Set (AcS), *AddWin* and $\mathcal{N}$, respectively, are used. Increasing $\mathcal{N}$ favours the UL, whereas too many cells in the AcS degrade the DL. Although several combinations of these parameters yield balanced capacity situations, only one of them also implies maximum capacity, which is the interesting one for the operator. In this sense, the service mix in the target area has the most relevant impact on this configuration, and the spatial distribution of the users has a lower effect.

Regarding *Monitoring*, KPIs have been defined taking into account 3GPP definitions on performance related data. In particular, the percentage of UEs that require more power than a certain threshold and a counter of the number of continuous frames in which the previous KPI is reached for 5% of users are defined. In fact, this acts as a standard time-to-trigger to avoid ping-pong effect, as it is usually done in other procedures in the resource management of wireless networks. From tests without implementing ATS, it is derived that the global DL transmission power is not representative enough of congestion in the cell.

Finally, the *Control* stage acts in a totally distributed manner, and parameters are modified on a cell by cell basis. Speed of reaction can be increased with a more conservative *Monitoring*, that means reducing the threshold in the first KPI. However, this is at the cost of increasing the number of reconfigurations.

Simulations with different service mix cases showed capacity gains in terms of simultaneous users around 30% when ATS was running.

10.5.4 Automated Tuning of Call Admission Control

Call Admission Control (CAC) procedure together with Congestion Control (CC) impacts the values and the mutual relation of two important indicators, namely call blocking and dropping rates (denoted *BR* and *DR*, respectively). These indicators are used for the quality assessment of serving the offered traffic. They are based on counters of events that occur relatively rare and thus cannot be directly used as KPIs in the tuning tight schedule due to poor reliability of their measurements [WJ08, NWS+09].

Hence, a new concept of decentralised online automated and measurement-based CAC tuning is developed [WL10]. The proposed solution is based on traffic monitoring, its analysis and predictive reconfigurations of the CAC procedure in order to increase the quality of serving traffic. It is assumed to pass over information about cell coupling and close the whole measurement and decision process in a single cell. Although it suggests to reorganise the scheme of measurement gathering and processing in UMTS, which would require a certain effort, the reward of opening new possibilities for managing the network seems to be tempting.

The optimisation of the CAC procedure can be based on the following objective function minimisation:

$$CF = BR + W \cdot DR, \qquad (10.1)$$

where $W > 0$ is a weight factor that is assumed as $W = 4$, since operators pay more attention to protect the network against dropping than blocking. It is realised by a certain reserve of load assumed in the CAC procedure, which means that the CAC load threshold for new calls (denoted as η_{max_ON}) is set below the CC one (η_{max}). The values of the η_{max_ON} and η_{max} thresholds impact the *BR* and *DR* indicators.

Predictive reconfigurations of the CAC load threshold are based on indicators that are related to *BR* and *DR* but measured with much better reliability during short periods of time. Performed simulations for voice traffic showed that the optimal load threshold $\eta_{max\_ON\_opt}$ depends on the traffic volume offered within a cell ($A_{offered}$) and the user mobility profile characterised by the easy to measure mean time of serving calls in a cell ($t_{average}$). When the traffic volume is greater, the cost function (10.1) reaches its minimum for smaller values of η_{max_ON}. On the other hand, when mobile stations are moving faster, $t_{average}$ is smaller and so is the optimal η_{max_ON} value. These relations, obtained empirically, were used as a basis for dynamic control of the η_{max_ON} threshold.

Defined tuning algorithm was compared to the static optimisation for the case of an unexpected traffic density disturbance. A certain increase of traffic volume was assumed in the central cell of a dynamic 7-cell network model. As a result of the dynamic CAC optimisation, a gain of approximately 12–15% in the cost function *CF* (10.1) value was reached for the central cell.

Dynamic CAC threshold configuration according to the traffic conditions variation enables increasing the network resilience for incidental disturbances. Simulation results showed that the proposed tuning method works properly, does not impact the stability of network operation and ensures measurable profits in the case of an atypical unplanned growth of generated traffic volume.

10.5.5 Automated Tuning of HSDPA Code Allocation

Although, HSDPA (and more generically HSPA) continues evolving through new 3GPP standard definitions, the RRM algorithms that are implemented in the vendor equipment are a key factor to its success and performance improvement. These algorithms are not defined by the standard, and that is why several investigations are being carried out to find the best possible implementations.

In this sense, the work presented in this subsection is aimed to assess to which extent it is worthwhile to make a dynamic management of HSPA three most important resources: devoted power, percentage of users assigned to Rel'99 or HSDPA and codes, when both technologies are deployed under the same carrier [GLRB08]. From the study, one of the first conclusions reached is that the benefits of HSDPA imply that in general there is no clear benefit in introducing an ATS to manage power or the percentage of UEs assigned to HSDPA; both can be handled by straightforward rules-of-thumb.

On the other hand, the management of the Orthogonal Variable Spreading Factor (OVSF) code tree is more complicated. Initially, three questions can be posed: First, how the codes should be assigned to meet QoS targets; second, if this assignment is dependent on changes in traffic patterns; and third, if code allocation should be considered for the inclusion in a UMTS ATS. By means of simulation, the cell throughput along with several collateral effects were studied for different codes allocations. This was done for different geographical UEs distributions, and several engineering rules were obtained:

- Effects on blocking probability: Blocking probability upper bounds the maximum number of codes to be considered by the ATS. Both HSDPA and Rel'99 blocking are proportional to the number of reserved codes. So, by favouring HS-PDSCHs too much, a negative effect also appears in HSDPA.
- Effects on SHO areas: An indirect effect of the previous point is that AcS sizes are reduced if the number of allocated codes surpass a certain threshold, which value depends on the Rel'99 TFs. This implies connections with Node-Bs that are not the best option in terms of DL power. As a consequence, DL interference increases with the number of HS-PDSCHs. Of course, when the number of fully blocked users is important, DL power is again reduced, but with a clearly inadequate performance of the network.
- Effects on Rel'99 throughput and dropping: Rel'99 throughput is maintained, but degradation and eventually dropping are proportional to the number of HS-PDSCHs.
- The optimum number of codes to be assigned to HSDPA is tightly related to UEs spatial distribution. For users concentrations far away from the Node-B, there is no point in reserving more than five codes, this value can even be decreased as the cell size increases. A higher value does not improve HSDPA performance, codes are wasted, and blocking probabilities are unnecessarily increased in both technologies. On the other hand, when users are close to the Node-B (distance below 150 m), the number of codes to be allocated is just limited by the maximum

allowable blocking probability. When UEs are homogeneously distributed, the optimum number depends on the cell size.

Given this, an ATS is proposed to make mid-term reservations based on the majority of RF conditions in HSDPA UEs. Consequently, reported Channel Quality Indicator (CQI) measurements are continuously monitored, and the corresponding histogram is computed. After a simulation based study, the first quartile and the standard deviation of the histogram are considered as inputs for the self-tuning of the system. Also, from this analysis, a decision Look-Up-Table is defined to connect these KPIs and the codes to be applied by the *Control*.

A post-processing of the KPIs is again necessary to avoid too frequent reallocations and an excessive ping-pong effect. The study revealed that a combination of a standard time-to-trigger with an FIR filter-based running average is a proper option. When comparing the performance of one network running the ATS against another with a fixed code allocation, important benefits are obtained. The ATS proposal improves the performance of HSDPA networks coexisting with UMTS Rel'99. An optimum number of codes is allocated for each technology, and, hence, the cell throughput can be optimised while minimising both Rel'99 and HSDPA blocking probabilities and degradation below specific thresholds.

10.6 Self-organising Networks

While a well thought-out planning of mobile radio networks and auto-tuning of parameters already assure good performance, it has been found in recent years that dynamic, autonomously operating processes carrying out substantial parameter changes that may even intervene with classical network planning and optimisation processes are required in order to effectively optimise mobile radio networks. These processes rely on accurate information concerning the state of the network and thus require constant and reliable collection of system measurements. The term self-organising network (SON) has emerged in order to identify a network exhibiting functionality of the discussed nature. Thoughtful and thorough investigation and development of self-organising functionality is required since important network parameters are affected to an extent making it critical for reliable and stable network operation.

Self-organising functionality is designed to carry out a variety of network operation tasks in an autonomous way. It thus reduces the need for costly human interaction in operation of the network and leads to a strong reduction of the operator's Operational Expenditures (OPEX). Furthermore, the optimisation gains expected by the introduction of self-organising functionality in mobile radio networks will increase efficiency and capacity of the network and reduce the required number of sites and thus Capital Expenditures (CAPEX) of the network operator [vdBLE+08]. The introduction of self-organising functionality is therefore envisioned as a key technology in achieving significant cost reductions without sacrificing performance.

This section discusses specific aspects of and requirements on self-organising functionality and presents approaches for the solution of typical use cases in the field of SONs.

10.6.1 Requirements and Assessment Criteria

Several technical and economical requirements have to be considered for the successful development of self-organising functionality. Concerning the technical requirements, self-organising functionality has to assure stable operation that is robust against erroneous measurements and has to provide a performance gain that is in balance with the additional complexity. Depending on the task that is carried out by self-organising functionality, different timing requirements can apply, depending on the goal and cause. All algorithms that carry out tasks in the field of self-organising functionality and that influence the same parameters have to interact in order to resolve conflicts and to coordinate actions. Finally, certain architectural requirements arise depending on the implementation of self-organising functionality [ALS$^+$08a].

On the economic side, cost efficiency requirements apply, meaning that self-organising functionality should reduce CAPEX and OPEX. Furthermore, so-called LTE deployment aspects require self-organising functionality to speed up the roll-out of LTE networks, to simplify operational processes, to ease the introduction of new services and to provide end-user benefit in terms of high service quality [3GP07, NGM06, ALS$^+$08a, ALS$^+$08b].

The measurement of the effectiveness of self-organising functionality is carried out using different metrics in a benchmarking approach. Relevant metrics are:

- performance metrics indicating Grade of Service (GoS), such as call blocking ratio and call dropping ratio, and QoS, such as packet delay statistics, packet loss ratio, transfer time statistics, throughput statistics, fairness,
- coverage metrics, such as service coverage and data rate coverage, which specify the fraction of area where a certain service or a minimum bit rate can be experienced,
- capacity metrics indicating the capacity of a cell in terms of the maximum number of concurrent calls, the maximum supportable traffic or the spectrum efficiency,
- CAPEX, where it has to be noted that costs or cost per unit may increase, depending on the complexity of the self-organising functionality,
- OPEX, where in extreme cases, all OPEX resulting from manual adjustment can be removed. Furthermore, OPEX can be reduced by minimisation of drive tests.

The metrics are evaluated for a fixed scenario and for different approaches or algorithms. In order to compare different algorithms, the values of the metrics together with an estimate of the optimisation effort can be compared. In general, some of the metrics will behave oppositely, meaning that most likely, there will be no algorithm that performs high in all metrics. In order to facilitate a strict ranking, a utility function can be established by weighting and summing the different metrics according to the operator's policy [ALS$^+$08a, ALS$^+$08b].

10.6.2 Self-organising Use Cases

Use cases describe particular problems arising in mobile communication networks and define the desired outcome of the problem-solving solution. According to the operational process of self-organising functionality, they are categorised into three areas:

- *self-configuration* denotes the process of automatically obtaining the initial configuration or a reconfiguration of a Base Station (BS),
- *self-optimisation* refers to the process of automatically adapting algorithms and parameters in order to adapt to changes in the network, traffic and environmental conditions,
- *self-healing* describes the process of automatically resolving possible coverage and capacity gaps resulting from failures.

Several use cases have been defined for self-organising networks [3GP09, NGM07, SOC08]. This section describes some of the use cases and introduces relevant self-organising approaches.

10.6.2.1 Call Admission Control Parameter Optimisation

CAC decides about the admission of call requests and is thus a central task in cellular mobile radio networks. The challenge with CAC is to admit as many calls as possible while guaranteeing QoS of ongoing calls and assuring successful HandOver (HO) to the cell. Performance metrics for CAC algorithms are thus GoS metrics like call blocking ratio and handover failure ratio and QoS metrics such as traffic loss ratio and call throughput.

The capability of a cell to admit additional calls is based on its capacity. The capacity $C(t)$ of a cell at time t can be estimated according to [SSB10] by

$$C(t) = \frac{1}{\eta} \sum_{m=1}^{N_R} r_m(t) \qquad (10.2)$$

with η the load factor, N_R the number of resource units and $r_m(t)$ the bit rate achieved by resource unit m. The division by the load factor is required to consider the potential capacity of not scheduled resource units.

The key parameter in CAC is the fraction of the capacity of the cell that is reserved for HO calls. For self-organising tuning of this parameter, thresholds for call blocking ratio, handover failure ratio, traffic loss ratio and the ratio of calls not fulfilling their throughput requirement are defined. In order to filter fluctuations, the performance metrics are averaged using exponentially decaying weights and then constantly compared against the defined thresholds. If any of the thresholds is exceeded, the fraction of the cell capacity reserved for HO calls is increased, and if all of the metrics fall below the thresholds, the fraction of the cell capacity reserved for HO calls is decreased. In order to avoid oscillations, a hysteresis value is considered [SSB10].

10.6.2.2 Handover Parameter Optimisation

In cellular mobile radio networks, the HO of users from one cell to another cell is required in order to support mobility of the users. Undesired effects such as call drops and ping-pong HOs, for example, can occur due to high cell load or shadowing, respectively. In order to avoid these effects, it is necessary to control HOs by constantly adapting the individual values of HO parameters for each cell.

An HO is taken into account if the Receive Signal Reference Power (RSRP) of the connected cell is smaller than the RSRP of another cell considering a hysteresis value. If this condition holds for a certain time, called Time-To-Trigger (TTT), the HO is initiated. Hysteresis value and TTT are thus the control parameters for this use case, each combination of values for the two control parameters is called an operating point. Relevant metrics for evaluation of the performance of self-organising HO parameter optimisation algorithms are HO failure ratio, ping-pong handover ratio and call dropping ratio.

In order to determine the influence of the control parameters on the performance metrics and in order to identify actions to be taken if HO performance drops, system level simulations are carried out. Hexagonal scenarios are widely used in simulation due to their convenience in modelling and run time efficiency. These scenarios, however, do not reflect irregularities and particularities of the real scenario and may thus be significantly inaccurate.

A realistic modelling of mobile radio scenarios requires a variety of detailed input data. Accurate propagation prediction can be obtained by applying ray-tracing based on detailed data on land-use, terrain and building height, shape and arrangement of buildings. User mobility can be modelled with high detail using vector data describing the road network and the Simulation of Urban Mobility (SUMO) package [KHRW02]. Finally, traffic data and the network layout can be obtained from a network operator; alternatively, traffic data could also be derived from land-use information. Note that besides the data on the network layout, several publicly available sources for the required information exist [JSK10].

Figure 10.16 shows system level simulation results of the call dropping ratio for different operating points. The effect and thus the importance of a realistic modelling is clearly visible.

Simulation results such as those from Fig. 10.16, for example, can be used to design rules for HO optimisation. Table 10.9 shows the adequate action for a drop of each of the performance metrics in dependence on the control parameter values [JBMK10]. Comparing the rules of the second row of Table 10.9 to Fig. 10.16(a), it can be seen how establishing and maintaining an operating point assuring reliable and stable operation of the network with respect to the call dropping ratio is aimed for.

In order to enable self-organising optimisation of the HO parameters, thresholds are defined for each of the three performance metrics. If a threshold is exceeded for a certain time duration, a new operating point is chosen according to Table 10.9. The choice of the values of the thresholds can be taken from a broad range and depends on the policy of the network operator [JBMK10].

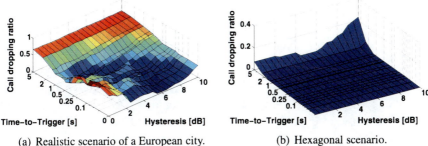

(a) Realistic scenario of a European city.   (b) Hexagonal scenario.

Fig. 10.16 Call dropping ratio for different operating points [JSK10]

Table 10.9 HO parameter optimisation rules [JBMK10]

Metric	TTT	Hysteresis	Action	
			TTT	Hysteresis
Call dropping ratio	≤ 0.6 s	< 3.5 dB	+	+
	≤ 0.6 s	3.5–6.5 dB		+
	≤ 0.6 s	> 6.5 dB	−	−
	> 0.6 s	≤ 6 dB	−	
	> 0.6 s	> 6 dB	−	−
HO failure ratio		< 5 dB	+	
		5–7 dB	+	+
		> 7 dB		+
Ping-pong HO ratio		< 2.5 dB	+	
		2.5–5.5 dB	+	+
		> 5.5 dB		+

10.6.2.3 Load Balancing

Different cell sizes and inhomogeneous spatial distribution of offered traffic may lead to load imbalances among the cells of cellular mobile radio networks. In order to maintain service quality, it may be beneficial to shift load from a highly loaded cell to surrounding lightly loaded cells and to obtain a more balanced distribution of the traffic load. Handing over users from a heavily loaded cell to a lightly loaded cell is thus the core of Load Balancing (LB). As a consequence, the LB use case is strongly related to the HO parameter optimisation use case.

Different parameters are suited to serve as control parameters for the LB use case, including the control parameters from the HO parameter optimisation use case. The goal of LB is the improvement of the GoS while yielding QoS parameters limits. GoS parameters and QoS parameters are thus performance metrics for the LB use case.

In order to be able to evaluate the load of a cell, the virtual cell load η' is defined by

$$\eta' = \frac{1}{N_R} \cdot \sum_{k=1}^{K} \frac{D_k}{R(\gamma_k)} \qquad (10.3)$$

with N_R the total number of resource units, K the number of users of the cell, D_k the average bit rate requirement of user k, γ_k the SINR of user k, and $R(\gamma_k)$ the achievable bit rate per resource unit. Virtual load η' can be used to detect heavily loaded or even overloaded cells and to identify lightly loaded cells and choose the cell that is best suited to accept users from the heavily loaded cells [LSJB10].

10.6.2.4 Self-optimisation of Home eNodeBs

In cellular mobile radio networks, indoor coverage is often an issue since the radio signal may be considerably attenuated by the walls of the building. At the same time, users often use services with high capacity requirements in indoor locations. In order to solve this discrepancy and to take load away from macro cells, a new type of network element, the so-called home eNodeB, also called femtocell, has been introduced, see e.g. Sect. 10.1.

Home eNodeBs are deployed by the user and are not accessible by the operator. Furthermore, large numbers of home eNodeBs with small coverage areas are expected to be deployed. As a consequence, a significant amount of self-configuration and self-optimisation capabilities is required in home eNodeBs.

A number of challenges arise from the application of home eNodeBs. Since home eNodeBs may restrict access to only certain users, the so-called Closed Subscriber Group (CSG) users, they can cause coverage holes for users not belonging to this group due to creating additional interference. In this context, the downlink transmit power of a home eNodeB has been identified to influence the portion of connected CSG users and connected non-CSG users while having only a small impact on the achievable throughput of users of both types. The downlink transmit power of home eNodeBs is thus a well-suited parameter for self-optimisation of home eNodeBs [ZST+10].

Another important issue with home eNodeBs is the selection of HO parameters. While the HO of users to home eNodeBs is in principle desired in order to reduce load on macro cells, it may cause inefficient operation in connection with fast moving users. Low values for hysteresis and TTT, for example, cause high signalling traffic and potentially many ping-pong HOs but at the same time increase the achievable throughput. Hysteresis value and TTT should therefore be set as low as possible but sufficiently high to avoid ping-pong HOs [ZST+10].

10.6.2.5 Interference Coordination and QoS Related Parameter Optimisation

In cellular mobile radio networks, inter-cell interference is inherently present. A frequency planning resulting in a rather inflexible allocation of resources to the cells can be carried out in order to control strength and effect of inter-cell interference.

Due to the variety of services and the mobility of the users, however, spatial distribution of offered traffic is inhomogeneous and dynamically changing. In order to assure efficient network operation, a dynamic and flexible allocation of resources to the cells regarding inter-cell interference is thus desirable. Channel adaptive schedulers already provide means of dynamic resource allocation and interference control. However, it is believed that further performance gains can be achieved by using self-organising techniques.

Considering propagation conditions and the spatial distribution of offered traffic and inter-cell interference, the distribution of the average spectrum efficiency of a cell can be determined. Alternatively, the distribution can be estimated based on SINR measurements from the UE, since the measurements reflect propagation conditions and the distribution of offered traffic and inter-cell interference. Based on the distribution of the average spectrum efficiency and considering the traffic mix, the distribution of the amount of bandwidth required by a cell in order to carry the offered traffic can be determined for a given maximum transmit power [HKS08].

Assuming a sufficient number of independent users in the cell, the distribution of the required cell bandwidth has Normal shape. Mean and standard deviation of the distribution depend on the maximum transmit power P_S and the average inter-cell interference power $\bar{P}_I$. The probability p_{QoS} that sufficient resources in terms of cell bandwidth B_{cell} and maximum transmit power P_S to provide the required QoS are allocated to a cell is thus given by

$$p_{QoS} = 1 - \Phi\left(\frac{B_{cell} - \mu_{cell}(P_S, \bar{P}_I)}{\sigma_{cell}(P_S, \bar{P}_I)}\right) \quad (10.4)$$

with $\Phi(\cdot)$ the CDF of the Normal distribution with zero mean and variance one and $\mu_{cell}(P_S, \bar{P}_I)$ and $\sigma_{cell}(P_S, \bar{P}_I)$ mean and standard deviation of the distribution of the amount of bandwidth required by a cell in order to carry the offered traffic [HKS08, HKG08].

The model of (10.4) is called Power-Bandwidth Characteristic since it is cell-specific and relates the cell bandwidth, maximum transmit power and a QoS metric. It considers propagation conditions and the distribution of offered traffic and inter-cell interference and can be used to evaluate the performance of a cell and to determine resource allocation for the cells. The application of the model to optimisation problems allows one to find optimum resource allocations for a group of cells or even for a whole network. In many cases, such optimisation problems are convex problems, depending on optimisation goal and optimisation variables, thus enabling a dynamic self-organising allocation of resources [HKG09a, HKG09b, HKG+09].

10.6.2.6 X-Map Estimation

The exploitation of UE measurements as proposed in Sect. 10.6.2.5 is taken one step further in the context of so-called X-Maps. X can be received power, throughput or any other measurement. While in Sect. 10.6.2.5, measurements are collected in order to obtain distributions of certain physical parameters or performance metric values, X-Maps relate the measurement values or the performance metric values of the

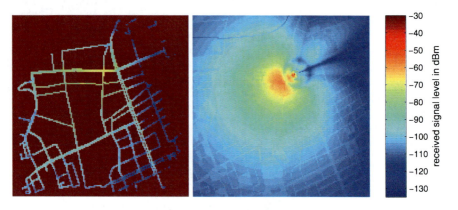

Fig. 10.17 X-Map estimation results using value mapping (*left*) and model calibration (*right*) [NKA11] (©2011 IEEE, reproduced with permission)

UEs with the geographic location of their recording. A geographic map with overlay performance information approximating the spatial characteristics of the network is thus obtained. X-Maps are therefore capable of continuously tracking the network state and reducing the costs involved in drive/walk tests. Furthermore, they are envisioned for application in network monitoring and self-organising optimisation of coverage, capacity and quality [NKA11].

For the estimation of X-Maps, measurement values or performance metric values can be directly mapped to the location of their recording. This leads to accurate but incomplete maps since information is only available from locations that are visited by UEs. Alternatively, the measurements can be used to calibrate the propagation model, increasing the accuracy of the model which can be used to obtain information for every location. Figure 10.17 shows the results of X-Map estimation for both approaches.

An important aspect in the context of X-Map estimation is the accuracy of the X-Map, which depends mostly on the estimation approach and on the accuracy of the location estimation. Different techniques for location estimation are relevant, in particular Global Navigation Satellite System (GNSS), such as Global Positioning System (GPS), for example, and Observed Time Difference of Arrival (OTDOA) are suited concerning accuracy and availability. Additionally, hybrid approaches, combining GNSS and OTDOA are of great relevance since both techniques complement one another, and in many cases one technique provides location information where the other is not applicable or highly inaccurate, as it may be the case for GPS in street canyons or for OTDOA in areas close to BSs [NKA11].

Table 10.10 summarises accuracy results of the different X-Map estimation approaches and shows the effect of location estimation accuracy on X-Map estimation accuracy. The results are given in terms of mean and standard deviation of the X-Map estimation error in dB [NKA11]. The first row, assuming perfect location estimation, shows the trade-off between accuracy and completeness of the two X-Map estimation approaches. In the second column, the percentage of valid user position estimates is listed, showing the gain of the hybrid location estimation approach.

Table 10.10 X-Map estimation error in dB [NKA11]

Location estim.	Valid positions	Value mapping		Model calibration	
		Mean	Std. dev.	Mean	Std. dev.
Perfect estim.		0.0	0.2	2.1	6.6
GPS	77.1%	0.1	2.3	2.6	6.6
OTDOA	66.4%	0.0	4.6	4.6	6.7
GPS + OTDOA	90.6%	0.0	2.3	2.9	6.7

10.6.2.7 Cell Outage Management

Failures in the network infrastructure may locally cause large performance degradation or even coverage holes in cellular networks. Detection of such failures can take much time and is sometimes only possible through customer feedback. A way of automatically and fast detecting failures and taking countermeasures is therefore important.

Cell Outage Management (COM) is an approach aiming to minimise the performance degradation resulting from failures in BSs. It is divided into Cell Outage Detection (COD) and Cell Outage Compensation (COC). COD is carried out by constantly monitoring measurements. These measurements can arise from the BS affected by the failure, from surrounding BSs, from UEs or other network elements. An algorithm that combines different measurements to detect operation failures is required. Assessment criteria for evaluation of COD algorithms are detection delay, detection probability and false detection probability.

COC can be achieved by adjusting the radio parameters that have an impact on coverage and capacity of cells surrounding the cell in outage. Furthermore, due to the adaptation of radio parameters, updates of secondary parameters may be required, such as the neighbour cell list, for example, since new neighbour relations can appear. In order to evaluate the performance of COC approaches, provided coverage, capacity and QoS metrics are suited. Furthermore, the deployment effort in terms of additional signalling traffic, for example, has to be taken into account in the evaluation of COM approaches.

The potential impact of COM approaches varies depending on the scenario in which COM is carried out. The following considerations apply:

- high BS density and traffic load will lead to higher effect of COM,
- different service types will provide different potential gain from COM,
- for a cell at the border of a group of LTE cells, less neighbour cells are available for COM,
- user mobility will effect the perceived outage and thus influence COM,
- spatial traffic distribution sets limits of the maximum effect of COM since the compensation is easier for users located at the cell borders than for users located in the cell centre.

While adapting the radio parameters of cells surrounding a cell in outage may reduce coverage holes and resolve capacity issues in affected areas, users belonging to well-functioning cells may be affected negatively. A trade-off between both effects thus has to be found. The trade-off is governed by the operator's policy, which may range from just providing coverage to providing high quality.

Note that there is no single policy for the whole network. Instead, different policies will be applied to different cells, different operation situations and different time of the day or week. In order to be able to quantitatively assess the operator's policy, the different targets have to be weighted and combined into a cost function. All cells that are affected by COC have to be considered in this cost function [AJK$^+$09a, AJK$^+$09b].

10.6.3 Multi-objective Self-organising Functionality

Section 10.6.2 presents self-organising approaches and algorithms for several use cases. Each of the use cases is treated separately while in practice, however, several use cases may occur in parallel. As a consequence, the integration of the separate algorithms into an overall self-organising concept is necessary.

The challenge of such an overall self-organising concept lies in inherently different goals of the use cases. To give an example, an LB algorithm might try to reduce the coverage area of a cell in order to avoid an overload situation, while another algorithm might try to avoid resulting coverage holes by increasing the coverage area of the cell. As a consequence, self-organising algorithms have to be coordinated in order to avoid oscillation and unstable system behaviour.

Based on the used control parameters, self-organising algorithms can be divided into several groups. Algorithms that affect the same parameter belong to the same group. Functional coupling thus exists exclusively among algorithms within the same group, but never between algorithms of different groups. Consequently, coordination is required only among algorithms of the same group.

Additionally, self-organising algorithms and the need for coordination can be classified according to the interaction of the algorithms. The following interactions exist:

- Self-organising algorithms may be *triggered* due to a different, previously carried out self-organising algorithm. The two algorithms are thus carried out sequentially, and no coordination is required.
- Several self-organising algorithms may be activated simultaneously due to the same system performance degradation. In this case, all activated algorithms have to *cooperate*, and coordination is required.
- If different algorithms that affect the same control parameter are simultaneously active, they have to *co-act*, and coordination is required in order to resolve conflicts and to avoid oscillation.

Two types of conflicts, namely parameter value conflicts and goal conflicts, may occur in the coordination of self-organising algorithms. Parameter value conflicts arise when different algorithms try to set different values for the same control parameter. Goal conflicts occur when the metric optimised by one algorithm is negatively affected by another algorithm with a different goal. The solution to these conflicts has to be carried out according to priorities that have been assigned to the different goals. The priorities depend on the operator's policy [JAT+09a, JAT+09b].

References

[25007]	ETSI TS 102 250-1. Speech processing, transmission and quality aspects (STQ); QoS aspects for popular services in GSM and 3G networks; part 1: Identification of quality of service aspects, v1.2.1, technical specification. Technical Report, 2007.
[25008]	ETSI TS 102 250-2. Speech processing, transmission and quality aspects (STQ); QoS aspects for popular services in GSM and 3G networks; part 2: Definition of quality of service parameters and their computation, v1.6.2, technical specification. Technical Report, 2008.
[3GPa]	3GPP. Evolved universal terrestrial radio access network (E-UTRAN); self-configuring and self-optimizing network (SON) use cases and solutions. Technical Report 3GPP TR 36.902.
[3GPb]	3GPP. Telecommunication management, study on management of evolved universal terrestrial radio access network (E-UTRAN) and evolved packet core (EPC). Technical Report 3GPP TR 32.816.
[3GP03]	3GPP. Services and service capabilities. Technical Report 3GPP TS 22.105 v6.20.0, 3GPP, 2003.
[3GP07]	3GPP. TR 32.816: Study on management of evolved universal terrestrial radio access network (e-UTRAN) and evolved packet core (EPC). Technical Report, Release 8, v1.3.1, 2007.
[3GP09]	3GPP. TR 36.902: Self-configuring and self-optimizing network use cases and solutions. Technical Report, Release 9, v9.0.0, 2009.
[AHK+03]	K. Aardal, S. V. Hoesel, A. M. C. A. Koster, C. Mannino, and A. Sassano. Models and solution techniques for frequency assignment problems. *Oper. Res. Quart.*, 1(4):261–317, 2003.
[AJK+09a]	M. Amirijoo, L. Jorguseski, T. Kürner, R. Litjens, M. Neuland, L. C. Schmelz, and U. Türke. Cell outage management in LTE networks. Technical Report TD-09-750, Braunschweig, Germany, February 2009.
[AJK+09b]	M. Amirijoo, L. Jorguseski, T. Kürner, R. Litjens, M. Neuland, L. C. Schmelz, and U. Türke. Cell outage management in LTE networks. Technical Report TD-09-941, Vienna, Austria, September 2009.
[AL03]	J. Anttila and J. Lakkakorpi. On the effect of reduced quality of service on multiplayer online games. *International Journal of Intelligent Games & Simulation*, 2(2):89–95 2003.
[ALS+08a]	M. Amirijoo, R. Litjens, K. Spaey, M. Doettling, T. Jansen, N. Scully, and U. Türke. Use cases, requirements and assessment criteria for future self-organising radio access networks. Technical Report TD-08-616, Lille, France, October 2008.
[ALS+08b]	M. Amirijoo, R. Litjens, K. Spaey, M. Doettling, T. Jansen, N. Scully, and U. Türke. Use cases, requirements and assessment criteria for future self-organising radio access networks. In *Third International Workshop on Self-Organizing Systems, IWSOS 2008*, December 2008.
[AMCV07]	N. Anastácio, F. Merca, O. Cabral, and F. J. Velez. QoS metrics for cross-layer design and network planning for B3G systems. Technical Report TD(07)225, COST 2100, Lisbon, Portugal, 2007.

[Ant04] J. Antoniou. *A system level simulator for enhanced UMTS coverage and capacity planning.* Master's thesis, Cyprus University, Cyprus, 2004.

[BCG+08] M. Barbiroli, C. Carciofi, P. Grazioso, D. Guiducci, and C. Zaniboni. Analysis of macrocellular and microcellular coverage with attention to exposure levels. In *COST 2100 TD(08)412*, February 2008.

[BCR09] M. Barbiroli, C. Carciofi, and G. Riva. Influence of propagation phenomena on wireless mesh networks performance. In *COST 2100 TD(09)845*, May 2009.

[BEGT09] M. Bohge, A. Eisenblätter, H.-F. Geerdes, and U. Türke. An interference coupling model for adaptive partial frequency reuse in OFDMA/LTE networks. Technical Report TD(09)757, COST 2100, Braunschweig, Germany, 2009.

[BHC07] S. Bieder, L. Häring, and A. Czylwik. Impact of imperfection in the array configuration on the capacity of UMTS-FDD based cellular networks with smart antennas. Technical Report TD(07)368, COST 2100, Vienna, Austria, 2007.

[Bud07] H. Buddendick. Investigation of 3G radio network topologies considering performance and exposure aspects. *Adv. Radio Sci.*, 259–264, 2007. [Also available as TD(07)212].

[CGV09] A. Carniani, L. Giupponi, and R. Verdone. Evaluation of spectrum opportunities in the GSM band. Technical Report TD(09)999, COST 2100, Vienna, Austria, 2009.

[CMLR10] S. Caban, C. Mehlfuhrer, G. Lechner, and M. Rupp. Testbedding MIMO HSDPA and WiMAX. Technical Report TD-10-10046, Athens, Greece, February 2010.

[CSV07] O. Cabral, A. Segarra, and F. J. Velez. Simulation of IEEE 802.11e in the context of interoperability. In *COST 2100 TD(07)328*, September 2007.

[CSV08] O. Cabral, A. Segarra, and F. J. Velez. Implementation of IEEE 802.11e block acknowledgement policies based in the buffer size. In *COST 2100 TD(08)474*, February 2008.

[CVP09] O. Cabral, F. J. Velez, and N. R. Prasad. Simulation of a weighted scheduling algorithm for IEEE 802.11e cross-layer design between PHY and MAC. In *COST 2100 TD(09)769*, February 2009.

[DECD04] I. Demirkol, C. Ersoy, M. U. Caglayan, and H. Delic. Location area planning and cell-to-switch assignment in cellular networks. *IEEE Trans. Wireless Commun.*, 3(3):880–890, 2004.

[dlRVLP+08] G. de la Roche, A. Valcarce, D. López-Pérez, E. Liu, and J. Zhang. Coverage prediction and system level simulation of WiMAX femtocells. Technical Report TD-08-617, Lille, France, October 2008.

[dlRZ09] G. de la Roche and J. Zhang. CWNetPlan project: combined indoor/outdoor wireless network planning. In *COST 2100 TD(09)810*, May 2009.

[ed03] J. Ferreira, editor. Final report on traffic estimation and services characterization. Technical Report Deliverable 34900/PTIN/DS/014/d6, IST SEACORN CEC, 2003.

[EG07] A. Eisenblätter and H.-F. Geerdes. Reconciling theory and practice: a revised pole equation for WCDMA cell powers. In *Proc. MSWIM'07*, Chania, Greece, 2007.

[EG08] A. Eisenblätter and H.-F. Geerdes. Capacity optimization for UMTS: bounds on expected interference. Technical Report TD(08)416, COST 2100, Wroclaw, Poland, 2008.

[EGGP08] A. Eisenblätter, H.-F. Geerdes, J. Gross, and O. Puñal. A two-stage approach to WLAN planning: detailed performance evaluation along the Pareto front. In *COST 2100 TD(08)535*, June 2008.

[EGS07a] A. Eisenblätter, H.-F. Geerdes, and I. Siomina. Integrated access point placement and channel assignment for wireless LANs in an indoor office environment. In *COST 2100 TD(07)256*, February 2007.

[EGS07b]   A. Eisenblätter, H.-F. Geerdes, and I. Siomina. Integrated access point placement and channel assignment for wireless LANs in an indoor office environment. In *IEEE International Symposium on a World of Wireless, Mobile and Multimedia Networks 2007*, Espoo, Finland, June 2007.

[EGT05]   A. Eisenblätter, H.-F. Geerdes, and U. Türke. Public UMTS radio network evaluation and planning scenarios. *Int. J. Mobile Network Design and Innovation*, 1(1):40–53, 2005. Scenarios available at momentum.zib.de.

[EU98]   ETSI-UMTS. Universal mobile telecommunications system (UMTS); selection procedures for the choice of radio transmission technologies for the UMTS. Technical Report ETSI TR 101 112 v3.2.0, ETSI, 1998.

[EU03]   ETSI-UMTS. Universal mobile telecommunications system (UMTS); quality of service (QoS) concept and architecture. Technical Report ETSI TR 123 107 v5.10.0, ETSI, 2003.

[FCFN07]   L. S. Ferreira, L. Caeiro, M. Ferreira, and M. S. Nunes. QoS performance evaluation of a WLAN mesh vs. a WiMAX solution for an isolated village scenario. In *EuroFGI Workshop on IP Quality of Service and Traffic Control*, 2007. [Also available as TD(08)409].

[FCX$^+$03]   L. Ferreira, L. M. Correia, D. Xavier, I. Vasconcelos, and E. R. Fledderus. Final report on traffic estimation and services chracterisation. Technical Report D1.4, IST-2000-28088 MOMENTUM, May 2003.

[FV05]   J. Ferreira and F. J. Velez. Enhanced UMTS services and applications characterisation. *Telektronikk*, 101(1):113–131, 2005.

[Gan]   GANDALF (monitoring and self-tuning of RRM parameters in a multi-system network). Technical Report. Eureka, Celtic (04.2005–12.2006).

[Gee08]   H.-F. Geerdes. *UMTS radio network planning: mastering cell coupling for capacity optimization*. Vieweg + Teubner, 2008. PhD thesis, Technische Universität Berlin, Germany.

[GJMS09]   A. Goldsmith, S. A. Jafar, I. Maric, and S. Srinivasa. Breaking spectrum gridlock with cognitive radio: an information theoretic perspective. *Proceedings of the IEEE*, 97(5):894–914, 2009. Digital Object Identifier: 10.1109/JPROC.2009.2015717.

[GLG10]   V. Garcia, N. Lebedev, and J.-M. Gorce. Selection of stations for multi cellular processing for MIMO Rayleigh channels. In *COST 2100 TD(12)074*, November 2010.

[GLRB08]   M. Garcia-Lozano and S. Ruiz-Boque. Study on the automated tuning of HSDPA code allocation. In *COST 2100 TD(08)410*, 2008.

[GLRBALOC07]   M. Garcia-Lozano, S. Ruiz-Boque, A. Andujar-Linares, and N. Oller-Camaute. Automatic tuning of soft handover parameters in UMTS networks. In *COST 2100 TD(07)344*, 2007.

[HAS96]   R. Händel, M. Anber, and S. Schröder. *ATM Networks, Concepts, Protocols, Applications*. Addison-Wesley, Reading, 1996.

[HK07a]   A. Hecker and T. Kürner. Application of classification and regression trees for paging traffic prediction in LAC planning. In T. Kürner, editor, *Proc. VTC 2007-Spring Vehicular Technology Conference, IEEE 65th*, pages 874–878, Dublin, Ireland, April 2007.

[HK07b]   A. Hecker and T. Kürner. Redistribution of cell measurement values for MTC paging traffic prediction in LAC planning. In T. Kürner, editor, *Proc. IEEE 18th International Symposium on Personal, Indoor and Mobile Radio Communications, PIMRC 2007*, pages 1–5, Athens, Greece, September 2007.

[HK07c]   A. Hecker and T. Kürner. Signal traffic estimation layout for measurement-based planning of location areas in mobile radio networks. In *SPECTS—10th Annual Symposium Performance Evaluation of Computer and Telecommunication Systems*, pages 459–466, San Diego, California, USA, July 2007. SCS—The Society for Modeling and Simulation International.

[HK09a] A. Hecker and T. Kürner. Merging the hierarchical information of performance statistics with the site data base of a planning tool. In *COST 2100 TD(09)984*, 2009.

[HK09b] A. Hecker and T. Kürner. Introduction of measurement-based estimation of handover attempts for automatic planning of mobile radio networks. In *Proc. IEEE International Conference on Communications ICC 2009—A Tradition of Innovation*, Dresden, Germany, June 2009.

[HKG08] P. P. Hasselbach, A. Klein, and I. Gaspard. Measurement based cell bandwidth allocation in cellular mobile radio networks using power-bandwidth characteristics. Technical Report TD-08-609, Lille, France, October 2008.

[HKG09a] P. P. Hasselbach, A. Klein, and I. Gaspard. Self-organising radio resource management for cellular mobile radio networks using power-bandwidth characteristics. In *Proc. ICT—MobileSummit 2009*, June 2009.

[HKG09b] P. P. Hasselbach, A. Klein, and I. Gaspard. Transmit power allocation for self-organising future cellular mobile radio networks. In *Proc. 2009 IEEE 20th International Symposium on Personal, Indoor and Mobile Communications (PIMRC09)*, September 2009.

[HKG$^+$09] P. P. Hasselbach, A. Klein, I. Gaspard, D. V. Hugo, and E. Bogenfeld. Performance analysis of dynamic radio resource allocation for self-organising cellular mobile radio networks. Technical Report TD-09-937, Vienna, Austria, September 2009.

[HKS08] P. P. Hasselbach, A. Klein, and M. Siebert. Interdependence of transmit power and cell bandwidth in cellular mobile radio networks. In *Proc. 2008 IEEE 19th International Symposium on Personal, Indoor and Mobile Communications (PIMRC08)*, 2008.

[HV02] A. Hoglund and K. Valkealahti. Quality-based tuning of cell downlink load target and link power maxima in WCDMA. In *Proc. VTC 2002 Fall—IEEE 56th Vehicular Technology Conf.*, vol. 4, pages 2248–2252, 2002.

[HVHL02] A. Hamalainen, K. Valkealahti, A. Hoglund, and J. Laakso. Auto-tuning of service-specific requirement of received EbNo in WCDMA. In *Proc. VTC 2002 Fall—IEEE 56th Vehicular Technology Conf.*, vol. 4, pages 2253–2257, 2002.

[JAT$^+$09a] T. Jansen, M. Amirijoo, U. Türke, L. Jorguseski, K. Zetterberg, R. Nascimento, L. C. Schmelz, J. Turk, and I. Balan. Embedding multiple self-organisation functionalities in future radio access networks. Technical Report TD-09-758, Braunschweig, Germany, February 2009.

[JAT$^+$09b] T. Jansen, M. Amirijoo, U. Türke, L. Jorguseski, K. Zetterberg, R. Nascimento, L. C. Schmelz, J. Turk, and I. Balan. Embedding multiple self-organisation functionalities in future radio access networks. In *IEEE 69th Vehicular Technology Conference VTC Spring 2009*, April 2009.

[JBMK10] T. Jansen, I. Balan, I. Moermann, and T. Kürner. Handover parameter optimization in LTE self-organizing networks. Technical Report TD-10-10068, Athens, Greece, February 2010.

[JJJ$^+$09] V. Jungnickel, S. Jaeckel, L. Jiang, L. Thiele, and A. Brylka. Measurements and evaluation results for cooperative multi-cell systems. Technical Report TD(09)730, COST 2100, Braunschweig, Germany, 2009.

[JSK10] T. Jansen, M. Schack, and T. Kürner. A universal realistic scenario for wireless communication network simulations. Technical Report TD-10-11074, Aalborg, Denmark, June 2010.

[KH05] T. Kürner and A. Hecker. Performance of traffic and mobility models for location area code planning. In A. Hecker, editor, *Proc. VTC 2005-Spring Vehicular Technology Conference 2005, IEEE 61st*, vol. 4, pages 2111–2115, Stockholm, Sweden, May 2005.

[KHRW02] D. Krajzewicz, G. Hertkorn, C. Rössel, and P. Wagner. SUMO (Simulation of Urban MObility); an open-source traffic simulation. In *4th Middle East Symposium on Simulation and Modeling*, September 2002.

[Kwo95] T. C. Kwok. A vision of residential broadband services: ATM-to-the-Home. *IEEE Network*, 9(5):14–28, 1995.
[LPGSZ+08] D. López-Pérez, F. Gordejuela-Sánchez, L. Q. Zhao, W. Huang, and J. Zhang. QoS performance evaluation of a WLAN mesh vs. a WiMAX solution for an isolated village scenario. Technical Report TD-08-442, Wroclaw, Poland, February 2008.
[LRTW10] S. Luna-Ramirez, M. Toril, and V. Wille. Performance analysis of dedicated signalling channels in GERAN by retrial queues. *Wireless Personal Communications*, 60(2):215–235, 2010. [Available as online-first, also available as TD(09)805].
[LRTWFN09] S. Luna-Ramirez, M. Toril, V. Wille, and M. Fernandez-Navarro. Optimal traffic sharing in GERAN. *Wireless Personal Communications*, 57(4):553–574, 2009. [Available as online-first, also available as TD(10)10022].
[LSJB10] A. Lobinger, S. Stefanski, T. Jansen, and I. Balan. Load balancing in downlink LTE self-optimizing networks. Technical Report TD-10-10071, Athens, Greece, February 2010.
[LWN02] J. Laiho, A. Wacker and T. Novosad, editors. *Radio Network Planning and Optimisation for UMTS*. Wiley, New York, 2002.
[MAR+07a] M. Marques, J. Ambrósio, C. Reis, D. Gouveia, D. Robalo, F. J. Velez, R. Costa, and J. Riscado. Design and implementation of IEEE 802.16 radio links with relays. Technical Report TD-07-026, Lisbon, Portugal, February 2007.
[MAR+07b] M. Marques, J. Ambrósio, C. Reis, D. Gouveia, D. Robalo, F. J. Velez, R. Costa, and J. Riscado. Design and planning of IEEE 802.16 networks. Technical Report TD-07-326, Duisburg, Germany, September 2007.
[MMSMC08] J. Matamales, D. Martin-Sacristan, J. F. Monserrat, and N. Cardona. Performance assessment of HSDPA networks from outdoor drive-test measurements. Technical Report TD-08-626, Lille, France, October 2008.
[MOMC09] J. Matamales, V. Osa, J. F. Monserrat, and N. Cardona. Automatic optimisation of operating UMTS networks. In *COST 2100 TD(09)830*, 2009.
[Mon05] V. Monteiro. Algorithm for dynamic radio resource allocation for 4G systems based on MC-CDMA. Technical report, Universidade de Aveiro, 2005 (in Portuguese).
[Mül10] C. M. Müller. On the importance of realistic traffic models for wireless network evaluations. In *COST 2100 TD(12)039*, November 2010.
[NGM06] NGMN. Next generation mobile networks beyond HSPA & EVDO. Technical report, White Paper, 2006.
[NGM07] NGMN. NGMN informative list of SON use cases. Technical Report, Annex Deliverable, 2007.
[NKA11] M. Neuland, T. Kürner, and M. Amirijoo. Influence of positioning error on X-map estimation. In *Proc. IEEE Veh. Tech. Conf. Spring (VTC Spring 2011)*, Budapest, Hungary, May 2011. [Also available as TD(10)11068].
[NPP10] A. Navarro, A. Pachon, and U. Garcia Palomares. Optimisation in fourth generation mobile networks using derivative-free optimisation in parallel. In *COST 2100 TD(10)12084*, November 2010.
[NWS+08] M. Nawrocki, M. Wagrowski, K. Sroka, R. Zdunek, and M. Miernik. On-line mobile network optimisation—network-in-the-loop approach—initial results. In *COST 2100 TD(08)614*, 2008.
[NWS+09] M. Nawrocki, M. Wagrowski, K. Sroka, R. Zdunek, and M. Miernik. On input data for the mobile network online optimisation process. In *COST 2100 TD(09)747*, 2009.
[OM08] J. Oostveen and O. Mantel. Requirements on ray-based propagation models for mobile network planning. In *COST 2100 TD(08)517*, June 2008.
[Pom09] A. J. Pomianek. Remarks on sector overlapping in CDMA networks. Technical Report TD(09)867, COST 2100, Valencia, Spain, 2009.

[PPC09]   C. N. Pitas, A. D. Panagopoulos, and P. Constantinou. Performance measurements and QoS analysis of high speed packet access networks. Technical Report TD-09-764, Braunschweig, Germany, February 2009.

[PSN07]   A. J. Pomianek, P. M. Slobodzian, and M. J. Nawrocki. Transformation function for analytical optimisation of CDMA networks. In *Proc. PIMRC'07*, Athens, Greece, 2007. IEEE. Presented as TD(07)342 at COST 2100 in Duisburg, Germany.

[RFMM10]   S. Robitzsch, J. Fitzpatrick, S. Murphy, and L. Murphy. An experimental evaluation of co-channel interference in IEEE802.11a multi-radio Mesh networks. In *COST 2100 TD(10)10060*, February 2010.

[SKK03]   K. Schloegel, G. Karypis, and V. Kumar. Graph partitioning for high-performance scientific simulations. In J. Dongarra, et al., editors, *The Sourcebook of Parallel Computing*, pages 491–541. Morgan Kaufmann, San Mateo, 2003.

[SOC08]   SOCRATES. Deliverable D2.1: use cases for self-organising networks. Technical Report, 2008.

[SSB10]   K. Spaey, B. Sas, and C. Blondia. Self-optimising call admission control for LTE downlink. Technical Report TD-10-10056, Athens, Greece, February 2010.

[STV+09]   P. Sebastião, R. Tomé, F. J. Velez, A. Grilo, F. Cercas, D. Robalo, A. Rodrigues, F. F. Varela, and C. X. P. Nunes. WLAN planning tool: a techno-economic perspective. In *COST 2100 TD(09)935*, September 2009.

[SVC+09]   P. Sebastião, F. J. Velez, R. Costa, D. Robalo, C. Comissário, and A. Rodrigues. Planning and deployment of WiMAX and Wi-Fi networks for health sciences education. *Telektronikk, Technical Architecture*, 105(2):173–185, 2009.

[TGGLRW10] M. Toril, P. Garcia-Guerrero, S. Luna-Ramirez, and V. Wille. Re-assignment of base stations to packet controllers by integer programming. Technical Report TD-10-10057, Athens, Greece, February 2010.

[TLRWS09] M. Toril, S. Luna-Ramirez, V. Wille, and R. Skehill. Analysis of user mobility statistics for cellular network re-structuring. In *Proc. 69th IEEE Vehicular Technology Conference*, pages 1–5, April 2009.

[TMWW10]   M. Toril, I. Molina, V. Wille, and C. Walshaw. Analysis of heuristic graph partitioning methods for the assignment of packet control units in GERAN. *Wireless Personal Communications*, 60(4):611–633, 2010. [Available as online-first].

[TW08]   M. Toril and V. Wille. Optimization of handover parameters for traffic sharing in GERAN. *Wireless Personal Communications*, 47(3):315–336, 2008.

[TWB06]   M. Toril, V. Wille, and R. Barco. Optimization of the assignment of cells to packet control units in GERAN. *IEEE Commun. Lett.*, 10(3):219–221, 2006.

[TWLRJ09] M. Toril, V. Wille, S. Luna-Ramirez, and K. Jarvinen. Performance analysis of location area re-planning in a live GERAN system. Technical Report TD-09-724, Braunschweig, Germany, February 2009.

[VaC+09]   F. J. Velez, P. Sebastião, R. Costa, D. Robalo, C. Comissàrio, A. Rodrigues, and A. H. Aghvami. WiMAX radio and network planning: limitations and challenges. Technical Report TD-09-774, Braunschweig, Germany, February 2009.

[VAH10]   F. J. Velez, H. Aghvami, and O. Holland. Basic limits for throughput computation and cost/revenue optimisation: a formulation for fixed WiMAX. Technical Report TD-10-11040, Aalborg, Denmark, June 2010.

[vdBLE+08] J. L. van den Berg, R. Litjens, A. Eisenblätter, M. Amirijoo, O. Linnell, C. Blondia, T. Kürner, N. Scully, J. Oszmianski, and L. C. Schmelz. SOCRATES: self-optimisation and self-configuRATion in wirelESs networks. Technical Report TD-08-422, Wroclaw, Poland, February 2008.

[Vel00]   F. Velez. *Aspects of cellular planning in mobile broadband systems*. PhD thesis, Instituto Superior Técnico, Lisbon, Portugal, 2000.

[VHPF02]	K. Valkealahti, A. Hoglund, J. Parkkinen, and A. Flanagan. WCDMA common pilot power control with cost function minimization. In *Proc. VTC 2002 Fall—IEEE 56th Vehicular Technology Conf.*, vol. 4, pages 2244–2247, 2002.
[VSC+09]	F. Velez, P. Sebastião, R. Costa, D. Robalo, C. Comissàrio, A. Rodrigues, and A. H. Aghvami. Field trials for a wireless PMP WiMAX network at 3.5 GHz. Technical Report TD-09-837, Valencia, Spain, May 2009.
[VTFDE11]	E. M. Vitucci, L. Tarlazzi, P. Faccin, and V. Degli-Esposti. Analysis of the performance of LTE systems in an interleaved F-DAS MIMO indoor environment. In *European Conference on Antennas and Propagation (EuCAP2011)*, Rome, Italy, April 2011. [Also available as TD(10)12021].
[WJ08]	M. Wagrowski and L. Janowski. Measurements of blocking and dropping rates in UTRAN. In *COST 2100 TD(08)607*, 2008.
[WL10]	M. Wagrowski and W. Ludwin. Decentralized call admission control autotuning for UTRAN. In *COST 2100 TD(10)11004*, 2010.
[WWWW05]	R. Wahl, G. Wölfle, P. Wertz, and P. Wildbolz. Dominant path prediction model for urban scenarios. In *14th IST Mobile and Wireless Communications Summit*, Dresden, Germany, June 2005.
[ZdlR10]	J. Zhang and G. de la Roche. *Femtocells: Technologies and Deployment*. Wiley, New York, 2010.
[Zha08]	Y. Zhang. *WiMAX Network Planning and Optimization*. Taylor & Francis, London, 2008.
[ZST+10]	K. Zetterberg, N. Scully, N. Turk, L. Jorguseski, and A. Pais. Controllability for self-optimisation of home eNodeBs. Technical Report TD-10-10061, Athens, Greece, February 2010.

Chapter 11
Resource Management in 4G Networks

Chapter Editor Silvia Ruiz Boqué, Chapter Editor Narcis Cardona, Andreas Hecker, Mario Garcia-Lozano, and Jose F. Monserrat

The chapter focuses on the resource management for Fourth Generation (4G) networks and is organized into five sections. The first section describes the results obtained in COST 2100 on Radio Resource Management (RRM) strategies; the second section deals with modeling, prediction and classification of traffic; the third section addresses radio resource management for 3GPP-Long Term Evolution (LTE) networks; the fourth section deals with Common Radio Resource Management (CRRM) in heterogeneous wireless networks; and, finally, the last section discusses trends and resource topics.

11.1 RRM Strategies in Cellular Networks

RRM optimization is an important issue in modern wireless systems due to the scarceness of resources and the growing demand for services requiring high reliability of transmission and/or high throughput along with low transmission delays. Most of the resources are limited, being then crucial to exploit efficiently the given spectrum in order to provide high spectral efficiency. It is also absolutely necessary to design proper CRRM strategies between different networks and to facilitate user network selection while optimizing global performance. Many techniques, completely different or designed as an evolution with respect Third Generation (3G) systems, have arisen:

- Use of multiple antennas increasing spectral efficiency without any increase in bandwidth or transmission power [BCC[+]07].

S. Ruiz Boqué (✉)
Universitat Politècnica de Catalunya, Barcelona, Spain

N. Cardona (✉)
Universidad Politécnica de Valencia, Valencia, Spain

- Resource allocation algorithms assigning resources to User Equipment (UE)s with best channel conditions and/or highest necessity according to their Quality-of-Service (QoS) requirements, through cooperation among different layers.
- Power control to reduce inter-cell interference minimizing the power consumption at the UE and Base Station (BS)s and ensuring an adequate QoS level. Initially designed to maintain a target Signal-to-Interference-plus-Noise Ratio (SINR), today it is combined with Adaptive Modulation and Coding (AMC) and focused in maximizing throughput and capacity.
- Scheduling algorithms applying different metrics to the users depending on their requirements. Classical algorithms are:
  - Throughput Maximization oriented: Maximal-rate and Proportional Proportional Fair Scheduling (PFS) considering only instantaneous channel condition of UEs [VTL02].
  - Delay minimization-oriented algorithms, such as Modified-Largest Weighted Delay First (M-LWDF), [AKR+01], and Exponential Scheduling (EXP), [SS00], considering queue status and traffic priority.
- Modifying existing algorithms to increase spectral efficiency:
  - Assigning, for example, the same resources to less correlated users in the multiuser transmission by multiple antennas.
  - Adapting M-LWDF and EXP algorithms by changing the weights according to instantaneous traffic load, urgency of the packets in the queue, and ratio between delay and throughput requirements.
  - Include user satisfaction measures to validate different scheduling approaches, [LPS09].

Through this section we will focus on the specific advances in resource allocation, scheduling, multiuser and cooperative scenarios, as well as metrics, testbeds, and trials developed inside the European Cooperation in Science and Technology (COST) action.

11.1.1 Resource Allocation

For most of the resource allocation algorithms given in the literature, it is essential that the Channel State Information (CSI) is available at the BS, and thus the resource manager entity can adaptively assign resources to users with best channel condition based on the chosen metrics.

Throughput Maximization-Oriented Algorithms Rate maximization-oriented algorithms achieve good performance in terms of overall system throughput.

- **Maximal-rate scheduler:** always schedules the UE with the highest achievable rate, i.e., the UE with the best channel:

$$k^*(n) = \arg\max_{k} \{r_k(n)\}, \qquad (11.1)$$

where $r_k(n)$ is the theoretically achievable rate for UE k. It exploits multiuser diversity due to independent variations of UEs' channels [KH95] but does not provide fairness resulting in starvation of some UEs.
- **PFS:** introduces certain fairness, assigning resources to the UE with the largest ratio between its instantaneous achievable rate and its transmission rate, averaged over a window of size t_c [VTL02]:

$$k^*(n) = \arg\max_k \left\{ \frac{r_k(n)}{T_k(n)} \right\}, \qquad (11.2)$$

where $T_k(n)$ is the average throughput of kth UE at time slot n:

$$T_k(n+1) = \left(1 - \frac{1}{t_c}\right) T_k(n) + \frac{1}{t_c} r_k(n) \cdot 1_{k=k^*(n)}, \qquad (11.3)$$

where $1_{k=k^*(n)}$ is 1 if UE k is selected and 0 otherwise. It enables UEs with poor channel conditions to transmit more frequently.

Delay Performance Oriented Algorithms Queue status and traffic priority should be considered if we want to design an efficient resource allocation algorithm, capable of meeting the QoS requirements of the heterogeneous applications.
- **M-LWDF:** tries to balance the weighted delays of UE's packets, selecting, at each scheduling cycle, UEs whose utility function is maximized [AKR+01]:

$$k^*(n) = \arg\max_k \left\{ a_k W_k(n) \frac{r_k(n)}{\overline{r}_k} \right\}, \qquad (11.4)$$

where $a_k = -\log(\delta_k)/D_{ik}$ is the required level of QoS in terms of delay bounds, D_{ik} the largest delay tolerated by UE k, δ_k is the largest probability that this delay can be exceeded, and $W_k(n)$ is the Head-Of-the-Line (HOL) packet delay (queue length can be used instead). M-LWDF is throughput optimal because it is able to keep all queues stable [KH95] but gives higher priority to delay bound of real-time services resulting in poor performance for best-effort traffic.
- **EXP:** modifies UE selection criterion compared to the M-LWDF algorithm as follows [SS00]:

$$k^*(n) = \arg\max_k \left\{ a_k \frac{r_k(n)}{\overline{r}_k} \exp\left(\frac{a_k W_k(n) - \overline{aW}}{\beta + \sqrt{\overline{aW}}} \right) \right\}, \qquad (11.5)$$

where the numerator of the exp( ) function is the weighted delay averaged over K UEs, and β is a positive constant to adjust the influence of the delay dependent term. It gives a slightly fairer treatment for delay tolerant traffic since the weight of the exp term only exceeds the channel state-dependent term when the UE's weighted HOL packet delay is significantly larger than the average weighted delay.

Resource allocation is not only related with the scheduling protocol: simultaneously to the decision of who is having permission to transmit, a set of parameters have to be assigned, like the modulation, the code, which subcarriers and how many of them, the transmitted power, etc.

	Average pr. cell capacity (bits/s/Hz) shown in (GP, Binary, Full) triplets		
Number of cells (N)	Suburban Macro	Urban Macro	Urban Micro
1	$(6.02, 6.02, 6.02)$	$(5.13, 5.13, 5.13)$	$(11.96, 11.96, 11.96)$
2	$(4.93, 4.93, 4.74)$	$(4.40, 4.40, 4.27)$	$(6.64, 6.64, 4.54)$
3	$(4.41, 4.40, 4.02)$	$(4.03, 4.03, 3.75)$	$(6.03, 6.03, 3.39)$
4	$(4.03, 4.01, 3.53)$	$(3.70, 3.69, 3.33)$	$(4.66, 4.65, 2.91)$
5	$(3.98, 3.95, 3.45)$	$(3.68, 3.67, 3.28)$	$(3.88, 3.85, 2.75)$
6	$(3.81, 3.78, 3.25)$	$(3.54, 3.53, 3.11)$	$(3.41, 3.36, 2.58)$
7	$(3.67, 3.64, 3.08)$	$(3.42, 3.41, 2.97)$	$(3.06, 3.00, 2.40)$

Fig. 11.1 Network capacity statistics

Binary Power Control for Multicell Scenarios It has been analytically shown that in a two-cell scenario a Binary Transmitted Power is the optimum possibility. Independently on the number of cells (being $N > 2$), the Binary Power Control is still the optimal solution when either a low SINR or a geometric–arithmetic mean approximation is applicable [AG07]. The mathematical framework of Geometric Programming (GP) [BKVH07] has been used in order to establish a sum capacity benchmark to compare with the Binary Power Allocation. An enormous simplification of the power allocation algorithm is given, while reducing the feedback rate is needed to communicate between nodes. Even in the general case (any SINR value), the loss considering the binary case is negligible with respect the possibility to transmit any power level. Moreover, the limited number of solutions facilitates a distributed resource allocation, especially if clusters of cells are considered [KØG07]. Clustering is a way of lowering the complexity, as well as reducing the required channel knowledge, at the cost of only a small reduction in network capacity. In Fig. 11.1 the average per cell capacity (R/N) in bits/s/Hz versus the number of cells (N) is given. Introducing Power Control (PCo) improves the throughput for $N \geq 2$, especially for the urban microcell environment, while there is only a marginal improvement when going from binary to optimal GP power control. Increasing N, the GP uses less transmit power than binary control, so GP reduces the average transmit power while achieving the same network capacity (Fig. 11.2).

Power Control for Downlink Multimedia Broadcast Multicast Service (MBMS)
An efficient power control for MBMS is described in [VP07] through an analysis of the effects of using different combinations of dedicated/shared channels, while considering that some of the channels automatically implement power control while others do not. Two approaches have been considered:

- Switching between a point-to-point channel (Dedicated Channel (DCH) with power control) up to a threshold number of UEs and a point-to-multipoint channel (Forward Access Channel (FACH) with power control disabled or Downlink Shared Channel (DSCH) with inner-loop power control enabled) for the whole cell. In DCH the Down-Link (DL) power increases linearly with the number of UE, while for the FACH, the power remains constant, so the point where both

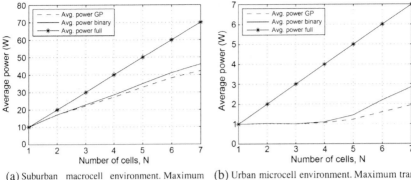

Fig. 11.2 Average sum transmit power for macro- and microcell versus the number of cells N

lines cross can be used as the threshold for switching. Switching between DCH and DSCH or FACH could cause a degradation of UEs QoS because the threshold is highly dependent on the UE position. Moreover, switching requires complex mechanisms over the Iub and Iur with a significant signaling overhead.
- Using only a point-to-multipoint channel per multicast group in each cell (FACH or DSCH). This is the most efficient solution when a large number of UEs have to be served from the beginning. Up to 1 km DSCH is better in terms of transmitted power, but it is appropriated only when the number of UEs is not very high, FACH being the best option for a high number of UEs.

Call Admission Control (CAC) for Downlink MBMS The goal of CAC is the management of new connections satisfying their QoS requirements while preserving the QoS of ongoing connections. For MBMS, the CAC should be based in power and throughput Downlink Power and Throughput based CAC algorithm (DPTCAC) considering traffic patterns for Voice, Video Telephony, and Multimedia Web Browsing. The studied algorithm, [VP], is based on an estimation of the required DL power level while knowing the currently used DL power for ongoing users, considering also the maximum transmission power of the BS. DL transmission power is taken into account for the initial establishment of the shared channel (FACH or DSCH) while connections running over this channel are admitted after evaluation if there is enough remaining capacity. The required BS transmission power level to support the new connection is estimated by

$$P_{DL}^* = SINR_t \cdot \left(P_N + \frac{\sum_{j=1}^{k} P_{DL,j}(r_j)}{N} + \frac{1-\alpha}{N} \cdot P_{DL,0}(r_0) \right). \quad (11.6)$$

Admission is made considering the sum of the estimated required power level and currently used power for ongoing connections against the maximum physical limit of target BS's transmission power [EC05, ECH04]:

$$P_{DL}^* + P_{DL,0} \leq P_0^{max}, \quad (11.7)$$

Fig. 11.3 Base station transmitted power with TCAC and DPTCAC [NP06] (©2006 IEEE, reproduced with permission)

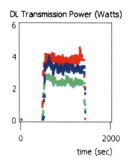
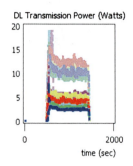

where: P_{DL}^* is the estimated required DL transmission power for the new UE, $SINR_t$ the target SINR to ensure QoS, P_N the background and thermal noise power, $P_{DL,j}(rj)$ the total DL transmission power of BS j perceived at UE's location (at distance r_j from BS j), N the service spreading factor, α the own cell DL orthogonality factor, $P_{DL,0}(r0)$ the total transmission power of the target BS perceived at UE location (at distance r_0 from target BS), k the number of BSs in the network, and P_0^{max} the maximum physical transmission power of the target BS. A scaling factor can be introduced enabling the implementation of a guard-power. Results compare the performance of a classical Throughput-Based CAC algorithm (TCAC) with the DPTCAC. Figure 11.3 shows how BS DL transmission power for DCHs is directly correlated to the number of accepted UEs. TCAC demonstrates low transmission power levels since the cell capacity is restricted to low values, while DPTCAC improves the cell capacity and forces increased transmission power levels. With TCAC there is much unused potential in terms of transmission power usage, while DPTCAC can unleash this potential achieving higher cell capacities as a trade-off. In addition to DPTCAC, during the admission process of an increasing number of UEs, transmission power does not increase constantly. When high power levels are reached and new connection requests are received, Power Control mechanism is invoked to regulate power levels. This process is made in order to provide room for additional connections given that minimum power requirements for all connections can still be met.

Cell throughput in Fig. 11.4 reveals the intuitive fact that throughput in the DL direction is directly correlated to cell capacity; the increased number of UEs allowed by DPTCAC is reflected in the increased throughput compared to reference TCAC. Simulations show that DPTCAC does not affect performance negatively at higher cell capacities, being the end-to-end delay and the jitter comparable.

Dynamic Resource Allocation (DRA)   DRA gives resources to cells according to the actual demand [HKG08], the purpose of the algorithm being a reduction in the outage while maintaining reuse distance, so that inter-cell interference remains lower than a threshold value. The algorithm to determine the number of required resource units considers QoS parameters. A modified demand vector fulfilled with the available number of resource units is generated, and later, a suboptimal policy-based resource assignment algorithm with very low complexity is tested. The problem can

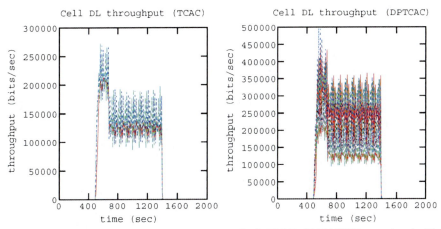

Fig. 11.4 Cell DL throughput with TCAC and DPTCAC [NP06] (©2006 IEEE, reproduced with permission)

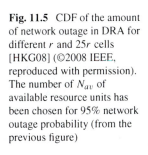

Fig. 11.5 CDF of the amount of network outage in DRA for different r and $25r$ cells [HKG08] (©2008 IEEE, reproduced with permission). The number of N_{av} of available resource units has been chosen for 95% network outage probability (from the previous figure)

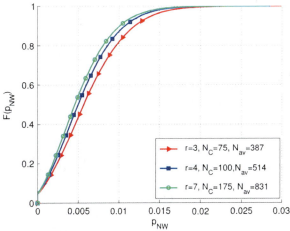

be expressed as a constrained combinatorial optimization problem:

$$\min_A p_{NW}$$
$$\text{s.t.} \quad a_{li} \cdot a_{lk} = 0 \qquad \forall l, i, k \in \mathbb{N}(i) \qquad (11.8)$$
$$A \in \{0, 1\}^{(n_{av} \times N_C)}, \quad n_{av} \leq N_{av},$$

p_{NW} being the amount of outage, and N_C the number of cells. The second equation is the interference constraint assuring that the maximum inter-cell interference is yielded. a_{li} are equal to 1 if resource unit l is assigned to cell i and 0 if not. The third equation assures that algorithm does not exceed the number of available resource units N_{av}.

Figure 11.5 shows the CDF of the amount of network outage in DRA for different r (neighborhood group sizes) and a number of cells of 25 times r. The number

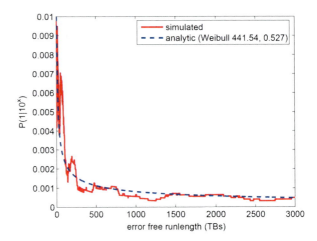

Fig. 11.6 Conditional link error probability [HKG08] (©2008 IEEE, reproduced with permission)

N_{av} of available resource units has been chosen for 95% network outage probability. DRA achieves an outage of zero in 5% of all cases. Although the network outage probability is high, the amount of real outage is below 3%. The amount of outage decreases with increasing r. With the suboptimal algorithm, results are achieved faster and with a small degradation. Suboptimal algorithm is based on trying to assign as many resource units as possible with the reuse distance d_r, meaning that two cells whose centers are at a distance of d_r will use as much as possible the same resource units. The N_C cells are divided into r cells groups such that the centers of all cells within each group are at a distance of multiples of the reuse distance and the available resource units N_{av} are divided into r resource groups.

Dynamic Frequency Allocation Up-Link (UL) Adaptive Resource Allocation for MultiCarrier CDMA (MC-CDMA) systems is based on selecting the best Group Of Frequencies (GOF) following some quality criterion, as, for example, the predictability of the link errors for each UE [CKV08]. The probability that the next Transport Block (TB) will be transmitted with error depends on the error-free runlength since the last error occurrence. For an enhanced cross-layer scheduling mechanism, e.g., for H.264/AVC video streaming over MC-CDMA, the scheduler needs to know about the expected error probability in the next Transmission Time Interval (TTI). The probability that the next TB will be transmitted with error based on the error-free run-length since the last error occurrence, can be calculated and fitted via a Weibull distribution.

Figure 11.6 represents the simulated conditional link error probability from the scenario with six pedestrian video users. The estimator perfectly predicts the transmission error probability based on the number of error-free TBs since the last error occurrence.

A binary decision becomes necessary, and thus a threshold is required for separating the areas of transmitting high- and low-priority packets. The proposed method is compared with the selection of the GOF using the average Signal-to-Noise Ratio (SNR) values of the last TTI in the GOF, where only the predictive nature of the

Fig. 11.7 Generic power profiles in static ICIC schemes: (**a**) full frequency reuse, (**b**) reuse factor 3, (**c**) soft frequency reuse, (**d**) partial frequency reuse [GGLRO10] (©2010 ICST, reproduced with permission)

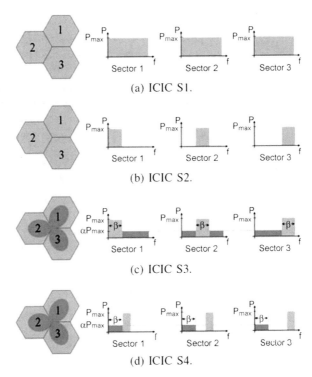

channel from the last to the current TTI is used. Moreover, a high average SNR in the GOF does not necessarily mean error-free transmission in case there is high fading in some subcarriers. As has been shown in Fig. 11.6, the error probability becomes smaller the higher the distance from the last transmission error. Regarding the selection of the GOF, UEs have to transmit on the GOF with the highest error-free run-length. Of course, in order to know about past error events in all GOFs in all TTIs, UE has to transmit a packet in all GOFs in all TTIs. However this is also required in the SNR-based GOF selection mechanism. The proposed method reduces the error ratio by one order of magnitude with respect the SNR method.

Static Intercell Coordination Inter-Cell Interference Coordination (ICIC)
Static ICIC for OFDMA-based realistic nonregular cellular layout is studied in [GGLRO10] giving special attention to the efficiency vs. fairness trade-off (see Fig. 11.7). ICIC gain has been obtained considering the constraints associated to the frequency domain scheduling (PFS is considered) and the effect of the Adaptive Modulation and Coding (AMC), because the impact of ICIC on the system throughput must be analyzed using a model in which the interactions with additional RRM functions are also considered. Both system- and UE-oriented metrics have been considered strategic to improve fairness while keeping spectrum efficiency as high as possible.

A constrained optimization technique based on the sum-rate maximization problem, to optimally allocate bandwidth and transmit power to Partially Frequency

Reuse networks, has been analyzed in [KWMM10]. To arrive to a convex optimization problem, equal power allocations for the Full Frequency Reuse users (inner) in all cells have been considered, and, in the case of fixed bandwidth allocation, the power allocation is based on a simple water-filling algorithm. It is also shown that most of the cell outer bandwidth can be reallocated as cell inner bandwidth whenever the outer user is idle, as this increases the rate of the inner users and also the total sum-rate. Simulations have been done considering two users per cell: one located in the inner region with reuse-1 and the other located at the outer region with reuse-3.

A first insight into frequency scheduling in a macrocell/femtocell combined scenario is addressed in [Y.H10], where the authors propose that any femtocell located in a given macrocell sector is allowed to reuse the bandwidth of the two adjacent sectors of the same overlaying macrocell. The spectrum is divided into three equal portions, one for each of the three macrocell sectors. Three types of femtocell reuse are analyzed: Full Reuse (the femtocell can reuse whatever channels are used by the two adjacent sectors independently of its location in the cell), Partial Reuse, and Mixed Reuse (defining some areas inside the macrocell and the channels associated to them in terms of femtocell reuse). The selection of the specific channel is based on a greedy algorithm selecting the less interfered ones. Simulations show that there is an advantage in terms of SINR with respect the scenario with only macrocell deployment, but no significant differences among the different reuse schemes have been appreciated.

11.1.2 Scheduling

In conventional cellular networks, the scheduling algorithm only schedules a single UE during a TTI, being not optimal. When channel quality is poor, it would be better to share resources among multiple UEs but limiting the interference level. Main reasons for introducing cross-layer approaches in wireless networks are:

- The wireless links with time-variant network topologies, high error rates, etc.
- The possibility of an opportunistic use of the channel by dynamically adapting transmission parameters.
- The broadcast nature of the channel allowing the reception of multiple packets simultaneously.

Optimized scheduling strategies for a mixture of different wireless multimedia services are currently investigated [FTV00, LJSB05]. Since most of the approaches heavily rely on cross-layer design principles, it is impossible to compare them via simple theoretical analysis. A software environment able to simulate different algorithms in the presence of mixed packet-base realistic traffic over wireless shared time variant channels with a cross-layer approach is required. This is deeply analyzed in [CV07b]. The optimum strategy will be the one combining Maximum Throughput (MaxTP) and Fair Service (FS) algorithms. In the Advanced Scheduling implemented, the metric for each UE considers the Time-to-Deadline, the Type of Service, and the Channel State. Simulations have been done considering Variable

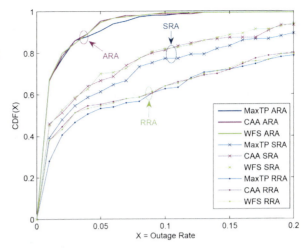

Fig. 11.8 CDF of the video users' outage rate depending on scheduling and resource allocation strategies [CTV07] (©2007 IEEE, reproduced with permission)

Bit Rate (VBR) video stream and File Transfer Protocol (FTP) data. An opportunistic distribution of resources is considered where, at each scheduling time instant, the resource allocation which is a completely air interface module, formulates independent allocation proposals for each UE with a nonempty radio link buffer based on the current available common resource budget in the system. Proposals contain the following information: Transport Block Size, number of required Resource Unit (RU)s, modulation and coding, and required transmit power. The scheduler selects the best according to the desired policy, considering the average SNR for each GOF with free RUs and two different thresholds. Time slot and spreading codes are selected randomly, the transmit power is equal to the maximum power divided by the number of allocated RU per slot, fixed Quaternary Phase Shift Keying (QPSK) modulation and Bose–Chaudhuri–Hocquenghem (BCH) codes with variable coding rate are considered. Then the resources are removed from the budget, and the resource allocator determines new proposals for the other UEs based on the remaining resource budget, until either all UEs have been allocated, resource budget is empty, or the maximum number of Resource Units that can be allocated per UE has been reached.

An evaluation of different Cross-Layer Resource Allocation and Scheduling strategies for MC-CDMA depending on traffic loads and interference level is performed in [CV07a], showing that this is especially important for video UEs under heavily loaded cells. The resource allocator is a completely air-interface-aware module while the scheduler is an air-interface-unaware module dealing with the application level. The performance of this Channel Adaptive Resource Allocation (ARA) scheme is compared with a Non-Adaptive Simple Resource Allocation (SiRA), where the modulation and coding format are fixed, and with the Random Resource Allocation (RRA), where the modulation and coding scheme are fixed, and the GOF are selected randomly. A Channel and Application Aware (CAA) scheduler is considered according to three parameters: Time to Deadline, Type of Service, and Channel State. It is compared with classical Maximum Throughput and Wireless Fair Service techniques. Outage rate is represented in Fig. 11.8, showing that the

Fig. 11.9 Impact of the traffic load on different scheduling policies [CTV07] (©2007 IEEE, reproduced with permission)

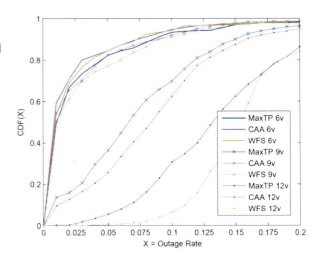

scheduling policy is not determinant because the traffic load is smaller than the system capacity and that the allocation procedure implementing link adaptation CAA is the best one.

Figure 11.9 shows the impact of video traffic load on the scheduling policies. With nine video UEs, the system behaves almost like with six for CAA and Wireless Fair Service (WFS) since the buffer management strategy discards unnecessary packets guaranteeing good performance, while MaxTP quickly degrades, since no mechanism considers the delay sensitiveness of the application. Only the CAA can cope with a very loaded system, being the worst technique in this case WFS since it misses UEs with the best channel quality trying to preserve fairness in very unfavorable conditions.

Another efficient cross-layer mechanism for Scheduling Video Streaming over a Universal Mobile Telecommunications System (UMTS) DCH is presented in [KNRW07], using link prediction at the Radio Link Control (RLC) protocol layer, while scheduling the more important data in the time intervals with lowest predicted error probability. The distortion is estimated according to the size of the lost information in bits and in picture blocks in a novel proposed model. Losses are considered at the RLC Protocol Data Unit (PDU) level, rather than at the frame/slice level, in two steps: Primary distortion within a frame, caused by a lost PDU, and Distortion propagation in the following frames until the end of the Group Of Pictures (GOP). The distortion is modeled empirically, an exponential function being at the end. Link errors are have been also modeled considering that due to the effects of jointly coded and interleaved TBs within one TTI, the radio link produces error bursts separated by short gaps (gap length < 12 for the 384 kbps bearer). These bursts lying within up to three consecutive TTIs form an error cluster. Power control separates successive clusters by long gaps to reach the target link quality (1% of block error rate). Transmitter receives the error feedback information after a certain delay (around 3 TTIs), so error prediction within one error cluster is not possible, but allows the estimation and prediction of the location of the error clusters or the

Table 11.1 Number of erroneous frames and mean PSNR for the presented scheduling methods [Rup09]

Method	Error-free frames [%]	Mean PSNR [dB]
Without scheduling	73.86	29.29
I sched. (d_{max} = 3600 TBs)	76.27	30.26
II sched. (d_{max} = 8400 TBs)	76.63	30.32
RLC sched. (3600/1200)	66.07	30.44
RLC sched. (8400/3600)	73.01	32.03
Scheduling limit	82.09	35.53

length of the long gaps. The start of intervals with very low error probability and also the length of these intervals are determined. The performance of the scheduler is controlled by the scheduler buffer length and the maximum acceptable delay. The link error model identifies the intervals with low error probability, delivering this information to the scheduler. An estimation of the cumulative distortion that would be caused by the loss of every RLC PDU in the buffer is provided. If an RLC PDU remains in the buffer longer than the maximum acceptable delay, it is scheduled immediately. Otherwise, at each step within the time interval, the RLC PDU with the lowest distortion is chosen and sent. Simulations show that this method reduces the number of frames with high distortion, while other methods, as, for example, the prioritized scheduling of the I frames [KNR06], globally reduce the number of erroneous frames since it moves the errors from the first frame in GOP to the later ones. A comparison between both techniques is given in Table 11.1, where it can be seen that the mean Peak Signal-to-Noise Ratio (PSNR) improves around 2 dB with the RLC scheduling.

A scheduling algorithm is proposed for the UL of a wireless communication system with several access points and frequency bands that is able to schedule simultaneously multiple users transmitting with equal power. This Group Proportional Fair Scheduling (GPFS) is analyzed in detail in [KCM07] and compared with PFS. In fact the algorithm is an extension of the PFS, allowing the simultaneous transmission of the users with best fading conditions.

A novel cross-layer algorithm for jointly optimizing the power spectrum and the target-rate allocation in time in multicarrier systems performing Dynamic Spectrum Management (DSM) is proposed in [WSN09]. Rate allocation has the goal of avoiding interference decreasing the power consumption. Assuming that the channel is slowly varying, periodically repeated finite-length rate-schedule can be precomputed together with the optimal power spectrum allocation and only updated when channel conditions change. Traffic variability cannot be exploited by this strategy, but it allows for energy reduction by interference avoidance in the time domain in addition to that in the frequency domain. It is repeatedly assumed in the literature that the Dual Optimization DSM algorithms reach near-optimal solutions. Bounds to the maximization of the weighted sum-rate and the minimization of the weighted sum-power are also derived from his work. Simulations show for all but the highest feasible target-rates that the energy savings by periodic rate-scheduling in addition to that of DSM are negligible in the Digital Subscriber Line (DSL).

Interference management in Femtocell Networks using Distributed Opportunistic Cooperation is considered in [PBS+10], where cooperating groups are automatically organized among severely interfered femtocells to avoid interference seeking to improve their performance by sharing spectral resources, minimizing the number of collisions and maximizing the spatial reuse. A hybrid access policy, based on SINR feedback and macro users' position estimation, is defined and compared with closed and open policies. The proposed algorithm allows the femtocells to cooperate and self-organize into disjoint coalitions while significantly improving their performance in terms of average pay-off per femtocell, relative to the noncooperative case, showing also that large pervasive cooperation can be detrimental when available resources are highly contended, a marginal selfish behavior of femtocells being preferable.

11.1.3 Space Division Multiple Access (SDMA) and Cooperative Multipoint Transmission

Multiple-Input Multiple-Output (MIMO) algorithms are grouped in:

- Beam-forming, implemented in both ends, improving the SNR and allowing an increase in data rate, coverage, and a reduction in errors. The adaptive interference suppression performed at the receiver allows more simultaneous transmissions and the simultaneous reception of different messages from multiple transmitters.
- Space diversity to mitigate fading effects. Gains are limited if the signal at the antennas are not sufficiently uncorrelated.
- Spatial multiplexing to increase data rates. Multipath is exploited transmitting different data streams on different antennas, which can be separated at the receiver.
- Combinations of the previous techniques can be used, and adaptive MIMO-systems that automatically choose the most favorable technique according to the channel are the best option.

Theoretical capacity gains that can be achieved in a reservation-based Ad Hoc Network when using MIMO techniques are studied in [RTNL07] giving indications on how different strategies could affect the Media Access Control (MAC) and routing algorithms. Results are given for two channel models: Spatial Channel Model Extended (SCME) defined by the WINNER project [Win] (32 nodes in a rectangular area of sides 1, 2, 4, 6, and 12 km length, testing 1×1, 2×2, 4×4, and 8×8 MIMO with antennas in a Uniform Linear Arrays (ULA)), and ray tracing for urban environments (2×2, 4×4, and 8×8 MIMO with antennas on a circle with 1-m radius Uniform Circular Array (UCA)). When comparing the mean routing lengths MIMO is better than Single-Input Single-Output (SISO) especially for the 2 and 4 km, the difference being small for 12 km because the difference in pathloss between alternative routes exceeds their difference in MIMO gains. For dense network

11 Resource Management in 4G Networks

Table 11.2 Median number of links

	1 km	2 km	4 km	6 km	12 km
SISO	95	54	39	37	36
2×2	98	58	40	37	36
4×4	101	61	41	37	36
8×8	102	64	43	38	36

(1 km), the average route lengths are short, and the used routes already contain high capacity links in the SISO case. Median number of links used in the network is given in Table 11.2, where it can be seen that MIMO increases the number of links carrying traffic, especially for the dense network.

MIMO systems performing beamforming can be used to generate constructive and destructive interference to restrict transmissions in beams toward a predefined direction combined with a scheduling scheme based on PFS [VS08]. For a given user k, its channel capacity (Shannon) is

$$C_k = BW \cdot \log_2 \left(I_{n_R} + \frac{SINR_k}{n_T} H_k \cdot H_k^H \right). \tag{11.9}$$

BW being channel bandwidth, n_T and n_R the numbers of transmit/receive antennas, H_k the channel matrix of UE k, and $(\cdot)^H$ the complex transpose conjugate operator. SINR of UE k is

$$SINR_k = \frac{\overbrace{\operatorname{tr}(H_k \cdot H_k^H) \cdot \frac{P_k}{d_k^\gamma}}^{S_k}}{N_k + \underbrace{\beta_k \cdot \operatorname{tr}(H_k \cdot H_k^H) \cdot \frac{P_k}{d_k^\gamma}}_{ISI_k} + \underbrace{\sum_{users\, i \neq k} G_{ik} \cdot \operatorname{tr}(H_i \cdot H_i^H) \cdot \frac{P_i}{d_k^\gamma}}_{MAI_k}}, \tag{11.10}$$

tr(.) being the matrix trace operator, P_k the power allocated to UE k, d_k^γ its pathloss, S_k the signal power of UE k, ISI_k the inter-symbol interference of UE k, and MAI_k the sum of the received other UEs' power considering only intra-cell interference, G_{ik} the fraction of the transmit power of UE i which is emitted in the direction of UE k, and β_k the fraction of the transmitted signal power which will be perceived as interference by UE k.

Simulations assume that BSs perfectly know the channel matrix, pathloss, angular position, and even the queue rate of each UE. PFS tries to maximize the relation between the potential throughput of UE k and the mean throughput. The algorithm sorts UEs according to the PFS criterion, evaluating the capacity of each UE's channel without considering interferences, comparing the effect of allocating all the resources to the first UE, or to the first two, and so on. Then a second algorithm loops on resources, finding the UE which benefits most. When allocating for the second UE takes into account the interference of the first beam half powered, and so on, this technique includes the Multi-Access Interference (MAI) avoiding allocating UE too close to each other. In Table 11.3 it can be seen that for low queue rates, the channel

Table 11.3 Throughput gain of multistream scheduling

Mean queue rate	Multistream cell throughput	Single beam throughput	Throughput gain of multistream
1 Mbs	1696 kb	1562 kb	1.10
2 Mbs	3109 kb	2163 kb	1.44
4 Mbs	4256 kb	2356	1.80
8 Mbs	4802 kb	2579	2.02

capacity is large enough to deal with UEs' queues, but when it becomes larger, the capacity of a single link is overwhelmed, and the multistream scheduling gain starts to increase.

Directional antennas combined with a coordination among neighboring BS to allocate resources are an effective way to reduce Inter-Cell Interference. In [KWBM11] the BSs are equipped with multiple antennas that are used in combination with main-lobe steering beamformers in both 3G-LTE and 802.16e WiMAX realistic scenarios. Keeping the resource assignment static for several subsequent frames gives the BS the possibility to measure the interference level that its UEs receive. The algorithm tries to find the optimal point in time for changing resource assignment trying to keep it over multiple frames as constant as possible. This is illustrated in Fig. 11.10 for TTIs 1 to 3, where the resource assignments do not change although the interference situation changes. As soon as the interference exceeds a certain level, the BSs changes the resource assignment of the affected UEs (it is the case of the purple—third—UE after TTI3).

Problems in applying traditional scheduling algorithms on SDMA are addressed in [CJK09b]. These algorithms assign each time slot to only one UE, the channel state being represented by UEs' instantaneous SNR and expressed as the maximal rate each UE can theoretically achieve. But in MIMO systems supporting SDMA multiplexing of UEs in the UL, UE's SNR depends not only on its link condition but also on the level of correlation of spatial signatures of UEs selected in the same time slot. Interference arising from channel correlations depend on the detection algorithm implemented at the receiver. A new idea optimized for the receiver with linear Zero Forcing (ZF) detection algorithm supporting adaptive adjustment of the utility function weighting factors based on the traffic type, queue status, and link conditions of each UE is tested. ZF receivers obtain the transmitted data through the

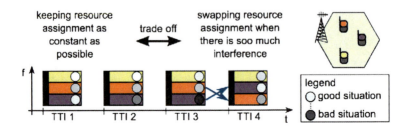

Fig. 11.10 Example of resource assignments variation [KWBM11]

multiplication of the received vector **y** with the matrix **G**, a pseudo inverse of the channel matrix **H** of size $M \times N$, M being the number of antennas at the BS, and N the number of spatially multiplexed UEs:

$$\mathbf{Gy} = \mathbf{G}(\mathbf{Hx} + \mathbf{n}) = \mathbf{x} + \mathbf{Gn}. \tag{11.11}$$

The decoding process enhances the distortion of the signal from the ith UE due to additive noise:

$$\hat{x}_i = x_i + g_{i1}n_1 + \cdots + g_{iN}n_N. \tag{11.12}$$

Assuming equal average noise power N_0 on all receive antennas and P_i being the received signal power, the SNR for the ith UE can be expressed by

$$SNR_i = \frac{P_i}{N_0} \frac{1}{\sum_{j=1}^{N} |g_{ij}|^2}. \tag{11.13}$$

The algorithm works as follows: First, the UE correlation matrix **R** is obtained by defining, for each UE, the set of UEs that are not violating the maximal correlation constraint and can be selected in the same slot. Second, the set of spatially multiplexed UEs using the incremental approach is selected. In each iteration the UE with the highest utility out of the set of UEs that are not violating the maximal correlation constraint relative to UEs selected in previous iterations is selected. Finally, optimal transmission modes are selected for each UE, using a recursive procedure at the BS. This algorithm performs better than PFS specially for Voice over Internet Protocol (VoIP) traffic and better than M-LWDF and EXP algorithms for VoIP and best-effort traffic.

A new utility-based scheduling metric for heterogeneous traffic is derived from the M-LWDF scheduling rule, adding adaptive priority weights that are periodically updated, based on past QoS level, which is calculated using Exponential Moving Average (EMA) function with forgetting factor α [CJK09a]. The objective is obtaining good fairness properties guaranteeing high level of UE satisfaction also at the cell edge. The utility function for the kth user in the nth frame is

$$u_k(n) = D_k(n)^{a_k(n)} \cdot \frac{r_k(n)}{\bar{r}_k} \cdot b_k(n), \tag{11.14}$$

which is the same as for the M-LWDF rule but replacing the fixed priority weight with two adaptive weights, $a_k(t)$ and $b_k(t)$, determining UE priority (delay- and rate-dependent, respectively), D_k is the HOL packet delay, and r_k is the theoretically achievable transmission rate for UE k. The weights for each UE are periodically adapted to optimize UE satisfaction level and resource utilization simultaneously. For delay-sensitive applications, it is important to keep the packet end-to-end delay under the defined deadline and the transmission rate above the threshold level. Delay-tolerant best-effort traffic has a strong adaptability to delay and bandwidth. Thus priority can be adapted on the QoS level of UEs with delay-sensitive traffic instead of the UE's own QoS level. The average throughput is updated in each frame, while the average waiting time is adapted only if the UE was scheduled in

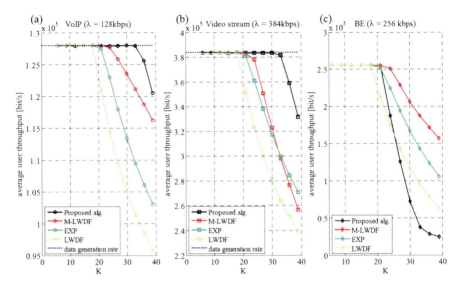

Fig. 11.11 Average UE throughput as function of the number of UEs for different scheduling algorithms for (**a**) VoIP traffic, (**b**) video streaming traffic, and (**c**) best-effort traffic and comparing Largest Weighted Delay First (LWDF), M-LWDF and exponential algorithm with the proposed one

the previous frame. The magnitude of the increment/decrement of the weight depends on the intra-application UE's QoS level, defined as the ratio between the UE's individual QoS level and the averaged QoS level of all UEs using the same application type. UEs with lower satisfaction level are given higher priority increment.

The algorithm is able to keep the queues of delay-sensitive UEs stable for higher number of active UEs, but there is a performance decrease (in throughput) for delay tolerant best-effort UEs, as is expected, which does not cause significant deterioration (see Fig. 11.11).

Multi-Cell Processing (MCP) allows the combination of signals from two or more BSs to improve UEs' SINR level due to the fact that the strongest interferers will be considered as useful signals. MCP can be used to achieve uniform capacity in full spectral reuse system with a Multiple-Input Single-Output (MISO) strategy, allowing the selection of up to four transmit antennas, oriented to improve the cell-edge UEs SINR [GLG09]. DL without CSI available at the BS is considered but the CSI from the analyzed BS to neighboring BSs is needed which is difficult to obtain in real time. To skip this, a central coordinator performs the scheduling scheme selection and resources allocation to the BSs, based on UE received power level made available via feedback for handover purposes. The MCP technique is designed to use the whole available spectral resources in each cell in a coordinated manner. All BS are assumed to transmit with a constant power level P. Macrodiversity can be used like a distributed MISO technique obtaining the SINR by an Alamouti scheme with Matrix A as defined in 802.16e standard [IEE05]. The two strongest received signals are selected to be active part of matrix A. The uniform capacity, also known

Fig. 11.12 Uniform capacity versus active users [GLG09] (©2009 SEE, reproduced with permission)

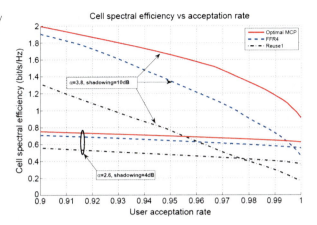

as ubiquitous capacity or maximum fairness, is given by

$$C_u = w(x)\log_2(1+\gamma(x)) \quad \forall x, \tag{11.15}$$

$$w(x) = \frac{W_{\text{tot}}}{\int_{\text{cell}} \frac{dx}{\log_2(1+\gamma(x))}} \cdot \frac{1}{\log_2(1+\gamma(x))}; \tag{11.16}$$

$w(x)$ represents the portion of bandwidth allocated to the position x, C_u the normalized constant capacity that any UE at any location x will achieve, W_{tot} the total available amount of resources of the cell. In a cooperating network the total resource need of a cell includes not only its own UEs' consumption but also the neighboring cells UEs' consumption, with which the cell performs MCP. Results show that the optimal MCP model performs slightly below Fractional Frequency Reuse 4 (FFR4) when no shadowing is present, due to the lack of diversity, while outperforms other techniques when shadowing standard deviation is beyond 4 dB (see Fig. 11.12).

When the requirements of the UEs in resources are considered, the previous capacity formula has to be modified. A frequently used fairness is the index of Jain [GLG10]. It is equal to $1/M$ for the last fair allocation and equal to one for the fairest case. Another possibility is the use of the α-fairness policy [MW00], where 0 offers the maximum throughput, 1 is the proportional fairness, 2 is the harmonic fairness, and infinity the maximum fairness or uniform capacity. Jain's index and the cell total capacity have been chosen as the two measures to balance, using the α-fair allocation of resources since it allows a practical allocation and is easy to solve [MW00]. The selection of a cooperation mode for a mobile defines its SINR. Once every mobile is attributed to one or several BS, the resource amount it will consume and so its capacity are computed. The trade-off that has to be tackled is the capacity gain of a given UE versus its resource consumption. Different criteria have been defined to decide whether the UE i employs Coordinated Multiple Point (CoMP) with N_i cooperating BSs or it keeps the Reuse 1 mode ($N_i = 1$).

Figure 11.13 shows the evolution of both total throughput and Jain index for different α-fairness. The objective is to improve either the Jain index, or the total

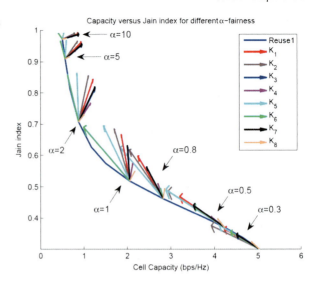

Fig. 11.13 Jain index versus total throughput. *Blue curve* is the baseline Reuse 1. Each point corresponds to a value of α. *Arrows* represent the evolution of both Jain index and throughput using CoMP

throughput, or even both when possible. For low values of α, the improvement provided by CoMP results in a small fairness increase while strongly decreasing the capacity. In this case the use of simple coordinated transmission is worthless. Using Reuse 1 with higher α is much easier and more efficient. For proportional fairness $\alpha = 1$, only two methods increase both capacity and fairness; for harmonic fairness, all methods except K_6 improve both. For α higher than 1, K_6 is the criterion offering the best fairness but does not improve well the capacity. For high values of α, CoMP cannot improve the fairness much since it is already close to 1, but it improves the capacity almost twice.

In addition to antenna arrays, the relay channel model enables the use of distributed antennas belonging to multiple relaying terminals. This form of space diversity is referred to as cooperative diversity because terminals share antennas and other resources to create a virtual array through distributed transmission and signal processing. In cooperative networks the mutual interference can be a limiting factor which needs to be characterized to optimize the allocation of power among source and relay nodes. The effect of relay position and power allocation strategy employing space–time codes under interference constraints in realistic scenarios is analyzed in [TZC10], other important issues addressed also in the work being the channel modeling, protocols, and resource management. The analysis of the outage probability drives the choice of power allocation criteria, studying how the geometry and links quality affect the performance, a key parameter of analysis being the appropriate reuse distance in cellular networks with relay.

An analysis of the interactions between data traffic characteristics and Coordinated Multipoint transmission algorithms is given in [Mue10], revealing that traffic characteristics, if not taken into account in the design of CoMP algorithms, can lead to largely reduced coordination gains or additional transmission delays with a possibly negative impact on user experience. As for CoMP algorithms, the work is focused on Coordinated Scheduling and Coordinated Beamforming in the downlink of an LTE-like system considering that all Evolved NodeB (eNB)s cooperate using the X2 interface. Available traffic models are reviewed making suggestions about how and which models to use in wireless network evaluations.

Heterogeneous load scenarios where data, voice, video, and other interactive services have to be delivered simultaneously to users is optimized by Context-Aware Resource Allocation (CARA) in [PKV10]. The approach combines two techniques: Context awareness providing the scheduler with information about the application environment focusing on the application foreground/background state and Transaction-based Scheduling managing traffic flows as transactions with attached finish time, QoS requirements, and context information. This helps the scheduler to extend in advance its allocation decision to multiple TTIs. Large utility gains have been obtained considering two different traffic types, the utility functions being derived directly from context information. To reduce the extra cost associated to the context information, efficient signaling protocols should be defined.

11.1.4 Metrics, Testbeds, and Trials

The increasing complexity of current and future mobile wireless technologies requires the implementation of adequate platforms to evaluate and optimize their performance. The implementation of such advanced simulation tools has become a very challenging task when investigating CRRM techniques since different Radio Access Technology (RAT)s need to be simultaneously emulated in a single platform. One of the platforms developed during the project is SPHERE [MSGC07], which integrates General Packet Radio Service (GPRS), Enhanced Data for Global Evolution (EDGE), High-Speed Downlink Packet Access (HSDPA), and Wireless Local Area Network (WLAN). The four RAT are emulated in parallel and at the packet level, which enables an accurate evaluation of the final user perceived QoS.

The cellular environment entity stores the location of each base station and the interfering relations among them considering 27 omnidirectional or sectored sites with wrap-around technique and with radius of 50 m for WLAN and 500 m for the rest. The radio link module characterizes the pathloss, shadowing, and fast fading. Mobility model is implemented at the mobile station unit. The BS is responsible for the MAC and RRM functions and controls the channel pool where the status of all channels per RAT is maintained. When a mobile terminal requests a channel from a given RAT, the serving BS is examined to search for an available channel. If it is not possible, a channel from a different RAT is assigned, or it is placed in a queue until a transmitting mobile releases its channel. GPRS and EDGE users will be served in a First Come First Served (FCFS) scheduling policy; in HSDPA they can be served either in Round Robin (RR), according to the Max C/I criterion, or with a PFS. In WLAN, real-time traffic is served through Hybrid Coordination Function Controlled Channel Access (HCCA) with FCFS policy, while best effort users contend to get the channel using the Enhanced Distributed Channel Access (EDCA) protocol. Link adaptation, multichannel operation for GPRS and EDGE, multicode allocation in HSDPA, and call admission control have been also implemented. User traffic is defined by knowing the session-arrival process and the traffic models for web browsing, real-time H.263 video transmission, and email. The link management module handles the radio transmission and emulates channel errors by means of a Look-Up-Table (LUT) mapping the emulated Carrier to Interference Ratio (CTIR) average value to a Block Error Rate (BLER).

The platform has been validated comparing the results with the maximum possible performance of each RAT to prove that obtained results are within the expected range.

A versatile emulation platform of cellular access networks capable to execute real-time emulation for various services which require moderate transfer rates is explained in [PVB09]. The hardware resources required by the emulator are one computer equipped with two network cards, the emulation being performed based on delay and error statistics generated by simulations or real measurements. The tool also includes a simulation platform which generates the required statistic tables and proposes an emulation/simulation strategy which significantly reduces the number of the required statistics tables by ensuring the reusability of these statistics.

Main operations performed in the emulation process are the operations related to the SNR computation (updated once every second), the UL connection, and the DL connection. Packet-delay and errors are inserted according to the appropriate statistics which characterize the emulated situation. Parameters used in the simulation are those characterizing: the cell by number of active users, multipath propagation profile of the channel, carrier frequency, Probability Density Function (PDF) of fast fading due to mobility; the users by main user and secondary users, speed, attenuation, services (priority, average bit rate, etc.); the Orthogonal Frequency Division Multiplexing (OFDM) modulation, Forward Error Correction (FEC) coding techniques, and multiuser access method. The allocation of resources and scheduling is done by considering the value of the instantaneous SNR (depends on fading and background noise/interference), priority of the service and average bit rate, and coding rate imposed by the Hybrid ARQ (H-ARQ) process.

A testbed for MIMO HSDPA and Worldwide Interoperability for Microwave Access (WiMAX) has been developed by [CMLR10] presenting an excellent trade-off between effectiveness and efficiency. To reduce cost and time, during the measures, only the physical layer is analyzed, considering one Tx and one Rx with a maximum of four antennas each, so only the UL/DL scenario from a single BS to a single user can be considered (the interference is included considering other measures with different Tx locations). All transmit-data blocks are pregenerated in Matlab and stored on parallel solid state hard disks for instantaneous access. Results for HSDPA and WiMAX measurements in urban and alpine scenario demonstrate the efficiency of the testbed. There are significant performance gains of Double Transmission Antenna Array (DTxAA) compared to Single-stream Transmit Antenna Array (TxAA), especially in the urban scenario. In the WiMAX system, Low Density Parity Check Code (LDPC) codes promises about 1-dB gain over Turbo Channel coding.

The Mobile Innovation Center (MIC) designed a test network focusing on mobile service and software development which integrates UMTS, IEEE 802.11, and

Ethernet networks providing the infrastructure for real-life experiments and developments for internetworking issues and a physical interface for investigating radio level questions [Faz07]. MIC structure contains the whole UMTS radio network with a NodeB with an indoor antenna as well as a Radio Network Controller (RNC) and the core network. The main areas of research are: Mobile radio technologies dealing with radio aspects of networks, radio coexistence and cooperation between technologies, and radio channel modeling; Techniques enabling an increase of the radio data transmission efficiency covering, among others, intelligent antenna systems, and Beyond 3G (B3G) detection techniques are also main points of research; Integration and management of heterogeneous mobile networks, traffic modeling, security, and real-time multimedia information transfer; Services and mobile applications, user's behavior.

11.2 Modeling, Prediction, and Classification of Traffic

Traffic modeling is an important part in many aspects of communications including the area of mobile radio networks. The knowledge about characteristics of user generated traffic and the generation of traffic values, either per user or aggregated (e.g., per radio cell), is strongly related to network simulations and emulations. In these contexts, the significance of traffic models is independent of the network generation, whereas the complexity increases.

Increasing complexity has been observed in radio network planning, in which prediction of traffic plays an important role, especially in planning of UMTS networks due to the influence of the cell load on the link budget. In contrast to Global System for Mobile Communications (GSM), the expected traffic has to be estimated not only according to volume,[1] but also according to location.[2]

The second characteristic of 3G traffic in contrast to Second Generation (2G) appears in the specification of the UMTS QoS classes [3GP07]. Although different services with different data rates have been in operation in 2G, simulation and planning have focused on voice traffic. This reasonable simplification was no longer valid for UMTS. Since then, multiservice traffic has had to be considered. Contributions in the topic of multiservice traffic were raised, e.g., by the investigations of the Models and Simulations for Network Planning and Control of UMTS (MOMENTUM) project [FCX+03] or the Simulation of Enhanced UMTS Access and Core Networks (SEACORN) project [AH04]. In general, the procedural method consists of the definition of a set of services based on the UMTS QoS classes and the building of a traffic model for each service.

---

[1] Traffic volume leads to the number of required channels.

[2] Location of users leads to required transmission powers that lead to the interference situation that leads to the size of the coverage area.

11.2.1 Simulation

In the context of B3G and the investigation of multiservice traffic validation, single-traffic models are combined to form a multiservice traffic model [JPV]. Concerning characteristics of traffic, the main difference between 3G and B3G is seen in higher data rates demanded by additional applications. Considered applications are based on the investigations of the SEACORN project that defines various services and environments including office, business city center, and vehicular scenarios. The representative applications for the chosen vehicular scenario are voice at 12.2 kbit/s, video telephony at 144 kbit/s, multimedia web browsing at 384 kbit/s, and assistance in travel at 1536 kbit/s, respectively. Dedicated traffic models were built with the most relevant activity/inactivity characteristics [Fer03].

11.2.2 Emulation

The operator Telefónica from Spain introduced its Traffic Loading Platform (TLP). Traffic loading is a mandatory stress test that mobile operators must perform to any network element before its deployment. The TLP is able to load mobile stations with real traffic from UEs and has been designed for repetitive and controllable traffic loading of Node Bs. The TLP is designed for UMTS/HSDPA/High-Speed Up-link Packet Access (HSUPA) testing, but it is also functional for any mobile standard between 800 MHz and 2500 MHz, including GSM/GPRS, WiFi, and MIMO connections with respect to 3rd Generation Partnership Project (3GPP)/LTE and mixed GSM/UMTS traffic loading.

The TLP is used to test any configuration of a base station before it is deployed in the real network and to assess the expected performance in a low-risk laboratory environment. Up to eight terminals can be connected to three base stations. The services which are controlled by the TLP and offered to the user terminals include voice, video calls, video streaming, and file download.

The example of TLP shows the typical characteristics of network evolution: several services with several different applications with several different data rates made available by several different but parallel operated system architectures. With respect to 4G, this description is not expected to change radically. In fact, the giant step in the development of traffic modeling is considered during the evolution from 2G to 3G.

The description of the UMTS Terrestrial Radio Access Network (UTRAN) testbed designed by the university of Namur, Belgium, is more detailed. In order to characterize the UMTS QoS classes, the focus has been set on four representative applications (see Table 11.4). To characterize the composite traffic of a single user, each representative application is fit in a three-level model consisting of a session level, a connection level, and a packet level.

The session level lasts as long as the application is running. Its statistics are mainly influenced by user behavior. The connection level describes the connection behavior of a single session. The pattern chosen for the sources is based on

11 Resource Management in 4G Networks

Table 11.4 3GPP traffic classes, applications and statistical distributions for emulation of UTRAN

3GPP Traffic Classes	Representative Applications	Statistical distributions		
		Session-Level	Connection Level	Packet-Level
Conversational	VoIP	Poisson	Two States Markovian	Constant
Interactive	Web browsing	Poisson	Geometric & Exponential	Pareto & Exponential
Streaming	Video streaming	Poisson	–	Pareto
Background	E-mail	Poisson	Lognormal & Pareto	Lognormal

a high/low model in which it could be possible to generate traffic on both states. The packet level represents the packet inter-arrival and size distribution for each state of the connection level.

Based on the three-level division, the four classes are described in terms of statistical distributions [KLL01, LS02, BS98, RLGPC+99] and 3GPP standards [3GP98, 3GP04a, 3GP04b] when applicable. These descriptions are summarized in Table 11.4.

This example shows a little bit more in detail the concept of multiservice traffic modeling. Due to the fact that every single model consists of statistical distributions, the increase of modeling complexity is limited. In summary, traffic modeling bas been mainly based on the multiservice concept since 3G. With respect to 4G, applications with higher data rates are expected, and the characteristics of packet switched data delivery have to be solely taken into account. Circuit switching is not expected to be included any more in any system specification beyond the specification of LTE.

11.2.3 Planning

The automation of network planning processes requires reliable input data for optimization in order to achieve reasonable results. Methods adopted by MOMENTUM to derive input data show poor results when verified against measurements of a real network, but also show a tendency to improvement when a traffic model becomes more complex [KH05]. However, a high modeling accuracy is only possible at the expense of a huge analysis time that is not available in radio network planning. Furthermore, the user behavior is likely to change during short periods of time, so that repeated and, therefore, fast modeling algorithms have to be adopted. Such algorithms are based on Linear Multiple Regression Analysis (LMRA) and Classification And Regression Trees (CART) as well as on the combination of the two methods [HK07].

Modeling itself is based on a learning data set as input data. Figure 11.14 depicts the process of input data extraction from geographic maps. Every cell is represented by its Cell Assignment Probability (CAP). The CAP is a geographic map as well.

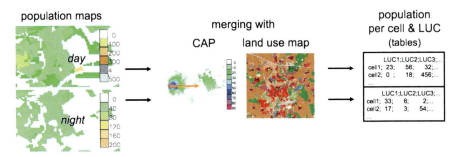

Fig. 11.14 Extraction of input data for traffic modeling [HK07] (©2007 IEEE, reproduced with permission)

Combining CAP and population distribution assigns a population to every cell. Considering the land use map, population figures can be extracted as histograms over Land Use Class (LUC)s for every cell. Therefore, the input data consists of land use dependent variables containing population figures. This procedure is similar to MOMENTUM.

The performance of the model is validated against performance measurement values from the Network Management System (NMS) (see Sect. 10.4.1). It could be shown that the combination of the two above-mentioned methods shows modeling results of arbitrary accuracy. However, the results are too optimistic, as modeling and estimation of traffic values with respect to a new network plan is not the same.

Furthermore, during COST 2100 a paradigm shift has been experienced. Network simulations and their corresponding models have been considered to be able to deliver reliable input data for optimization. For that reason, many corresponding investigations were related to UMTS radio network planning. However, this promise has not been fulfilled. Instead of building general models calibrated with measurement data, investigations have dealt with direct processing of measurement data, afterward (cf. Sect. 10.4). Still, the conclusions drawn from the planning aspect do not reduce the importance of traffic modeling for simulation and emulation of networks.

11.3 3GPP LTE Radio Resource Assignment

Existent 3G and 3.5G networks are currently responding to the rapid expansion of data services in mobile environments. This new demand is caused by factors such as an increased technological awareness among users, phones becoming "smarter," and the continuous development of wireless networks. In the near future, LTE and WiMAX networks are expected to establish a dominant position worldwide, and in both cases their radio access network (DL for LTE) is based on OFDM.

Among the different advantages of OFDM, its use to multiplex users in the form of Orthogonal Frequency-Division Multiple Access (OFDMA) implies an almost negligible intra-cell interference because of the inherent orthogonality among users in a cell. Despite this fact, efficient transmission of multimedia traffic over OFDMA networks in general, and in the 3GPP LTE in particular, requires the design of strategies to assign radio resources smartly. In LTE, assignments are performed in resource blocks of 180 kHz (12 subcarriers at a 15-kHz spacing) with a temporal resolution of 1 ms. Nevertheless, along this section, some investigations use other values but without invalidating its applicability in LTE systems at all.

Radio resource assignment is going to be governed by scheduling algorithms executed in the eNode-B and by ICIC strategies that will complement them. These last are of special importance in order to control inter-cell interference at the cell edge. ICIC basically restricts the resources to be used by the scheduler so that inter-cell collisions in the assignments of subcarriers are minimized and the QoS is leveraged among all users in the cell no matter of their position.

Given this, some researches investigate scheduling-focused approaches and deal with the resource assignment problem from a single-cell perspective. On the other hand, a second group of studies deal with multicell environments and mostly cope with the ICIC problem.

11.3.1 Resource Assignment in Single-Cell Scenarios

Following the previous taxonomy and starting with scheduler centric studies, the importance of addressing the resource assignment from a cross-layer viewpoint is outlined in [GKC09]. Physical layer (PHY)-centric approaches tend to solve the maximization of the cell throughput subject to constraints on minimum user rates and maximum available power; however, information from the MAC layer is paramount to guarantee the desired QoS.

With this cross-layer objective in mind, the authors show that the PHY-centric resolution is indeed a nonlinear integer optimization problem and optimum resolution methods require prohibitive execution time. Thus, they are not applicable in real systems. An investigation performed by means of system-level simulations shows that, among the existent alternatives, the proposal by Zhang [ZL04] is of special interest. This scheme is an iterative algorithm with low complexity that assumes equal power allocation for all subcarriers and that reduces capacity in just a 0.28%. This algorithm is the start point in the definition of an improved cross-layer schema.

Zhang's strategy is extended in order to take into account QoS information from the MAC layer. From this information the minimum rate that every user requires is computed dynamically, i.e., one of the constraints in the optimization problem is updated along time. As a second improvement, the assignment of resources based on the frequency-selective behavior of the channel is complemented taking into account the temporal evolution of the frequency-averaged channel gain (SNR gain). Users are also prioritized following this second metric. Table 11.5 shows the simplified flowchart of the scheme.

Strategies derived from the permutation of the mentioned improvements have been assessed and compared with the original Zhang's algorithm. In particular, Ta-

Table 11.5 Flowchart of resource assignment algorithm proposed in [ZL04]

	START
	⇓
	SNR Calculation
	⇓
(STEP 1)	User Prioritization according to Frequency-averaged SNR
	⇓
(STEP 2)	Update rate constraints based on instantaneous queued packets
	⇓
(STEP 3)	Perform radio resource assignment following Zhang's scheme (considering the adaptation imposed by step 1 and step 2)

Table 11.6 Resource assignment schemes considered for evaluation in [GKC09]

Algorithm's name	Step 3	Step 2	Step 1
Static RA	[ZL04]	–	–
SNR Enhanced Static RA	[ZL04]	–	Yes
Dynamic RA	[ZL04]	Yes	–
SNR Enhanced Dynamic RA	[ZL04]	Yes	Yes
Dynamic FDM	Basic	Yes	–
SNR Enhanced Dynamic FDM	Basic	Yes	Yes

ble 11.6 relates the names of the strategies with the features that each scheme is considering.

Results show a relatively low utilization of "static RA", around 60% of the total cell capacity. Furthermore, this scheme introduces great inefficiency and large packet delays even at low cell loading. On the other hand, the "dynamic RA" option improves performance up to an 83% under high cell loading, and delay is also visibly improved. The best results in terms of throughput are obtained when the two proposed extensions are used, with a system utilization of 96% for the "SNR Enhanced Dynamic RA" case.

This way, and given the gains introduced when multiuser diversity is exploited in frequency and time domains, the same authors propose in [GDC09] a heuristic joint time-frequency allocation algorithm that outperforms frequency domain only schemes. Besides, unlike the majority of previous investigations, multiple traffic classes requiring different QoS are considered.

The foundation of this heuristic algorithm is as follows: at the beginning of every temporal frame, the Constant Bit Rate (CBR) user with worst average rate is selected and allocated the best subchannel that is simultaneously the worst over Best Effort

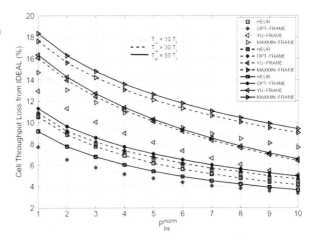

Fig. 11.15 Comparative performance of the algorithm proposed in [GKC09]

(BE) users. Once all CBR users are assessed, the remaining resources are assigned to BE ones. This second allocation is done following a proportional fair policy and ensuring that the maximum rate between the best and worst user is below a certain threshold. Regarding power, it is equally distributed among all subcarriers, the main reason for this is the considerable reduction in the complexity of the problem. In particular, the proposal can be solved with polynomial-time complexity.

This algorithm outperforms other existent options; in particular, it is compared against these approaches:

- HEUR: This is the actual new proposal.
- IDEAL: Assuming a perfect knowledge of the achieved rates over a certain temporal window (note that this implies noncasuality), the problem can be posed as an Integer Linear Programming one. IDEAL is the quasi-optimum solution after having relaxed the integrality constraint.
- OPT-FRAME: Optimization only performed in the frequency dimension. Each frame allocation is independent from others.
- YU-FRAME: A second per-frame optimization; in particular, this is the proposal in [YZQ07].
- MAXMIN-FRAME: Third per-frame optimization, proposal in [RC00].

The curves in Fig. 11.15 compare the average cell throughput of the different approaches with respect to the IDEAL case. This is represented as a function of the DL transmitted power, normalized to the minimum needed power to satisfy the CBR users and avoid outage. Since the scenario only has one cell, this value is directly related to SNR conditions in the cell. Finally, note that HEUR works on a per-frame basis but with the objective of optimizing the average throughput computed in a certain temporal window. For this reason, different window times (T_w) are included in the study.

From the graph it is clear that HEUR outperforms existing per-frame schemes. For example, paying attention to the case in which the temporal window is of 50 frames ($T_\mathrm{w} = 50 T_\mathrm{f}$), YU is improved by 7.2% at low SNR (4.8% on average) and

MAXMIN by 9.2% at low SNR (7.2% on average). Moreover, it is even better than the per-frame upper bound, the OPT-FRAME case, with gains between 1.3 and 2.2%. The average loss with respect to the IDEAL case is just 5.6% on average.

The work in [GC10] follows the development of algorithms that distribute the power evenly among all subcarriers and that consider single-cell scenarios with heterogeneous service mix. Two new techniques are proposed with a close-to-optimum performance and low execution complexity. These strategies are motivated by the resemblance of the problem with the *generalized assignment combinatorial optimization problem* [MT90], and indeed its development is mainly inspired by ideas utilized in the related literature.

The proposed algorithms are evaluated in a comparative manner among the following four schemes:

- IP: Exact optimal case, which corresponds to the Integer Linear Programming solution. The problem is NP-Hard and so computationally very complex.
- LP: Solution extracted after relaxing the integrality of the previous scheme. This simplification allows finding the allocation under polynomial execution complexity.
- HEUR1: First proposed heuristic. The procedure is inspired by the solution approach given in [MT90]. It is composed of two phases:
  - HEUR1 Phase 1: BE users are ignored, and the best CBR sub-channels are selected. The remaining unallocated resources are assigned to BE users using a best-rate user policy.
  - HEUR1 Phase 2: This is a round of comparisons and subchannels swaps between pairs of users to improve the allocation.
- HEUR2: Second proposed heuristic, in this case inspired by the works of [Wil] and [FW97]. Again the solution is approached on two phases with the same computational complexity:
  - HEUR2 Phase 1: The optimal solution of the unconstrained problem is extracted by simply allocating each subchannel to the user that experiences the maximum data-rate on it.
  - HEUR2 Phase 2: Reallocations aiming at rendering the solution feasible, subject to the required constraints.

Given this, Fig. 11.16 represents the accumulated throughput for BE users under different loading conditions of CBR users.

From the figure the interest on the two new heuristic techniques rely on:

- Both perform close to the optimal scheme but with a significantly lower execution time.
- As the number of CBR users increases, the losses with respect to the optimal solution tend to be larger.
- The first heuristic option performs better than the second, with losses of 3.1% under good SNR conditions and 6.5% otherwise. The second option records losses of 6.9% and 15%, respectively; in this last situation, if the CBR loading is increased, losses can surpass 23%.

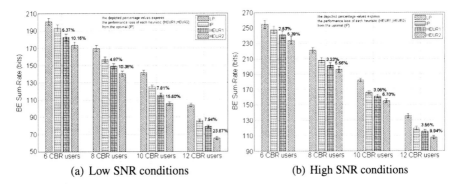

Fig. 11.16 Comparative performance of the algorithms proposed in [GC10]

All previous investigations consider single-cell scenarios. However, as previously stated, when considering multicell environments, some type of coordination among cells is required to avoid collisions in the assignment of resources, and this is responsibility of ICIC schemas. This is the context of the second group of research works concerning resource allocation in LTE networks.

11.3.2 Resource Assignment in Multicell Scenarios

ICIC strategies can be classified in multiple manners. A first possibility is between static and dynamic. In this sense, it is important to note the difference in the temporality of scheduling and ICIC decisions: whereas the scheduler acts in a scale of milliseconds, the allocation of resources to cells by ICIC can last much more or be even fixed. On the other hand, depending on whether the cells have to exchange information and take actions in a centralized manner, one can talk about distributed or (semi-)centralized strategies, as it is detailed later on.

11.3.2.1 Static ICIC

The great advantage of static schemes is twofold: first, the lack of signaling overhead and the low complexity involved; secondly, they allow making interference predictable, and therefore scheduling techniques such as those in Sect. 11.3.1 can easily be extended to multicell environments without dramatic changes in their basis.

Static proposals tend to split the available bandwidth into an inner and an outer part, and intermediate ones can also be considered. The two basic static strategies are *Fractional Reuse* [SOAS03] and *Soft Reuse* [Hua05]. In both cases users are categorized according to their position, and resources are assigned depending on it. Whereas fractional reuse applies reuse 1 only for inner users and a higher value

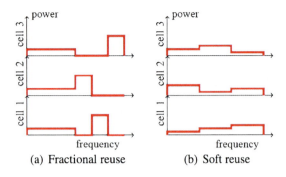

Fig. 11.17 Example of static ICIC strategies for a regular pattern of three cells

for outer ones (Fig. 11.17a), soft reuse allows sharing the overall bandwidth by all eNodes-B but restricts the power on each subcarrier (Fig. 11.17b).

The work in [RHG+09] compares fractional and soft reuse with two classic hard reuse deployments with a reuse factor of 1 and 3, respectively. The study is done over a scenario with 21 cells where the bandwidth has been divided in three parts, as in Fig. 11.17. Two possible power levels are considered, P_{MAX} and εP_{MAX}, and for the fractional case, the inner band occupies βBW_{TOT}, the total available bandwidth. The comparison is done by means of system-level simulations in which the link-to-system level mapping has been done using the LTE PHY layer simulator from [OSR+09], from which more details can be found in Sect. 6.4.2.

Concerning the Cumulative Density Function (CDF) of the SINR, hard reuse 3 is the one that provides the best values, while hard reuse 1 is the worst. Soft and fractional reuse strategies fall in between, being closer to hard reuse 1 and 3 respectively. Improvements in the obtained CDF generally imply cell-edge users with higher throughput values. However, this does not imply a higher global throughput, and so the final values have to be adjusted accordingly.

Results indicate that β is one of the parameters with a higher impact on throughput, and its choice must be carefully done for each particular scenario. On the other hand, similar CDFs can be obtained with an appropriate tuning of ε and β, and so no clear rules can be derived to choose between soft or fractional reuse.

An improvement to the basic soft reuse model is proposed in [CGV]. The authors propose a procedure to compute the power per subcarrier or power mask to maximize, as in most of the previous works, the summation of individual user rates. Since the approach is completely static, the mechanism can be executed off-line, during the radio planning process. The resulting power masks are provided to the cell schedulers that act accordingly, so the operation is not only static but also fully distributed.

The proposed mechanism organizes the available bandwidth in K groups of subcarriers, K being the size of a group of adjacent cells, similar to the concept of *cluster* in traditional geometric frequency planning. Then the algorithm aims at finding the K power levels that maximize the global capacity. Different permutations of this power vector are assigned among the cells of the same cluster and the same permutation is reused between correspondent cells of different clusters. Two models are presented for the computation of the power mask:

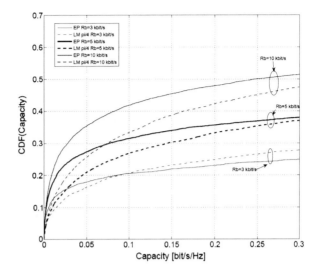

Fig. 11.18 CDF of the capacity for different target rates [CGV], ©2009 IEEE, reproduced with permission

- Linear Model: The K power levels lay on a straight line forming an angle ϑ with the average power value. Discretization of ϑ and a brute force algorithm are used to find the best angle between 0 and $\pi/4$.
- Iterative Procedure: Without loss of generality this case is particularized for $K = 3$. A gradient ascent method is used to optimize the parameters for a set of different scenarios, with different users positions and channel realizations. The average value of these individual solutions is proposed as the final one.

System-level simulations have been performed to compare the performance of these schemes with the case in which all cells and subcarriers are assigned with equal power levels. Results indicate that no substantial gain is obtained in case a maximum sum rate scheduler is used. On the other hand, if fairness-oriented schedulers are used, a significant improvement is reached for the linear model. This is graphically supported by Fig. 11.18 that represents the outage network capacity for different user target rates R_b^* for the linear model (LM) and the equal power case (EP).

The work in [PC10] is claimed to be an extension of the previous approach. In particular, the power profiles structure is now defined depending on the position of the users, which are organized in three groups according to their distance to the eNode-B.

Concerning the scheduling part, the authors pursue to guarantee fairness to some degree. Keeping this in mind, several schedulers have been tested in conjunction with the proposed ICIC strategy:

- Distance-Based: Users are served in order, following their distance to the eNode-B (farthest first).
- Ascendent SINR-based: Resource blocks are ranked according to a measure of SINR that does not consider pathloss or shadowing. Then, for each resource, the user with the higher SINR is selected. This approach favors inner users.

- Distance-descendent SINR-based: Users are ranked according to their distance (farthest first), and they are assigned with the resource block having the highest SINR. This approach favors cell-edge users.

The proposed schedulers are also compared to a power control algorithm which guarantees a received power equal to −82 dBm.

Throughput and fairness have been quantified, and results show that the distance-based approach is the worst one. This is logical because it does not consider propagation or interference, so even the baseline power control performs better. The other two options imply much better results, with similar average values but with almost opposite variations, which is logical considering their opposite behavior favoring inner or outer users. The authors conclude that these preliminary results show that channel aware scheduling in conjunction with power planning outperforms power control and unaware scheduling.

11.3.2.2 Dynamic ICIC

In dynamic ICIC, the frequency/power allocation that is provided to the scheduler evolves along time, for example, to adapt the system to different traffic loads or variations in the long-term channel conditions. Although this process could be executed in a distributed manner, most of the proposals in the existent state-of-the-art require an exchange of information among cells and a certain level of centralized decision making. Thus, a central unit is required. This entity determines the best suited resources and distributes them to the cells so that schedulers use them accordingly. In the context of LTE, this central element could be a *master* eNode-B that could interchange information with other nodes through the X2 interface. This new signaling overhead is the main drawback of dynamic approaches.

Following this architecture and considering the approaches explained in Sect. 11.3.2.1, two basic types of dynamic strategies can be defined: firstly, those derived from the static fractional frequency reuse; secondly, those that dynamically change the power mask, i.e., derived from the static soft frequency reuse.

The approach in [GRO+09] is an example of the first group. In this work the authors evaluate a mechanism that decides new frequency masks for a group of cells repeatedly in time. Once, the frequency masks are reported back to the cells, the particular schedulers administer them with a proportional fair policy. Note that unlike classic fractional reuse, the bandwidth is not divided in subparts; now each mask has been computed evaluating all the available subcarriers. That is why, a structure of three frames is defined, and each one serves a particular user class. Then, synchronization among eNodes-B is assumed.

Simulations have been run to test this mechanism in terms of throughput. Two scenarios have been considered, SISO and MIMO 2 × 2 with spatial multiplexing. Figure 11.19 shows the histogram of the average cell throughput for the SISO case. It can be clearly observed that the bar graph is divided in three subgraphs, each one corresponding to one class of users. It is noticeable how the quality doubles when leaving the cell edge and entering an intermediate area of cells. This deviation in

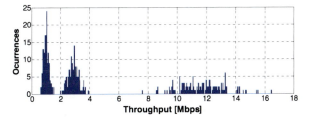

Fig. 11.19 Histogram of the average cell throughput for the proposal in [GRO+09] (©2009 IEEE, reproduced with permission)

quality is aggravated when considering inner cell throughput, which can be up to 16 times higher than that from outer areas. So even though the scheduler serves each class with a proportional fairness strategy, the technique should be complemented to improve inter-class fairness.

Regarding dynamic strategies derived from static soft frequency reuse, the work in [Smo08] aims at finding the optimum power mask in a similar way as [CGV] and [PC10] but adding the adaptation feature. The available bandwidth is again divided into K groups, from which different permutations are defined and used in different cells. Two algorithms are proposed:

- Noncooperative algorithm: This falls into the category of static approaches. As in [PC10], the groups of subcarriers are assigned to different distances in the cell, with different permutations among different cells. The concept of reuse factor for the permutations is defined.
- Cooperative algorithm: In this case some kind of cooperation between eNodes-B is required, and powers are computed adaptively. The procedure is as follows: First, the T weakest users per cell are assigned to a subcarrier group, and this is done sequentially user-by-user. After this first assignment, the required transmitted powers are computed, and these values are used as interference to compute the best assignment of the remaining users.

Both algorithms are tested in the context of a WiMAX system-level simulator; nevertheless, the scenario is completely generic, and results can be directly extrapolated to the 3GPP LTE standard.

Significant results are shown in Fig. 11.20 where the Signal-to-Interference Ratio (SIR) is plotted for different transmission powers and configurations of the proposed ICIC strategies. A random assignment algorithm is also simulated as a lower-bound reference. According to the curves, a higher number of subcarrier groups implies better SIR values. The option with six groups outperforms others. In particular, for a given required SIR, the proposed noncooperative approach can improve the required power in up to 5 dB, and up to 3 dB when compared with the cooperative and noncooperative options with a smaller set of groups. For the same number of groups, noncooperative and cooperative perform in a similar manner in the loaded scenario.

Another investigation on the dynamic update of power masks can be found in [BEG09]. This work generalizes to LTE systems the cell coupling model for WCDMA systems that was previously used in investigations such as [Tur06]. In particular, since subcarriers are orthogonal in frequency, now there are as many coupling equations as resource blocks, and their resolution yields the power values

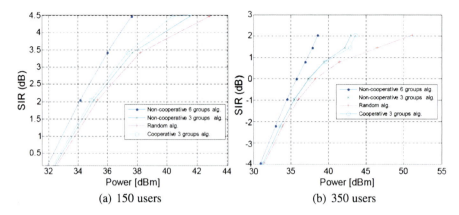

Fig. 11.20 Algorithms performance for different levels of load in the network

to be provided to the scheduler. Note that, unlike [CGV] or [Smo08], now the power mask is smoothly adapted to available resource blocks, and there is no cluster-based cell organization.

Considering the resolution of this equation system, the task of finding optimal power masks is represented by a nonlinear optimization model that pursues the maximization of the overall throughput. So, as in other previously commented works, the problem is addressed from a heuristic viewpoint. In this case the authors propose a method that combines the principles of binary search and greedy algorithms. Note that again a central unit is required, responsible for receiving the average path loss values of the different active links and finding the final solution. Since the procedure implies adopting solutions as long as they are better than the current, the final proposal is expected to be suboptimal. Nevertheless, results show that the generated power allocations improve system capacity significantly when compared with some previously known schemas.

Figure 11.21 shows the average cell throughput (Fig. 11.21a) and the average throughput of the weakest user (Fig. 11.21b) for the proposed schema and the following four existent methods:

- A: Frequency reuse 3.
- B: Frequency reuse 1.
- C: Soft reuse. Three power levels are interleaved across cells (recall Fig. 11.17b). The mask is given by [1; 0.5; 0.3], which means that the lowest power level equals one third of the higher one.
- D: Soft reuse with mask [1, 0.1; 0.01].
- E: This is indeed the new dynamic heuristic proposal.

Total cell throughput results show that an appropriate and dynamic choice of the power mask clearly impacts on the system performance. It can be observed how the proposal more than doubles the capacity achieved by a classical hard frequency reuse with reuse factor of 3. In addition, Fig. 11.21(b) shows how this increased throughput is not detrimental to users difficult to serve, who also improve their in-

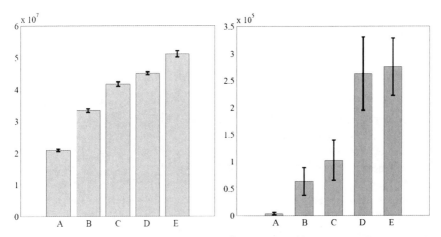

(a) Cell throughput (bps), 99.9% confidence intervals (b) Weakest user (bps), 95% confidence intervals

Fig. 11.21 Comparison of average throughput obtained by the proposal in [BEG09] and other existent alternatives

dividual rate. Note that in this case the confidence intervals are much larger because only the throughput of one single user is considered.

Fractional or soft reuse is not the only possibility to achieve coordination among several cells. If the geographical position of users is known to some extent, this new degree of freedom might allow deploying a "coordinated unitary reuse," as it is proposed in [KWBM11]. In this case, once a predefined interference and channel quality threshold is exceeded, the resource allocation for a certain mobile is changed with a certain probability. If cells are assumed to be fully loaded, this change would correspond to a swap of resources between two mobile stations.

The research work compares several swap strategies, requiring different levels of signaling:

- Without signaling: Random swap.
- Mobile stations report the SINR on all resource blocks: the change can be done in three ways, prioritizing the swapper, the swapee, or both.
- Neighboring eNodes-B communicate their scheduling decision and an estimation of the geographical position (subsector) of each mobile. This information, combined with radio planning calculations, allows predicting the interference a mobile would get after a change in the resource allocation. This strategy is named "altruistic." Note that it is considered that each-sector is equipped with a main-lobe steering beamformer, consisting of a linear array of four sector antennas. For each resource block, the main lobe is pointed at the receiving mobile station, which is assumed to the perfectly tracked.

By means of simulation it is computed the coordination gain that is introduced by each strategy. A lack of coordination implies lower SINR values, either because resources are static, or because they are changed with high probability. Swap probabilities of about 17% to 0.2% give the best results. The comparison of strategies yields that those based on SINR measurements perform better, with gains up to 1.2 dB when both swapper and swapee are jointly considered in the decision change. On the other hand, their weak point is a higher amount of signaling over the radio channel. If this is restrictive, then the altruistic case is the best option, with gains up to 0.7 dB.

In order to close this section, it is worth outlining that in all the previous research works, simulation is the essential tool to obtain performance metrics. An analytical evaluation of some of the complex heuristic techniques that have been commented would be very complex, besides the new multimedia services cannot be easily characterized as it happened with the voice service in 2G networks. The main drawback of this methodology is that realistic simulations are very time consuming and they have to be repeated if the network setup is changed. In this respect, the work in [FKWK09] proposes as alternative a semi-analytical method that gives a much faster estimate of the average cell throughput.

For this purpose, the PDF of the SIR from a pathloss prediction tool is needed assuming that all base stations transmit at its maximum power. Note that this input does not consider short-term fading or scheduling. These two factors are included by the proposed analysis. In particular, the authors find the PDF of the SIR gain due to fair reuse (round robin) and proportional fair scheduling under fast fading conditions. Given this, the final PDF of the SIR is found by combining the two previous random variables. From here, the probabilities for the utilization of the different modulation and coding techniques can be computed, and so the average cell throughput can be calculated.

A comparison with analytical results shows that the difference is around 2%. The authors indicate that for proportional fair scheduling, this difference enlarges as the number of users increases. In particular, the approximation tends to lower the average throughput. Nevertheless, for 30 active users, this difference reaches just a 10%.

11.4 CRRM in Heterogeneous Networks

Technology evolution points to an always best connected wireless paradigm [GJ03] with different technologies coexisting in the same area. The so-called seamless connectivity, and the resulting benefit for the user Quality of Experience (QoE), can be only accomplished through the coordination of the coexistent wireless technologies. This trend toward cooperation among heterogeneous RAT is also known as beyond 3G communications. Its main goal is to serve each user with the RAT best-suited to the running application.

In order to highlight the relevance of the concept of RATs cooperation in beyond 3G networks in the framework of European research, note that this issue has been deeply studied in other European Information and Communications Technologies (ICT) Projects, like Ambient Networks [Amb04], Aroma [Aro06], and Daidalos [Dai04], with the common objective of improving the end user experience.

11 Resource Management in 4G Networks

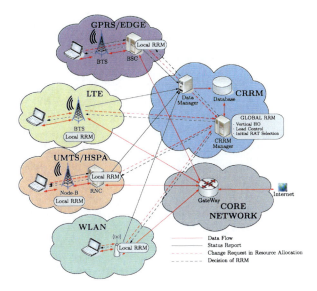

Fig. 11.22 CRRM in heterogeneous networks

This complex heterogeneous system requires an underlying architecture that allows user handovers. In the normal functioning, the CRRM entity should decide on which is the optimal network and resource for each user and service demand, provided that users should be able to switch transparently from one RAT to another. An example of architecture of CRRM with several 3GPP and Institute of Electrical and Electronics Engineers (IEEE) technologies is shown in Fig. 11.22.

This section summarizes the main contributions from COST 2100 action concerning the CRRM concept. Given its special interest, the RAT Selection topic has been differentiated from the rest in Sect. 11.4.2.

11.4.1 General Aspects About CRRM

In any Radio Access Network (RAN), users share a set of available resources being the RRM entity who decides on the distribution policy. An additional functional unit, the CRRM entity, is responsible for the interworking of the RANs not only of the same RAT but also of different RATs. The RRM most important functions are: initial RAT and cell selection, CAC, congestion control, power control, scheduling or resource allocation, HandOver (HO), and Vertical HandOver (VHO). These functions must be distributed into the RRM and the CRRM entities as illustrated in Fig. 11.22. The CRRM/RRM interaction degree defines which entity manages each function. The different interaction degrees are [PRSADG05]:

- Low interaction degree: the RRM entities provide all the functionalities, and the CRRM entity only establishes some policies that configure the resource management behavior.

- Intermediate interaction degree: the CRRM entity manages the initial RAT selection and VHO functions. The local RRM entities provide some RRM measurements, such as the list of candidate cells for the different RATs and cell load measurements, so that the CRRM can take into account the resource availability in each RAT.
- High interaction degree: in this case, the CRRM entity is involved in most of the functionalities, leaving only the power control and scheduling for the RRM entities. Thus, CRRM is involved in each intra-system HO procedure requiring a more frequent measurement exchange. Similarly, joint congestion control mechanisms could be envisaged to avoid overload situations in any of the underlying RANs.
- Very high interaction degree: this approach introduces the joint scheduling in the CRRM entity. The RRM entities only manage the power control functions. This solution requires that the CRRM entity make decisions at a very short time scale in the order of milliseconds, with the possibility of executing frequent RAT changes for a given terminal.

In order to make proper decisions, the RRM and CRRM entities have access to a huge amount of spatial and temporal data. This data consists of counters and performance indicators, mainly generated by each BS. Since the number of these parameters is increasing, and the CRRM entity requires a high-level view of network performance, a common and integrated parameter that can evaluate radio resources availability and network conditions is required. In order to implement this important task, namely to provide a desired QoS, a Cost Function (CF) model was proposed in [SC07a]. The main success of this work was firstly the integration of a joint Cost Function (CF) for wireless heterogeneous networks, capable of being used by RRM and CRRM algorithms. Many, if not all, CRRM algorithms, policies, and strategies can be based on this CF as demonstrated in [SC07b], since all BSs and MTs will be marketed by their own cost on the network. Besides, the formulation can be also extended to encompass MIMO and Location Area (LA) processes [SKC08]. Thus, it is easy to compare and classify the most relevant nodes in the radio network, enabling the creation of candidate lists for a given criterion. The service priority scheme has an important impact on results, since this mechanism switch UE services to a given RAN, therefore, being responsible for the load distribution factor within the CRRM domain. The results based on this CF are promising, since CRRM policies based on it can enhance CRRM capabilities and sensibilities.

Another aspect that must be taken into account is that, with the concept of heterogeneous networks, users not only should be connected anywhere, anytime, but also they should be served with the best available connection, what can be only accomplished with the interworking of the different technologies. For that reason, the standardization bodies are doing their best to make the interworking possible. For instance, the 3GPP not only allows UMTS to interwork with GPRS (two 3GPP RATs) but also establishes the basis for a WLAN interworking (a non-3GPP RAT). In addition, IEEE association has recently published IEEE standard 802.21 Media Independent Handover (MIH). Although this standard has grabbed a lot of attention, few works have addressed the integration of an 802.21-enabled solution with existing 3GPP and non-3GPP networks. To fill this gap, [MES09] thoroughly investigated a 802.21-based realization of a representative multiradio management framework. For each functional block, the authors evaluated how 802.21 can be used and how it integrates with different RATs. The conclusions are that current WLAN and WiMAX networks (including the respective amendments) provide sufficient support for 802.21 functions. However, in 3GPP access networks, support for 802.21 services is still very limited. A certain lack of functionality or mismatch of services primitives can be observed with respect to measurement reporting and the support of discovery and registration. A provisioning of information about neighbor networks based on the 802.21 Information Service cannot be integrated in 3GPP RATs as easily as it is the case for WLAN or WiMAX. The most severe problem arises with respect to the exchange of signaling messages between MIH functions on the terminal side and in the access network, where currently no appropriate transport solution exists.

11.4.2 RAT Selection

In heterogeneous networks RAT selection is, without any doubt, the most important function implemented by the CRRM entity. RAT selection encompasses two additional functionalities, namely, the initial RAT selection and the VHO management. User mobility may make the best RAT or cell to which a user is connected change over time. Therefore, the initial selection might not be optimum in the future. In general, HO (or intra-system HO) is understood as the process of change of the cell the UE is transmitting to, but considering the same RAT. Otherwise, the VHO (or inter-system HO) also implies a RAT change.

Current cellular system standards include the necessary mechanisms to perform intra-system HOs. An HO occurs when the signal quality of the current cell decreases under the quality perceived from a contiguous cell. Usually a hysteresis margin is used to avoid continuous changes near the cell boundaries. VHO are similar but including RAT changes. The new RAT can be selected with the help of an initial RAT selection algorithm, just like if a new call was asking for admission.

In [SC09, SC10] the authors updated a simple analytical model, initially used for intra-system HOs, being proposed to be used in the VHO problematic, by assuming the existence of RRM and CRRM entities. The results confirmed the validity of this analytical model. With its usage, it was shown that high bit-rate RAN is the RAN group that produces more impact on other groups, by generating more signaling traffic related with CRRM functionalities, triggering and managing VHOs. Results also demonstrated that data oriented services have more impact on the overall CRRM performance varying by a factor from 2 up to 5, depending on the output parameter under study.

However, VHO can be also executed to perform load balancing. Indeed, although there is an efficiency charge for each RAT to handle different services, balancing the load between multiple systems allows a better utilization of the available radio resources, and more importantly, maintains the QoS provided to the end users. This

Fig. 11.23 Suitability for the load balancing selection algorithm [CVR+08] (©2009 IEEE, reproduced with permission)

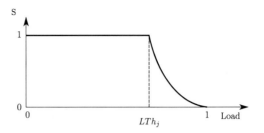

multiplexing gain, due to combined systems, in contrast to disjoint ones, can be achieved if cross-system information pertaining to each radio access technology is taken into account along the system functioning, not only in the initial attachment of users. In [MCR+08, CVR+08] the authors proposed a RAT selection based on load suitability where the systems are HSDPA and WLAN, under delay constraints services. The concept of suitability was used in terms of preferred access system to accommodate the service, but this concept suitability can change as load increase, in order to maintain the QoS of delay constraints services. Thus, the goal was to optimize the load in each RAT, without loss of QoS guarantee, or with a gain in QoS provisioning. The rationale behind the proposed algorithm was the following: a preferable RAT is selected by default to handle a service, assuming in this case that the service traffic is flexible and can be handled by more than one RAT. Besides, an empirical algorithm for load balancing among cells of different RATs was proposed when a new call is requested. The algorithm is targeted to flexible traffic and imposes certain flexibility on the system, meaning that the service can be held by each RAT. The algorithm for the suitability, S, is expressed by the following equation and depicted graphically in Fig. 11.23:

$$S(L(cell_{i,j})) = \begin{cases} 1 & \text{if } L(cell_{i,j}) \leq LTh_j, \\ (\frac{1-L(cell_{i,j})}{1-LTh_j})^2 & \text{if } L(cell_{i,j}) > LTh_j, \end{cases} \quad (11.17)$$

where $cell_{i,j}$ represents the cell or access point i belonging to the RAT j, $L(cell_{i,j})$ is the normalized load in the $cell_{i,j}$, LTh_j is the load threshold for RAT j, and $S(L(cell_{i,j}))$ is the suitability value for accepting a new user in the $cell_{i,j}$.

Figure 11.24 compares the overall QoS goodput as a function of the offered load in the absence and presence of CRRM with the Suitability algorithm (where the preferable RAT is HSDPA). The QoS goodput increases with offered load until the maximum HSDPA system capacity is reached. The QoS goodput does not exceed 1.7 Mbps and starts to tail-off at around 1.5 Mbps. This effect is due to the use of the MaxCIR scheduler (it will always provide services to users in the near vicinity of the base station). The authors mapped the 1.7-Mbps goodput to a $Load_{\text{Threshold}} = 0.6$. Using this threshold and comparing the goodput "with" and "without" CRRM the observable gain with 60 users is $2700/1500 = 80$.

It is also worth noting that in the framework of COST 2100 action, some practical work has been carried out concerning heterogeneous network. For instance, in [Nav08] a project is described that develops a seamless connection system for mobile users that allows the reselection between WLAN/GPRS/CDMA2000/WiMAX

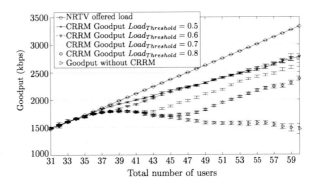

Fig. 11.24 CRRM in heterogeneous networks [CVR+08] (©2009 IEEE, reproduced with permission)

networks. The same authors presented in [GEN+10] an analysis of vertical handover using a simulation tool, NCTUns, in order to compare different algorithms and approaches for vertical handovers proposed for 4G technologies. The proposed algorithm evaluates some parameters included in a specific cost function, such as the network bandwidth, economic cost of the connection, and the mobile node mobility, that allows selecting the optimum candidate network.

11.4.3 Joint Dynamic Resource Allocation

Two main approaches to the resource allocation exist, namely, Fixed Resource Allocation (FRA) and DRA. The first one is the most extended in the classic circuit-switched networks conceived for voice conversations, which is the case of GPRS. The second one is the best suited for present and future packet-switched networks, where the bursty nature of new services traffic makes the Fixed Resource Allocation (FRA) schemes underutilize the available resources. Nevertheless, the DRA approach requires a smart scheduling algorithm to guarantee the users' satisfaction in terms of QoS. The objective of the DRA algorithm is to select, for each user, the optimal amount of radio resources to be allocated. Several DRA algorithms for a unique RAT have been proposed in the literature. From low to high CRRM/RRM interaction degrees, the DRA is performed separately in each RAT, but within a very high interaction degree, a Joint Dynamic Resource Allocation (JDRA) is also possible.

In [CMMSC07] the authors proposed an extremely efficient JDRA algorithm to decide on the best RAT and resource quantity to allocate to users. The complexity of the optimization problem requires the usage of advanced techniques to find, at least, a suboptimum solution. For example, linear programming or game theory are two of these techniques. Nevertheless, Hopfield Neural Networks (HNNs) have been identified as fast hardware optimizers that can obtain a valid solution in a few microseconds. This fast response is a consequence of the simplicity of each individual neuron and their parallel interworking. Therefore, problems that are more complex need more neurons, i.e., more hardware, but maintain the fast response of simpler problems. This feature makes HNNs be the best candidates for suboptimal and real-time schedulers. In [CMMSC07] a new HNNs-based algorithm to jointly distribute among users the set of resources of all RATs available in a heterogeneous wireless system is proposed. This algorithm takes simultaneously into account the kind of service and specific QoS requirements of each user and the resource availability and characteristics of each RAT. The algorithm was evaluated through simulations in a basic WLAN scenario with two access points and mobile users. As compared with other static and dynamic strategies, the HNNs-based JDRA algorithm was always preferred since it was able to adapt the resource distribution to the variable scenario conditions.

11.5 Trends and Open Research Topics in RRM

The majority of the new proposals regarding Radio Resource Management in Access Networks deal somehow with breaking the formal layered architectures and take decisions based on the knowledge at the MAC and upper layers of both the channel and the spectrum usage status. These ideas are developed around the concepts of Cognitive Radio, Cross-Layer Systems, and Hybrid Networks.

11.5.1 Cross-Layer Systems

Several publications have reviewed the literature of this subject and attempted to classify the approaches to it found therein. After a review of the definitions written in recent literature, in [Bur07] it is assumed that the "protocol design by the violation of a reference layered communication architecture is cross-layer design with respect to the particular layered architecture." The meaning of layered architecture in this context firstly is that there can be no direct communication between nonadjacent layers, and secondly that a higher layer calling on the services of a lower one can take no account of how these services are provided.

More precisely, four main aspects on protocols design are identified as cross-layer techniques [Bur07]:

- Cross-layer design: the protocol of one or more layers is designed taking into account the detailed design of other layers.
- Cross-layer linkage: layers communicate with one another, rather than by the conventional mechanism of passing information up or down the stack.
- Cross-layer integration: complete merging of one or more layers.
- Cross-layer analysis, in which the performance of a given layer is evaluated taking into account. The performance of other layers and without necessarily violating the architecture at all.

The fundamental advantage, of course, is that cross-layer techniques remove restrictions on the implementation of communication systems. Any architecture represents a set of constraints on the system; allowing violations of it introduces further degrees of freedom. It is well known that the joint performance of many functions can be improved by implementing them jointly, and of course the layered architecture forbids this if they reside on different layers.

The effect of this is particularly important in wireless systems, while layered architectures like the OSI architecture, or TCP/IP, were originally developed for wired networks. In wireless systems the channel is much more variable, and it is important for systems to be able to adapt to the nature of the channel, which often involves passing information between layers. Moreover, the channel is often more hostile, and the performance gains available from joint implementation of functions may be more important to achieve good performance overall. Further, the wireless channel provides opportunities for interactions between layers due to the broadcast nature of the medium which are not usually present in wired systems: for example, cooperation between different nodes in a network is much easier, because direct wired links do not have to be provided between them.

There are also disadvantages to cross-layer design—inevitably, since the rules of a layered architecture were introduced for a reason, and their violation clearly may have negative consequences. Layered architectures were introduced to allow independent design of different layers and to ensure stability when different parts of a protocol were designed by different suppliers. Hence, if cross-layer techniques are applied, much more care must be taken to ensure stability, especially that unintended coupling (as well as the intended linkages) does not occur between functions on different layers.

11.5.1.1 Cross-Layer Resource Allocation and Scheduling in Cellular Systems

Joint implementations of scheduling and resource allocation (RA) algorithms have been proposed to offer optimal distribution of resources among multiple users in cellular systems. However, the practical implementation of such an approach is unfeasible since the optimization space is too large to be explored in the inter-scheduling intervals (in the order of milliseconds). Some approximations to fully joint distribution of resources based on splitting the scheduling and the RA are defined, like in [CV07a], where the resource allocator is a completely air-interface aware module, and the scheduler is an air-interface unaware module. This strategy is suboptimal due to the functional split, but it allows a cross-layer approach due to the tight interaction between the scheduler, which also deals with the application level, and the resource allocator, which interacts with the physical layer.

In MC-CDMA, resource allocation deals with assigning the amount of bits that could be transmitted (i.e., the transport block size), the number of RUs required, the modulation and channel coding scheme, and the transmit power. This decision can be taken either from a cross layer fully optimized to very simple approaches. The Adaptive Resource Allocation (ARA) is an example of an optimized strategy where all allocation parameters are jointly considered [CV07a]. This strategy identifies for each user the best "group of frequencies" (GOF), every GOF having at least one slot and spreading code available in the resource budget, and

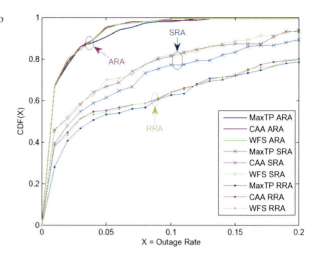

Fig. 11.25 CDF of the video users outage rate depending on scheduling and resource allocation policies [CV07a] (©2007 IEEE, reproduced with permission)

taking the decisions based on the estimation of the "normalized SNIR" and the channel quality. In Fig. 11.25 the ARA strategy is compared with a Simple RA algorithm (same as ARA but without link adaptation) and a Random RA algorithm (GOF is randomly selected), showing that ARA outperforms the rest of approaches, despite the scheduling strategy, in particular for the MC-CDMA video users case.

As for the scheduler, several strategies have been proposed in cross-layer systems, based on the knowledge of the channel state: the Channel- and Application-Aware (CAA) scheduler [CV07a], defined according to three parameters, the Time-to-Deadline (TD), the Type-of-Service (TS), and the Channel State (CS); the Maximum Throughput (MaxTP), where the flow of the user with the best receiving conditions is selected at any time, maximizing the overall system throughput; and the Wireless Fair Service (WFS), whose goal is to reach long-term fairness among users. As seen in Fig. 11.25, MaxTP is less efficient compared to the other two approaches, but differences are not so significant. Nevertheless, in Fig. 11.26 [CV07a] it is shown that for heavy loaded systems, it starts to make sense to the use of the cross-layer CAA scheduler compared to the other approaches that degrade very rapidly.

11.5.1.2 Cross Layer in Wireless Ad Hoc Networks

In the case of Ad Hoc Wireless Networks, some works have proposed to use cross-layer algorithms, mainly for scheduling [CTV08]. In the particular case of an Ad Hoc Network, the Sink nodes play the role of collecting, elaborating, and sending the packets to a control unit that takes decisions accordingly. Some results have been obtained for hierarchical ad hoc video networks, like in [CTV08], where an emergency environment with nodes equipped with camera devices is analyzed. In the mentioned work, a cross-layer strategy is used, so the scheduling algorithm takes

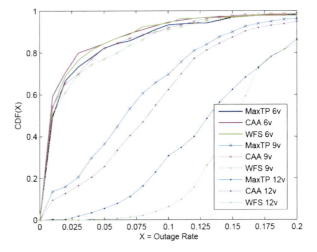

Fig. 11.26 Impact of the traffic load on different scheduling policies [CV07a] (©2007 IEEE, reproduced with permission)

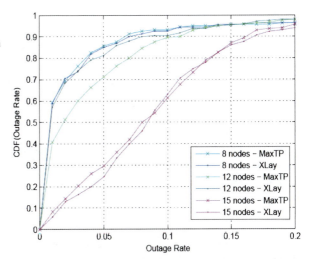

Fig. 11.27 CDF of the outage rate depending on the number of nodes: comparison with the maximum throughput

into account information coming from both the physical layer and the application level. In particular, firstly the scheduler identifies the channel state (CS) according to the estimated average channel gain coming from the physical layer. Then, it selects the node to be allocated, also taking into account the time to deadline (TD) of the video packet considered, computed as the difference (in number of frames) between the deadline of the HOL packet in the radio link buffer and the current system time. This approach is evaluated in Fig. 11.27 for different numbers of video nodes, showing that cross-layer strategy outperforms the Maximum Throughput scheduler in case of heavily loaded system, but performance is similar in case of fully loaded system.

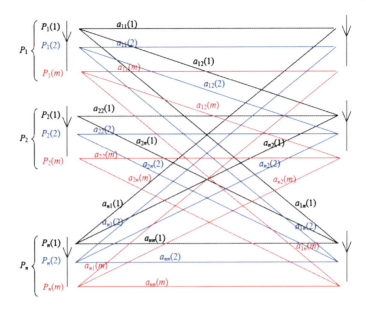

Fig. 11.28 System model (*colors* indicate separate non-interfering channels)

11.5.2 Cognitive Radio Resource Management

Cognitive radio has been attracting a great deal of attention as a means of making available large amounts of currently underused spectrum for wireless communications. The concept allows "cognitive users" to transmit in spectrum which is currently unused (even though it might be assigned by a regulator), using spectrum sensing to detect other users (including "incumbent users" that have explicit usage rights to the spectrum).

One of the focuses of application of cognitive radio in Resource Management is the dynamic power allocation between channels. In the conventional view of cognitive radio channel assignment is a "hard decision"; that is, a user is either assigned a channel (usually a single channel) or not and is expected to transmit all its power in the assigned channel. However there may be multiple channels available, subject to varying degrees of interference, and optimum capacity may then be obtained by distributing the power between them. In [Bur09a, Bur09b] the insights from information theory are used to compare the total capacity available using a fully distributed power allocation scheme with an optimum scheme based on perfect knowledge of all links. The expressions for optimum allocation assuming a system model like in Fig. 11.28 can be obtained as an extension of the water-filling power allocation problem, which can be solved iteratively.

The optimization for a certain user l consists in finding the power to be allocated $Pl(k)$ for all channels of index k with nonzero power such that

$$\frac{N + A_l(k)}{a_{ll}(k)} + P_l(k) = \frac{1}{c_l + B_l(k)}. \tag{11.18}$$

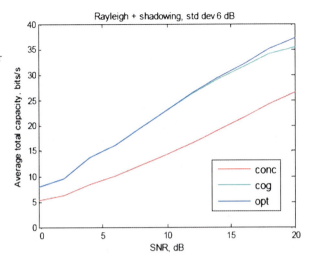

Fig. 11.29 Average total capacity versus signal-to-noise ratio for optimum and cognitive power allocation and for conventional channel allocation ('conc') [Bur09a] (©2009 IEEE, reproduced with permission)

This is equivalent to water-filling, with $(N + Ai(k)/a_{ll}(k))$ taking the role of the nonwhite noise-plus-interference term in Shannon's water-filling approach. If the term B is ignored, the power allocation for the lth user is optimized by water-filling, taking into account the interference from other users in the m channels. However, to maximize the overall capacity, it has also to be taken into account the effect of the power allocation in the lth channel on the capacity of the other users. The B term provides this allowance, being a sort of sensitivity to interference from the lth user. The right-hand side, $1/(c_l + B_l(k))$, takes the role of the "water level": however it is no longer constant across the channels, which rather detracts from the water-filling analogy.

The application of this approach to a practical case [Bur09a] relays on the fact that the term B is most sensitive to the product of terms $a_{li}(k)$ (gain of the channel from transmitter l to receiver i) and $P_i(k)$ (power allocated to transmitter i). On the assumption of reciprocity, the use of time-division duplexing, and that the power transmitted on the return channel is the same as on the outward (also assuming reciprocity of the wanted link), this product can be estimated from the interference received during the return transmission. Hence, for cognitive modified water-filling, we will assume that this is known to user l and hence that B can be estimated by multiplying this by a constant value related to the signal-to-interference-plus-noise ratio at the other users.

Numerical results [Bur09a] show the capacity available in a cognitive radio system through cognitive allocation of power between channels available to a user, compared to an optimum power allocation algorithm and more conventional cognitive channel allocation, where each user transmits only in their best channel. Figure 11.29 shows that a significant capacity increase, in excess of 50%, is available using the power allocation algorithm and moreover that a cognitive power allocation algorithm is possible, operating in a distributed fashion, which approaches very closely to the optimum capacity.

A second aspect of Cognitive Radio (CR) applied to Access Networks is the Dynamic Spectrum Access (DSA). DSA is aimed to facilitate accessing and exploiting preallocated, but unused, radio spectrum dynamically.

One of the most important factors in CR and DSA networks is how to detect the available spectrum. Sensing techniques can be divided into three major categories [GPE09]. First, Matched filter is the optimal way which maximizes Signal-to-Noise Ratio (SNR), demodulates the primary signal already known to the CR, and is then able to read detailed information about the primary system. Matched filter therefore requires detailed information about the primary systems, which might be stored in the CR memory. The CR needs to achieve coherency with the primary system to be synchronized in both time and frequency. In OFDMA transmissions with Mobile WiMAX as the primary system, the secondary system could decode the Preamble, FCH, and MAPs in the OFDMA frame to obtain knowledge about the available resource elements. In cases with horizontal striping or rectangular scheduling, the FCH could be used to survey the subcarrier to subchannel mapping to be able to operate in white holes in the frequency domain. When vertical striping is used, the secondary system could simply utilize the available time period.

Second, Energy detection is a simpler approach and is considered to be more unreliable in that it is susceptible to uncertainty in noise power. If the probability of false alarm is high, the spectrum band will often be detected as occupied by the PUs although it actually is not. This will reduce spectrum utilization rate. The delay between detection of the primary signal to vacation of the spectrum will cause interference to the primary system. Energy detection could be utilized in the time domain for vertical striping but seems to be a difficult task for horizontal striping due to hardware requirements on the amount of sensors and processing power required.

Third, Cyclostationary Feature detection analyzes the periodicity of the statistics, mean, and autocorrelation of the primary signal. Available periods in the time domain can be easily detected when vertical striping is used. Variable consecutiveness, defined as a measure of difference in occupancy among consecutive OFDMA frames over time, might be important to model when defining medium access schemes for secondary systems, especially when considering sensing interval. Consecutiveness of subcarrier usage could be detected when horizontal striping is used, but this would require huge amounts of sensors when 1024 subcarriers are used in a 10-MHz channel.

A common problem for the three sensing methods mentioned above is that the signal might be too weak to be detected, and cooperative sensing [GPE09] between several sensing nodes might then be a more optimal solution in terms of sensing reliability. Operator assistance would otherwise be the optimal way for a secondary system to utilize white holes, either giving real time information about scheduling and spectrum usage over a wired connection or using beaconing, which means that information is sent when the spectrum is occupied and available.

These concepts have been tested for OFDMA radio access technologies [GPE09] by simulating the primary spectrum usage in the OFDMA system Mobile WiMAX. Figure 11.30 shows the DL occupancy distribution map in both time and frequency domain represented by temperature map.

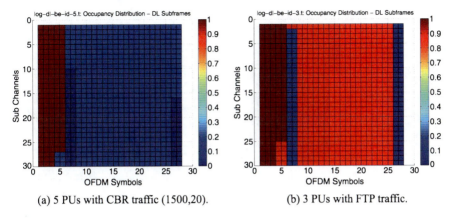

(a) 5 PUs with CBR traffic (1500,20). (b) 3 PUs with FTP traffic.

Fig. 11.30 DL subframe: the average occupancy distribution as the temperature map

These results [GPE09] will be useful as guidelines when designing medium access control (MAC) protocols and schemes for secondary systems that exploit white holes in primary OFDMA networks.

11.5.3 Radio Resource Management in Hybrid Hierarchical Architectures

The Hybrid Hierarchical Architecture (HHA) represents a particular case of Wireless Hybrid Network where sensor nodes transmit their samples to an infrastructure network through multiple hops. In the HHA, gateway terminals implementing both cellular and infrastructureless air interfaces allow integration of the two separate paradigms characterizing the Wireless Sensor Network (WSN) and the cellular network. This type of Architecture requires both the characterization of the traffic canalized by the mobile gateway and provided to the infrastructure network and the design of the scheduling techniques implemented at the infrastructure side.

A scenario of such a hierarchical network where an IEEE 802.15.4 WSN, organized in a tree-based topology, is connected, through a mobile gateway, to an infrastructure network using a cellular UMTS air interface has been studied in [BV08]. The scenario is composed of three levels, namely levels 0, 1 and 2: an IEEE 802.15.4 compliant WSN (level 2), which has to periodically transmit data taken from the environment to a sink, that is, the mobile gateway (level 1); the latter must forward data received to a UMTS radio access port (level 0). In this scenario a specific issue arises: the UMTS scheduler needs to allocate radio resources to the mobile gateway that generates data according to the inputs received from the WSN.

Figure 11.32 shows some statistics of the traffic generated at the gateway in terms of the WSN parameters, namely:

- Superframe Order (SO), which defines the interval of time in which there can be arrivals at the gateway;

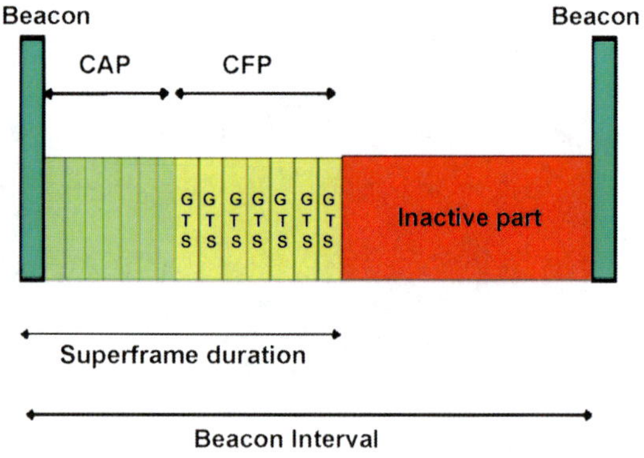

Fig. 11.31 Superframe structure used in the IEEE 802.15.4 Wireless Sensor Network [BV08] (©2008 VDE, reproduced with permission)

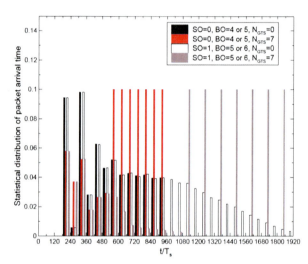

Fig. 11.32 Statistical of the packet arrival time for $N1 = 10$, for different values of SO, BO, and $NGTS$ [BV08] (©2008 VDE, reproduced with permission)

- Beacon Order (BO) of the gateway superframe (which depends on the number of routers N_R) and defines the Beacon Interval (BI), that is, the periodicity of the arrivals at the gateway (a sequence of arrivals every BI sec);
- N_1, which is the number of nodes that generate traffic;
- N_{GTS}, the number of Guaranteed Time Slots that can be allocated by the coordinator to specific nodes within a superframe (see Fig. 11.31).

References

[3GP98]   3GPP. Selection procedures for the choice of radio transmission technologies of the UMTS. TS 30.03, 3rd Generation Partnership Project (3GPP), March 1998.
[3GP04a] 3GPP. Feasibility study for enhanced uplink for UTRA FDD. TR 25.896, 3rd Generation Partnership Project (3GPP), April 2004.
[3GP04b] 3GPP. IP transport in UTRAN. TR 25.933, 3rd Generation Partnership Project (3GPP), January 2004.

11 Resource Management in 4G Networks

[3GP07] 3GPP. Quality of service QoS concept and architecture. TS 23.107, 3rd Generation Partnership Project (3GPP), September 2007.

[AG07] G. E. Øien and A. Gjendemsjø. Binary power control for multi-cell capacity maximization. In *COST 2100 2nd MCM*, Lisbon, Portugal, 2007. [Available as TD(07)218].

[AH04] J. Antoniou and G. Hadjipollas. Final report on simulation of enhanced UMTS. Technical report, IST SEACORN CEC Deliverable 34900/UCY/DS/046/f1, IST Central Office, Brussels, Belgium, March 2004.

[AKR+01] M. Andrews, K. Kumaran, K. Ramanan, A. Stoylar, and P. Whiting. Providing quality of service over a shared wireless link. *IEEE Commun. Mag.*, 39(2):150–154, 2001.

[Amb04] Network context management: Concepts, scenarios and analysis of state of the art. Technical Report AN-R6.1, Ambient Networks Project, 2004.

[Aro06] First report on AROMA algorithms and simulation results. Technical Report D09, Aroma Project, 2006.

[BCC+07] E. Biglieri, R. Calderbank, A. Constantinides, A. Goldsmidth, A. Paulraj, and H. V. Poor. *MIMO Wireless Communications*. Cambridge University Press, Cambridge, 2007.

[BEG09] M. Bohge, A. Eisenblatter, and H. Geerdes. An interference coupling model for adaptive partial frequency reuse in OFDMA/LTE networks. Technical report, COST 2100, Braunschweig, Germany, February 2009. [Available as TD(09)757].

[BKVH07] S. Boyd, S. J. Kim, L. Vandenberghe, and A. Hassibi. A tutorial on geometric programming. Optimization and Engineering 2007. [Online] http://www.stanford.edu/~boyd/papers/gp_tutorial.html, 2007.

[BS98] S. A. Brian and B. Smith. An experiment to characterize videos stored on the web. In *Proceedings of the 5th Multimedia Computing and Networking (MMCN)*, pages 166–178, USA, January 1998.

[Bur07] A. G. Burr. Cross-layer issues in COST 2100—a WG1 view. In *COST 2100 Duisburg Meeting*, pages 1–7, September 2007. [TD(07)317].

[Bur09a] A. G. Burr. Capacity of cognitive channel and power allocation. In *Information Theory Workshop, 2009. ITW 2009. IEEE*, pages 510–514, 11–16 October 2009. [Also available as TD(07)872].

[Bur09b] A. G. Burr. Cognitive channel and power allocation: information theoretic bounds. In *COST 2100 Braunschweig Meeting*, pages 1–13, February 2009. [TD(08)746].

[BV08] C. Buratti and R. Verdone. A hybrid hierarchical architecture: from wireless sensors to the fixed infrastructure. In *Proc. 14th European Wireless Conference (EW 2008)*, pages 1–7, June 2008. [Also available as TD(08)469].

[CGV] V. Corvino, D. Gesbert, and R. Verdone. A novel distributed interference mitigation technique using power planning. In *Proc. IEEE WCNC 2009*.

[CJK09a] T. Celcer, T. Javornik, and G. Kandus. Fairness oriented scheduling algorithm with QoS support for broadband MIMO systems with heterogeneous traffic. In *COST 2100 9th MCM*, Vienna, Austria, 2009. [Available as TD(09)927].

[CJK09b] T. Celcer, T. Javornik, and G. Kandus. Resource allocation algorithms applying multiuser spatial multiplexing in MIMO systems. In *COST 2100 8th MCM*, Valencia, Spain, 2009. [Available as TD(09)849].

[CKV08] V. Corvino, W. Karner, and R. Verdone. Error prediction based frequency selection in MC-CDMA resource allocation. In *COST 2100 4th MCM*, Wroclaw, Poland, 2008. [Available as TD(08)475].

[CMLR10] S. Caban, C. Mehlführer, G. Lechner, and M. Rupp. Testbedding MIMO HSDPA and WiMAX. In *COST 2100 10th MCM*, Athens, Greece, 2010. [Available as TD(10)10046].

[CMMSC07] D. Calabuig, J. F. Monserrat, D. Martin-Sacristan, and N. Cardona. Joint dynamic resource allocation for coupled heterogeneous wireless networks. A new Hopfield neural network-based approach. In *COST 2100 Duisburg Meeting*, pages 1–4, September 2007. [TD(07)382].

[CTV07]   V. Corvino, V. Tralli, and R. Verdone. Cross-layer resource allocation for MC-CDMA. In *Proc. ISWCS 2007*, pages 267–271, Trondheim, Norway, October 2007.
[CTV08]   V. Corvino, V. Tralli, and R. Verdone. Cross-layer scheduling for multiple video streams over a hierarchical emergency-deployed network. In *COST 2100 Wroclaw Meeting*, pages 1–7, February 2008. [TD(08)401].
[CV07a]   V. Corvino and R. Verdone. Cross-layer resource allocation for MC-CDMA. In *Proc. 4th IEEE ISWCS Conf.*, pages 267–271, Trondheim, Norway, October 2007. [Also available as TD(07)300].
[CV07b]   V. Corvino and R. Verdone. Scheduling techniques of mixed traffic over MC-CDMA under varying load and channel conditions. In *COST 2100 2nd MCM*, Lisbon, Portugal, 2007. [Available as TD(07)201].
[CVR+08]   O. Cabral, F. Velez, J. Rodriguez, V. Monteiro, A. Gameiro, and N. R. Prasad. Optimal load suitability based RAT selection for HSDPA and IEEE 802.11e. In *Proc. Wireless VITAE 2009*, pages 1–8, May 2008. [Also available as TD(10)11095].
[Dai04]   QoS architecture and protocol design specification. Technical Report D321, Daidalos Project, 2004.
[EC05]   S. E. Elayoubi and T. Chahed. *Admission Control in the Downlink of WCDMA/UMTS*. Springer, Berlin 2005.
[ECH04]   S. E. Elayoubi, T. Chahed, and G. Hébuterne. Connection admission control in UMTS in the presence of shared channels. *Computer Communications*, 27, June 2004.
[Faz07]   P. Fazekas. Testbed and research at the mobile innovation center at the Budapest university of technology and economics. In *COST 2100 2nd MCM*, Lisbon, Portugal, 2007. [Available as TD(07)249].
[FCX+03]   L. Ferreira, L. M. Correia, D. Xavier, I. Vasconcelos, and E. R. Fledderus. Final report on traffic estimation and services charcterisation. Technical Report D1.4, IST-2000-28088 MOMENTUM, May 2003.
[Fer03]   J. Ferreira. Final report on traffic estimation and services characterization. Technical report, IST SEACORN CEC Deliverable 34900/PTIN/DS/014/d6, IST Central Office, Brussels, Belgium, September 2003.
[FKWK09]   A. Fernekeß, A. Klein, B. Wegmann, and D. Karl. Semi-analytical cell throughput estimation of OFDMA systems with scheduling based on SINR distributions. Technical report, COST 2100, Valencia, Spain, May 2009. [Available as TD(09)806].
[FTV00]   M. Ferraccioli, V. Tralli, and R. Verdone. Channel based adaptive resource allocation at the MAC layer in UMTS TD-CDMA systems. In *52nd Vehicular Technology Conference, 2000*. IEEE VTS-Fall VTC 2000, vol. 6, pages 2549–2555, September 2000.
[FW97]   L. Foulds and J. Wilson. A variation of the generalized assignment problem arising in the new Zealand dairy industry. *Annals of Operations Research, Springer Netherlands*, 69:105–114, 1997.
[GC10]   A. Gotsis and P. Constantinou. Adaptive multi-user OFDM resource allocation for heterogeneous data traffic: the uniform power loading case. Technical report, COST 2100, Athens, Greece, February 2010. [Available as TD(10)10079].
[GDC09]   A. Gotsis, D. Domnakos, and P. Constantinou. Joint resource allocation and user scheduling for single-cell OFDMA networks supporting multiple data traffic classes. Technical report, COST 2100, Valencia, Spain, May 2009. [Available as TD(09)828].
[GEN+10]   A. Garcia, L. Escobar, A. Navarro, A. Arteaga, and A. Vasquez. Analysis of vertical handoff algorithms with NCTUns. In *COST 2100 Aalborg Meeting*, pages 1–8, June 2010. [TD(10)11035].
[GGLRO10]   D. Gonzalez, M. Garcia-Lozano, S. Ruiz, and J. Olmos. On the performance evaluation of static inter-cell interference coordination in realistic cellular layouts. In *Proc. MONAMI Conf.*, 2010. [Also available as TD(10)11053].
[GJ03]   E. Gustafsson and A. Jonsson. Always best connected. *IEEE Wireless Commun. Mag.*, 10(1):49–55, 2003.

[GKC09] A. Gotsis, N. Koutsokeras, and P. Constantinou. Dynamic subchannel and slot allocation for OFDMA networks supporting mixed traffic: upper bound and a heuristic algorithm. *IEEE Commun. Lett.*, 13(8):576–578 2009. [Also available as TD(07)384].

[GLG09] V. Garcia, N. Lebedev, and J. M. Gorce. Multi-cell processing for uniform capacity improvement in full spectral reuse system. In *Proc. COGIS 2009*, Paris, France, 2009. [Also available as TD(09)989].

[GLG10] V. Garcia, N. Lebedev, and J. M. Gorce. Capacity-fairness trade-off in small cells networks using coordinated multi-cell processing. In *COST 2100 10th MCM*, Athens, Greece, 2010. [Available as TD(10)10004].

[GPE09] P. Gronsund, H. N. Pham, and P. E. Engelstad. Towards dynamic spectrum access in primary OFDMA systems. In *2009 IEEE 20th International Symposium on Personal, Indoor and Mobile Radio Communications*, pages 848–852, 13–16 September 2009. [Also available as TD(07)875].

[GRO$^+$09] D. Gonzalez, S. Ruiz, J. Olmos, M. Garcia-Lozano, and A. Serra. Link and system level simulation of downlink LTE. Technical report, COST 2100, Braunschweig, Germany, February 2009. [Available as TD(09)868].

[HK07] A. Hecker and T. Kürner. Application of classification and regression trees for paging traffic prediction in LAC planning. In T. Kürner, editor, *Proc. VTC2007-Spring Vehicular Technology Conference IEEE 65th*, pages 874–878, Dublin, Ireland, April 2007. doi:10.1109/VETECS.2007.189.

[HKG08] P. P. Hasselbach, A. Klein, and I. Gaspard. Dynamic resource assignment DRA with minimum outage in cellular mobile radio networks. In *Prov. IEEE Veh. Tech. Conf. (VTC2008-Spring)*, May 2008. [Also available as TD(08)429].

[Hua05] G. Huawei. 3GPP TSG RAN WG1 Meeting #41. *R1-050507—Soft Frequency Reuse Scheme for UTRAN LTE*. 3GPP, 2005.

[IEE05] IEEE 802.16e-2005: Air interface for fixed and mobile broadband wireless access systems, 2005.

[JPV] J. M. Juárez, R. R. Paulo, and F. J. Velez. Modelling and simulation of B3G multi-service traffic in the presence of mobility.

[KCM07] K. Kansanen, P. Chaporkar, and R. R. Müller. Group proportional fair scheduling. In *COST 2100 3rd MCM*, Duisburg, Germany, 2007. [Available as TD(07)385].

[KH95] R. Knopp and P. Humblet. Information capacity and power control in single cell multi user communications. In *1995 IEEE International Conference on Communications, 1995*. ICC '95 Seattle, 'Gateway to Globalization', vol. 1, pages 331–335, June 1995.

[KH05] T. Kürner and A. Hecker. Performance of traffic and mobility models for location area code planning. In A. Hecker, editor, *Proc. VTC 2005-Spring Vehicular Technology Conference 2005 IEEE 61st*, vol. 4, pages 2111–2115, Stockholm, Sweden, May 2005. doi:10.1109/VETECS.2005.1543707.

[KLL01] A. Klemm, C. Lindemann, and M. Lohmann. Traffic modeling and characterization for UMTS networks. In *Proceedings of the 44th Global Telecommunications Conference, 2001. GLOBECOM '01. IEEE*, vol. 3, pages 1741–1746, San Antonio, TX, USA, November 2001. doi:10.1109/GLOCOM.2001.965876.

[KNR06] W. Karner, O. Nemethova, and M. Rupp. Link error prediction based cross-layer scheduling for video streaming over UMTS. In *Proc. 15th IST Summit on Mobile and Wireless Commun.*, June 2006.

[KNRW07] W. Karner, O. Nemethova, M. Rupp, and C. Weidmann. Distortion minimizing network-aware scheduling for UMTS video streaming. In *COST 2100 3rd MCM*, Duisburg, Germany, 2007. [Available as TD(07)304].

[KØG07] S. G. Kiani, G. E. Øien, and D. Gesbert. Maximizing multi-cell capacity using distributed power allocation and scheduling. In *Proc. WCNC 2007—IEEE Wireless Commun. and Networking Conf.* 2007.

[KWBM11]   M. Kaschub, T. Werthmann, C. M. Blankenhorn, and C. F. Mueller. Interference mitigation by distributed beam forming optimization. *Journal of RF-Engineering and Telecommunications*, September/October 2010. [Also available as TD(10)11036].

[KWMM10]   B. Krasniqil, M. Wolkerstorfer, C. Mehlfuhrer, and C. F. Mecklenbrauker. Weighted sum-rate maximization for two users in partial frequency reuse cellular networks. In *COST 2100 11th MCM*, Aalborg, Denmark, 2010. [Available as TD(10)11022].

[LJSB05]   G. Liebl, H. Jenkac, T. Stockhammer, and C. Buchner. Joint buffer management and scheduling for wireless video streaming. In *Proc. ICN 2005*, April 2005.

[LPS09]   B. G. Lee, D. Park, and H. Seo. *Wireless Communication Resource Management*. Wiley, New York 2009.

[LS02]   F. Y. Li and N. Stol. QoS provisioning using traffic shaping and policing in 3rd-generation wireless networks. In *Proceedings of the 3rd Wireless Communications and Networking Conference, 2002. WCNC2002. 2002 IEEE*, vol. 1, pages 139–143, Orlando, FL, USA, March 2002. doi:10.1109/WCNC.2002.993478.

[MCR+08]   V. Monteiro, O. Cabral, J. Rodriguez, F. Velez, and A. Gameiro. HSDPA/WiFi RAT selection based on load suitability. In *Proc. ICT-MobileSummit 2008—ICT Mobile and Wireless Summit*, pages 1–8, June 2008. [Also available as TD(08)672].

[MES09]   C. M. Mueller, H. Eckhardt, and R. Sigle. Realization aspects of multi-radio management based on IEEE 802.21. In *Proc. ICWWC 2009—7th International Conference on Wired/Wireless Communications*, pages 1–10, May 2009. [Also available as TD(09)825].

[MSGC07]   J. F. Monserrat, D. M. Sacristán, D. Gozálvez, and N. Cardona. SPHERE—a simulation tool for CRRM investigations. In *COST 2100 2nd MCM*, Lisbon, Portugal, 2007. [Available as TD(07)244].

[MT90]   S. Martello and P. Toth. *Knapsack Problems: Algorithms and Computer Implementations*. Wiley, New York, 1990.

[Mue10]   C. Mueller. On the importance of realistic traffic models for wireless network evaluations. In *COST 2100 12th MCM*, Bologna, Italy, 2010. [Available as TD(10)12039].

[MW00]   J. Mo and J. Walrand. Fair end-to-end window-based congestion control. *IEEE/ACM Trans. Networking*, 8(5):556–567, 2000.

[Nav08]   A. Navarro. A crosslayer system for automatic reselection between WiFi, UMTS, GPRS and WiMax. In *COST 2100 Trondheim Meeting*, pages 1–10, June 2008. [TD(08)508].

[NP06]   M. Neophytou and A. Pitsillides. Hybrid CAC for MBMS-enabled 3G UMTS networks. September 2006.

[OSR+09]   J. Olmos, A. Serra, S. Ruiz, M. Garcia-Lozano, and D. Gonzalez. Link level simulator for LTE downlink. Technical report, COST 2100, Braunschweig, Germany, February 2009. [Available as TD(09)779].

[PBS+10]   F. Pantisano, M. Bennis, W. Saad, R. Verdone, and M. Latva-aho. Interference management in femtocell networks using distributed opportunistic cooperation. In *COST 2100 12th MCM*, Bologna, Italy, 2010. [Available as TD(10)12064].

[PC10]   F. Pantisano and V. Corvino. Distance-based power planning and resource allocation for LTE. Technical report, COST 2100, Athens, Greece, February 2010. [Available as TD(10)10101].

[PKV10]   M. Proebster, M. Kaschub, and S. Valentin. A motivation for context-aware scheduling in wireless networks. In *COST 2100 12th MCM*, Bologna, Italy, 2010. [Available as TD(10)12045].

[PRSADG05]   J. Perez-Romero, O. Sallent, R. Agusti, and M. A. Diaz-Guerra. *Radio Resource Management Strategies in UMTS*. Wiley, New York, 2005.

[PVB09]   Z. A. Polgar, M. Varga, and V. Bota. Real time software emulation platform for packet based cellular transmission systems. In *COST 2100 8th MCM*, Valencia, Spain, 2009. [Available as TD(09)857].

[RC00]   W. Rhee and J. Cioffi. Increase in capacity of multiuser OFDM system using dynamic subchannel allocation. In *Proc. VTC 2000 Spring—IEEE 51st Vehicular Technology Conf.*, pages 1085–1089, Tokyo, Japan, May 2000.

[RHG+09] S. Ruiz, E. Haro, D. Gonzalez, M. Garcia-Lozano, and J. Olmos. Comparison of 3G-LTE DL scheduling strategies. Technical Report, COST 2100, Valencia, Spain, May 2009. [Available as TD(09)868].

[RLGPC+99] A. Reyes-Lecuona, E. González-Parada, E. Casilari, J. C. Casasola, and A. Diaz-Estrella. A page-oriented WWW traffic model for wireless system simulations. In *Proceedings of the 16th ITC, Teletraffic Engineering in a Competitive World*, pages 1271–1280, Edinburgh, June 1999.

[RTNL07] J. Rantakokko, O. Tronarp, J. Nilsson, and E. Löfsved. Ad hoc network capacity utilizing MIMO-techniques. In *COST 2100 2nd MCM*, Lisbon, Portugal, 2007. [Available as TD(07)223].

[Rup09]   M. Rupp. *Video and Multimedia Transmissions Over Cellular Networks: Analysis, Modelling and Optimization in Live 3G Mobile Networks*. Wiley, New York, 2009.

[SC07a]   A. Serrador and L. M. Correia. A cost function for heterogeneous networks performance evaluation based on different perspectives. In *Proc. ISTMWC 2007—16th Mobile and Wireless Communications Summit*, pages 1–5, July 2007. [Also available as TD(07)208].

[SC07b]   A. Serrador and L. M. Correia. Policies for a cost function for heterogeneous networks performance evaluation. In *Proc. PIMRC 2007—IEEE 18th Int. Symp. on Pers., Indoor and Mobile Radio Commun.*, pages 1–5, September 2007. [Also available as TD(07)313].

[SC09]   A. Serrador and L. M. Correia. A model to evaluate CRRM QoS parameters. In *COST 2100 Valencia Meeting*, pages 1–6, May 2009. [TD(09)809].

[SC10]   A. Serrador and L. M. Correia. A model to evaluate vertical handovers on JRRM. In *COST 2100 Aalborg Meeting*, pages 1–6, June 2010. [TD(10)11002].

[SKC08]   A. Serrador, B. W. M. Kuipers, and L. M. Correia. Impact of MIMO systems on CRRM in heterogeneous networks. In *Proc. WCNC 2008—IEEE Wireless Commun. and Networking Conf.*, pages 2864–2868, April 2008. [Also available as TD(08)404].

[Smo08]   L. Smolyar. Interference reduction for downlink multi-cell OFDMA systems. Technical Report, COST 2100, Lille, France, October 2008. [Available as TD(08)633].

[SOAS03] M. Sternad, T. Ottoson, A. Ahlen, and A. Svensson. Attaining both coverage and high spectral efficiency with adaptive OFDM downlinks. In *Proc. VTC 2003 Fall—IEEE 58th Vehicular Technology Conf.*, pages 2486–2540, October 2003.

[SS00]   S. Shakkottai and A. Stoylar. Scheduling for multiple flows sharing a time-varying channel: the exponential rule. Bell Laboratories Technical Report, December 2000.

[Tur06]   U. Turke. *Efficient methods for WCDMA radio network planning and optimization*. PhD thesis, Universitat Bremen, Bremen, Germany, 2006.

[TZC10]   V. Tralli, L. Zuari, and A. Conti. Effects of power allocation and reuse distance in relay-assisted wireless communications with mutual interference. In *COST 2100 11th MCM*, Aalborg, Denmark, 2010. [Available as TD(10)11077].

[VP]   V. Vassiliou and A. Pitsillides. Call admission control for MBMS-enabled 3G UMTS networks.

[VP07]   V. Vassiliou and A. Pitsillides. Power control for multicasting in IP-based 3G mobile networks. In *COST 2100 3rd MCM*, Duisburg, Germany, 2007. [Available as TD(07)363].

[VS08]   J. Vanderpypen and L. Schumacher. Multistream proportional fair scheduling applied on beamforming technologies. In *COST 2100 6th MCM*, Lille, France, 2008. [Available as TD(08)663].

[VTL02]   P. Visawanath, D. N. C. Tse, and R. Laroia. Opportunistic beamforming using dumb antennas. *IEEE Trans. Inform. Theory*, 48(6):1277–1294, 2002.

[Wil]   J. Wilson. A simple dual algorithm for the generalized assignment problem. *Journal of Heuristics*, 2(4):303–311, 1997. doi:10.1007/BF00132501

[Win] WINNER (wireless world initiative new radio) project. http://www.ist-winner.org.
[WSN09] M. Wolkerstorfer, D. Statovci, and T. Nordström. Duality-gap bounds for multi-carrier systems and their application to periodic scheduling. In *COST 2100 9th MCM*, Vienna, Austria, 2009. [Available as TD(09)991].
[Y.H10] Y. Haddad. Femtocell SINR performance evaluation. In *COST 2100 11th MCM*, Aalborg, Denmark, 2010. [Available as TD(10)11014].
[YZQ07] G. Yu, Y. Zhang, and P. Qiu. Adaptive subcarrier and bit allocation in OFDMA systems supporting heterogeneous services. *Wireless Personal Commun.*, 43(4):1057–1070, 2007.
[ZL04] Y. Zhang and K. Letaief. Multiuser adaptive subcarrier-and-bit allocation with adaptive cell selection for OFDM systems. *IEEE Trans. Wireless Commun.*, 3(5):1566–1575, 2004.

Chapter 12
Advances in Wireless Ad Hoc and Sensor Networks

Chapter Editor Paolo Grazioso, Velio Tralli, Pawel Kulakowski, Andrea Carniani, and Lubomir Dobos

Traditionally, until the end of 20th century, wireless systems could be subdivided in three broad categories, namely point-to-point, point-to-multipoint and area systems. The three categories are rather distinct from each other.

Point-to-point links are generally bidirectional and symmetrical, providing the same transmission capacity at both ends of the link, while the two terminals usually are identical.

Point-to-multipoint systems, on the other hand, are intrinsically hierarchical, inasmuch as there is one central terminal which is in charge of communication with any other peripheral terminal. The individual links can be either symmetrical or asymmetrical, and they can also be unidirectional.

Finally, area systems (such as mobile radio and broadcasting) are characterized by one station providing service to all the terminals within its coverage area. Mobile radio links are generally bidirectional, and they can be either symmetrical or asymmetrical depending from the application. On the contrary, broadcasting systems are intrinsically unidirectional, with the terminals only able to receive the signal emitted by the transmitter. Area systems in general may also support terminal mobility, allowing the terminal to interact with the central station from whichever location of the coverage area and even to move among coverage areas of different stations.

The development of new, advanced concepts and standards brought along distributed network architectures that cannot be encompassed in the above taxonomy. The familiar one-to-one and one-to-many relationships cease to apply, and each terminal is immersed in a web-like network of similar elements, exchanging information and/or co-operating with each other.

In this chapter we address some of the hottest topics related to the evolution of advanced wireless network systems and architectures.

We start presenting the main contributions of COST 2100 to the theory of wireless networks. Then we present a thorough overview of activities regarding Wireless Sensors Networks and Mesh Networks, while the subsequent section deals

---

P. Grazioso (✉)
Fondazione Ugo Bordoni (FUB), Bologna, Italy

with Cognitive Radio. Finally, the possibilities offered by Mobile Ad Hoc Network (MANET) are discussed.

12.1 Contributions to Wireless Networks Theory

In this first section we start discussing the main contributions to the theory of wireless networks. Work in this field has the aim of finding and understanding fundamental behavior of wireless networks and general methods and rules for using and organizing network and radio resources to optimize capacity and performance. The work aimed at deriving analytical tools and model for performance evaluation in general conditions is also encompassed.

The contributions of COST 2100 in this area are related to capacity limits in both wireless ad hoc and sensor networks and cognitive radio systems and to analytical tools for performance evaluation of multiple access protocols for ad hoc networks. We illustrate and discuss them starting from cognitive radio system and ending to ad hoc and sensor networks.

Cognitive radio [MJ99] allows "cognitive users" to transmit in spectrum which is currently unused (even though it might be assigned by a regulator), using spectrum sensing to detect other users (including "incumbent users" which have explicit usage rights to the spectrum). Information theory literature (see [DVT08] for an overview) provides upper bounds on the total capacity of such a system, allowing for the interference between the users, both cognitive and incumbent.

In the conventional view of cognitive radio channel, a user is expected to transmit all its power in a single assigned channel. However there may be multiple channels available, subject to varying degrees of interference, and optimum capacity may then be obtained by distributing the power between them. The work presented in [Bur09] addressed multi-channel power allocation in a cognitive radio system and used insights from information theory to compare the total capacity available using a fully distributed power allocation scheme with an optimum scheme based on perfect knowledge of all links. The radio channel includes both shadow and Rayleigh fading.

In the system model a set of n point-to-point wireless users share a set of m channels. Different users on the same channel are subject to interference: the gain between the transmitter of user i and the receiver of user j on channel k is given by $a_{ij}(k)$. There is no interference between different channels. User i has a fixed maximum power P_i to be divided between the m channels: the kth channel is allocated a power $P_i(k)$. It is assumed that the interference from other users is treated as noise which, denoted with N, is the same on all the channels: no joint detection or transmission is attempted. The power allocation problem considered is the following:

$$\max_{P_i(k), \forall i, k} \sum_{i=1}^{n} \sum_{k=1}^{m} f\left(a_{ij}(k)\frac{P_i(k)}{N}, \sum_{j \neq i} a_{ji}(k)\frac{P_j(k)}{N}\right)$$

$$\text{s.t.} \quad \sum_{k=1}^{m} P_i(k) = P_i, \quad P_i(k) \geq 0,$$
(12.1)

where $f(x, y) = \log(1 + x/(1 + y))$.

The work in [Bur09], after the discussion of algorithms that achieve the optimal solution in most cases, describes a practically realizable power allocation scheme for cognitive radio, which is based on an iterative modified water-filling algorithm suited to distributed implementation (referred to as cognitive power allocation). More specifically, some parameters of the algorithm can be estimated from the interference received during the return transmission by each user separately.

Figure 11.29 shows a result obtained for the average capacity as a function of Signal-to-Noise Ratio (SNR): we observe again that the cognitive algorithm achieves the optimum average capacity except at very high signal-to-noise ratio. It is also observed that at moderate SNR the capacity of the power allocation schemes exceeds that of channel allocation by more than 50%.

In ad hoc wireless networks the network topology may be a priori unknown, and data from source nodes may reach destination nodes by means of a relaying process involving other nodes of the networks. In the last few years, starting from the work in [GK00], the research has addressed the issue of how network capacity or user throughput scale with the number of nodes n. The authors in [GK00] focused on "multihop" wireless networks where the wireless channel is shared by a suitable set of point-to-point links which can reliably coexist if each receiver is not interfered (or the interference level is below a given threshold) by other transmitters. They showed that for a network in a planar region of area A with arbitrary fixed node locations and traffic pattern, the transport throughput of the network in bit-meter per second scales as $\Theta(\sqrt{An})$. From another point of view, if we define the node density $\delta = n/A$, we can say that the transport throughput scales as $\Theta(A\sqrt{\delta})$, i.e., it grows as $\sqrt{\delta} = \sqrt{nA}$ in networks with a spatial constraint or as $A = n\delta$ in networks with a density constraint.

More recent information-theoretic results for density constrained wireless networks have shown that the *transport capacity*, i.e. the maximum value of the feasible transport throughput in bit-meter per seconds, scales as $O(n)$ regardless the presence or absence of fading in the channel model, and this scaling law is achievable with multihop networks with nearest neighbor communication. This leads to the conclusion that multihop communication strategy is optimal with respect to capacity scaling. However, only recently, a similar result for spatially constrained wireless networks has been derived in [OLT07] showing that the aggregate throughput is with high probability bounded by $O(n \log(n))$ in a network with unit area. Therefore, in this class of networks, multihop communication strategy results as not optimal with respect to capacity scaling law. To fill this gap, [OLT07] proposed a hierarchical architecture based on distributed MIMO communication which is able to achieve with high probability an aggregate throughput which scales as n^ε where ε can be made arbitrarily close to 1.

In the work presented in [Tra08] linear wireless networks with variable number of nodes are considered and a basic cut-set bound for transport capacity is derived highlighting the functional dependence on both node density and network length. Moreover, transport throughput scaling with respect to node density δ is investigated in a wireless network with multihop communication showing that a capacity increase up to $\log \delta$ is achieved in the presence of fading by using, instead of a simple nearest-neighbor communication, suitable algorithms that adaptively schedule source-destination links in the networks.

With more details, a linear wireless network with length $L = nr$ is considered, where r is the minimum distance between nodes. The node density is $\delta = n/L = 1/r$. The signal received by each node is the sum of useful signal, interference from other transmitters and noise, and includes the effects of path-loss (modeled as $L_{1m} r_{ij}^{\beta}$, where L_{1m} is the channel attenuation at 1 meter from transmitter, and $r_{ij} = |i-j|r$ is the distance between nodes i and j). All the signals are bandlimited with bandwidth B, and the transmitted signal power is upper-bounded by P_M.

The main results obtained are the following:

Result 1 Consider a regular linear wireless network. In a space constrained network the capacity scales with δ as $O(BL^2 \alpha \delta \log \delta)$, whereas in density constrained networks the capacity scales with L as $O(BP_M/(L_{1m}\sigma^2)\delta^{\beta} L)$ if $\beta \geq 2$ and no CSI at the transmitter (or if $\beta \geq 3$ with CSI). Therefore, capacity scales in a different way with respect to the number of nodes n, and only in density constrained networks it is limited by the available power.

Result 2 Consider a linear multihop wireless network with an arbitrary configuration of $L\delta$ fixed nodes with Rayleigh fading without CSI at the transmitters. There exists a regular configuration of nodes and an algorithm which adaptively schedules multiple overlapped source-destination links such that the network can achieve a transport throughput greater than

$$\frac{LB}{1+\Delta} \sum_{j=1}^{l} I_j \left[m, \frac{L_{1m}\sigma^2}{P_M r_C^{-\beta}} + 2 \left(\frac{1}{1+\Delta-\mu} \right)^{\beta} \frac{2\beta-1}{\beta-1} + \frac{l}{(1-\mu)^{\beta}} \right], \quad (12.2)$$

where $\Delta + 1$ is the minimum normalized distance between receiving and interfering nodes, the term $(1-\mu)^{-\beta}$ is a bound for intra-cluster interference, and $I_j(m,z) = \int_0^{\infty} \log(1+x/z)(1-e^{-x})^{m-j} e^{-jx} m!/[(m-j)!(j-1)!] dx$.

The result is obtained by considering a scheme with hops of fixed length r_C small enough to have interference limited communication. The nodes are grouped in $n_C = L/r_C$ clusters of $r_C \delta$ nodes and $m = \mu r_C \delta$ (here, μ is arbitrarily chosen and tuned to keep m integer) source-destination pairs per cluster are adaptively scheduled to use the radio channel. This adaptive multihop scheme selects l out of m overlapped source-destination links in the cluster (those with the largest gain). It is implicitly assumed at the cluster level a distributed or centralized control algorithm which selects and schedules the routes, which requires a cross-layer interaction between physical layer and upper layers. It can be shown that asymptotically, as $m = \mu r_c \delta \longrightarrow \infty$, the sum of integral functions becomes $\sum_{j=1}^{l} I_j(m, z) \geq (1 - \mu)^\beta \ln(m)$ if l is chosen with a logarithmic law, i.e. $l = \lambda \log(m)$ (here, $\lambda < 1$ is an arbitrary parameter tuned to have l integer). Parameter μ, which appears in both m and z, may be chosen to optimize throughput. The transport throughput asymptotically scales as $\log(\delta)$.

The last result indicates that in the presence of Rayleigh fading the wireless network adopting the underlined multihop scheme can achieve a transport throughput which increases with δ. This can never be achieved in a multihop network without fading. To obtain this result, a channel adaptive routing/scheduling algorithm is needed. The comparison with a scheme based on distributed MIMO communication shows that multihop communication is still competitive if node density is not very large, despite the scaling law is still far from δ^ε with ε approaching 1 of the latter scheme.

In the area of ad hoc wireless networks some effort has also been spent on derivation of analytical tool for performance evaluation of Media Access Control (MAC) protocols. In [MOV09] a mobile ad hoc network where packets are transmitted to their intended destinations using single-hop communication over a nonfading channel is considered. The MAC protocols investigated are ALOHA and CSMA. Packets are located randomly in space and time according to a 3-D Poisson point process (PPP), which consists of a 2-D PPP of transmitter (TX) locations in space and a 1-D PPP of packet arrivals in time. All multi-user interference is treated as noise, and the signal-to-interference-plus-noise ratio (SINR) is used to evaluate the performance of the communication system. The main objective of the network is correct reception of packets, the metric used for performance evaluation is outage probability. A packet transmission is counted to be in outage either if it is backed off from transmission a given maximum number of times (only in the case of CSMA) or if it is received erroneously after a given maximum number of retransmission attempts (in both ALOHA and CSMA).

Approximate expressions are derived for the outage probability of ALOHA (slotted and unslotted) and CSMA (with transmitter sensing and with receiver sensing), as functions of the transmission density and the number of backoff M and retransmission attempts N. These are all reported in [MOV09] and verified by simulation. It is also found that the outage probability is a decreasing function of M and N, although the decrease is rather insignificant for high values of M and N. Moreover, significant gain can be provided by increasing the number of backoffs. In fact, for large number of backoffs, CSMA with receiver sensing can outperform slotted ALOHA.

12.2 Wireless Sensors Networks and Mesh Networks

The idea of ambient and pervasive intelligence, realized as a distributed wireless network being able to sense the environment and sometimes also react properly, increasingly attracts the attention of telecommunication world. As this idea remains in a clear compliance with the topic of European Cooperation in Science and Technology (COST) 2100 Action, it should not be surprising that this scientific area is intensively investigated by COST 2100 researchers.

Distributed wireless networks sensing their environment are usually described as Wireless Sensor Networks (WSNs). While this notion is very broad, WSN research is usually concentrated on low-cost and low-power simple devices gathering some data from the vicinity and transmitting it by wireless medium to one or more network centers, called sinks.

The WSN topic is also supported by very interesting considerations regarding Wireless Mesh Networks (WMNs), giving some general suggestions concerning Quality-of-Service (QoS), radio propagation, channel throughput and interferences.

12.2.1 Applications

There exists a tremendous diversity of WSN applications, both predicted and already implemented, like monitoring an environment in dangerous regions, controlling traffic in streets, controlling an inventory in storehouses, tracking patients in hospitals or monitoring enemy forces in a battlefield. A classification of WSN applications can be found in [BVL08], where they are divided into two groups: (a) high-density large-scale networks with massive data flows, dynamic routing and multihop communication, (b) low/medium-density networks supporting transaction-based data flows, static routing and single-hop communication. The application areas of these two network types are presented in Figs. 12.1 and 12.2 respectively, dividing them into two additional sub-categories depending on user requirements: event detection (ED) and process estimation (PE) networks.

An additional survey of wireless sensor networks applications is given in [FBVL07]. The paper concentrates on three application areas: automotive industry, precision agriculture and intelligent clothes, introducing the examples of situations where sensors can enhance the functionality of diverse tools in numerous domains of human life.

First, a detailed system that could aid a person when driving a car is described. The system is composed of sensors measuring tire pressure, acceleration, light, temperature and driver body parameters. Then, the potential of wireless sensor networks applied in agriculture is considered, with sensor motes serving to gather the data about humidity, temperature and carbon dioxide level and controlling the agriculture facilities. Finally, sensor networks integrated with clothes worn by hospital patients or sportsmen are discussed. Such networks can perform many functions, like patients monitoring, detecting their diseases, checking the body parameters during sport trainings, etc. The related issues of Body Area Networks (BANs) are treated with more details in Sect. 14.9.

An example of a specific WSN application related to animals tracking is discussed in [GSGSL$^+$10]. A stationary WSN-based system suitable for surveillance and monitoring of a certain area in a national park, e.g. an entrance to a wildlife

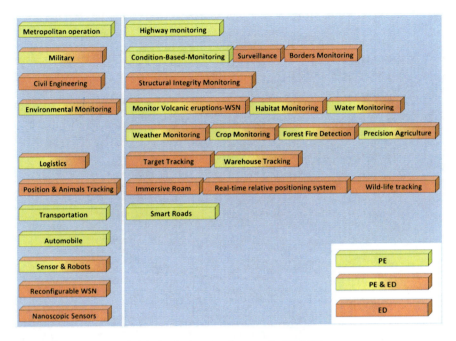

Fig. 12.1 Applications of high-density large-scale networks [BVL08]

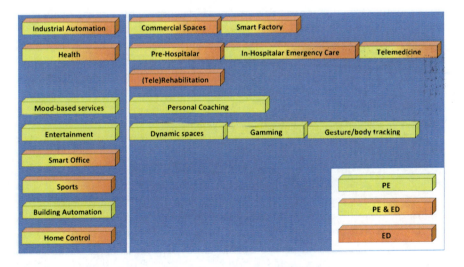

Fig. 12.2 Applications of low/medium-density networks [BVL08]

passage, is introduced. The network consists of several dozen simple motion sensors and few cameras, mounted on popular and commercially available Imote2 platforms (Fig. 12.3). Despite the network simplicity, the whole system is able to detect

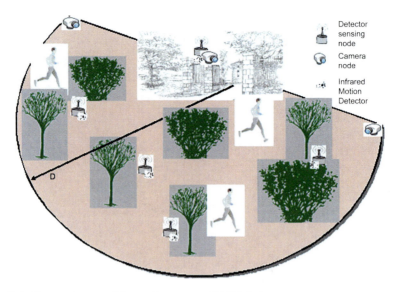

Fig. 12.3 The architecture of the tracking system [GSGSL+10]

an animal entering into the monitored zone, track its direction and speed and finally identify it with one of the cameras.

The extensive computer simulations of system performance are reported. The most important simulation result is the detection failure probability, i.e. the percentage of the cases where the network was unable to detect and register the speed and direction of the tracked animal. As shown in [GSGSL+10], for a given animal speed, the failure probability can be minimized and kept below 0.1%, but it is always a trade-off with the power consumption and network cost related to its density. On the basis of measurements that are also reported, the worst-case life time of each node is assessed as about 1 month. However, the network can work properly even if some nodes are unusable because of energy depletion.

Another environmental application, a Wireless Sensor and Actuator Network (WSAN) intended to detect and extinguish the fire in a forest is analyzed in [KCM10]. Forest fire detection is a classical application of WSNs, but there the research topic is extended, taking into account actuators extinguishing the fire and focusing on sensors and actuators cooperation. The spreading of the fire is illustrated using a well-known model based on percolation theory and explaining its relations with epidemics propagation models. Then it is shown how the temperature data gathered by sensors can be used to create a fire map and to make automatic decisions where the actuators should perform their actions. The system performance is validated through the computer simulations, documenting the correctness of the decisions taken by the system and the efficiency of fire-fighting actions related to the sensors density.

As WSNs are also designed to monitor dangerous and hardly accessible zones, it can happen that the access to a network is restricted to its perimeter only. In such a situation, the problem of dissemination and gathering the sensors data can be solved

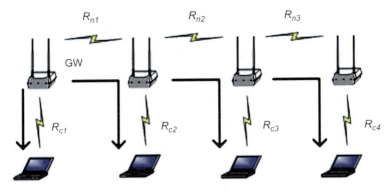

Fig. 12.4 An example of wireless mesh network [HBEH08]

using random linear coding schemes [VSC+09, SVCS10, SCV+10]. It is described with more details in Sect. 9.1.

Finally, application discussions can be concluded by an example of an Institute of Electrical and Electronics Engineers (IEEE) 802.11-based WMN providing the Internet service for an isolated village. While Wireless Local Area Network (WLAN) mesh network cannot compete with more complex technologies, like Worldwide Interoperability for Microwave Access (WiMAX), it is still a viable and cost-attractive solution for small networks having not more than 20 users. The detailed QoS simulation results can be found in [FCFN08].

12.2.2 Physical Channel and Network Throughput

An analysis of network capabilities and potential applications should be supported very strongly with studies of its throughput and capacity of its radio channels. The research considering fundamental issues for mesh networks, an impact of propagation parameters on network performance, is reported in [BCMR09]. The included simulation results show the influence of the path loss exponent on the system throughput and packet delay. The research is focused on the classical Rayleigh channel model with Additive White Gaussian Noise (AWGN).

On the other hand, the analysis how the throughput scales depending on network topology is performed in [HBEH08]. A mesh network consisting of client devices and access points is considered. All the nodes are connected wirelessly, while the capacity of each link can be different, as described in Fig. 12.4. The authors derive analytical formulas showing how the throughput obtainable by each node depends on the number of nodes. The transmission with no inter- or intra-network interferences is assumed. Numerous scenarios (symmetrical and asymmetrical links, single frequency or with re-use of some frequency bands) are taken into account. The theoretical results are also confronted with measurements conducted with a real IEEE 802.11 b/g mesh network. In the area with limited interferences, the analytical formulas show a very good compliance with performed tests.

The topic of channel interferences in IEEE 802.11 mesh networks is covered in [RFMM10]. In order to simplify the deployment of 802.11 network and avoid cable infrastructure between access point nodes, the promising strategy of using multi-radio interface nodes is analyzed. The first radio interface serves network users, while the second one provides the connection between access points. A well-known problem of this strategy is the fact that two radio-interfaces (and two antennas) mounted at one node create the interferences to each other, even if working on different frequency channels.

The reported measurements provide a very detailed analysis of the network performance in such adverse conditions. Not only the channel interferences are investigated, but also the influence of the channel separation and physical antenna separation. At the same time, some useful hints are given how to design 802.11 mesh networks.

Regarding WSNs, the primary goal of deploying a sensor network is to gather data from the environment. But, how much of data can be delivered by a network? An analysis of this problem is performed in two companion papers [VFB08, BFV08].

Multi-sink scenario with sensors and sinks deployed according to uniform random distribution is studied [VFB08]. The problem is stated as the calculation of area throughput defined as the amount of samples per unit of time successfully transmitted to the sinks from the given area. Without loss of generality, a certain finite square shape area is considered. Sinks send queries to sensors asking for the environmental data. Each sensor aggregates its data, grouping D samples into a packet. Then, it chooses the sink with the strongest signal and responds once per D queries sending the packet.

The probability that a sensor is able to deliver its packet is calculated considering two issues. First, the sensor must have at least one sink in its range, taking into account the path loss and shadowing effects. Second, the packet transmission from the sensor to the sink must occur without failures and collisions with packets of other competing sensors, i.e. the access to the medium must be properly resolved by the MAC protocol. While the IEEE 802.15.4 MAC scheme is assumed in the paper, the results are valid for other CSMA protocols as well. The formulas and numerical results presented in the paper give the overall view of the strategies that could be applied to get the maximum data from a certain area covered by a sensor network.

In [BFV08] these studies are continued, while concentrating more on medium access details. Both beacon-enabled and non-beacon-enabled modes of IEEE 802.15.4 standard are investigated. Comparing these two modes, the advantage of beacon-enabled scheme is shown, especially when the number of guaranteed time slots is chosen carefully.

Related studies, but more concerned with assessing the data traffic, can be found in [BV08]. A hybrid network architecture—a sensor network connected with a UMTS of WLAN gateways—is analyzed. The work by Buratti and Verdone stresses the importance of having statistics about the sensor data traffic received at the mobile gateway. It shows how the parameters of MAC protocol (like number of guaranteed time slots, beacon and super-frame orders) affect the traffic incoming to the gateway. Thus, as we can see, these results can be helpful when designing the gateway.

On the other hand, a MAC protocol designer can "shape" these statistics depending on the traffic requested in the gateway.

12.2.3 Energy Efficiency Issues

As the concept of WSNs is concentrated on the idea of a network consisting of simple, low-cost low-power devices, the energy efficiency is a key feature. However, the existing standard that could be applied to WSNs, IEEE 802.15.4, has some drawbacks strictly related to energy issues.

In [RSO+09], the energy efficiency in WSNs is considered, and the detailed analysis of IEEE 802.15.4 is undertaken. The protocol is able to work in two modes, with or without beacons. Yet, with beacons, the protocol suffers from several issues that reduce its energy performance. First, the beacons can collide, as there is no Carrier Sense Multiple Access/Collision Avoidance (CSMA/CA) scheme used for them. Second, the hidden terminal problem can result in collisions in star networks. Also, the 802.15.4 super-frame does not scale well in multihop networks. Finally, some nodes, called full functional devices (FFD), must be continuously kept active and have very short lifetime.

Taking into account these issues, the research is focused on non-beacon-enabled mode. A new routing algorithm, called V-Route, is proposed that limits the energy expenditure. V-Route disables the beacons and the super-frame structure. Reduced functional devices are removed from a network, all the nodes have FFD functionality, and they can communicate with all their neighbors according to CSMA/CA competition rules. The whole network is divided into so-called timezones showing a node distance to the central node, called coordinator. The procedure of creating the timezones starts when nodes are broadcasting beacon requests in order to find the coordinator node. When the coordinator is localized, first, its neighbors are associated to the network as belonging to the 1st timezone. Their neighbors that are not in the coordinator range are classified as the nodes from the 2nd timezone, etc., creating the particular network hierarchy (see Fig. 12.5).

The proposed protocol also takes into account the possible topology dynamics, e.g. caused by nodes energy depletion or their mobility. All the nodes have the timezone expiration time defined. If no update messages are received before that time, the timezone is considered outdated, and the node broadcasts a timezone update request to be classified to a new timezone.

Between the nodes, three types of the transmissions are possible: (a) upstream, the transmission to the direct neighbor located in the timezone with lower number; (b) downstream, the transmission to the one of the neighbors from the timezone with higher number; and (c) broadcast, to all the neighbors. All the transmissions between the nodes are arranged according to the schedule presented in Fig. 12.6 (an example for 4 timezones is shown). To limit the collisions caused by simultaneous broadcast transmissions, the nodes from the same timezone can use the broadcast mode only once for each of 4 frames.

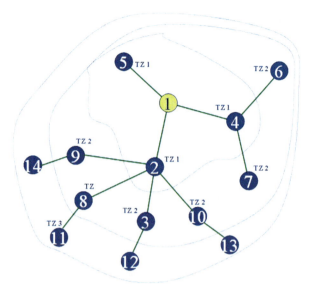

Fig. 12.5 A network divided into timezones [RSO+09] (©2009 Springer, reproduced with permission)

Together with the V-Route protocol, numerous suggestions how to enhance the performance of the communication scheme are presented. First, it is proposed to turn off transmitter radios as soon as possible, i.e. just after finishing the transmission of the last packet. Second, an additional bit should be added to the transmitted packets, showing if the packet is the last one. It allows one also to turn off the receiver radio more quickly. Next, idle listening can be successfully limited by decreasing the maximum number of backoffs in the collision avoidance mechanism. Instead of that, in a case of a channel access failure, the data packet is re-scheduled in the next frame. Finally, node micro-controllers can be occasionally put into stand-by mode. The suggested enhancements were tested in practice with a network consisting of Philips AquisGrain sensors (CC2420 radio transceivers and AT-Mega 128L microcontrollers). It was shown that up to 85% of the energy spent by sensor nodes can be saved if the proposed changes were applied.

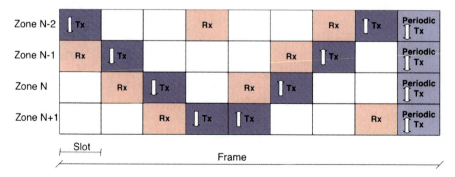

Fig. 12.6 The transmission schedule for V-Route protocol [RSO+09] (©2009 Springer, reproduced with permission)

The energy efficiency issues in the broader range of networks (ad hoc, WSNs) are also considered in [CCLR08]. The authors validate non-regenerative relays as a technique of energy saving. Two-hop and three-hop scenarios are analyzed, and the network parameters (noise figure, nodes positions and efficiency) are shown where using the relays can be profitable. The authors also stress the profit of sharing the energy consumption between many nodes.

12.2.4 Large-Scale Networks

In the framework of COST 2100 cooperation, some presented works were related to wireless sensor networks consisting of large number of nodes, e.g. hundreds or thousands, possibly randomly deployed in an outdoor area. Because of their scale and not-fully controlled deployment, such networks should have communication protocols implemented, meeting following criteria. First, the protocols should be highly scalable, i.e. be able to work properly independently of the number of nodes in a WSN. Second, they should be able to react for network topology changes. Any topology variations, due to sensor energy depletion (some nodes may disappear from the network), wireless channel fading or node movements, should not interrupt the network smooth running. Finally, the communication protocols should be robust in the sense that they could operate in irregular network topologies.

The well-known solution for large-scale WSNs is a family of geographic routing protocols, broadly described in open literature. Geo-routing is based on the assumption that instead of maintaining large routing tables, wireless nodes exploit the knowledge about their geographic positions. Messages are routed, hop by hop, to the nodes geographically closer to the destination. In case of a data packet that is stuck somewhere in the network, some recovery mechanisms are applied, also based on geographical position awareness. Thus, it is crucial to have not only geo-routing, but also localization algorithms.

Among the localization schemes suitable for large-scale WSNs, there is still no well-accepted approach on how to solve the localization issue, especially for outdoor sensor networks. Because of large number of sensor nodes and their desirable low cost, it is not feasible to mount a GPS receiver at each node. Numerous localization techniques were proposed so far, like Ultra-WideBand (UWB), acoustic, Time of Arrival (ToA) or Received Signal Strength (RSS) ones. However, all of them suffer from similar weaknesses: they require expensive additional hardware increasing the system cost (UWB, acoustic), very exact measurements (ToA) or do not guarantee the appropriate precision (RSS).

Localization in WSNs can be also performed using Azimuth of Arrival (AoA) measurements. In [KVAEL+10], an AoA localization system concept suitable for small-size sensor nodes is described. It is proposed that few reference nodes in a considered WSN are equipped with an array of four antennas arranged in a square. By changing the direction of maximum radiation of the antenna array, a rotating beacon is created. In order to do that, the phases of the signals at all four antennas are adjusted to have four radio waves interfering constructively at a specific direction, according to the well-known scanning phased array (beamforming) technique. The chosen antenna parameters guarantee the whole antenna array would have dimensions appropriate for tiny wireless devices.

The reference nodes emit their own positions and the beacons with rotating antenna patterns. Each sensor registers the time when the power of a beacon is the strongest. On this basis, sensor can calculate the angles to the reference nodes and then their own coordinates.

The conducted Monte Carlo simulations consider random positions of sensors and reference nodes, the radio propagation channel model specific for outdoor WSNs, the radio noise and transceiver parameters of real sensor motes. Also, to the best of the author knowledge, [KVAEL+10] is the first paper where the influence of SNR on the AoA localization accuracy is taken into account in computer simulations. The obtained results show very good localization accuracy. Additionally, as the whole algorithm is working in a decentralized manner and no signals need to be transmitted from sensors to reference nodes, its scalability seems to be especially promising for large-scale WSNs.

Geo-routing protocols usually combine two approaches. Greedy forwarding, which selects the neighbor whose distance to the sink is minimal, is used as long as it is possible. If it fails, because a node has no neighbors closer to the sink (there is a hole in the network), the recovery procedure is applied. The most common solution is face routing, which consists of two operations. First, the network is planarized, i.e. the crossing links are removed. Then, the packet is routed along the borders of so-called faces that arise in planarization process. While it is proved that face routing guarantees packet delivery, it is still very energy and time consuming procedure.

In [KGH09b], another approach to geo-routing is adopted. It is proposed to omit the planarization process at all. If the greedy routing fails, the packet is forwarded along the border of the encountered hole using the information about the angles between the nodes (Fig. 12.7). The routing decisions are taken in a distributed way with very limited data: own node position, sink node position and two previous hop positions.

The details of the routing procedure can be found in [KGH09b]. The provided computer simulation results show also that the proposed algorithm allows one to reach the sink in a smaller number of hops than the schemes based on face routing approach.

The proposed routing algorithm is further developed in [KELGH09], where it is supplemented by a forwarder resolution procedure. It enables sensors being neighbors of a transmitting node to choose the best forwarder in a distributed way. The whole procedure starts when a node has a packet to forward or is the source of a message. The node monitors the radio channel, and, if the channel is idle, it sends the packet with its own position data. Each neighbor node then checks how good forwarder it is, i.e. it calculates its routing metric according to the rules of geo-routing protocol. Then, the neighbors respond to the initial packet in a time-slotted scheme. In the first slot, the best possible forwarders respond, and, later, the worst do. For the initial node, only the first slot with a single response is relevant: that neighbor is the

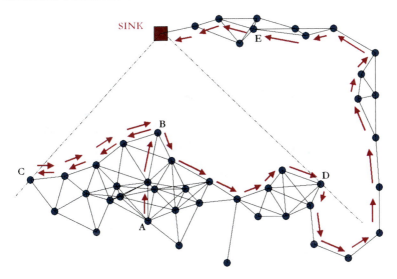

Fig. 12.7 Geo-routing: a packet forwarded along the network border [KGH09b]

next forwarder. Then, the initial node sends a short confirmation packet, so as to let non-involved nodes go to sleep, and immediately the data packet. If there is a collision in the first occupied slot (the packet cannot be decoded), the initial node sends another short packet indicating the number of the collided slot. In this case, only the nodes previously collided are allowed to reply. Their responses are re-mapped onto the whole pool of time slots, proportionally to their routing metrics. If there is another collision, this process is repeated, until the first occurrence of a slot with a single reply. Finally, in all the cases, the one neighbor is chosen as a forwarder.

Similar algorithms were previously proposed in the open literature. However, it is the first procedure which considers not only greedy routing, but can also cooperate with the recovery schemes, like those introduced in [KGH09a]. Moreover, it guarantees that the best neighbor is chosen as the forwarding node.

It is also shown and proved by computer simulations that the overhead of the proposed forwarder resolution procedure is very limited. Furthermore, the procedure performs quite well, even if the radio channel parameters are not known and cannot be correctly assessed by the sensor nodes.

12.3 Cognitive Radio

Radio is a broadcast medium, and thus all users coexisting in the same frequency band interfere with each other. As the number of wireless systems and services has grown exponentially over the last two decades, the availability of prime wireless spectrum has become severely limited. This limited available spectrum and the inefficiency in the spectrum usage necessitate a new communication paradigm to exploit the existing wireless spectrum opportunistically. This new networking paradigm is referred to as *Cognitive Radio Networks* and provides the capability to use or share the spectrum in an opportunistic manner; thus, Dynamic Spectrum Access (DSA) techniques allow the cognitive radio to operate in the best available channel [ALVM08]. Cognitive Radio is currently attracting a great deal of attention as means of making available large amounts of under-used spectrum for wireless communications, so it is not a surprise that this scientific area is investigated by European Cooperation in Science and Technology (COST) 2100 researchers. In particular, the problems of interference management [FOKK09], spectrum opportunities in different bands [GPE09, CGV10], and the capacity that can be achieved in a novel class of data channel [PU10] are investigated.

12.3.1 Interference Management

The main constrain for secondary (unlicensed) users (SUs) is seek their signals with those of the primary (licensed) users (PUs) without impacting the transmission of primary system. In particular, licensed transceiver/primary user is interfered when a cognitive radio/secondary user, located within a certain range, transmits on the same frequency.

In [FOKK09] the concept of sensor network aided cognitive radio is proposed to manage the interference that secondary users cause to primary system, where a fixed sensor network deployed firstly localizes both the licensed radio/primary user and the cognitive radio/secondary user with a certain given accuracy. Then the distance between them is estimated and, provided that they use the same frequency, used to determine the probability of the SU causing interference to the PU. If this probability is high, the SU should be instructed to adjust its transmit power or frequency such that the SU interference experienced by the PU becomes acceptable.

In this work the interference experienced by a receiving PU (who is known to be somewhere within a region of known size and shape) caused by an SU, which is generally located in a different, but also known region, is considered. In particular, the probability of this kind of interference is studied, since knowledge of it can be used for SU power control, outage capacity analysis, and sensor design. Under certain assumptions, the interference probability can be found by means of geometrical considerations.

In particular, this work has analyzed interference probabilities for sensors with ranging capacity and for Boolean sensors: the results show that with range estimates available, the resolution is higher and the resulting interference probability provides more information than if the sensors are Boolean (which only report whether a transmitter is within the maximal sensing range or not). To keep interference probability under a certain desired threshold, the authors propose to use a feedback on transmit power from the sink of sensors network to the secondary transmitter.

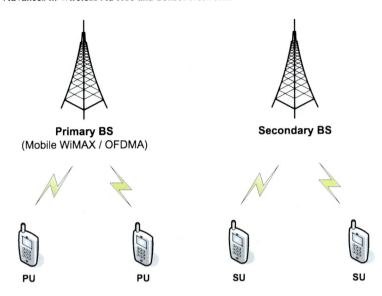

Fig. 12.8 OFDMA system scenario [GPE09]

12.3.2 Spectrum Opportunities

In this section we propose two works performed within COST 2100 about the evaluation of opportunities for secondary usage of spectrum for DSA in a Primary OFDMA System [GPE09] and in GSM band [CGV10].

Orthogonal Frequency Division Multiple Access (OFDMA) is the major transmission and access technology for future mobile wireless broadband systems such as Mobile WiMAX and 3GPP Long Term Evolution (LTE). It is therefore of great interest to survey the opportunities for secondary systems to utilize spectrum in primary OFDMA systems. Since capacity is allocated as frequency-time resource elements within the OFDMA frame, it is necessary to study OFDMA spectrum in both the frequency and time domain to characterize the opportunities for DSA operation in primary OFDMA systems. In [GPE09], for secondary communications, the authors suppose to use the interweave paradigm for cognitive radio, which is based on the idea of opportunistic communications. The basic idea is to exploit the temporary space–time frequency voids, referred to as spectrum holes, which are not in constant use in both licensed and unlicensed bands, for a secondary use of spectrum.

In this work, the usage of spectrum in systems that use OFDMA transmission and give directions on how to derive schemes for OFDMA occupancy useful for secondary systems that aim to operate in primary OFDMA spectrum is characterized. In particular, the Mobile WiMAX is considered as the primary OFDMA system, while the secondary system consists of a Secondary BS (Base Station) and SUs that may use Cognitive Radio to detect white holes in the primary spectrum, as shown in Fig. 12.8.

In order to obtain strong schemes, the well known ns-2 simulator with a Mobile WiMAX implementation, developed by the WiMAX Forum, and different traffic models (web browsing, file transferring, voice data, video streaming) are used to simulate the primary OFDMA system.

This work shows that it should be possible to utilize white holes in a mobile OFDMA broadband system, but that the OFDMA scheduling technique and the traffic models used by primary users will have significant impact on the characterization of available spectrum. Operator assistance is also considered as important to be able to utilize the available capacity within primary OFDMA systems with high activity.

In [CGV10] the possibility of exploiting underutilized channels in the GSM bands is explored. The focus is on the feasibility of the proposed approach, so that we consider a system-level simulator based on a realistic GSM network deployed in the city of Bologna. In this context we evaluate the impact that the operation of multiple secondary users has on the performances of the primary system, as a function of the number of secondary users in the scenario and their transmission power. In addition, we study the performances that can be obtained by the secondary system under the condition of marginal interference to the primary system.

In particular, for secondary communications, the underlay paradigm for cognitive radio is considered: this paradigm encompasses techniques that allow concurrent primary and secondary transmissions, as long as the interference generated by the SUs is below some acceptable threshold. In this work, the authors address the spectrum decision problem and propose an algorithm to select the most appropriate frequency channel for the secondary communications: the algorithm proposed in this paper takes advantage of the traditional frequency planning of the GSM network, which is based on the concept of spatial reuse. In addition, the algorithm proposed takes advantage of some fuzzy logic concepts in order to make the most suitable decision in a scenario characterized by high uncertainty.

The GSM network of an Italian operator in the Bologna area is considered as the primary system, where secondary users want to exploit the GSM band for secondary ad hoc communications, to provide an alternative way of communication during, e.g. an event with a high concentration of people (fair or conference) where the participants want to communicate with each other, as shown in Fig. 12.9.

Preliminary simulation results, obtained by means of a realistic C simulator of the GSM network of an Italian operator in the Bologna area, show that by means of appropriate spectrum decisions, the spectrum can be reused taking advantage of the traditional GSM frequency planning, without causing harmful interference to the primary system.

12.3.3 Secondary Wireless Communication Through a Novel Class of Channel

In [PU10], the authors consider secondary communication channels, defined via utilization of the protocol overhead information in legacy communication systems, here referred to as primary systems. The secondary communication is carried out

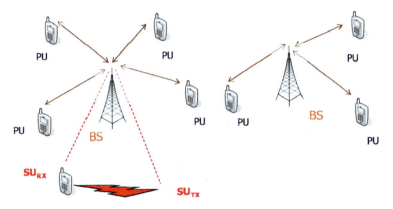

Fig. 12.9 GSM system scenario

by reordering the packets used by the primary system. In particular, the authors introduce a new framework for computing the capacity of secondary communication using this novel class of channels: in fact, the fundamental element of any packet-based communications is the overhead carried in the packet, used to determine how packet should be delivered. This overhead be suitably manipulated in order to provide transfer of additional information, thus providing a secondary communication channel, by minimally affecting the original primary communication channel. The data rates that can be sent are relatively modest, but using this approach, the secondary messages are sent via the header of the primary system, which is robustly encoded and can be received in a large area of systems that are already deployed; hence, the applications that require low, but robust data rate, such as M2M communication, are the ideal application for the approach presented in this work. To evaluate the capacity that can be achieved, the authors here develop a new framework to find the capacity of the combinatorial model under packet erasures without using Shannon's result: in particular, they take advantage of the specific state structure for the channel at hand in order to limit the number of necessary input symbols to a number lower than $2F$ (where F is the number of packet transmissions in one frame), using the Markov chain. The challenge is to find practical methods for encoding the additional bits provided by the reordering phase, as well as assessment of the additional capacity offered by the secondary communication by accounting of the constraints (e.g. scheduling) of the specific primary systems.

12.4 Ad Hoc Networks and MANET

This section will deal with the possibilities and the possible improvements offered by the MANET (*Mobile Ad Hoc Network*) concept. The MANET presents very important part of wireless networks, and it belongs to next generation of mobile radio networks. It is a collection of communication devices or nodes that communicate without any fixed infrastructure, such as base stations and pre-determined organization of available links [PCD08b]. Therefore, it is an attractive networking option for connecting mobile devices quickly and spontaneously. The MANET have several salient characteristics, such as *dynamic changes of topologies, bandwidth-constrained, links capacity, energy-constrained operation, limited physical security* [PCD09a]. In this section we focus some problems and advances in MANET, namely *Security and QoS, Cross layer design, Location of MANET terminals, IEEE 802.11 routing* and *Simulation of the stochastic channel model*.

12.4.1 Security and QoS in MANET

Security and QoS are very important and dynamically developing areas of research for MANET. Especially in MANET, specific services have specific QoS and security needs. Although the ongoing trend is to adopt MANET for commercial uses due to their unique properties, the main challenge is to provide ability to configure their own level of security and QoS [PCD08a]. The basic idea of QoS is to offer additional service classes on top of the standard best-effort variant. The QoS is defined as a set of service requirements to be met by the network while transporting a packet stream from source to destination. Intrinsic to the notion of QoS is an agreement or a guarantee by the network to provide users with a set of measurable pre-specified service attributes in terms of delay, jitter, available bandwidth, packet loss, and so on [PCD09a]. Security presents very complex problems, and there are many types of attacks that can affect MANET functionality. There are five main security issues considered in MANET: *confidentiality, authentication, availability, integrity, non-repudiation*. In order to achieve these goals, the security mechanisms and solutions should provide complete protection spanning the entire protocol stack. There is no single mechanism that will provide all the security services in MANETs [PCD08b].

Article [PCD08a] introduces the new concept of how to integrate security and QoS into one common parameter in MANET. In former literature, security is interpreted as a dimension of QoS, but integration of both concepts had not been studied before. The main idea is to provide QoS and security in parallel and to minimize negative mutual influences. Also the new concept of modified Security Service Vector (SSV) was introduced and described. Modified SSV is based on the SSV designed for wired IP networks. The SSV was proposed to determine a number of customizable Security Services with choices of customizable Service Degrees. Authors introduce how the modified SSV was implemented into MANET and describes how the dynamic routing protocol (DSR) was modified. This modification is needed for cooperation between modified SSV and MANET [PCD08b]. The process of integration of security and QoS is described as follow: At the beginning (*first phase*), the user (*source node*) specifies security and QoS requirements (*SSV data*). The modified SSV collects these data and stores them into the main memory. After this process, the modified SSV adds data to the routing packet [PCD08b]. In the second phase, a user who wants to activate a security-related service with another user (*destination node*) or server, sends request packets that contain the SSV data to the adjacent nodes, i.e. those located near the source node. If the intermediate node is not the destination node, it compares its own SSV data with SSV date stored in the request packet. If the node can provide the requested service, it sends the routing packet to the next adjacent node. This process is repeated until the destination node is reached. Then the destination node sends a replay packet to the source node, fulfilling the requested SSV data, and the service is activated [PCD09a]. Integration of security and QoS presents new research areas and is still at the beginning. The process of integration provides new abilities to specify service requirements, and users have the new concept of interaction with routing and system processes. The MANET can also provide new types of services with specified security and QoS levels. Simulation results show that the activation of modified SSV does not excessively increase the total delay of MANET [PCD08a, PCD08b, PCD09a].

12.4.2 Cross Layer Design in MANET

The Cross Layer Design (CLD) approach is a new dynamic area of research for MANET networks. This approach provides new possibilities to increase the performance and adaptability of MANET. Research on cross-layer networking is still at a very early stage, and no consensus exists on a generic cross layer infrastructure or architecture. The research carried out so far reflects the diversity of the problems caused by the system dynamics in ad hoc networks. Cross-layering is not a simple replacement of a layered architecture, nor it is a simple combination of layered functionality. Cross-layering tries to share information among different layers that can be used as input for algorithms, for decision processes, for computations and adaptations. The specific characteristics of MANET lead to benefits and problems that the CLD approach aims to solve. Solutions can be divided into the following areas: *adaptation, mobility, energy control and power control, different QoS requirement* and *security*.

In [PCD09b], the new approach to CLD is introduced. The article shows how CLD can improve cooperation between security and QoS. This article is closely linked with [PCD09b]. In the latter a new CLD model between non-adjacent layers was introduced. The CLD model is used for collecting QoS and security related data that are necessary for the modified SSV and modified routing protocol. The model also provides an interface between user and system (routing protocol and modified SSV). The task of CLD is to enable transferring and collecting data from application layer to the routing protocol operating on network layer [PCD09c]. The collected SSV data consists of information about security and QoS parameters that a node is able to provide. The new CLD model provides information about SSV collected on application layer to each node in the network (also to nodes that only forward packets and act as router-hop nodes). In the case of routing and destination nodes, the CLD provides additional algorithms used to activate the process of collecting information from routing packet. If the SSV data are not stored in memory, CLD also activates the process of collecting SSV data on the node [PCD09c, PCD09a]. The individual CLD algorithms on source, routing and destination nodes are tested. Next simulations were oriented to cooperation between CLD and SSV for MANET networks consisting of 10 to 50 nodes. The obtained results show that this cooperation is effective to improve MANET operation.

12.4.3 Location in MANET

The location process in MANET is the determination of the mobile nodes position. Localization methods are divided into two groups: (i) *direct approaches*, also called the methods of absolute localization (position of mobile nodes is manually configured, or all the mobile nodes are equipped by GPS receivers) and (ii) *indirect approaches*, also called the methods of relative localization (position of mobile nodes is calculated) which have two forms: *methods based on propagating signal parameters measurement (called Range-based)* and *methods without signal parameters measurement (called Range-free)*. The localization methods are also divided into two groups according to whether or not the beacon nodes (with known coordinates) help the localization process: *Beacon-free location methods* and *Beacon-based location methods*.

Authors of [DC09b] introduces a new beacon-based location method for MANET terminals, based on ring triangulation. The proposed algorithm consists of several phases. Distances between nodes are calculated on the basis of: (i) *measurement of the signal propagation time between two nodes* and (ii) *measurement of the received signal power level*. The algorithm is iterative. In the first phase all mobile nodes which are in the range of at least three steady beacons are located. In the subsequent iterations, all mobile nodes already localized in previous phase become virtual beacons [DC09a]. Every phase of proposed algorithm consists of following stages: *selection of a mobile node suitable for localization process, the calculation of the distances between localized mobile node and all beacons in its range, selection of appropriate beacons to the unknown node, triangulation and position calculation for location mobile node* [DC09a]. This algorithm is used for locating mobile nodes in urban environment. The accuracy of the algorithm increases when the number of phases decreases [DC09a]. In [DS09], the authors are focused on determining the relative position of nodes for networks without a presence of beacons (nodes knowing their absolute positions) and nodes equipped with a built-in GPS module. The localization protocol is based on Self-Positioning Algorithm (SPA) and consists of two parts. First, the distance between nodes is measured and computed with the method described in the paper. In the second phase, the localization algorithm uses measured distances as input data for positioning. The accuracy of the algorithm was computed as a mean of differences between the actual and computed distance of two nodes, evaluated for all possible combination of nodes. The error was presented as a percentage of the radio range size. The simulation show that the positioning errors range from 20 to 70%, depending on various network parameters (number of nodes, network size, type of environment, etc.) [DS09].

12.4.4 IEEE 802.11 Routing in MANET

Packet forwarding is a very important aspect of MANET. Stations that generate the packets must know where they should forward their data and this must be done efficiently. Since the family of standards IEEE 802.11 only regulates the Medium Access Control (MAC) and Physical (PHY) layers, the routing principles are not covered by it, which describes only how the direct communication between two stations is performed.

In [FCV10] authors present a new cross-layer approaches that inject values read from the Physical and MAC layers into the routing component, in a search for the best possible path. The experiences were performed by using the IEEE 802.11e simulator that was developed at Instituto de Telecomunicações, Portugal. In this work some tests were performed in custom made IEEE 802.11e simulator, by introducing 12 new different routing techniques. Authors directly compared them with a simple routing algorithm (called standard), which selects the paths with the minimal number of hops. From the results it was verified that the traditional approach is outperformed by all cross-layer proposals. In fact, these ones can deliver up to three times more packets than the standard one and decrease the latency per packet. The overall best performance was achieved by a proposal which uses links that have the highest signal to interference-plus-noise ratio. It generates paths with a larger number of hops, whose links have a baud rate near the maximum one and reliably deliver the packets from the source to the destination.

12.4.5 A Stochastic Channel Model for Simulation of Mobile Ad Hoc Networks

This section presents a new stochastic channel model used for simulation of MANET. In MANET, the channel model needs to be both accurate and computationally efficient. Inaccuracies in the channel model can seriously affect various network performance measurement [ELW+08]. The channel model has to provide realistic time and spatial variability as a terminal move. However, to meet the necessary low-complexity constraints, commonly used channel models for network simulations are very simplified.

Authors in [EFW10] show that, when developing and evaluating channel models, it is important to consider the correlation between different channels parameters. Authors are focused on describing the structure of a stochastic channel model, based on estimated channel parameters from measurements. The parameter generation of the model was described. For evaluation, results were obtained from a peer-to-peer measurement campaign taken at 300 MHz in a urban environment in Linköping (Sweden) [EFW10]. The proposed model includes different channel parameters as the distribution functions of large-scale fading, Ricean K-factor and delay spread. It also considers the correlation between the channel parameters, and an AR-filter is used to maintain accurate spatial correlation.

References

[ALVM08] I. F. Akyildiz, W.-Y. Lee, M. C. Vuran, and S. Mohanty. A survey on spectrum management in cognitive radio networks. *IEEE Communication Magazine*, 7:40–48, 2008.

[BCMR09] M. Barbiroli, C. Carciofi, A. Masini, and G. Riva. Influence of propagation phenomena on wireless mesh networks performance. Technical Report TD-09-845, Valencia, Spain, May 2009.

[BFV08] C. Buratti, F. Fabbri, and R. Verdone. Area throughput for an IEEE 802.15.4 based wireless sensor network. Technical Report TD-08-676, Lille, France, October 2008.

[Bur09] A. Burr. Capacity of cognitive channel and power allocation. In *IEEE Information Theory Workshop*, pages 510–514, October 2009.

[BV08] C. Buratti and R. Verdone. A hybrid hierarchical architecture: from wireless sensors to the fixed infrastructure. Technical Report TD-08-469, Wroclaw, Poland, February 2008.

[BVL08] L. M. Borges, F. J. Velez, and A. S. Lebres. Taxonomy for wireless sensor networks services characterisation and classification. In *Proc. of Conftele' 2009—7th Conference on Telecommunications*, Vila da Feira, Portugal, May 2009. [Also available as TD(08)532].

[CCLR08] J. Chen, L. Clavier, C. Loyez, and N. Rolland. Energy considerations for non-regenerative relays in heterogeneous ad hoc networks. Technical Report TD-08-625, Lille, France, October 2008.

[CGV10] A. Carniani, L. Giupponi, and R. Verdone. Evaluation of spectrum opportunities in the GSM band. In *2010 European Wireless Conference (EW)*, pages 948–954, April 2010. [Also available as TD(09)999].

[DC09a] L. Dobos and V. Cipov. Beacon based localization algorithm for MANET terminals. In *International Conference on Applied Electrical Engineering and Informatics 2009 (AEI2009)*, vol. 1, pages 25–38, September 2009. [Also available as TD(09)915].

[DC09b] L. Dobos and V. Cipov. Beacon based location algorithms for MANET terminals. Technical Report TD-09-915, Wien, Austria, September 2009.

[DS09] L. Dobos and B. Svoboda. GPS and beacon free location algorithms for MANET. Technical Report TD-09-916, Wien, Austria, September 2009.

[DVT08] N. Devroye, M. Vu, and V. Tarokh. Achievable rates and scaling laws for cognitive radio channel. *EURASIP Journal on Wireless Communication Networking*, 2008(2):1–12, 2008.

[EFW10] G. Eriksson, K. Fors, and K. Wiklundh. A stochastic channel model for simulation of mobile ad hoc networks. Technical Report TD-10-11084, Aalborg, Denmark, June 2010.

[ELW+08] G. Eriksson, S. Linder, K. Wiklundh, P. D. Holm, P. Johansson, F. Tufvesson, and A. F. Molisch. Urban peer-to-peer MIMO channel measurements and analysis at 300 MHz. In *Military Communications Conference, IEEE (MILCOM2008)*, pages 1–8, San Diego, CA, USA, November 2008.

[FBVL07] J. M. Ferro, L. M. Borges, F. J. Velez, and A. S. Lebres. Applications of wireless sensor networks. Technical Report TD-07-327, Duisburg, Germany, September 2007.

[FCFN08] L. S. Ferreira, L. Caeiro, M. Ferreira, and M. S. Nunes. QoS performance evaluation of a WLAN mesh vs. a WIMAX solution for an isolated village scenario. Technical Report TD-08-409, Wroclaw, Poland, February 2008.

[FCV10] J. M. Ferro, O. Cabral, and F. J. Velez. Looking for an efficient routing method for IEEE 802.11e ad hoc networks. Technical Report TD-10-10011, Athens, Greece, February 2010.

[FOKK09]　J. T. Flam, G. Oien, A. N. Kim, and K. Kansanen. Sensor requirements and interference management in cognitive radio. Technical Report TD-09-749, Braunschweig, Germany, February 2009.

[GK00]　P. Gupta and P. R. Kumar. Capacity of wireless networks. *IEEE Transactions on Information Theory*, 46:388–401, 2000.

[GPE09]　P. Gronsund, H. N. Pham, and P. E. Engelstad. Opportunities for dynamic spectrum access in primary OFDMA systems. Technical Report TD-09-875, Valencia, Spain, May 2009.

[GSGSL+10]　A. J. Garcia-Sanchez, F. Garcia-Sanchez, F. Losilla, P. Kulakowski, J. Garcia-Haro, A. Rodriguez, J. V. Lopez-Bao, and F. Palomares. Wireless sensor network deployment for monitoring wildlife passages. *Sensors J.*, 10(8):7236–7262, 2010. [The initial version available as TD(10)10089].

[HBEH08]　V. Hassel, T. O. Breivik, P. E. Engelstad, and L. Henden. Analysis and measurements of the throughput in IEEE 802.11 mesh networks. Technical Report TD-08-551, Trondheim, Norway, June 2008.

[KCM10]　P. Kulakowski, E. Calle, and J. Marzo. Sensors-actuators cooperation in WSANs for fire-fighting applications. In *Proc. 6th IEEE Int. Conf. on Wireless and Mobile Computing, Networking and Communications (WiMob)*, pages 726–732, October 2010. [Also available as TD(10)12098].

[KELGH09]　P. Kulakowski, E. Egea-Lopez, and J. Garcia-Haro. Competitive MAC protocol for azimuth routing in large-scale wireless sensor networks. Technical Report TD-09-869, Valencia, Spain, May 2009.

[KGH09a]　P. Kulakowski and J. Garcia-Haro. Alternatives to face routing for localized nodes in wireless sensor networks. Technical Report TD-09-775, Braunschweig, Germany, February 2009.

[KGH09b]　P. Kulakowski and J. Garcia-Haro. Azimuth routing for large-scale wireless sensor networks. In *VIII Jornadas de Ingenieria Telematica (JITEL)*, pages 177–182, September 2009. [Also available as TD(09)775].

[KVAEL+10]　P. Kulakowski, J. Vales-Alonso, E. Egea-Lopez, W. Ludwin, and J. Garcia-Haro. Angle-of-arrival localization based on antenna arrays for wireless sensor networks. Technical Report 6, November 2010. [The initial version available as TD(08)544].

[MJ99]　J. Mitola and G. Q. Maguire Jr. Cognitive radio: making software radios more personal. *IEEE Personal Communications*, 6:13–18, 1999.

[MOV09]　M. Kaynia, G. E. Oien, and R. Verdone. Analytical assessment of the effect of backoffs and retransmissions on the performance of ALOHA and CSMA in MANETs. Technical Report TD(09)922, COST 2100, 2009.

[OLT07]　A. Ozgur, O. Leveque, and D. Tse. Hierarchical cooperation achieves linear capacity scaling in ad-hoc networks. In *IEEE INFOCOM*, pages 382–390, May 2007.

[PCD08a]　J. Papaj, A. Cizmar, and L. Dobos. Integration security as QoS parameter via security service vector in MANET. Technical Report TD-08-434, Wroclaw, Poland, February 2008.

[PCD08b]　J. Papaj, A. Cizmar, and L. Dobos. Security service vector in MANET. In *International Conference on Applied Electrical Engineering and Informatics 2008 (AEI2008)*, vol. 1, pages 130–138, September 2008. [Also available as TD(08)434].

[PCD09a]　J. Papaj, A. Cizmar, and L. Dobos. Integration of modified security service vector to the MANET. *Journal of Electrical and Electronics Engineering*, 2(3):135–140, 2009. [Also available as TD(09)917].

[PCD09b]　J. Papaj, A. Cizmar, and L. Dobos. Using cross layer design to collecting of SSV data in MANET. Technical Report TD-09-917, Wien, Austria, September 2009.

[PCD09c]　J. Papaj, A. Cizmar, and L. Dobos. Using cross layer design to collecting of SSV data in MANET. In *International Conference on Applied Electrical Engineering and Informatics 2009 (AEI2009)*, vol. 1, pages 66–73, September 2009. [Also available as TD(09)917].

[PU10] P. Popovski and Z. Utkovski. Secondary wireless communication through reordering of user resources. Technical Report TD-10-11096, Aalborg, Denmark, June 2010.

[RFMM10] S. Robitzsch, J. Fitzpatrick, S. Murphy, and L. Murphy. An experimental evaluation of co-channel interference in IEEE 802.11a multi-radio mesh nodes. Technical Report TD-10-10060, Athens, Greece, February 2010.

[RSO+09] A. G. Ruzzelli, A. Schoofs, G. M. P. O'Hare, M. Aoun, and P. Van der Stok. Coordinated sleeping for beaconless 802.15-based multi-hop networks. In *Proc. Intern. Conf. on Sensor Systems and Software (S-CUBE)*, Pisa, Italy, September 2009. [Also available as TD(08)437].

[SCV+10] C. Stefanovic, V. Crnojevic, D. Vukobratovic, L. Niccolai, F. Chiti, and R. Fantacci. Contaminated areas monitoring via distributed rateless coding with constrained data gathering. Technical Report TD-10-12016, Bologna, Italy, November 2010.

[SVCS10] C. Stefanovic, D. Vukobratovic, V. Crnojevic, and V. Stankovic. A random linear coding scheme with perimeter data gathering for wireless sensor networks. Technical Report TD-10-11010, Aalborg, Denmark, June 2010.

[Tra08] V. Tralli. Linear wireless networks with variable length and density: scaling laws and design considerations. In *IEEE Wireless Communications and Networking Conference (WCNC)*, pages 1763–1768, April 2008.

[VFB08] R. Verdone, F. Fabbri, and C. Buratti. Area throughput for CSMA based wireless sensor networks. Technical Report TD-08-526, Trondheim, Norway, June 2008.

[VSC+09] D. Vukobratovic, C. Stefanovic, V. Crnojevic, F. Chiti, and R. Fantacci. A packet-centric approach to distributed rateless coding in wireless sensor networks. Technical Report TD-09-752, Braunschweig, Germany, February 2009.

Part IV
Broadcasting, Body and Vehicle Communications

Chapter 13
Hybrid Cellular and Broadcasting Networks

Chapter Editor David Gomez-Barquero, Chapter Editor Peter Unger, Karim Nasr, Jussi Poikonen, and Kristian Nybom

Mobile multimedia broadcasting (i.e., delivering mass multimedia services to portable devices as cell phones or PDAs) is a fast emerging area with potential economic and societal impact. Digital broadcasting networks especially designed for mobile services are currently under deployment, DVB-Handheld (DVB-H) being the most representative technology in Europe with commercial services in several countries [FHST06]. On the other hand, Multimedia Broadcast Multicast Service (MBMS) extensions to the existing cellular networks are currently under standardization [HHH$^+$07, HHH$^+$]. Unfortunately, neither the broadcasting nor cellular systems alone can provide a cost effective service provision under all possible scenarios. Obviously, broadcasting over large areas makes sense only in scenarios with a lot of users consuming the same content, whereas cellular systems can deal much better with service personalization, but the existing infrastructure and currently allocated spectrum is too limited to hope for a mass market deployment.

New systems based on interworking/integration of cellular and broadcasting networks are currently under investigation in an effort to provide affordable mass multimedia services. The goal of this approach is to join the advantages of both networks. The benefits shown in several publications are manifold [Bri09, GB09, Ung10], such as reducing the system cost [GBBZC07], reducing the overall necessary data rate [UK06], improving the perceived area coverage [SLKAM09], reducing the electromagnetic exposure [USK07], etc.

This chapter covers Radio Resource Management (RRM) aspects for 3G MBMS services, coverage and capacity optimization aspects of DVB-H networks, as well as and network planning issues of hybrid cellular 3G and broadcasting DVB-H systems. The chapter is concluded with a brief overview of the second generation digital terrestrial TV standard DVB—Second Generation Terrestrial (DVB-T2).

D. Gomez-Barquero (✉)
Universidad Politecnica de Valencia, Valencia, Spain

P. Unger (✉)
Technische Universität Braunschweig, Braunschweig, Germany

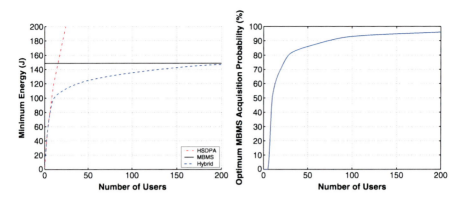

Fig. 13.1 Hybrid 3G MBMS-HSDPA file download results without time constraints. File size 512 KB. Acquisition probability 99%. MBMS reference case without macro-diversity [GBFAC09] (©2009 IEEE, reproduced with permission)

13.1 Cellular Multicasting/Broadcasting MBMS

In order to efficiently transmit the same content to several users simultaneously, the 3G standard has been enhanced with MBMS [HHH[+]07], introducing new point-to-multipoint radio bearers and multicast support in the core network. MBMS will lead to a better utilization of the existing radio resources provided by 3G, enabling the provision of new multimedia multicast services. MBMS can be deployed in the unpaired time-division duplex (TDD) 3G spectrum, which is currently unused, without impacting current 3G voice and data services (which employ the paired frequency-division duplex, FDD, 3G spectrum). Since its introduction in 3G release 6, MBMS is under continuous enhancement. Release 8, also known as iMB (integrated Mobile Broadcast), provides improved coverage and capacity performance, including single-frequency network operation mode [WDA10]. MBMS is also being considered in the 3G long-term evolution (LTE) standardization process [HHH[+]].

Reference [GBFAC09] focuses on the efficient multicast delivery of files in future 3G mobile networks with HSDPA and MBMS. Delivery configurations studied include: only p-t-p transmissions with HSDPA (one for each active user), a single p-t-m transmission with MBMS, and using both jointly in a hybrid approach employing HSDPA for error repair of the MBMS transmission. The optimum HSDPA and MBMS transmission configurations are investigated as functions of the time to deliver the file when used separately, and the optimum trade-off between the initial MBMS file transmission and the HSDPA error repair for the hybrid delivery. Similar issues but in LTE are investigated in [GBMFA[+]09].

As an example of the results obtained, Fig. 13.1 shows the minimum energy that can be achieved with a hybrid delivery for the MBMS reference case without macro diversity to obtain a 99% acquisition probability of a 512 KB file as a function of the number of users compared to using MBMS and HSDPA separately [GBFAC09]. The optimum MBMS acquisition probability increases as a function of the number of active users, as shown in the figure. This is simply because the efficiency of

MBMS compared to HSDPA increases with the number of users. For few users per cell, it equals to zero percent, meaning that MBMS is not used and all users are served with HSDPA. For more users than this threshold, the optimum delivery configuration corresponds to a hybrid MBMS-HSDPA delivery. The gain increases until the crossing point of the two reference curves using HSDPA and MBMS separately. This is the point where the highest energy reduction with the hybrid delivery is achieved (about 30% energy reduction in this case). After this point, the gain decreases with the number of users per cell. For very large number of users, there is very little gain using HSDPA as a complement of the MBMS transmission, and the optimum MBMS acquisition probability is close to the overall target. The gain brought by the hybrid delivery increases for higher acquisition probabilities targets, as it becomes very costly to serve the last percentage of users using only MBMS [GB09].

In [VP07a], a novel hybrid Connection Admission Control (CAC) scheme combining downlink transmission power and aggregate throughput, in the case of dedicated and shared connection setup respectively, is presented. The motivation for introducing this hybrid CAC approach is twofold: the need to use a representative resource metric on which to base CAC decisions and the necessity to take advantage of the particularities of the resource allocation procedure in true multicast environments with connection sharing. The proposed algorithm has been compared against a reference throughput-based CAC algorithm. Simulation results show a beneficial effect on cell capacity in terms of the number of users without observable degradation in the offered Quality of Service (QoS). In [VP07b], new approaches for power control for MBMS 3G services are proposed, demonstrating simultaneously the trade-off between the number of users and the average distance from the Node B when transmitting to a multicast group.

In [HRT07], it is investigated the use of adaptive relays in LTE in order to improve the capacity and coverage of MBMS.

13.2 Mobile Broadcasting DVB-H

13.2.1 Introduction

DVB-H (Handheld) is the European standard for terrestrial mobile TV. Italy was the first country to launch commercial services in summer 2006. Currently it is also on air in the main urban areas of Finland, Switzerland, Austria, the Netherlands, Hungary and Albania. Although alternative technologies exist, such as T-DMB (Terrestrial Digital Multimedia Broadcasting) in South Korea, ISDB-T (Integrated Services Digital Broadcasting—Terrestrial) in Japan, and Media FLO (Multimedia Forward Link Only) in USA; DVB-H seems today the most relevant mobile broadcasting technology worldwide, as it is also being deployed in South-East Asia and Africa. A key to the success of DVB-H is that it is introduced almost exclusively as a link layer on top of DVB-T (Digital Video Broadcast—Terrestrial), the European standard for digital terrestrial TV (DTT), which is the most widely adopted system in the world. This way it is possible to share the same network infrastructure (e.g., transmitters, antennas, multiplexers).

DVB-T was primarily designed for fixed rooftop reception, and the reception quality is degraded in mobile channels without additional processing such as receiver antenna diversity. Reception of DVB-T signals is not practical for battery-powered handheld devices with a single built-in antenna. The main features introduced with DVB-H are: discontinuous transmission technique known as time-slicing where data is periodically transmitted in so-called bursts, which reduces the power consumption of terminals and enables seamless hand-overs between cells with different frequencies; optional FEC mechanism at the link layer called MPE-FEC (Multi-Protocol Encapsulation FEC), which ensures more robust transmissions under mobility and impulsive interference conditions; and the use of IP. Multiprotocol encapsulation (MPE) is the adaptation protocol used to transport IP streams over MPEG-2 DVB-T transport streams.

With time-slicing data is periodically transmitted in bursts of maximum size 2 Mb. Each burst contains information of the time difference to the next burst of the same service. Terminals synchronize to the bursts of the desired service and switch their receivers front-end off when bursts of other services are being transmitted. When MPE-FEC is employed, each IP datagram burst is encoded with a Reed–Solomon (RS) code at the link layer to generate parity data which is transmitted with the source IP data to compensate for potential transmission errors [Him09]. The maximum percentage of errors per burst that can be corrected is equal to the rate of parity data transmitted. For example, a code rate 3/4 can cope with up to 25% errors. Therefore, MPE-FEC can cope with errors that represent a fraction of the burst, but it cannot recover from complete lost bursts.

Lost bursts have a major impact on the service coverage because they are perceived as a coverage discontinuity. This situation may be especially evident in the initial phases of the DVB-H network deployment, where mobile users experience temporary lack of coverage (outage areas) when moving across the service area. But mobile users may experience errors in covered areas as well because of the vulnerability of the physical layer to signal fading (fast fading and shadowing) and impulse noise.

13.2.2 Simulation Aspects

Conventional network planning for both analogue and digital TV broadcasting networks is based on a static approach that targets to guarantee a certain coverage level, or percentage of locations in which the average signal strength exceeds a given value with a target high probability. However, mobile broadcasting networks require also dynamic analysis over time, since the actual service quality perceived by the users depends on the distribution of transmission errors over time [Him09]. Therefore, QoS issues cannot be studied only from performance indicators that reflect an average within the service area, such as the coverage probability commonly used in network planning. A relevant question is then how to implement more detailed broadcast network analysis through simulations in a computationally feasible manner.

In simulations aiming at performance evaluation of contemporary wireless communication systems, significance of the communication channel is emphasized, since the degradation of a signal propagating from a transmitter to a receiver is strongly dependent on their locations relative to the external environment. Wireless mobile communications, where the receiver is in motion, presents additional challenges to channel modeling, as it is necessary to account for variation in the signal distortion as a function of time for each transmitter–receiver pair. Furthermore, in mobile broadcasting systems such as DVB-H thousands of mobile receivers may attempt to receive information simultaneously from a single transmitter. In developing and analyzing such systems, modeling the variation of the transmitter–receiver link quality on the network scale is a complicated task both computationally and descriptively.

In reference [Poi09], a computationally efficient, algorithmically simple simulation model for transmission of OFDM (Orthogonal Frequency Division Modulation) signals over time-variant, frequency-selective wireless propagation channels is presented. The considered model is based on applying a set of finite-state models to approximate modulation symbol errors occurring over the subcarriers of an OFDM signal. Parameters for the finite-state models are derived explicitly from propagation channel characteristics and the OFDM configuration used in the transmission. The validity of the proposed model is studied using theoretical capacity analysis and by comparing results with simulations performed using a reference propagation model. This kind of simulation models are directly applicable in performing system-level analyses of the physical level performance of mobile broadcasting systems such as DVB-H. In [Poi09], it is shown that considerable gain in computational efficiency can be achieved using this kind of finite-state modeling approach without significant loss in simulation accuracy.

The model outlined above is specified for simulating bit or byte errors in transmitting OFDM signals over time-variant wireless channels. For purposes of system simulation, it is beneficial to have the option of models with higher levels of abstraction, which include effects of digital processing such as error correction coding in the system. Since DVB-H is designed as a delivery system for MPEG-2 format Transport Stream (TS) packets, it is natural to model the performance of a single link in the DVB-H system in terms of TS packet errors. Finite-state packet error models and their application in a system simulator for DVB-H are described in [PPI09].

The structure of the proposed DVB-H system simulator is illustrated in Fig. 13.2 [GBPPC10]. The simulator consists of modules for modeling user mobility, radio coverage, physical layer performance—including multipath propagation effects—and receiver Quality-of-Service (QoS) measures. These modules can be specified independently according to the data available on the transmission scenario and the target application of the simulation. The underlying objective in the development of this simulator was that it could be used for modeling the time-variant reception quality of a large number of users moving within the DVB-H coverage area. Refer-

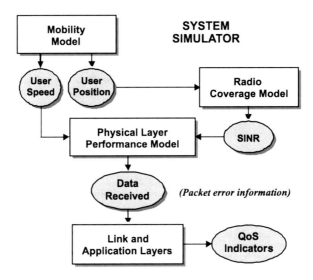

Fig. 13.2 Dynamic DVB-H system-level simulator structure [GBPPC10]

ence [GBPPC10] presents an overview of this simulator, its applications, and validation against field measurements. Recommended implementations for the simulator modules are the mobility models presented in references [KBW06, BB03], correlated log-normal [Gud91] or site-specific [SAZ03] radio coverage models, and the parameterized Markov models for physical layer performance estimation presented in reference [PPI09].

Application examples for such DVB-H simulator are: quality of service estimation for streaming services, coverage estimation for file delivery services, and RRM in hybrid DVB-H/cellular Networks, see references [GBPPC10] and [GB09]. For hybrid networks, the simulator can be used to investigate the potential cost delivery savings that can be achieved if a cellular network is used to repair errors of the DVB-H transmissions, as well as to dimension the post-delivery cellular repair phase. Another application would be to solve the problem of predicting the optimum configurations beforehand. As an illustrative example, Fig. 13.3 shows the evolution of the acquisition probability of a 16 Mb file in a hybrid network as a function of the time for different numbers of transmitted bursts using application layer FEC (AL-FEC) for DVB-H (see Sect. 13.2.10) and for different numbers of active users [GBPPC10]. The optimum transmission configuration will depend on the relative cost of transmitting one additional repair burst in DVB-H compared to using the cellular network to serve a given percentage of users. In this type of analysis, both the mobility and the number of users within the hybrid network are crucial simulation parameters, and the computationally efficient dynamic system simulator is of great benefit by reducing both the descriptive and the computational complexity of the simulation task.

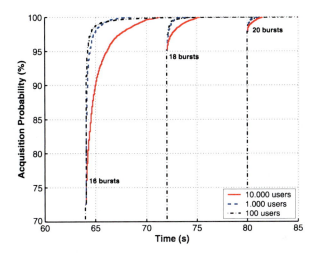

Fig. 13.3 File acquisition probability of a 16 Mb file in a hybrid DVB-H/HSDPA network. Vehicular rooftop users [GBPPC10] (©2010 IEEE, reproduced with permission)

13.2.3 Network Planning Algorithms for SFNs

Radio network planning and optimization represent a very important part of the network design due to the significant influence to the network cost and signal quality. In general, planning and optimization of radio networks is a highly complex task. Sophisticated algorithms and methods are needed in order to achieve efficient solutions in a reasonable amount of time.

In a Single Frequency Network (SFN) all transmitters are synchronized and utilize the same frequency and transmit exactly the same data. The advantages of SFNs are manifold: better coverage, less interference, less power, higher reliability, and an efficient use of the spectrum [Mat05]. An SFN has to be carefully synchronized in order to avoid inter-symbol interference (ISI), and the guard intervals of the OFDM system limits the distance between the transmitters. Therewith, static delays can be adopted for each transmitter, which influences the ISI of the network coverage area.

The delay of signals originated from different transmitters has to be analyzed in order to estimate the ISI. As a reference model for estimating the amount of interference or useful signal power Eq. (13.1), which has been defined in [BH03], is typically used. If $w(\Delta t) = 1$, no inter-symbol interference and a value of zero stands for full contribution to interference, since the signal arrives outside the guard interval. The weighting value depends on the length of the useful part of the OFDM symbol T_u, the length of the guard interval T_g, and the relative time of arrival Δt, which is compared to the selected position of the FFT window. Signals arriving within the guard interval contribute to the useful signal power. Due to the pilot carriers pattern employed, the total loss of constructive signal components occurs beyond a relative delay of $T_u/3$,

$$w(\Delta t) = \begin{cases} 0 & \text{if } \Delta t \leq T_g - T_u/3, \\ (\frac{T_u + \Delta t}{T_u})^2 & \text{if } T_g - T_u/3 < \Delta t \leq 0, \\ 1, & \text{if } 0 \leq \Delta t \leq T_g, \\ (\frac{T_u + T_g - \Delta t}{T_u})^2 & \text{if } T_g < \Delta t < T_u/3, \\ 0, & \text{if } T_u/3 < \Delta t. \end{cases} \quad (13.1)$$

In radio network planning the signal-to-interference-and-noise ratio (SINR) has to be determined for each location of the selected scenario in order to estimate the signal quality at the receiver's side. A signal propagation model is used to calculate the receiving signal power. For SFNs, the contributing signals with the power sum of correlated signals and interfering signals have to be considered. In order to optimize an SFN, several parameters are to be included; for example: the transmitter position, antenna hight, antenna sectorization and orientation, transmitting power and the static delay. Since the number of parameters of an SFN is high, the search for an optimal solution is highly complex. The computational effort increases exponentially with the number of transmitters. Therefore, operation research methods have to be used in order to find a good solution with an acceptable amount of time.

Heuristic methods have been widely excepted and are very prominent approaches in the field of network optimization, such as Simulated Annealing and Genetic Algorithms.

Two examples for optimizing the network's coverage by using the simulated annealing approaches have been described in [GLRBL[+]10] and [RLE[+]09]. In [GLRBL[+]10] the static delay is optimized for DVB-T networks, and the applicability is shown by means of three different scenarios, which have been analyzed including a digital terrain model for realistic signal propagation predictions. The optimization objective is to minimize the ISI by searching the optimal set of static delays for the involved transmitters.

The paper [RLE[+]09] deals with DVB-T2 network optimization also using the Simulated Annealing approach. A realistic network scenario has been analyzed, and the coverage of an initial network setup has been improved applying SA optimization.

Another promising heuristic approach has been described in [MAA10]. The Particle Swarm Optimization (PSO) algorithm is used to optimize the coverage area of SFNs by adapting the transmission parameters, e.g., the static signal delay, antenna gain, and sector orientation. The PSO approach is a modern heuristic optimization technique similar to Genetic Algorithm or Simulated Annealing. As shown in Fig. 13.4, the presented approach comprises a signal propagation prediction model for determining the signal strength and the signal delay at each location of the scenario using a realistic scenario and transmitter setup. Furthermore, the PSO approach including a fitness function is used in order to find an optimal solution for the transmitter parameters.

Different PSO schemes and different antenna options (omnidirectional and sectorized antennas) have been considered for the optimization process. The presented approach has been applied for an SFN with 18 transmitters. The results show that the approach is able to reduce the interference and to increase the overall coverage.

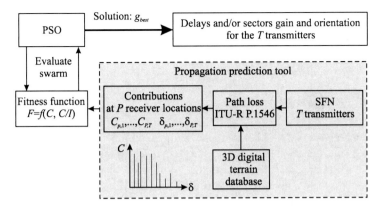

Fig. 13.4 Block diagram of the Particle Swarm Optimization approach [MAA10]

13.2.4 Network Planning for SFNs with Local Service Areas

Two different approaches are currently being applied when deploying broadcast networks: Multi Frequency Networks (MFN) and Single Frequency Networks (SFN). In an MFN, each transmitter utilizes a dedicated frequency for transmission, and thus, the transmitters operate independently and may transmit different services. Frequency planning has to be performed to apply an appropriate frequency reuse factor in order to avoid inter-cell interference. The SFN approach uses a single frequency layer, whereas all transmitters are synchronized and transmit exactly the same data. The so-called SFN gain allows for more cost efficient network deployment, especially for large-scale broadcast networks.

In some scenarios, the constraint of transmitting the same data within the entire coverage area may be inefficient. Some services are then provided in locations, where they are requested by the users rarely or not at all. A location-dependent service provision would be more suitable and less resource demanding. This is one of the main benefits of hybrid cellular and broadcasting networks. But for the broadcasting network, an ideal deployment approach would be a combination of an MFN and an SFN, which provides global coverage with services which are consumed within the entire coverage area, and local services, which are distributed in specific regions only. In [MU07] a new approach has been presented, which enables local services within an SFN mode for the DVB-H standard. This approach makes use of the time slicing technology, whereas some slices are used for global and the others for local services. The advantage is that a significant amount of capacity can be saved, since the reserved slices for local services can be reused in the different local service areas (LSA). Interference is caused between these areas. A feasibility study in terms of evaluating the achievable coverage sizes of local service areas has been performed in [UM07] with theoretic models and realistic scenarios. The coverage areas of the global services still have the traditional SFN advantages. Since the same transmitters are used for both local and global services, the respective coverage areas are interdependent. This issue is investigated in [UK08], which discusses methods for a joint optimization process.

13.2.5 Use of Gap-Fillers

On-channel repeaters (gap-fillers) are receiving increasing attention in digital broadcasting systems due to the advantages they can offer for single frequency networks deployment and coverage. The main purpose of a gap-filler is to extend the coverage in a cost-effective manner. The gap-filler receives the signal off air, amplifies it, and then retransmits it on the same frequency as received. There are several issues and trade-offs related to the design of such systems. These issues include echo cancellation and channel estimation for stable operation of the repeater, pre-distortion to improve amplifier linearity, and Multiple Channel Power Amplification (MCPA) that allows multiple radio channels to use a single amplifier hence reducing the costs of the overall network deployment. The general requirements are a high gain (usually above 80 dB), large dynamic range, good selectivity, and good isolation between the transmitter and receiver antennas.

The main problem with gap-fillers is the effect of strong feedback signals from the transmitting antenna and surrounding environment toward the receiving antenna. These feedback signals can interfere with the weak received signal of the repeater causing oscillations and system instability. The unwanted feedback signal is also known as coupling loop interference and arises when the gain of the amplifier is larger than the isolation between the antennas at the two ends of the repeater as well as from unwanted reflections (or echoes) from the surrounding environment. Carefully positioning the antennas, increasing the physical separation distance between the transmitter and the receiver antennas of the gap-filler is necessary but generally not sufficient. Effective cancellation can be obtained by combining the physical separation of the antennas together with digital signal processing (DSP) techniques to model and remove the feedback channel between the antennas. Another main requirement is an effective technique for unwanted channel estimation. The possible approaches for echo cancellation using DSP techniques fall in three main categories time domain, space domain, or a hybrid of both.

Figure 13.5 shows the block diagram of a gap-filler based on an open-loop channel estimator, [NCBG07]. This approach guarantees the stability of the system and allows a low processing delay, as well as an operation with comparatively poor isolation between receive and retransmission antennas.

A low power training sequence is buried in the transmitted OFDM signal for unwanted channel estimation based on the correlation principal. At start-up, only the training sequence is fed to the transmitting antenna for quick initial channel estimation by opening the switch. After the first channel estimates are acquired, the switch is closed for normal operation. This approach is selected since the applied CAZAC (Constant Amplitude Zero Autocorrelation) training sequence is uncorrelated with the transmitted OFDM signal and hence ensures good estimation. The update rate of

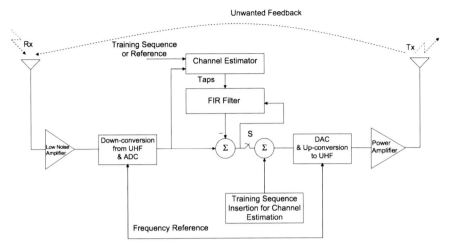

Fig. 13.5 Block diagram of the on-channel repeater with echo canceller and channel estimator [NCBG07]

the channel estimator should match the rate of change of the channel. The drawback of this approach is that the training sequence will act as an unwanted interference and would slightly degrade the OFDM signal. The training sequence level should hence be as low as possible. Once the channel complex coefficients are estimated, the taps of an FIR (Finite Impulse Response) filter are updated, and the output of the filter is subtracted from the input ideally canceling the unwanted echoes and ensuring the stability of the on-channel repeater, which is the main purpose of the echo canceller. This on-channel repeater architecture results in a universal design that can be used for any DVB or DAB standard, since it does not rely on standard specific information like scattered pilots to estimate the unwanted channel. Details of the performance of this gap-filler design can be found in [NCBG07].

13.2.6 Soft Handover

DVB-H operates using time-slicing, which is a form of time division multiplexing, where data is periodically transmitted in bursts. The main goal is to reduce the battery power consumption of the terminals, but it also enables the receivers to use the intervals without data transmission for monitoring the radio environment. If the receiver detects a strong signal in an adjacent cell during a burst interval, it can switch to the center frequency of the alternative signal and continue receiving the service if the cells are suitably synchronized. This kind of abrupt cell switching is called *hard handover*. Another approach is *soft handover*, where the receiver switches between the adjacent cell transmissions, receives both signals, and combines them. With synchronized DVB-H transmissions, this combining can be performed on the packet level, using correctly received packets from either service. In principle, this results in considerable diversity gain in the reception, since the received signals differ in transmission time, frequency, and spatial propagation route. In [May08], the results of a measurement campaign studying soft handover in a DVB-H network are presented, and a significant gain in decrease of packet errors in the reception is reported.

13.2.7 Interference Issues

In situations where different systems operate in the same area and in the same frequency band, inter-system interference issues (i.e., interfering signals from one system to the other) may occur. In a DVB-H system, two inter-interference system issues can be considered:

- A DVB-H transmitter is the source of disturbing signals.
- The DVB-H receivers are disturbed by the other radio system.

In the first case, a DVB-H transmitter interferes other wireless communication system deployed in the same area. Interference issues between DVB-T and cellular systems was investigated in [BGGR09]. In the second case, other radio systems can interfere the operation of a DVB-H network. The immunity of DVB-H against external interferences was investigated in [Moz09b, Moz09a], and [Moz10], by means of simulations and laboratory measurements. Figure 13.6 shows the simulation platform, which includes both physical and link DVB-H layers [Moz09b]. The simulation platform was validated with laboratory measurements [Moz09a].

Two different interfering signals were considered. One signal was a narrowband signal characteristic of wireless remote control systems [Moz10]. Figure 13.7 shows an example of the results obtained. The narrowband signal interferes only some part of the DVB-H signal in the frequency domain, and hence only some subcarriers are disturbed. We can note how the bit error rate can be limited by the interference instead of noise. However, the frequency interleaving and forward error correction performed at the DVB-H physical layer can efficiently correct this type of errors.

The other disturbing signal considered was a DVB-T2 signal. In this case, the interfering signal has flat spectrum, which could be compared to Gaussian noise. However, the interference issues should take into account different parameters that may affect the signal in the time and frequency domain.

13.2.8 Transmit Diversity

Transmit diversity techniques with multiple antennas have long been proposed to improve the performance and capacity of wireless systems. In transmit Delay Diversity (DD) [ZCL$^+$07], the same information is transmitted from two or multiple spatially separated antennas but with a delay of several OFDM symbol intervals, as

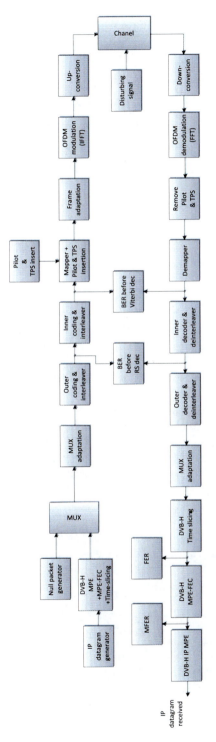

Fig. 13.6 Block diagram of the DVB-H simulator

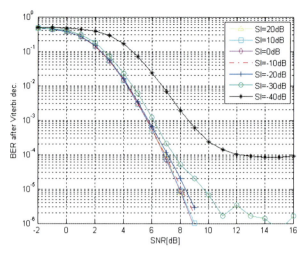

Fig. 13.7 Influence of a narrowband signal in DVB-H. Bit error rate before Viterbi vs. Signal-to-Interference (SI) ratio. DVB-H transmission mode: FFT 8K, GI 1/4, modulation QPSK, code rate 1/2. AWGN channel

illustrated in Fig. 13.8. The effect of diversity is reducing the probability of observing deep fades at the receiver and hence enhancing reception quality. A diversity gain will always be observed in Non-Light-Of-Sight (NLOS) situations provided that the observed signals have uncorrelated fading and the receiver is able to exploit it. The result is that the number of transmitter sites and the total network radiated power required to achieve effective coverage are reduced.

In [BBZ+08], laboratory measurements were conducted to investigate the potential benefits of transmit diversity for different DVB-H configurations and propagation scenarios. Several channel models, cross correlation values, and Doppler shifts were tested for different DVB-H parameters. The overall transmitted power was normalized for all measurements to ensure fair comparisons (i.e., either 100% of the power was transmitted through a single antenna, or 50% of the power was transmitted by each of the two antennas). The Bit Error Rate (BER) metric was found to be impractical as a figure of merit of the system due to the bursty nature of the errors, and hence the ESR (Erroneous Second Ratio) criterion was used. The ESR is the percentage of erroneous seconds observed during a defined integration period. A transmission duration of 20 minutes was found to be adequate for most scenarios and resulted in a stable and accurate assessment. The ESR was measured

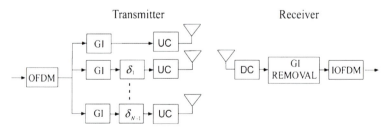

Fig. 13.8 Transmit delay diversity applied to an OFDM system [ZCL+07] (©2007 IEEE, reproduced with permission)

13 Hybrid Cellular and Broadcasting Networks

Table 13.1 Comparison of transmit and receive diversity gains at different Doppler frequencies

Doppler Frequency	MISO Gain	SIMO Gain
10 Hz	3 dB	9.5 dB
170 Hz	7 dB	9.8 dB

as a function of the Carrier-to-Noise Ratio (C/N) at the receiver input at a Threshold of Visibility (ToV) of 3%. This ToV is equivalent to two erroneous seconds every minute. It should be pointed out that the same generic transmit diversity techniques can be applied to other OFDM-based broadcast systems (e.g., DVB-T/SH, DAB, DMB, MediaFLO).

The results obtained in [BBZ+08] showed that transmit DD significantly improves receiver performance until the impact of Doppler on inter-symbol interference becomes unacceptably high for a particular receiver's implementation. It has also been confirmed through practical measurements that the gain predicted form theoretical simulations can be realized or exceeded using actual equipment under realistic propagation conditions. It was also shown that the diversity gain obtained from Maximal Ratio Combiner (MRC) receive diversity is higher than transmit diversity gains, especially at low Doppler. The gap between MISO and SIMO gains reduces considerably as the Doppler frequency increases.

Table 13.1 compares SIMO and MISO gains for different Doppler frequencies for QPSK targeting high-speed mobile and handheld applications. The selected environment for this test is Typical Urban (TU6), and the transmit delay diversity value is 1 s. Tests were conducted for low and high Doppler values and for different channel cross correlations. The diversity gain was evaluated for a reference ESR value of 5%.

The results were further validated with DVB-T/H field trials [BNB+10]. It was observed that performance predicted previously through theoretical investigations and laboratory measurements can be realized in real-world situations. For DVB-T, the gain was found to be 4 dB in terms of CNR reduction and 1 Mbps in terms of data bitrate improvement. For DVB-H, the gain was considerably smaller (1 dB), because the DVB-H chipset implementation was already optimized for high mobility reception and because of the time interleaving effect of the MPE-FEC encoding.

13.2.9 File Delivery

Apart from audiovisual streaming services, also other file-type services can be offered over DVB-H. These services involve downloading files from a dedicated broadcast server, such as video and audio clips, software updates, etc. Since these services typically have a zero Bit Error Rate (BER) tolerance, additional FEC coding is required to ensure robustness in the transmission. In order to improve the DVB-H robustness for file delivery, Raptor codes have been adopted as Application Layer FEC (AL-FEC) [Sho06] to be used instead of MPE-FEC. The key with

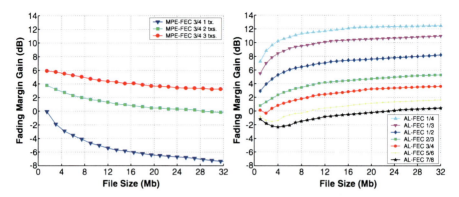

Fig. 13.9 Fading margin gain for 95% file acquisition probability vs. File size [GB09]. FFT 8K GI 1/4 16-QAM 1/2. TU6 $f_d = 10$ Hz. Uncorrelated shadowing $\gamma = 5.5$ dB, $d_{corr} = 0$ m

AL-FEC is that it can provide protection across the whole file (i.e., across multiple bursts), rather than across one burst.

Figure 13.9 shows the fading margin gain as a function of the file size for MPE-FEC when the file is transmitted once, twice, and three times, and for different code rates with Raptor AL-FEC [GB09]. The fading margin gain is defined as the difference between the reference CNR threshold for conventional streaming services (i.e., CNR requirement for 5% burst error rate with MPE-FEC code rate and burst size 2 Mb) and the required CNR to achieve a 95% file acquisition probability.

In the figure we can notice that MPE-FEC suffers a penalization for large files. The fading margin gain is negative for the case with one single file transmission for files spanning more than one burst. This means that the area coverage is reduced compared to conventional streaming services. In contrast, with AL-FEC the area coverage for file download services is higher than streaming services with MPE-FEC. The reason is that it is possible to benefit from the time diversity of the mobile channel during the transmission of the files that span several bursts. The gain for large files is thus twofold, because MPE-FEC performs worse and multi-burst FEC performs better.

Since AL-FEC codes typically can be implemented in software, this allows for the introduction of new codes. In [NV07] different sparse-graph AL-FEC codes are compared when decoded iteratively. It is shown that sparse-graph FEC codes can quite easily be designed to achieve small reception overheads at medium to high Signal-to-Noise Ratio (SNR) levels. Typically, the performance of AL-FEC codes is measured in terms of the reception overhead, which is defined as the ratio of the amount of received data to the amount of source data. In [NV07], the AL-FEC codes are compared in terms of the reception overhead vs. SNR. In [VSB07] two subclasses of Low Density Generator Matrix (LDGM) codes suitable for large content distribution applications are analyzed. The subclasses of the codes are staircase and triangle LDGM codes, and the parity-check matrices are designed at random, using the Progressive Edge Growth (PEG) algorithm [HEA05] and the Approximated Cycle Extrinsic Message Degree (ACE) constrained code construction algorithm

[TJVW04]. It is shown through Monte Carlo simulations that short codes are highly dependent on the code design, but when the code length is increased, the average reception overhead factor decreases quickly without respect to the design algorithm. Therefore, in the limit of long code lengths, the design of the codes is reduced to the selection of suitable degree distributions.

When designing communication standards, the coding gain is traditionally used as a metric for determining the reduction in SNR that a certain FEC code provides to the system for achieving a certain BER. Nybom [Nyb09] introduces a new metric for AL-FEC codes in broadcasting scenarios, called the *receiver coding gain*. The receiver coding gain determines the gain in energy consumption for using a certain FEC code as compared to uncoded transmissions, and it is consequently implementation dependent. However, it can give guidelines to how expensive it is to use an FEC code from the receiver point of view. The receiver coding gain is defined as

$$G_{dB} = 10 \log_{10}\left[\frac{\varepsilon_0}{\varepsilon_c + \frac{T_b E_c}{L_0 \bar{P}}}\right], \qquad (13.2)$$

where ε_0 and ε_c are the reception overheads for uncoded and coded transmission, respectively, T_b is the transmission rate, E_c is the total energy used by the Forward Error Correction (FEC) decoder, L_0 is the amount of source data, and $\bar{P}$ is the average power consumed by the receiver.

13.2.10 Multiburst FEC for Streaming Delivery

Multiburst FEC is currently only standardized in DVB-H for file delivery services as AL-FEC with Raptor coding. But Raptor codes can be applied to streaming services in DVB-H as well. This can be done either at the application layer with AL-FEC or at the link layer with MPE-iFEC [GB09], achieving practically the same performance (only some implementation and signaling specific aspects differ). It is also possible to provide a multiburst protection of the transmission with the same Reed–Solomon code adopted in MPE-FEC employing a sliding window approach [SLK+09]. However, existing DVB-H terminals are not specified to handle the increased memory requirements.

Compared to the conventional intra-burst MPE-FEC mechanism, for streaming services, multiburst FEC can improve the transmission robustness by delivering the content as a succession of larger source blocks spanning several bursts. The potential gain is very significant, as it is possible to partially cope against slow fading (shadowing). Reference [GBU08] quantifies the reduction in the fading margin required to cope with shadowing that can be achieved in a time-sliced system with multiburst FEC as a function of the code rate and the number of jointly encoded slots. Performance evaluation results of multiburst FEC with DVB-H field measurements with Raptor and sliding Reed–Solomon MPE-iFEC encoding can be found in [GBGC09] and [GGBC08], respectively. It should be pointed out that the use of Raptor coding with AL-FEC can be also beneficial to provide mobile DVB-T

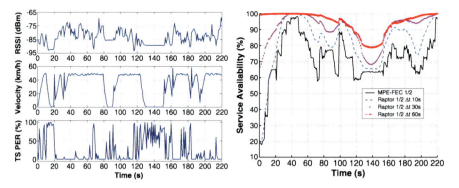

Fig. 13.10 Example of simulated trajectory (*left*) and service availability with multiburst FEC (*right*) [GBUKC10] (©2010 IEEE, reproduced with permission)

services, providing time interleaving at the application layer. Interested readers are referred to [GGBSL09] and [LGGBC09].

The gain of multiburst FEC compared to MPE-FEC depends on several factors, such as the code rate, the interleaving depth in number of bursts jointly encoded, the FEC scheme employed, and the target residual error rate. However, the actual gain is very difficult to quantify in real life, as it depends on the time evolution of the transmission errors experienced by the users and hence on the velocity and trajectory of the users. One important parameter is, for example, the coverage level perceived by the users. As if it is too high, there will not be any error to correct, whereas if it is too low, it will not be possible to recover any data. The gain is thus not constant across the service area, and it cannot be directly included in the link budget.

For the conventional intra-burst MPE-FEC scheme, coverage estimation is a semi-static process due to the reduced burst duration of few hundreds of milliseconds. In a realistic scenario, a coverage map can be easily derived directly mapping the signal availability in each small area (coverage map grid resolution) into service availability level. This cannot be done with multiburst FEC, as the probability of correctly receiving a burst depends on previous and/or future reception conditions. A valid approach for coverage estimation for multiburst FEC is to perform dynamic system-level simulations, as proposed in [GBPPC10] and outlined above in Sect. 13.2.2. As the performance of multiburst FEC depends on the mobility and trajectory of the users, accurate received signal strength predictions and realistic mobility patterns are required, together with an accurate and computationally efficient DVB-H performance model.

Figure 13.10 shows an example of the received signal level (RSSI), velocity and TS packet error rate at the physical layer of one simulated trajectory in the city of Braunschweig (Germany) [GBUKC10] and the service availability for multiburst FEC 1/2 with Raptor coding and different latencies (that correspond to different number of bursts jointly encoded). The reference case with MPE-FEC 1/2 is also shown for comparison.

In the figure one can see the increasing gain obtained with multiburst FEC as a function of the latency. But more importantly here, shown results can be directly translated into space, and thus a service coverage map can be plotted. It is worth mentioning that results are valid for both forward and backward directions, but they are specific for the particular velocity trace considered.

13.3 Network Planning and RRM for Hybrid Cellular and Broadcasting Networks

The combination of the 3G mobile network UMTS and a broadcast network such as DVB-H has shown high potential to efficiently serve the user's requests. By using both technologies, the advantages of both network types can be exploited. UMTS offers individual services on the one hand and is expected to be deployed with full service coverage. On the other hand, a high data rate broadcast downlink is provided by DVB-H, which enables the combination of several user's requests for a very efficient service delivery.

In a hybrid service delivery mode both networks are interdependent, and the overall performance depends on the cooperation of both networks. This makes an optimized network planning method and radio resource management indispensable. Both radio network planning and radio resource management are nontrivial tasks for single network, and so a combination of two networks is even more challenging. This section deals with approaches for efficient planning and radio resource management for hybrid networks.

13.3.1 Hybrid Network Planning

In this section the methods to efficiently deploy a hybrid network structure, which consists of both technologies UMTS and DVB-H, are summarized. These methods are described in detail in [Ung10]. The performance of both technologies is interdependent since requested traffic can be switched from one network. The influence of the load switching on the network performance is discussed in [UK09].

The optimized cooperation of both networks, i.e., an optimized load balancing approach, has to be considered, which is responsible for an optimized allocation of the network resources. One possible solution has been proposed in [Heu09]. It focuses on two service types, streaming multimedia and file downloaded, and describes the dynamic and average performance of hybrid networks while optimizing the overall end-to-end delay. One of the key measures is to collect requests for popular content and serve the relevant users by a single transmission by the broadcast network.

For an optimized network planning process, the selected load balancing scheme has to be considered. In the case of network planning, besides the overall service

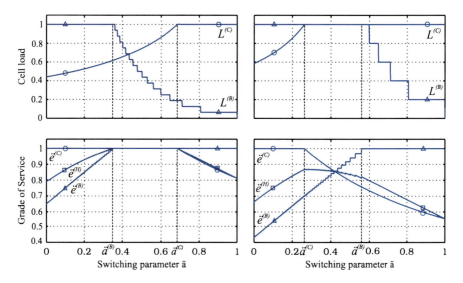

Fig. 13.11 Performance indicators cell load and grade of service in a hybrid network for a single cell depending on the switching ratio δ. The *left side* shows an ideal case, the *right side* shows a problematic case

delivery speed, the grade of service is the key performance criteria which describes the amount of traffic served compared to the overall requested traffic. A model has been developed which includes the grade of service estimation for both the UMTS and the DVB-H network. Using this model, criteria for deployment strategies and network parameter can be determined.

The performance of the UMTS and the broadcast network can be described by means of an equation-based model. As a basis, the existing performance model of UMTS, the interference coupling system [Gee08], describes the performance and the interdependencies of UMTS cells and can be efficiently used for simulating also for large-scale UMTS networks. This model has been used and extended by a broadcast system model and the combination of both to develop a hybrid network system model [UK09]. Key performance parameters are the cell load L and grade of service λ of the UMTS cells and the broadcast cells. A combined criterion for the grad of service of hybrid networks has been defined and denoted as $\lambda^{(H)}$. Furthermore the load switching parameter δ has been introduced, which describes the ratio of traffic switched between the networks. Due to the network capabilities, the switching of content is not possible for the entire range but limited. In [UK09] the so-called switching bound concept has been introduced which defines the constraints of load switching based on the performance of UMTS and broadcast cells. The switching bound concept has been identified as a strong tool to define the necessary network structure and parameters. Depending on the user requirements, the planning and network specifications, an optimized hybrid network can be developed.

Figure 13.11 shows an example of the performance of a hybrid network. Both performance indicators, cell load and grade of services are shown. In both cases the curve for the UMTS ($L^{(C)}$ and $\lambda^{(C)}$) and the broadcast network ($L^{(B)}$ and $\lambda^{(B)}$) are presented for different load switching values δ. Furthermore, the overall grade of service ($\lambda^{(H)}$) for the hybrid network is presented. The cellular switching bounds $\delta^{(C)}$, which is defined by the UMTS cell parameter, and the broadcast bound $\delta^{(B)}$, which is defined by the DVB-H cell parameters, are shown representing the operating range of the load switching parameter.

Both left-hand figures show an ideal case. Load switching can be applied between both bounds in order to reach a perfect grade of service ($\lambda = 1$). The cell load of both network types shows that all traffic can be transmitted by both network layers if the load switching parameter value is in between the switching bounds. Otherwise, the cells are overloaded, and some users cannot be served. At the right-hand side, a problematic case is shown. The amount of traffic is too high, so that the entire traffic load cannot be served by both cells independent on the grade of switching. Nevertheless, tt is shown that the grade of service of the hybrid network can be increased compared to the single network case.

Using these information, the network parameter can be adapted in order to increase the network performance. As one of the hybrid network planning rules, it has been identified that the broadcast bound needs to be the lower bound and the UMTS bound the upper one. In this case perfect grade of service can be guaranteed.

13.3.2 Hybrid RRM

In this subsection, we present illustrative results on the potential cost delivery savings that can be achieved if an evolved 3G cellular network with HSDPA capabilities is used to repair errors of the broadcast transmissions of a DVB-H network [GB09]. In a realistic scenario, there may be users that experience significantly worse DVB-H reception conditions than the majority, so that it may be more efficient to serve them through the cellular network. We will not address the decision on whether to transmit the service via DVB-H or the cellular network by applying a load balancing algorithm. We will consider that the service is popular enough so that it should be transmitted through DVB-H. We focus on file download services because they exhibit the largest potential from a hybrid cellular-broadcast delivery point of view (for streaming delivery, a progressive deployment of the broadcast network as a complement of the cellular network where and when needed should be the preferred option). File delivery requires an error-free reception of the files, which cannot be guaranteed for every user after a DVB-H transmission because some users might have experienced too bad reception conditions. Moreover, it does not have very stringent time constraints such as streaming delivery. To enable an easy and efficient implementation of the repair mechanisms, we adopt the use of forward error correction at the application layer (AL-FEC) with Raptor coding. With AL-FEC, specific source packets are not required to be retransmitted, and repair transmissions consist of additional parity packets that can be used for all users. Hence, repair data can be efficiently delivered with p-t-m broadcast/multicast transmissions with DVB-H or MBMS.

From an RRM point of view, the problem of efficient file delivery in hybrid cellular and broadcasting systems is more complex than when each network is operated separately. The reason is that an additional variable is needed to relate the importance of the resources of both radio accesses since they cannot be directly compared. We adopt a cost-based RRM framework where the delivery cost is proportional to the amount of radio resources employed in each network [Bri09]. The broadcast cost is proportional to the amount of data transmitted in DVB-H. The cellular cost is proportional to the transmission energy, defined as the product of the transmit power, P_{tx}, and the active transmission time, T_{tx}:

$$E_c = P_{tx} \cdot T_{tx}. \qquad (13.3)$$

Note that by minimizing the energy, the radio resource usage is minimized, and the cell capacity is maximized.

Assuming that the original source file is initially transmitted through DVB-H (with a given burst size and cycle time), the optimum repair configuration will be the one that completes the delivery in due time and minimizes the delivery cost:

$$C_R = c_b \cdot n_b + c_c \cdot \sum_i^K E_{c,i}, \qquad (13.4)$$

where n_b is the number of repair bursts transmitted in DVB-H, and c_b and c_c are the costs per transmitted DVB-H burst and per energy unit (J) of the cellular network, respectively.

The optimum amount of parity data that should be transmitted in DVB-H will depend on the number of active users and the number of cells within the service area, K. Note that the broadcast cost is independent of these values. However, the optimum configurations will heavily depend on the ratio c_b/c_c, which is difficult to quantify in real-life implementations. Moreover, prices may vary over time. For example, according to the congestion level experienced by each radio access (the cost of delivering certain amount of data over the cellular network during peak hours is higher compared to off-peak hours).

Figure 13.12 shows the cost delivery savings that can be achieved combining DVB-H and HSDPA in the hybrid network depicted in Fig. 13.3, compared to the reference case where the DVB-H network and the HSDPA network are used separately [GB09]. The cost saving is pictured as a function of the ratio c_b/c_c for different numbers of active users in the system. Savings of up to 55% are achieved. In the situations where there are no savings, the optimum configuration consists on using only DVB-H or HSDPA.

We can observe that the savings initially increase as a function of the ratio c_b/c_c. In this area the broadcast cost is smaller than the reference case using only HSDPA. The maximum gain occurs for the c_b/c_c ratio which provides the same cost using HSDPA and DVB-H separately. Larger c_b/c_c values than this threshold imply that cellular network is more efficient than the DVB-H network (when operated independently). The maximum gain also depends on the number of active users in the system. The lower the number of users, the higher the maximum gain. However, the differences are rather small. For a given number of users, there is a c_b/c_c value which minimizes the delivery cost employing jointly DVB-H and HSDPA.

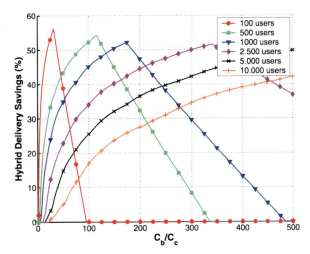

Fig. 13.12 Hybrid DVB-H/HSDPA delivery cost savings vs. Ratio c_b/c_c [GB09]

13.4 Second Generation Terrestrial Broadcasting DVB-T2

DVB-T2 is the second generation standard for terrestrial television broadcasting [VBT$^+$09]. The standardization process was launched in 2006 with the study on commercial requirements. In 2007, a call for technologies was issued and finally in 2009 DVB-T2 was standardized. As compared to its predecessor DVB-T, DVB-T2 has a more efficient physical layer using state-of-the-art technologies to achieve close to optimal performance in terms of true bit-rate in Quasi Error Free (QEF) transmissions. The use of Low Density Parity Check Code (LDPC) and Bose–Chaudhuri–Hocquenghem (BCH) coding instead of convolutional and Reed–Solomon (RS) coding gives a significant increase in robustness. Due to the more efficient FEC coding, DVB-T2 supports both higher-order Quadrature Amplitude Modulation (QAM) constellations and larger Fast Fourier Transform (FFT) sizes. Smaller guard intervals are supported with fewer pilots required, and more flexible pilot structures are available, resulting in a notable increase in capacity. Moreover, DVB-T2 uses rotated QAM constellations combined with cyclic Q-delay, effectively giving receivers the possibility to cope with erasures. DVB-T2 supports Multiple-Input Single-Output (MISO) antenna reception, has a more efficient interleaving scheme than DVB-T, and provides service-specific robustness through a concept called the Physical Layer Pipe (PLP). Each service is assigned to a PLP, which is a virtual data stream going through the DVB-T2 physical layer, and each PLP can have a code rate, modulation, and time interleaving length of its own. Because of the PLPs and the flexible interleaving possibilities in DVB-T2, either time diversity or power saving can be obtained, simply by distributing QAM cells belonging to a specific PLP in different ways inside the Orthogonal Frequency Division Multiplexing (OFDM) symbols. DVB-T2 also includes two Peak-to-Average Power Ratio (PAPR) techniques to combat disadvantages with OFDM signals resembling Gaussian noise. The time and frequency synchronization is also more efficient in DVB-T2 through the use of a preamble, that allows for a much wider choice of transmission

Table 13.2 Comparison of CNR for QEF BER with code rate 1/2 (Ricean channel with ideal channel estimation)

Modulation	RS-Viterbi	BCH-LDPC	Difference
QPSK	3.6 dB	1.7 dB	1.9 dB
16-QAM	9.6 dB	4.7 dB	4.9 dB
64-QAM	14.7 dB	8.2 dB	6.5 dB

parameters without increasing the overall synchronization time [VBT[+]09]. Due to the many improvements in DVB-T2, the overall throughput of the system is close to the theoretical channel capacity. DVB-T2 gives an increase in bit-rate of at least 30% as compared to DVB-T, and for some configurations up to 70%.

Table 13.2 shows some simulation results over a Ricean channel with ideal channel estimation for the required Carrier-to-Noise Ratio (CNR) for achieving QEF Bit Error Rate (BER) using either the DVB-T or the DVB-T2 FEC codes[1] [RBEL[+]07, RBLE[+]08]. As can be seen from the table, there is a notable gain in the CNR in DVB-T2 as compared to DVB-T, which increases with increasing modulation orders.

DVB-T2 includes a highly flexible time interleaving scheme, allowing for configurations supporting different use cases. In [GVGBC10], the performance of time interleaving in DVB-T2 in the context of mobile reception is analyzed. Through the use of time interleaving, it is possible to provide different trade-offs in terms of time diversity, latency, and power saving. In this way, in the same frequency channel, it is possible to accommodate different use cases: fixed, portable, and mobile.

In DVB-T2, FEC blocks belonging to the same PLP are grouped into interleaving frames for time interleaving purposes, and time interleaving is performed on TI (Time Interleaving) blocks. Each TI block contains a dynamically varying number of FEC blocks that are interleaved before transmission. DVB-T2 supports three different TI schemes that are illustrated in Fig. 13.13. As can be seen, multiframe interleaving corresponds to interleaving the interleaving frames over multiple T2 frames. Multiframe interleaving increases the time diversity at the expense of additional latency and longer channel switching times. Frame hopping corresponds to mapping interleaving frames onto nonconsecutive T2 frames, allowing power saving by having the receiver front ends switched off between the transmissions of the relevant PLPs. Frame hopping results in an increase in the average channel switching time, which is proportional to the frame interval. Sub-slicing is used when each PLP is transmitted inside several subslices per T2 frames. Subslices from different PLPs are placed regularly one after another inside the T2 frames. The maximum time diversity is achieved when the number of subslices is equal to or higher than the number of OFDM symbols in the T2 frame and PLPs are continuously transmitted over time.

Figure 13.14 illustrates simulation results for different TI configurations over the TU-6 channel. Three TI configurations are considered: *reference*, *handheld*, and

---
[1] The QEF BER corresponds to a BER of 10^{-4}.

13 Hybrid Cellular and Broadcasting Networks

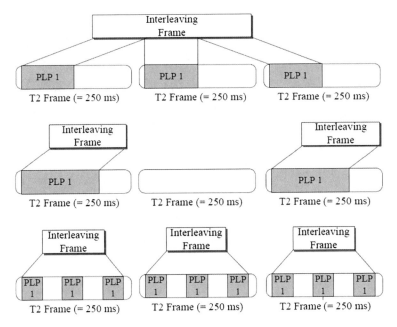

Fig. 13.13 Frame mapping options in DVB-T2: multiframe interleaving (*top*), frame hopping (*center*), and subslicing (*bottom*) [GVGBC10] (©2010 IEEE, reproduced with permission)

maximum diversity. The reference configuration does not employ multiframe interleaving, frame hopping, or subslicing. This configuration provides the shortest channel switching times and the best power saving levels at the expense of reduced time diversity. The handheld configuration employs both frame hopping and subslicing. The number of subslices is configured to the maximum value of 270, which results in continuous transmission inside the T2 frames. The maximum diversity configuration employs multiframe interleaving combined with 270 subslices. This represents the upper bound in time diversity that can be provided in DVB-T2.

The QoS criterion selected for the simulations is BBFER (Baseband Frame Error Rate) 1%. Figure 13.14(a) shows the performance gain achievable when using multiframe interleaving over 1, 2, and 3 T2 frames for the Doppler frequencies 10 Hz and 80 Hz. The results show an improvement of up to 2.6 dB for 10 Hz Doppler and up to 0.8 dB for 80 Hz Doppler when using multiframe interleaving. Figure 13.14(b) shows the results for 1, 5, 9, and 270 subslices. Similarly to multiframe interleaving, subslicing achieves a greater improvement in low Doppler frequencies.

In Fig. 13.14(c), the results for the performance of the *reference* configuration with different number of FEC blocks per T2 frame are illustrated. These results show the impact of the TI memory utilization on the system performance. The utilization of the full TI memory (129 FEC blocks) achieves a gain up to 1.2 dB with 10 Hz of Doppler and up to 1 dB with 80 Hz of Doppler. Higher utilizations of the TI memory for a certain data rate can be achieved by means of frame hopping. If a frame interval of 2 is employed, twice as much information must be transmitted every 2 T2 frames.

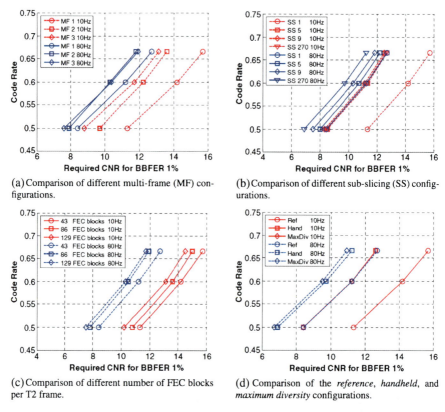

Fig. 13.14 Simulation results of time interleaving configurations over the TU-6 channel [GVGBC10] (©2010 IEEE, reproduced with permission)

In this case, frame interval values of 1, 2, and 3 result in 43, 86, and 129 FEC blocks transmitted per T2 frame. Another way of increasing the TI memory usage is to multiplex several services in the same stream in order to increase the data rate of the PLP. Since the transmission parameters cannot be particularized for each one of the services transmitted inside the same PLP, it is recommended to multiplex services that belong to the same use case.

In Fig. 13.14(d) the *reference*, *handheld*, and *maximum diversity* configurations are compared. As can be seen, the handheld and maximum diversity configurations achieve practically the same performance. The utilization of multiframe interleaving provides little improvement when the number of subslices is 270 and the information is continuously transmitted inside the T2 frame. Nevertheless, the maximum diversity configuration achieves a gain up to 3.1 dB in the case of 10 Hz and up to 1.8 dB in the case of 80 Hz of Doppler with respect to the reference configuration.

Based on these simulation results, it can be concluded that by means of the different TI scenarios provided in DVB-T2, it is possible to achieve gains of 3.1 dB with 10 Hz of Doppler and up to 1.8 dB with 80 Hz of Doppler. It should also be noted that, due to limitations in the available TI memory, DVB-T2 may not provide sufficient time diversity in shadowing scenarios [GVGBC10].

References

[BB03] P. I. Bratanov and E. Bonek. Mobility model of vehicle-borne terminals in urban cellular systems. *IEEE Trans. Veh. Technol.*, 52(4):947–952, 2003.

[BBZ+08] R. Bari, M. Bardm, Y. Zhang, K. M. Nasr, J. Cosmas, K.-K. Loo, R. Nilavalan, H. Shirazi, and K. Krishnapillai. Laboratory measurement campaign of DVB-T signal with transmit delay diversity. *IEEE Transactions on Broadcasting*, 3(54):532–541, 2008.

[BGGR09] M. Barbiroli, A. Guidotti, P. Grazioso, and G. Riva. Coexistence and mutual interference between mobile radio and broadcast systems—preliminary simulation results. Technical Report TD-09-709, Braunschweig, Germany, February 2009.

[BH03] R. Brugger and D. Hemingway. OFDM receivers—impact on coverage of intersymbol interference and FFT window positioning on OFDM receivers. Technical Report EBU Technical Review, European Broadcasting Union (EBU), BPN059, 3rd edition, July 2003.

[BNB+10] R. Bari, K. M. Nasr, M. Bard, J. Cosmas, R. Nilavalan, K.-K. Loo, and K. Krishnapillai. A field measurement campaign for a DVB-T/H system with transmit delay diversity. Technical Report TD-09-923, Vienna, Austria, September 2010.

[Bri09] A. Bria. *Mobile multimedia multicasting in future wireless systems—a hybrid cellular and broadcasting system approach*. PhD thesis, Royal Institute of Technology, Stockholm, Sweden, 2009.

[FHST06] G. Faria, J. Henriksson, E. Stare, and P. Talmola. DVB-H: digital broadcast services to handheld devices. *Proceedings of the IEEE*, 94(1):194–209, 2006.

[GB09] D. Gómez-Barquero. *Cost efficient provisioning of mass mobile multimedia services in hybrid celular and broadcasting systems*. PhD thesis, Universidad Politécnica de Valencia, Valencia, Spain, 2009.

[GBBZC07] D. Gómez-Barquero, A. Bria, J. Zander, and N. Cardona. Affordable mobile TV services in hybrid cellular and DVB-H systems. *IEEE Network*, 21(2):34–40, 2007.

[GBFAC09] D. Gómez-Barquero, A. Fernández-Aguilella, and N. Cardona. Multicast delivery of file download services in evolved 3G mobile networks with HSDPA and MBMS. *IEEE Transactions on Broadcasting*, 55(4):742–751, 2009.

[GBGC09] D. Gómez-Barquero, D. Gozálvez, and N. Cardona. Application layer FEC for mobile TV delivery in IP datacast over DVB-H systems. *IEEE Transactions on Broadcasting*, 2(55):396–406, 2009.

[GBMFA+09] D. Gómez-Barquero, J. F. Monserrat, A. Fernández-Aguilella, J. Calabuig, and M. Cardona. Efficient multicast and broadcast services in long term evolution. Technical Report, Joint COST 2100/NEWCOM++ Workshop on RRM for LTE, Vienna, Austria, September 2009.

[GBPPC10] D. Gómez-Barquero, J. Poikonen, J. Paavola, and N. Cardona. Development and applications of a dynamic DVB-H system-level simulator. *IEEE Transactions on Broadcasting*, 56(3):358–368, 2010.

[GBU08] D. Gómez-Barquero and P. Unger. On the fading margin gain due to multi-slot FEC in TDMA broadcast systems. Technical Report TD-08-641, Lille, France, September 2008.

[GBUKC10] D. Gómez-Barquero, P. Unger, T. Kürner, and N. Cardona. Coverage estimation for multi-burst FEC mobile TV services in DVB-H systems. *IEEE Trans. Veh. Technol.*, 59(7):3491–3500, 2010.

[Gee08] H.-F. Geerdes. UMTS Radio Network Planning: Mastering Cell Coupling for Capacity Optimization. Technische Universität Berlin, 2008.

[GGBC08] D. Gozálvez, D. Gómez-Barquero, and N. Cardona. Performance evaluation of the MPE-iFEC sliding RS encoding for DVB-H streaming services. In *Proc. PIMRC 2008—IEEE 19th Int. Symp. on Pers., Indoor and Mobile Radio Commun.*, Cannes, France, September 2008. [Also available as TD(08)538].

[GGBSL09] D. Gozálvez, D. Gómez-Barquero, T. Stockhammer, and M. Luby. AL-FEC for improved mobile reception of MPEG-2 DVB-T transport streams. *International Journal of Digital Multimedia Broadcasting*, 2009.

[GLRBL+10] M. Garcia-Lozano, S. Ruiz-Boque, M. Lema, E. Torras, J. J. Olmos, and F. Minerva. Metaheuristic proposal to minimize self-interference in single frequency networks. Technical Report TD-10-11032, Aalborg, Denmark, June 2010.

[Gud91] M. Gudmundson. Correlation model for shadow fading in mobile radio systems. *Elect. Lett.*, 37(23):2145–2146, 1991.

[GVGBC10] D. Gozálvez, D. Vargas, D. Gómez-Barquero, and N. Cardona. Performance evaluation of DVB-T2 time interleaving in mobile environments. In *Proc. IEEE Vehicular Technology Conference (VTC) Fall*, Ottawa, Canada, September 2010. [Also available as TD(12)001].

[HEA05] Y. Hu, E. Eleftheriou, and D.M. Arnold. Regular and irregular progressive edge growth tanner graphs. *IEEE Trans. Inform. Theory*, 51(1):386–398, 2005.

[Heu09] C. Heuck. *Optimierung hybrider (Rundfunk-/Mobilfunk-)Netze durch Steuerung der Lastverteilung*. Dissertation, Technische Universität Braunschweig, 2009, in German.

[HHH+] F. Hartung, U. Horn, J. Huschke, M. Kampmann, and T. Lohmar. MBMS—IP multicast/broadcast in 3G networks. *International Journal of Digital Multimedia Broadcasting*.

[HHH+07] F. Hartung, U. Horn, J. Huschke, M. Kampmann, T. Lohmar, and M. Lundevall. Delivery of broadcast services in 3G networks. *IEEE Transactions on Broadcasting*, 53(1):188–199, 2007.

[Him09] H. Himmanen. *On transmission system design for wireless broadcasting*. PhD thesis, University of Turku, Turku, Finland, 2009.

[HRT07] R. Höckmann, E. Reetz, and R. Tönjes. Adaptive relaying for MBMS in the 3GPP LTE system environment. Technical Report TD-07-330, Duisburg, Germany, September 2007.

[KBW06] D. Krajzewicz, M. Bonert, and P. Wagner. The open source traffic simulation package SUMO. In *Proc. RoboCup*, 2006.

[LGGBC09] J. López, C. Garcí, D. Gómez-Barquero, and N. Cardona. Planning a mobile DVB-T network for Colombia. In *Proc. IEEE Latin-American Conference on Communications (LATINCOM)*, Medellin, Colombia, September 2009.

[MAA10] J. Morgade, P. Angueira, and A. Arrinda. Coverage optimization for single frequency networks using a particle swarm based method. Technical Report TD-10-10077, Athens, Greece, February 2010.

[Mat05] A. Mattsson. Single frequency networks in DTV. *IEEE Transactions on Broadcasting*, 4(51):413–422, 2005.

[May08] G. May. Analysis of DVB-H soft handover gain. Technical Report TD-08-515, Trondheim, Norway, June 2008.

[Moz09a] P. Mozola. Comparison of the simulation and the measurement physical layer of the DVB-T/H system. Technical Report TD-09-842, Valencia, Spain, May 2009.

[Moz09b] P. Mozola. Modelling DVB-H system for compatibility analysis. Technical Report TD-09-755, Braunschweig, Germany, February 2009.

[Moz10] P. Mozola. Compatibility analyses of the DVB-H and wireless narrow band systems. Technical Report TD-10-1170, Aalborg, Denmark, June 2010.

[MU07] G. May and P. Unger. A new approach for transmitting localized content within digital single frequency broadcast networks. *IEEE Transactions on Broadcasting*, 4(53):732–737, 2007.

[NCBG07] K. M. Nasr, J. Cosmas, M. Bard, and J. Gledhill. Performance of an echo canceller and channel estimator for on-channel repeaters in DVB-T/H networks. *IEEE Transactions on Broadcasting*, 3(53):609–618, 2007.

[NV07] R. Nybom and D. Vukobratović. A survey on application layer forward error correction codes for IP datacasting in DVB-H. Technical Report TD-07-355, Duisburg, Germany, September 2007.

[Nyb09] K. Nybom. *Low-density parity-check codes for wireless datacast networks*. PhD thesis, Åbo Akademi University, Turku, Finland, 2009.

[Poi09] J. Poikonen. *Efficient channel modeling methods for mobile communication systems*. PhD thesis, University of Turku, Turku, Finland, 2009.

[PPI09] J. Poikonen, J. Paavola and V. Ipatov. Aggregated renewal Markov processes with applications in simulating mobile broadcast systems. *IEEE Trans. Veh. Technol.*, 58(1):21–31, 2009.

[RBEL+07] S. Ruiz-Boqué, C. Enrique, C. Lopez, G. Martorell, L. Alonso, M. Garcia, and F. Minerva. Physical layer simulation platform for DVB-T and DVB-T2. Technical Report TD-07-343, Duisburg, Germany, September 2007.

[RBLE+08] S. Ruiz-Boqué, C. Lopez, C. Enrique, L. Alonso, J. Olmos, M. Garcia, and F. Minerva. Performance of DVB-T2 improvements over low Doppler mobile channels. Technical Report TD-08-451, Wroclaw, Poland, February 2008.

[RLE+09] S. Ruiz, C. Lopez, C. Enrique, M. Garciam, and F. Minerva. DVB-T2 network optimization with simulated annealing. Technical Report TD-09-993, Vienna, Austria, September 2009.

[SAZ03] S. R. Saunders and A. Aragón-Zavala. *Antennas and Propagation for Wireless Communication Systems*. Wiley, Hoboken, NJ, 2003.

[Sho06] A. Shokrollahi. Raptor codes. *IEEE Trans. Inform. Theory*, 52(6):2551–2567, 2006.

[SLK+09] B. Sayadi, Y. Leprovost, S. Kerboeuf, M. Alberi-Morel, and L. Roullet. MPE-IFEC: an enhanced burst error protection for DVB-SH systems. *Bell Labs Techn. J.*, 14(1):25–40, 2009.

[SLKAM09] B. Sayadi, Y. Leprovost, S. Kerboeuf, and M. Alberi-Morel. Efficient repair mechanism of real-time broadcast services in hybrid DVB-SH and cellular systems. *Bell Labs Techn. J.*, 14(1):41–54, 2009.

[TJVW04] T. Tian, C. Jones, J.D. Villasenor, and R.D. Wesel. Selective avoidance of cycles in irregular LDPC code construction. *IEEE Trans. Commun.*, 52:1242–1247, 2004.

[UK06] P. Unger and T. Kürner. Radio network planning of DVB-H/UMTS hybrid mobile communication networks. *Wiley European Transactions on Telecommunication*, 17(2):193–201, 2006.

[UK08] P. Unger and T. Kürner. Optimizing the local service areas in single frequency networks. Technical Report TD-08-467, Wroclaw, Poland, February 2008.

[UK09] P. Unger and T. Kürner. Switching bounds for hybrid (DVB-H/UMTS) IP Datacast networks. Technical Report TD-09-841, Valencia, Spain, May 2009.

[UM07] P. Unger and G. May. Coverage estimation for localized service areas in single frequency networks. Technical Report TD-07-215, Lisbon, Portugal, February 2007.

[Ung10] P. Unger. *Radio access network planning and optimization of hybrid cellular and broadcasting systems*. PhD thesis, Technical University of Braunschweig, Braunschweig, Germany, 2010.

[USK07] P. Unger, M. Schack, and T. Kürner. Minimizing the electromagnetic exposure using hybrid (DVB-H/UMTS) networks. *IEEE Transactions on Broadcasting*, 53(1):418–424, 2007.

[VBT+09] L. Vangelista, N. Benvenuto, S. Tomasin, C. Nokes, J. Stott, A. Filippi, M. Vlot, V. Mignone, and A. Morello. Key technologies for next-generation terrestrial digital television standard DVB-T2. *IEEE Commun. Mag.*, 47(10):146–153, 2009.

[VP07a] V. Vassiliou and A. Pitsillides. Call admission control for MBMS-enabled 3G UMTS networks. Technical Report TD-07-364, Duisburg, Germany, September 2007.

[VP07b]   V. Vassiliou and A. Pitsillides. Power control for multicasting in IP-based 3G mobile networks. Technical Report TD-07-363, Duisburg, Germany, September 2007.
[VSB07]   D. Vukobratović, V. Senk, and D. Bajic. On LDPC codes for content distribution applications. Technical Report TD-07-240, Lisbon, Portugal, February 2007.
[WDA10]   Z. Wang, P. Darwood, and N. Anderson. End-to-end system performance of IMB. In *Proc. IEEE International Symposium on Broadband Multimedia Systems and Broadcasting (BMSB2010)*, March 2010.
[ZCL$^+$07]   Y. Zhang, J. Cosmas, K.-K. Loo, M. Bard, and R. Bari. Analysis of cyclic delay diversity on DVB-H systems over spatially correlated channel. *IEEE Transactions on Broadcasting*, 1(53):247–255, 2007.

Chapter 14
Vehicle-to-Vehicle Communications

Chapter Editor Christoph Mecklenbräuker, Laura Bernadó, Oliver Klemp, Andreas Kwoczek, Alexander Paier, Moritz Schack, Katrin Sjöberg, Erik G. Ström, Fredrik Tufvesson, Elisabeth Uhlemann, and Thomas Zemen

This chapter discusses major results and conclusions from Special Interest Group C bringing together various aspects of mobile-to-mobile communication from all working groups. Vehicle-to-vehicle communication scenarios are emphasized.

Traffic telematics applications are currently under intense research and development for making transportation safer, more efficient, and cleaner. Communication systems which provide "always on" connectivity at data rates between 1 and 10 Mb/s to highly mobile surface traffic (cars and trains) are urgently required for developing traffic telematics applications and services. Currently much attention is given to advanced active safety, but the application area also ranges to improved navigation mechanisms and infotainment services. Mobile-to-mobile communications need to be reliable and trusted: Drivers in cars which are equipped with vehicle-to-vehicle communications need to rely on the accuracy and timeliness of the exchanged data.

Automotive manufacturers, road authorities, broadcast companies, and telecom providers are the key players in the value chain for such future systems. These communication systems provide an extended information horizon to warn the driver or the vehicular systems of potentially dangerous situations in an early phase.

14.1 Performance Metrics

Most traditional performance metrics for digital communications are relevant for mobile-to-mobile communications as well. The objective with communications is to reproduce the transmitted messages at the receiver with the required quality at the lowest cost. What needs to be specified is therefore the notion of quality and cost; both can be considered as performance metrics.

C. Mecklenbräuker (✉)
Vienna University of Technology (VUT), Vienna, Austria

Cost can, e.g., be measured in required transmit power, total power consumption, transmitted signal bandwidth, or transceiver complexity. Shannon's channel capacity describes the optimal trade-off between power and spectral efficiency for very low error rates. For practical systems, channel capacity is a useful metric when sufficiently long messages are to be transmitted.

What is a relevant quality measure depends on the application. Nevertheless, to allow for generic designs of communication systems, it is useful to specify quality measures that are applicable for a large range of applications. Such a quality measure is the probability of error for a randomly selected bit or block of bits in the bit stream that make up the message. If errors are correlated in time, it might be necessary to characterize the error process in more detail. That is, suppose that $e(n)$ is the error process, which is 1 if the nth bit (or block of bits) is in error and 0 otherwise. The bit error probability is then $\mathscr{E}\{e(n)\}$ (assuming that $e(n)$ is wide-sense stationarity), which, of course, is not a complete description of the error process. A complete description is the joint pdf for $e(n)$, but we can also settle for second order statistics (e.g., the autocorrelation function).

Latency (delay) is another quality measure of importance. We can measure delay between any two points in the transmission chain, but it normally makes sense to measure delay with respect to a certain layer in the protocol stack. For instance, latency can be defined as the delay from when a frame transmission request is issued to the transmitter Media Access Control (MAC) layer until the frame is delivered from the receiver MAC layer. The latency is a random process, which can be characterized in the same way as $e(n)$ above. For real-time data traffic, we wish to deliver messages before a certain deadline. A suitable metric is the probability that the latency exceeds the deadline (the so-called missed deadline ratio), which can be found from the cumulative distribution function of the latency.

Numerous other metrics have been proposed and studied, such as the delivery ratio, throughput, and delay jitter. The relevance of these metrics in safety-related vehicle-to-vehicle applications requires much further discussion.

14.2 Channel Characterization

It is vital to understand the characteristics of the radio channel, before sophisticated communication systems can be developed. Notably in vehicular communications, the radio channel is challenging, due to its high variability in time and the resulting Doppler shifts. In this section the vehicular radio channel is described by its time-delay and time-Doppler characteristics, delay spreads, small-scale fading statistics, and spatial correlations. The presented results and examples are taken from three channel measurement campaigns, [PKZ+08, RKVO08, PBK+10] that are described in Sect. 2.1.3. The channel sounders used in these measurement campaigns provide the discrete time-variant Impulse Response (IR) $h^{(p)}[k,l]$ and the discrete time-variant Transfer Function (TF) $H^{(p)}[k,q]$, where k is the time index, l the delay index, q the frequency index, and p the Multiple-Input Multiple-Output (MIMO) link index.

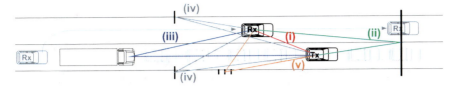

Fig. 14.1 Scatterers distribution of a traffic congestion scenario [PBK+10] (©2010 IEEE, reproduced with permission)

14.2.1 Time-Delay Domain and Time-Doppler Domain

In vehicular communications the observed fading process does not fulfill the Wide-Sense Stationary (WSS) Uncorrelated Scattering (US) property [Bel63]. Since it can be assumed that the process is stationary for a given time period, it is meaningful to represent its power spectral density as a function of time. In this sense, the Local Scattering Function (LSF) $C_H^{(p)}[k,q;l,m]$, which is defined for non-Wide-Sense Stationary Uncorrelated Scattering (WSSUS) channels in [Mat05], can be used in order to calculate the time-variant Power-Delay Profile (PDP)

$$PDP[k;l] = \sum_{p=1}^{N_t \times N_r} \sum_{q=0}^{L-1} \sum_{m=-M/2}^{M/2-1} C_H^{(p)}[k,q;l,m] \qquad (14.1)$$

and time-variant Doppler Spectral Density (DSD)

$$DSD[k;m] = \sum_{p=1}^{N_t \times N_r} \sum_{q=0}^{L-1} \sum_{l=0}^{L-1} C_H^{(p)}[k,q;l,m], \qquad (14.2)$$

where m is the Doppler index. The way of estimating the LSF from vehicular channel measurement data is described in [PZB+08].

14.2.1.1 Significant Scatterers

In this subsection important scatterers from highway measurements are analyzed by the time-variant PDP and DSD [PBK+10]. In the considered scenario the Transmitter (Tx) vehicle is stuck in a traffic congestion on the right lane, whereas the Receiver (Rx) vehicle overtakes the Tx vehicle on the left lane, see Fig. 14.1. The significant scattering contributions in the PDP, see Fig. 14.2(a), and DSD, see Fig. 14.2(b), are labeled from (i) to (v). The propagation paths from each contribution are shown in Fig. 14.1. Contribution (i) corresponds to the Line-Of-Sight (LOS) path with slightly increasing delay at the beginning, where the Tx is going faster than the Rx (negative Doppler shift). After some seconds the Rx vehicle accelerates (decreasing delay and positive Doppler shift) and overtakes the Tx vehicle at 14.3 s (shortest delay and Doppler shift zero). Path (ii) corresponds to a single bounce reflection produced by a big traffic sign placed ahead of both vehicles. Since both vehicles

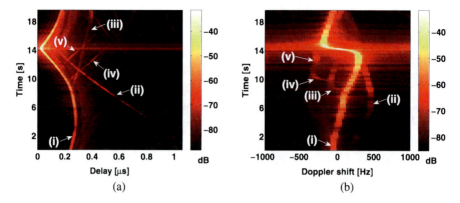

Fig. 14.2 PDP (**a**) and DSD (**b**) of a traffic congestion scenario [PBK+10] (©2010 IEEE, reproduced with permission)

Table 14.1 RMS delay for different environments [RKVO08]

	Urban	Highway	Campus	Suburban
Mean delay spread, $\bar{\sigma}_\tau$	373 ns	165 ns	112 ns	104 ns

are approaching the traffic sign, the delay of this path is decreasing and its Doppler shift is positive. The large truck standing in front of the Rx in the beginning of the measurement causes multipath contribution (iii) after the Rx overtakes the truck at 8.7 s. Multipath contributions (iv) and (v) correspond to temporary traffic signs from a construction site at both sides of the road, which contribute to the received signal as soon as they are left behind. The other cars present in the measurement do not significantly contribute as scatterer, and they do not shadow the LOS path.

14.2.1.2 Delay Spread

The delay spread σ_τ is an important parameter for the characterization of a radio channel. In [RKVO08] the Root-Mean-Square (RMS) delay spread is analyzed for different scenarios and summarized in Table 14.1. As expected, the highest delay spreads are found for the urban scenario, because of its high density of scatterers. The highway scenario shows the second highest delay spread followed by similar delay spreads for the suburban and campus scenarios.

14.2.2 Large-Scale and Small-Scale Fading

In [RKVO08] different composite models for large- and small-scale fading are compared with the vehicular channel measurement results. For the comparison, a Rice-lognormal model [CV94], a Suzuki model [Suz77], a Nakagami-lognormal model, and a Weibull-lognormal model are considered. These four distributions were fitted against the measured V2V narrowband channel

$$h_{\text{narrow}}[k] = \sum_{p=1}^{N_t \times N_r} \sum_{l=0}^{L-1} h^{(p)}[k,l], \qquad (14.3)$$

where $N_t = 30$ and $N_r = 30$. The Cramer–von Mises criterion and the Kolmogorov–Smirnov test were applied, in order to evaluate the empirical distributions. The resulting goodness-of-fit values were used to estimate how often each of the composite distributions have the best fit, the second best fit, etc., as described in [RAMB05]. The Weibull-lognormal and the Rice-lognormal distributions showed the best fit. Since there is no physical intuition behind the Weibull distribution, the Rice distribution is suggested in [RKVO08], in order to model the V2V narrowband channel.

In [BZK+on] also a Rice distribution, but only for the first delay bin, is observed. The later delay bins are mostly Rayleigh distributed. The variation of the K-factor, of the first delay bin, in time, frequency, and space is investigated in [BZK+on]. For this, a crossing scenario with buildings only in one quadrant was chosen. During the measurement run, there is a transition between Non-Line-Of-Sight (NLOS) and LOS situation. The investigation of the first delay bin is done separately for 24 frequency subbands of 10 MHz (frequency variation), for each time instance (time variation), and for 16 MIMO links (space variation). A strong variation of the K-factor over time (e.g., $K_{t=6.5s} = 0.6$, $K_{t=11.5s} = 7.5$, and $K_{t=17.5s} = 48.4$ for frequency subband 1, link 10, and an overall measurement run duration of 20 s) is found. The time-dependent K-factor occurs because both vehicles are moving, and therefore the LOS and NLOS situations and the scatterer distribution are varying with time. Unexpectedly, the K-factor varies significantly in frequency, which is not the common assumption for channels with a relative bandwidth below 10%. One possible reason for this frequency variation is the frequency varying contribution of small scattering elements, such as foliage from trees surrounding the crossing (the considered wavelength is 54 mm). Another cause of the frequency variation of the K-factor is the frequency-dependent antenna pattern gains throughout the 240 MHz measurement bandwidth. The different patterns for each single antenna element are the explanation of the space-varying K-factor, which is analyzed by observing the different MIMO links. Given these observations, a stochastic approach could be followed to model the multidimensional variability of the K-factor for such communication scenarios.

14.2.3 Spatial Correlation

Spatial diversity can be used, in order to increase the reliability of communication links. This is of special interest for V2V communications (e.g., safety-related messages). Spatial diversity is related with a general characteristic of the MIMO radio channel, the spatial correlation. In [NBP05] the diversity measure is defined by the rank of the covariance matrix $\mathbf{C}$ if the Rician part $\bar{\mathbf{H}}$ of the MIMO channel is lying in the subspace spanned by the Rayleigh part $\tilde{\mathbf{H}}$ of the channel. The overall radio channel is defined by the sum of the Rician part and the Rayleigh part $\mathbf{H} = \bar{\mathbf{H}} + \tilde{\mathbf{H}}$. If there is a component of $\bar{\mathbf{H}}$ orthogonal to the subspace spanned by $\tilde{\mathbf{H}}$, there exists an Signal-to-Noise Ratio (SNR)-dependent critical data rate for Rician fading MIMO channels below the diversity goes to infinite. This means that up to this critical data rate reliable transmissions with zero outage are achievable.

In [PZK+09] the spatial correlation and the spatial diversity over time are investigated for a highway scenario, where the two vehicles are passing during they are going in opposite directions. The spatial correlation of the MIMO links with LOS show a high correlation, in contrast to the NLOS links, which show low correlation. In these measurements 4×4 MIMO circular antenna arrays, where the four antenna elements are equally distributed over the whole circle, were used. In this case always four links have LOS character and about two links partly LOS character. Of course the links with LOS behavior are changing over time, especially when the two vehicles are passing. The four LOS links also show the highest K-factors compared to the other links. For the investigation of the spatial diversity, the covariance-based diversity measure from Nabar [NBP05] is compared with the diversity measure defined by Ivrlac [IN03]

$$\Psi_{\mathbf{C}}[k, f] = \left(\frac{\text{tr}(\mathbf{C})}{\|\mathbf{C}\|_F} \right)^2. \qquad (14.4)$$

The two diversity measures show the same behavior with the main difference that the Nabar measure is integer-valued and the Ivrlac measure not. Both diversity measures are frequency- and time-variant. The spatial diversity is much lower when the vehicles are closer and is roughly inversely proportional to the K-factor of the channel.

14.3 Channel Modeling

For system- and link-level simulations, it is required that the channel model is capable of mimicking the properties described in the previous section in a realistic way. It is important to remember that the specific behavior of the mobile-to-mobile channel is generally different from the behavior of the "conventional" cellular channel. There are three major approaches to the modeling of the mobile-to-mobile channel: (1) ray tracing, (2) geometric stochastic approaches, and (3) purely stochastic approach.

With the purely stochastic approach the statistics of the Doppler shift, the expected power for the various delays, and, for multiantenna models, also the angle-of-arrival and angle-of-departure are specified. Based on this information, different channel samples are generated, often by a tapped-delay-line approach. Usually different Doppler spectral densities are specified for the various taps, and such a model is used as a reference model in the IEEE 802.11p standard. The major advantage of

Fig. 14.3 Example of ray-tracing simulation for mobile-to-mobile communication

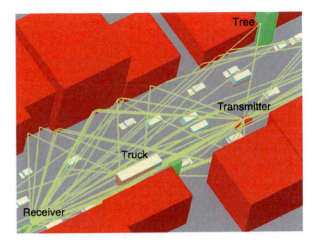

this modeling approach is that the complexity of the model is reasonably low, but, on the other hand, it might be hard to model the nonstationary nature of the mobile-to-mobile channel in a realistic way. Therefore we mainly focus on the two alternative approaches, the ray-tracing approach and the geometric stochastic approach.

14.3.1 Ray-Tracing Approach

When using ray tracing for channel modeling, we start by creating a realistic model of the environment, including houses, other cars, road signs. Then, by a high-frequency approximation for the iteration between the radio waves and the objects in the virtual environment created, we aim to predict every possible path for the radio wave between the Tx and the Rx. If we have moving objects in the environment, those movements can easily be incorporated in the model. Usually we use a so-called microscopic model [MFP$^+$08] for the surrounding traffic, where each car creates one or several specular components, multipath components, that may contribute to the total impulse response when all possible paths between Tx and Rx are evaluated. In this paper, six different types of cars (van, compact car, truck, etc.) are represented, each with its unique size and material parameters. In this way it is also possible to study the influence of shadowing effects in a realistic way, a phenomena that often is neglected but may be important in situations where, e.g., the line-of-sight between two communicating cars is broken by a larger car. In Fig. 14.3 an example of how the simulated environment may look like is given together with the estimated 50 strongest paths between the Tx and Rx. Often the agreement with measurements is really good, given that a realistic 3D description of the environment is available, and nonstationary scenarios can easily be modeled. The drawback is that the method is computationally very demanding and also that it might be hard to create detailed enough descriptions of the environment.

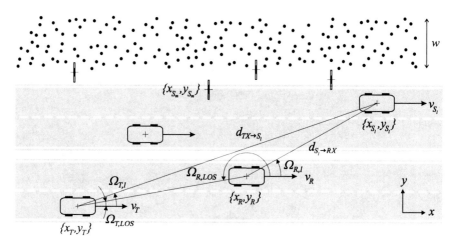

Fig. 14.4 Example of the geometry of a GSCM [KTC+09] (©2009 IEEE, reproduced with permission)

14.3.2 Geometric Stochastic Approach

For the geometric stochastic channel models Geometry-Based Stochastic Channel Model (GSCM), a simplified description of the environment is created. This description might or might not try to resemble the physical reality, but the positions of the possible scatterers are chosen according to a statistical distribution in both cases. A simplified ray tracing, or rather ray launching, is performed to calculate the contributions from the different scatterers and thereby the impulse response between the Tx and Rx. In the same way as for the ray-tracing models, the technique does naturally include multiantenna modeling and modeling of the temporal evolution of the channel. Movements of the Tx, Rx, and the scattering objects can be described in a similar way as in the ray-tracing approach. There are basically two main categories of GSCM, the reference models, as described in [CWL08], where pure geometric distributions of the scatterers are considered, and the GSCMs where the aim is to distribute the scatterers at positions more similar to a real scenario, see, e.g., [KTC+09]. In the reference models the scatterers usually are prescribed to be located on circles around the Tx or Rx cars (for double scattering processes between the two circles) or on an ellipse (for single scattering processes), where the Tx and Rx cars are located in the foci of the ellipse. With those reference models, it is usually possible to theoretically calculate, e.g., correlation functions for the temporal and spatial behavior of the model. By proper statistical description of, e.g., Angle Of Arrival (AOA) and Angle Of Departure (AOD), the output of those models may be adjusted to reflect the behavior of measured channels. Using the other approach, a virtual road is, e.g., created, and then scatterers are positioned according to certain distributions with reference to the road. Figure 14.4 shows an example of such a GSCM with both diffuse scattering and discrete scatterers along the road.

14.4 Vehicular Networks Simulation

The need for integrated vehicular network simulators which employ realistic channel modeling and appropriate mobility models becomes more and more apparent. This is motivated by the continuous development of Vehicle-to-X (V2X) communication protocols and corresponding system architectures. Moreover, in order to analyze the impact of V2X applications on the behavior of vehicles, an interconnection of different simulation tools is needed. On the one hand, mobility models in current network simulation tools like linear mobility, mass mobility, etc. are relatively simple in comparison to the real behavior of vehicles in urban areas. Therefore, these approaches are not able to appropriately model the dynamic behavior of vehicular networks with high relative velocities of wireless hosts and the complex behavior of road traffic, which is characterized by independent mobile nodes. For this purpose, microscopic road traffic simulation tools are available that accurately model vehicular mobility regardless of V2X communication. On the other hand, in current V2X communication simulations the radio channel is often regarded as the less critical part compared to the whole system. In some publications, the maximum propagation range of a transmitter is even set to a fixed value, particularly with regard to higher layer investigations. In reality, the maximum allowed distance between transmitter and receiver, within which a connection is feasible, can vary considerably due to multipath propagation. Particularly in dense urban environments, the frequency-selective and time-variant character of the propagation channel is visible. Hence, an accurate modeling of the radio channel is indispensable for V2X investigations. The vehicular communication channel is often investigated using ray optical, pure stochastic or even geometry-based stochastic approaches. However, the computation time of ray optical propagation tools for V2X communication is an important issue since the number of vehicles in a simulation scenario can be very large.

The architecture of the integrated simulator proposed in [SSK09] is depicted in Fig. 14.5. The basis for the simulation is the scenario data which provide the input to both, the road traffic and the radio channel simulator. The scenario comprises an urban area of 1.5 km × 1.5 km in the city of Braunschweig, Germany, including residential zones, small industrial areas, and also parks with vegetation. Different streets like one-way and two-way streets with a different number of lanes and different speed limits of 30 or 50 km/h are part of the scenario. In order to use such a scenario for simulations, different types of data are required. For the input of the road traffic simulation, street data with lane and speed information, the number of incoming and outgoing vehicles, traffic light phases and turning probabilities at crossroads are needed. For the description of the environment, which is an important input for the radio channel simulation, 3D building and elevation data are required. Furthermore, models of the vehicles and models of the surrounding traffic signs have to be included in the scenario data.

The core part of the integrated simulation environment is a network simulator. This block interacts with three other simulators via dedicated interfaces. Basically, the mobility data in terms of speed and positions of the vehicles are generated by the road traffic simulator. These data are fed to the network simulator that requests

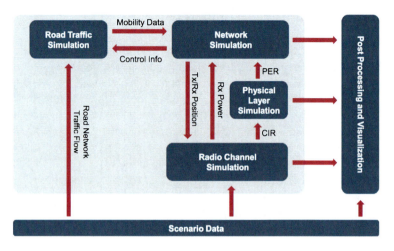

Fig. 14.5 Integrated simulation environment for V2X communications

the channel properties for the current Tx–Rx link from the radio channel simulator. Based on the calculated channel impulse response obtained from the ray-optical-based model of the radio channel simulator, the resulting power at the Rx is determined, and the corresponding Packet Error Rates (PERs) are derived for a specific SNR. However, the specific properties of the V2X channel in terms of time variance and frequency selectivity have a great impact on the system performance. Therefore, the network simulator can also be fed by performance metrics obtained by a Physical layer (PHY) simulator that is compliant to IEEE 802.11 [NSK10]. In this PHY simulator, both deterministic and stochastic channel models have been implemented. On the one hand, deterministic channel modeling by means of ray-optical techniques can be used to analyze the system behavior in a specific situation, e.g., two cars are approaching a certain intersection. In this way, the performance of the system regarding a certain application like collision avoidance can be investigated for a very realistic traffic scenario. On the other hand, the use of stochastic channel models enables the system analysis in various traffic scenarios, e.g., a highly mobile highway scenario. The latter approach is very useful to evaluate different signal processing algorithms in the receiver.

In V2X communications, the applied antennas that spatially filter the transmission paths on the transmitter and receiver sides, respectively, have a great impact on the resulting propagation link and, therefore, have to be taken into account when V2X systems are simulated. In most simulations, the influence of the antennas is neglected, and only simple isotropic radiators instead of realistic antenna models are used. In order to obtain realistic simulation results, the investigation of the channel thus needs to consider an antenna model which accounts for several nonideal effects such as mutual coupling or the influence of a finite ground plane size. In [SKSK10], a comparison of a roof-top module and a distributed antennas configuration for future V2X systems is carried out in different scenarios. The results of this investigation verify the strong influence of the antenna placement and the antenna configuration on the channel parameters, which shows the need for the integration of realistic antennas in the simulation environment.

14.5 Antenna Issues

Vehicle-to-vehicle (V2V) communication systems focus on the improvement of driver safety and efficiency of road traffic. In the context of Intelligent Transportation Systems (ITS), wireless systems of this kind aim at sharing safety- and efficiency-related traffic information between moving vehicles to enable sustainable vehicle traffic with a largely decreased number of road fatalities. With the recent bandwidth allocation of 30 MHz at 5.9 GHz for ITS in Europe, V2V communication systems are close to market introduction. Due to the time-frequency selective fading behavior of the vehicular communication channel as described in Sect. 14.2, antenna-related effects impacted by the vehicular mounting position and mutual coupling play an integral role in the performance of V2V communications. Particularly the directional behavior of the antenna defines the reliability of safety-related communications between moving vehicles and, as such, the end-to-end arrival success of exchanged message types. Impairments in the directional behavior of automotive antenna equipment caused by vehicular mounting effects may degrade the link performance between transmitting (Tx) and receiving (Rx) vehicles as described in [KTP+09].

One of the major applications of ITS is envisioned in the area of collision avoidance systems. Such applications shall prevent from traffic accidents by periodically resolving the geographical positions of surrounding vehicles in a local map. Exchanging Cooperative Awareness Messages (CAM) between individual vehicles is thus targeted to reduce the number of road fatalities by applications for safety-critical traffic situations such as, e.g., intersection assistance. Since the quality of the local map depends on the behavior of each individual communication link between Tx and Rx, i.e., vehicle dynamics and propagation conditions ultimately define the reliability and robustness of vehicular communication systems in the 5.9-GHz range. The performance assessment of current implementations for vehicular communications like IEEE 802.11p and the design of future systems requires a profound understanding of the underlying propagation channels and related communication system aspects. Even though the antenna design methodology for V2V antennas is already well explored, a little is known about the system-related effects of V2V antenna systems that are impacted by conventional automotive mounting concepts [KTP+09]. Also, the effect of antenna mounting positions on the vehicle shell plays an integral role by limiting the signal-to-noise conditions between Tx and Rx as described in [RFZ09].

This section shall help to assess the importance of the antenna interface for safety-relevant V2V communications by reviewing some of the principal system impairments that are caused by vehicular antenna integration. To do so, Sect. 14.5.1 focuses on the analysis of some of the characteristic antenna effects that are subject to vehicular mounting and therefore impact on the reliability of safety-related V2V communication systems.

In addition, multiple antenna techniques have gained considerable attention in the field of V2V communication systems (e.g., [PBK+10]). Those provide appropriate means for adaptive network scalability, interference mitigation, and link reliability for safety-related communication applications in cooperative ITS.

With the increased redundancy by introducing multiple antennas into the V2V radio frontend, some of the impairments caused by in situ vehicular antenna integration can be improved as described in [KT09, RFZ09, RPZ09, KTP+09]. This may help meeting the requirements for robust system definition in safety-relevant communication applications for ITS. As an example, Multi-Element Antenna (MEA) topologies are considered in Sect. 14.5.2 to improve the reliability of the wireless link. The proposed diversity combining scheme between multiple antennas may help fulfilling the requirements for real-time communications by statistically improved link conditions. However, improved system reliability comes at the expense of additional hardware costs for antennas and signal processing components.

Therefore, the use of multiple antenna equipment for wireless vehicular communications is in conflict with stringent cost constraints driven by parameters in the electrical vehicle architecture and during automotive batch production.

Section 14.5.3 exemplifies the performance metrics for a realistic two-element V2V antenna configuration that is integrated into an automotive-compliant roof-top antenna module with additional antennas for Global Postioning Systems (GPS) and multiband cellular radio.

14.5.1 Challenges of Automotive Antenna Integration

V2V communications is restricted to the horizontal xy-plane, thus requiring terrestrial coverage of the V2V antenna frontend. Due to the plethora of different use cases for ITS (e.g., highway vs. traffic intersection) and the relative movement and direction of vehicles in the xy-plane, an omnidirectional coverage of the V2V antenna beam pattern for all azimuth angles φ with maximum gain at $\vartheta = 90°$ is highly desired.

This set of requirements is in conflict with the conventional mounting conditions that are defined by the automotive industry. A relevant position for automotive-compliant V2V antenna equipage is given by the conventional roof-top antenna module. This one is centered and located at the back of the metallic vehicle roof. The vehicle roof itself provides some properties that negatively influence on the radiation pattern of V2V antennas as follows: It represents a metallic surface with finite dimensions $x_{\text{roof}} \times y_{\text{roof}}$ with, e.g., potential insets for nonmetallized sun roofs or railings oriented in parallel to the driving direction. In addition to the railings that cause shading to the left and right sides of the vehicle, the roof provides a finite bent angle φ_{roof} at the mounting position of the roof-top antenna. φ_{roof} is typically dependent on the vehicle type and can be in the range of $\varphi_{\text{roof}} \in [10, \ldots, 15]°$ for sedan cars. Dependent on the bent angle, the embossing of the vehicle roof may therefore cause significant shading in driving direction.

Additionally, the roof-top antenna module also provides functionality for a couple of different broadcasting and telecommunications services. Those typically include antennas for cellular communications and Global Positioning System (GPS), as well as for satellite radio services (e.g., US Satellite Digital Audio Radio Systems (SDARS)). Generally, the entire roof-top antenna equipment is enclosed by a dielectric housing that shall prevent from environmental influence and which is subject to aesthetic design considerations. Summarizing the above-mentioned effects that impact on the performance for V2V antennas, roof-top mounted equipment is subject to:

- Beam tilt in ϑ-direction due to finite vehicle roof-top dimensions $x_\text{roof} \times y_\text{roof}$;
- Shading of radiation pattern due to bent angle φ_roof;
- Mutual coupling with additional antenna elements being enclosed in the same roof-top antenna module;
- Mutual coupling with the dielectric housing enclosing the antenna arrangement.

Besides the traditional mounting position on the vehicle roof, vehicle manufacturers explore alternative mounting sites for V2V antenna equipment, as surveyed in [RFZ09]. Such alternative mounting sites are the front and rear bumpers, the left- and rear-side mirrors, and the bottom of the vehicle itself. Resulting performance of multiple-antenna systems is compared in [RSZ09, KSS+10]. It must be noted that choices for alternative mounting concepts are driven not only by performance, but also by cost and aesthetic design considerations. The impact of the aforementioned mounting conditions will be analyzed for a V2V antenna design, which is based on the concept of a short-circuited circular patch, driven in its fundamental mode in agreement with [GPSVRI+06, KTP+09]. Antenna elements of this kind generally support the requirement of a monopole-like radiation pattern for almost omni-directional, terrestrial coverage. They also provide a certain but limited headroom with respect to beam shaping in elevation ϑ. The typical antenna layout (top-view and side-view) including an eccentric feeding pin and a centered, metallic post is shown in Fig. 14.6. According to [TKP+10], an antenna prototype is manufactured on *Rohacell* dielectric material with a sheet thickness of $t_\text{Patch} = 1$ mm and a relative permittivity of $\varepsilon_{r,\text{Patch}} \simeq 1.0$. The outer diameter of the circular patch and the shorting post amount to $d_\text{Patch} = 10.0$ mm and $d_\text{sp} = 1.0$ mm. Figure 14.7 highlights the effects of the vehicular antenna integration on a metallic roof-top with finite dimensions $x_\text{roof} \times y_\text{roof}$. Figure 14.7(a) represents the vertical antenna gain of the short-circuited patch antenna which is centered on the rear part of the vehicle roof-top. As a consequence from the mutual coupling with the evoked electrical currents on the roof-top, the main beam direction of the configuration is shifted from $\vartheta = 90°$ to an angular range at $\vartheta \simeq 72°$. The beam tilt results in a degradation of the underlying Signal-to-Noise Ratio (SNR) performance with respect to vehicular transmitters located in the horizontal xy-plane.

As a consequence of the finite size of the vehicle roof, the average gain degradation is in the order of magnitude of $\overline{\Delta G} \simeq 4$ dB at $\vartheta = 90°$. The impact of

Fig. 14.6 Geometry of the short-circuited, circular patch antenna [TKP+10] (©2010 EurAAP, reproduced with permission)

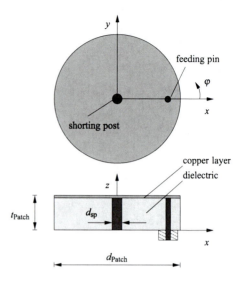

this mounting-based degradation of the V2V antenna performance in the horizontal plane is increased even further in case of a realistic tapering of the vehicle roof with a bent angle of $\vartheta_{\text{roof}} = 10°$ as depicted in Fig. 14.7(b). This particular mounting situation results in a drop of directive antenna gain to approx. -6.5 dBi in driving direction ($\vartheta = 90°$ and $\varphi = 90°$), thus yielding a significant sensitivity loss of the associated receiver hardware.

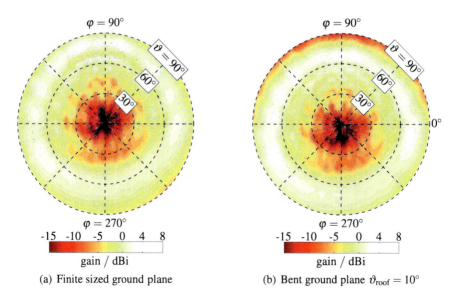

Fig. 14.7 Antenna radiation patterns on finite sized ground plane with (**a**) $\vartheta_{\text{roof}} = 0°$ and (**b**) $\vartheta_{\text{roof}} = 10°$ of elevation [Kle10] (©2010 IEEE, reproduced with permission)

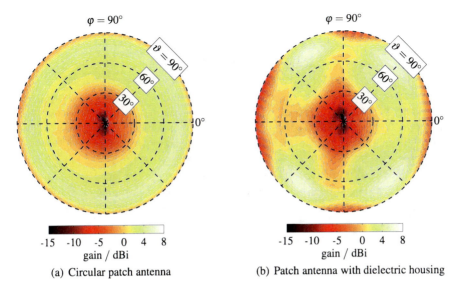

Fig. 14.8 Measurement of antenna radiation pattern with custom design cover (a) antenna without cover and (b) antenna with cover [Kle10] (©2010 IEEE, reproduced with permission)

The effect of the dielectric antenna housing can be seen from Fig. 14.8. Calibrated radiation pattern measurements of the short-circuited patch antenna were taken at 5.9 GHz in a spherical nearfield chamber on a ground plane (GP) with diameter $d_{GP} = 1$ m and rolled edges in order to reduce diffraction effects of the electromagnetic field. Figure 14.8(a) shows the results of the standalone mounted at the center of the GP, whereas Fig. 14.8(b) represents the radiation pattern including an automotive-compliant, dielectric design cover.

It can be seen that the dielectric design cover of the roof-top module leads to a significant deterioration of the related antenna radiation patterns at the considerer frequency of operation. Due to the physical dimensions of the roof-top module that are typically in the order of $(w \times l) \simeq (60 \times 110)$ mm, Fig. 14.8(b) shows the geometry- and frequency-dependent interference pattern resulting from reflection and refraction effects inside of the antenna housing. Due to symmetry properties of the arrangement, the scattering processes within the antenna compartment yield an azimuthal modulation of the associated far field patterns with a periodicity of $180°$ in azimuth φ. The resulting peak-to-peak variation amounts to approximately 5 dB in the horizontal plane ($\vartheta = 90°$). It is thus shown that the performance of V2V antenna elements may significantly depend on the mounting conditions on the vehicular roof-top. The finite size of the vehicle roof and its bent angle lead to a severe degradation of azimuthal antenna coverage due to beam tilting and shadowing of the V2V antenna under consideration. Also, the impact of any metallo-dielectric surroundings of the antenna configuration must be carefully reviewed for proper beam-pattern layout and link performance in V2V communications. The effect of mutual coupling with additional antennas enclosed in the same roof-top compartment will be highlighted in Sect. 14.5.3.

14.5.2 Multiple-Element Antennas for V2V Communications

V2V channel measurements as in [MTKM09, PBK+10] have shown that the vehicular radio channel is subject to multipath propagation of the electromagnetic field as described in Sect. 14.2. The presence of spatial multipath components depends on the channel conditions and is defined by the particular automotive use case. Multipath propagation can be effectively exploited by using MEA configurations to improve the quality of the wireless link incorporating adaptive combining schemes as in [KT09, RFZ09, RPZ09, KTP+09]. The impact of MEA array orientation is discussed in [KT09]. Multiple-element antennas can furthermore compensate for radiation pattern related impairments resulting from the particular mounting conditions as described in Sect. 14.5.1. The MEA diversity performance in spatio-temporal fading conditions depends on the mean effective antenna gain $G_{\text{MEG},\nu}$ and the power correlation ρ_{ij} between antenna elements i and j of the configuration [Tag90]. In this section, we will refer to the simplified AOA characteristic $p(\vartheta, \varphi)$ which is given by a stochastic formulation [PMF98] based on a two-dimensional power density in azimuth φ and elevation ϑ as shown in [RFZ09]. Here, a Laplacian density with a mean angle of wave incidence m_φ and an angular spread σ_φ is applied in azimuth. In elevation ϑ, the channel is modeled by a mean angle of incidence m_ϑ and a respective angular spread σ_ϑ defined by a Gaussian power density. Using this two-dimensional angular power density $p(\vartheta, \varphi)$, the mean effective antenna gains $G_{\text{MEG},\nu}$ of antenna element ν are derived as follows:

$$G_{\text{MEG},\nu} = \int_{\varphi=0}^{2\pi} \int_{\vartheta=0}^{\pi} \left[\frac{1}{1+\text{XPD}} G_{\vartheta,i}(\vartheta,\varphi) \frac{\text{XPD}}{1+\text{XPD}} G_{\varphi,i}(\vartheta,\varphi) \right] \times p(\vartheta,\varphi) \sin\vartheta \, d\vartheta \, d\varphi.$$

Here, XPD determines the cross-polarization ratio of the considered channel scenario. This quantity is set to XPD = 0 subsequently, thus neglecting potential polarization conversion by means of multipath propagation in the channel scenario throughout. In (14.5), $\mathbf{G}(\vartheta, \varphi) = [G_\vartheta, G_\varphi]^T$ describes the polarization-variant gain distribution of the antenna elements. The mean effective antenna gains $G_{\text{MEG},\nu}$ provide proportionality with the respective branch SNRs of a diversity antenna configuration. Equation (14.5) determines the antenna covariance R_{ij} between radiation elements i and j:

$$R_{i,j} = K_0 \int_0^{2\pi} \int_0^{\pi} \left[\text{XPD}\, G_{\varphi,i}(\vartheta,\varphi) G^*_{\varphi,j}(\vartheta,\varphi) + G_{\vartheta,i}(\vartheta,\varphi) G^*_{\vartheta,j}(\vartheta,\varphi) \right]$$
$$\times e^{j\mathbf{k}\mathbf{r}_{ij}} p(\vartheta,\varphi) \sin\vartheta \, d\vartheta \, d\varphi. \tag{14.5}$$

The antenna variances σ_i^2, σ_j^2 can be computed from (14.5) replacing $i = j$. In (14.5), K_0 is a normalization constant. The phase term $j\mathbf{k}\mathbf{r}_{ij}$ in (14.5) employing the wave vector $\mathbf{k}$ accounts for the spatial correlation between the antenna elements i and j. Subsequently, we will assume a spatial channel characteristic featuring a narrow angular spread in elevation $\sigma_\vartheta = 5°$ and a mean angle of incidence in elevation of $m_\vartheta = 90°$ as described in [KT09]. In azimuth, a mean angle of incidence is

Fig. 14.9 Two-dimensional angular power density $p(\vartheta, \varphi)$ [TK08b] (©2008 EuMA, reproduced with permission)

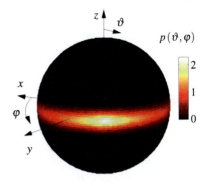

given by $m_\varphi = 120°$, and the azimuthal angular spread is assumed to be $\sigma_\varphi = 80°$. Figure 14.9 depicts a spatial diagram of the angular power density $p(\vartheta, \varphi)$ in terms of the considered AOA scenario. The considered channel scenario will represent the spatial distribution of incoming waves at the receiver location for a specific use case with multiple vehicles and scatterers located between the transmitting and receiving vehicles. Here, $\sigma_\varphi = 80°$ is applied to model the angular dispersion of scatterers in the xy-plane.

14.5.3 An Example Multistandard Antenna Module

Starting with the mounting-dependent impact on the V2V antenna performance as discussed in Sect. 14.5.1, we will furthermore investigate the effects of mutual coupling for V2V antennas for a realistic automotive-grade multistandard antenna module, including antenna elements for GPS and multiband cellular coverage, as well as a two-element MEA configuration for V2V communications as in [TK08a, KT09]. Conventionally, those antennas for cellular and navigation services will be co-located with the antenna frontend for V2V communications in the same roof-top compartment. The two 5.9 GHz V2V antennas V2V$_{\#1}$ and V2V$_{\#2}$ are spatially separated by approx. $|\mathbf{r}_{ij}| \simeq 44$ mm. This is shown in Fig. 14.10, which demonstrates the general antenna layout for multiple radio services jointly combined in a single roof-top antenna compartment. For laboratory measurements, the feeding point of each V2V antenna is directly matched to a 50 Ω SMA connector. Impedance matching at 50 Ω provides a return loss better than 16 dB for each V2V antenna within the considered frequency band of operation between 5.75 GHz and 5.95 GHz. Transmission coefficients S_{21} and S_{12} between the two V2V antennas range below -15 dB in the entire frequency range of operation and guarantee a sufficient degree of power decoupling.

Figure 14.11 depicts the measured antenna gain patterns of the two V2V antennas in a spherical near field chamber. Following Fig. 14.11(a), V2V$_{\#1}$ provides an almost rotational symmetric radiation pattern due to the large spatial separation to the GPS patch antenna on the one hand and to the cellular antenna on the other

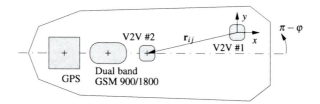

Fig. 14.10 Antenna placement of V2V$_{\#1}$ and V2V$_{\#2}$ relative to the GPS and cellular antennas in an automotive-grade multistandard antenna module. Vehicle driving direction at $\varphi = 180°$ [TK08b] (©2008 EuMA, reproduced with permission)

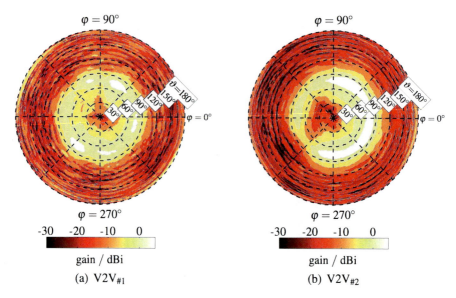

Fig. 14.11 Gain patterns $G_\vartheta(\vartheta, \varphi)$ of V2V$_{\#1}$ and V2V$_{\#2}$ at 5.9 GHz [TK08b] (©2008 EuMA, reproduced with permission)

hand. The directive antenna gain at 5.9 GHz amounts to approximately 6.9 dBi. Figure 14.11(b) shows the radiation pattern of V2V$_{\#2}$ at 5.9 GHz. This V2V antenna exhibits a spatially selective radiation pattern in the xy-plane, thus emphasizing the azimuthal region $|\varphi| \leq 90°$. The beam steering of antenna element V2V$_{\#2}$ results from mutual coupling with the cellular antenna located in close proximity to the V2V antenna. The cellular antenna acts as an electromagnetic director at 5.9 GHz due to its comparatively large physical size. The interaction with the cellular antenna leads to an enhanced directive antenna gain of approximately 8.7 dBi for antenna element V2V$_{\#2}$, which largely reduces the antenna gain in vehicle driving direction $\varphi = 0°$. To generally improve the beam pattern shape of V2V equipment, reflector–director techniques with discrete scatterers enclosed in the roof-top antenna compartment were successfully applied as reported in [KTP$^+$09]. Thus, mutual coupling with the dielectric antenna cover as shown in Sect. 14.5.1 and

Table 14.2 $G_{\text{MEG},\nu}$ of V2V$_{\#1}$ and V2V$_{\#2}$ at 5.9 GHz

Antenna #	$G_{\text{MEG},\nu}/\text{dB}$
V2V$_{\#1}$	-0.61
V2V$_{\#2}$	0.54

with additional antennas for different services in the same compartment may therefore strongly impact on and limit the achievable system performance of ITS in the 5.9 GHz range. Table 14.2 represents the mean effective antenna gains $G_{\text{MEG},\nu}$ of elements V2V$_{\#1}$ and V2V$_{\#2}$ in the considered channel scenario as in Sect. 14.3. Due to the slightly increased antenna gain of antenna element V2V$_{\#2}$ at angular positions $|\varphi| \leq 90°$, this antenna provides increased mean effective gain performance yielding $G_{\text{MEG}} = 0.54$ dB. This leads to an improved SNR behavior in comparison to antenna element V2V$_{\#1}$ with a mean effective antenna gain of $G_{\text{MEG}} = -0.61$ dB. The power correlation between the two V2V antenna elements amounts to $\rho_{ij} = 0.31$ in the considered channel scenario. Following the comparable mean effective antenna gains for V2V$_{\#1}$ and V2V$_{\#2}$ and the reasonably low value of antenna correlation, effective diversity combining can be achieved. The considered MEA performance metrics therefore point at improved communication performance and increased robustness of the radio link in safety-relevant V2V scenarios. Thus, in order to compensate for multipath fading in vehicular communication channels and to relax mounting-based coverage impairments, multiple-element V2V antennas are beneficial. Those can provide a better reliability of the wireless link and may add scalability features in vehicular networks by means of adaptive antenna techniques.

14.6 Wireless Access in Vehicular Environments

The IEEE standard 802.11p is known as Wireless Access in Vehicular Environments (WAVE). The WAVE physical layer specification is based on the previously published standard 802.11a in the 5 GHz band. Its main characteristics are: orthogonal frequency division modulation (OFDM), dedicated frequency band: 5.850–5.925 GHz, 10 MHz bandwidth, and 8 different operational regimes (coding-modulation schemes) achieving various data rates.

The 802.11p is rooted in the 802.11a standard for Wireless Local Area Network (WLAN) coverage in indoor environments. The vehicular scenario on roads and highways is markedly different from the one indoors. Hence substantial efforts are required to characterize the physical layer performance of 802.11p and propose improvements mainly for channel estimation. These efforts are described in Sect. 14.6.1. Also, the topology changes of a vehicular ad hoc network demand new ideas for the MAC layer to keep the channel access delays upper bounded, which is addressed in Sect. 14.6.2.

14.6.1 Physical Layer Performance and Proposed Improvements

A physical layer simulator model is implemented in [Kem07, NSK10]. In [Kem07], two kinds of channel impulse responses are used for evaluation: synthetically generated ones with a tapped-delay line channel model (as proposed in WAVE) and measured channel impulse responses. Whereas in [NSK10] a ray-tracing approach and a stochastic channel model are used for evaluation.

In [Kem07], the packet error rate (PER) results are presented for three different vehicular scenarios: rural, urban, and highway. The SNR for both channel models is chosen as 20 dB. The PER increases with packet length, data rate, and decreasing SNR. Comparing the performance results for both channel impulse responses, the simulation using the proposed WAVE channel model delivers more pessimistic results. The threshold for a correct system operation is set to PER < 0.1. In this case it is not possible to send a frame with length of 1000 bytes at an SNR of 20 dB for vehicles driving with 140 km/h. The packet payload impact on PER is evaluated in [NSK10] for vehicles driving on a highway. For the same simulation setting, it is shown that by using a soft-input Viterbi decoder the system performance improves about 5 dB.

PER results are also presented in [BCZB10], where a nonstationary vehicular channel model for highway scenario is used [KTC[+]09]. Here, the evaluations are done for a coding and modulation scheme using QPSK and a convolutional code with rate 1/2 that achieves 6 Mbps data rate. The length of the frame is set to 200 bytes. The simulated scenario consists of a car that drives by a fixed access point, where the access point acts as transmitter, and the car is the receiver. The model differentiates static, mobile, and diffuse scattering objects. Classical channel estimators rely on constant statistical channel properties (WSSUS assumption). If such estimators are applied for channels showing nonstationary behavior, they get more and more mismatched to the actual channel properties with longer frame duration. Hence the Bit Error Rate (BER) toward the end of the frame increases, leading to an increase in PER with packet length. The results in [BCZB10] also point out the strong influence of the diffuse scatterers, which makes the channel estimation more difficult using the pilot pattern defined in the IEEE 802.11p draft standard.

Experimental results from a vehicular measurement campaign are presented in [PTA[+]10]. The authors defined a performance metric, the frame success ratio as FSR = 1 − PER. Based on it, the achievable communication ranges are defined: FSR > *threshold* determines the distance over which the FSR remains above this *threshold*. The achievable range decreases for longer frame lengths. However, it is only applicable for thresholds above 0.5. Figure 14.12 shows the achievable range, which decreases with increasing data rate. This corroborates the results shown by computer simulations in [Kem07, BCZB10].

In order to cope with the nonstationarity of the channel, new techniques for channel estimation have to be investigated. A first step the characterization of this nonstationarity. Reference [MPZ[+]08] analyzes the temporal evolution of the signal subspace by means of the principal angle between subspaces at two times t and Δt. It

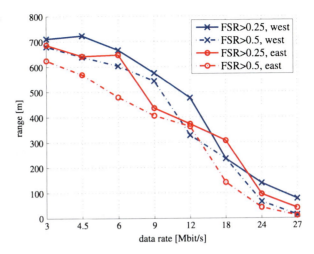

Fig. 14.12 Achievable range vs. data rate (120 km/h, 200 Byte) [PTA+10] (©2010 IEEE, reproduced with permission)

measures the dependency of two random vectors. The principal angle is larger when there is a stronger temporal variation of the dominant singular value.

The effect of using outdated statistical channel information is investigated in [BZPK09]. The spectral distance metric named spectral divergence [Geo07b] is used to evaluate the performance degradation of a temporal mismatched Linear Minimum Mean Square (LMMSE) filter. Further, the spectral divergence concept introduced in [Geo07a] for the continuous, infinite, and noise-free case is extended to a more realistic case, where noise is taken into account. By allowing a higher Mean Square Error (MSE), channel statistics could be used longer in time. Results show that the spectral divergence measure and the MSE degradation are correlated metrics.

The results presented until now are based on the IEEE 802.11p standard. Reference [ZCCR09] evaluates the prediction algorithm introduced in [ZMFK07] on real vehicular measurements at 2.5 GHz. The minimum-energy band-limited prediction method is used on a synthetically generated channel, using the Clarke's channel model with 30 paths, and on the measurements showing a good match between the results. The prediction MSE increases with mobile speed and with prediction horizon length. Further, in non-LOS situation, the predictor performs slightly worse than in LOS conditions.

14.6.2 Medium Access Control Layer Performance and Proposed Improvements

The MAC protocol is a key component in data communication, responsible for scheduling transmissions on the shared communication channel. The decision on which MAC protocol to use depends on, e.g., network topology and deployment. Two commonly used MAC protocols are Carrier Sense Multiple Access (CSMA) and Time-Division Multiple Access (TDMA). The latter offers a fixed number of time slots for each user such that each node knows when it is allowed to transmit and will then have exclusive right to the channel during the specific slot. However, TDMA does not support variable packet lengths; it requires synchronization among nodes and is usually deployed in centralized networks containing an Access Point (AP) or a Base Station (BS) responsible for sharing the resources, the time slots, among the users. CSMA, on the other hand, can be used in decentralized ad hoc networks, it is simple, supports arbitrary packet lengths, and does not require synchronization. When nodes want to transmit in CSMA, they start by listening to the channel and, if there are any ongoing transmissions, the node(s) will perform a backoff procedure, i.e., defer channel access by a random (backoff) time before transmission attempts begins anew.

CSMA has had enormous success through the WLAN standard IEEE 802.11 for wireless Internet, where high throughput is of great importance. In [KAH08] transmission between an AP and mobile vehicles is considered. However, in the common Intelligent Transportation Systems (ITS) communications architecture, currently under development in Europe, an ad hoc network structure is suggested for low-delay traffic safety applications. European Telecommunications Standards Institute (ETSI) proposes to use a European variant of Institute of Electrical and Electronics Engineers (IEEE) 802.11p called ITS-G5, which unlike traditional IEEE 802.11 requiring access points, uses a decentralized network topology. The ad hoc structure is advantageous since it eliminates the problem of guaranteeing coverage, but it also implies that scalability problems become imminent. In order to avoid this problem, the current suggestion within ETSI is to keep the network load below 25%.

Since a Vehicular Ad Hoc Network (VANET) is distributed, the corresponding MAC method must be decentralized. The MAC method must also allow for self-organization to cope with rapid network topology changes when nodes enter and leave the network. Furthermore, it must have support for all nodes within the radio range, which means coping with overloaded situations in terms of a high number of nodes and/or a high amount of injected data traffic. In addition, certain applications may have further requirements on fair or predictable channel access. Many traffic safety applications are based on the existence of time-triggered position messages, broadcasted periodically by all vehicles. These messages, denoted Cooperative Awareness Messages (CAM) by ETSI, contain position, speed, heading, etc. of each vehicle. There are also event-driven messages, denoted Decentralized Emergency Notification Message (DENM), broadcasted as a response to a hazard or an event. A MAC method used for CAM and DENM must be predictable, fair, and scalable. However, it has been pointed out in several articles [BSBU07, BUS09, BUSB09, PSN10] that since CSMA contains a random backoff procedure, channel access cannot not be provided within a predictable upper-bounded time. During certain traffic situations, periodic messages are dropped by the sending node before they are even transmitted. This is due to the fact that periodic messages have a deadline because as soon as updated position information is available, it is better to transmit the new information. Therefore the old message becomes useless and should be thrown away. In addition, the evaluations show that some nodes were forced to drop several consecutive position messages [BUSB09], resulting in vehicles being invisible to its surroundings for up to ten seconds. Consequently, with CSMA channel access is not fair, since some nodes are forced to drop more messages than others. At low penetration rates, i.e., few ITS equipped vehicles, CSMA works satisfactorily [PSN10], since the probability of long channel access delays is low. However, as more and more vehicles are equipped and can communicate, there will be situations where periodic channel access is not possible due to network overload. The randomness in CSMA causes problems also once channel access has been granted. Since the number of discrete backoff values to choose from is limited, concurrent transmissions from several closely located nodes are possible. If longer backoff periods are allowed, similar to what is done for multiple backoffs in acknowledged communication, the problem is mitigated, [KØV09]. The problems with scalability therefore become more prominent.

One potential remedy to the problems with CSMA could be the use of Self Organizing Time Division Multiple Access (STDMA) proposed in [BSBU07, BUS09] and [BUSB09]. This scheme is already in commercial use within the shipping industry for decentralized networks. In STDMA all nodes always get periodic access to the channel regardless of the number of nodes within radio range. This means that STDMA is predictable, fair, and scalable. However, STDMA does require synchronization and position messages such as CAM to be present in the system. Hence, STDMA uses a cross-layer approach for MAC by using the position information to schedule concurrent transmissions to be as far apart as possible to exploit spatial channel reuse. Even if the first generation of VANETs will be based on the IEEE 802.11p standard and its MAC method, CSMA–STDMA constitutes an interesting option when penetration rates increases and scalability issues of mature systems emerge.

14.7 Car-to-Car Applications

Let us briefly review the recent history of traffic safety, traffic management, and commercial applications and the corresponding research initiatives in Europe. In the beginning of the 1970s there was a peak of fatal accidents. About 18500 people died due to traffic accidents in Germany. International Transport Research Documentation (ITRD) [ITR] is internationally tracking these numbers since 1972. The first steps to lower the numbers of fatal accidents were the use of seat belts, airbags, ABS, ESP, etc. It helped to decrease the numbers in Germany to around 5000 in 2008. In the 1980s the first research programs on how traffic safety and traffic efficiency could be increased by wireless technologies were launched. The main goal of all these programs and applications was to improve the active safety and traffic efficiency by widening the driver's horizon with information on how to avoid accidents and navigate around congestions. Today we have traffic broadcast services with delay times which are in the range of several minutes to half an hour. Car-2-Car communication will generate and give us much faster access to the information needed for different applications. There are three major groups of applications addressing the different tasks with a variety of approaches.

Traffic safety applications are in need of real time information, i.e., in the range of 100 ms to 2–3 seconds depending on the situation for early warning or to inform the driver of an accident, road conditions, emergency vehicle, obstructions, such as broken down vehicles or road construction. Applications operate at intersections with vehicles, bikers, and pedestrians and also, for example, as pass assistant or wireless road sign information. Technological challenges for safety applications are to assure not only low latency by direct (ad hoc) communication between vehicles and vehicles to infrastructure, but also high position accuracy, reliability, and communication security.

Traffic management applications are the ones that drivers are most aware of, because the benefit in reduced travel time is quite evident. The main goals are to keep the traffic fluent by avoiding congestions and to inform the driver about the traffic situation. Additionally, applications dealing with different aspects of "finding available parking spaces" belong also to this group. These mobility applications can be realized with or without an infrastructure-based solution, e.g., WiFi communication to roadside units or cellular communication. In infrastructure-based solutions vehicles could deliver floating car data, e.g., speed, position, parking space, temperature, etc. to the traffic management center. A traffic light assistant, for instance, could supply speed recommendations to the vehicle advising the optimum speed to pass the traffic light (Travolution) [Tra]. Without a dedicated infrastructure Car-2-Car communication could help to organize traffic or predict the time delay to the destination. Self-Organizing Traffic Information System Self-Organizing Traffic Information System (SOTIS) [WER04], for instance, is such an application. In SOTIS, where a segment-oriented approach is adopted, vehicles send their driving times of (congested) road segments to approaching vehicles by Car-2-Car communication. Receiving vehicles aggregate the information in their knowledge base and generate new information for the next send event. The so-called hop-by-hop technique via ad hoc communication by vehicles is named multihop communication.

Commercial applications range from home entertainment server connected via WiFi to the car, car-to-mobile device, and broadband connections, as well as telematic services, e.g., for fleet management. Complete new service concepts to improve costumer relations like an extranet or a portal for car data, parking space reservation, booking and e-payment for services, and tolling are within this application range. Commercial applications are probably the group of applications which could help to deploy the Car-2-Car communication [SR05].

Such applications and the associated technologies are investigated in order to agree on an open international standard for a cooperative Intelligent Transport System. The TD(09)759 [MHL[+]09] describes the Vehicle Information System Program in the Kainuu region in North-East Finland, where a whole region teams up to supply a test area for various car applications. The following overview is by far not complete, but it will show the trends of what was done and where we are heading to.

One of the first pan European programs was EUREKA PROMETHEUS [EUR] from 1987 to 1995, where autonomous driving and Car-to-Car (Car-2-Car) communication, as well as convoy driving with vehicle tracking was shown. A French–German Project was the Inter Vehicle Hazard Warning from 2001–2002 which concentrated on Car-2-Car communication in the 869 MHz frequency range with an advantage of a relative wide range of about 1 km compared to WiFi with 300–500 m. At this point the US Federal Communications Commission (FCC) had already allocated a spectrum of 75 MHz bandwidth in the 5.9 GHz region for communication named Dedicated Short Range Radio (Dedicated Short Range Communication (DSRC)). This was investigated in the US programme named Vehicular Security Communication (VSC) [VSC] from 2002 onwards. Main interest was in the technology and traffic safety and also in the applications for traffic efficiency. Communication was mainly between car and Road Side Unit (RSU), for example a traffic light, but toll collection and commercial communication were of interest, too. Out of this program the working group for the new Car-2-Car communication standard IEEE 802.11p, based on the Wifi standard IEEE 802.11a, was established in 2004, and according to the IEEE project timeline [IEE], a final version was completed in July 2010. The German project FleetNet (2000–2003) concentrated on the communication aspects with different media such as UMTS and Wifi using multihop network protocols to overcome the short communication range of a single Wifi link. In addition, they also looked into communication by 24 GHz car radar. On the application side the Internet access as part of a business case was looked at. From 2001 to 2005, InVent [InV] researched traffic efficiency and assistant applications by means of cellular communication. PReVent [PRe], as a European Community funded Project, had its main goal set by the European commission to half the number of dead people by 2010. Active security applications where applied to reduce the dangers by estimation of time and gravity of a situation and first inform, then warn and at last assist the driver. To achieve this, the projects made use of new sensors and Car2X communication to increase the information range and prevent accidents. The subprojects concentrated on the applications like safe speed, safe following, lateral support, intersection safety (InterSafe), Vulnerable Road users, Collision Mitigation, wireless local danger warnings (WILLWARN), etc. During the same time from 2004–2008 the national German project Network on Wheels (NoW) [Net] worked on the communication links. NoW was able to base its work on the results derived by FleetNet. The main targets were the specification of geo-based routing and forwarding protocols, adaptation of the communication protocols to the changing communication channel, and security in ad hoc car networks. Beside the safety applications the project NoW also covers deployment applications. All previous studies had shown that the deployment needs to be supported by other applications than safety to cover the infrastructure costs. NoW delivered input to the standardization on the European level and to the Car-to-Car Communication Consortium (C2C CC) [Car]. This consortium of European car manufacturers and suppliers works on the standardization, harmonization, and promotion of the V2X communication technology. Due to this activity, an European wide exclusive 30 MHz frequency band with the option for another 20 MHz was granted 5.875 MHz to 5.905 MHz and optionally up to 5.925 MHz. Supported by COMeSafety [COM] which is a platform for exchanging information and results, but on an international basis. It supports the ongoing European activities of CVIS [Cooc], Safespot [Saf], and COOPERS [Coob], as well as the German AKTIV [Ada] program. CVIS concentrates its activities on the definition of a standardized architecture (ISO CALM) of the communication platform for the safety and efficiency application with a strong bond to the Infrastructure, by using not only Wifi with 802.11p, cellular 2G/3G, but the 5.8 GHz Toll system too. Whereas COOPERS has a stronger affiliation with the infrastructure operators and tries to minimize the gap between them and the telematics applications of the car industries. COMeSafety helps to consolidate research results from related projects also from the USA, for example, IntelliDrive (formerly VII) [Int] and CICAS [Cooa], and from Japan.

Most of these projects demonstrated applications on controlled test sites with a very limited number of interacting traffic partners. One of the major demonstration events was organized by the C2C CC [Car]. In October 2008 nine OEMs using communication hardware from several suppliers showed the interoperability while realizing four different safety related applications: Broken down vehicle warning, approaching emergency vehicle warning, roadworks warning, and warning on a motorcycle passing an intersection [Car]. Single applications and interoperability between the partners were well addressed, and the results are a good input for what has to follow now. The next step will be field operational tests (FOT) [FOT], where the communication and the application will have to work in the field with a lot of other partners in addition to the common day-to-day traffic which creates the scenario to prove the usefulness and what is needed on a statistical base to evaluate the complete ecosystem. This will give a better feedback on questions like: What type of congestion control is necessary if not only 5% but more than 90% of the participating cars are using the communication channel? What will be the real potential in increasing the safety, mobility, and efficiency?

The German project SIM-TD [SIM] has its focus on the whole variety of applications to be tested on an integrated communication platform in the field and on secluded test sites. The main difference to other projects is the number of test vehicles. It is planned that 100 *controlled* vehicles with dedicated test drivers exercising test scenarios and about 300 *uncontrolled* vehicles with drivers using the cars in their normal day-to-day life in the field test area near Frankfurt/Main, Germany. This includes the infrastructure and infrastructure communication in a large scale. Out of all applications, the most efficient and effective where derived in such a way that with the V2X functions about 50% of the deadly accidents should be addressed.

The CVIS project has six different test sites in France, Germany, Italy, Netherlands–Belgium, Sweden, and the United Kingdom, each operating a selection of the CVIS applications [Cooc]. The test site in Dortmund, e.g., is for demonstrating Enhanced Driver Awareness in an interurban environment. All those should provide results on a broader base for communication in fully populated environment, i.e., channel congestion, in case of traffic jams etc. and the acceptance and usefulness of the applications.

14.8 Standardization

In 1999, the FCC in the US allocated a 75-MHz band at 5.850–5.925 GHz dedicated to ITS and asked the American Society for Testing and Materials (ASTM) to define an ITS standard emanating from the IEEE standard 802.11 for WLAN. The standard was termed DSRC, a somewhat misleading name, since the term DSRC has been referring to a type of Radio Frequency Identification (RFID) systems for, e.g., Electronic Toll Collection (ETC) in Europe and Japan since the 1980s. IEEE has now taken over the work of ASTM and has developed IEEE 802.11p which amends the 802.11 both at the physical layer and the medium access control layer. To complement IEEE 802.11p, IEEE develops an entire protocol stack, called Wireless Access in Vehicular Environments (WAVE), which includes an application layer, IEEE 1609.1, security, IEEE 1609.2, network and transport layers, IEEE 1609.3, and a medium access layer based on IEEE 802.11p, IEEE 1609.4.

In Europe it was not possible reserve a 75-MHz band for ITS due to previous allocations. At 5.795–5.805 GHz there is a band, called Road Transport and Traffic Telematics (RTTT), which is dedicated for the European Committee for Standardization (CEN) standard DSRC, i.e., used for ETC. However, 30 MHz at 5.875–5.905 GHz has been set aside in a first step toward a harmonized European frequency band for ITS applications. ETSI has standardized a profile of IEEE 802.11p tailored to the European 30-MHz band. The ETSI Technical Committee (TC) on ITS, formed in December 2007, consists of five Working Groups (WG): WG1: User and Application Requirements, WG2: Architecture, Cross Layer, and Web Services, WG3: Transport and Network, WG4: Media and Medium Related, and WG5: Security. WG1 has defined two types of traffic safety messages, event-driven DENM and time-triggered CAM. CAMs are broadcasted periodically by all nodes and contain information about vehicle speed, position, and direction. In Europe the CAMs will initially have an update rate of 2 Hz and packet lengths of 800 bytes, whereas the current suggestion in the US is packet lengths in the order of 100–300 bytes updated with a frequency of 10 Hz.

The International Organization for Standardization (ISO) is developing a more general framework termed Communication Architecture for Land Mobiles (CALM). Note that CALM also includes a profile of IEEE 802.11p, denoted CALM M5. The idea of CALM is to use all types of already existing access technologies such as GSM, UMTS, LTE, WiFi, WiMAX, Infrared, CEN DSRC, CALM M5, etc., to provide seamless connections to end ITS users. CALM uses Internet Protocol version 6 (IPv6) in a bridging layer above the different wireless access technologies such that ITS applications can be developed without knowledge of the underlying access technology.

14.9 Summary and Outlook

For V2X channels, key properties are pathloss, fading statistics, temporal variance, and delay spread. The propagation conditions are influenced by the antennas and their placement on the vehicle. The vehicle's roof can strongly influence the antenna pattern; The details of intersections and lane merging should be carefully modeled, as they impact safety-critical applications.

The region over which the transmitter provides coverage is smaller than the region in which it creates interference. Due to the high speeds involved, V2X channels show strong time variance (the channel *state* changes) and nonstationarity (the channel *statistics* change).

The channels are simulated efficiently by both GSCM and TDLs. Possible nonstationarities are implicitly modeled by GSCMs, while TDLs have to be modified to provide time-varying tap locations and statistics. Ray-based models, on the other hand, provide accurate results for specific locations and surrounding structures.

Currently, the dominating standard for vehicular communications is IEEE 802.11p, which is derived from the popular 802.11a (WiFi) standard. Simulations showed an error floor, which motivates improved receiver architectures using advanced channel estimators. The experiments motivated to place access points as high as possible on gantries. The site-specific deployment of the APs will have a major impact on the coverage and reliability of infrastructure-based communications.

We note the following open issues: Applications and security mechanisms for V2V communication still need much standardization to enable inter-operability among vehicles from different manufacturers and countries, and the scenario for their introduction is still open.

The following questions need to be answered: How can the data be filtered and processed so that the driver is not irritated? How will scenarios work in different countries? For example, a warning received from a motorcycle in Germany is seen as a useful safety application, but how are motorcyclist warnings handled in other countries, e.g., France or Italy? Among the technological challenges, we mention positioning to distinguish between different lanes for road crossing assistance in dense urban environments and the use of MIMO for scalable and dependable communication over time-variant channels, to name but a few. Overall, the field of vehicular propagation channels and system design continues to provide fascinating and worthwhile challenges.

References

[Ada] Adaptive and Cooperative Technologies for the Intelligent Traffic (AKTIV). http://www.aktiv-online.org/.

[BCZB10] L. Bernadó, N. Czink, T. Zemen, and P. Belanović. Physical layer simulation results for IEEE 802.11p using vehicular non-stationary channel model. In *IEEE International Conference on Communications*, Cape Town, Sud Africa, June 2010. Also available as TD(10)11029.

[Bel63] P. A. Bello. Characterization of randomly time-variant linear channels. *IEEE Transactions on Communications Systems*, 11(4):360–393, 1963.

[BSBU07] K. Bilstrup, E. Ström, U. Bishop, and E. Uhlemann. Medium access control schemes intended for vehicle communication. Technical Report TD(07)369, Duisburg, Germany, September 2007.

[BUS09] K. S. Bilstrup, E. Uhlemann, and E. G. Ström. Performance of IEEE 802.11p and STDMA for cooperative awareness vehicle applications. Technical Report TD(09)938, Wien, Austria, September 2009.

[BUSB09] K. Bilstrup, E. Uhlemann, E. G. Ström, and U. Bilstrup. On the ability of the 802.11p MAC method and STDMA to support real-time vehicle-to-vehicle communications. *EURASIP Journal on Wireless Communications and Networking*, 2009 (Article ID 902414):13, 2009. This article has been published in an Open Access Journal and is therefore freely available to all. Follow the link the full text.

[BZK+on] L. Bernadó, T. Zemen, J. Karedal, A. Paier, A. Thiel, O. Klemp, N. Czink, F. Tufvesson, A. F. Molisch, and C. F. Mecklenbräuker. Multi-dimensional K-factor analysis for V2V radio channels in open sub-urban street crossings. In *International Symposium on Personal, Indoor and Mobile Radio Communications (PIMRC 2010)*, Istanbul, Turkey, submitted for publication. Also available as TD(10)10015.

[BZPK09] L. Bernadó, T. Zemen, A. Paier, and J. Karedal. Performance degradation of a mismatched wiener filter for non-WSSUS processes. Technical Report TD(09)858, Valencia, Spain, May 2009.

[Car] Car-to-Car communications consortium. http://www.car-2-car.org.

[COM] COMeSafety. http://www.comesafety.org.

[Cooa] Cooperative intersection collision avoidance systems (CICAS). http://www.its.dot.gov/cicas/index.htm.

[Coob] Co-operative systems for intelligent road safety, contract No. FP6-2004-IST-4 No. 026814. http://www.coopers-ip.eu/.

[Cooc] Cooperative vehicle infrastructure systems (CVIS), contract No. FP6-2004-IST-4-027293-IP. http://www.cvisproject.org/en/cvis_subprojects/test_sites.

[CV94] G. E. Corazza and F. Vatalaro. A statistical model for land mobile satellite channels and its application to nongeostationary orbit systems. *IEEE Trans. Veh. Technol.*, 43:738–742, 1994.

[CWL08] X. Cheng, C.-X. Wang, and D. I. Laurenson. An adaptive geometrical-based stochastic model for space–time-frequency correlated MIMO mobile-to-mobile channels. In *COST 2100 TD(08)472*, Wroclaw, Poland, February 2008.

[EUR] EUREKA PROMETHEUS. http://en.wikipedia.org/wiki/EUREKA_Prometheus_Project.

[FOT] Field operational test network (FOT-Net). http://www.fot-net.eu.

[Geo07a] T. T. Georgiou. Distances and Riemannian metrics for spectral density functions. *IEEE Trans. Signal Processing*, 55(8):3995–4003, 2007.

[Geo07b] T. T. Georgiou. Distances and Riemannian metrics for spectral density functions. *IEEE Trans. Signal Processing*, 55(8):3995–4003, 2007.

[GPSVRI+06] V. Gonzalez-Posadas, D. Segovia-Vargas, E. Rajo-Iglesias, J. L. Vazquez-Roy, and C. Martin-Pascual. Approximate analysis of short circuited ring patch antenna working at TM_{01} mode. *IEEE Transactions on Antennas and Propagation*, 54(6):1875–1879, 2006.

[IEE] IEEE 802.11 working group project timelines. http://www.ieee802.org/11/Reports/802.11_Timelines.htm.

[IN03] M. T. Ivrlac and J. A. Nossek. Quantifying diversity and correlation in Rayleigh fading MIMO communication systems. In *3rd International Symposium on Signal Processing and Information Technology (ISSPIT)*, pages 158–161, December 2003.

[Int] IntelliDrive. http://www.intellidriveusa.org.

[InV] InVent. http://www.invent-online.de.

[ITR] ITRD—International Transport Research Documentation. http://www.oecd.org/statisticsdata/0,3381,en_2649_34351_1_119656_1_1_1,00.html.

[KAH08] P. Kulakowski, J. V. Alonso, and J. G. Haro. Performance analysis of outdoor IEEE 802.11 WLAN with mobile sensors/vehicles. Technical Report TD(08)453, Wroclav, Poland, September 2008.

[Kem07]	T. Kempka. Physical layer simulations for vehicular environments. Technical Report TD(07)345, Duisburg, Germany, September 2007.
[Kle10]	O. Klemp. Performance considerations for automotive antenna equipment in vehicle-to-vehicle communications. In *Proc. URSI International Symposium on Electromagnetic Theory (EMTS)*, pages 934–937, 2010.
[KØV09]	M. Kaynia, G. E. Øien, and R. Verdone. Improving the performance of MAC protocols through channel sensing and retransmissions. Wien, Austria, September 2009. [Also available as TD(09)922].
[KSS$^+$10]	D. Kornek, M. Schack, E. Slottke, O. Klemp, I. Rolfes, and T. Kürner. Effects of antenna characteristics and placements on a vehicle-to-vehicle channel scenario. In *Proc. ICC 2010—IEEE Int. Conf. Commun.*, Cape Town, South Africa, 23–27 May 2010.
[KT09]	O. Klemp and A. Thiel. Diversity performance assessment of multielement antenna configurations in vehicle-to-vehicle communications. In *COST 2100 TD(09)743*, Braunschweig, Germany, February 2009.
[KTC$^+$09]	J. Karedal, F. Tufvesson, N. Czink, A. Paier, C. Dumard, T. Zemen, C. F. Mecklenbräuker, and A. F. Molisch. A geometry-based stochastic MIMO model for vehicle-to-vehicle communications. *IEEE Trans. Wireless Commun.*, 8(7):3646–3657, 2009.
[KTP$^+$09]	O. Klemp, A. Thiel, A. Paier, L. Bernadó, J. Karedal, and A. Kwoczek. In-situ vehicular antenna integration and design aspects for vehicle-to-vehicle communications. In *COST 2100 TD(09)982*, Wien, Austria, September 2009.
[Mat05]	G. Matz. On non-WSSUS wireless fading channels. *IEEE Trans. Wireless Commun.*, 4:2465–2478, 2005.
[MFP$^+$08]	J. Maurer, T. Fügen, M. Porebska, T. Zwick, and W. Wiesbeck. A ray-optical channel model for mobile to mobile communications. In *COST 2100 TD(08)430*, Wroclaw, Poland, February 2008.
[MHL$^+$09]	P. Mikkonen, M. Hämäläinen, T. Lehikoinen, K. Lindberg, O. Rajaniemi, and A. Kananen. Introduction to the vehicle information system program ongoing in Kainuu region in North-East Finland. In *COST 2100 TD(09)759, 7th Management Committee Meeting*, Braunschweig, Germany, 2009.
[MPZ$^+$08]	C. Mecklenbräuker, A. Paier, T. Zemen, G. Matz, and A. Molisch. On temporal evolution of signal subspaces in vehicular MIMO channels in the 5 GHz band. In *Joint Workshop on Coding and Communications (JWCC 2008)*, St. Helena (CA), USA, October 2008.
[MTKM09]	A. F. Molisch, F. Tufvesson, J. Karedal, and C. F. Mecklenbräuker. Propagation aspects of vehicle-to-vehicle communications—an overview. In *Proceedings of the 2009 IEEE Radio and Wireless Symposium*, vol. 1, pages 179–182, 18–20 January 2009.
[NBP05]	R. U. Nabar, H. Bölcskei, and A. J. Paulraj. Diversity and outage performance in space–time block coded Ricean MIMO channels. *IEEE Trans. Wireless Commun.*, 4:2519–2532, 2005.
[Net]	Network on Wheels. http://www.network-on-wheels.de.
[NSK10]	J. Nuckelt, M. Schack, and T. Kürner. Deterministic and stochastic channel models implemented in a physical layer simulator for Car-to-X communications. In *Advances in Radio Science, Kleinheubacher Berichte 2010*, vol. 9, pages 1–6, 2010. [Also available as TD(10)12013].
[PBK$^+$10]	A. Paier, L. Bernadó, J. Karedal, O. Klemp, and A. Kwoczek. Overview of vehicle-to-vehicle radio channel measurements for collision avoidance applications. In *71st IEEE Vehicular Technology Conference (VTC2010-Spring)*, Taipei, Taiwan, 2010. [Also available as TD(09)928].
[PKZ$^+$08]	A. Paier, J. Karedal, T. Zemen, N. Czink, C. Dumard, F. Tufvesson, C. F. Mecklenbräuker, and A. F. Molisch. Description of vehicle-to-vehicle and vehicle-to-infrastructure radio channel measurements at 5.2 GHz. In *COST 2100 TD(08)636*, Lille, France, October 2008.

[PMF98] K. I. Pedersen, P. E. Mogensen, and B. H. Fleury. Spatial channel characteristics in outdoor environments and their impact on BS antenna system performance. In *48th IEEE Vehicular Technology Conference*, vol 2, pages 719–723, 1998.
[PRe] PReVent. http://www.prevent-ip.org/en/home.htm.
[PSN10] S. Papanastasiou, E. G. Ström, and P. Nordqvist. The effect of security overheads on spatial congestion in vehicular networks. Technical Report TD(10)10080, Athens, Greece, September 2010.
[PTA+10] A. Paier, R. Tresch, A. Alonso, D. Smely, P. Meckel, Y. Zhou, and N. Czink. Average downstream performance of measured IEEE 802.11p infrastructure-to-vehicle links. In *IEEE International Conference on Communications*, Cape Town, Sud Africa, May 2010. [Also available as TD(10)014].
[PZB+08] A. Paier, T. Zemen, L. Bernadó, G. Matz, J. Karedal, N. Czink, C. Dumard, F. Tufvesson, A. F. Molisch, and C. F. Mecklenbräuker. Non-WSSUS vehicular channel characterization in highway and urban scenarios at 5.2 GHz using the local scattering function. In *International Workshop on Smart Antennas (WSA 2008)*, pages 9–15, 26–27 February 2008.
[PZK+09] A. Paier, T. Zemen, J. Karedal, N. Czink, C. Dumard, F. Tufvesson, C. F. Mecklenbräuker, and A. F. Molisch. Spatial diversity and spatial correlation evaluation of measured vehicle-to-vehicle radio channels at 5.2 GHz. In *13th DSP Workshop & 5th SPE Workshop*, Marco Island, Florida, USA, January 2009. [Also available as TD(09)751].
[RAMB05] M. Riback, H. Asplund, J. Medbo, and J. E. Berg. Statistical analysis of measured radio channels for future generation mobile communication systems. In *61st IEEE Vehicular Technology Conference (VTC2005-Spring)*, Stockholm, Sweden, 2005.
[RFZ09] L. Reichardt, T. Fügen, and T. Zwick. Effect of antenna placement on car-to-car communication channels. In *COST 2100 TD(09)829*, Valencia, Spain, May 2009.
[RKVO08] O. Renaudin, V.-M. Kolmonen, P. Vainikainen, and C. Oestges. First results of 5.3 GHz car-to-car radio channel measurements. In *COST 2100 TD(08)510*, Trondheim, Norway, June 2008.
[RPZ09] L. Reichardt, J. Pontes, and T. Zwick. Performance improvement using multiple antenna systems for car-to-car communications in urban environments. In *COST 2100 TD(09)966*, Wien, Austria, September 2009.
[RSZ09] L. Reichardt, C. Sturm, and T. Zwick. Performance evaluation of SISO, SIMO and MIMO antenna systems for car-to-car communications in urban environments. In *Int. Conf. Intelligent Transport Systems Telecommunications (ITST)*, pages 51–56, 20–22 October 2009. doi:10.1109/ITST.2009.5399385.
[Saf] Safespot (Integrated Project), Contract No. IST-4-026963-IP. http://www.safespot-eu.org/.
[SIM] Safe and intelligent mobility—test field Germany (SIM-TD). http://www.simtd.de.
[SKSK10] M. Schack, D. Kornek, E. Slottke, and T. Kürner. Analysis of channel parameters for different antenna configurations in vehicular environments. In *2010 IEEE 72nd Vehicular Technology Conference Fall (VTC 2010-Fall)*, pages 1–5, September 2010. [Also available as TD(10)12036].
[SR05] W. Specks and B. Rech. Car-to-X communication—the integration of the vehicle into our connected world. In *Elektronik im Kraftfahrzeug*, Baden-Baden, 6–7 October 2005.
[SSK09] H. Schumacher, M. Schack, and T. Kürner. Coupling of simulators for the investigation of car-to-x communication aspects. In *IEEE Asia-Pacific Services Computing Conference, 2009. APSCC 2009*, pages 58–63, December 2009. [Also available as TD(09)773].
[Suz77] H. Suzuki. A statistical model for urban radio propagation. *IEEE Transactions on Communications*, 25:673–680, 1977.
[Tag90] T. Taga. Analysis for mean effective gain of mobile antennas in land mobile radio environments. *IEEE Trans. Veh. Technol.*, 39:117–131, 1990.

[TK08a] A. Thiel and O. Klemp. Initial results of multielement antenna performance in 5.85 GHz vehicle-to-vehicle scenarios. In *Europ. Microwave Conf. (EuMC)*, pages 1743–1746, 27–31 October 2008. doi:10.1109/EUMC.2008.4751813.

[TK08b] A. Thiel and O. Klemp. Initial results of multielement antenna performance in 5.85 GHz vehicle-to-vehicle scenarios. In *Proceedings of the 2008 European Conference on Wireless Technology*, vol. 1, Amsterdam, The Netherlands, 2008, pages 322–325, 2008.

[TKP+10] A. Thiel, O. Klemp, A. Paier, L. Bernado, J. Karedal, and A. Kwoczek. In-situ vehicular antenna integration and design aspects for vehicle-to-vehicle communications. In *Proc. 4th European Conference on Antennas and Propagation*, Barcelona, Spain, Vienna, Austria, April 2010. [Also available as TD(09)982].

[Tra] Travolution. http://www.vt.bv.tum.de/index.php?option=com_content&task=view&id=69&Itemid=1.

[VSC] Vehicle safety communications. http://www.sae.org/events/ads/krishnan.pdf.

[WER04] L. Wischhof, A. Ebner, and H. Rohling. Self-organizing traffic information system based on car-to-car communication: prototype implementation. In *1st International Workshop on Intelligent Transportation (WIT 2004)*, Hamburg, Germany, 2004.

[ZCCR09] T. Zemen, S. Caban, N. Czink, and M. Rupp. Validation of minimum-energy band-limited prediction using vehicular channel measurements. In *17th European Signal Processing Conference (EUSIPCO)*, Glasgow, Scotland, August 2009.

[ZMFK07] T. Zemen, C. F. Mecklenbräuker, B. H. Fleury, and F. Kaltenberger. Minimum-energy band-limited predictor with dynamic subspace selection for time-variant flat-fading channels. *IEEE Trans. Signal Processing*, 55(9):4535–4548, 2007.

Chapter 15
Body Communications

Chapter Editor Arie Reichman, Chapter Editor Jun-ichi Takada,
Dragana Bajić, Kamya Y. Yazdandoost, Wout Joseph, Luc Martens,
Christophe Roblin, Raffaele D'Errico, Carla Oliveira, Luis M. Correia,
and Matti Hämäläinen

The chapter describes the activity carried out by COST 2100 on Body Area Networks (BANs) and is organised into three sections: the first section discusses the applications and the requirements; the second section presents the activity on channel measurements and modelling; finally, the third section addresses the existing transmission technologies and describes the results obtained for PHY, MAC and NET layers; low power design is also addressed.

15.1 Applications and Requirements

The idea of body area network (BAN) has been originally proposed by Zimmerman in 1996 [Zim96]. Among several means of BAN, where a set of communicating devices are located inside, on or around the (predominantly) human body, such as electric near-field communications [Zim96] or molecular communications [ABB08], recent advances in wireless technology have led to the development of wireless body area networks (WBAN). These devices comprise miniature interconnected sensor/actuator nodes, each one with its own energy supply for autonomous operation. The collected data are used either in situ or transmitted from the body to an access point and then forwarded to a clinic, trainer or elsewhere, in a real-time manner [Dru07, OC08, GVH+06].

Body area networks address a technology segment that is still in its early stage and uncovered by standards. Wide research efforts aim to provide a broad range of data rates at lower power consumption. Transmission distance is limited—immediate environment around and inside a person's body. Some of challenges

---

A. Reichman (✉)
Ruppin Academic Center, Kfar Saba, Israel

J.-i. Takada (✉)
Tokyo Institute of Technology, Tokyo, Japan

and opportunities do rely upon general wireless sensor networks. However, with the antenna–body interaction being an integral part of the channel and with the transmitter and receiver that almost collocate to each other, the features of WBAN channels require unique solutions [YSP10, HHN+07, FDD+06, GLP+03, Pro]. This technology segment is currently in standardisation process of IEEE 802.15 Task Group 6 (802.15.6 TG). It develops guidelines for wireless technologies and outlines a range of user scenarios such as in hospital, home and gym.

This technology is expected to be a breakthrough invention primarily in medicine, enabling the concepts like telemedicine and m-health to become real and facilitating highly personalised and individual care. For that purpose, three types of nodes with distinct propagation features can be defined [Jov05, Rei09, ZLK07, PKaAP+07, YSP10]:

- In-body (implant) nodes, placed inside the body;
- On-body (body surface) nodes, placed on the skin surface or at most a few centimetres away;
- Off-body (external) nodes: between a few centimetres and up to a few metres away from the body.

The first signals to be thought of when considering WBANs are blood pressure and electrocardiograph (ECG) waveforms, or, alternatively, their derivates: heart rate, QT interval, systolic and diastolic blood pressure. Their importance emerges from the fact that cardiac diseases are worldwide leading factor of mortality and morbidity. Other physiological parameters suitable for on-body sensing are temperature, oxygen level, respiration, muscle tension, electroencephalograph (EEG) signals and, to some extent, electromyography (EMG) signals. Constant monitoring during the daily routine gives an opportunity to spot episodic abnormalities that are not likely to occur in hospital bed, achieving better diagnostic and prognostic values [otESoCtNASoPE04]. Another important application is prediction of involuntary microsleep and falling asleep events while driving or performing dangerous but monotonous tasks [BF10].

Here node design comes into the focus, since it must be tightly but imperceptibly attached to the body and therefore it must be small-sized, implying smaller battery and less power. Since transmission is a major energy consumer, a substantial reduction could be made by intelligent processing, compressing the signal or extracting only its significant features while removing the characteristic artefacts caused by (inevitable) body-sensor friction and sensor displacement. A careful trade-off between communication and computation is crucial for an optimal design node [Jov05]. Considering the implant nodes, pacemakers that control commands for adjusting the heart rhythm and cardioverter defibrillator, programmed to detect cardiac arrhythmia and correct it by delivering a jolt of electricity are the most important application examples. Similarly, measuring the glucose level automatically releases the injection of proper amount of insulin. Monitoring the signals from a combination of kinetic sensors that measure acceleration and angular rate of body movement EMG and muscle tension sensors envisages advances in neurodegenerative and myodegenerative diseases [BLT09].

Data rates required for each one of the nodes that transmits signals aligned with heart-beats or slow signals like temperature are less than 100 b/s. Raw signal waveforms (EEG, ECG, EMG, pressure) are more consuming, but would not exceed 3 kb/s. Eventually, the technology development will increase the medical demands: contrary to ordinary monitoring, complex scientific investigation requires high sampling frequencies in order to extract the tiniest signal features. If, in parallel, a video signal is transmitted in order to correlate the signal pathology with patient's behaviour, data rates increase dramatically.

Other type of WBAN traffic is event-driven. If some abnormality in signal is detected, an alarm will be transmitted to doctor or to ambulance. This is of particular importance for cardiovascular diseases. Consider a scenario where thousands of amateur marathon runners are provided with sensors, with a possibility to spot and alert a particular person before he/she collapses. Alarm might be also raised if an asthmatic patient with an ambient sensor enters an environment full of allergens. Fall detector is valuable during the rehabilitation process, but also in another major application, i.e. geriatric care. Elderly people will continue to live alone and unattended much longer if a fall detector, coupled with several other vital monitoring sensors, is implemented.

Finger language interpreter for speech-impaired persons will be designed with sensors attached to fingers and hands that can detect the movements and relative positions between fingers and hands. WBAN can collect these data and send them to a signal processor in real time, where it would be converted to vocal language. Similar applications are interactive games where the body movements are transferred to movements of computer hero at video screen.

Sport monitoring research can also considerably benefit from wireless connectivity. A close study of body functions can be performed under physiological states during active training. Observing heart-rate, breathing, movements and function of muscles and joints, performances of elite sportsmen could be improved. This is not restricted to humans: trainers of horses and dogs could also improve the performance of their racing animals. Other species require adjusting data rates according to the body size, while the sensing signals remain the same. Subsets of sporting applications are fitness and recreation, with similar but less demanding range of signals.

Personal audio systems are focused to a personal music player device. The challenge in this use case arises as soon as a second audio source is added to the network and the requirement to potentially switch dynamically between sound sources is introduced, e.g. listening to the music while a phone call comes. The application scenario would then require that the music be muted and the call be activated on the wireless headset. Once the phone call is terminated, the headset should resume audio playback.

The range of WBAN applications is plentiful and colourful, emphasising its vital feature, i.e. interdisciplinarity. Without the respect and compromises of designers' constraints and end users necessities, no successful application can be made. In a case of a positive outcome, positive loops, analogous to Moore's law, are expected;

still modest transmission rates will grow following the increasing demands of consumers, thus verifying the success of technology.

15.2 Channel Measurements and Modeling

During the last few years there has been a significant increase in the number and variety of on/in-body health monitoring devices, ranging from simple pulse monitors, activity monitors, and portable Holter monitors, to sophisticated and expensive implantable sensors. WBAN for sensing and monitoring of vital signs is one of most rapidly growing wireless communication systems because patient cares and diagnosis day by day and more and more is connected with and dependent upon concepts and advances of electronics and electromagnetics.

The number of available radio-frequency (RF) medical devices is increasing every year, and the complexity and functionality of these devices is also increasing at a significant rate. Propagation model plays a very important role in designing wireless communication systems. Therefore, a reliable and efficient communication link is necessary to guarantee the best connection from/to in-body and on-body devices. In this section we provide outline and possible scenarios of our radio channel model and measurement setup for body area communications.

The use of in-body or on-body medical devices is not without many significant challenges, particularly, the increasing of propagation losses in biological tissue and effect of human body on RF signal. Therefore, to ensure the efficient performance of BAN wireless communication, the channel model needs to be characterised and modelled for reliable communication system with respect to environment and antenna.

Most of the WBAN devices are involved in the electromagnetic coupling into and/or out of the human body. This coupling usually requires an antenna to transmit a signal into a body or to pick up a signal from a body. The antenna operating environment for on-body and in-body antennas are different from the traditional free-space communications. An antenna can be designed in either air or the dielectric of the body. If the antenna is designed in air, the antenna's best performance will be achieved when air surrounds the antenna. If an antenna is designed in or near the dielectric of the body, the best performance from the antenna will be achieved when the antenna is actually inside the body cavity or placed on body surface. Therefore, to design an antenna for body area network, it is necessary to place the antenna in the medium in which it will be expected to operate.

The human body is not an ideal medium for radiowave transmitting. It is partially conductive and consists of materials of different dielectric constants, thickness, and characteristic impedance. Therefore, depending on the frequency of operation, the human body can lead to high losses caused by power absorption, central frequency shift, and radiation pattern destruction. The absorption effects vary in magnitude with both frequency of applied field and the characteristics of the tissue which is largely based on water and ionic content. It is very difficult to determine the absorption of electromagnetic power radiated from an implanted source by the human

15 Body Communications

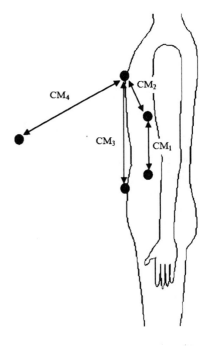

Fig. 15.1 Possible communication links for WBAN (CM_1: in-body, CM_2: in-body to on-body, CM_3: on-body, CM_4: on-body to off-body)

body. Quite a few investigations have been done to determine the effect of human body on radiated field, and almost all of these studies have been based on external sources.

Understanding the effect of human body on radiowave propagation is complicated by the fact that the body consists of components that each offer different degrees, and in some cases different types, of RF interaction. The liquid nature of most body structures gives a degree of RF attenuation, whereas, the skeletal structure introduces wave diffraction and refraction at certain frequencies. Prior to taking into consideration any BAN data communications, the effect of the human body on the RF signal must be understood. Unlike the standard communication all the way through constant air, the various tissues of body have their own unique conductivity, dielectric constant, and characteristic impedance.

The BAN devices are operated while close to, attached to or inside a human body, a fact which affects the communication performance. Reliable operating range is difficult to predict for these systems. BAN devices in hospitals are often used to replace hard wiring, so when similar performance is expected, the limitations of radio propagation compared to wires must be accounted for each application. The channel models for BAN could be categorised into different scenarios. Possible communication links for WBAN are shown in Fig. 15.1 [YSP10]. The scenarios are determined based on the location of the communicating nodes, i.e. in-body, on-body and off-body.

A detailed knowledge of the BAN channel (related to antenna/body interactions) is required to analyse and design properly systems at the physical (PHY) layer, media access control (MAC) layer and networking layers.

15.2.1 In-body Implants and Channel

15.2.1.1 Introduction

Only limited research deals with in-body implants or channels. In COST 2100, this topic is discussed only in [YK09, YH09, YH10, KJVM10]. Knowledge of the propagation characteristics in the body is important for successful transceiver design in dielectric media and for assessing potential health effects of electromagnetic radiation. The degree to which antennas can communicate with each other in a medium, characterised using the concept of path loss, is an important aspect. Information regarding RF propagation is typically obtained by physical experiments. This is difficult for medical implants but possible for homogeneous tissues as shown in [KJVM10].

Literature focuses mainly on two frequency bands for in-body communication: the 400 MHz Medical Implant Communication Services (MICS) band and the license-free industrial, scientific and medical (ISM) band at 2.4 GHz.

The Federal Communication Commission (FCC) [FCC] and European Telecommunications Standards Institute (ETSI) [ETS02] have allocated the frequencies in the 402–405 MHz range to be used for MICS. The maximum emission bandwidth to be occupied is 300 kHz. The maximum power limit is set to 25 μW equivalent radiated power (ERP) by ETSI, while FCC defined in the USA a maximum equivalent isotropic radiated power (EIRP) of 25 μW. Among primary reasons for selecting these frequencies, one can point to better propagation characteristics for medical implants, reasonable sized antennas and worldwide availability.

Also the 2.4 GHz ISM frequency band, with a maximal permitted EIRP of 100 mW could be selected. Signal propagation in the body will be easier for the MICS band than for the higher-frequency band of 2.4 GHz. However, the maximal permitted radiated power is lower in the MICS band than in the 2.4 GHz ISM band. This frequency band can be chosen as there are no licensing issues in this band and the higher frequency allows the use of a smaller antenna.

15.2.1.2 Implant Antennas

Medical Implant Communication Service (MICS) Band The analysis of [YK09, YH10] is only based on simulations. A three-dimensional full-wave electromagnetic field simulator, High Frequency Structural Simulator (HFSS), is used in [YK09, YH10] and integrated in a 3D immersive & visualisation platform. This virtual environment allows for more natural interaction between experts with different backgrounds such as engineering and medical sciences. Designing an efficient antenna for implantable devices is an essential requirement for reliable MICS operation.

Figure 15.2 shows the implant antenna used in the simulations. The antenna is composed of a single metallic layer and is printed on a side of a D51 (NTK) substrate with dielectric constant $\varepsilon_r = 30$, loss tangent $\tan\theta = 0.000038$ and thickness

Fig. 15.2 Implant antenna layout [YK07] (©2007 EuMA, reproduced with permission)

1 mm. A copper layer with thickness of 0.036 mm has been used as a metallic layer. The dimension of the antenna is $8.2 \times 8.1 \times 1$ mm, which is quite appropriate for some medical applications. The metallic layer is covered by RH-5 substrate with dielectric constant $\varepsilon_r = 1.0006$, loss tangent $\tan\theta = 0$ and thickness of 1 mm. Good impedance matching is obtained using the simulated return loss of this antenna for the MICS frequency band [YK09, YH10].

Industrial, Scientific and Medical (ISM) Band In [KJVM10] wave propagation at 2.457 GHz is investigated in human muscle tissue with relative permittivity $\varepsilon_r = 50.8$ and conductivity $\sigma = 2.01$ S/m [FCC01] by measurements as validation and simulations for insulated dipoles.

Two identical insulated dipoles (Fig. 15.3) are designed where the dipole arms are perfect electric conductors (PEC) surrounded with an insulation made of polytetrafluoroethylene with $\varepsilon_r = 2.07$ and $\sigma = 0$ S/m. Dipole antennas are used in [KJVM10] as they are the best understood antennas in free space and have a simple structure. At length $\ell_1 = 3.9$ cm, resonance is obtained for the frequency of 2.457 GHz. The resonance appears when the antenna is equal to half the wavelength in a homogeneous medium equivalent to the combination of the insulation and the muscle tissue medium. Hence $\lambda_{res} = 7.8$ cm, where λ_{res} is the wavelength at which resonance occurs, and we can derive the equivalent permittivity

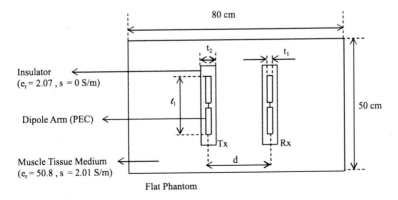

Fig. 15.3 Simulation setup and parameters [KJVM10]

$\varepsilon_{\text{r,equiv}} = 2.45$, which is closer to the permittivity of the insulation than to the muscle tissue. The dipole arms have a diameter $t_1 = 1$ mm with the diameter of the insulation $t_2 = 5$ mm. Insulated dipoles are selected instead of bare dipoles because the insulation prevents the leakage of conducting charges from the dipole and reduces the sensitivity of the entire distribution of current to the electrical properties of the ambient medium. This property makes insulated dipoles valuable for communication [KSOW81].

15.2.1.3 In-body Channel Models

Path Loss Path loss (PL) is defined as the ratio of input power at port 1 (P_{in}) to power received at port 2 (P_{rec}) in a two-port setup. PL in terms of transmission coefficient is defined as $1/|S_{21}|^2$ with respect to 50 Ω when the generator at the Tx has an output impedance of 50 Ω and the Rx is terminated with 50 Ω. This allows us to regard the setup as a two-port circuit for which we determine $|S_{21}|_{\text{dB}}$ with reference impedances of 50 Ω at both ports:

$$\text{PL}|_{\text{dB}} = \frac{P_{\text{in}}}{P_{\text{rec}}} = -10\log_{10}|S_{21}|^2 = -|S_{21}|_{\text{dB}}. \quad (15.1)$$

In this case, it is no longer possible to separate the influence of the antenna from the influence of the body. Therefore, in [YH09, KJVM10] the antenna is included as a part of the channel model.

Path Loss in 400 MHz MICS Band Simulations have been performed for four near-surface implants and two deep-tissue implants in a typical male body. The near-surface scenarios include applications such as Pacemaker (located below the left pectoral muscle), Vagus Nerve Stimulation (Right Neck & Shoulder) and two Motion Sensor applications located in right hand and right leg. The deep tissue implant scenarios consider endoscopy capsule applications for upper stomach (95 mm below body surface) and lower stomach (118 mm below body surface). For each transmitter location, the received power was calculated for a grid of points within a cylindrical area around the body. Then, the resulting data was partitioned into three sets: in-body to in-body, in-body to on-body, and in-body to off-body propagation sets. The in-body to in-body set includes all of the sample points that completely reside inside the body. Likewise, the in-body to body surface set includes all points that reside within a definable distance (i.e., 2 mm, 10 mm and 20 mm) from the body surface; and finally the in-body to out-body propagation set distinguishes all of the points that reside further away from the body surface.

The path loss PL in dB at some distance d can be statistically modelled by the following equation:

$$\text{PL}(d) = \text{PL}(d_0) + 10n\log\left(\frac{d}{d_0}\right) + S, \quad d \geq d_0, \quad (15.2)$$

where d_0 is the reference distance (e.g. 50 mm), and n is the path loss exponent which depends on the environment where RF signal is propagating. The human body

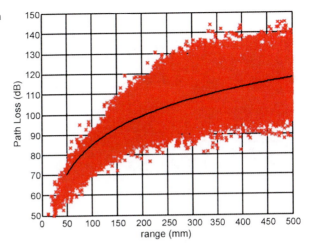

Fig. 15.4 Scatter plot of path loss versus distance for near surface implants to another implant [YH09]

is an extremely lossy environment, and therefore, much higher values than $n = 2$ (free space) for the path loss exponent are expected. S is the random scatter around the mean and represents the shadow fading in dB. The path loss is modelled by considering separate sets of parameters for deep tissue implants versus near-surface implants. Figure 15.4 shows a scatter plot of PL as a function of Tx–Rx separation for near-surface implants to another implant. The mean value of the random PL is displayed by a solid line. Shadowing effects are modelled by a random variable with a normal distribution with zero mean and standard deviation σ_s i.e., $\mathcal{N}(0, \sigma_s^2)$. The parameters of the model (15.2) can be found in [YH09]. The results of this model are based on simulation data only and should be validated using physical experiments.

Path Loss in 2.4 GHz ISM Band Measurements are executed using a vector network analyser NWA (Rohde & Schwarz ZVR) and the scattering parameters $|S_{11}|_{dB}$ and $|S_{21}|_{dB}$ (with respect to 50 Ω) between Tx and Rx for different separations are determined. The path loss is then calculated from $|S_{21}|_{dB}$ as shown in Eq. (15.1). A flat phantom, representing the trunk of a human body and recommended by CENELEC standard EN50383 [CEN02] (dimensions $80 \times 50 \times 20$ cm^3), is filled with muscle tissue simulating fluid (relative permittivity $\varepsilon_r = 50.8$ and conductivity $\sigma = 2.01$ S/m at 2.45 GHz [FCC01]). Simulations are performed using a 3D electromagnetic solver SEMCAD-X (SPEAG, Switzerland), a finite-difference time-domain (FDTD) program and FEKO (EMSS, South Africa), a method of moment (MoM) program. In [KJVM10] only PL in homogeneous tissues are investigated.

Figure 15.5 shows the simulated and measured PL in human muscle tissue as a function of distance d for the insulated dipole. The measured and the simulated values show excellent agreement up to 8 cm. The deviations between the measurements and the simulations are very low; average deviations up to 8 cm were 0.8 dB with SEMCAD-X and 1.3 dB with FEKO. High PL is obtained in human muscle tissue and PL increases of course with respect to distance.

Fig. 15.5 Path loss vs. separation between the Tx and the Rx of measurement and simulations [KJVM10]

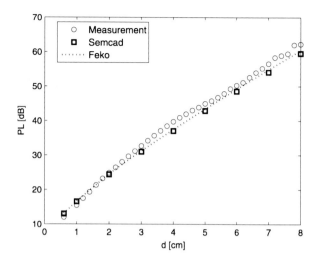

The measurement results of Fig. 15.5 are used to develop a PL model as a function of distance in human muscle tissue at 2.457 GHz. The PL is modelled as follows [KJVM10]:

$$\text{PL}|_{\text{dB}} = \begin{cases} (10\log_{10} e^2)\alpha_1 d + C_1|_{\text{dB}} & \text{for } d \leq d_{\text{bp}}, \\ (10\log_{10} e^2)\alpha_2 d + C_2|_{\text{dB}} & \text{for } d \geq d_{\text{bp}}, \end{cases} \quad (15.3)$$

where the parameters α_1 and α_2 are the attenuation constants [cm^{-1}], $C_1|_{\text{dB}}$ and $C_2|_{\text{dB}}$ are constants, $d_{\text{bp}} = 2.78$ cm is the breakpoint where the mutual coupling between the transmitter and the receiver ends, and d is in cm. The Tx and the Rx are close to each other, and this causes the antennas to interact with each other and alter the impedances due to mutual coupling. It is observed that the input impedance of the Tx keeps changing up to a certain separation between the Tx and the Rx, after which the input impedance becomes constant. This variation in the input impedance due to mutual coupling ceases to exist for separations larger than d_{bp}.

The parameter values in (15.3) are obtained by using a least square-error method and are provided in [KJVM10]. The low values of the maximum deviations of 0.56 dB and 1.22 dB suggest excellent agreement between the measurement and the derived models.

The model consists of two regions, *Region* 1 and *Region* 2. *Region* 1, $d \leq d_{\text{bp}}$, is defined as the region which is very close to the Tx dipole and extends from 0 cm to 2.78 cm. In this region mutual coupling appears between the Tx and the Rx. In *Region* 2, $d \geq d_{\text{bp}}$, the mutual coupling between the Tx and the Rx disappears.

The parameters α_1 and α_2 are the attenuation constants of *Region* 1 and *Region* 2, respectively. α_1 is equal to 0.99 in *Region* 1, and the value is higher than α_2 due to the Tx and Rx being close to each other, and hence α_1 depends on both the Tx and Rx and on the dielectric properties of the human muscle tissue and insulation. α_2 is lower and equals to 0.66 in *Region* 2 and depends on the dielectric properties of the human muscle tissue. The parameters $C_1|_{\text{dB}}$ and $C_2|_{\text{dB}}$ are constants with values 7.18 dB and 15.79 dB, respectively, and they depend on the dielectric properties of the medium and insulation. The proposed model is provided to validate the performance of in-body wireless communication systems.

In [KJVM10] models are provided for relative permittivity ε_r and conductivity σ combinations spanning a range of human tissues at 2.4 GHz.

15.2.2 On-body Antennas and Channel

As components and devices for WBANs are integrated in embedded systems—here body worn—the design constraints are particularly stringent: low-power-consumption devices are needed for demanding applications regarding autonomy (e.g. medical surveillance), surety and safety are mandatory for vital applications, radiations are subject to regulatory limits for public health and coexistence reasons, and size, aspect ratio and weight should be carefully dimensioned; all these constraints are notably intimately related to the acceptability by the future consumers. In particular, the size constraint on sensors and terminals is stringently reported on the antenna design.

15.2.2.1 On-body Antennas

The strong influence of the proximity of a human subject on the behaviour of on-body antennas—considering the frequency range for which the energy does not penetrate deeply in the body (say, for simplicity, above 1 GHz)—appears significantly differently in the narrowband and ultrawideband (UWB) cases. For both, the near field coupling to the body modifies antenna currents, hence the input matching, and induces energy absorption, often significantly. However, in the narrowband case, the dominant effect, and major drawback, is the shift of the resonance frequency causing mismatch, in addition to the losses inside the body, resulting in the unacceptable collapse of the total efficiency. Conversely, in the UWB case, the proximity of the body may improve the matching (often increasing the bandwidth) for two reasons: first—but this is a very general effect—losses favour the matching, lowering the $|S_{11}|$ more or less as a whole; second, the high permittivities of human tissues, acting as a sort of additional substrate (in particular for planar tangent antennas without *screening isolation*), tend to shift-down the band. The consequence in the narrowband case is that much attention should be paid first to the matching aspects. In the UWB case, even though impedance matching is important, the true performance indicator for antenna comparison purpose—aiming eventually the channel characterisation and radio link performance—should be mainly found directly in the transmission characteristics. However, although less sensitive than narrowband antennas, UWB antennas can couple strongly to the body, resulting notably in energy losses. That is why *desensitisation techniques* can be also interesting in UWB. One option is the field screening with either a (floating) ground plane or a ferrite sheet. Another

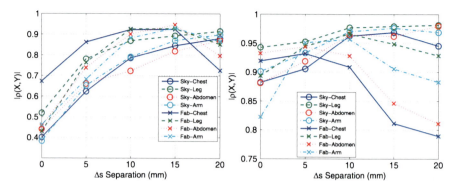

Fig. 15.6 Correlation coefficient of isolated and on-body antennas S_{11}, without (*left*) and with (*right*) absorbing material [TCAB10]

one, proposed in [TCAB10], is to use absorbing materials; the objective is twofold: first, to master the matching thanks to a controlled energy absorption (allowing a desensitisation with respect to the main sources of variability: population dispersion, tissues inhomogeneity and variable antenna distance from the body), and second, to reduce the radiation inside the body (public health aspect).

Two antennas are chosen for demonstration purpose: a commercial (SkyCross) and a fabric one. The isolated and on-body reflection coefficients S_{11} are compared for different distances from the body (Fig. 15.6) thanks to the analysis of the magnitude of the complex correlation coefficient between the isolated and on-body S_{11} given by

$$\rho(X,Y) = \frac{\mathbb{E}[(X-\mu_X)(Y-\mu_Y)]}{\sigma_X \sigma_Y}. \tag{15.4}$$

It is observed that the correlation coefficient values exceed 0.8 (and 0.9 most of the time) with the absorbing sheet, whereas it varies from 0.4 to 0.9 without. Furthermore, a reduction of up to 4.7 dB of radiated power into the body is observed.

15.2.2.2 On-body Static Channel (Statistical Analysis)

In the very context of the on-body channel, the physics of the propagation along or around the body has been addressed specifically, showing without any ambiguity that the dominant mechanisms of propagation are very specific, mainly combining free space waves, creeping waves (travelling at the speed of light in air [ZMAW06, ROHD07]), on-body reflections (from arms and shoulders) and off-body MPCs (reflection, diffraction and scattering coming from the environment), resulting in very particular channels, depending on the studied scenario. It is not worthwhile to recall that creeping waves are waves diffracted around a body of smooth convex surface in the shadow region. It should be underlined again at this occasion that, because of the high complex permittivities of most biological tissues over a very large frequency range, almost no energy propagates through the body (for example, attenuation *through the head* of the order of 200 dB or more are reported [ZMAW06]). As well, it has been shown that the path loss mechanism is dominated by energy absorption from body tissues. It has been notably underlined in, e.g., [Tak08, RDG+09] that: "The body area channel is very different from other wireless channels in the sense that the antenna–human body interaction is an integral part of the channel..." and "... it is impossible to separate the antennas and propagation channel in the same way as the double directional channel modelling, but the Tx and Rx antennas are the integral part of the channel" (cited from [Tak08]) and that: on-body "propagation in WBAN is indeed confined within a limited range around human body where transmitter and receiver are almost co-located to each other. The propagation channels in WBAN therefore have limited number of Multi-Path Components (MPCs), ...", and "Contrary to the definition of fast fading and slow/shadow fading in large area propagation, on-body channel fading is mainly due to the body shadowing effects and the body movements." (cited from [LDO08b]).

Various contributions have been produced worldwide in the very last years. In particular, interesting contributions have been provided by IEEE 802.15 Working Groups: in November 2006, UWB channel models have been proposed by the IEEE 802.15.4a Task Group, comprising a first WBAN model [MCC+06, MBC+07]; the IEEE 802.15 Task Group 6 (TG6) has been established in November 2007[1] to standardise PHY and MAC layers for BANs and has already provided channel models in the BAN context in April 2009 [YSP10]. In Europe, contributions have been notably provided by ETH Zurich [ZAS+03, ZMAW05, ZMAW06], UCL (Leuven), IMEC (Leuven, Eindhoven), and ULB (Brussels) [FDD+06, FDR+05, FRD+06, ROHD07], the University of Birmingham and Queen Mary University of London [HHN+07, HH06], and Queen's University of Belfast [CS06, CS07b]. Worldwide, contributions have come from Japan (NICT and Tokyo Tech) [ATT+08a, ATT+08b, KTM+08, Tak08], Korea, Singapore, Australia (NICTA) [MHS+08] and the USA.

An overview of NICTs work achieved notably in the framework of its contribution to the TG6 channel working group is proposed in [YTA+09]. An uncommon situation, interesting for medical applications, is considered: measurements are performed for various scenarios (notably "S3" and "S4" of TG6) with a subject playing the role of a patient lying on a bed in a hospital room (Fig. 15.7). Detailed results are also provided in [ATT+08a, ATT+08b, YSP10].

A detailed state-of-the-art including the most recent IEEE 802.15.6 contributions has been published in [RDG+09]. It is reported that a very large dispersion of the results, notably as regards PL data and models are observed, both in anechoic chamber (Fig. 15.8) and in indoor premises (Fig. 15.9). The reasons are manifold. First, it points out a lack of standard protocols regarding measurement procedures, analyses and modelling. Second, it underlines the large variability of the BAN channel with regard to population (human subjects), postures and movements (including involuntary micro-movements, breathing, etc.), used antennas (antenna type, distance

---

[1] All TG6 documents are downloadable at: https://mentor.ieee.org/802.15/documents?is_group =0006.

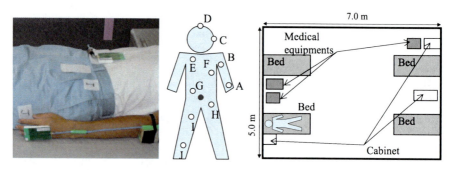

Fig. 15.7 NICT's measurement scenarios in a hospital room [ATT+08a] (©2008, reproduced with permission)

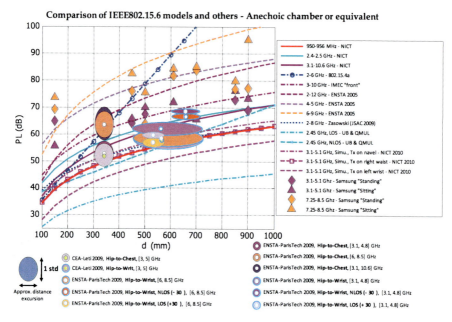

Fig. 15.8 Comparison of PL models assessed from measurements in anechoic chambers [RDG+09]

from the body and polarisation), positioning on the body and surrounding environment. Third, representativeness of the proposed models is not established, as it is not clear whether they are specific or nonspecific: in particular, the numerous above-mentioned sources of variability should be more widely investigated, which would found the statistical approach more rigorously; precisely, the collected statistical sets often seem either too small or too specific. Typically, three types of approach have been proposed:

1. A *classical* approach—which could be quoted as *physical*—considering well-defined situations (such as *around* or *in front of the torso*, or *along arm* or

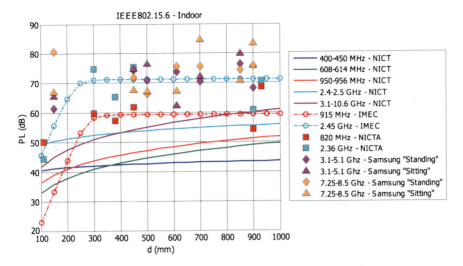

Fig. 15.9 Comparison of 802.15.6 PL models in indoor environments [RDG+09]

leg, etc.), and e.g. modelling the PL as a function of the inter-antenna distance [ZAS+03, RDM+04, FDR+05, FDD+06, FRD+06, MCC+06, MBC+07, HHN+07, ROHD07, TTJI09, JRVM09]. Note that in several of these works, antennas are considered as point sources.

2. An approach—which could be quoted as *hybrid*—considering *scenarios* related to potential applications, but still modelling the PL as a function of the distance, either gathering all these configurations in a unique PL(d) model (sometimes distinguishing the radio link nature—LOS or soft/hard NLOS) [HHN+07, Tak08, ATT+08a, ATT+08b, YTA+09, TTT+09, BCNT09, OLMC10], or providing PL(d) models according to the reference node location [YHSP+10].

3. And a *scenario-based* approach considering application-dependent configurations (such as *hip to chest* or *hip to wrist* or *ear to ear*, etc.) completely separately, the PL depending only on the scenario and not on the distance [CS06, CS07a, KCPW08, MHS+08, DO09, RDG+09, DO10b, DO10a, RM10, DO11, Rob10].

Note that all these three approaches have been considered in the standard IEEE 802.15.6 (see [YSP10] and related contributions). The characterisation and modelling of the BAN channel following a purely scenario-based approach is proposed in the TG6 channel models "C" [YTA+09] for various links between antennas and body postures, at 820 MHz and 2.36 GHz [MHS+08], over the UWB sub-bands 3.1–5.1 GHz and 7.25–8.5 GHz in both anechoic chamber and office room environments [KCPW08], at 868 MHz in [CS06], at 2.45 GHz in [HHN+07], over the 2.4–2.5 GHz ISM band and 3–5 GHz UWB sub-band in [DO09, DO10b], and over a 2–12 GHz UWB in [RM10], all considering various postures and/or movements in indoor premises (or in anechoic chambers for last two). The scenario-based approach presents the advantage of not merging very different radio link situations which could be near field (NF) or far field (FF), quasi-LOS, pseudo-periodic (from LOS to NLOS), or completely NLOS (such as *hip to chest*, or *hip to wrist* or *front to back*). Indeed, depending on the configuration, the *distance effect* can be completely hidden by other dominant effects such as strong shadowing or fading due to movements.

In *classical channels* (indoor, urban, etc.) antennas are usually far away enough to be in their respective FF zones and (at most) weakly perturbed by their environment. In this context the power law model is valuable and physically founded, and the constant term PL_0 is primarily fixed by the chosen reference distance. This is particularly true when the deembedding/deconvolution of the antenna responses is achieved leading to the "intrinsic" channel. Considering the "ideal" case of the free space channel between point sources, the Friis formula holds, and the PL_0 term depends uniquely on the reference distance (and the frequency if not averaged over the band). However, most often, channel models include the contribution of antennas, affecting the PL_0 value. For the BAN channel, the physical interpretation of this term is completely different: of course, it depends on the chosen reference distance—as the increasing trend of the PL with the distance is physically founded and practically observed for well-defined particular configurations, e.g. for radio links around the trunk—but the main contribution resides in the absorbed energy by the body notably in the vicinity of antennas, which is actually related to the in situ total efficiency (i.e. including losses into the body in addition to return losses) of each antenna. It consequently also strongly depends on the used antenna type, e.g. narrowband or broadband, well-grounded—and *screened* from the body—or not, well-balanced or not, etc., and, depending on this choice, significantly (or not) on the antenna distance to the body. Actually, for the BAN channel, most researchers consider that antenna deembedding/deconvolution is impossible to achieve properly both practically and conceptually.

In addition, measurements have shown that for most of the indoor premises, on-body cluster(s) and off-body clusters (multipath components (MPCs) coming from surrounding scatterers) can be resolved in the UWB context, supporting the approach proposed by IMEC since 2005 [FRD+06] to separate on-body and off-body contributions in the models structure; this position was also approved in [MCC+06], following the 15.4a final report on UWB channel models [MBC+07]. This approach can bring a simplification of the model elaboration and of its structure and a significant reduction of the required measurement campaigns effort for analysing the numerous sources of variability of the BAN channel.

In the statistical analysis of variability, an alternative approach to measurement is numerical simulation. It has been notably used to elaborate the 15.4a model, and more recently, a 3D virtual reality platform has been developed by NICT (Japan) and NIST (USA) [YHSP+10] (Fig. 15.10).

Path Loss Models Both exponential models with typical attenuation coefficients ranging from 1 to 2 dB/cm at 2.4 GHz and power law models (with PL exponents ranging from 1.5 to 5 depending on the frequency range, LOS/NLOS configurations, and environments, with values up to more than 7 for UWB measurements in anechoic chambers) have been used for the PL. Typical PL ranges from 30 to 110 dB,

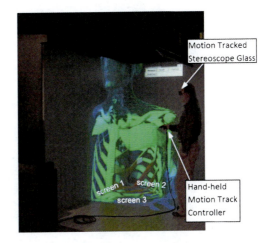

Fig. 15.10 3D virtual reality visualisation environment [SPYH+09] (©2009, reproduced with permission)

depending on the frequency, the radio link configuration (LOS/NLOS) including body postures or motion, the environment type, the antenna type and its distance to the body for most of them. General trends are the following:

- The higher the frequency, the more the PL values.
- The main origin of medium scale and/or slow fading, is the body shadowing effect, contrary to the usual "exogeneous" shadowing.
- The shadowing effect of the body is strong, e.g. typically ranging between 5 and 15 dB (on average from max to min) over the UWB low band, typically 3–5 GHz [DO09].
- Because of the off-body MPCs incoming from the surrounding environment, PL values are higher in anechoic chamber.
- Antenna types, and even more, antenna distance to the body, may cause PL variations as high as 15 dB [RM10].
- Both in UWB and narrowband contexts, the shadowing statistics have been found to be lognormally distributed by most authors [YSP10, YTA+09, Tak08, LDO08b, LDO08a, JRVM09, TTT+09, BCNT09, TTJI09, DO09, RDG+09, OLMC10].

The PL has been proved to depend on the body posture (Fig. 15.11 [OLMC10]), on the frequency range [YSP10, RM10], on the antenna types and, for most of them, on their distance to the body. Parametric analyses have been proposed in [RM10] and [Rob10] (Figs. 15.12, 15.13, and Table 15.1).

On-body Channel Impulse Response (CIR) and Power Delay Profile (PDP) Models for the UWB The CIR/PDP model proposed by NICT [YSP10] is a single cluster tapped delay line/Poisson process model (Table 15.2), where a_l, t_l and ϕ_l are respectively the path amplitude, the path arrival time and the phase of the lth path, L the number of the arrival paths, $\delta(t)$ the Dirac distribution, Γ an exponential decay with a Rician factor γ_0, S a normal distribution with zero-mean and standard deviation σ_S, λ the path arrival rate, and $\bar{L}$ the average number of the arrival paths L. It is modified in [DO10a] to model the time-varying channel.

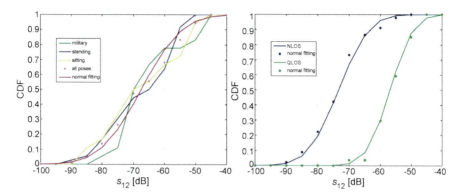

Fig. 15.11 (*Left*) Path gain CDF for various postures and (*right*) clustered according to their nature (quasi-LOS or NLOS), obtained from measurements with patch antennas at 2.45 GHz [OLMC10]

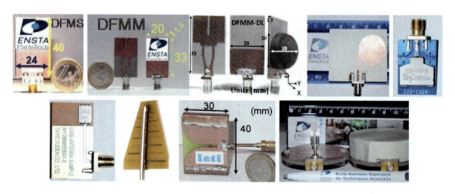

Fig. 15.12 Nine analysed UWB antennas [Rob10]

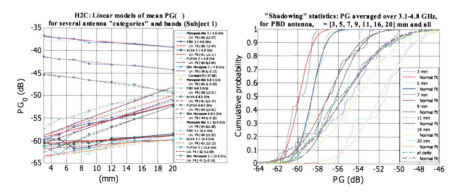

Fig. 15.13 *Hip to chest* scenario; (*left*) Linear models of PG(δ) per "antenna type" and sub-bands; (*right*) PG empirical CDF and models with Planar Balanced Dipoles and δ as a parameter (3.1–4.8 GHz) [Rob10]

Table 15.1 *Hip to chest* scenario—mean path gain linear fits $PG_0(\delta) = PG_{00} + K\delta$ [Rob10]

Subject[†]	Monopole-like[‡]		PBD		ALVA		PLPDA		Staircase M.	
Band (GHz)	PG_{00}	K (mm^{-1})	PG_{00}	K (mm^{-1})	PG_{00}	K (mm^{-1})	PG_{00}	K (mm^{-1})	PG_{00}	K (mm^{-1})
3.1–4.8	−64.26	0.2706	−60.66	0.4434	−61.08	0.1110	−60.79	0.6385	−36.56	−0.1339
6–8.5	−64.31	0.2288	−58.23	0.5090	−61.58	0.1388	−64.49	0.6128	−44.56	−0.2238
3.1–10.6	−64.78	0.2577	−60.47	0.5280	−62.26	0.1259	−62.41	0.5859	−41.06	−0.1601

[†]Male: $h = 1.68$ m, $m = 65$ kg, $BMI = 23.03$ kg/m^2. [‡]"Monopole-like" = {DFMS, DFMM, SkyCross, Taiyo Yuden}

Table 15.2 UWB CIR/PDP model summary from IEEE 802.15.6 [YSP10]

PDP Model		
	$h(t) = \sum_{l=0}^{L-1} a_l \exp(j\psi_l)\delta(t - t_l)$	
	$10\log_{10}|a_l|^2 = \begin{cases} 0 & l = 0 \\ \gamma_0 + 10\log_{10}(\exp(-\frac{t_l}{\Gamma})) + S & l \neq 0 \end{cases}$	
	$p(t_l|t_{l-1}) = \lambda \exp(-\lambda(t_l - t_{l-1}))$	
	$p(L) = \frac{\bar{L}^L \exp(\bar{L})}{L!}$	
	ψ_l is modelled by a uniform distribution over $[0, 2\pi)$	
a_l	γ_0	−4.60 dB
	Γ	59.7
	σ_S	5.02 dB
t_l	$1/\lambda$	1.85 ns
L	$\bar{L}$	38.1

Delay Spread   In anechoic chamber, reported values are in the range of a few nanoseconds, with typical average values of 1–2 ns. In indoor premises, typical values depend strongly on the environment, ranging from 10–15 ns for a small office up to more than 40 ns for either large rooms or highly diffusive environments [RDG+09]. It has been found to depend on the inter-antenna distance, for example in [JRVM09] in which fitted models of τ_0 and τ_{rms} are provided for radio links along the torso, the back, the arm and the leg in a modern office environment at 2.45 GHz (Fig. 15.14). Both quantities are found to be lognormally distributed around the means.

Analytical or Semi-analytical Models   An analytical 3D on-body propagation model is derived in [LKD+10b] (see also [FKR+10] and [RFCO07]). The model describes the mutual scattering interference from different body components, i.e. trunk and arms represented with three vertical homogeneous lossy cylinders, to the on-body propagation with a polarised point source. The full field solution is derived from a polarised point source for the propagation with any desired direction of polarisation. It is computed along ($z'z$ direction) and around (φ direction) the *trunk cylinder*. Preliminary measurements were conducted in anechoic chamber to measure the propagation in vertical polarisation on the trunk with dipole antennas at 2.45 GHz. The results showed similar channel power loss along the trunk with the model, and similar trends of the scattering interference from the body was observed (Fig. 15.15).

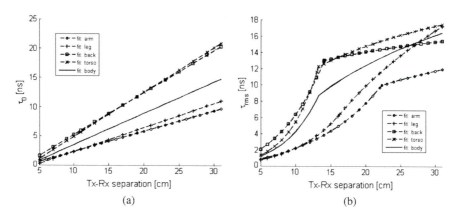

Fig. 15.14 (a) Mean excess delay τ_0 and (b) RMS delay spread τ_{rms} vs. antenna separation [JRVM09]

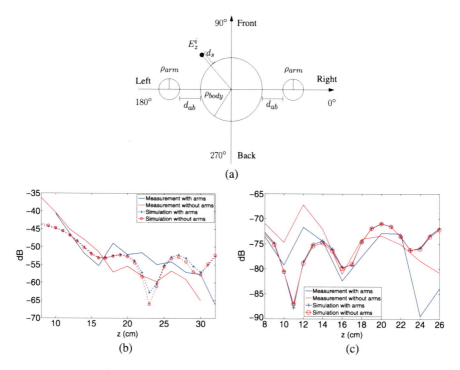

Fig. 15.15 (a) Human body and three cylinders modelling. Measurement and simulation comparison on channel power along z-direction at (b) $\phi = 135°$ and (c) $\phi = 225°$ [LKD+10a] (©2010 IEEE, reproduced with permission)

15 Body Communications

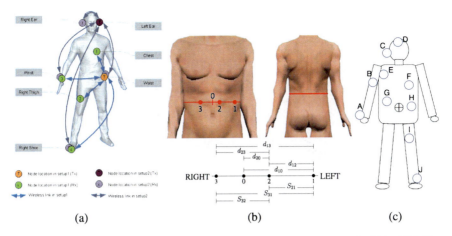

Fig. 15.16 On-body channel measurement setup in (**a**) [DO09, DO10b], (**b**) [LDO08a, LDO08b, LDO09b] and (**c**) [KT10]

15.2.2.3 On-body Dynamic Channel

The on-body dynamic channel has been investigated in [DO09, DO10b, DO10a, LDO08a, LDO08b, LDO09b, KT10] in different bands.

A measurement campaign has been carried out in the time domain at CEA-Leti in both anechoic chamber and indoor environments [DO09, DO10b]. The measurement test-bed is able to collect simultaneously up to four CIRs, corresponding to different locations on the body. The sounded bandwidth being very wide, the channel characterisation is carried out in both ISM band around 2.45 GHz and in the UWB low band 3–5 GHz. Seven subjects of various morphologies were involved to account for the population variability. Two measurement sets were carried out with central node on *hip* or on *left ear*, the sensors being placed on different body positions (Fig. 15.16a). For both anechoic chamber and indoor environments, three cycles of 4 seconds have been recorded in the *still* and *walking* scenarios, whereas one cycle of 20 seconds has been collected in the *running* scenario.

In [LDO08a, LDO08b, LDO09b] measurements were performed at 2.4 GHz in a quasi-anechoic environment with off-body MPCs assumed negligible, except for the floor. Three vertically polarised antennas were placed on the abdomen at the same height (Fig. 15.16b). To limit the impact of the body on the antennas, 5-mm thick dielectric spacers were inserted. The torso was kept immobile while the arms were swinging (assuming a periodicity around 1 Hz) to mimic the walking movement.

Measurements of Tokyo Tech and NICT were performed with a channel sounder at 4.5 GHz [KT10] with 120-MHz bandwidth. The Tx antenna is placed on the navel, whereas ten positions are considered for the Rx sensors (Fig. 15.16c). Walking on the spot (Action I) and repetitive standing up/sitting down (Action II) are considered.

In [DO09] the time-dependent power transfer function $P(t_n)$ is expressed as

$$P(t_n) = G_0 S(t_n) F(t_n), \tag{15.5}$$

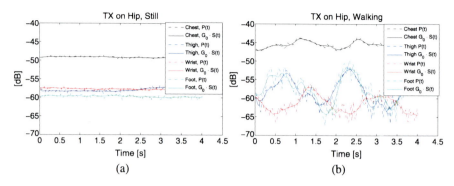

Fig. 15.17 Slow fading component in indoor for Tx on hip at 2.45 GHz band: (**a**) still and (**b**) walking subject [DO09] (©2009 IEEE, reproduced with permission)

where G_0 represents the mean channel gain, and $S(t_n)$ and $F(t_n)$ the slow and fast fading components. The slow component is basically due to the *shadowing* by the body, which mainly depends on its movements, as shown in Fig. 15.17. When the subject does not move, the slow component is very moderate, since the shadowing condition basically remains unchanged. However, the mean channel gain values vary significantly from one subject to another. This is due to morphological differences of subjects and to the proprieties of their tissues, even if antenna positioning on each subject is carefully reproduced. The most appropriate model for G_0 is found from measurements to be lognormal:

$$G_0|_{dB} \sim \mathcal{N}(\mu_{0_\mathbb{S}}, \sigma_{0_\mathbb{S}}), \tag{15.6}$$

where the mean value and standard deviation depend on the *scenario* $\mathbb{S}$ (representing a given on-body radio link, movement and environment) as described above [DO09, DO10b]. The statistical analysis of G_0 is performed on the mean channel gain of different subjects, so that the normal distribution accounts for the dissimilarities between human bodies. Some local micro-variations of the antenna emplacement, due to dissimilar morphologies, produce also differences in the mean channel gain, and their effect is taken into account by Eq. (15.6).

As shown in Fig. 15.18, the channel gain variance can be considerable especially in anechoic chamber (large $\sigma_{0_\mathbb{S}}$), where the propagation occurs mainly by on-body creeping waves, line of sight and a small reflection from the ground. As expected, the PL is higher in anechoic chamber.

The statistical analysis of the slow component has been carried out on the time dependent samples $S(t_n)|_{dB}$ of the whole population (7 subjects); A lognormal distribution has been also found:

$$S(t_n)|_{dB} \sim \mathcal{N}(0, \sigma_{s_\mathbb{S}}), \tag{15.7}$$

where the standard deviation $\sigma_{s_\mathbb{S}}$ dB accounts for the slow variations of the power transfer function given by shadowing from human body, in a specific scenario $\mathbb{S}$. In Table 15.3 we list the values of $\sigma_{s_\mathbb{S}}$ in the 2.45-GHz band. As expected, we have a large standard deviation in moving scenario, especially in the walking case. The

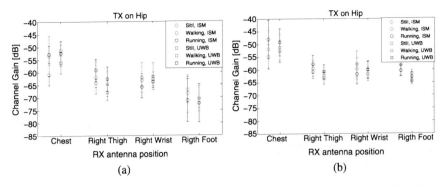

Fig. 15.18 Channel gain for Tx on the hipx: (**a**) anechoic chamber, (**b**) indoor [DO09] (©2009 IEEE, reproduced with permission)

Table 15.3 Shadowing standard deviation σ_{sg} in dB

Tx on Hip/Rx on		Chest	Thigh	R Wrist	R Foot
Anech.	Still	0.61	0.41	0.95	0.41
	Walking	2.15	5.27	4.31	4.97
	Running	2.39	2.47	3.49	2.69
Indoor	Still	0.60	0.24	0.26	0.24
	Walking	1.52	3.27	2.66	2.57
	Running	2.00	1.98	2.37	1.80
Tx on L Ear/Rx on		R Ear	Hip	R Wrist	R Foot
Anech.	Still	0.23	0.45	1.15	0.56
	Walking	0.54	2.07	3.35	4.21
	Running	0.52	2.82	2.98	1.57
Indoor	Still	0.21	0.28	0.31	0.30
	Walking	0.90	2.20	1.80	2.02
	Running	0.70	2.19	1.88	1.24

standard deviation of the running subject is smaller than the walking one, since people tend to keep arms close to the body, limiting their oscillation. By the way, when the subject is running on a spot, his legs' movement is faster but less oscillating than in the walking scenario. Slow fading effect is smaller in indoor because MPCs incoming from the surrounding environment somewhat mitigate the shadowing by the human body.

In [AIK+10] the authors take two parallel approaches, i.e. numerical simulation and experiment, to analyse the channel gain fluctuation during the movement. In the simulation, dynamic motion of the human body is modelled by Poser, and FDTD is applied for the channel response calculation. The simulation results show that the antenna orientation can strongly impact the channel gain, but a simple polarisation

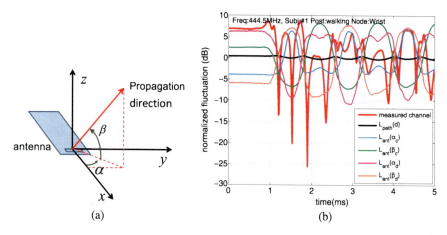

Fig. 15.19 (**a**) Definition of antenna rotation and (**b**) fluctuation of path gain and rotational antenna motion measured in walking postures [AIK+10]

diversity scheme can improve it by +28 dB in the worst case. In the experiment, dynamic motion of the human body is captured by the motion capture system, and the dynamic channel response at 444.5 MHz has been synchronously measured with the motion capture. Analysing the variation of the channel gain with the distance, it has been shown that large variations occur even for small variations of distance. It can be concluded from these observations that the distance is not a major factor of the BAN channel fading. As a consequence, the IEEE path loss exponential model [YSP10] is not suited to describe the channel gain variation in dynamic context.

Besides the masking effect of the body, MPCs and variations of antennas orientation due to body motion are also potential causes of fading. However, the comparison between the fluctuation of the channel gain and the antenna orientation cannot be done directly, since these quantities have different dimensions. Considering that the antenna variable orientation will change the path gain according to its radiation pattern (on-body), the loss $L_{\mathrm{ant,dB}}(\varphi)$ due to the antenna rotation is assumed to be a linear function of the angle of antenna rotation relative to the propagation direction φ. In Fig. 15.19 the normalised fluctuations of measured channel gain, LOS gain calculated from the transmission distance, and antenna loss due to the antenna rotation are depicted for the wrist to navel channel in walking scenarios. The trend of variation and numerous ripples suggest that the channel gain variation has closer similarity to pattern or inverse pattern of antenna rotation than to the distance variation.

One of the main issues in dimensioning a BAN is the knowledge of the channel status at any time. This is of practical importance to adopt cooperative approaches between nodes which undergo different shadowing effects. In [DO09] it was shown that scenarios submitted to strong shadowing (generally when one antenna is on a limb) present the highest correlation. For instance the *hip to wrist*, *hip to thigh* and *hip to foot* are strongly correlated during walking. It is noteworthy that *thigh and foot* are correlated whereas both *wrist and thigh* and *wrist and foot* are anti-correlated because during walking human beings tend to alternate arms and legs movement.

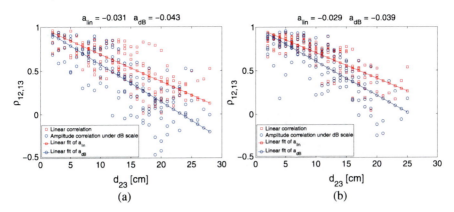

Fig. 15.20 Measurement results and linear fit of $\rho_{12,13}^{dB/lin}$ for (**a**) $d_{12} = 12$ cm and (**b**) $d_{12} = 14$ cm [LDO09a, LDO08a] (©2009 EurAAP, reproduced with permission)

The fading correlation properties of *hip to wrist* channels have been investigated in [LDO08a, LDO08b, LDO09b]. Two kinds of correlation coefficients are defined. The first one is the correlation coefficient of the channel fading amplitude in dB scale according to the lognormal distribution of fading found in [LDO08a]:

$$\rho_{12,13}^{dB} = \frac{\mathbb{E}[(|\overline{S_{12}}|_{dB} - \mu_{\overline{|S_{12}|}_{dB}})(|\overline{S_{13}}|_{dB} - \mu_{\overline{|S_{13}|}_{dB}})]}{\sigma_{\overline{|S_{12}|}_{dB}} \sigma_{\overline{|S_{13}|}_{dB}}}. \quad (15.8)$$

The second is the direct correlation coefficient of the complex channel fading in linear scale:

$$\rho_{12,13}^{lin} = \frac{\mathbb{E}[(|\overline{S_{12}}| - \mu_{\overline{|S_{12}|}})(|\overline{S_{13}}| - \mu_{\overline{|S_{13}|}})]}{\sigma_{\overline{|S_{12}|}} \sigma_{\overline{|S_{13}|}}}. \quad (15.9)$$

S_{13} and S_{12} are the S-parameters measured according to the scheme shown in Fig. 15.16(b). The distance difference d_{23} is assumed to bring the decorrelation effect to the channel fading correlation. It is therefore expected to establish a relationship between the correlation coefficient and the distance difference of the channels if the other parameters d_{12} and d_{10} are fixed. The natural boundary condition $\rho_{12,13} = 1$ should be obviously fulfilled when $d_{23} = 0$. Both correlation coefficients computed from the measurement for d_{12}, equal to 12 and 14 cm, respectively, are presented in Fig. 15.20 as well as linear regression models by least square. The results in Fig. 15.20 display a clear decreasing trend of $\rho_{12,13}^{dB/lin}$ against d_{23}. Also, the impact of the decorrelation effect from the noise get amplified as the propagation distance increases and the SNR decreases. Measurement results show a decorrelation distance which ranges generally from 10 to 15 cm.

The three-cylinder model previously mentioned (Fig. 15.15) has been also used to represent the time-variant channel fading caused by the movement of arms (Fig. 15.21). The analytical solution of the EM field in Fig. 15.21(a) is related to the general multiple cylinder scattering problem [LRDO09]. The aforementioned azimuth on-body propagation model is evaluated by comparing the simulation with

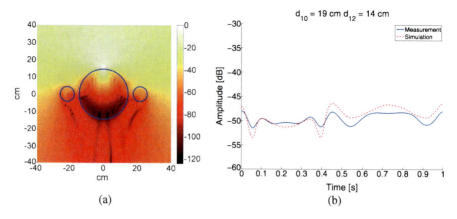

Fig. 15.21 (a) Normalised field power between the single- and multiple-cylinder scattering simulations, $d_s = 5$ mm. The white $+$ is the position of the line source, the *black circles* indicate the body trunk and arms [LDO09b]. (b) Wave pattern comparison over time [LDO09b, LKC+11] (©2011 Lingfeng Liu et al., reproduced with permission)

the measurements. The average wave patterns of the channel fading of S_{12} with $d_{12} = 14$ cm and $d_{10} = 19$ cm are compared in Fig. 15.21(b). Preliminary simulations show a promising similarity with the measurement results. The measurement sample and simulation fit on each local minimum and maximum. The deviation of the simulation to the measurement sample is under 3 dB throughout the period. This model is a rather general propagation channel model in WBAN and can be extended easily to other purposes of applications.

An alternative solution to reproduce the channel space-time correlation proprieties, is to model the slow fading as an autoregressive (AR) process [DO10b]. According to its posture, the body acts as an obstacle between the Tx and the Rx antennas. The consequential effect on the power delay profile is a variation of the delay dispersion characteristic. In [DO10b] it has been shown that when the antennas are in LOS, the energy is concentrated in a small number of paths, which yields a less dispersive channel, whereas in NLOS the power decreases, and delay spread increases. For instance the arm swinging for the wrist to hip link causes a delay spread variation from 194 ps to 23 ns. The mean number of MPCs $\widetilde{K}(t_n)$ follows a certain regularity according the shadowing conditions [DO10b]:

$$\widetilde{K}(t_n) = \exp(m \cdot S(t_n)|_{dB} + m_0). \tag{15.10}$$

From the generated correlated slow fading components with the AR model [DO10b], the mean value of MPCs $\widetilde{K}$ can be computed as a function of $S(t_n)|_{dB}$ with (15.10); a Poisson-distributed random variable $K(t_n)$ is then generated with this mean value. Several PDP models have been given in [DO10b] for well-defined scenarios and shadowing conditions, extending the IEEE model [YSP10] to the dynamic context adopting a compatible model structure. For the UWB channel, the small amplitude variations around the mean value given by the exponential decay of the PDP follow a Nakagami-m distribution. The fast fading component can also

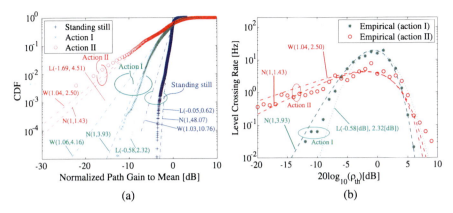

Fig. 15.22 (a) Cumulative distribution function of the normalised gain and (b) level crossing rate function of the threshold level [KT10]

be expressed as $F(t_n) = |\chi(t_n)|^2$. Measurement results show that $|\chi(t_n)|$ follows a Rice distribution at 2.45 GHz.

An alternative approach blending slow and fast fading is proposed in [KT10]. The CDFs of the fluctuations of the normalised PG to mean for various actions and sensor locations (Fig. 15.16c) are shown in Fig. 15.22. The dotted, dashed and dash-dotted lines denote Log-normal, Nakagami-m and Weibull distributions, respectively. Weibull distribution shows a better fit than Nakagami-m distribution for severe fading in dynamic scenarios such as Action II. The empirical level crossing rates (LCR) of Actions I and II (with computed maximum Doppler frequencies of respectively $f_{D,I} = 19.38$ Hz and $f_{D,II} = 4.08$ Hz) are shown in Fig. 15.22, as well as models based on closed-form expressions of the Log-normal [CS07b], Nakagami-m [CS07a] and Weibull [SZKT04] fading processes, fitted by LS method.

The Doppler spectra of various scenarios were investigated in [DO11]. It is shown that on-body channels can be assumed, as a first approximation, as fixed radio links with moving scatterers producing the channel time-variance. This assumption is quite correct when both antennas are placed on the torso, on the same limb, or, more generally, when their relative motion remains low, e.g. *hip to foot* and arm swinging. Contrary to Jakes/Clarke model, the Doppler spectrum is consequently centred around $\nu = 0$ Hz (Fig. 15.23) as in fixed wireless systems. It is hence more appropriate to assess the Doppler bandwidth for a given threshold below the peak value at 0 Hz instead of the maximum Doppler shift: at -20 dB it does not exceed 10 Hz in the worst scenario. A model of Doppler spectrum in anechoic chamber and indoor is proposed, by separating the on-body and off-body scatterers.

15.2.3 On-body to Off-body Antennas and Channel

On-body to off-body (on–off body) communications relate to a communication link established between a body-worn device and a remote access point located in the

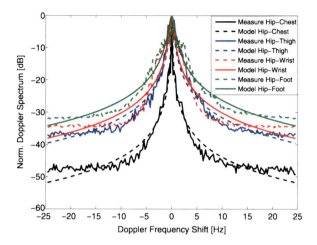

Fig. 15.23 Normalised Doppler spectra: human walking in anechoic chamber [DO11] (©2011, reproduced with permission)

range of a personal area network. An application of these systems is, for instance, in hospitals, health care facilities or at home, to transmit patient measurement data to a nearby receiver, or, for the patient to receive therapeutic instructions. Firemen and rescue workers communications are another important application of on–off body communications.

The operating environment of on–off body communications is quite different from traditional wireless networks, as one end of the radio link is on the body. Nevertheless, most of the channel is off the body and in the surrounding space, with the body dictating the non-stationary behaviour of the radio channel. It is worthwhile noticing that the parameters describing the channel are system independent, concerning only multipath components features (i.e., spatial, temporal and power). The Nakagami-m distribution, among many standard models, was found to describe best the on–off body channels fading phenomenon caused by realistic actions of human beings in various typical environments [CS07a]. Also, Rayleigh-like fading caused by multipath propagation is reported in some indoor environments [TRV[+]09].

The reliability of on–off body communications is conditioned by the influence of the body, whose impact on propagation of nearby electromagnetic signals demands a careful analysis. Some theoretical investigation was done in [OC10], under the simplistic approach of considering the human body as a circular homogeneous dielectric cylinder, which is impinged by a normal incident TE or TM plane wave. Results from this work confirm the well-known phenomenon of reflection and diffraction of the incoming plane wave when it approaches the body. The extent of these phenomena will depend on the frequency, relative dimensions of the body, and composition of human tissues.

Comparison of different size body parts and tissues exposed to a plane wave shows little fluctuations of electromagnetic fields for the smaller body parts, thus, being concluded that, if body dimensions are not large compared to the wavelength, a small impact on propagation is verified. Accordingly, harder patterns of reflections

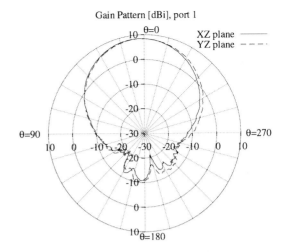

Fig. 15.24 Measured radiation pattern of a textile antenna for on–off body communications [VTH+10] (©2010 IEEE, reproduced with permission)

are verified for larger body elements, and narrower shadow areas become visible for body tissues with large penetration depths. No huge differences are expected between TE and TM incidences.

The presence of the human body leads to changes in antenna radiation pattern and input impedance, causing shifts of the resonant frequency and also reduction of its efficiency. Antennas with maximum radiation patterns away from the body are required, minimising radiation into the user. Also, in on–off body communications, characterisation of the far-field radiation pattern is appropriated. Figure 15.24 depicts the radiation pattern of a textile antenna designed for on–off body communications.

Even though antennas are deterministically describable components, their performance on on–off body communications has a stochastic nature, far from being deterministic. Accordingly, some investigation has been done on the concept of statistically characterising user's influence in on–off body communications, endeavouring to provide useful information in a link budget estimation for channel modelling. Reference [OC10] points out a simple approach for a parametric analysis on the variations of the total electromagnetic fields nearby a body impinged by a plane wave, on a squared area of observation. Maximum likelihood estimators for a 95% confidence interval are presented, assuming a Normal distribution (as a first approximation) for the field values. This statistical analysis allows a general overview and quantification of mean field variations for on–off body communications and also of standard deviations. For instance, at 2.45 GHz, the standard deviation reaches up to 2 dB, but up to 5 dB deviations are found at 900 MHz. Very simple geometrical-optics-based models might be used to have a glimpse on the possible range of changes in the radiation pattern of an antenna located in the vicinity of the human body, as it was done in [MC10]. This approach is applied under the drawback that, for the frequencies of interest (0.915, 2.45 and 5.8 GHz), body dimensions are not much larger than the wavelength, as required for the validity

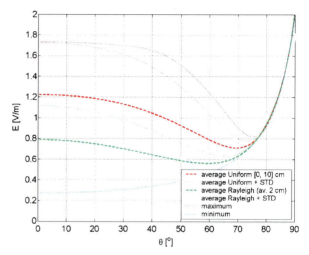

Fig. 15.25 Average radiation pattern of an antenna located over a 2/3 muscle tissue (infinite dielectric half-space), at 915 MHz, for parallel polarisation [MC10] (©2010 IEEE, reproduced with permission)

of geometrical optics approximation. An isotropic antenna, with parallel or perpendicular polarisations, was considered over an infinite lossy dielectric half space or ellipse (representing body tissues), and calculations of the reflection coefficients resulting from wave incidence on these tissues have been performed. The statistics of radiation pattern, i.e. the average, minimum, maximum and standard deviation, have been calculated for uniform or Rayleigh distributed antenna-body distances. An example of the study performed is presented in Fig. 15.25, which depicts the average radiation pattern of an antenna located nearby a 2/3 muscle tissue, at 915 MHz, for parallel polarisation.

The shape of the radiation pattern strongly depends on the distance between antenna and the body. Results in [MC10] show much higher variations for Rayleigh-distributed antenna-body separations than for uniformly distributed ones, which can go as far as 60% and 30%, respectively. When modelling various tissues, results show that the average radiation pattern is changing no more than 15%, but changes of standard deviation can reach 70%. Likewise, the standard deviation fluctuates significantly over observation angles and also for various frequencies and tissues. In general, the standard deviation of average radiation pattern is higher for perpendicular polarisations than for the parallel ones.

Results from a similar statistical study using a female voxel body model [MOLC10] show that the influence of the human body is higher when the antenna is attached to the body, due to the stronger coupling and reduction of radiation efficiency. The main relative variations of the radiation pattern to the case of the isolated antenna are observed in the backward direction, where the radiation level is very low. For the antenna located near the chest, considering a Rayleigh distribution of the distance from the patch to the body, the average pattern difference can reach 11%. In general, moving the antenna away from the body results in smaller pattern differences; however, this is not a monotonous trend, especially in the case of antenna located on the chest.

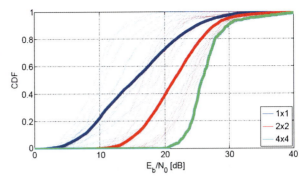

Fig. 15.26 Cumulative distribution for different system configurations, with $N \times N$ receiving/transmitting antennas for $N = 1, 2$ and 4 [TVH+] (©2010 IEEE, reproduced with permission)

A key point to characterise the radio channel and assess the quality of prediction models is to have reliable measurements that capture the dynamic radio environment around the end points of the on–off communication link, with singular interest near the body. Namely, radio channel measurements with real test persons provide good statistics on the time-variant channel, as far as appropriated channel sounding is concerned. A measurement campaign, at 2.45 GHz, targeting a fire fighter working indoor environment is presented in [TRV+09]. Wearable dual-polarised patch antennas, made of highly conductive textile materials, were integrated into the front and back sides of a jacket of a fire fighter walking on a corridor. This configuration evokes *front to back* diversity, which is an important issue in body centric communications, since the front and back antennas virtually cover two complementary half spaces, diminishing the human body impact on propagation. Also, two dual-polarised patch antennas have been used as the remote (off-body) access point, setting up a multiple-input multiple-output (MIMO) scheme. The signals for the different channels are approximately Rayleigh distributed, but with important differences in average received signal levels. Those differences result from inevitable shadowing and correlation between signals. Figure 15.26 shows the cumulative distribution of the signal levels, the higher performance of the 4×4 MIMO system, 16 dB better than the single-input single-output (SISO) scenario. Moreover, the bit error rate (BER) curves displayed in Fig. 15.27 illustrate that the performance of the system improves significantly when compared to a SISO system. Results from [TRV+09] confirm that the use of MIMO in on–off body communications might enhance the reliability of radio links, even though there are some limitations imposed by the available space and somehow inevitable mutual coupling. The use of cross-polarised antennas, for instance, can overcome the small space problem. Correlated signals, resulting essentially from mutual coupling, can still be however constructive if correlation is low enough (e.g. [Ala98] reports a substantial gain if the envelope correlation coefficient is lower than 0.7). MIMO channel models including this correlation can be successfully applied to measurement data, as far as the random numbers needed for their generation follow the same distribution as the original signals.

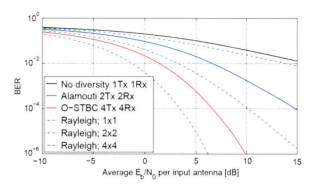

Fig. 15.27 BER curves showing diversity and array gain [TVH+] (©2010 IEEE, reproduced with permission)

15.3 Transmission Technology

15.3.1 Existing WBAN Radio Technologies

As shown in the previous sections, wireless body area networks (WBAN) are used to transfer information originated from sensors in a close proximity of a human body, or even inside a human body, first to a local (body) access point, and then further away. In addition to medical, welfare- or health-related applications, entertainment applications such as a music transmission from MP3 player to wireless headset can be seen as a WBAN network operation. In general, WBAN can be build using different kinds of existing radio technologies, such as IEEE 802.11a/b/g/n [HDS+10], IEEE 802.15.1 Bluetooth or Bluetooth low-energy technology (previously known as Bluetooth Lite, Wibree, Low-Power Bluetooth) [IEE05, SIG09], IEEE 802.15.3 [IEE03], IEEE 802.15.3a/WiMedia/ECMA-368 [ECM08], IEEE 802.15.4 ZigBee [Whe07], 802.15.4a [KP10], to list some of the most popular options. Each of the listed technologies has its own peculiarities, including different modulation, data rate, transmission power, power consumption, etc., which defines their major operational environment and prior application. Moreover, the number of individual channels at each of these standards has impacts on the maximum number of simultaneous users that can access the channel at the same time without collisions in the channel and thus, reduction in the system performance. In addition to the above technologies, there are several other short-range communication technologies targeted for automation or processing industry, such as WirelessHart [SHM+08], Z-Wave [Wal06] or HomeRF [Wir], but they have a clear deviation from the targeted end-user WBAN application.

The application range where all these listed radio technologies can be utilised is numerous, and the WBAN is only one option. For example, IEEE 802.11 family is more directed toward wireless local area or personal area networks (WLAN and WPAN, respectively) than WBAN, but still it can be used as a link from WBAN access point to room access point, and therefore it can be part of a personal wearable node. However, combining WLAN radio into the WBAN system typically increases the number of radios used in the system, which directly reflects to the implementation complexity and signal processing needs. Currently there is no explicit standard for WBAN, but in November 2007 IEEE established a study group IEEE 802.15.6 to define one. During the March 2010, the IEEE 802.15.6 successfully managed to merge all the proposals into one, which is a positive sign for the future WBAN technological development [KW10]. As pointed out in [OC08], there are still several open questions for WBAN research, such as which radio technology utilises in most efficient way the radio channel, or selection of the frequency range in the penetration depth point of view for in-body communication. Distinctive of the WBAN communication is a data rate demand which is not targeting to those ones needed to transfer, e.g. high-definition television signal. For peer-to-peer WBAN data links, it is sufficient to transfer rather small amount of data, and the data rate is more around hundreds of kilobits than megabits per second. The coming IEEE 802.15.6 standard still defines the required data rate between 10 kb/s and 10 Mb/s [Rei10]. However, the total network traffic is a cumulative sum of the traffic in independent links. According to the same standard, the 10 kb/s minimum data rate should be supported between two BAN nodes within a range up to 3 m with packet error rate (PER) less than or equal to 10% for 256 octet payload with a link success probability of 95% with all the channels defined by the standard document [Rei10]. WBAN traffic can be divided in different ways. For example in [GGS$^+$09], depending on the final application, WBAN operation is based on three different communication categories: streaming with a continuous data flow and a constant bandwidth; monitoring with periodic data; and alarm, when a communication is done only when necessary. The streaming-type communication conveys data from a source to a destination or multiple targets. Applications are mainly related to entertainment and data upload/download operations. The measured data in a monitoring category can be reported further based on pre-defined scheduling, on event-based decisions, or the transmission can be continuous. Typically data rates are smaller than in the streaming case. Mutual property for these two applications is a need for reliable communication between the source and destination, and the delay is not so critical. However, the alarm-type transmission typically reports critical changes in the measured parameter, and therefore the transmission delay has more importance, and the data amounts to be transferred are quite small [GGS$^+$09].

One aspect that should also be kept in mind is the requirement to allow mobility when using WBAN. Even if the network is attached to human body, it does not mean a stable transmission environment and line-of-sight links between the communicating nodes. For example, during walking cycle, arms and legs are moving and consequently affecting the link quality between the nodes. This could mean changes in the throughput, but still, especially in medical applications, the information is not allowed to vanish anyway. Moreover, the movement changes the absolute position of the WBAN within a space. Therefore the WBAN and its supporting network need to adapt to the changes in interference and coexistence environments, as well as support handovers between adjacent access points [Rei10].

15.3.2 PHY Layer

The physical (PHY) layer defines the interface between transmission media and the device. In wireless communication, this means how the signal is using the electromagnetic frequency spectrum. The most interesting physical layer technologies to be used in WBAN applications are Bluetooth LE, ZigBee and UWB. As a power consumption point of view, Bluetooth, though being a short-range technology, is losing the competition. Within this COST Action, UWB physical layer approaches have been studied in more details, and the performances of direct sequence UWB (DS-UWB) and UWB frequency modulation (UWB-FM) are considered, for example in [VHI10b]. Both of these technologies have also been selected by the IEEE 802.15.6 to be included in the coming WBAN standard [KW10].

Typically, the radio regulations and standards are defining how the communication system accesses the transmission media. However, all the regulations are not globally accepted or prevailing. For example, the FCC in the USA liberated a frequency band from 3.1 GHz to 10.6 GHz for unlicensed UWB transmission, but in Europe, the corresponding frequency range is only from 6 GHz to 8.5 GHz. The maximum power spectral density for transmission is limited to -41.3 dBm/MHz in both regions.

If detect and avoid (DAA) mechanism is used, also the band between 3.1 GHz and 4.8 GHz is allowed for European use. In Japan, the frequency band from 7.25 GHz to 10.25 GHz is allowed without restrictions with the same power spectral density limitation. The use of frequency band 3.4–4.8 GHz is allowed if DAA is obtained. These examples show that the global UWB regulation is not solid and there are lots of regional variations. However, it is possible to identify a global band from 7.25 GHz to 8.5 GHz for unlicensed UWB applications. Moreover, there are designated frequency bands for UWB radar imaging systems and vehicular radars which are omitted here due to their evident separation from the WBAN use. In a DS-UWB, a Gaussian monocycle or its high-order derivatives, or some other short waveforms which are easy to generate are options for the signalling. A centre frequency and bandwidth of a transmitted signal can be modified to fit the spectrum mask by adjusting a pulse length and shape. In UWB-FM [GKvdM$^+$05], the information signal is spread over a UWB band by applying a double frequency modulation (FM). A constant envelope UWB signal is created by a low-modulation index digital frequency shift keying (FSK) followed by a high-modulation index analog FM. By properly selecting the system parameters, it is possible to allocate both DS-UWB and UWB-FM spectra overlapping in the frequency domain as shown in Fig. 15.28 [VHI10b]. The UWB RF aspects have been discussed in Chap. 6 in more details.

Bit error rate (BER) performances of these two different UWB physical layer solutions have been studied through simulations using two independently developed WBAN radio channel models, IEEE 802.15.6 model [YSP10] and WBAN model developed at the Centre for Wireless Communications, University of Oulu, Finland [TPRI$^+$]. Thus both models being WBAN on-body models, their delay profiles are deviating from each other, and hence, the UWB systems' performances when using these models are different. The CWC's WBAN radio channel measurements have

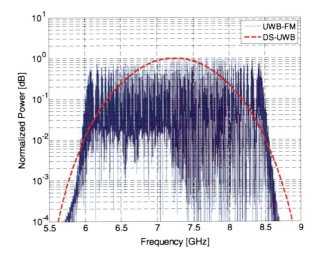

Fig. 15.28 Spectra of the studied DS-UWB and UWB-FM signals [VHI10a] (©2010 IEEE, reproduced with permission)

been carried out at the real hospital environment, accordingly at the Oulu University Hospital. Hence, they are reflecting to the real environment; operation room, ward and corridor. The CWC's hospital radio channel models are illustrating an on-body link of a laying down patient and also a body-to-external access point link of a standing patient in a regular ward environment. The frequency band of the models covers the range from 3.1 GHz to 10 GHz. In both on-body and on-body to off-body models, the channel impulse responses at CWC's measurements are more concentrated to shorter delays with stronger power gains. This is the most significant difference between these two models, and its impact on the transceivers' performances can be seen from the simulation results as shown, for example, in [VHIT09]. In Fig. 15.29, the DS-UWB and UWB-FM systems' performance results are shown for the IEEE 802.15.4a and CWC's channels. In DS-UWB, coherent maximum ratio combining (MRC) and non-coherent square law combining (SLC) are used at the receiver. The number of rake fingers obtained in the DS-UWB is 1, 3 and 10. Correspondingly,

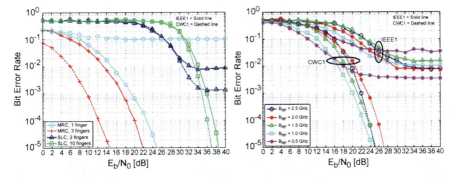

Fig. 15.29 Performances of DS-UWB and UWB-FM in IEEE and CWC on-body channels [VHI10a] (©2010 IEEE, reproduced with permission)

Table 15.4 Parameters for DS-UWB and UWB-FM simulations

Parameter	Value, DS-UWB	Value, UWB-FM
Pulse length (T_p) [ns]	1.5	–
Pulse type	Eigenpulse	–
Modulation method	BPAM, OOK	–
Processing gain (PG_{dB})	20.63, 31.42	–
Combining technology	MRC, SLC	–
Type of rake receiver	partial rake	–
Number of rake fingers (N_f)	1, 3, 10	–
Center frequency [GHz]	7.25	–
−20 dB bandwidth (B_{-20dB}) [GHz]	2.5	–
RF bandwidth (B_{RF}) [GHz]	–	0.5, 1.0, 1.5, 2.0, 2.5
Carrier frequency (f_c) [GHz]	–	7.25
Modulation index of FSK (β_{FSK})	–	1

the UWB-FM receiver is utilising different RF bandwidths. As can be seen, the channel models have great impact on the system performance. Both channel models used in the simulations are targeted to WBAN applications, but still the UWB systems' performances are different. As a general rule, the system design is a trade off between targeted quality of service level and an implementation complexity. The system parameters used in the simulations are shown in Table 15.4 [VHI10b].

15.3.3 MAC and NET Layers

The medium access control (MAC) layer is responsible to share media (channel) between different users in coordinated fashion. The channel itself can be cable, fibre, body channel, and so on. In wireless communication, the propagation media is a radio channel. There are several possible MAC layer protocols that can be utilised in WBAN applications to allow channel access. For example, slotted Aloha (S-Aloha) and the preamble sense multiple access (PSMA) proposed in [HRG+09] are reasonable candidates for WBAN MAC. A carrier-less nature of IR-UWB signal is also causing additional requirements for the receiver to maintain a specified quality of service.

According to the IEEE 802.15.4 standard [IEE06], time is constructed by a superframe structure. The network management is carried out by a network coordinator. A superframe is composed by an inactive part and an active part, which consists of the contention access period (CAP) and the contention-free period (CFP). The channel access is managed by slotted carrier sense multiple access with collision avoidance (CSMA/CA) algorithm. Personal area network (PAN) coordinator can allocate seven guaranteed time slots (GTS) at maximum, and one GTS can occupy more than one slot inside a superframe. Each superframe starts with a beacon

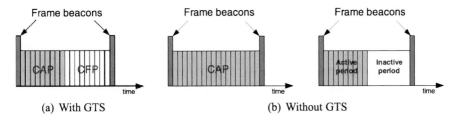

Fig. 15.30 The IEEE 802.15.4 superframe structure [IEE06]

packet, which is transmitted by the coordinator. The superframe structure defined by [IEE06] for guaranteed and non-guaranteed time slots is presented in Figs. 15.30(a) and 15.30(b), respectively. The standard itself gives a detailed description for the frame structure.

The channel access can be based on competition during a CAP period, or MAC can provide CFPs to channel access to guarantee data transmission [KGT+10]. In addition, the different physical layer requirements reflect to MAC layer through, for example, various amounts of data to be transferred within a certain time by different applications. This requires that MAC layer is capable to adapt to these changes in service requirements. MAC layer has also its own impact on transceiver's power consumption through the mechanism it allows users to get channel access. How effectively MAC can handle handshaking procedures or retransmission rules, gives estimates of the energy that is needed in the devices from MAC viewpoint. In addition, radios, screens, etc. will consume power, which is needed to take into account when calculating the total power consumption. Thus keeping the transceivers in a sleeping mode, a certain amount decreases the power consumption if compared to the situation where the transceivers are always in active states. To guarantee a quality of transmission and equal possibilities for channel access, MAC is following a superframe structure where each user can have its own, dedicated time slot (based on competition or not).

A performance of beacon-enabled mode of the IEEE 802.15.4 network, which is utilising a star topology, is studied in [SBV09]. The results show unambiguously that the increase in the amount of WPANs in the network impacts on the packet error rate and delays of the adjacent WPANs in a harmful fashion due to the increased traffic. Both PER and delay are increasing linearly when the number of PAN coordinators, and thus the number of adjacent WBANs in the same room, increases. This performance degradation should be taken into account when installing several WBANs in the same environment. Typically it is assumed that the collision of a packet in a channel causes complete loss of the transmitted information. As can be seen from Fig. 15.31(a) (no inactive period within a beacon interval, duty cycle $\delta = 100\%$), the system throughput is increasing if the partial energy from the colliding packets can be utilised at detection. This phenomenon is called a capture effect. The throughput is affected in a different way if the inactive period is increased, as shown in Fig. 15.31(b) ($\delta = 25\%$). This is due to the increased competition while CAP is smaller. During the inactive period, there is no traffic in the channel.

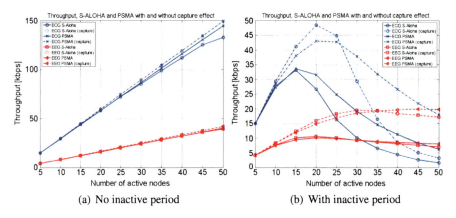

Fig. 15.31 Throughput for S-Aloha and PSMA with and without capture effect [SBV09]

For wireless sensor or body area networks, three cooperative transmission mechanisms have been studied in [GGS+09]: A relay channel, multihopping, and virtual MIMO. A relay channel utilises one or more relaying nodes to retransmit the message from source to sink. Cooperation can be done at PHY layer, so routing algorithms are not needed in higher layers. In WBAN, a beacon-enabled star topology is typically used, but in a cooperative mode, each node can relay its all neighbours. In a star network topology, all the nodes are connected to the PAN coordinator, which makes the channel access mechanism simpler to handle. In a large mesh network, relaying can, however, benefit more than in a smaller-scale WBAN networks. Relaying can improve the transmission reliability by forwarding the transmitted packet if the direct path is broken [DRM11]. On the other hand, using relaying instead of increasing transmission power, the link does not cause additional interference to the adjacent WBAN networks nor cause harmful radiation to human organs.

For relaying link studies in [GGS+09], the MIMO type on-body radio channels were measured using two men and one woman as a subject. The study focused on signals having 2.45-GHz centre frequency and 10-MHz bandwidth. The antennas were located in different parts of a human body as shown in Fig. 15.32, and the subjects were walking inside a room during the recordings. The studied performance of the relayed WBAN network is utilising Bluetooth Low-Energy technology [SIG09].

Due to the complex structure of a human body, the propagation paths between the transmitting and receiving antennas are often obstructed. As mentioned earlier, the body movement during the walking, for example, will change regularly the link goodness between two communicating nodes, and therefore it has an impact on the quality of the transmission. If the link quality is studied through outage probability, i.e. the probability that the packet error rate is higher than a certain threshold, it can be found out that using a relay node instead of one hop link can improve the transmission reliability. An additional relaying node can solve the communication problem, e.g. between the nodes locating in a back and in front of human torso. In Fig. 15.33, the cumulative distribution functions (CDF) of PER for wrist to hip link are shown for three persons. The different walking style has an impact on propaga-

Fig. 15.32 Measured on-body links [DRM11] (©2011 IEEE, reproduced with permission)

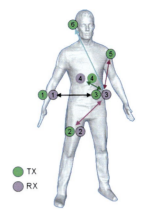

Fig. 15.33 CDF of PER for different persons. Link is from wrist to hip [DRM11] (©2011 IEEE, reproduced with permission)

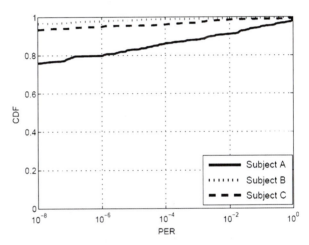

tion characteristics, which influences PER [DRM11]. By analysing the CDFs of the measured PER from the walking persons, it is possible also to distinguish different persons. The paper shows also results for different links when a relay node is used. For relaying on-body links, a hip is a natural location for a BAN coordinator node. The following comparison is carried out from links from hip to other parts of the body, as shown in Fig. 15.32. In Figs. 15.34 and 15.35, the CDFs for different links from hip are shown without a use of relaying node and with a relay node locating at a shoulder, respectively.

As shown in Fig. 15.35, the double hop link outperforms the quality of one hop link, since it is able to mitigate the fading caused by the body movement. If all the information provided by the channel can be combined, the performance is the best. The latter case in Fig. 15.35 utilises both direct path and relayed information. The performance evaluation when utilising cooperative transmission schemes showed that in the case where human is moving, the impacts of fast and slow fadings can be mitigated. If the combined path mechanism is obtained at the source node, the received PER can be significantly improved if compared to corresponding one hop

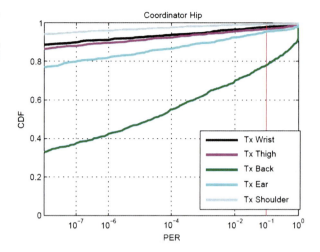

Fig. 15.34 CDF of PER for different links from hip [DRM11] (©2011 IEEE, reproduced with permission)

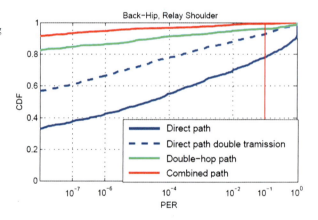

Fig. 15.35 CDF of PER for different links from hip using relay at shoulder [DRM11] (©2011 IEEE, reproduced with permission)

or two hop links, as shown in Fig. 15.36 using ear, thigh and shoulder as a relay node location [DRM11].

The performance in a network layer (NET) can be a bottleneck if the network load increases suddenly, such as after an emergency event. The quality of NET functionality is studied through parameters such as outage, throughput and achievable transmission rate in [DRHD09]. If the NET throughput suffers, there is no guarantee that all the critical information is going through the network. In mesh networks, the amount of delivered data through network depends on various circumstances: channel condition, number of hops, energy of the nodes which impacts on routing, etc. In WBAN network the traffic is typically bursty, which is caused by temporally and spatially unsynchronised sensor nodes, i.e. their reporting times depend on the parameter the sensor is monitoring. This causes notable changes to network traffic. Together with the low power consumption and computation constrain requirements set to the sensor nodes, the network coordination function cannot be heavy. A closed-form expression for the probability that transmission is successful

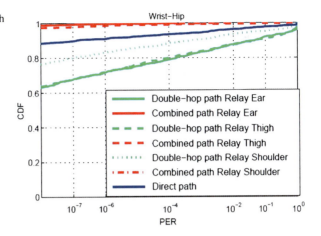

Fig. 15.36 CDF of PER with cooperation [DRM11] (©2011 IEEE, reproduced with permission)

in interference-limited BANs with large-scale fading (i.e., shadowing) has been derived in [DRHD09]. The received power follows zero-mean Gaussian distribution in dB scale (log-normal in a linear scale). Using a centralised network topology, it has been shown that a duplex transmission capability of the nodes does not play a critical role, especially for the limited transmission rates. In [DRHD09], it has also been shown that a maximum achievable transmission rate exists and it depends on the amount of deployed nodes. In addition, the interference from adjacent WBANs decreases the achievable transmission rate.

For BANs with a mesh topology, the transmission strategy depends on the traffic profile. When the transmission probability of each node is limited, thus being such as a passive monitoring of a patient or deep sleep of the nodes, long-range transmissions can be used in order to save energy and avoid multiple relays (multihops). On the other hand, when the sensors have a substantial activity, the performance decreases exponentially with the number of relaying hops. The shortest possible hop strategy should be used, but it might come at the cost of numerous relaying. However, even though the maximum sustainable number of hops is small, relaying is still suitable approach for BAN based applications as shown in [DRM11]. It has also been observed that it is extremely difficult to reach a security level of 90% without any transmission control protocol [DRHD09]. The achievable transmission rate of a link T [bit/s/Hz] can be calculated as

$$T = \tau \log_2\left(1 + \frac{P_0(d)}{N + P_{\text{int}}}\right) = \tau \log_2(1 + \text{SINR}), \qquad (15.11)$$

where τ depends on the duplex mode of the transmission; in a full-duplex communications, $\tau^{\text{full}} = pP_s$, and in a half-duplex mode, $\tau^{\text{half}} = p(1-p)P_s$, where p is the transmission probability, $P_0(d)$ is the received power as a function of distance, N is the ambient noise power, P_{int} is the total received interfering power from all the undesired transmitters, SINR is the signal-to-interference-plus-noise ratio, and P_s is the outage probability, i.e., the probability that the received SINR is below a given threshold. It can be seen that T is a product of the probabilistic channel throughput and the link capacity.

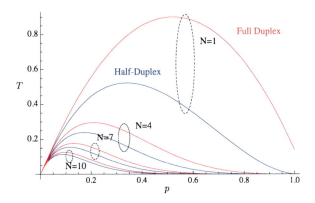

Fig. 15.37 Achievable channel transmission rate T [DRHD09]

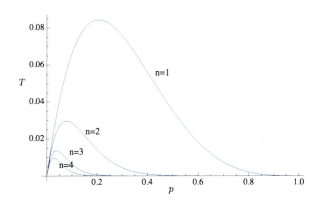

Fig. 15.38 Achievable channel transmission rate in mesh network [DRHD09]

In Fig. 15.37, the achievable channel transmission rate T is presented as a function of a transmission probability p for different number of nodes forming a star network. The parameters to calculate the curves are: SINR limit for outage probability and required link SINR are both 10 dB, and standard deviation is 4 dB [DRHD09]. As the results show, comparing full- and half-duplex systems, T is quite similar if transmission probability is low. If the full duplex mode is used, the achievable channel transmission rate increases considerably. Corresponding results using the same parameters with different numbers of hops, n, in the mesh network are presented in Fig. 15.38. Using a peer-to-peer communication without relaying gives the best channel rate, and the rate is decreasing when the number of hops is increasing [DRHD09]. Total number of nodes forming the studied network is 21.

Using link adaptation (LA) technique to estimate the momentary link quality and then accordingly changing the modulation scheme can be used to improve the reliability of the transmission. For example, the WBAN nodes that are transmitting periodically information from sensors to WBAN coordinator can adapt the changes within the WBAN links. An LA method presented in [MVB10a] and [MVB10b] is based on the signal-to-noise ratio (SNR) calculation. In the case of high SNR, higher order modulation schemes can be used if the system allows variable modulation schemes, but when the SNR decreases, more robust modulation is used, which

Table 15.5 Parameters for energy calculations [KGT+10]

Parameter	Comment	Value
M_{Tx}	Tx power consumption	20 mW
M_{Rx}	Rx power consumption	116 mW
M_{sleep}	Low-power mode power consumption	0.2 mW
I	Battery current (typ. commercial value)	1100 mA
V	Battery voltage (typ. commercial value)	1.2 V
B_h	Battery value at observation time	$I \cdot V$

means that the WBAN links do not provide constant data rate all the time. At the end, the use of LA improves the received packet error rate and therefore the overall reliability of the WBAN.

15.3.4 Low Power Design

A WBAN can include in implanted medical or status sensors that cannot be replaced or recharged without a surgery operation, so energy efficiency of the nodes is a fundamental issue. Another important requirement is latency if real-time communication between the sensors and the final user is needed. Throughput S, energy consumption E and expected battery lifetime B_{lt} are the most common parameters when the comparison between different technologies and architectures are done. The maximum acceptable energy consumption of the node depends on the application needs. In some cases, the requirement for battery life is 5–10 years, but some applications can tolerate changes of energy source monthly, weekly or even daily. This indicates a typical battery life time variation for WSN nodes. The metrics for the WBAN usability can be presented as [KGT+10]

$$\begin{cases} S = \dfrac{N_{pkt\_Rx} N_{bits\_per\_pkt}}{T_{sim}}, \\ E = \dfrac{E_{data} + E_{ack}}{N_{bits\_Rx}}, \\ B_{lf} = \dfrac{B_h}{E} T_{sim}, \end{cases} \qquad (15.12)$$

where N_{pkt_Rx} and $N_{bits\_per\_pkt}$ are the total number of received packets and the size of the data packet in bits, respectively, T_{sim} is a simulation time, E_{data} and E_{ack} are the total energies spent for data transmission and receiving an acknowledgement, respectively, N_{bits_Rx} is the number of correctly received bits at MAC layer, and B_h is an instantaneous battery power. Using the values from Table 15.5 and typical parameters from biomedical measurement devices as presented in Table 15.6, the comparative analysis for energy consumption with different MAC protocols can be done.

Table 15.6 Parameters for biomedical measurements [KGT+10]

Biomedical measurement	Sample rate (samples/s)	Resolution (bits/sample)	Information rate (bit/s)
ECG	1250	12	15000
EEG	350	12	4200
Blood oxygenation	50	16	800
Heart rate	25	24	600
Body temperature	5	16	80

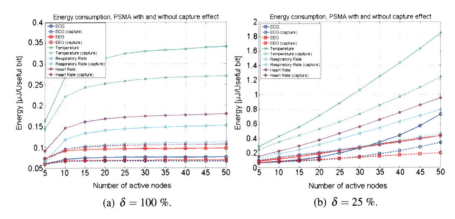

Fig. 15.39 PSMA energy consumption with and without capture effect [KGT+10] (©2010 IEEE, reproduced with permission)

The results shown in Figs. 15.39(a) and 15.39(b) for several applications unambiguously indicate the usefulness of the capture effect [KGT+10]. If the partial energy is utilised at the reception, the amount of retransmissions can be decreased, which then has an influence on the node's energy consumption. Figure 15.40 translates the previous energy consumption results into the expected battery life. If the amount of transferred data is small, the battery life is longer if compared to the applications which are transferring data more frequently. Again, the capture effect extends the battery life.

The energy consumption calculations for WBAN links using cooperative communication instead of one hop link are presented in [DRM11]. The calculations assume 7.5-mA current when transmitting at -12 dBm power level and 11.8-mA current when receiving at 1-Mbps data rate. The transmissions are formed between source (S), relay (R) and destination (D) nodes as using SD, SR and RD links. The data packet is assumed to be transmitted within a 1-ms time slot. The transceiver utilised aggregate data transmission in the relaying mode when it is able to transmit within a same time slot its own data and the relayed data (Data S and Data R, respectively).

Fig. 15.40 PSMA expected battery life with and without capture effect, $\delta = 100\%$ [KGT+10] (©2010 IEEE, reproduced with permission)

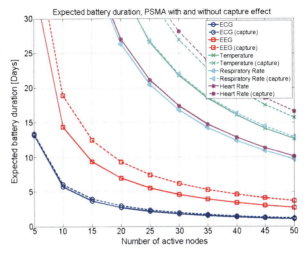

Fig. 15.41 PER vs. transmission power (link: back to hip) [DRM11] (©2011 IEEE, reproduced with permission)

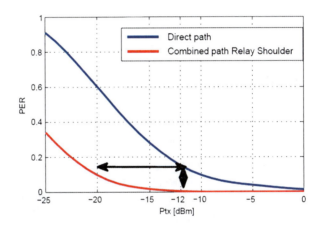

As can be seen from the Table 15.7, the energy consumption when both direct path and relayed path information is used exceeds the ones from only one or two hop links. However, the profit of the increased energy consumption is a better system performance, as shown in Fig. 15.41 [DRM11]. When looking at the overall power consumption at PER level of 0.1, it can be seen that the power saving when combined mechanism is used is 33% for the studied Bluetooth Low-Energy chip. The energy saving is due to the lower transmission power and diversity that can be utilised at the reception which then leads to the decreasing amount of retransmissions. Based on the results, it can be seen that the use of cooperation schemes in moving WBAN application is beneficial in energy consumption point of view when studying a certain target PER level.

Table 15.7 Energy consumptions for different relaying mechanisms [DRM11]

Node	Direct [mA]	Double Hop [mA]	Combined [mA]
D	Rx(Data S) + Rx(Data R) = 23.6	Rx(Data R + Data S) = 11.8	Rx(Data S) = 23.6 + Rx(Data R + Data S) = 23.6
S	Tx(Data S) = 7.5	Tx(Data S) = 7.5	Tx(Data S) = 7.5
R	Tx(Data R) = 7.5	Rx(Data R) + Tx(Data R + Data S) = 19.3	Tx(Data R + Data S) + Rx(Data S) = 19.3
Total	38.6 mA/packet	38.6 mA/packet	50.4 mA/packet

References

[ABB08]   I. F. Akyildiz, F. Brunetti, and C. Blazquez. Nanonetworks: a new communication paradigm. *Computer Networks*, 52(12):2260–2279, 2008.

[AIK+10]   T. Aoyagi, Iswandi, M. S. Kim, J. Takada, K. Hamaguchi, and R. Kohno. Study of relation between body motion and channel response of dynamic body area channel. In *COST 2100*, Aalborg, Denmark, June 2010. [TD(10)11060].

[Ala98]   S. M. Alamouti. A simple transmit diversity technique for wireless communications. *IEEE J. Select. Areas Commun.*, 16(8):1451–1458, 1998.

[ATT+08a]   T. Aoyagi, J. Takada, K. Takizawa, N. Katayama, T. Kobayashi, K. Y. Yazdandoost, H.-B. Li, and R. Kohno. Channel models for wearable and implantable WBANs. IEEE 802.15 Working Group Document, July 2008. IEEE P802.15-08-0416-02-0006

[ATT+08b]   T. Aoyagi, J. Takada, K. Takizawa, H. Sawada, N. Katayama, K. Y. Yazdandoost, T. Kobayashi, H.-B. Li, and R. Kohno. Channel models for WBANs—NICT. IEEE 802.15 Working Group Document, November 2008. IEEE P802.15-08-0416-04-0006.

[BCNT09]   L. Betancur, N. Cardona, A. Navarro, and L. Traver. A statistical ultra wide band—body area network channel model for body propagation environments. In *COST 2100*, Valencia, Spain, May 2009. [TD(09)819].

[BF10]   A. Baharav and G. D. Furman. Sleepiness on task can be detected using heart rate fluctuations. In *6th Conference of the European Study Group on Cardiovascular Oscillations*, April 2010.

[BLT09]   D. Bajic and T. Loncar-Turukalo. Cardiovascular data transmission: end user experience. In *COST 2100*, Braunschweig, Germany, February 2009. [TD(09)741].

[CEN02]   CENELEC. Basic standard for the calculation and measurement of electromagnetic field strength and SAR related to human exposure from radio base stations and fixed terminal stations for wireless telecommunication systems (110 MHz–40 GHz). Technical Report EN50383, European Committee for Electrotechnical Standardization CENELEC, September 2002.

[CS06]   S. L. Cotton and W. G. Scanlon. A statistical analysis of indoor multipath fading for a narrowband wireless body area network. In *Proc. PIMRC 2006—IEEE 17th Int. Symp. on Pers., Indoor and Mobile Radio Commun.*, Helsinki, Finland, September 2006.

[CS07a]   S. L. Cotton and W. G. Scanlon. Characterization and modeling of the indoor radio channel at 868 MHz for a mobile bodyworn wireless personal area network. *IEEE Antennas Wireless Propagat. Lett.*, 6:51–55, 2007.

[CS07b]	S. L. Cotton and W. G. Scanlon. A higher order statistics for lognormal small-scale fading in mobile radio channels. *IEEE Antennas Wireless Propagat. Lett.*, 6:540–543, 2007.
[DO09]	R. D'Errico and L. Ouvry. Time-variant BAN channel characterization. In *Proc. PIMRC 2009—IEEE 20th Int. Symp. on Pers., Indoor and Mobile Radio Commun.*, Tokyo, Japan, September 2009. [Also available as TD(09)879].
[DO10a]	R. D'Errico and L. Ouvry. Analysis and modeling of delay profile in time-variant on-body channels. In *COST 2100*, Athens, Greece, September 2010. [TD(10)10090].
[DO10b]	R. D'Errico and L. Ouvry. Delay dispersion of the on-body dynamic channel. In *Proc. EuCAP 2010—4th Euro. Conf. on Antennas and Propagat.*, Barcelona, Spain, April 2010. [Also available as TD(10)10090].
[DO11]	R. D'Errico and L. Ouvry. Doppler characteristics and correlation proprieties of on-body channels. In *Proc. EuCAP 2011—European Conf. Antennas Propagat.*, Rome, Italy, April 2011. [Also available as TD(10)12091].
[DRHD09]	J.-M. Dricot, S. Van Roy, F. Horlin, and P. De Doncker. Outage, local throughput, and achievable transmission rate of narrowband body area networks. In *COST 2100*, Wien, Austria, September 2009. [TD(09)939].
[DRM11]	R. D'Errico, R. Rosini, and M. Maman. A performance evaluation of cooperative schemes for on-body area networks based on measured time-variant channels. In *Proc. ICC 2011—IEEE Int. Conf. Commun.*, Kyoto, Japan, June 2011. [Also available as TD(10)11083].
[Dru07]	S. Drude. Requirements and application scenarios for body area networks. In *Proc. 16th IST Summit on Mobile and Wireless Commun.*, pages 1–5, July 2007.
[ECM08]	ECMA. High rate ultra wideband PHY and MAC standard. ECMA International, 2008. Standard ECMA-368.
[ETS02]	ETSI. Electromagnetic compatibility and radio spectrum matters (ERM), radio equipment in the frequency range 402 MHz to 405 MHz for ultra low power active medical implants and accessories; part 1: Technical characteristics, including electromagnetic compatibility requirements, and test methods. Technical Report EN 301 839-1, European Telecommunications Standards Institute ETSI, 2002.
[FCC]	FCC. Fcc: Wireless services: medical device radiocommunications service: Medradio. [Online]. Available: http://wireless.fcc.gov/services/index.htm?job=service_home&id=medical_implant.
[FCC01]	FCC. Revised supplement C evaluating compliance with FCC guidelines for human exposure to radiofrequency electromagnetic fields. Technical Report OET Bulletin 65, Federal Communication Commission, Office of Engineering and Technology, 2001.
[FDD+06]	A. Fort, C. Desset, P. De Doncker, P. Wambacq, and L. Van Biesen. An ultra-wideband body area propagation channel model—from statistics to implementation. *IEEE Trans. Microwave Theory Tech.*, 54(4):1820–1826, 2006.
[FDR+05]	A. Fort, C. Desset, J. Ryckaert, P. De Doncker, L. Van Biessen, and S. Donnay. Characterization of the ultra wideband body area propagation channel. In *Proc. ICUWB 2005—IEEE Int. Conf. on Ultra-Wideband*, Zurich, Switzerland, September 2005.
[FKR+10]	A. Fort, F. Keshmiri, G. Roqueta, C. Craeye, and C. Oestges. A body area propagation model derived from fundamental principles: analytical analysis and comparison with measurement. *IEEE Trans. Antennas Propagat.*, 58(2):503–514, 2010.

[FRD+06] A. Fort, J. Ryckaert, C. Desset, P. De Doncker, P. Wambacq, and L. Van Biessen. Ultra-wideband channel model for communication around the human body. *IEEE J. Select. Areas Commun.*, 24(4):927–933, 2006.

[GGS+09] J.-M. Gorce, C. Goursaud, C. Savigny, G. Villemaud, R. d'Errico, F. Dehmas, M. Maman, L. Ouvry, B. Miscopein, and J. Schwoerer. Cooperation mechanisms in BANs. In *COST 2100*, Valencia, Spain, May 2009. [TD(09)862].

[GKvdM+05] J. F. M. Gerrits, M. H. L. Kouwenhoven, P. R. van der Meer, J. R. Farserotu, and J. R. Long. Principles and limitations of ultra-wideband FM communications systems. *EURASIP J. Applied Signal Processing*, 2005(3):382–396, 2005.

[GLP+03] S. K. S. Gupta, S. Lalwani, Y. Prakash, E. Elsharawy, and L. Schwiebert. Towards a propagation model for wireless biomedical applications. In *Proc. ICC 2003—IEEE Int. Conf. Commun.*, vol. 3, pages 1993–1997, June 2003.

[GVH+06] B. Gyselinckx, R. Vullers, C. Van Hoof, J. Ryckaert, R. F. Yazicioglu, P. Fiorini, and V. Leonov. Human++: emerging technology for body area networks. In *Proc. of IFIP International Conference on Very Large Scale Integration*, pages 175–180, October 2006.

[HDS+10] G. Hiertz, D. Denteneer, L. Stibor, Y. Zang, X. P. Costa, and B. Walke. The IEEE 802.11 universe. *IEEE Commun. Mag.*, 48(1):62–70, 2010.

[HH06] P. S. Hall and Y. Hao, editors. *Antennas and Propagation for Body-Centric Wireless Communications*. Artech House, Norwood, MA, USA, 2006.

[HHN+07] P. S. Hall, Y. Hao, Y. I. Nechayev, A. Alomalny, C. C. Constantinou, C. Parin, M. R. Kamarudin, T. Z. Salim, D. T. M. Hee, R. Dubrovka, A. S. Owadally, W. Song, A. Serra, P. Nepa, M. Gallo and M. Bozzetti. Antennas and propagation for on-body communication systems. *IEEE Antennas Propagat. Mag.*, 49(3):41–58, 2007.

[HRG+09] J. Haapola, A. Rabbachin, L. Goratti, C. Pomalaza-Ráez, and I. Oppermann. Effect of impulse radio-ultra wideband based on energy collection on MAC protocol performance. *IEEE Trans. Veh. Technol.*, 58(9):4491–4506, 2009.

[IEE03] IEEE Computer Society. Wireless medium access control (MAC) and physical layer (PHY) specifications for high rate wireless personal area networks (WPANs). IEEE Standard for Information Technology, 2003. IEEE Std 802.15.3-2003.

[IEE05] IEEE Computer Society. Wireless medium access control (MAC) and physical layer (PHY) specifications for wireless personal area networks (WPANs). IEEE Standard for Information Technology, 2005. IEEE Std 802.15.1-2005.

[IEE06] IEEE Computer Society. Wireless medium access control (MAC) and physical layer (PHY) specifications for low rate wireless personal area networks (WPANs). IEEE Standard for Information Technology, 2006. IEEE Std 802.15.4-2006.

[Jov05] E. Jovanov. Wireless technology and system integration in body area networks for m-health applications. In *Proc. of 27th Annual International Conference of the IEEE Engineering in Medicine and Biology Society*, pages 7158–7160, September 2005.

[JRVM09] W. Joseph, E. Reusens, G. Vermeeren, and L. Martens. Experimental determination of on-body channel for two humans in multipath environment. In *COST 2100*, Braunschweig, Germany, February 2009. [TD(09)702].

[KCPW08] N.-G. Kang, C. Cho, S.-H. Park, and E. T. Won. Channel models for WBANs—Samsung Electronics. IEEE 802.15 Working Group Document, November 2008. IEEE P802.15-08-0781-00-0006.

[KGT+10] L. Kynsijärvi, L. Goratti, R. Tesi, J. Iinatti, and M. Hämäläinen. Design and performance of contention based MAC protocols in WBAN for medical ICT using IR-UWB. In *Proceedings of Body Area Networks-Enabling Technologies for Wearable and Implantable Body Sensors (BAN2010)*, Istanbul, Turkey, September 2010. [Also available as TD(10)11056].

[KJVM10] D. Kurup, W. Joseph, G. Vermeeren, and L. Martens. In-body path loss model for homogeneous human tissues. In *COST 2100*, Aalborg, Denmark, June 2010. TD(10)11017 (presented as TD(10)12023 in Bologna, Italy).

[KP10] E. Karapistoli and F.-N. Pavlidou. An overview of the IEEE 802.15.4a standard. *IEEE Commun. Mag.*, 48(1):47–53, 2010.

[KSOW81] R. W. P. King, G. S. Smith, M. Owens, and T. T. Wu. *Antennas in Matter Fundamentals, Theory and Applications*. MIT Press, Cambridge, 1981.

[KT10] M. S. Kim and J. Takada. Statistical characterization of 4.5 GHz narrowband on-body propagation channel. In *COST 2100*, Athens, Greece, February 2010. [TD(10)10031].

[KTM+08] M. S. Kim, J. Takada, L. Materum, T. Kan, Y. Terao, Y. Konishi, K. Nakai, and T. Aoyagi. Statistical property of dynamic BAN channel gain at 4.5 GHz. IEEE 802.15 Working Group Document, July 2008. IEEE P802.15-08-0489-00-0006.

[KW10] R. Kohno and E. T. Won. WiBAN-SMA merger announcement. IEEE 802.15 Working Group Document, March 2010. IEEE P802.15-10-0215-02-0006.

[LDO08a] L. Liu, P. De Doncker, and C. Oestges. Fading correlation analysis for front abdomen propagation in body area networks. In *COST 2100*, Trondheim, Norway, June 2008. [TD(08)522].

[LDO08b] L. Liu, P. De Doncker, and C. Oestges. Fading correlation measurement and modeling on the front and back side of a human body. In *COST 2100*, Lille, France, October 2008. [TD(08)642].

[LDO09a] L. Liu, P. De Doncker, and C. Oestges. Fading correlation measurement and modeling on the front side of a human body. In *Proc. EuCAP 2009—European Conf. Antennas Propagat.*, pages 969–973, March 2009. [Also available as TD(08)522].

[LDO09b] L. Liu, P. De Doncker, and C. Oestges. Time-variant on-body channel fading characterization and modelling with dynamic human body. In *COST 2100*, Wien, Austria, September 2009. [TD(09)919].

[LKC+11] L. Liu, F. Keshmiri, C. Craeye, P. De Doncker, and C. Oestges. An analytical modeling of polarized time-variant on-body propagation channels with dynamic body scattering. *EURASIP J. Wireless Commun. Networking*, 2011, Article ID 362521, 2011. doi:10.1155/2011/362521.

[LKD+10a] L. Liu, F. Keshmiri, P. De Doncker, C. Craeye, and C. Oestges. 3-d body scattering interference to vertically polarized on-body propagation. In *Proc. AP-S 2010—IEEE Antennas Propagat. Soc. Int. Symp.*, Ontario, Canada, June 2010. [Also available as TD(10)11013].

[LKD+10b] L. Liu, F. Keshmiri, P. De Doncker, C. Craeye, and C. Oestges. 3-D body scattering interference to on-body propagation with polarized point source. In *COST 2100*, Aalborg, Denmark, June 2010. [TD(10)11013].

[LRDO09] L. Liu, S. V. Roy, P. De Doncker, and C. Oestges. Azimuth radiation pattern characterization of omnidirectional antennas near a human body. In *Proc. ICEAA 2009—Int. Conf. Electromag. in Advanced Appl.*, Torino, Italy, September 2009.

[MBC+07] A. F. Molisch, K. Balakrishnan, C.-C. Chong, S. Emami, A. Fort, J. Karedal, J. Kunisch, H. Schantz, U. Schuster, and K. Siwiak. IEEE 802.15.4a channel model—final report. IEEE 802.15 Working Group Document, October 2007. IEEE P802.15-04-0662-04-004a.

[MC10] M. Mackowiak and L. M. Correia. A statistical approach to model antenna radiation patterns in off-body radio channels. In *Proc. PIMRC 2010—IEEE 21st Int. Symp. on Pers., Indoor and Mobile Radio Commun.*, Istanbul, Turkey, September 2010. [Also available as TD(10)10041].

[MCC+06] A. F. Molisch, D. Cassioli, C. Chong, S. Emami, A. Fort, A. Kannan, J. Karedal, J. Kunish, and H. G. Schantz. A comprehensive standardized model for ultrawideband propagation channels. *IEEE Trans. Antennas Propagat.*, 54(11):3151–3165, 2006.

[MHS+08] D. Miniutti, L. Hanlen, D. Smith, A. Zhang, D. Lewis, D. Rodda, and B. Gilbert. Narrowband channel characterization for body area networks. IEEE 802.15 Working Group Document, July 2008. IEEE P802.15-08-0421-00-0006.

[MOLC10] M. Mackowiak, C. Oliveira, C. Lopes, and L. M. Correia. A statistical analysis of the influence of the human body on the radiation pattern of wearable antennas. In *COST 2100*, Bologna, Italy, November 2010. [TD(10)12041].

[MVB10a] F. Martelli, R. Verdone, and C. Buratti. Link adaptation in IEEE 802.15.4-based wireless body area networks. In *Proc. PIMRC 2010—IEEE 21st Int. Symp. on Pers., Indoor and Mobile Radio Commun.*, Istanbul, Turkey, September 2010.

[MVB10b] F. Martelli, R. Verdone, and C. Buratti. Link adaptation in wireless body area networks. In *COST 2100*, Bologna, Italy, November 2010. [TD(10)12043].

[OC08] C. Oliveira and L. M. Correia. Exploiting the use of MIMO in body area networks. In *COST 2100*, Trondheim, Norway, June 2008. [TD(08)543].

[OC10] C. Oliveira and L. M. Correia. A statistical model to characterize user influence in body area networks. In *COST 2100*, Athens, Greece, February 2010. [TD(10)10091].

[OLMC10] C. Oliveira, C. Lopes, M. Mackowiack, and L. M. Correia. Characterisation of on-body communications at 2.45 GHz. In *COST 2100*, Bologna, Italy, November 2010. [TD(10)12069].

[otESoCtNASoPE04] Task Force of the European Society of Cardiology, the North American Society of Pacing, and Electrophysiology. Heart rate variability: standards of measurement, physiological interpretation and clinical use. *Circulation*, 93(5):1043–1065, 2004.

[PKaAP+07] C. S. Pattichis, E. C. Kyriacou, M. S. Pattichis, A. Panayides, S. Mougiakakou, A. Pitsillides, and C. N. Schizas. A brief overview of m-health e-emergency systems. In *Proc. of 6th International Special Topic Conference on ITAB*, pages 53–57, September 2007.

[Pro] FP7 Project. WiserBAN—smart miniature low-power wireless microsystem for body area networks. http://www.wiserban.eu/. [Page available at April 16, 2011].

[RDG+09] C. Roblin, R. D'Errico, J. M. Gorce, J. M. Laheurte, and L. Ouvry. Propagation channel models for BANs: an overview. In *COST 2100*, Braunschweig, Germany, February 2009. [TD(09)760].

[RDM+04] J. Ryckaert, P. De Doncker, R. Meys, A. de Le Hoye, and S. Donnay. Channel model for wireless communication around human body. *Elect. Lett.*, 40(9):543–544, 2004.

[Rei09] A. Reichman. Body area networks: applications and challenges. In *COST 2100*, Braunschweig, Germany, February 2009. [TD(09)717].

[Rei10] A. Reichman. UWB PHY for body area networks. In *COST 2100*, Aalborg, Denmark, June 2010. [TD(10)11018].

[RFCO07] G. Roqueta, A. Fort, C. Craeye, and C. Oestges. Analytical propagation models for body area networks. In *IET Seminar on Antennas and Propagat. for Body-Centric Wireless Commun.*, vol. 24, pages 90–96, April 2007.

[RM10] C. Roblin and N. Malkiya. Parametric and statistical analysis of UWB BAN channel measurements. In *COST 2100*, Athens, Greece, September 2010. [TD(10)10098].

[Rob10] C. Roblin. Parametric modeling of antennas influence on the path loss for UWB on-body WBAN scenarios. In *COST 2100*, Bologna, Italy, November 2010. [TD(10)12093].

[ROHD07] S. V. Roym, C. Oestges, F. Horlinm, and P. De Doncker. On-body propagation velocity estimation using ultra-wideband frequency-domain spatial correlation analysis. *Elect. Lett.*, 43(25):1405–1406, 2007.

[SBV09] A. Sanchez, C. Buratti, and R. Verdone. Performance analysis of body communication networks based on IEEE 802.15.4 with star topologies. In *COST 2100*, Valencia, Spain, May 2009. [TD(09)880].

[SHM+08] J. Song, S. Han, A. K. Mok, D. Chen, M. Lucas, and M. Nixon. WirelessHART: Applying wireless technology in real-time industrial process control. In *Proc. RTAS '08—IEEE Real-Time and Embedded Tech. and Appl. Symp.*, St. Louis, MO, USA, 2008.

[SIG09] Bluetooth SIG. Bluetooth specification version 4.0. Bluetooth SIG Standard, 2009.

[SPYH+09] K. Sayrafian-Pour, W. B. Yang, J. Hagedorn, J. Terrill, and K. Y. Yazdandoost. A statistical path loss model for medical implant communication channels. In *Proc. PIMRC 2009—IEEE 20th Int. Symp. on Pers., Indoor and Mobile Radio Commun.*, pages 2995–2999, September 2009. [Also available as TD(10)12008].

[SZKT04] N. C. Sagias, D. A. Zogas, G. K. Karagiannidis, and G. S. Tombras. Channel capacity and second-order statistics in Weibull fading. *IEEE Commun. Lett.*, 8(6):377–379, 2004.

[Tak08] J. Takada. Static propagation and channel models in body area. In *COST 2100*, Lille, France, October 2008. [TD(08)639].

[TCAB10] W. Thompson, R. Cepeda, S. Armour, and M. Beach. Improved antenna mounting method for UWB BAN channel measurement. In *COST 2100*, Aalborg, Denmark, June 2010. [TD(10)11048].

[TPRI+] A. Taparugssanagorn, C. Pomalaza-Raez, A. Isola, R. Tesi, M. Hämäläinen, and J. Iinatti. Preliminary UWB channel study for wireless body area networks in medical applications. *International Journal of Ultra Wideband Communications and Systems (IJUWBCS)*, 2(1):14–22, 2011.

[TRV+09] P. Van Torre, H. Rogier, L. Vallozzi, C. Hertleer, and M. Moeneclaey. Application of channel models to indoor off-body wireless MIMO communication with textile antennas. In *COST 2100*, Vienna, Austria, September 2009. [TD(09)963].

[TTJI09] R. Tesi, A. Taparugssanagorn, M. Hämäläinen, and J. Iinatti. UWB channel measurements for wireless body area networks. In *COST 2100*, Lille, France, October 2009. [TD(08)649].

[TTT+09] L. Traver, C. Tarin, D. Toledano, C. Roblin, A. Sibille, and N. Cardona. Head to body UWB-BAN channel measurements. In *COST 2100*, Valencia, Spain, May 2009. [TD(09)816].

[TVH+] P. Van Torre, L. Vallozzi, C. Hertleer, H. Rogier, M. Moeneclaey, and J. Verhaevert. Indoor off-body wireless MIMO communication with dual polarized textile antennas. *IEEE Trans. Antennas Propagat.*, 59(2):631–642, 2011.

[VHI10a] H. Viittala, M. Hämäläinen, and J. Iinatti. Impact of difference in WBAN channel models on UWB system performance. In *Proc. ISSSTA 2010—11th Int. Symp. on Spread Spectrum Tech. Appl.*, pages 175–180, October 2010.

[VHI10b] H. Viittala, M. Hämäläinen, and J. Iinatti. UWB system performance in two different WBAN channels. In *COST 2100*, Athens, Greece, February 2010. [TD(10)10032].

[VHIT09] H. Viittala, M. Hämäläinen, J. Iinatti, and A. Taparugssanagorn. Different experimental WBAN channel models and IEEE 802.15.6 models: comparison and effects. In *Proc. ISABEL 2009—2nd Int. Symp. on Appl. Sci. in Biomed. and Commun. Tech.*, Bratislava, Slovakia, 2009.

[VTH+10] L. Vallozzi, P. Van Torre, C. Hertleer, H. Rogier, M. Moeneclaey, and J. Verhaevert. Wireless communication for firefighters using dual-polarized textile antennas integrated in their garment. *IEEE Trans. Antennas Propagat.*, 58(4):1357–1368, 2010. [Also available as TD(09)963].

[Wal06] J. Walko. Home control. *Computing Control Engineering Journal*, 17(5):16–19, 2006.

[Whe07] A. Wheeler. Commercial applications of wireless sensor networks using ZigBee. *IEEE Commun. Mag.*, 45(4):70–77, 2007.

[Wir] P. Wireless. HomeRF: overview and market positioning. http://www.palowireless.com/homerf/homerf.asp. [Page available at March 2, 2010].

[YH09] K. Y. Yazdandoost and K. Hamaguchi. Channel models for implant communications. In *COST 2100*, Vienna, Austria, September 2009. [TD(09)904].

[YH10] K. Y. Yazdandoost and K. Hamaguchi. Antennas for body area network communications. In *COST 2100*, Athens, Greece, February 2010. [TD(10)002].

[YHSP+10] K.Y. Yazdandoost, K. Hamaguchi, K. Sayrafian-Pour, W. B. Yang, J. Hagedorn, and J. Terrill. Study of on-body propagation using a 3D virtual reality platform. In *COST 2100*, Bologna, Italy, November 2010. [TD(10)12008].

[YK07] K. Y. Yazdandoost and R. Kohno. An antenna for medical implant communications system. In *Proc. EuMC 2007—European Microwave Conf.*, pages 968–971, October 2007.

[YK09] K. Y. Yazdandoost and R. Kohno. An antenna for medical implant communications system. In *COST 2100*, Valencia, Spain, May 2009. [TD(09)808].

[YSP10] K. Y. Yazdandoost and K. Sayrafian-Pour. Channel model for body area network (BAN). IEEE 802.15 Working Group Document, November 2010. IEEE P802.15-08-0780-12-0006.

[YTA+09] K. Y. Yazdandoost, K. Takizawa, T. Aoyagi, J. Takada, and R. Kohno. An overview of NICT's channel model planning for wireless body area network. In *COST 2100*, Braunschweig, Germany, February 2009. [TD(09)754].

[ZAS+03] T. Zasowski, F. Althaus, M. Stäger, A. Wittneben, and G. Tröster. UWB for noninvasive wireless body area networks: channel measurements and results. In *Proc. UWBST 2nd IEEE Conf. on Ultra Wideband Systems and Tech.*, Reston, VA, USA, November 2003.

[Zim96] T. G. Zimmerman. Personal area networks (PAN): near-field intra-body communication. *IBM Systems J.*, 35(3–4):609–617, 1996.

[ZLK07] B. Zhen, H.-B. Li, and R. Kohno. IEEE body area networks for medical applications. In *Proc. of 4th IEEE Int. Symp. on Wireless Comm. Systems*, pages 327–331, October 2007.

[ZMAW05] T. Zasowski, G. Meyer, F. Althaus, and A. Wittneben. Propagation effects in UWB body area networks. In *Proc. ICUWB 2005—IEEE Int. Conf. on Ultra-Wideband*, Zurich, Switzerland, September 2005.

[ZMAW06] T. Zasowski, G. Meyer, F. Althaus, and A. Wittneben. UWB signal propagation at the human head. *IEEE Trans. Microwave Theory Tech.*, 54(4):1836–1845, 2006.

Chapter 16
Wrapping Up and Looking at the Future

Chapter Editor Luis M. Correia

16.1 Wrapping Up

COST 2100 spanned over a number of different areas, dealing with mobile radio communications, not only from the techniques themselves, but addressing some applications as well. This book is structured in a such way that this is clearly seen, starting with channel modelling and measurements, then, looking into antennas in terminals, after which ultrawideband, MIMO, cooperative and distributed systems, and coding, modulation and signal processing, are addressed; networks aspects are also included, from radio resource management to deployment, optimisation and operation. The application areas encompass several kinds of systems and networks, i.e., ad hoc and sensor, hybrid cellular and broadcasting, mobile-to-mobile, and body area ones. In what follows, a brief overall view of the main results is presented, with an effort to link and integrate them.

The modelling of radio channels plays a key role in any wireless communications system, being even more important when mobile radio ones are at stake. The complexity of the propagation environments, and the increased efficiency that a system takes in the transmission of information, together with the impact that the radio channel has on a system (due to the constraints that it imposes), makes this area as one that has experienced continued work for many years. On the other hand, the increased complexity of the environments, together with the availability of equipment to measure channel parameters, has created the conditions to use channel measurements for the test, and even development, of models. With the new paradigms of radio channels, new figures of merit appeared, like frequency selectivity, directivity, correlation, and polarisation, among others. Additionally, much wider bands (up to ultrawideband) and much higher bands (up to sub-millimetre waves) emerged as well. This also means that new methods and approaches for performing the measurements need to be investigated and perfected.

L.M. Correia (✉)
Istituto Superior Técnico (IST), Universidade Técnica de Lisboa, Lisbon, Portugal

The specifications of a number of sounders are given, which enables their comparison, as well as the influence of several parameters on the measurement results. Also, the calibration of the measurement set-up is addressed. Besides this, measurement procedures should be clearly defined, and a set of very useful recommendations for those wishing to be involved in this type of activity is given. On the top of this, estimation of channel parameters continues to play a major role, and several methods and algorithms are analysed and compared.

Quite many measurement campaigns were held within COST 2100, leading to a large number of sets of data. Measurements can be taken from three different approaches: single-sounder sequential, in which a single channel sounder is used sequentially along multiple routes to mimic multiple users in a sufficiently static scenario, or along the same route sequentially with different base stations; single-sounder multi-node, where a single channel sounder is used, with multiple nodes connected to the sounder by RF cables; and multi-sounder, by using more than one equipment, typically a single transmitter and multiple receivers are used. Measurement scenarios can be split into multi-node (multi-user) and multiple base stations. This is clearly an area that crosses with the applications one, since the measurement scenarios need to be adjusted, i.e., when considering mobile-to-mobile systems or body area networks, obviously, measurements need to be performed and adapted to the corresponding environments.

Channel models have been updated and further developed for a number of environments (indoor, outdoor-to-indoor, and outdoor, in several different types for each one), concerning path loss and wideband parameters. Both deterministic and stochastic approaches were taken, serving different purposes on the modelling perspective, namely complex environments and MIMO systems, for both static and time variant scenarios. Modelling, together with measurements, addresses mainly the "new" bands above 2 GHz, mostly up to the 5 GHz one. For a given frequency band, angular spread can range in more than one order of magnitude (between $0.7°$ and $24°$), while delay spread varies much less (within [110, 540] ns). Polarisation aspects are important as well, since, in general, antennas do not keep the polarisation ratio in the full three-dimensional radiation pattern, hence, allowing for coupling in between orthogonal polarisations, which needs to be considered by channel models.

The COST 2100 Multi-Link MIMO Channel Model is presented, describing the physical wideband radio propagation in various scenarios, ranging from pico- to macro-cells, in terms of delay and direction domains at both transmitter and received ends, via physical clusters, i.e. groups of multipath components. This model is described by parameters associated to polarisation, diffuse multipath, and multi-link features, being either external (deterministic) related to the environment, or stochastic linked to the random nature of propagation; it requires also the carrier frequency and the bandwidth. One of the key parameters of this model, the number of clusters of scatterers, can basically be taken as 3 or 4, regardless of the environment being indoor or outdoor and whether line-of-sight exist or not. The model was implemented in both MatLab and C++, and validated by measurements.

Channel models are very much connected to antennas, namely their radiation patterns (but also impedance bandwidth, efficiency and mutual coupling), hence, the latter need to be properly analysed and taken into account. The influence of the user on the antenna is an "old problem", but MIMO brought a new attention to it, not only because of the location of the multiple antennas on mobile terminals, but also due to the changes of their performance coming from the stochastic behaviour (not only from user to user, but time wise for the same user as well). This also implies that propagation environments should be taken into account.

The location of multiple antennas in terminals puts challenges that cannot be minimised, namely because of the coupling created by their close-spaced location. A spacing of half wavelength seems to be enough to guarantee a low correlation in between antennas, but this can still put a problem for the lower frequency bands (e.g., around the 800 to 900 MHz), due to the size of the phones, although this problem is not critical for larger terminals, like laptop computers. The presence of the user can have quite an impact, leading to a decrease of the Mean Effective Gain up to 15 dB, which halves the ergodic Shannon capacity for a 2×2 MIMO configuration. User influence can be mitigated by the design of proper (planar) antennas in mobile terminals, which can even decrease the mutual coupling in between them. Additionally, a statistical modelling approach can be taken for the ensemble of antennas and channel variability, or for just the antenna itself. These problems may not be acute for mobile-to-mobile systems, since antennas are to be placed on cars, hence, the variability comes only from the change of direction of the vehicle (which even may not be that important for a quasi-omnidirectional antenna), but the results of these studies find a strong application in ad hoc, sensor, and body area networks.

The standardisation efforts of COST 2100 have been performed via the work on Over-the-Air measurement of antennas integrated in mobile terminals, including MIMO handsets. Many setups were examined and compared with different levels of cost/complexity trade-offs, and several lab evaluations of different setups were investigated, building the foundations for a round robin measurement campaign that was held, together with a comparison between real environment references and MIMO Over-the-Air setups. Four major findings were extracted from this work: some results indicate that the devices under test cannot reach the expected Over-the-Air throughput; anechoic-chamber based methods show good repeatability and reproducibility across different labs; anechoic-chamber based and two-stage test methods show similar results; and reverberation and anechoic methods produce different results.

Ultrawideband systems find application in the transmission of very high data rates at very short distances, using frequencies in the order of a few GHz. The design of antennas, considering radiation pattern, impedance bandwidth, and polarisation, has been addressed, not only exploring the mathematical modelling, but also by looking into implementations in printed antennas technology. The antenna gain is of the order of a few dB, exhibiting a polarisation discrimination of at least 10 dB.

Ultrawideband channel measurements were performed in a variety of indoor environments (office, residential, industrial, car, aircraft, road tunnel, and even a body area network), looking at parameters like path loss, delay spread, and Ricean K-factor. The range of variation of parameters related to path loss and the K-factor is quite large, due to the variety of scenarios, but delay spread is in the range of tens of nanoseconds. The use of ultrawideband for localisation and tracking has also been explored. The mean error of the estimation can be 0.5 m (or even lower), when using more than 5 nodes in line-of-sight, which can be considered very promising.

The implementation of MIMO requires quite some changes in both mobile terminals and base stations, both on the transmitting and receiving ends. On the one hand, it is desirable to use the best algorithms and techniques, in order to increase the data rate, but on the other, efficiency aspects need to be taken into consideration, e.g. power consumption. Therefore, there is a trade-off that needs to be achieved, which requires study. One of the main reasons for the increase of consumption in terminals is the use of a transceiver analogue component in the architecture composed of a front-end stack-up, each of the processing chains being dedicated to an antenna contribution; in order to limit this, the terminal architecture can be based on different multiplexing or transmission techniques, allowing the use of a single front-end and proving a good performance vs. power consumption trade-off. In here, the dependence on the application scenario is quite strong: the constraints on MIMO antenna solutions are quite different when mobile-to-mobile systems or body area networks are considered. In a car, there is not too much restriction to the number of antennas, and power consumption is just a "minor drawback", whether in a body network the number of antennas/sensors that can be used is under several restrictions, and clearly power efficiency is a must. This shows, once again, that specific solutions need to found for each case and application.

Adaptive MIMO and precoding can be used to increase performance. For the latter, spatio-temporal techniques are used. Concerning the former, two types of gain can be considered: diversity gain, measured by the rate of decrease of BER with SNR at a constant link throughput, and multiplexing gain, measured by the increase in link throughput with SNR at a constant BER. For example, for two antennas, the coding gain with the full LTE codebook is 3 dB better than for MMSE detection. A multiple terminal analysis in MIMO has to consider: multiple users, multiple user diversity, multiple base stations, and multiple virtual users. Other techniques, like Transmit Diversity and Space-Time Block Coding, prove also to increase system performance. Application areas may not play a key role in this topic, as they play in previous ones, but still the matter of the number of antennas, and power consumption, needs to be taken into account when examining the usage of these techniques.

Cooperative and distributed communication systems represent an important shift of the classical historic role of the physical layer functionality. Traditionally, the physical layer was viewed from a point-to-point communications perspective, all communication tasks related to the information transfer in a more complicated multiple-source and multiple-node structure being delegated to upper layers. Cooperative and distributed systems can be viewed as the physical layer coding and processing algorithms being aware, actively using the knowledge of the network structure/topology, and even the simplest scenario of the bidirectional relay communication can show a substantial throughput improvement. Three different scenarios can be considered: single-source and multiple-node or relay; multiple sources and multiple interim nodes or relays; distributed signal processing and coding. The design of the networks, being ad hoc and sensor, mobile-to-mobile, cellular, or body area ones, will again be of key importance. The number of relay nodes that one may use, the way that hops are established having power constraints into consideration, or the distribution of the processing among the various nodes definitely need to take the type of networks into account.

Advanced coding and signal processing issues are addressed as well, e.g., advanced coding techniques and the design of physical layer Forward Error Correction schemes based on LDPC codes (coding over bits is replaced by coding over packets, packet-based application layer Forward Error Correction techniques being presented), iterative coding techniques (iterative demodulation and decoding, and turbo equalisation are discussed), multicarrier and adaptive schemes, channel estimation and synchronisation, and network interference. These techniques are applied to a number of cases, LTE (with an OFDMA downlink access, and SC-FDMA being used in the uplink) being one of the most used examples. Performance gains vary, depending on the technique and on other parameters (e.g., scenario and system, among others).

The deployment, optimisation and operation of mobile radio networks include three key aspects: coverage, which is largely independent of the specific system; optimisation, which critically depends on the standard; and exploitation of data, taken from actual measurements, in a structured way. Hot topics currently under discussion on coverage are strategies to reduce the level of electromagnetic exposure, while still providing a high level of coverage, especially indoors. In order to achieve that goal, new concepts, like femto-cells, are being introduced in various networks, but they can also contribute to bring problems to operators, since this may imply that operators loose the complete control of the network structure.

The two most widespread standards families used in mobile and wireless networks are being developed by 3GPP and IEEE, both being address here. Optimisation techniques are presented for both 3GPP's UMTS, IEEE's 802.16 (WiMAX) and 802.11 (WLAN), with two different approaches, using input data available from radio network planning tools and using real network measurement values. An operator's effort in the initial configuration and in the optimisation of mobile radio networks consumes a considerable amount of their Operational Expenditures (OPEX); therefore, currently developed solutions and standardisation activities in this area are focused on the automation of the network organisation process.

Different levels for the integration of measurement data exist: off-line processing of measurement data and feeding these data back to the planning process, focusing on traffic and mobility issues; introduction of auto-tuning processes, which, although using a higher degree of automation, still rely on the traditional planning process; introduction of self-organising networks (SONs), where an entity inside the network collects the measured data in the live network and changes its configuration in an optimised way. This last concept of SONs is already being standardised in 4G networks (LTE), enabling a more dynamic and flexible network operation and supporting a wider range of services and business areas.

Although, traditionally, the aspects mentioned in the last three paragraphs are associated to cellular related networks, one can easily envisage that they will be extended to other types of networks. The concept of wireless mesh network, addressed as one of the techniques to be used in cellular networks, can be easily applied to mobile-to-mobile communications, or even multiple body area networks. Ad hoc networks will also play a key role bridging with cellular ones, as femto-cells somehow already indicate. Furthermore, with the trend to include sensor networks into larger scale ones, e.g., cellular, the concept of network will really encompass all types of nodes that route information from one point to another. This means that many of the algorithms developed for a specific network will be extended to allow for the heterogeneity of networks, whatever their access technique is, or any other characteristic of the network.

Spectrum became a scarce asset since many years, and lately, with the auctioning of frequency bands, the pressure to use it in the most efficient way possible became even stronger. In mobile and wireless systems, this usually means to increase the efficiency of the radio resource management algorithms. Furthermore, with the larger variety of services being offered, with different transmission requirements (concerning reliability and/or throughput, among others), the problem gained an augmented complexity. The heterogeneity of the networks is an additional constraint to be taken into account, which, in turn, allows for a better optimisation of the usage of the resources. The design of optimum Common Radio Resource Management strategies among different networks needs to be performed, in order to facilitate user network selection while optimising global performance. Many techniques have been proposed: algorithms assigning resources to mobile terminals with best channel conditions and/or highest need according to their QoS requirements, through cooperation among different layers; power control to reduce inter-cell interference, minimising also the power consumption at both mobile terminals and base stations, ensuring also an adequate QoS level; scheduling algorithms applying different metrics to users, depending on their requirements; and modifying existing algorithms to increase spectral efficiency. The outcome of these algorithms and strategies can really lead to significant improvement in performance, e.g., the number of VoIP or video stream users can double, with the same reference quality.

Traffic modelling is an important component as well. The knowledge about the characteristics of user-generated traffic is usually related to network simulation and emulation. This is an area that is not dominated by operators, i.e., users generate at their own will traffic coming from very different services and applications, with different characteristics, with different mixes depending on location or time of day, and many other aspects not controlled by network operators. Adequate traffic modelling enables a better planning and optimisation of networks, hence, being of great value. Statistical distributions have been established for the four service classes established by 3GPP, at the session, connection, and packet levels.

When addressing cross-layer issues, the fact that a layered architecture is still maintained means that there cannot be direct communication between non-adjacent layers and that a higher layer calling on the services of a lower one cannot take account of how these services are provided. Four main aspects on protocol design are identified as cross-layer techniques: design—the protocol of one or more layers is designed taking the detailed design of other layers into account; linkage—layers communicate with one another other than by the conventional mechanisms of passing information up or down the stack; integration—complete merging of two or more layers; analysis—the performance of a given layer is evaluated accounting for the performance of other layers, and without necessarily violating the architecture at all. The fundamental advantage, of course, is that cross-layer techniques remove restrictions on the implementation of communication systems.

The theory of wireless networks is addressed as well, aiming at finding and understanding the fundamental behaviour of wireless networks, as well as establishing general methods and rules for using and organising networks, and radio resources to optimise capacity and performance. The derivation of analytical tools and models for performance evaluation in general conditions is also encompassed. The contributions are related to capacity limits in both wireless ad hoc and sensor networks, as well as cognitive radio systems.

Distributed wireless networks, with the capability of sensing the environment and even actuating on it, are becoming increasingly important, with many foreseen applications, ranging from industrial scenarios to environmental ones and encompassing so different areas from urban to body ones. These types of networks, i.e., wireless sensor networks, are usually characterised by low-cost and low-power simple devices, gathering data from the vicinity and transmitting them to one or more sinks; in many cases, a mesh structure is taken. Additionally, cognitive radio networks are addressed as well, where the existing frequency spectrum is exploited opportunistically. This approach may hugely increase the availability of large amounts of underused spectrum for wireless communications. For example, simulations show that by means of appropriate spectrum decisions, GSM spectrum can be reused for other systems, without causing harmful interference to GSM.

MANETs (Mobile Ad Hoc Networks) are also an important part of the next generation of wireless networks, being a collection of communication devices or nodes that communicate without any fixed infrastructure (such as base stations) and predetermined organisation of available links. These networks have several characteristics, such as dynamic changes of topologies, bandwidth-constrained, links capacity, energy-constrained operation, limited physical security, which are addressed.

Although the heterogeneity of wireless networks is usually taken as a mixture of cellular and local area ones, or even of several cellular systems, the emergence of mobile multimedia broadcasting has enabled a new perspective into the area. Neither broadcasting nor cellular systems alone can provide a cost effective service provision under all possible scenarios; therefore, the goal is to join the advantages of both networks. The benefits shown encompass the reduction of the system cost, the decrease of the required overall data rate, the improvement of the perceived area coverage, and the reduction of the electromagnetic exposure, among others. The combination of UMTS with DVB-H has shown a high potential to efficiently serve users' requests. As one of the hybrid network planning rules, it has been identified that the broadcast bound needs to be the lower bound and the cellular bound the upper one, an optimum grade of service being guaranteed.

Mobile to mobile communications, especially in vehicle-to-vehicle communication scenarios, represent a very promising and certain area of future networks. Traffic telematics applications enable transportation safer, more efficient, and cleaner, namely concerning cars, buses, and trains; additionally, the areas of navigation mechanisms and infotainment services need to be included in the set. One of the major differences to other systems lies on the radio channel, since its behaviour is quite different from the usual cellular networks, due to its fast time variation and to the resulting Doppler shifts. Mobility models need to be taken into account for different scenarios (e.g., roads and urban structures) and several communication types (vehicle-to-vehicle and vehicle-to-infrastructure). Results show a strong influence of antenna placement and configuration on channel parameters, implying the need for the integration of realistic antennas in the simulation environments.

Finally, body area networks address a technology segment that is still in its early stage and uncovered by standards. Several characteristics involve a broad range of data rates at lower power consumption and limited transmission distance in the immediate environment around and inside a person's body, namely from a wireless sensor networks perspective. Three types of nodes with distinct propagation features can be defined: in-body (implant) nodes, placed inside the body; on-body (body surface) nodes, placed on the skin surface or at most a few centimetres away; and off-body (external) nodes, between a few centimetres and up to a few metres away from the body. The range of body area networks applications is plentiful and colourful, emphasising its vital feature, i.e., interdisciplinarity. The path loss in between two nodes can vary quite a lot (ranging in a couple of dB for the same scenario), depending on the location of the sensors, and the mobility of the body needs also to be accounted for. A possible approach is to take a stochastic view on the radiation pattern of the antennas. Moreover, transmissions aspects do need to be considered, namely low-power constraints, since many nodes may not be replaced or recharged without surgery.

16.2 Looking at the Future

The area of mobile and wireless communications is evolving in directions that enable one to discuss some of research topics that are going to be addressed in the near future. Networks of the Future, or Future Internet, became a "hat" under which a number of topics are being investigated, but besides this, other topics gained relevance by themselves. Some of these areas are not in the scope of COST 2100, like user interfaces and three-dimensionality of information in user's terminals, user-aware applications development, and augmented reality. Even in the area of networks and protocols, many topics are being researched today, like new architectures, information-centric networks, cloud networking, virtualisation, and "post-IP" protocols. But the area of radio continues to have quite many topics to be looked at, especially because the trend to go wireless (as opposed to wireline) in communications is increasing, and will not stop; this encompasses propagation and channels, radio interfaces, and radio networks, among other aspects. In what follows, some objectives are identified, along with the way they relate to the research areas and application areas that, on the one hand, serve as a host for the developments obtained in these areas and, on the other, establish requirements and constraints that need to be taken into account.

Currently, one of the major research directions is still to increase the data rate in the air interface. Although this quest will slow down in some years from now, after it reaches a value that enables "instantaneous" information, i.e., transfer at speeds with a waiting time below user's perception, it seems clear that it will still play a role in the years to come. This increase in data rate has many implications, ranging from larger bandwidths (hundreds of MHz) to higher carrier frequencies (up to THz) and including new modulation and coding schemes, new access techniques, and parallel transmission in heterogeneous networks (i.e., transporting information in parallel channels from different systems).

Very much related to the previous area is the one on spectrum sharing and the many techniques that enable an increased efficiency of spectrum usage. Due to the physical constraints, coming essentially from antennas and propagation, some of the frequency bands are already simultaneously "over-populated" with many systems and desired for the implementation of many others. Associated to this matter, is the decreasing size of cells, which went from tens of kilometres in the beginning of cellular networks to the current tens of metres (and is further decreasing). The minimisation of interference is definitely a goal, but on the other hand, one can envisage the possibility of systems that take advantage of interference. The opportunistic, cooperative, and cognitive use of a slice of the spectrum for a given communication needs to be considered. Work on transmission/reception techniques has to continue, namely considering multiple antennas and new transceiver architectures.

Heterogeneous networks need to work in a cooperative way, in a self-organised perspective, so that resources (in a broader view, ranging from radio channels to energy) are used in the most efficient way. Cooperation among different networks and among network nodes will be essential; one has to consider both the link and network levels, enabling, e.g., the maximisation of end-to-end throughput and of global capacity through network coding or other forms of cooperation. Cooperation also involves the proper detection, mitigation and management of inter-network interference, arising from the possibility of autonomous networks merging and splitting. Multi-hop relaying is another approach that needs to be considered along these lines. Still, a holistic approach, across the layers, of the whole communication network has to be accomplished, so that additionally flexibility is obtained, and the efficiency goal is properly achieved.

Energy efficiency is definitely another major topic for work. For years, all the efforts in this area have concentrated exclusively on the side of the mobile terminal, and no attention was paid to the fixed part of the system. Since a few years, environmental concerns on energy usage, together with foreseen high energy prices, created an enormous focus into this area on the network side, which obviously includes the radio component. Again, more efficient radio interface aspects need to be developed, together with radio resource algorithms (in homogeneous and heterogeneous networks) that take energy efficiency into account, without degrading the requested quality of service. Also, one needs to revisit beamforming and MIMO from the energy efficiency viewpoint, as these techniques may play a role in this matter. Such an approach of decreasing power consumption and, hence, the transmitted one as well will also ease the problem of people's exposure to electromagnetic fields.

Another area in mobile and wireless communications where developments will be progressing is security, as a natural consequence of the increase of mobile commerce and other mobile terminals based on business and usage applications. Capturing the "signature of a location" via channel characteristics (in space, frequency or time) can be a way to bring security aspects into the radio component.

Body Area Networks will continue to be one of the major application areas, being related to many sectors, e.g., health, entertainment, commerce, and even communications. Concerning the last one, one needs to address the problems associated to short-range links coming from a user carrying a personal "RF SIM card" that will allow his/her identification by any device in the surroundings, which will then establish the connection to the network. Furthermore, aspects of propagation in-, on- and off-body scenarios need to looked at, as well as the join consideration of antennas and propagation in channel modelling, considering the effects of the body proximity and movement.

Machine-to-Machine communication networks (or the Internet of Things), where a user has no intervention in the communications system, requires still a lot of work. This area encompasses car-to-car and car-to-infrastructure communications, but it includes also transmission in environments like building walls, and inside public transports. Additionally, low power consumption requirements and communication to control or information networks need to be considered as well.

The last two paragraphs address networks that are closely related to Wireless Sensor Networks, hence, many of the problems are common. Additionally, one can envisage the extension of the MIMO concept to the use of different multiple antennas, located at somehow random locations (e.g., buttons on clothes on a body) or at fixed ones (on or inside a car), and a virtualisation of MIMO antennas. One can even consider that architects and designers, and civil and mechanical engineers, will incorporate the location of sensors/base stations/access points in streets, houses, offices, cars, and many other environments.

The extension of Location Based Services taking advantage of a very high accuracy, enabling the distinction of devices on the body will be just a matter of time; in this case, the user is using the network for location purposes and then applying this information for a variety of services for his/her own purpose. But, one can invert this concept, i.e., the network knowing where users are, and take advantage of it. For example, by using users' terminals (and other devices) as channel sensors, the network can establish a geographical map of channel conditions and, hence, of channel quality, and with that information, forecast (for each user) in an efficient way the possibilities of services availability, implementing the concept of User-Aware Networks.

Context Awareness is another area that is expected to witness major developments in the future. In here, one is not only referring to the previously mentioned areas (ranging from location based services to sensors), but also to a lot of additional capabilities, which include interactive games in the mobile terminal with the real surroundings of the user. This will require some short-range radar characterisation and a channel mapping of surroundings.

An integration of many of the previous areas will certainly be done by the implementation of Smart Environments, ranging from the human body to an urban area.

In these environments, sensors, actuators, embedded systems, user terminals, and many other types of communication devices will cooperatively pursue given tasks and exchange information, sharing resources, such as spectrum or energy. Areas of application include Health, Transport, Energy, Entertainment, and e-Government, among others.

Finally, the impact of Social Networks on communication systems and networks needs to be envisaged. We are becoming more and more prosumers (producers/consumers) of information (which implies that up- and down-link data rates need to be paired), sharing information of all kinds, including games and entertainment (which implies the support of many different systems and devices), assuming an "instantaneous" exchange of data (which implies data rates to be very high at the air-interface), and relying on electronic media instead of paper-based one (which implies dealing with a capacity several orders of magnitude larger than the one that is available today), just to mention a few aspects. Additionally, requirements coming from Point-to-Multi-Point or Multi-Point-to-Multi-Point systems, associated to real-time geographically based ad hoc or mesh networks (i.e., networks established just for a period of time, based on the interest of users, in a given location, enabling the communication and the sharing of information among the users of a group), must be put together with the needs identified above. After all, enabling all sorts of communication among people is the ultimate goal of all the work we do in mobile and wireless communications.

The new COST Action IC1004 will, for sure, address many of these challenges.

Index

0–9
60 GHz antennas, 164
60 GHz broadband links, 27
60 GHz-UWB real-time channel sounder, 33
802.11p, 601, 602

A
α-Stable interference, 357, 394
Absorption, 173
Accuracy of beam pattern, 52
Adaptive modulation, 389
AF, 348–350, 366
AGC, 51
Aircraft cabin, 36
Alamouti, 368
Ambiguity function, 42
Amplify-and-forward, 342
Antenna, 263, 266
    Multi mode, 266
Antenna deembedding, 624
Antenna Pattern, 91
Antenna selection, 173, 177, 178
AOA, 287
AZFML, 321

B
Back-2-back system calibration, 51
Bandwidth, 149, 173
Bayesian estimation, 45
Beam pattern, 52
Beamforming, 268
Beampattern, 50
BFC, 353
Block erasure channels, 365
Block Fading Channel (BFC), 353
Body surface node, 610

BS Power Constraint, 347

C
CAC, 549
CAC algorithm, 445
Capacity maximisation, 415
Capacity under interference, 107
Capital Expenditures (CAPEX), 443
Cell Outage Management (COM), 451
Channel, 270
    Capacity, 286
Channel eigenmodes, 155, 156
Channel estimation, 390
Channel interferences, 528
Channel Modeling, 278
    Deterministic, 281
    Deterministic-stochastic, 282
    Hybrid deterministic, 282
    Hybrid Ray Tracing/FDTD, 282
    Ray Tracing, 282
    Statistical, 279
Channel Quality Indicator (CQI), 211
Channel sounding, 23
Cluster, 53
Clustering, 46
Cognitive radio, 186
Common clusters, 125
Common scatterers, 110
Compactness, 151
Compress-and-forward, 342
Condition numbers, 106
Connection Admission Control, 549
Context Awareness, 670
Cooperative Diversity, 341, 342, 356
COOPERS, 601
Coordinated Multi Point (CoMP), 418
Correlation, 151, 152, 158, 160

COST 2100 channel model, 125
COST 2100 channel model implementation, 135
COST 2100 channel model parameters, 128
CPR, 98
Cross Polarization Ratio, 86
CVIS, 601, 602
CWNetPlan, 410

D

Decode and Forward (DF), 342, 343, 348, 366
Dense multipath component, 42
Deterministic Channel Models, 78
DF, 348, 349, 356, 357
DFMM, 185
Dielectric loading, 173
Dielectric properties, 163
Dielectric resonator antenna, 150, 179
Differential feeding, 263
Diffuse multipaths, 123
Diffuse scattering, 42, 85
Digital signal processing (DSP), 556
Direct sequence spread spectrum, 23
Distributed systems, 6
Diversity, 152, 197, 224, 560
Diversity combining, 345
Diversity gain, 177
DMC, 42
Drop-based simulations, 104
DSP, 556
Dual-link model, 112
DVB-H, 547, 549–552, 555, 557, 558, 560, 561, 563, 565–568, 667
DVB-H/HSDPA, 553
DVB-T, 549, 550, 554, 558, 561, 569, 570
DVB-T/H, 561
DVB-T2, 569–572

E

E-band links, 39
ECC, 249
Effective gain, 186
Effective Roughness approach, 85
Efficiency, 149, 173
EKF, 45
Electromagnetic Models, 80
Energy efficiency, 529
Environment database, 85
Error Vector Magnitude (EVM), 212
ESPAR, 177
ESPRIT, 43
Exposure to electromagnetic fields, 408
Extended Kalman Filter, 45

F

F-DAS (Fiber-Distributed Antenna System), 418
Fading emulator, 204, 215, 223
FCC, 23, 249
Femtocell, 411
FF, 254
Field trials, 422
Finite scattering channel model, 313
Finite-difference methods, 80
FleetNet, 601
Frame synchronization, 392
FWHM, 262

G

Giga-bit/s, 27
Graph partitioning problem, 434

H

Hand grip, 163
Hand grip position, 157, 163
Handset MIMO, 198, 199
Handsets, 187
Hard handover, 557
Hearing Aid Compatibility (HAC), 199
High resolution parameter estimation, 50
High-Definition TV, 27
HO parameter optimisation, 446
Home eNodeBs, 448
Hospital environments, 37
HSDPA, 548, 549, 568

I

ICIC, 487
IEEE 802.11p, 20, 22
IEEE 802.15.4a, 23
IFDMA, 388
Imaging, 289
Impedance mismatch, 173
Implant antenna, 614
Implant node, 610
Impulse Radio Ultra-WideBand (IR-UWB), 356
In-body channel model, 616
In-body node, 610
Infotainment, 38
Inter-cell correlations, 110
Interference, 393
Interference coordination, 448
InterSafe, 601
Intersymbol interference, 386
IQHA, 150
ISM band, 614
ISO CALM, 602
Isolation, 173–176, 180–182

Index

K
Keller's cone, 90
Kirchhoff's scattering theory, 29
Kronecker model, 42

L
LA re-planning, 435
Large-scale networks, 531
Large-scale parameters, 109, 110
Light-Of-Sight (NLOS), 560
Linear precoders for MU MIMO, 314
Link adaptation (LA), 650
Load balancing, 447
Local cluster, 120
Localization, 287, 531
Lognormal, 187
LTE, 198, 224, 231, 235, 243, 368, 407, 413, 444, 486, 549, 664, 665

M
MAC, 23
MAC layer, BAN, 644
MAC multi-user MIMO capacity, 313
MAS, 255
Matching network, 153, 154, 168
Matrix collinearity, 106
Maximum length binary sequence, 24
MBMS, 547–549, 567
MBMS-HSDPA, 548, 549
Mean gain, 262, 265
Measurement system, 50
Measurement-based planning, 430, 432
MEG, 159, 160, 200, 201, 206
Mesh networks, 426
MICS band, 614
Millimeter-waves, 27
MIMO, 149, 164, 173, 188
MIMO BC, 314
MIMO capacity, 159
Mobility modelling, 433
Modelling user influence, 163
MOMENTUM, 416
MORANS, 416
MPC, 39
MU diversity, 317
Multicasting/Broadcasting MBMS, 548
Multihop, 396
Multipath component distance, 46
Multiple access, 393
Multiple antenna systems, 149–151, 158
Multiple antenna terminals, 150
Multiple BS MIMO, 318

Multiple virtual-user MIMO, 319
Mutual coupling, 151–153, 159, 164, 173, 176, 182, 183

N
Network coding, 342
Network Lifetime, 355
Network Management System, 431
Network on Wheels, 601
Network planning, 547
Network Power Constraint, 347
Neutralisation line, 150
NodeB emulator, 226
Non-stationarity, 20
NPC, 347

O
OFDM, 51, 251, 274, 387
OFDMA, 486
Off-body node, 610
On-body node, 610
Operational Expenditures (OPEX), 407, 443, 665
Optimisation, 430
Optimisation of 3GPP RAN, 430
Orthogonal Space–Time Block Code, 351
OSTBC, 351, 352
OTA, 163, 171, 172

P
Parameter estimation, 39
Particle Filter, 45, 46
Path loss, 30, 272
PCS correlation, 433
PDF, 184
PDP, 41, 239
Peer-to-peer channels, 114
Performance counter statistics, 431
Phase modes, 251
Phase noise, 52
Phaseless antenna patterns, 164
PHY layer, BAN, 642
Picocell, 411
PIFA, 174
Planning tool, 423, 432
Polarization, 40, 86, 96, 122
Polarization diversity, 263
Power-Delay Profile, 41, 239
Precoding, 202
Precoding Matrix Indicator (PMI), 211
Propagation
 in tunnels, 69
 in vegetation, 68
 outdoor-to-aircraft, 74

Propagation (cont.)
  outdoor-to-indoor, 73
  outdoor-to-outdoor, 71
Propagation channel, 150
Propagation path, 39
Pseudo noise sequence, 23
PSMA, 644
Public transportation scenarios, 30

Q
QHA, 178
QoS, 550, 551, 666
Quality-of-Service (QoS), 198, 549

R
Radar, 289
Radiation pattern, 97, 149, 251
Radiative Transport Theory, 89
Rank Indicator (RI), 211
RARE, 44
Ray Launching Models, 82
Ray Tracing, 282
Ray Tracing Models, 82
RCS, 260
Realistic scenarios, 430
Receive diversity, 151
Received-Signal Strength Indicator, 237
Relay selection, 397
Resource Assignment, 486
Reverbertion Chambers, 204
RIMAX, 44
Ringing, 262
Routing, 529, 531, 532

S
Safespot, 601
SAGE, 44, 277
SAM, 174
SC-FDMA, 665
Scheduling, 202, 487
Self-configuration, 445
Self-healing, 445
Self-optimisation, 445
Self-organising network (SON), 408, 443
Self-organising networks (SONs), 665
SEM, 185, 256
SFN, 554, 555
SFNs, 553
Shadowing by human bodies, 30
Shadowing correlation, 108
Shannon capacity, 176
Short-Term Dynamic (STD) system-level simulations, 417
SIC, 41

Similarity measures, 105
Similarity regions, 103
Single carrier transmission, 386
Single Frequency Network (SFN), 553
Single Frequency Networks, 555
Single-bounce cluster, 121
Slepian modes, 252
Smart Environments, 670
SMEM, 185, 256
Soft handover, 557
SOSF, 116
Space–Time Block Coding, 664
Spatial multiplexing efficiency, 165
Specific Absorption Rate (SAR), 199
Spectral divergence, 107
Spectroscopy, 29
Spheroidal coupler, 164
Stationarity, 75
Stochastic, 184, 186, 189
Sub-millimeter-waves, 27
Super-antenna, 164
Superframe, 644
Switched parasitics, 173, 176, 177

T
TDOA, 287
Terahertz, 29
Terminal antenna, 149, 150
Ternary code, 25
Throughput, 527
TOA, 287
Tracking, 524
Traffic modelling, 433
Traffic models, 418
Travolution, 600
Twin cluster, 121
Two-ray model, 30

U
UCA, 54
ULA, 53
Ultra fast file transfer, 30
Ultra-wideband, 23
Ultrawideband, 663
UMTS, 173, 175, 566, 567
User interaction, 150
User terminal, 150
UWB, 164, 182, 183, 185, 186
  Antenna, 263, 266
  Antenna Array, 267
  Antenna impulse response, 262, 266
  Antenna quality measures, 262
  Beamforming, 269
  Channel, 286
  Imaging, 289

UWB (cont.)
    Localization, 287
    Pulser, 268
    Radar, 289
    Regulations, 249
    Time Domain, 270

V
Vector network analyzer, 23, 228
Vehicle-to-infrastructure, 20
Vehicle-to-vehicle, 20
VHP, 157, 163
Virtual antenna array, 342
Virtual MIMO, 341, 342
Visibility region, 90, 119
Visibility region assignment table, 127
Visibility region group, 127
VoIP, 666

W
Wave Propagation
    Aircraft, 274
    In-car, 274
    Indoor, 272
    Industrail environment, 275
    Tunnel, 275
WBAN, 609
WBAN channel, 612
Wideband array model, 42
WILLWARN, 601
WiMAX, 211, 419
WINNER, 187
WINNER channel model, 77
Wireless Mesh Networks, 426, 523
Wireless Network Coding, 358, 361
Wireless relay channel, 342
Wireless relay network, 343
Wireless Sensor Networks, 523
WLAN, 425

X
X-Map estimation, 450
XPD, 98